Lecture Notes in Electrical Engineering

Volume 285

For further volumes:
http://www.springer.com/series/7818

Tutut Herawan · Mustafa Mat Deris
Jemal Abawajy
Editors

Proceedings of the First International Conference on Advanced Data and Information Engineering (DaEng-2013)

 Springer

Editors
Tutut Herawan
Faculty of Computer Science
  and Information Technology
University of Malaya
Kuala Lumpur
Malaysia

Jemal Abawajy
School of Information Technology
Deakin University
Burwood, VIC
Australia

Mustafa Mat Deris
Faculty of Computer Science
  and Information Technology
Universiti Tun Hussein Onn Malaysia
Batu Pahat, Johor
Malaysia

ISSN 1876-1100   ISSN 1876-1119   (electronic)
ISBN 978-981-10-1340-9   ISBN 978-981-4585-18-7   (eBook)
DOI 10.1007/978-981-4585-18-7
Springer Singapore Heidelberg New York Dordrecht London

Preface

We are honored to be part of this special event in the First International Conference on Advanced Data and Information Engineering (DaEng-2013) together with Data Engineering Research Center. The Data Engineering Research Center is a non-profit organization devoted to promoting research and publication in fields of science and technology

The First International Conference on Advanced Data and Information Engineering which held in Kuala Lumpur, Malaysia during the period of December 16–18, 2013 has attracted 187 papers from 15 countries from all over the world. Each paper was peer reviewed by at least two members of the Program Committee. Finally, only 81 (43 %) papers with the highest quality were accepted for oral presentation and publication in these volume proceedings.

The papers in these proceedings are grouped into eight sections and two in conjunction workshops:

- Database Theory
- Data Warehousing and Mining
- Image and Video Processing
- Information Processing and Integration
- Information Retrieval and Visualization
- Information System
- Mobile, Network, Grid and Cloud Computing
- Web Data, Services and Intelligence
- Workshop on Recent Advances in Information Systems
- Workshop on Mining Educational Data for Effective Educational Resources.

On behalf of DaEng-2013, we would like to express our highest appreciation to be given the chance to do cooperation with Data Engineering Research Center for their support. Our special thanks go to the Steering Committee, General Chairs, Program Committee Chairs, Organizing Chairs, all Program and Reviewer Committee members, and all the additional reviewers for their valuable efforts in the review process that helped us to guarantee the highest quality of the selected papers for the conference.

We also would like to express our thanks to the two keynote speakers, Jason J. Jung from Yeungnam University and Palaiahnakote Shivakumara from University of Malaya. Our thanks also go to Xiuqin Ma from Northwest Normal University for her willingness to deliver tutorial of world-class standard.

Our special thanks are due also to Ramesh Nath Premnath for publishing the proceeding in Lecture Notes in Electrical Engineering of Springer. We wish to thank the members of the Organizing and Student Committees for their very substantial work, especially those who played essential roles.

We cordially thank all the authors for their valuable contributions and other participants of this conference. The conference would not have been possible without them.

Tutut Herawan
Mustafa Mat Deris
Jemal H. Abawajy

Acknowledgments

This work is supported by Ministry of Higher Education (MOHE) and Research Management Centre (RMC) at the Universiti Teknologi Malaysia (UTM) under Research University Grant Category (VOT Q.J130000.2528.02H99).

Conference organisation

Steering Committee

Jemal H. Abawajy   Deakin University
Prabhat K. Mahanti   University of New Brunswick

Chair

Tutut Herawan   University of Malaya
Noraziah Ahmad   Universiti Malaysia Pahang

Publication Chair

Tutut Herawan   University of Malaya
Mustafa Mat Deris   Universiti Tun Hussein Onn Malaysia
Jemal H. Abawajy   Deakin University

Program Committee Chair

Mohamed Ariff Ameeden   Universiti Malaysia Pahang
Zailani Abdullah   Universiti Malaysia Terengganu

Organizing Chair

Prima Vitasari   Institut Teknologi Nasional
Mazlina Abdul Majid   Universiti Malaysia Pahang
M. Ikhmatiar Sibghotulloh   Data Engineering Research Centre

Student Committee

Ng Liang Shen	Universiti Malaysia Pahang
M. Z. Rehman	Universiti Tun Hussein Onn Malaysia
Habib Shah	Universiti Tun Hussein Onn Malaysia
Younes Saadi	Universiti Tun Hussein Onn Malaysia
Iwan Tri Riyadi Yanto	Universitas Ahmad Dahlan
Ardiansyah Salleh	Universiti Malaysia Pahang
Rofildc Hasudungan	Universiti Malaysia Pahang

Program Committee

Abderrahmane Lakas	United Arab Emirates University, United Arab Emirates
Abul Hashem Beg	Charles Sturt University, Australia
Adina Magda Florea	University Politehnica of Bucharest, Romania
Ahmad Nazari Mohd Rose	UNISZA, Malaysia
Ahmad Shukri Mohd Noor	Universiti Malaysia Terengganu
Ahmed N. Abd Alla	Universiti Malaysia Pahang
Aldo Humberto Romero	CINVESTAV-IPN Unidad Querétaro, México
Ainul Azila Che Fauzi	Universiti Malaysia Pahang
Amina Bouraoui	Ecole Supérieure des Sciences et Techniques de Tunis
Angela Guercio	Kent State University at Stark, USA
Anisur Rahman	Charles Sturt University, Australia
Aristotel Tentov	The Ss. Cyril and Methodius University, Macedonia
Azizul Azhar Ramli	Universiti Tun Hussein Onn Malaysia
Azlinah Mohamed	UiTM, Shah Alam Selangor, Malaysia
Bana Handaga	Universitas Muhammadiyah Surakarta, Indonesia
Ching-Cheng Lee	California State University at East Bay, USA
Chang-Woo Park	Korea Electronics Technology Institute, Korea
Constantin Volosencu	University of Timisoara, Romania
Dino Isa	University of Nottingham, Malaysia Campus
Elrasheed Ismail Sultan	Universiti Malaysia Pahang
Feng Feng	Xi'an University of Posts and Telecommunications, China
Gang Zhang	Guangdong University of Technology, China
Habib Shah	Universiti Tun Hussein Onn Malaysia
Hai Tao	Universiti Malaysia Pahang
Hairulnizam Mahdin	Universiti Tun Hussein Onn Malaysia
Hsing-Wen Wang	National Changhua University of Education, China
Ibrahim Kamel	University of Sharjah, United Arab Emirates
Irina Mocanu	University Politehnica of Bucharest, Romania
Iwan Tri Riyadi Yanto	Universitas Ahmad Dahlan, Indonesia
Kamaruddin Malik Mohamad	Universiti Tun Hussein Onn Malaysia
Ke Gong	Chongqing Jiaotong University, China
Kwang Baek Kim	Silla University, South Korea

Laxmisha Rai	Shandong University of Science and Technology, China
Ma Xiuqin	Northwest Normal University, China
Mamta Rani	Central University of Rajastha, India
Md. Geaur Rahman	Charles Sturt University, Australia
Mieczyslaw Drabowski	Cracow University of Technology, Poland
Mohamed Ariff Ameedeen	Universiti Malaysia Pahang
Mohd Farhan Md Fudzee	Universiti Tun Hussein Onn Malaysia
Mohd Helmy Abd Wahab	Universiti Tun Hussein Onn Malaysia
Mohd Fadzil Hassan	Universiti Teknologi Petronas, Malaysia
Mohd Najib Mohd Salleh	Universiti Tun Hussein Onn Malaysia
Mohd Isa Awang	UNISZA, Malaysia
Mohd Zainuri Saringat	Universiti Tun Hussein Onn Malaysia
Mokhtar Beldjehem	University of Ottawa, Canada
Muh Fadel Jamil Klaib	Jadara University, Jordan
Nazri Mohd Nawi	Universiti Tun Hussein Onn Malaysia
Nawsher Khan	University of Malaya
Noor Aida Husaini	Universiti Tun Hussein Onn Malaysia
Noorhaniza Wahid	Universiti Tun Hussein Onn Malaysia
Noraini Ibrahim	Universiti Tun Hussein Onn Malaysia
Noraziah Ahmad	Universiti Malaysia Pahang
Norhalina Senan	Universiti Tun Hussein Onn Malaysia
Norhamreeza Abdul Hamid	Universiti Tun Hussein Onn Malaysia
Noriyani Mohd Zin	Universiti Malaysia Pahang
Noryusliza Abdullah	Universiti Tun Hussein Onn Malaysia
Nureize Arbaiy	Universiti Tun Hussein Onn Malaysia
Nurul Azma Abdullah	Universiti Tun Hussein Onn Malaysia
Qin Hongwu	Northwest Normal University, China
Pabitra Kumar Maji	B. C. College, Asansol, West Bangal, India
Ping Zhu	Beijing University of Posts and Telecommunications, China
R. B. Fajriya Hakim	Universitas Islam Indonesia
Rabiei Mamat	Universiti Malaysia Terengganu
Rathiah Hashim	Universiti Tun Hussein Onn Malaysia
Roslina Mohd Sidek	Universiti Malaysia Pahang
Rozaida Ghazali	Universiti Tun Hussein Onn Malaysia
Sambit Bakshi	National Institute of Technology Rourkela, India
Siddhivinayak Kulkarni	University of Ballarat, Australia
Shahreen Kasim	Universiti Tun Hussein Onn Malaysia
Tuncay Ercan	Yasar University, Turkey
Ventzeslav Valev	Bulgarian Academy of Sciences, Sofia, Bulgaria
Wan Maseri Wan Mohd	Universiti Malaysia Pahang
Yap Bee Wah	UiTM, Shah Alam Selangor, Malaysia
Yongfeng Huang	Tsinghua University, Beijing, China
Zafril Rizal M. Azmi	Universiti Malaysia Pahang
Zarina Dzolkhifli	Universiti Malaysia Pahang

Workshop on Recent Advances in Information Systems

Maizatul Akmar Ismail   Universiti Malaya

Workshop on Mining Educational Data for Effective Educational Resources

Henda Chorfi   King Saud University
Hend Al-Khalifa   King Saud University
Muna Al-Razgan   King Saud University

Contents

Part VI Information System

Part VII Mobile, Network, Grid and Cloud Computing

Contents

Part I
Database Theory

A Log File Analysis Technique using Binary-Based Approach

Sallam Osman Fageeri, Rohiza Ahmad, Baharum Baharudin

Department of Computer and Information Sciences, Universiti Teknologi PETRONAS
Bandar Seri Iskandar, 31750 Tronoh, Perak, Malaysia
sallam_fageeri@hotmail.com, {rohiza_ahmad, baharbh}@petronas.com.my

Abstract. Log files are an important by product of any computing systems including database systems. They contain a huge amount of historical data. Although many algorithms have been designed to utilize the information stored in such files, many of them can still be further improved in terms of execution time and memory usage. In this research paper, a binary-based approach for mining frequency of data items in database transaction log files is introduced. Both the data structures and the algorithms used will be presented according to the sequence of the methodology stages carried out in this research work. The stages are pre, during and post scanning of the log file. Initial experimentation of the approach reveals a significant improvement in terms of the execution time taken to perform frequency analysis of a database transaction log file. To validate the approach, performance comparison was also done against the popular Apriori algorithm. Initial result has shown enhancement in terms of execution time using the binary-based approach.

Keywords. Log File, Frequency Mining, Binary-based Algorithm, Data Structure

1 Introduction

Transaction log files are often kept in order to trace executions or activities performed on a computing system such as database systems, e-commerce systems, e-business websites, etc. As for database systems, the transaction information in the files can be used, for example, to recover any database transaction failure as well as rolling the database back into its previous consistent state. Log files are usually flat files that contain main components such as, timestamp, information about the executed events along with the event identifier [1]. As an example, a database log file may contain the following information about a transaction: the SQL statement, the queried database, the table involved and the selected attributes. In other words, a log file provides the transaction history and user events [2]. Log files are useful sources of information that can be mined to get hidden information such as identifying the frequent items queried to be recommended to customers and decision makers, as well as ranking the most popular search results.

T. Herawan et al. (eds.), *Proceedings of the First International Conference
on Advanced Data and Information Engineering (DaEng-2013)*, Lecture Notes
in Electrical Engineering 285, DOI: 10.1007/978-981-4585-18-7_1,
© Springer Science+Business Media Singapore 2014

Furthermore, the more data saved in log files usually means the more accurate results might be obtained from them. For instance, no significant information could be concluded from a log file which contains two lines of data as compared to a log file which has million records of transactions. However, the massive size of log files does incur costs in terms of execution time when they are being analyzed. Hence, decreasing the latency of user requests has gained the interest of numerous researchers over the past few years[3-6]. With the increasing demand for further user information, it has become very important to discover hidden information from a large and a huge data repository such as log files [7]within a short period of time interval.

In this research paper, a binary-based approach which is able to analyze database log files, in particular, mining of frequent data items, in a short timeframe is introduced. The discussion on the approach will include both the data structures as well as the algorithms used for the processing. To validate the new approach, a well-known algorithm in data mining, the Apriori algorithm [8] was used for execution time performance comparison.

The remaining of this paper will present some related works in Section 2, the proposed approach in Section 3, initial experimental result in Section 4 and conclusions in Section 5

2   Literature Review

This section reviews some of the previous research works which are based on the log file usage as well as the approach utilized in analyzing them. The work presented in [9], for instance, describes the usage of different log files for managing bandwidth and server capacity. For that purpose, the authors have discussed different types of log files including transfer logs, agent logs, error logs and referrer logs. The article describes the whole works that need to be done on the IIS web server log, from the step of obtaining raw log files until the step before the mining process being initialized. However, within the paper, the authors did not discuss the approach utilized in analyzing those log files in detail.

In addition to the above, in[10], a different method is introduced to assist system administrators in tuning the performance of web based applications. The paper discusses the analysis of web log data of the NASA website using an in-depth analysis approach which can capture valuable information such as topmost errors and potential website visitors. By recognizing the most and the least active days of the server, suitable days for scheduling the server shut down can be determined. Furthermore, the authors in [11] have derived user interests through the analysis of server log file and meta data of the page content accessed. The result of the analysis is a graph of files that provides a view page of the user interest. The authors combined two analysis methods in their work: statistical as well as graph analysis. In analyzing the log file, the weight factor (amount of time spent on a page) is used instead of the frequency (number of times a page is accessed). This was due to their assumption that time spent by a user on a specific web page is more useful in finding links to other interesting pages than the number of times a page is being accessed.

In terms of algorithms used for mining frequent items in log files, Apriori algorithm [8] uses breadth-first approach. The algorithm performs multiple passes over the log files. Firstly, the frequency of individual available items is obtained. Secondly, the most frequently accessed items are selected and paired with all available items and the frequency of access is again counted and compared. In the third pass, the most frequent accessed pair(s) are selected and combined with a third item and again the frequency of access is counted and compared. This process continues until the prune step is reached. The Apriori algorithm has been used by many researchers including those reported in [12-14].

Other than the above, in [15-16] the authors have used a data mining algorithm called FP-Growth which uses a data structure called FP-tree for mining relevant access patterns. The access patterns are again generated using the web log file. The authors also claimed that the technique used is efficient for mining both short frequent patterns as well as long frequent patterns. Based on further readings of the technique, it could be said that FP-growth is probably the most popular algorithm besides the well-known Apriori. Both have been constantly referred by most researchers in frequent pattern mining of log files.

Based on the literature survey of existing algorithms, some areas which can be improved have been identified. The areas are either in terms of the execution time as general performance, the memory usage for storing intermediate results or the way of analyzing the log files itself. At the moment, most approaches require multiple scan of the log file which obviously incurs unnecessary additional cost in terms of time and memory space. For all the mentioned drawbacks, we believe that the binary-based approach has the potential to take care of most of these issues.

3 Proposed Binary-Based Log File Analysis Technique

3.1 Log File Structure and Content

As mentioned in the introduction section, log files can be generated from many sources and for various purposes. For example, a log file which is generated based on customers' purchasing activities over the Internet can be used to help an online store to decide which products to sell. Another log file which contains users' status of login attempts to a corporate network can be used to balance the load of the network. In our work however, we are focusing on log files which contain database transactions. For the purpose of simplifying our discussion on the proposed technique, in this paper we will highlight database log files which contain user queries in the form of database tables only. However, using similar technique, there should be no problem to include attributes as well in the log file. Table 1 shows sample content of a database log file that is used in this study (Note: *TID* stands for transaction id and *item used* stands for the database tables queried).

Table 1. Sample content of a database log file

TID	Item Used
100	A C D
200	B C E
300	A B C E
400	B E

For example, in the above table, the first transaction queries data from three database tables which are A, C and D.

3.2   Log File Items Relationship

Consider the following assumption to find the association between the database tables requested: Let $I = \{i_1, i_2 \ldots i_n\}$ be a set of database table names. Let D be a database transaction log file and T be a set of database transactions, such that $\forall\, T \in D$, $T \subseteq I$. A transaction T contains a set of items X if and only if $X \subseteq T$. An association rule is an implication of the form $X \Rightarrow Y$, where $X \subseteq I$, $Y \subseteq I$, and $X \cap Y = \phi$. A rule $X \geq Y$ holds in the transaction set D with confidence c if c% of the transactions in D that contain X also contain Y. The rule $X \Rightarrow Y$ has support s in the transaction set D where s% of transactions in D contains X also contain Y. The problem of finding association rule can be decomposed as

- Find all frequent set of items in the log file.
- Use the frequent item to generate the possible combination and association between the other related items.

3.3   Technique Framework

Basically, the binary-coded technique proposed has three stages of implementation. As depicted in Fig. 1, the first stage, which is called the pre-scanning stage, sets the required data structures for supporting the later log file analysis. The data structures used are a vector for storing the name of all database tables in the database and a single dimensional array which will keep the frequency of query terms occurrences in the log file. The second stage is referred to as the scanning stage. In stage 2, the log file will be scanned line per line and the corresponding frequency counter will be incremented accordingly. In stage 3, which is named as the post-scanning stage, associated items will be grouped together to show ranking of access popularity. Fig. 1 shows both the data structures and algorithm(s) involved in each stage.

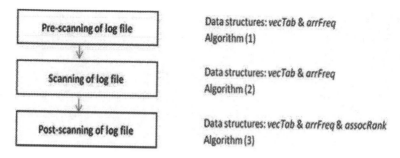

Fig. 1. Main stages of binary-based log file analysis technique

3.4   Binary-based Technique

Stage 1: Pre-scanning of log file.

As shown in earlier Fig. 1, two data structures are created in stage 1: a vector *vecTab* and an array *arrFreq*. vecTab is used to assign a numeric identifier to each of the database tables in the database. For example, the first database table read from the database will be kept in position 0 of the vector, hence, has 0 as its numeric identifier. As for the array *arrFreq*, its size will be determined by the number of tables in the database. For instance, if the number of tables n, is 3, then the number of rows will be $2^n = 2^3 = 8$ with indices of 0 to 7. The number of rows corresponds to the number of possible combinations of the 3 tables. This idea is adopted from the truth table concept in Boolean algebra. If we have 3 elements, e.g., A, B and C, then the potential combinations will be $2^3 = 8$ which are: null, A, B, C, AB, AC, BC and ABC. Fig. 2 shows the truth table representation of A, B and C, and its meaning in our technique.

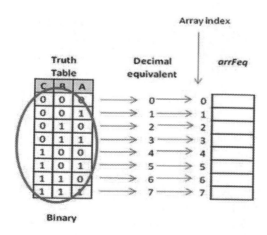

Fig. 2. Truth table relationship with *arrFreq* (1=True, 0=False)

Algo. 1 below presents self explanatory steps which are involved in populating *vec-Tab* and creating *arrFreq*.

```
Begin
    Query table names from database
    Set index to 0
    While not end of result set
        vecTab[index] ← table
        Increment index by 1
    end while
    Create arrFreq[2ⁿ]
End
```

Algo. 1. Algorithm for populating *vecTab* and creating *arrFreq*

Stage 2: Scanning of log file.

Having *arrFreq* in such a way eases the process of incrementing relevant frequency. For example if we have three tables named as A, B and C which are kept in *vecTab* locations 0, 1 and 2 respectively, then a query containing A and C will increment the frequency at index 5 of *arrFreq* by 1. This index is determined as follows: $2^0 + 2^2 = 1 + 4 = 5$ (Note: the first exponent, 0, is the location at which A is stored in *vecTab* and the second exponent, 2, is the location of C in *vecTab*). Looking back at the truth table in Fig. 2, we can see that index 5 is for TFT or 101 which is the binary for 5. Algo. 2 shows the algorithm used for stage 2.

```
Begin
    while not Eof (log)
        read transaction T from log
        for each table in T
            exp   ← map(table)
            index ← index + 2ᵉˣᵖ
        end for
        arrFreq[index] ← arrFreq[index] + 1
    end while
End
```

Algo. 2. Algorithm for scanning log file and populating *arrFreq*

   The algorithm starts by reading the first record in the log file. For each table name mentioned in the record, its numeric code will be found by mapping it to the index where it is stored in *vecTab*. The code is then used in a power formula where 2 is the base and the code is the exponent. As shown in Formula (1), summing up each table's power formula will give the index of *arrFreq* for which the frequency is to be incremented.

$$\text{Index} = \sum_{i=1}^{n} 2^{\text{map(table}_i)} \quad \text{where map(table}_i) \text{ is the index of table}_i \text{ in } vecTab \quad (1)$$

Table 2 gives few more examples on the application of Formula (1) in determining which frequency in *arrFreq* to increase based on the database tables queried in a transaction (Note: Assume the database queried has four tables, i.e., A, B, C and D; and numerical identifier for A is 0, B is 1, C is 2 and D is 3).

Table 2. Samples of Formula (1) application

Tables Queried	Binary Representation	Formula	Resulting *arrFreq* index
A C D	$1\ 1\ 0\ 1_2$	$2^0+2^2+2^3$	13
B C	$0\ 1\ 1\ 0_2$	2^1+2^2	5
A B C D	$1\ 1\ 1\ 1_2$	$2^0+2^1+2^2+2^3$	15

Stage 3: Post-scanning of log file.

Once *arrFreq* has been populated based on the content of the log file, ranking of popular associations for a table can be derived. For example, if a database has three tables, and table A has numerical identifier of 0, then combinations involving A in *arrFreq* will be located at odd indices from 1 till 7. Hence, the frequency values at those indices can be compared to see the most frequently accessed combinations involving A. Fig. 3 shows a sample of how the association information of each table can be extracted.

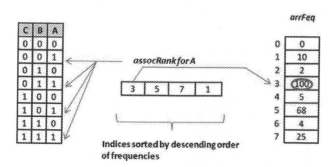

Fig. 3. Sample association ranking

After that, finding which combination(s) is the most accessed will just require a conversion of the first value in *assocRank* to its binary equivalent. For example, the value 3 in *assocRank* is 0011_2 which is BA. In other words, the most frequently accessed term is the combination of B and A. Due to space limitation, Algorithm (3) is not provided here but will be presented in our future publication.

4   Experimental Results

To validate the binary-based approach, experiments have been conducted using a synthetic dataset with different transaction weights to get meaningful response time. The experiments were performed on Intel® core™ 2 Duo CPU, 1.57GHz, and 01 GB of RAM computer, under visual studio.net. The obtained results are plotted as in Fig. 4. Using the mentioned dataset, which is provided by the QUEST generator from IBM's Almaden lab, the binary-based approach when compared to the well-known Apriori algorithm managed to outperform Apriori on the different transaction weights, in terms of the number of database passes as well as execution time taken to process the log file for providing the most frequent accessed items.

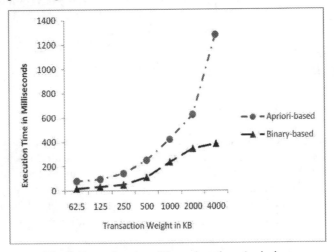

Fig. 4. Execution time: Binary-based vs. Apriori

5   Conclusion

In this research paper, a binary-based approach for frequency itemsets mining of database transaction log files is introduced. The approach which has three main stages of implementation manages to outperform the execution time performance of Apriori algorithm. This is due to the approach's direct identification of which frequency to be incremented using the binary-based formula given, as well as the requirement for a single pass over the log files. At the moment, the approach uses an array data structure for keeping the frequency counts. For future work, we intend to use other data structures such as vector or binary tree in order to cater for memory usage especially for handling sparse data set. In addition, more experiments will be conducted to compare the binary-based approach with other popular algorithms.

References

1. Nagappan, M., Robinson, B.: Creating Operational Profiles of Software Systems by Transforming their Log Files to Directed Cyclic Graphs. In: Proceedings of the 6th International Workshop on Traceability in Emerging Forms of Software Engineering, pp. 54–57. ACM, New York, NY, USA (2011).

2. Fronza, G.I., Sillitti, A.: Failure Prediction based on Log Files using Random Indexing and Support Vector Machines. Journal of System and Software, Elsevier B.V. (2012).
3. Gao, L., Zhang, Z., Towsley, D.: Catching and Selective Catching: Efficient Latency Reduction Techniques for Delivering Continuous Multimedia Streams. In: Proc. 1999 ACM Multimedia Conf, pp. 203–206. ACM, NY, USA (1999).
4. Jeswani, D., Gupta, M., De, P., Malani, A., Bellur, U.: Minimizing Latency in Serving Requests through Differential Template Caching in a Cloud. In: Proc. 2012 IEEE Fifth International Conference on Cloud Computing, pp. 269–276. IEEE (2012).
5. Sharifi, A., Kultursay, E., Kandemir, M., Das, C.R.: Addressing End-to-End Memory Access Latency in NoC-Based Multicores. In: Proc. MICRO '12 Proceedings of the 2012 45th Annual IEEE/ACM International Symposium on Microarchitecture, pp. 294–304. IEEE Computer Society Washington DC, USA (2012).
6. Murayama, D., Oota, N., Suzuki, K-I., Yoshimoto, N.: Low-Latency Dynamic Bandwidth Allocation for 100 km Long-Reach EPONs. Journal of Optical Communications and Networking, vol. 5, no. 1, p. 48 (2012).
7. Yu, X.: A Novel Approach to Mining Access Patterns. In: Proc. 3rd International Conference on Awareness Science and Technology (iCAST), pp.350-355. IEEE, (2011).
8. Agrawal, R.: Fast Algorithms for Mining Association Rules. In: Proc. 20th Int. Conf. Very Large Data Bases (VLDB), vol. 1215, pp. 1–32 (1994).
9. Wahab, M. H. A., Mohd, M. N. H., Hanafi, H. F., & Mohsin, M. F. M. : Data Pre-processing on Web Server Logs for Generalized Association Rules Mining Algorithm. World Academy of Science, Engineering and Technology, vol. 48 (2008).
10. Suneetha, K. R.: Identifying User Behavior by Analyzing Web Server Access Log File. IJCSNS International Journal of Computer Science and Network Security, vol. 9, no. 4, pp. 327–332 (2009).
11. Stermsek, G., Strembeck, M., Neumann, G.: A User Profile Derivation Approach based on Log-File Analysis. In: Proc. of IKE, pp. 258–264 (2007).
12. Ilayaraja, M.: Mining Medical Data to Identify Frequent Diseases using Apriori Algorithm. In: Proceedings of the 2013 International Conference on Pattern Recognition, Informatics and Mobile Engineering, pp. 194–199. IEEE (2013).
13. Singh, J., Ram, H.: Improving Efficiency of Apriori Algorithm Using Transaction Reduction. International Journal of Scientific and Research Publications, vol. 3, no. 1, pp. 1–4. (2013).
14. Luan, R., Sun, S., Zhang, J., Yu, F., Zhang, Q.: A Dynamic Improved Apriori Algorithm and its Experiments in Web Log Mining. In: 2012 9th International Conference on Fuzzy Systems and Knowledge Discovery, pp. 1261–1264. IEEE (2012).
15. Mishra, R.: Discovery of Frequent Patterns from Web Log Data by using FP-Growth algorithm for Web Usage Mining. International Journal of Advanced Research in Computer Science and Software Engineering, vol. 2, no. 9, pp. 311–318. IJARCSSE, India (2012).
16. Vyas, Z.V., Ganatra, A.P., Kosta, Y.P., Bhesadadia, C.K.: Modified RAAT (Reduced Apriori Algorithm Using Tag) for Efficiency Improvement with EP(Emerging Patterns) and JEP(Jumping EP). In: Proc. 2010 International Conference on Advances in Computer Engineering, pp. 238–240. IEEE (2010).

An Application of Oversampling, Undersampling, Bagging and Boosting in Handling Imbalanced Datasets

Bee Wah Yap[1], Khatijahhusna Abd Rani[2], Hezlin Aryani Abd Rahman[1],
Simon Fong[3], Zuraida Khairudin[1], Nik Nairan Abdullah[4]

[1,2] Faculty of Computer and Mathematical Sciences, Universiti Teknologi MARA,
Selangor,Malaysia
[3] Faculty of Science and Technology, University of Macau, China
[4] Faculty of Medicine, Universiti Teknologi MARA, Selangor, Malaysia
[1]{beewah, hezlin, zuraida_k}@tmsk,.uitm.edu.my, [2] ejahhusna@gmail.com,
[3]ccfong@umac.mo, [4]niknairan@yahoo.com

Abstract. Most classifiers work well when the class distribution in the response variable of the dataset is well balanced. Problems arise when the dataset is imbalanced. This paper applied four methods: Oversampling, Undersampling, Bagging and Boosting in handling imbalanced datasets. The cardiac surgery dataset has a binary response variable (1=Died, 0=Alive). The sample size is 4976 cases with 4.2% (Died) and 95.8% (Alive) cases. CART, C5 and CHAID were chosen as the classifiers. In classification problems, the accuracy rate of the predictive model is not an appropriate measure when there is imbalanced problem due to the fact that it will be biased towards the majority class. Thus, the performance of the classifier is measured using sensitivity and precision Oversampling and undersampling are found to work well in improving the classification for the imbalanced dataset using decision tree. Meanwhile, boosting and bagging did not improve the Decision Tree performance.

Keywords- Bagging, Boosting, Oversampling, Undersampling, Imbalanced data

1 Introduction

In recent years, there have been great interests in mining imbalanced datasets. In data mining classification problems, most classifiers such as logistic regression, decision tree and neural network work well when the class distribution of the categorical target or response variable in the dataset is balanced. However, for real problems such as document classification [1], loan default prediction [2], fraud detection [3] or medical classification [4] which involve a binary response variable, the dataset are often highly imbalanced. For a binary response variable with two classes, when the event of interest (eg: 'Died' due to a certain illness) is underrepresented, it is referred to as the positive or minority class. Thus, the number of cases for the negative or majority class is very much higher than the minority cases. When the percentage of the minority class is less than 5%, it is known as a rare event [5]. When a dataset is

T. Herawan et al. (eds.), *Proceedings of the First International Conference
on Advanced Data and Information Engineering (DaEng-2013)*, Lecture Notes
in Electrical Engineering 285, DOI: 10.1007/978-981-4585-18-7_2,
© Springer Science+Business Media Singapore 2014

imbalanced or when a rare event occurs, it will be difficult to get a meaningful and good predictive model due to lack of information to learn about the rare event. There are three approaches to handling imbalanced datasets: data level, algorithmic level and combining or ensemble methods. The data level approach involves resampling to reduce class imbalance. The two basic sampling techniques include random oversampling (ROS) and random undersampling (RUS). Oversampling randomly duplicates the minority class samples, while undersampling randomly discards the majority class samples in order to modify the class distribution. It has been reported that oversampling may lead to overfitting as it makes exact copies of the minority samples while undersampling may discards potential useful majority samples [6-10]. The algorithmic level approach is when machine learning algorithms are modified to accommodate imbalanced data while combining methods involve mixture-of-experts approach [6]. Meanwhile, [11] categorized the approaches as algorithm level, data level, cost-sensitive approach [12] and ensemble methods. Cost-sensitive methods combine algorithm and data approaches to incorporate different misclassification costs for each class in the learning phase. The two most popular ensemble-learning algorithms are boosting and bagging. Bagging stands for "**Bootstrap Aggregating**" whereby bootstrap samples are drawn randomly with replacement. Meanwhile, Boosting algorithms tries to improve the accuracy of a classifier by a reweighting of misclassified samples ([5], [13-14]).

This study examined the predictive performance of three decision tree (DT) algorithms: CART (Classification and Regression Tree), C5 and CHAID (Chi-Square Automatic Interaction Detection) after using oversampling, and undersampling techniques for a cardiac surgery imbalanced dataset. The DT performances are also compared using the bagging and boosting technique.

The rest of this paper is structured as follows: Section 2 reviews some past studies on comparison and applications of methods in handling imbalanced datasets. The ROS, RUS, Bagging and Boosting methods are explained in Section 3. The results are presented in Section 4 and Section 5 concludes the paper.

2   Literature Reviews

The class imbalance problem has been reported as a major obstacle to the induction of a good classifier in Machine Learning algorithms [15]. Most studies on comparisons of methods for handling imbalanced datasets used several different data sets, several different approaches and several classifiers such as Logistic Regression, C4.5, neural network and SVM (Support Vector Machine). This section reviews some of these studies.

In a study by [16] on nosocomial infection risk, the dataset comprises of 683 patients, whereby only 75 (11%) were infected or positive and 89% were negative cases. The difficulty to recognize the minority class took them to propose resampling techniques. They used a new resampling approach in which both oversampling of rare positives and undersampling of the noninfected majority rely on synthetic cases (prototypes) generated via class-specific subclustering. They reported that their novel resampling approach performs better than classical random resampling. The predictive performance of Support Vector Machine (SVM) Decision tree, Naïve Bayes, Adaptive Boosting (Adaboost) and Instance-Based Learner (IB1) improved with their new sampling approach. Their results also shows that support vector algorithm in which asymmetrical margins are tuned to improve recognition of rare positive cases are effective for nosocomial infection detection.   [17] implemented three different algorithms, namely, Logistic Regression (LR), Neural Network (NN) and Chi-squared Automatic Interaction Detection (CHAID) to a marketing dataset which consist of 2826 (17%) who bought the product (positive examples) and 14130 (83%) who did not buy the product (negative examples). The three classifiers performance were based on accuracy, hit rate and AUC and were compared for various imbalance datasets generated from the original dataset. They reported that hit rate (precision) is a better measure of classifier performance for imbalanced dataset and CHAID can be used to develop marketing models. Meanwhile, [1] implemented undersampling and cost sensitive learning in handling imbalanced data in biomedical document classification. They concluded that both undersampling and cost sensitive learning can improve the performance of Bayesian Network classifier. The measures of performance used were sensitivity rate, precision rate, F-score and false positive rate (FPR).   The Synthetic Minority Oversampling Technique (SMOTE) was proposed by [18] and involves generation of synthetic samples. Their experiment involves nine different imbalanced datasets and three classifiers, which are decision tree classifier, Ripper classifier and a Naïve Bayes Classifier. They found that combination of SMOTE and undersampling performs better than only undersampling the majority class. The methods were evaluated using area under Receiver Operating Characteristics Curve (AUC), accuracy of minority class and accuracy of majority class.

Several studies have compared the Bagging and Boosting methods. Boosting has been shown to be promising in handling imbalanced data. The case study by [5] on predicting customer attrition risk showed that combination of boosting and case sampling can improve logistic regression performance. A good explanation on bagging and boosting algorithm is given by [19]. They implemented these techniques on two datasets and showed the significant performance of boosting. Recently, the hybrids of bagging and boosting techniques such as RUSBoost and UnderBagging are reported to achieve higher performances than many other complex algorithms [11]. Meanwhile [14] also investigated four boosting and bagging techniques: SMOTEBoost, RUSBoost, EBBBag and RBBag. Their experiments showed that bagging generally outperforms boosting for noisy and imbalanced data. They recommended bagging without replacement techniques for handling imbalanced data. Recently, [20] reported that combining under-sampling, classification threshold selection, and using an ensemble of classifiers can improve the Naïve Bayes classifier to overcome the imbalance problem

Although there have been various developments for handling imbalanced data especially in the ensemble methods, the new variants or hybrid approach are quite complex, not yet available in data mining software and may be difficult for practitioners. Besides, there is still no conclusive evidence as to which is the best approach although undersampling and oversampling remain popular as it is much easier to implement. The next section explains the application to a real dataset using the sampling, bagging and boosting techniques.

3   Methodology

3.1 Cardiac Surgery Data

In this study, we only focus on the binary (or two classes) classification problems. The positive instances belong to the minority class and the negative instances belong to the majority class. The Cardiac Surgery data were obtained from a local hospital. The data contain cases from a study on prediction of survival of cardiac surgery patients. The response variable has two classes: alive and died. The cardiac surgery dataset comprises of 4976 cases with 4.2% who had 'Died' after surgery and 95.6% 'Alive' cases. For this study, eight independent variables were selected: gender (f,m); Age Group (18-40, 40-60, above 60); Comorbidities (Hypertension, Diabetes, Both, None); Surgery Type (CABG only, CABG and Valve Surgery, Others); Chest Reopen (Yes, No); Atrial Fibrillation (Yes, No), Wound Infection (Yes, No); EUROScore. There were no problems of imbalanced data for the categorical predictors.

3.2 Undersampling and Oversampling

IBM SPSS Modeler 15.2 was used for random undersampling and oversampling of the imbalanced data. The supernode was used to perform these sampling techniques. First, we need to determine the distribution of two classes before we proceed to balance out the data. In undersampling, the majority classes are eliminated randomly to achieve equal distribution with the minority class.   On the other hand, in oversampling the minority classes are replicated to achieve equal distribution with the majority class. Thus for undersampling the class distribution of minority to majority cases is 209:209 while for oversampling it is 4767:4767.

3.3   Bagging and Boosting

The bagging method proposed by [21] is a bootstrap ensemble method that can be applied to enhance model stability. In the Bagging approach, all instances in the training dataset have equal probability to be selected. All samples were replicates based on bootstrap approach. The replicates are samples drawn with replacement and with the same size as the training sample. For each bootstrap set, one model is fitted. The final predictions of the cases are produced using the voting approach.

Consider a training dataset with N samples belonging to two classes. The two classes are labeled as $y \in \{0,1\}$. The steps involved in the Bagging process ([13], [19-21]).

Are as follows:

1.   For iterations $t = 1, 2, \ldots, T$: # by using T=10

a)   Randomly select a dataset with N samples from the original training with replacement.

b)   Obtain a learner, $f(x)$ *(predictive model or classifier)* from the resample dataset

c)   By using the model, $f(x)$ predicts the cases.

2.   Combine all predicted model $f^t(x)$ into an aggregated model $f^A(x)$

3.   By using voting approach, return class that has been predicted most often.

The adaptive Boosting algorithm, named AdaBoost is available in IBM SPSS Modeler 15.0. Consider a training dataset with N samples belonging to two classes. The two classes are labeled as $y \in \{0,1\}$. The steps involved in the Boosting process are as follows by [19, 22-23]:

1.   Assign initial equal weights to each samples in the original training set:

$$w_i^1 = 1/N, i = 1,2,...,N$$

2.   For iterations $t=1, 2, \ldots, T$: # by using T=10
a)   Randomly select a dataset with N samples from the original training set using weighted resampling. The chance for a sample to be selected is related to its weight. A sample with a higher weight has a higher probability to be selected.
b)   Obtain a learner, $f(x)$ *(predictive model or classifier)* from the resampled dataset.
c)   Apply the learner $f(x)$ to the original training dataset. If a sample is misclassified, its error=1, otherwise=0.
d)   Compute the sum of the weighted errors of all training samples.

$$error^t = \sum_{i=1}^{N}(w_i^t \times error_i^t)$$

e)   Calculate the confidence index of the learner $f(x)$:

$$\alpha^t = \frac{1}{2}\ln\left(\frac{1-error^t}{error^t}\right)$$

The confidence index of the learner $f(x)$ depends on the weighted error.
f)   Update the weights of all original training samples:

$$w_i^{t+1} = w_i^t \exp(-\alpha^t * error_i^t)$$

If samples are correctly classified, the weights are unchanged, while the weights for misclassified samples are increased.

g)   Then, renormalize weight, $w_i^t = \dfrac{w_i^t}{\sum_i^N w_i^t}$ so that, $\sum_i^N w_i^{t+1} = 1$

h) $T=t+1$, if error<0.5, and $t<T$, repeat steps (a)-(g); otherwise, stop and T=t-1.
i)   After T iterations, $t=1,2...T$, there are T predicted model $f^t(x), t = 1,2,....,T$. The final prediction for case j, is obtained by the combined prediction of the T models using voting approach:

$$y_j = sign \sum_{t=1}^{T} \alpha^t f^t(x)$$

Figure 1 displays the modelling flow using IBM SPSS Modeler 15.0. The original data set is connected to the TYPE node which is connected the PARTITION node for splitting the data into Training (70%) and Testing (30%) samples. The CART model nodes are then connected to the PARTITION node. The diamond shaped gold nuggets are the generated models. The performance measures are then obtained for the training and testing samples. The process is repeated for C5 and CHAID algorithms.

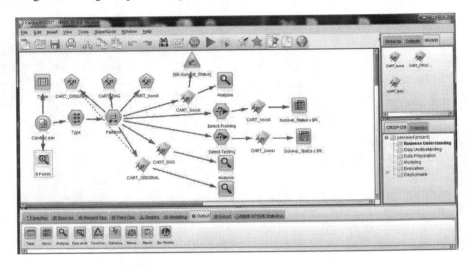

Fig 1. Bagging and Boosting using CART as classifier

3.4   Model Performance Measures

The classification accuracy rate (Acc), sensitivity (Sen), specificity (Spec) and precision rate (Pre) were chosen as the criteria in measuring the performance of the Decision Tree model.

Table 1. *Confusion Matrix*

Actual Class	Predicted Class	
	Positive ('Died')	Negative ('Alive')
Positive('Died')	True Positive (TP)	False Negative (FN)
Negative ('Alive')	False Positive (FP)	True Negative (TN)

Based on Table 1, the calculations are as follows:

$$Acc = \frac{TP + TN}{TP + FP + TN + FN} \qquad Sen = \frac{TP}{TP + FN} \; ,$$

$$Spec = \frac{TN}{TN + FP} \; , \qquad Pre = \frac{TP}{TP + FP}$$

4 Results

The first column in Table 2 shows the performance measures for the original dataset. As expected the specificity is high (100%) and sensitivity is 0%. The results in the second and third column shows that with oversampling and undersampling, the sensitivity for the testing set has increased to 69.4% and 68.7% respectively. Oversampling has been reported to be prone to overfitting but in this study there was no problem of overfitting. The CART_Bagg results are similar to CART model for the original data set and CHAID. Meanwhile, CART_Boost improves with testing sensitivity (27.9%) and precision (42.2%). Taking into consideration that the small sample of minority class will result in much smaller number of minority cases in the training and testing samples, the CART, CHAID AND C5 algorithms were applied to the original data without any data partitioning. Both CART and CHAID classified all 209 minority cases into the majority group (sensitivity=0%) while C5 correctly classified 28 (13.4%) minority cases.

In Table 2, the results for CHAID are similar with CART for Original dataset with 0% sensitivity. Results for C5, C5_Boost and CHAID_Bagg are also similar with testing sensitivity 25% and precision 48.6%. Bagging is not available for C5 in IBM SPSS Modeler. The results in Table 2 and Table 3 show that sampling approach

Bee Wah Yap et al.

performs better than bagging and boosting methods. Boosting and bagging did not
improve the sensitivity of the decision tree classifiers.

Table 2. Results for CART as base classifier

		CART_Original	Os	Us	CART_Bagg	CART_Boost
Acc	Training	95.9	79.1	81.9	95.9	96.2
	Testing	95.5	76.7	71.1	95.5	95.1
Sen	**Training**	**0.0**	**71.4**	**76.7**	**0.0**	**34.7**
	Testing	**0.0**	**69.4**	**68.7**	**0.0**	**27.9**
Spe	Training	100.0	86.4	87.2	100.0	98.8
	Testing	100.0	84.5	73.5	100.0	98.2
Pre	**Training**	**0.0**	**83.6**	**85.8**	**0.0**	**55.7**
	Testing	**0.0**	**82.8**	**71.9**	**0.0**	**42.2**

Notes: **Acc**: Accuracy, **Sen**: Sensitivity, **Spe**: Specificity, **Pre** : Precision,
Os: Oversampling, **Us**: Undersampling, **Bagg**: Bagging, **Boost**: Boosting

Table 3. Results for C5 and CHAID as base classifiers

		CART_Original	C5	C5_Boost	CHAID	CHAID-Bagg	CHAID_Boost
Acc	Training	95.9	96.2	96.2	95.9	96.2	96.5
	Testing	95.5	95.4	95.4	95.5	95.4	94.6
Sen	Training	**0.0**	**31.9**	**31.9**	**0.0**	**31.9**	**35.5**
	Testing	**0.0**	**25.0**	**25.0**	**0.0**	**25.0**	**23.5**
Spe	Training	100.0	98.9	98.9	100	98.9	99.1
	Testing	100.0	98.7	98.7	100	98.7	97.9
Pre	Training	**0.0**	**56.9**	**56.9**	**0.0**	**56.9**	**63.3**
	Testing	**0.0**	**48.6**	**48.6**	**0.0**	**48.6**	**34.8**

Notes: **Acc**: Accuracy, **Sen**: Sensitivity, **Spe**: Specificity, **Pre** : Precision,
Os: Oversampling, **Us**: Undersampling, **Bagg**: Bagging, **Boost**: Boosting

5 Conclusion

Sampling approaches are much easier to implement for improving prediction of the minority case of a two-class classification problem. The random undersampling advantage is that all the minority cases are maintained as replication of minority case in oversampling will cause overfitting since it makes duplicates copy of the existing data. Besides, most classifiers assume that all cases are independent. The application of bagging and boosting in this study shows that they do not perform better than the random sampling strategies. For future research, a simulation study should be carried out whereby data are generated and then the different approaches are compared so as to obtain a conclusive decision on the best strategy to handle imbalanced data. The simulation study could investigate the effect of different methods of handling imbalanced data with different percentage of imbalance and for different classifiers. It is also important to note that the classifiers performance depend on data quality All datasets should be cleaned and imbalanced problems in categorical predictors (or features) should be determined so as to obtain a good predictive model with results that can be generalized.

Acknowledgments We thank the Research Management Institute (RMI) Universiti Teknologi MARA and the Ministry of Higher Education (MOHE) Malaysia for the funding of this research under the Malaysian Fundamental Research Grant, 600-RMI/FRGS 5/3 (16/2012).

References

1. Laza, R., Pavón, R., Reboiro-Jato, M., Fdez-Riverola, F.: Evaluating the effect of unbalanced data in biomedical document classification. Journal of integrative bioinformatics, 8(3):177, (2011). Doi:10,2390/biecoll-jib-2011-177
2. Brown, I., & Mues, C. An experimental comparison of classification algorithms for imbalanced credit scoring data sets. *Expert Systems with Applications, 39*(3), 3446-3453, (2012). doi: 10.1016/j.eswa.2011.09.033
3. Wei, W., Li, J., Cao, L., Ou, Y., Chen, J.: Effective detection of sophisticated online banking fraid on ectremely imbalanced data. World Wide Web (2013) 16:449–475. doi: 10.1007/s11280-012-0178-0
4. Rahman, N.N., Davis, D.N.: Addressing the Class Imbalance Problems in Medical Datasets. International Journal of Machine Learning and Computing, 3(2), 224-228, (2013).
5. Au, T., Chin, M.-L., & Ma, G.: Mining Rare Events Data by Sampling and Boosting: A Case Study. In S. Prasad, H. Vin, S. Sahni, M. Jaiswal & B. Thipakorn (Eds.), Information Systems, Technology and Management (Vol. 54, pp. 373-379): Springer Berlin Heidelberg, (2010).
6. Kotsiantis, S. B., Pintelas, P. E., Kanellopoulus, D.: Handling imbalanced datasets: A review. GESTS International Transactions on Computer Science and Engineering, Vol.30, (2006).
7. Drummond C., Holte, R. C.: C4.5, Class Imbalance and Cost-Sensitivity: Why Undersampling beats Oversampling, Workshop on Learning from Imbalanced Datasets II, ICML, Washington DC, (2003).

8.   Drummond C., Holte, R. C.: Severe Class Imbalance: Why Better Algorithms Aren't the Answer. Proceedings of 16th European Conference of Machine Learning, LNAI 3720, 539-546, (2005).
9.   Weiss, G. M.:Mining with rarity: a unifying framework. Sigkdd Explorations, 6(1), 7-19 (2004).
10.   Chawla, N. V.: Data mining for imbalanced datasets: An overview Data mining and knowledge discovery handbook (pp. 853-867): Springer, (2005).
11.   Galar. M.,   Fern´andez, A., Barrenechea, E., Bustinc, H., Herrera, F.: A review on Ensembles for Class Imbalanced Problems: Bagging-, Boosting- and Hybrid Based Approaches. IEEE Transactions on Systems. Man, .and Cybernetics-Part C. Applications and Reviews. Vol.42, No.4, 463-484 (2012).
12.   Chawla, N. V., Cieslak, D. A., Hall, L. O., Joshi, A.: Automatically countering imbalance and its empirical relationship to cost. Data Mining and Knowledge Discovery 17, 2, 225-252 (2008).
13.   Kotsiantis, S., Pintelas, P.: Combining bagging and boosting. International Journal of Computational Intelligence, 1(4), 324-333 (2004).
14.   Khoshgoftaar, T.M., Van Hulse, J., Napolitano, A.: Comparing Boosting and Bagging techniques with Noisy and Imbalanced Data, IEEE Transactions on Systems. Man, .and Cybernetics-Part A. Systems and Humans. Vol.41,No.3, 552-568 (2011).
15.   Batista, G. E., Prati, R. C., Monard, M. C.: A study of the behavior of several methods for balancing machine learning training data. *ACM Sigkdd Explorations Newsletter, 6*(1), 20-29,(2004).
16.   Cohen, G., Hilario, M., Sax, H., Hugonnet, S., Geissbuhler, A.: Learning from imbalanced data in surveillance of nosocomial infection. Artificial Intelligence in Medicine, 37(1), 7-18 (2006).
17.   Duman, E., Ekinci, Y., Tanriverdi, A.: Comparing alternative classifiers for database marketing: The case of imbalanced datasets. Expert Systems with Applications, *39*(1), 48-53 (2012).
18.   Chawla, N. V., Bowyer, K. W., Hall, L. O., Kegelmeyer, W. P.: SMOTE: synthetic minority over-sampling technique, Journal of Artificial Intelligence Research, 16, 321-357 (2002).
19.   Cao, D.-S., Xu, Q.-S., Liang, Y.-Z., Zhang, L.-X., Li, H.-D.: The boosting: A new idea of building models. Chemometrics and Intelligent Laboratory Systems, *100*, 1-11(2010). doi: http://dx.doi.org/10.1016/j.chemolab.2009.09.002
20.   Klement, W., Wilk, S., Michaowski, W., Matwin, S.: Classifying severely imbalanced data. C. Butz and P. Lingras (Eds.): Canadian AI 2011, LNAI 6657, pp. 258–264 (2011).
21.   Breiman, L.: Bagging predictors. Machine learning, 24(2), 123-140 (1996).
22.   Freund, Y., Schapire, R. E.: A desicion-theoretic generalization of on-line learning and an application to boosting Computational learning theory (pp. 23-37): Springer,(1995).
23.   IBM SPSS Modeler 15 Algorithms Guide. IBM Corporation (2012).

Excel-Database Converting System using Data Normalization Technique

Rohiza Ahmad[1], Prum Saknakosnak[2], Yew Kwang Hooi[3]

Department of Computer and Information Sciences, Universiti Teknologi PETRONAS
Bandar Seri Iskandar, 31750 Tronoh, Perak, Malaysia
{ [1]rohiza_ahmad, [3]yewkwanghooi} @petronas.com.my, [2]nakprum@gmail.com

Abstract. This paper is intended to present an application system which can facilitate users in converting from Excel spreadsheet to database structure by using data normalization technique. It provides users with an automated system that can develop a database in its third normal form (3NF) from an Excel spreadsheet. There are two major problems of Excel spreadsheet that are emphasized in this project which are data redundancy and update anomaly. Prototyping-based methodology is used in order to develop this system. The system just creates the SQL codes for creating the database structure and populating the database; however, how the users make use of it such as interface for database interaction is not covered at this stage of project. Based on its functionalities, accuracy and performance, this system is able to bring more benefit and convenience for its users.

Keywords. SQL commands, Excel spreadsheet, Data normalization, Normal Forms, Functional dependency

1   Introduction

Data and information are important assets and valuable resources for individuals, companies, businesses and organizations. They can be used to help run a business as well as to make good decisions in achieving the goal of the business or organization. Therefore, proper data management is really needed.

There are a number of data management software that people always use to store their data and information. Among all of them, for certain reasons, Excel spreadsheet has been considered by many to be one of the most familiar software for users to key in their data. However, using Excel spreadsheet to store data and information has certain drawbacks. Among them, the spreadsheet will become unmanageable when the amount of data stored is large. Even more important than that are the problems of data redundancy and update anomaly.

Data redundancy refers to having information that is repeated in two or more places. In Excel, the same data will need to be entered in several different places if the data is needed there. As for update anomaly, in database terminology, it refers to problems which can occur when insertion, modification or deletion is done to the data

T. Herawan et al. (eds.), *Proceedings of the First International Conference on Advanced Data and Information Engineering (DaEng-2013)*, Lecture Notes in Electrical Engineering 285, DOI: 10.1007/978-981-4585-18-7_3,

[1]. For example, due to the existence of multiple copies of the same data, any modification to one of the copies may lead to data inconsistency since some of them might be missed out during the update process.

Therefore, this research project intends to help Excel spreadsheet users in converting their stored data into a third normal form (3NF) database. According to Connolly and Begg [1], a 3NF database is minimal in redundancy and free from update anomaly. Hence, there will be two objectives to be achieved by the project which are to construct algorithms for automating the manual process of database normalization starting from First Normal Form (1NF) up until 3NF and also to develop an application system that can facilitate in converting an Excel spreadsheet to a database structure using the data normalization algorithms developed. As for the scope of the project, the output of the application system will only be SQL commands for the users to create and populate the database. How the users make use of the codes such as executing the codes in certain DBMS or querying the data once the database has been created is not covered at this stage of the project.

The remaining of this paper will present some related works in Section 2, the methodology that has been opted to carry out this project work in Section 3, some results from the project in Section 4 and conclusions in Section 5.

2   Related Work

Due to the importance of database normalization in refining and ensuring high quality of relational database, many research works have been conducted in this area. Most of these research works were focusing on approaches to find functional dependencies, which are the bases for normalization steps, among attributes. For example, Yao and Hamilton [2] proposed an efficient rule discovery algorithm called FD_Mine for mining functional dependency from data. FD_Mine searches for functional dependencies by using equivalence and nontrivial closure. Besides, identifying the problem of discovering functional dependencies in a relation, the authors have also provided some results obtained from a series of experiments conducted on FD_Mine. Using both synthetic and real data, the experiments have proven the effectiveness of FD_Mine.

Apart from the above, it was mentioned by Beaubouef et al. [3] that rough relational database model can be developed in order to manage the uncertainty in relational database. In their work, they concentrated on rough functional dependencies when conducting the normalization process. In other words, normal forms are constructed based on roughly identified functional dependencies.

Chen et al. [4] on the other hand, demonstrated the deficiencies of normal form definitions based on functional dependency. According to them, the traditional way of solely defining normal forms based on functional dependency is not effective enough in removing common data redundancies and data anomalies. Instead, normalization process can yield better database designs with the addition of functional independency. For that, the authors have also introduced a new normal form concept which is based on functional independency. The new normal form was shown to be able to deliver additional criteria to further remove data redundancies and data anomalies.

Regarding algorithms to automate the process of data normalization, few works can also be found in the literature. Among them is the work done by Bahmani et al. [5]. The work presents a new complete automatic relational database normalization method which covers the data normalization technique up to Boyce-Codd Normal Form (BCNF). The approach uses dependency matrix and directed graph matrix to represent dependencies.

Much similar to the above approach is the work of Verma [6]. Verma has conducted a comparative study between manual and automatic normalization techniques using sequential as well as parallel algorithms. The technique for automatic normalization used was also depended upon the generation of dependency matrix, directed graph matrix, as well as determinant key transitive dependency matrix. From the study, it was found that the main benefit of the automatic approach as compared to the traditional one is the primary key. The key is automatically identified for each final table generated.

Another approach of normalizing relational database schemas is by using Methematica modules [7]. According to Yazici and Karakaya [7], Methematica provides a straightforward platform as compared to Prolog based tools which require complex data structures. The Methematica modules are put together into a window based system called JMath-Norm so that it can provide interaction between the user and Methematica. Apart from this, the interactive tool of the prototype is also a handy tool for teaching normalization theory in database management course.

3   Methodology

This project has adopted Prototyping-based methodology for the development of algorithms and the application system. As shown in Fig. 1, the methodology covers planning until system stages. Analysis, design and implementation phases are repeatedly executed until the prototype has been completed with enough functionalities. Once completed, the prototype can be implemented as a working system.

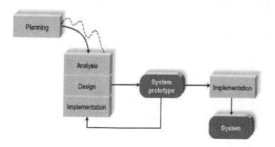

Fig. 1. Prototyping Development Methodology [8]

In this project, for the development of the prototype, the hardware used was a notebook, i.e., ACER Aspire 4736, and the software were Netbeans IDE 7.3 and Wamp-

Server 2.2D which includes Apache 2.2.21, MySQL 5.5.20 and phpMyAdmin 3.4.10.1. As for the programming languages, Java and SQL (MySQL) were chosen.

4   Results and Discussions

4.1   Data Gathering - Interview

Before initiating this project, internal interviews had been conducted to five staffs from the organization the authors are in. The staffs were from different departments and they use Microsoft Excel Spreadsheet in their everyday work. The main purpose for the interview was to get information about the experience of the real Microsoft Excel spreadsheet's users in doing their tasks in relation to the problem statement of the project. From the interview feedbacks, it can be concluded that the users did not have much knowledge about database. They could not differentiate between database and normal data storage such as Excel spreadsheet. From their points of view, as long as the software can be used to store certain amount of data, then it will become their database. However, regarding data redundancy and update anomaly, they did agree that these did occur in their data storage and a solution which can help them overcome the problems will be much welcomed.

4.2   System Flow

Fig. 2 shows the flow of operations for the system which has been developed. First, users need to import an Excel spreadsheet file into the system. The content of the spreadsheet will be displayed on the interface of the system. Next, the users will be guided to identify one functional dependency at a time. Each functional dependency identified will invoke one of the various forms of data normalization including 1NF, 2NF and 3NF. For easier understanding, 1NF refers to database table which has single data in all of its cells. As for 2NF, it refers to database tables which have no partial dependencies. In other words, the whole primary key is needed in order to tell the content of non-key attributes. 3NF on the other hand, refers to database tables which have no transitive dependencies. A 3NF table's non-key attributes can be identified solely based on the table's key, not from other non-key attribute. Upon completion of the normalization process, the system will create SQL codes according to the final data of tables from normalization. After that, the users just need to run the SQL codes in any DBMS selected. In this project, phpMyAdmin was used to show example of how to create database by importing the SQL file which has been created by the system.

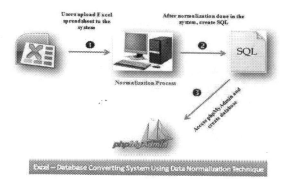

Fig. 2. System Flow for System

4.3 Normalization Algorithms

Based on the flow of the system, the normalization process basically begins with the submission of the Excel data to the conversion system. Hence, at this point of time, the data is not in any normal form or in other words, it is in the zero normal form (0NF). In order to normalize the data two steps have been taken: normalizing 0NF to 1NF and normalizing 1NF to 2NF and 3NF.

Zero Normal Form to First Normal Form - For this step, two functions were coded. The first is called generateInput(). The function returns an array in Java containing the data from the imported Excel file. The second, the outputArray() function, is used to generate array of 1NF by removing repeating groups from the original array. Fig. 3 shows the algorithms for both functions.

```
generateInput():
Begin
   SourceArray = data imported from Excel file
   Return SourceArray
End
outputArray():
Begin
   Generate two arrays namely tempArray and firstArray
   For all the entries in a row
     If any entry is not empty then
       tempArray = firstArray
     Else
       Fill with the nearest upper row of the same column
     Endif
Copy each row of tempArray into firstArray
Return firstArray
End
```

Fig. 3. Algorithms for generateInput() and outputArray()

From 1NF to 2NF and 3NF - In this part, the firstArray will become the main input in order to process to 2NF and 3NF. Two functions are used at this stage which are arrangeCompositeTable() and eliminateRedundancy(). The algorithm for the first function is given in Fig. 4 and the working of the second function is presented in Fig. 5. Basically, the first function is used to select the whole data columns from firstArray which have been selected by users as representing a functional dependency.

```
arrangeCompositeTable():
Begin
    If length of key is less or equal to zero
    Then return null
    End if
    For all the entries of firstArray
        Select the entries whose index is the same as the index
        of key and dependency selected by user
        Store these entries in output array
    End loop
    Return output
End
```

Fig. 4. Algorithm for arrangeCompositeTable()

As for the second function, it is used to eliminate data redundancy by generating a matrix and removing repeated data in each table set. As shown below, first, the same set of data result from the arrangeCompositeTable() will be compared. (Note: A to F are data, e.g., A is a course code CSC101, B is a lecturer named John, etc.)

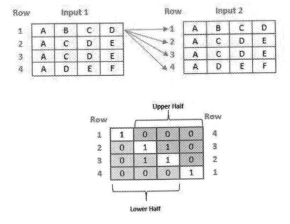

Fig. 5. Working of eliminateRedundancy()

Each row of input1 is compared to each row of input2 row by row. If the rows do not contain similar data, 0 will be placed in the resulting matrix underneath; otherwise, it will be 1. For example, in the above diagram, row 1 of input1 contains ABCD, while row 2 of input2 contains ACDE. Comparing the two will result in 0 being placed in cell "row 2, column 1"of the resulting matrix. After the resulting matrix has been fully filled up, the next step will be removing the rows in the matrix which appear to have number 1 (excluding the diagonal since this algorithm is to compare the same set of data. By default, the diagonal of the matrix will appear as number 1). There are two ways of removing the row either though upper half of the diagonal or lower half of the diagonal. Removing through the upper half will make the last row of the matrix as row number 1 in relation to data from arrangeCompositeTable(). On the other hand, removing through the lower half will end up with the first row being row number 1 in relation to data from arrangeCompositeTable(). However, both methods work accurately the same.

4.4   User Interface

For the convenient of users, an interface was developed for the system. The interface has three main functions for the users to interact with. The functions are for importing the Excel file, normalizing the data and creating SQL code. In case some users may have limited knowledge in database, the system also provides a HELP button which displays various steps for users to follow when identifying functional dependency.

4.5   Testing

The developed system was tested for validation. Regarding load testing, the system was tested with both versions of Excel 2003 and 2007 with different increasing number of columns and rows. According to the result, it shows that regardless of number of columns and rows of the Excel table, the system is still capable of performing the intended functionalities effectively. However, in terms of efficiency, slight increase in processing speed was observed when the number of rows was increased. As for usability testing, three testers with varying knowledge of database were asked to evaluate the system. They were found to be satisfied with the system's ability to maintain the consistency between the data stored in the Excel file in both versions and the data read to the application. According to them, overall, all functions are very good and processing of data is very fast.

In order to enhance the system, they also gave some suggestions for improvement such as the system should be able to determine functional dependency without input from users as well as to import Excel files with many sheets. Based on the feedbacks, future plan will be to improve the system if feasible.

5   Conclusion

In conclusion, Excel – Database Converting System Using Data Normalization Technique has been developed successfully as a prototype that covers the scopes which have been determined for the project. Basically, the prototype is able to import and process both versions of Excel file – 2003 and 2007 version into the system. It is also able to automatically normalize the data up to 3NF by minimizing data redundancy and update anomaly. The system is also capable of creating SQL file which will be used to create database in phpMyAdmin.

As for future enhancement, even though the system has been completely developed according to its scope and objectives, there are also certain additional features that can be enhanced. One of them is the identification of functional dependency which is currently identified by the users manually. Even though a HELP button is provided to guide the users, there are still possibilities that the users will not be able to identify the correct functional dependency based on the rule of normalization. Another possible enhancement is in terms of the interface for database interaction. The interface will be more useful and convenient for the users if it has update, edit and delete functionalities added. And lastly, future enhancement can also be done to the normalization process. Higher Normal Form such as BCNF, Forth Normal Form (4NF) and Fifth Normal Form (5NF) can be implemented to see if the solution can further reduce redundancy and update anomaly.

Acknowledgments. The authors would like to extend our gratitude and appreciation to Mr. Pan Sophoana for his strong support and technical assistance throughout the project.

References

1. Connolly, T., Begg, C.: Database Systems – A Practical Approach to Design, Implementation, and Management, 5th ed. Pearson Addison-Wesley, MA, USA (2010)
2. Yao, H., Hamilton, H.: Mining Functional Dependencies From Data. J. Data Mining and Knowledge Discovery, vol. 16, no. 2, pp. 197-219. Kluwer Academic Publishers Hingham, MA, USA (2008)
3. Beaubouef, T., Petry, F., Ladner, R.: Normalization in a Rough Relational Database. In: Slezak, D. et al. (eds.) RSFDGrC 2005. LNAI, vol. 3641, pp. 275-282. Springer, Heidelberg (2005)
4. Chen, T., Liu, S., Meyer, M., Gotterbarn, D.: An Introduction to Functional Independency in Relational Database Normalization. In: ACM-SE Proceedings of the 45th annual southeast regional conference, pp. 221-225. ACM Press, New York (2007).
5. Bahmani, A. H., Naghibzadeh, M., Bahmani, B.: Automatic Database Normalization and Primary Key Generation. In: Proceedings of Canadian Conference on Electrical and Computer Engineering 2008 (CCECE 2008), pp.11-16. IEEE Press, New York (2008)
6. Verma, S.: Comparing Manual and Automatic Normalization Techniques for Relational Database. International Journal of Research in Engineering & Applied Sciences (IJREAS), vol. 2, no. 2, pp. 59-67 (2012)
7. Yazici, A., Karakaya, Z.: Normalizing Relational Database Schemas Using Mathematica. In: V.N. Alexandrov et al. (eds.) ICCS 2006. Part II, LNCS, vol. 3992, pp. 375-382. Springer, Heidelberg (2006)
8. Dennis, A., Wixom, B., Tegarden, D.: Systems Analysis and Design with UML Version 2.0.- An Object-Oriented Approach, 2nd Ed. John Wiley & Sons, USA (2005).

Model for Automatic Textual Data Clustering in Relational Databases Schema

Wael M.S. Yafooz , Siti Z.Z. Abidin, Nasiroh Omar *and* Rosenah A. Halim

Faculty of Computer and Mathematical Sciences, UiTM Shah Alam, Selagor, Malaysia

Waelmohammed1@hotmail.com, {zaleha,nasiroh,rosenah}
@tmsk.uitm.edu.my

Abstract. In the last two decades, unstructured information has become a major challenge in information management. Such challenge is caused by the massive and increasing amount of information resulting from the conversion of almost all daily tasks into digital format. Tools and applications are necessary in organizing unstructured information, which can be found in structured data, such as in relational database management systems (RDBMS). RDBMS has robust and powerful structures for managing, organizing, and retrieving data. However, structured data still contains unstructured information. In this paper, the methods used for managing unstructured data in RDBMS are investigated. In addition, an incremental and dynamic repository for managing unstructured data in relational databases are introduced. The proposed technique organizes unstructured information through linkages among textual data based on semantics. Furthermore, it provides users with a good picture of the unstructured information. The proposed technique can rapidly and easily obtain useful data, and thus, it can be applied in numerous domains, particularly those who deal with textual data, such as news articles.

Keywords- relational databases, unstructured data, document clustering, query efficiency, textual data

1   Introduction

Unstructured information presents significant challenges in information management. Given their unorganized form, rapid management and retrieval of such information, which is essential in providing users with knowledge, is difficult[1, 2]. In addition, no rule or constraint exists in handling unstructured information. The amount of unstructured information increases as a result of the high reliance of users on digital data in almost all forms of daily tasks because this format is more secure, less storage space, and easy to retrieve as compared to hard copies[3]. Unstructured data can be found in relational databases such as news articles, personal data and textual documents. In spite of, Relational Database Management Systems (RDBMS) are powerful and robust data structures used in managing, organizing, and retrieving data [4, 5]. However, such systems contain massive amount of unstructured data, which are mea-

T. Herawan et al. (eds.), *Proceedings of the First International Conference on Advanced Data and Information Engineering (DaEng-2013)*, Lecture Notes in Electrical Engineering 285, DOI: 10.1007/978-981-4585-18-7_4,
© Springer Science+Business Media Singapore 2014

ningless and difficult to deal without proper organization if left unorganized. As a result, retrieval of significant information or pertinent knowledge becomes a challenge.

Few attempts have been made to deal with unstructured data in RDBMS. These attempts focus only on named entity and on extracting structured information often hidden in unstructured data. Such methods managed unstructured data in the database schema itself as an incremental repository, in databases structure (schema) [5-7], to answer a structured query [8-11] or retrieved such data by keyword search [11-15]. Such methods fail to represent the entire collection of textual documents (corpus) in meaningful clusters, which can help users acquire knowledge regarding the corpus. These methods also do not consider the semantic relation among unstructured data stored in relational databases. Moreover, the aforementioned methods require extra scripting and programming to manage and represent unstructured data in appropriate formats. Thus, these methods are time consuming and labor intensive.

This paper is an extension of our previous work [16]. The most common methods used for managing unstructured data in relational databases are investigated. In addition, an incremental and dynamic model for managing and clustering unstructured (textual) data is introduced. This model automatically processes data when the user loads textual documents into a relational database. The proposed technique is performing automatic textual data clustering and linking. By applying this concept, frequent term, which is used in document clustering, and named entity, which is used in information extraction, are presented. In addition, the semantic of the words in clustering process is included using WordNet database[17]. In this manner, the user can obtain the knowledge and useful data clusters based on the semantic relation among unstructured data. In addition, the efficiency of query processing is improved when retrieved the clustered data. Furthermore, the user is not required to develop extra programming for executing textual data the clustering on desktop application due to clustering process is performed automatically in database schema.

The rest of this paper is organized as follows. Section 2 presents related studies. Section 3 describes the proposed technique, that is, an incremental and dynamic repository for managing unstructured data in relational databases. Section 4 provides the conclusion.

2   Related studies

Relational databases contain massive amounts of unstructured information. Few attempts have been made to handle such information using information extraction techniques [18]. This challenge is addressed using the canonical link among named entities often found in articles because such entities exist in different formats or varieties in database records proposed by [7]. The proposed technique is executed in two processes. First, the named entities are recognized in the articles. Second, matching is performed by introducing the canonical link as the foreign key, which matches database records.

In [5], another method for extracting and storing structured information into a table is introduced. The table can be used for keyword search. In addition, three operators, namely, extract, cluster, and integrate, are presented. These operators can be used by the database administrator to manage extracted information. All the aforementioned methods focus on dealing with unstructured data in the same database schema. However, a method for storing extracted information is introduced in [6] by developing an intermediate database, called parse tree database (PTDB), and a query language, called parse tree query language (PTQL).

PTDB stores immediate data from the extracted information and works as intermediate between user and relational database. This process is the initial step in the proposed technique. PTQL is the query language used to retrieve extracted information from PTDB. This language decomposes retrieved queries into keyword-based queries and Structured Query Language (SQL). Recently, statistical technique based classification [19] and integration time series analytical data based on forecasting [20] are introduced for managing such data. Majority of aforementioned approaches deal with unstructured information in a low-level database schema. Another technique for dealing with such information is using query as a top-level database. This method is called answering the structured query, and the examples include SCORE[8, 9] , EXDB[21, 22] , Avatar [11],and SCOUT [10, 23]. Keyword search technique [12-15] is another method to manage unstructured data.

In a commercial database, such as Oracle®, the introduced oracle text [24, 25] which is consists of two main types of classification and clustering. Classification is based on a predefined class. Meanwhile, k-means and its variants clustering algorithms are used for clustering [26]. The numbers of clusters are needed to parameterize. In addition, text index on all textual data is required to create. This method is time consuming and cannot produce good quality data clusters. In addition, this is required users to enter the number of clusters in order to perform the clustering. Thus, user' should have a prior knowledge is on the textual data collection. Furthermore, several tedious steps are required before the clustering process. Table 1 summarizes the research works that focus on managing unstructured data in RDBMS in database schema or in query level.

Table 1. summarize of unstructured data management in RDBMS

Approach	Style	Strategy	Research work
Database	Inside Schema	Named Entity	[7]
Schema	Intermediate	Named Entity	[6]
	RDBMS		
	Inside Schema	Structured Information	[5]
	Inside Schema	All words	[24, 25]
Query Based	Decompose		[11-15]
	SQL	Common Words	

Most of aforementioned methods are based on extracting the structured from un-structured information by using named entity techniques in databases schema (Low level). Such methods are used only for building information extraction architecture within relational databases, while query based (high level) methods are running on top of the relational database. Such methods do not concern about organizing the actual data. These methods focus on retrieving textual documents from text databases. The keyword search techniques simplify the process of structured query language over relational databases only. However, they are time consuming and need extra tools or expert user to perform further processing. In addition, such methods do not concern in managing unstructured data from the start of the storing process.

3   Incremental and Dynamic Repository and Semantic Query

In this section, the proposed technique for managing unstructured information in rela-tional databases is presented. This technique includes an incremental and dynamic repository and semantic query.

3.1   Incremental and Dynamic Repository

Incremental and dynamic repository is a method for organizing and clustering textual data. This technique is embedded within the database schema. The technique pro-posed in this paper is divided into three stages, namely, data loading, data filtering, and data clustering.

Data loading.
    The first stage is data loading. During this stage, the user enters textual data into a database table. Before the data is stored, several steps are performed to rearrange the textual data. These steps include stemming that converts words into their source, stop words remove and noise cleaning, such as HTML or XML tags. In this step, there are two versions are original and cleaning of textual data. In original version, textual data which is stored in a specific data record while cleaning version is hold the textual document after rearrangement process. This version used to extract and stored such information in the incremental repository (database table) to be continuously used further processing. When all processes are completed, information filtering begins.

Data filtering.
    During this stage, concepts from information extraction and textual document clus-tering [27, 28] are applied. Information extraction converts textual data into structured information by extracting the named entity. Furthermore, in certain cases when named entity is ambiguous, tools such as IRC-names [29] and gazetteers are applied. Tex-tual document clustering adopts the clustering approach in which common words are

based on textual documents. Extracted information is stored in the incremental repository for usage in data clustering.

Data clustering.

Clustering is automatically executed based on common words and the named entity [30, 31]. In this manner, the process is faster when compared to common techniques used in traditional textual clustering methods. In addition, the clustering process is dynamic and its relies on some parameters entered by user at the first time such as minimum support of words

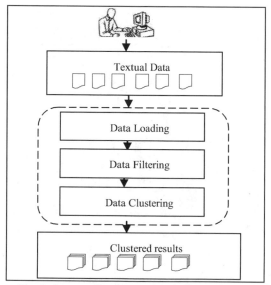

Fig. 1. Stages of incremental and dynamic repository

Minimum support of words used for disjoint data clusters to prevent the overlapping between them due to the representation of first cluster in hierarchical view. Hierarchical representation for textual document is preferable compare to partitonal representation[3]. Thus, textual document can be viewed in the form of topic and subtopic structure. Therefore, the user can obtain useful information when retrieving the textual document into meaningful clusters. Such data clusters can be used in many text domains such as text summarization, topic detecting and tracking, personal information management and managing extracted information.

Fig.1 demonstrates the three main stages of incremental and dynamic repository. The data clustering is performed automatically and incrementally in the database schema of RDBMS. Thus, the user does not need to another application for performing data clustering. In contrast, the traditional methods of clustering that execute in batch mode that gather all textual data before perform the clustering process. However, such methods are labor intensive and time consuming. Fig. 2 (a and b) show the comparison between   traditional methods and incremental and dynamic technique.

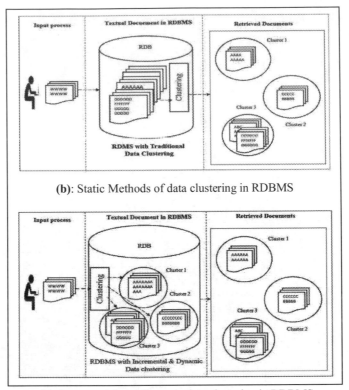

(b): Static Methods of data clustering in RDBMS

(b): Incremental and dynamic data clustering in RDBMS

Fig. 2. Clustering Process

Fig.2 demonstrates the method of data clustering in RDBSMS. Both methods consist of three stages. First, the user enters textual documents to RDBMS in input process. In the second stage, in static method the textual data is directly stored into database in the dynamic method, the data clustering is performed automatically based on the extracted information as discussed in section 3.1. In third stage, the traditional method is performed after whole texts are collected. However, in the proposed techniques the data cluster retrieved automatically without the need to do the clustering process. Thus, the time consumed is reduced. In addition, the searching process can be minimized by creating text index of the extracted keywords.

The primary benefits of the proposed technique are as follows. (1) The relationship among unstructured data within massive relational databases is determined. The discovered relationship is not only based on separate words, but also on semantic relation, which can be extracted from the WordNet database. (2) The user can obtain useful information and knowledge from the extracted information representing the textual documents. These extracted information known as the shortcut for textual documents. (3) Retrieval precision is improved because data are already clustered based on meaning. Furthermore, the proposed technique does not need user intervention except only at the first time of implementing the software package.

The software package is simple to be execute by a normal user and it can be included by database administer at the design or later. The proposed techniques provide structured query language (SQL) operator. The SQL operator can be used to retrieve the data clusters based on the semantic meaning of words as will be discussed in next section.

3.2   Semantic Operator in SQL

The semantic operator, which can be used in SQL query, is introduced to increase precision and recall. These factors are important in achieving good quality results. By using the semantic operator, the required word in a user query is obtained. The synonyms of the word are searched in the WordNet database[17]. In this manner, a list of words (required word + its synonyms) used to retrieve textual data cluster are represented by keywords are already extracted and stored in the incremental repository. These keywords act as the shortcut to retrieve specific textual document for example Table 2, presents table named "Articles," in database that contains four columns, namely "id," "articles," , "extracted_info" and "data_cluster". The "id" column holds the identifier name of the articles or their serial number. The "articles" column contains actual textual data. The "extracted_info" column contains data extracted from the original file. The "data_cluster" column holds cluster identifier. For example, a user needs to select data from the "Articles" table by issuing a query on extracted information under the word "perform." However, the user uses a different keyword, that is, "execute" (typed in normal case) instead of "perform," therefore, the retrieved result is null. In the proposed technique, the semantic operator will obtain the synonyms of "perform" from the WordNet database, and then, it will search for these synonyms. In this case, the results will show existing database records and the textual document will be retrieved.   The SQL format is as follows:

SELECT * FROM articles where SEMANTIC ('perform');

Table 2. Example of Semantic Query (Articles database table)

Id	Article	Extracted_Info	Data_Cluster
1	Microsoft announces about... Bill Gates.... in conference .. USA... perform	Microsoft, USA, Bill Gates, perform	C1
2	Oracle introduce.... IBM.....in united kingdom.... introduce	Oracle, united kingdom, IBM, introduce,	C2
3	Bill Gates said that...... evaluate	Bill Gates, evaluate, achieve	C1

The SEMANTIC operator can be used in the procedural structured query language which is provided in any relational databases. By using such language function or procedure can develop with passing parameter from the user. The user parameter is actual words that need to find its relevant in the relational database .The parameter can be presented not only in one word but it can be in many forms of words. In addition, the operator can be used to perform semantic integration between the unstructured information and keywords from user or applications. In this manner, the user can obtain a full picture about the content of textual documents. Fig. 3 shows the semantic query processes which begin with the user query with include the "SEMANTIC" operator in SQL command.

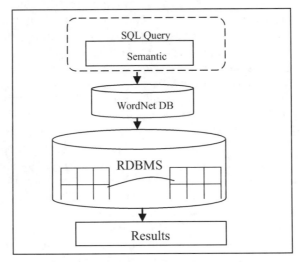

Fig. 3. Semantic Query Structure

The words are entered in semantic operator and sent to WordNet in order to obtain its synonyms. The synonyms used to look for the words that stored in the extracted columns. Thus, the query retrieves the required data in data clusters or only the ordinary requested data.

4   Conclusion

In this paper, an incremental and dynamic repository for managing unstructured data in relational databases is introduced. The system architecture of the proposed technique is described. This repository is created by using information extraction and textual document clustering techniques. Information extraction is conducted incrementally, whereas clustering is conducted dynamically. This method of retrieving data clusters is extremely fast as compared to traditional methods, in which an entire collection of textual documents (corpus) is clustered at one time. Such methods are time consuming. Furthermore, traditional methods are performed on top of a database using other applications with extra scripting and programming. Each time the user requests to see a certain data cluster, the entire clustering process needs to be repeated by extract the same collection of textual documents. In contrast, the proposed technique is embedded within the schema of relational databases. This technique presents the relationship among textual documents based on semantics. Moreover, it improves query efficiency when retrieving data clusters of textual documents. It proposed is useful for text domain researchers and developers, such as those who involve in on-line news services.

Acknowledgment

The authors wish to thank Universiti Teknologi MARA (UiTM) for the financial support. This work was supported in part by a grant number 600-RMI-/DANA 5/3/RIF (498/2012).

References

1. Doan, A., et al., Information extraction challenges in managing unstructured data. SIGMOD Record, 2008. Vol. 37, No. 4.
2. Doan, A., et al., The case for a structured approach to managing unstructured data. arXiv preprint arXiv:0909.1783, 2009.
3. Li, Y., S.M. Chung, and J.D. Holt, Text document clustering based on frequent word meaning sequences. Data & Knowledge Engineering, 2008. 64.1: p. 381-404.
4. Blumberg, R. and S. Atre, The problem with unstructured data. DM REVIEW, 2003. 13: p. 42-49.
5. Chu, E., et al., A relational approach to incrementally extracting and querying structure in unstructured data. Proceedings of the 33rd international conference on Very large databases, 2007. VLDB Endowment.
6. Tari, L., et al., Parse Tree Database for Information Extraction. IEEE TRANSACTIONS ON KNOWLEDGE and DATA ENGINEERING, 2010.
7. Mansuri, I.R. and Sarawagi, Integrating unstructured data into relational databases. Data Engineering, ICDE'06. Proceedings of the 22nd International Conference on. IEEE, 2006.
8. Roy, P., et al., Towards Automatic Association of Relevant Unstructured Content with Structured Query Results. Proceedings of the 14th ACM international conference on Information and knowledge management. ACM, 2005.
9. Roy, P. and M. Mohania, SCORE: symbiotic context oriented information retrieval. Advances in Data and Web Management. Springer Berlin Heidelberg, 2007: p. 30-38.
10. Jain, A., A. Doan, and L. Gravano, Optimizing SQL Queries over Text Databases. Data Engineering, . ICDE . IEEE 24th International Conference on. IEEE, 2008.
11. Kandogan, E., et al., Avatar Semantic Search: A Database Approach to Information Retrieval. SIGMOD , Chicago, Illinois,USA, 2006: p. 790-792.
12. Agrawal, S., S. Chaudhuri, and G. Das, DBXplorer: A System for Keyword-Based Search over Relational Databases. Data Engineering. Proceedings. 18th International Conference on. IEEE, 2002.
13. Hristidis, V. and Y. Papakonstantinou, Discover: Keyword search in relational databases. Proceedings of the 28th international conference on Very Large Data Bases. VLDB Endowment, 2002.
14. Li, G., et al., EASE: An Effective 3-in-1 Keyword Search Method for Unstructured, Semi-structured and Structured Data. Proceedings of the ACM SIGMOD international conference on Management of data, 2008.

15. Luo, Y., W. Wang, and X. Lin, SPARK: A Keyword Search Engine on Relational Databases. Data Engineering. ICDE. IEEE 24th International Conference on. IEEE, 2008.

16. YafoozA, W.M.S., S.Z. Abidin, and N. Omar, Towards automatic column-based data object clustering for multilingual databases. Control System, Computing and Engineering (ICCSCE), IEEE International Conference on. IEEE, 2011.

17. Miller, G., WordNet: A Lexical Database for English. Communications of the ACM 1995. 38.11: p. 39-41.

18. Sarawagi, S., Information Extraction. Foundations and Trends in Databases, 2008. Vol. 1, No. 3 (2007): p. 261–377.

19. Koc, M.L. and C. R´e, Incrementally Maintaining Classification using an RDBMS. Proceedings of the VLDB Endowment, 2011. Vol. 4, No. 5.

20. Fischer, U., et al., Towards Integrated Data Analytics: Time Series Forecasting in DBMS. Datenbank Spektrum 2013. 13.

21. Cafarella, M.J., et al., Structured querying of Web text. 3rd Biennial Conference on Innovative Data Systems Research (CIDR), Asilomar, California, USA, 2007.

22. Cafarella, M.J., Extracting and Querying a Comprehensive Web Database. Proc. of the 4 th Biennial Conference on Innovative Data Systems Research, Asilomar, CA, USA., 2009.

23. Jain, A., A. Doan, and L. Gravano, SQL Queries Over Unstructured Text Databases. Data Engineering. ICDE, IEEE 23rd International Conference on. IEEE, 2007.

24. Text, O., 11g Oracle Text Technical White Paper. 2007.

25. Text, O., an oracle technical white paper. 2005.

26. Jain, A.K., N. Murty, and P.J. Flynn, Data Clustering: A Review. ACM computing surveys (CSUR), 1999. 31.3: p. 264-323.

27. Su, C., et al., Text Clustering Approach Based on Maximal Frequent Term Sets. Proceedings of the IEEE International Conference on Systems, Man, and Cybernetics San Antonio, TX, USA, 2009.

28. Vishal Gupta, G.S.L., A Survey of Text Mining Techniques and Applications. JOURNAL OF EMERGING TECHNOLOGIES IN WEB INTELLIGENCE, 2009. VOL. 1, NO. 1.

29. Steinberger, R., et al., RC-NAMES: A Freely Available, Highly Multilingual Named Entity Resource. In RANLP 2011: p. pp. 104-110.

30. YafoozB, W.M.S., S.Z. Abidin, and N. Omar, Challenges and issues on online news management. Control System, Computing and Engineering (ICCSCE),IEEE International Conference on., 2011.

31. Fung, B.C.M., K. Wangy, and M. Ester, Hierarchical Document Clustering Using Frequent Itemsets. Proceedings of the SIAM international conference on data mining, 2003. 30. No. 5.

Sampling Semantic Data Stream: Resolving Overload and Limited Storage Issues

Naman Jain[1], Manuel Pozo[2], Raja Chiky[2], and Zakia Kazi-Aoul[2]

[1] VIT University, Vellore, TN 632014, India
namanjain2009@vit.ac.in
[2] ISEP - LISITE, Paris 75006, France
{manuel-jesus.pozo-ocana, raja.chiky, zakia.kazi}@isep.fr

Abstract. The Semantic Web technologies are being increasingly used for exploiting relations between data. In addition, new tendencies of real-time systems, such as social networks, sensors, cameras or weather information, are continuously generating data. This implies that data and links between them are becoming extremely vast.

Such huge quantity of data needs to be analyzed, processed, as well as stored if necessary. In this paper, we propose sampling operators that allow us to drop RDF Triples from the incoming data. Thereby, helping us to reduce the load on existing engines like CQELS, C-SPARQL, which are able to deal with big and linked data. Hence, the processing efforts, time as well as required storage space will be reduced remarkably.

We have proposed *Uniform Random Sampling*, *Reservoir Sampling* and *Chain Sampling* operators which may be implemented depending on the application.

Keywords: Big Data, Linked data-stream, Processing time, Sampling

1 Introduction

The semantic web handles many systems, such us Twitter, Facebook or Google, which generate increasing volumes of semantic data everyday. The problem of "too much (streaming) data but not enough (tools to gain and derive) knowledge" was tackled by [7]. They envisioned a Semantic Sensor Web (SSW), in which sensor data are annotated with semantic metadata to increase interoperability and provide contextual information essential for situational knowledge. CQELS[6], SPARKWAVE[5], C-SPARQL[3] etc. are existing technologies to exploit these semantic and streaming (continuous and infinite) data, and are based on recommended standard RDF, as the format of representation.

CQELS[6] is a native approach in an RDF environment based on 'white-boxes'. It provides its own processing model and its own operators to deal with streams, for example, window operators or query semantic operators. C-SPARQL[3] on the other hand, uses a 'black-box' approach which delegates the processing to other engines such as stream/event processing engines and SPARQL query processors by translating to their provided languages.

T. Herawan et al. (eds.), *Proceedings of the First International Conference on Advanced Data and Information Engineering (DaEng-2013)*, Lecture Notes in Electrical Engineering 285, DOI: 10.1007/978-981-4585-18-7_5,
© Springer Science+Business Media Singapore 2014

Although almost all the engines are based on the SPARQL Language, there are only a few systems which are able to process big quantity of data on the fly. Moreover, these engines do not feature any tool that would allow them to reduce the processing efforts and improve the processing time. For many applications, we must obtain compact summaries of the stream. These summaries could allow accurate answering of queries with estimates, which approximate the true answers over the original stream [4].

Thus, we propose the implementation of such sampling operators that could be used in conjunction with other existing real-time engines. These sampling operators will allow us to deal with the requested population by applying heuristic methods. *Uniform Random Sampling*, *Reservoir Sampling* and *Chain Sampling* were implemented on the data streams and were then compared with the error percentage, storage requirements and the load suffered by the engine. Thus, these sampling methods will help reduce processing time and the required memory space.

2   Extension to Existing Systems

We propose to extend existing semantic data stream querying engines by creating an external abstraction of the sampling operator. The extension acts as follows:

First, we recognize the different operators of the language and split the initial query according to them. This will allow us to use the same operator at different levels of the query: in the input, for sampling static data or streaming data and in the output to sample the result of the query, or both simultaneously.

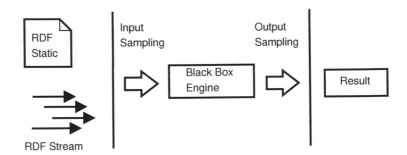

Fig. 1. Block Diagram

After splitting the query, we make syntax correction. If we identify any errors in our operator, we stop the execution of the query. Hence, if everything is correct, we create a key-value map where we store all the resources and all the requested sampling actions. Thus, we can apply the sampling operator independently by creating a thread for each of them.

We start the abstraction of the operator by taking out all the sampling instances in the query. This will allow us to avoid the correction of the rest of operators and leave the task to the specific engine itself.

When we send this query to the engine, as shown in Figure 1, it will run all the threads in charge of each source. By adapting those threads, we apply the sampling method just before the data comes to the engine, thus discharging the processing tasks. In a similar way, we adapt the output of the engines, in order to apply sampling methods at the output and lighten storage space.

3 Sampling Methods

The query would contain the information of sampling type and the sampling percentage for each unique stream. We apply the appropriate sampling method for each stream in different threads. We used *Uniform Random Sampling, Reservoir Sampling* and *Chain Sampling* to compare their advantages and disadvantages. We may choose an appropriate sampling method depending on the application used for.

We have implemented our sampling operators with CQELS[6] and C-SPARQL [3] engine. Example below shows simple type of sampling at 50% i.e. taking one triple and dropping the next one, using CQELS as:

```
PREFIX lv: <http://deri.org/floorplan/>
SELECT ?person ?locName
FROM NAMED <C:/floorplan.rdf>
WHERE {
STREAM <C:/rfid.stream> [NOW] [SAMPLING %50]
{?person lv:detectedAt ?loc }
GRAPH <C:/floorplan.rdf> {?loc lv:name ?locName } }
[OUTSAMPLING %80]
```

We implemented *Output Sampling* using
Operator: [**OUTSAMPLING %** {Sampling Percentage}]

As shown in Figure 1, the output of the engine goes through this operator and is given as final result only if the operator permits. Normal sampling algorithm has been used for this i.e. if %80 is specified, then only 4 out of 5 results are termed as final results. If *Output Sampling* operator is not specified, then the percentage is assumed to be 100 and no result is left out after the processing is done by the engine.

The sampling operators can also be implemented on similar systems like Sparkwave[5], ETALIS[1] after changing few configurations. In this paper, we use CQELS engine for the experimentation and the query is about computing average of *8-hour daily maximum ozone concentrations* as measured by *Clean Air Status and Trends Network* (CASTNET) in United States of America. The RDF data contains 2,159,133 triples consisting of 308,380 number of entries.[3]

---

[3] http://data-gov.tw.rpi.edu/wiki/Dataset_8/

3.1   Uniform Random Sampling

In this sampling method, incoming triples are sent to the engine, only if uniform random generation of true/false, with a probability equal to sampling percentage divided by 100, returns true. This helps to keep number of samples in proportion to total incoming data.

Operator: [**UNISAMPLING %**{Sampling Percentage}]

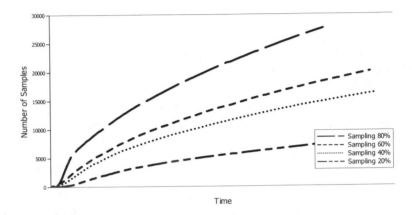

Fig. 2. Sample size variation/ Number of triples processed by engine in Uniform Random Sampling (During time span of 1 minute)

The query[4] is of the form

```
WHERE {
STREAM <C:/data-8.stream> [NOW][UNISAMPLING %60]
{?location vocab:ozone_8hr_daily_max ?value} }
```

Figure 2 shows sample size variation and number of triples processed by the engine. They both are same in this type of sampling as there is no removal of elements from the sample. This sampling method is less preferred as the sample size keeps on growing, which may lead to shortage of storage space, and even the outdated data element will be existing.

3.2   Reservoir Sampling

The problem of maintaining a sample of specified size 'k' is overcome by *Reservoir Sampling*[8]. In this, we add the triple 'i' to the sample with probability[k/i] and discarding a randomly chosen element from the reservoir (the sample) to make room for the new element.

Operator: [**RESSAMPLING** {Reservoir Size}]

---

[4] PREFIX & SELECT operators are omitted here to avoid repetition

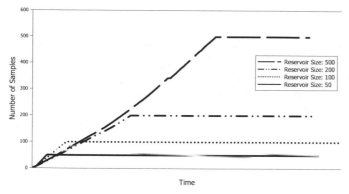

(a) Sample Size variation after using Reservoir Sampling

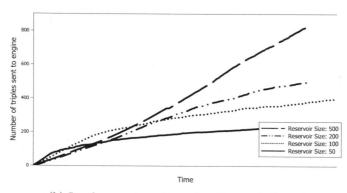

(b) Load on engine after using Reservoir Sampling

Fig. 3. Plot of sample size variation and load on engine after using Reservoir Sampling (All plots are for the same time duration of 1 minute)

For instance, here reservoir size of 500 ensures that the number of samples at any point may not exceed 500. Initially, triples are added to the reservoir until it is full. Then incoming triples, with stream number 'i' are added with probability 500/i, after replacing a randomly chosen element from the reservoir. The query becomes:

```
WHERE {
STREAM <C:/data-8.stream> [NOW][RESSAMPLING 500]
{?location vocab:ozone_8hr_daily_max ?value} }
```

Figure 3a shows that the sample size becomes constant after the reservoir is full and Figure 3b shows number of triples processed by the engine over the time. This sampling method helps us to limit the maximum storage space of the samples, but is not favorable when recent data is more important, as the sample data may have 'expired'. A significant reduction in load on the engine can also be noted in Figure 3b when compared to Figure 2.

3.3   Chain Sampling

Babcock et al. [2] proposed the *Chain Sampling* method for sequence based windows. In this method, if the window size is 'n' we add each new element 'i' in the sample with probability min(i,n)/n. As each element is added to the sample we choose an index, in the range (i+1,i+n), of the element that replaces it when it expires. Once the element with that index arrives, we store it and choose the index which replaces it when it expires, thus forming a chain of replacements. STEP operator is used to specify the size of steps that the sequence based window takes.

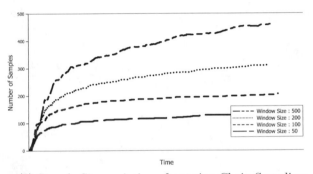

(a) Sample Size variation after using Chain Sampling

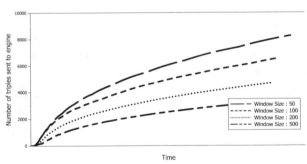

(b) Load on engine after using Chain Sampling

Fig. 4. Plot of sample size variation and load on engine after using Chain Sampling (All plots are for the same time duration of 1 minute)

Operator: [**CHNSAMPLING**{WindowSize} **STEP**{StepSize}]
and the query is of the form

```
WHERE {
STREAM <C:/data-8.stream> [CHNSAMPLING 500 STEP 2]
{?location vocab:ozone_8hr_daily_max ?value} }
```

The above query shows that window size 'n' is 500 and window moves in steps of 2. Figure 4a shows that the sample size increases as the window size is increased but Figure 4b shows that number of triples being processed by the engine is lower for higher window sizes. We also note from Figure 3b and Figure 4b that load on the engine is much higher for *Chain Sampling*.

Chain Sampling ensures to keep sample elements which are present only in the window and eliminates the problem of 'expired element' in the sample. But, the act of storing replacements does not reduce the memory space requirements and load on the engine as effectively as *Uniform Random Sampling* and *Reservoir Sampling*

Table 1. Variation of Sample size, Load on engine and Error due to different sampling methods

Method	Para-meter	Sample Size	Triples Processed	Effective Sampling %	% Triples Processed	% of Error
Uniform	20	10034	10034	20.06	20.06	0.1819
Random	40	19993	19993	39.98	39.98	0.1540
Sampling	60	29998	29998	59.99	59.99	0.0793
[Sampling %]	80	40002	40002	80.00	80.00	0.0494
	500	500	2807	1.00	5.61	1.3532
Reservoir	1000	1000	4925	2.00	9.85	0.8041
Sampling	2000	2000	8445	4.00	16.89	0.6047
[Reservoir Size]	5000	5000	16490	10.00	32.98	0.3123
	10000	10229	42589	20.45	85.17	0.4410
Chain	20000	13183	36332	26.36	72.66	0.5774
Sampling	30000	14857	31260	29.71	62.52	2.5181
[Window Size]	40000	16116	27298	32.23	54.59	0.6559

4 Experimentation

The queries in Section 3 were first executed without any sampling, and then after incorporating different sampling operators. After computing 50,000 entries of *8-hour daily maximum ozone concentrations*, without any sampling operator, their average was found to be **49.9161**. Also, the average of results obtained by 20 iterations (after applying each sampling operator) is demonstrated in the Table 1. All results are for 50,000 entries. The % of error was also computed for all sampling operators with a specific parameter as shown in Table 1. It was done by finding average of relative difference, of experimental value from actual value (49.9161, in this case), for all 20 iterations of the same query.

We may note that in *Uniform Random Sampling* probability of error reduces as the sampling percentage is increased while load on the engine is in coalition with the sampling percentage specified. In *Reservoir Sampling*, we see that the probability of error reduces with increase in reservoir size and a significant reduction in sample size as well as load on the engine is observed.

In *Chain Sampling* percentage of error depends on the elements in the current window and an improvement in sample size is observed in comparison to *Uniform Random Sampling* but load on the engine is comparatively high as sample replacements are also required to be processed by the engine.

5   Conclusion

The growing generated data from web applications is becoming a problem for the processing systems, and the relation between data is causing troubles when attempting to exploit data repositories. Therefore, In this paper we have proposed an extension of a real-time request system that allow us to reduce processing tasks and memory space requirements.

Different sampling methods suggested are useful for different applications. For example, if accuracy is required then we may go for *Uniform Random Sampling* with appropriate sampling percentage, and if memory size is fixed, we should go for *Reservoir Sampling. Chain Sampling* with suitable window size is favourable only when the samples need to be 'new'. In near future, we will build the sample according to the importance of incoming data elements, rather than randomly choosing them.

References

1. Anicic, D., Rudolph, S., Fodor, P., Stojanovic, N.: Stream reasoning and complex event processing in etalis. Semantic Web, 3(4): 397–407 (2012)
2. Babcock, B., Datar, M., Motwani, R.: Sampling from a moving window over streaming data. In Proceedings of the thirteenth annual ACM-SIAM symposium on Discrete algorithms, pp. 633–634. Society for Industrial and Applied Mathematics (2002)
3. Barbieri, D.F., Braga, D., Ceri, S., Della Valle, E., Grossniklaus, M.: C-sparql: Sparql for continuous querying. In: Proceedings of the 18th international conference on World wide web, pp. 1061–1062. ACM (2009)
4. Cohen, E., Cormode, G., Duffield, N.: Structure-aware sampling on data streams. In: Proceedings of the ACM SIGMETRICS joint international conference on Measurement and modeling of computer systems, pp. 197–208. ACM (2011)
5. Komazec, S., Cerri, D., Fensel, D.: Sparkwave: continuous schema-enhanced pattern matching over rdf data streams. In Proceedings of the 6th ACM International Conference on Distributed Event-Based Systems, pp. 58–68. ACM (2012)
6. Le-Phuoc, D., Dao-Tran, M., Parreira, J. X., Hauswirth, M.: A native and adaptive approach for unified processing of linked streams and linked data. In: The Semantic Web–ISWC 2011, pp. 370–388. Springer (2011)
7. Sheth, A., Henson, C., Sahoo S. S.: Semantic sensor web. Internet Computing 12(4), pp. 78–83. IEEE (2008)
8. Vitter, J. S.: Random sampling with a reservoir. ACM Transactions on Mathematical Software (TOMS), 11(1):37–57 (1985)

Shared-Table for Textual Data Clustering in Distributed Relational Databases

Wael M.S. Yafooz , Siti Z.Z. Abidin , Nasiroh Omar *and* Rosenah A. Halim

Faculty of Computer and Mathematical Sciences, UiTM Shah Alam, Selagor, Malaysia

Waelmohammed1@hotmail.com, {zaleha,nasiroh,rosenah}
@tmsk.uitm.edu.my

Abstract. High-performance query processing is a significant requirement of database administrators that can be achieved by grouping data into continuous hard disk pages. Such performance can be achieved by using database partitioning techniques. Database partitioning techniques aid in splitting of the physical structure of database tables into small partitions. A distributed database management system is advantageous for many businesses because such a system aids in the achievement of high-performance processing. However, massive amount of data distributed over network nodes affect query processing when retrieving data from different nodes. This study proposes a novel technique based on a shared-table in a relational database under a distributed environment to achieve high-performance query processing by using data mining techniques. A shared-table is used as a guide to show where the data should be saved. Thus, the efficiency of query processing will improve when data is saved at the same location. The proposed method is suitable for news agencies and domains that rely on massive amount of textual data.

Keywords- database clustering, relational database, distributed environment

1   Introduction

A relational database management system (RDBMS) is the backbone of numerous businesses for its robust data structure in managing, organizing, and retrieving data. The digital work of any firm can be processed in a same place based on a centralized database or at different areas based on a distributed database. The workload or "database transaction" in a centralized database is based on one machine. By contrast, a distributed database consists of more than one machine, thus improving system performance because the workload is divided [1], which is the concern of this study. However, the increasing amount of massive data in different network nodes will affect the query processing efficiency. The efficiency of database query can be better when data resides in same node thus the transportation cost in the network is also reduce [2]. Transportation cost, which is the time it takes to retrieve data from different nodes, is an important factor that must be considered in distributed databases. Therefore, database partitioning, clustering and fragmentation is used exchangeable,

T. Herawan et al. (eds.), *Proceedings of the First International Conference on Advanced Data and Information Engineering (DaEng-2013)*, Lecture Notes in Electrical Engineering 285, DOI: 10.1007/978-981-4585-18-7_6,
© Springer Science+Business Media Singapore 2014

which is employed to reduce the process of searching and retrieving for the required data by grouping data into continuous hard disk pages. In this way, the number of disk access is reduced, and the number of pages transfers from the secondary to primary storage, is minimized [3].

Database partitioning techniques aid in the splitting of the physical structure of database tables into small partitions based on the collected workload of database transactions. Thereafter, such workload is clustered based on the most accessed attributes or on the most similar records that may be accessed together. The physical structure of table is then divided in a process called splitting. Numerous attempts at database partitioning have been made. These attempts can be categorized into vertical [1, 4-8], horizontal [2, 9, 10], and mixed partitioning [11, 12].

Vertical partitioning is a method that groups the most frequent attributes (columns), which are accessed together into same network node. In vertical partitioning, response time will be reduced when searching for the required data because transportation cost decreases and most frequent attributes are stored in same network node or in a close network node. Users often do not need all the data on attributes; they only need specific tuples (records). Therefore, horizontal partitioning groups database records based on predicate or statistical analysis of the workload. Mix partitioning, which employs both vertical and horizontal clustering, is a good choice because records hold more data that are irrelevant to the requested information (more attributes).

All the aforementioned techniques can be static or dynamic. For static techniques, the data position never changes, depending on the base design. However, queries for any business change over time. All database partitioning methods employ meta-based statistical analysis (database workload information). Many studies focus on vertical clustering because it is more difficult to implement than horizontal clustering. However, all these techniques lack content similarity among data, which can be advantageous for many online news agencies or relational databases that store a large amount of textual data such as news articles, conference papers, and libraries.

This study proposes a novel technique based on shared-table. Shared-table holds frequent terms and named entities that can be found in any textual data that require storage in a relational database. In addition, shared-table conation columns hold the network node that stores such data. Thus, utilizing the shared-table conation column can be a guide for the next textual document to be stored if the documents bear any similarity with textual data in the network nodes. The similar textual data will then be stored in the same location (network node). Thus, the query will retrieve such data when requested from one network node or close network. Therefore, the efficiency of query processing will improve because data is clustered in same location or in a closed location.

The remainder of this paper is organized as follows: Section 2 presents some related works on database clustering. Section 3 explains the distributed database environment architecture. Section 4 discusses the idea of shared-table. Section 5 shows a comparison between existing methods and shared-table. Finally, Section 6 presents the conclusion of this paper.

2 Related Studies

This section presents several studies on database clustering techniques. These studies can be categorized into three: vertical, horizontal, and mix database clustering.

Vertical clustering approaches can be categorized into four: affinity-based [1, 5, 13-15], graph-based [15, 16], genetic algorithm-based [8, 12, 17], and transaction-based [3]. Most of these approaches are implemented in centralized [14, 15] or distributed environments [1, 2, 7, 11, 16, 18].The idea of vertical (attributes) clustering in an affinity-based approach begins with a constructed matrix called Attribute Usage Matrix (AUM) based on the Bond Energy Algorithm (BEA) [19]. BEA rearranges rows (attributes) and columns (transactions) into a two-dimensional matrix called AUM. Thus, the relationship between rows and columns can be informative. Most frequently accessed attributes are grouped together in one diagonal block. However, a clustering process can be evaluated based on user intervention to determine the similarity between attributes and the occurrence of overlapping [20]. AUM is extended to determine the degree of similarity between attributes based on the developed Affinity Attribute Matrix (AAM). However, AAM can only measure the similarity between two attributes in a column or in a row. Therefore, Navathe et al., [14] extended the work in [13] by proposing Navathe Vertical Clustering (NVP). NVP consists of two phases. The first phase involves the construction of the AAM, which is used as an input to perform clustering in the Clustered Affinity Matrix (CAM). Subsequently, CAM is used as an input in the second phase. The second phase performs iterative binary partitioning for existing clusters. However, time complexity is high because of the number of iterations in binary partitioning. In addition, overlapping between clusters is allowed. Therefore, S. B. Navathe & Ra [15] converts AAM into a graph-based approach to reduce time complexity. However, this method can only be applied in a small or centralized database.

Affinity-based methods lack the objective function [21]. Therefore, Muthuraj and Chakravarthy [20, 21] proposed an objective function called Partition Evaluator to assess clusters in a vertical database partitioning algorithm. In addition, the affinity concept only measures similarity between two attributes. Therefore, a transaction-based approach has been introduced by [3], where an efficient vertical clustering algorithm called Optimal Binary Partition Algorithm (OBP) was proposed. OBP clusters attributes based on a set of transactions. Important attributes often appear together in transactions. However, a large number of transactions in OBP affect the time execution of a query. Thus, time complexity increases. Therefore, Song and Gorla [8] proposed a genetic algorithm for database partitioning in a distributed database to generate more attribute partitions. However, genetic algorithms (GAs) are time-consuming. Several attempts have been made to achieve vertical partitioning.

All the aforementioned methods collect workloads, conduct statistical analyses, and partition tables based on the results of analysis. However, Abuelyaman [1] introduced "StatPart" based on a study of the environment and future tasks as a basis for attribute clustering at the initial stage of database design. Dynamic Vertical Partitioning Partitions (DYVEP) data based on the automatic monitoring of query frequencies introduced [7]. However, the communication cost within a distributed database is not considered. Therefore, Li and Gruenwald [4] proposed AutoClust, AutoClust as extended of [22] a previous version, AutoClust utilize a query optimizer to measure performance based on estimated cost in distributed database which lack in DYVEP.

Current studies focus on partitioning multimedia databases. Thus, Rodriguez & Li [6] introduced the Multimedia Adaptable Vertical Partitioning (MAVP).The MAVP static and database administrator shave to provide the information that MAVP needs to establish a vertical partitioning scheme. By contrast, [23] proved that Dynamic Multimedia ON line Distribution (DYMOND) docs not need a database administrator because it automatically obtains all input information.

Predicate or affinity-based methods are the two main approaches in horizontal (record) clustering. In the predicate approach, [10] terms that are homogeneously accessible from different applications are mined and grouped together. However, this method suffer requires the completion of the set of terms to perform record partitioning. Therefore, Ozsu & Valduriez [18] introduced efficient iterative algorithms called COMMIN. COMMIN is used to identify a minimal and complete predicate min term set from a predicate set of transaction queries. However, the min term predicate approach can hardly produce a complete predicate set.   The affinity-based method, which is used in vertical partitioning, was this introduced for horizontal clustering. [9] used two matrixes: Usage Predicate Matrix (UPM) and Predicate Affinity Matrix (PAM). Another research work utilizing PAM as input to GAs was introduced by [24], where a horizontal fragmentation was considered a traveling salesman problem. Khan & Hoque [2] recently introduced a new method for horizontal fragmentation and allocation in a distributed database. The method focused on the locality of the data recorded in a table called Attribute Locality Table precedence. This approach is implemented at the initial stage of distributed database system design by collecting information from a Create, Read, Update, and Delete matrix [25]. Mixed partitioning is the process of implementing vertical partitioning first, followed by horizontal partitioning, or vice versa. S. Navathe [11] Proposed mix database partitioning for a distributed database system. The idea is to construct grid representations based on data stored in attributes and records. The intersection between attributes and records is presented in grid form. Another approach utilizes GAs for mix partitioning in a relational database [12, 17].

3   Distributed Database Architecture

A distributed environment consists of numerous connected machines. The machines can either be in the same or different geographical areas [1]. The workload in a distributed environment is divided among all machines to enhance system performance. Thus, all machines run a parallel mechanism.

Fragmentation and allocation are two main elements of distributed database management systems. Fragmentation is a method by which to divide the physical structure of database tables into sub-tables. Figure 1 shows the environment structure of a distributed database.

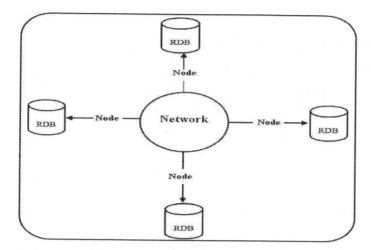

Fig. 1. Structure of the distributed database.

Allocation is a method by which to transfer sub-tables to one site or more over the network. In this way, if the user makes a query, the requested data will be retrieved from one network node. Thus, data retrieval improves when data are retrieved from more than one network node. Therefore, the transportation cost of data from the network node to the user is less than that of retrieving data from different nodes, which increases transportation cost.

4 Shared-Table

The proposed technique is designed for a relational distributed database. The technique, which deals with textual data, is called the Shared-Table Technique (STT). STT can be implemented in all network nodes that have a terminal for input data. STT will keep the database scheme as it is without partitioning. STT consists of a table and two main procedures: Mining Term and Allocation. The table referred to is a database table consisting of four columns: ID, Extracted_Data, Target_Node, and Time-stamp_ID as shown in Table 1. The ID column provides the identifier number, whereas the Extracted_Data column provides the extracted information comprising named entities [26-28] and frequent terms [29, 30]. The node address is stored in the Target_Node column, whereas the time of transaction is stored in the Time-stamp_ID column . Thus, when the user enters textual data directly, the mining term procedure runs.

The mining term procedure is the process of extracting the named entities and frequent terms, which are then stored in the Extracted_Data column. Such information will serve as a guide on where the subsequent textual data should be stored (in which node). Thus, the storing process matches the mining terms of the new textual data in the existing shared-table. Figure 2 shows a general view of the SST.

Table 1. Structure and simple information of STT

ID	Extracted_Data	Target_Node	Timestamp_ID
01	Malaysia,Najeeb, Ahmed, cairo, performance, electricity,	N1	12/5/20013 01:20
02	TNG,Malaysia, Najeeb, electricity	N1	12/5/20013 02:20
03	John,UK, Microsoft,Oracle	N4	12/5/20013 02:25
04	IBM,Oracle, USA	N4	15/5/20013 21:20
05	Football, Japan and China, team, Player	N5	16/5/20013 15:20

Table 1 demonstrates the structure of the STT and simple information about textual data and networks nodes that hold such data. In ID column, the identifier number
of textual document is "1". Identifier "1" hold information of specific textual data
such as the most frequent terms and named entities of textual documents which
stored in network node "N1", In the next record, identifier "2" , its hold information
about textual data which are similar to textual documents that hold identifier number
"1" . Thus, both textual data stored in network node "1". Each network node holds
textual documents that have similarity between its content. Such content is
represented by the named entities and most frequent terms of each textual document.
This method can be used as a guide on where data should be stored for easier retrieval. Therefore, retrieving process such data can be retrieved very fast rather than access
another network nodes over distributed environment. Furthermore, more other manipulation for such data can be done very efficiently. For instance, deleting data that
belong to specific action or specific month (s), it can be remove, backup or update
without any additional transportation cost over such environment. In case the data are
fully or partially similar based on content between two network nodes, the re-
allocation procedure will transform the small-sized data from network node to another
network node that has large-sized data

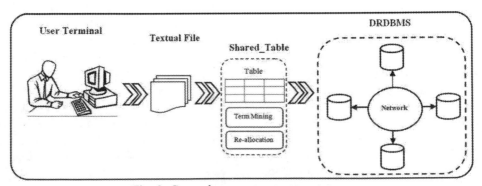

Fig. 2. General view of shared-table technique.

The re-allocation will consider and calculate the transportation cost to identify close network nodes that also have small data that are similar to that of the current node. This process is performed because the data that have to be transferred are often similar to that of another network node. In this case, the data that have to be transferred may be integrated with a far network node, thus affecting the transportation cost when retrieving data from these nodes. Therefore, the re-allocation procedure facilitates calculation and provides a better solution for the re-allocation of data between network nodes. In each database transaction, all tables will be replicated among all mining terms as new data are entered by the user.

Table 2. Transportation time between network nodes

Network Node	Node1	Node2	Node3	Node4
Node1	0	2	1	3
Node2	2	0	2	1
Node3	1	2	0	2
Node4	3	1	2	0

The example of query processing improvement is shown in Table. The measurement of transportation time between all network nodes is in seconds. If data is requested in the user query where query is issued from node 1 and data is allocated in different network nodes such as node number 2 and 3, *a transportation time* will occur in this case. There is also a *processing time* where it is a time taken to retrieve data from the secondary storage. Thus, the total *consuming time* (TC) to retrieve such data is the summation of transportation and processing time. TC= TC (Node1 to Node2) +TC (Node1 to Node3) + PT. Suppose that the *processing time* is 1 second, therefore, TC=2 +1+1=4. However, for data cluster that resides in the same network node, the total consuming time (TC) is equal 1, which is only the processing time.

5   Conclusion

The SST, which can reduce transportation cost over distributed relational databases, was introduced in this study. Query efficiency was improved by reducing the transport cost in a distributed database management system. Enhances efficiency can be achieved by grouping data that are likely to be accessed and retrieved together, specifically textual data that often have similar content. The SST serves as a guide on where data should be stored for easier retrieval. By utilizing the data mining concept, terms along with the location in which they are stored (in which network node) are contained in the shared-table.

Acknowledgment

The authors wish to thank Universiti Teknologi MARA (UiTM) for the financial support. This work was supported in part by a grant number 600-RMI-/DANA 5/3/RIF (498/2012).

References

1. Abuelyaman, E.S., An Optimized Scheme for Vertical Partitioning of a Distributed Database. IJCSNS International Journal of Computer Science and Network Security, 2008. VOL.8 No.1: p. 310-316.
2. Khan, S.I. and D.A.S.M.L. Hoque, A New Technique for Database Fragmentation in Distributed Systems. International Journal of Computer Applications, 2010. Volume 5– No.9: p. 0975 – 8887.
3. Chu, W.W. and I.T. Ieong, A Transaction-Based Approach to Vertical Partitioning for Relational Database Systems. Software Engineering, IEEE Transactions on, 1993. VOL. 19, NO. 8.
4. Li, L. and L. Gruenwald, Autonomous Database Partitioning using Data Mining on Single Computers and Cluster Computers. Proceedings of the 16th International Database Engineering & Applications Sysmposium. ACM, 2012.
5. Ma, H., K.-D. Schewe, and M. Kirchberg, A Heuristic Approach to Vertical Fragmentation Incorporating Query Information. Databases and Information Systems, 2006. 7th International Baltic Conference on. IEEE: p. 69-76.
6. Rodriguez, L. and X. Li, A vertical partitioning algorithm for distributed multimedia databases. . In e. a. A Hameurlain, editor, Proceedings of DEXA, . Springer Verlag, 2011. Vol 6861 (544—558).
7. RodríguezA, L. and X. Li, A dynamic vertical partitioning approach for distributed database system. Systems, Man, and Cybernetics (SMC), IEEE International Conference on. IEEE, 2011.
8. Song, S. and N. Gorla, A genetic Algorithm for Vertical Fragmentation and Access Path Selection. The Computer Journal, 2000. vol. 45, no. 1: p. 81-93.
9. Zhang, Y., On horizontal fragmentation of distributed database design. in M. Orlowska & M. Papazoglou, eds, Advances in Database Re- search, 1993. World Scientific Publishing: p. 121-130.
10. Ceri, S., M. Negri, and G. Pelagatti, Horizontal data partitioning in database design. in Proc. ACM SIGMOD, 1982.
11. S. Navathe, K.K., Minyoung Ra, Amixed fragmentation methodology for initial distributed database design. Journal of Computer and Software Engineering 1995. 3.4 (1995): p. 395- 426.
12. Gorla, N., V. Ng, and D.M. Law, Improving database performance with a mixed fragmentation design. J Intell Inf Syst (2012) 39, 2012. 39: p. 559–576.
13. Hoffer, H.A. and D.G. Severance, The Use of Cluster Analysis in Physical Database Design. Proceedings First Internutionul Conference on Vety Large Data Bases, 1975.
14. Navathe, S., et al., Vertical partitioning algorithms for database design. ACM Transactions on Database Systems (TODS) 9.4, 1984: p. 680-710.
15. Navathe, S.B. and M. Ra, Vertical Partitioning for Database Design: A Graphical Algorithm. ACM SIGMOD Record 18.2, 1989.
16. Ra, M., Horizontal partitioning for distributed database design. In Advances in Database Research, World Scientific Publishing, 1993: p. 101–120.

17. Ng, V., et al., Applying genetic algorithms in database partitioning. SAC '03 Proceedings of the ACM symposium on Applied computing, 2003: p. 544-549.
18. Ozsu, M.T. and P. Valduriez, Principles of Distributed Database Systems. 2nd ed., New Jersey: Prentice-Hall, 1999.
19. McCormick, W.T., P.J. Schweitzer, and T.W. White, Problem decomposition and data reorganization by a clustering technique. 1972. Operations Research 20.5: p. 993-1009.
20. Chakravarthy, S., et al., An objective function for vertically partitioning relations in distributed databases and its analysis. Distributed and parallel databases 2.2 1994. 183-207.
21. Muthuraj, J., et al., A formal approach to the vertical partitioning problem in distributed database design. Parallel and Distributed Information Systems, Proceedings of the Second International Conference on. IEEE, 1993.
22. Guinepain, S. and L. Gruenwald, Using Cluster Computing to Support Automatic and Dynamic Database Clustering. Cluster Computing, 2008 IEEE International Conference on. IEEE, 2008.
23. Rodríguez, L., et al., DYMOND: An Active System for Dynamic Vertical Partitioning of Multimedia Databases. Proceedings of the 16th International Database Engineering & Applications Sysmposium. ACM, 2012., 2012.
24. Cheng, C.-H., W.-K. Lee, and K.-F. Wong, A Genetic Algorithm-Based Clustering Approach for Database Partitioning. Systems, Man, and Cybernetics, Part C: Applications and Reviews, IEEE Transactions on 2002. VOL. 32, NO. 3: p. 215-230.
25. Surmsuk, P. and S. Thanawastien, The Integrated Strategic Information System Planning Methodology. 11th IEEE International Enterprise Distributed Object Computing Conference, 2007.
26. Montalvo, S., F. Víctor, and M. Raquel, NESM: a Named Entity based Proximity Measure for Multilingual News Clustering. Procesamiento de Lenguaje Natural, 2012. 48: p. 81-88.
27. Cao, T.H., T.M. Tang, and C.K. Chau, Data Mining: Foundations and Intelligent Paradigms Springer Berlin Heidelberg, 2012: p. 267-287.
28. YafoozB, W.M.S., S.Z. Abidin, and N. Omar, Challenges and issues on online news management. Control System, Computing and Engineering (ICCSCE),IEEE International Conference on., 2011.
29. Krishna, S.M. and S.D. Bhavani, An Efficient Approach for Text Clustering Based on Frequent Itemsets. European Journal of Scientific Research, 2010. ISSN 1450-216X Vol.42 No.3: p. 399-410.
30. Beil, F., M. Ester, and X. Xu, Frequent Term-Based Text Clustering. Proceedings of the eighth ACM SIGKDD international conference on Knowledge discovery and data mining. ACM, 2002.

Part II
Data Warehousing and Mining

A Graph-based Reliable User Classification

Bayar Tsolmon, Kyung-Soon Lee*

Division of Computer Science and Engineering, CAIIT, Chonbuk National University,
567 Baekje-daero, Deokjin-gu, Jeonju-si, Jeollabuk-do, 561-756 Republic of Korea
bayar_277@yahoo.com, selfsolee@chonbuk.ac.kr

Abstract. When some hot social issue or event occurs, it will significantly increase the number of comments and retweet on that day on Twitter. However, as the amount of SNS data increases, the noise also increases synchronously, thus a reliable user classification method is being required. In this paper, we classify the users who are interested in the issue as "socially well-known user" and "reliable and highly active user". "A graph-based user reliability measurement" and "Weekly user activity measurement" are introduced to classify users who are interested in the issue. Eight of social issues were experimented in Twitter data to verify validity of the proposed method. The top 10 results of the experiment showed 76.8% of performance in average precision (P@10). The experimental results show that the proposed method is effective for classifying users in Twitter corpus.

Keywords: Graph-based user metric, User classification, Timeline analysis

1   Introduction

With the rapidly increasing amount of data on the internet lately, the research on social user classification attracts more and more attention. Compared with the news and blog data, the Social Network Service (SNS) data are more widely used in the real-time event extraction and recommendation system. However, as the amount of SNS data increases, the noise also increases synchronously, thus a reliable user classification method is being required.

Since the existing user classification methods [1,2,3] only depend on the statically behavior of the user, there was a problem that some important user who has fewer numbers of followers might be missed. Especially the frequency of a retweet of tweets that is irrelevant to the event such as rumor or advertisement is high in the SNS environment.

There are recent works for social user classification based on user behavior analysis and timeline analysis. Social media has become indispensable to users recently. A rich set of studies has been conducted in various forms of social media. T. Tinati et al. [4] developed a model based upon the Twitter message exchange which enables us to analyze conversations around specific topics and identify key players in a conversation. Kwak et al. [5] compared three different measures of influence-

---

* Corresponding author.

T. Herawan et al. (eds.), *Proceedings of the First International Conference on Advanced Data and Information Engineering (DaEng-2013)*, Lecture Notes in Electrical Engineering 285, DOI: 10.1007/978-981-4585-18-7_7,
© Springer Science+Business Media Singapore 2014

number of followers, page-rank, and the number of retweets-finding that the ranking of the most influential users differed depending on the measure.

In this paper, we propose a graph based reliable user classification based on timeline and social user behavior analysis, and a user activity measurement method to extract reliable and highly active users who are needed for recommendation system or event extraction. The proposed method classifies the Twitter users as socially well-known user, reliable and highly active user, normal user and low active user. *Reliable and highly active user* is defined as the "user who writes a lot" about the issue, and that is "user who is being re-tweeted for several times". When the writing about an issue is being mentioned and being retweeted by people, it's become a reliable user.

The rest of the paper is organized as follows: Section 2 presents user classification; Section 3 describes the proposed method of a graph-based reliable user extraction method. Section 4 shows our experimental results on a Korean tweet collection. We conclude the paper in Section 5.

2   User Classification

1) User classification using follower and following ratio. It could be seen that the characteristic that makes Twitter a social network service is following and follower. Measuring follower and following ratio (*FFRatio*) that shows how much does the user do Twitter activity by using a Twitter user`s number of following and follower.

$$FFRatio(p) = \frac{\# \, of Follower(p)}{\# \, of \, Following(p)} \tag{1}$$

where, *Follower(p)* represents the number of followers of user p and *Following(p)* is the number of following of the user p.

2) User classification using a retweet and tweet ratio. The retweet ratios tend to be most meaningful when they are used to compare users within the same issue. The retweet ratio (*RTRatio*) of Twitter user who mentioned about an issue is calculated as follows.

$$RTRatio(p) = \frac{Total \, RT(p)}{Total \, Tweet(p)} \tag{2}$$

where, *TotalTweet(p)* represents how many tweets have been posted by the user p about the issue. *TotalRT(p)* is the number of retweets for all the tweets posted by the user p. The formula shows the average retweet ratio when user p writes a tweet about an issue.

A graph based on *FFRatio* and *RTRatio* is as in the Figure 1. The left side of Figure 1 shows the distribution of followings and followers. Following and follower ratio is almost same. The higher the number of the follower can be considered as a socially well-known user. Based on the *FFRatio* value of each user, the Twitter users can be categorized into three groups as follows (Table 1).

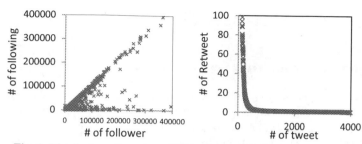

Fig. 1. Distribution of number of followings and followers and *RTRatio*

Table 1. User distribution based on the *FFRatio* value

User group	FFRatio	Description
CLB: Celebrity	Greater than 1.5	Follower >> Following
NRM: Normal	0.5 ~ 1.5	Follower ≈ Following
SPM: Spam	Less than 0.5	Follower << Following

The description of each group users is follows.
- CLB user: As a user who has many followers, classify as a celebrity
- NRM user: The case that number of following and follower is similar, classify as a normal
- SPM user: As a user who has more following than followers, classify as a spam or user with low activity

The right side of Figure 1 represents the retweet distribution (*RTRatio*). There is not much users with higher retweet frequency compared as entire users. From this figure it can be seen that users are divided by 4 big groups as follows (Table 2).

Table 2. Table of user distribution through the *RTRatio* value

User group	RTRatio	Description	
		Total RT	Total Tweet
A: Popular	Greater than 80	High	Low
B: Active	2.5 ~ 80	High	High
C: Normal	0.5 ~2.5	Low	High
D: Inactive	Less than 0.5	Low	Low

In this paper, we classify a Twitter user who is an important property for the event extraction based on *FFRatio* and *RTRatio* to four groups as follows.
User classification:
1) Socially well-known user (A-CLB & B-CLB)
2) Reliable and highly active user (A-NRM & B-NRM)
3) Normal user (C-NRM)
4) Low active user (D-SPM)
If the socially well-known users mention about the certain event, it indicates that a big social event happened. The reliable and highly active users are valuable users because they post important information every time an event occurs. The user who belongs to A-CLB (combination of group A and CLB) is the user who has a number of followers than following, and has 80 or more of retweet of the tweet wrote about an issue.

3   A graph-based reliable user extraction method

3.1   Extracting socially well-known users

It is hard to analyze user reliability based on the number of Twitter user's followers and tweets. However, highly active users tend to have a lot number of tweets and retweets. To extract socially well-known users, we adapted a HITS (Hyperlink-Induced Topic Search) [6] algorithm to extract "a user who is being retweeted several times" and "a user who is active in Twitter" by analyzing the social network among the Twitter users.

The directed graph is constructed as G = (V, E) on the each issue: V represents a user group and E represents a linkage group. The directed edge (p, q) $\in$ E is created when the tweet of a user p mentions a user q. Additionally applying mention, RT, Retweet value with edge weight between nodes to the existing HITS algorithm is as the same as in figure 2.

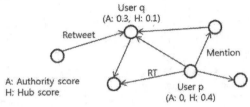

Fig. 2. Example of the weighted HITS graph illustration: Edge creation for Twitter users

Each user has an authority score and a hub score in HITS algorithm [6]. When out-link from the high authority node, it becomes higher hub, and when it is in-linked from high hub nodes, it becomes high authority. The formula that calculated by giving additional an edge weight on the original HITS algorithm to classify users by analyzing network among Twitter users in this paper is as follows.

$$HubScore^{(0)}(p) = FFRatio(p) \tag{3}$$

$$AuthScore^{(0)}(p) = RTRatio(p) \tag{4}$$

The weighted Hub score and Authority score are calculated as formula 5, 6.

$$HubScore^{(T+1)}(p) = \sum_{p \to q} w_{pq} \times AuthScore^{T}(q) \tag{5}$$

$$AuthScore^{(T+1)}(p) = \sum_{q \to p} w_{qp} \times HubScore^{T}(q) \tag{6}$$

The edge weight w_{qp} is as follows:

$$w_{qp} = \sum_{q \to p} FreqRT(q, p) + \sum_{q \to p} Mention(q, p) \tag{7}$$

The effectiveness of the HITS algorithm depends on the initial value and edge weight. The top ranked 100 users with high *AuthScores* are selected as socially well-known users.

3.2   Extracting reliable and highly active users

In the HITS algorithm, the generally reliable users in society are extracted without considering the timeline of each user. Since there is a user who writes a lot of tweets about the issue and actively write whenever the event related to the issue occurs, these users regularly have relatively higher activity than other users. In figure 3, user 2 and user 3 write tweets about the issue every week. User 1 and user 4 do Twitter activity on the issue only in a certain period.

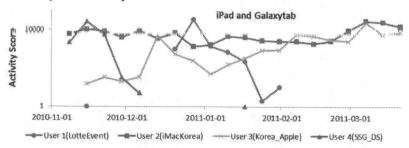

Fig. 3. A Graph of a user who has a high activity score

For cases like user 2 and user 3, the following formula calculates the average weekly activity score with user's weekly tweets and frequency of the retweet for measuring "user who writes often" as an average value.

$$Activity\ Score(u) = \frac{1}{w} \sum_{i=1}^{W} TweetFreq(u, d_i) \times RTFreq(u, d_i) \qquad (8)$$

where w shows the number of weeks; $TweetFreq$ shows the sum of tweets d that a user u wrote in the each ith week; $RTFreq$ represents the number of retweets d of the tweets written by a user u in the each ith week.

4   Experiments and Evaluation

We have evaluated the effectiveness of the proposed method on tweet collection. Eight issues are chosen and tweet documents for the issues are collected, spanning from November 1, 2010 to March 26, 2011 by Twitter API (all issues and tweets are written in Korean). Table 3 shows the number of users who wrote tweets on each issue.

Table 3. Twitter user set

Category	Issue	# of user
Product	Canon & Nikon	21,369
	iPad & Galaxy tab	115,022
People	Park Ji-Sung	29,568
	Kim Yu-Na	10,563
Company	Apple	71,878
	Samsung	108,800
Natural Disaster	Earthquake	110,345
Terrorism	Chonanham	19,473

The methods of comparative experiment to extract reliable user is as in the following. In this paper, the method based on Twitter user's *FFRatio* and *RTRatio* is used as a baseline for the extraction of reliable users.

Table 4. Comparative Methods & User classification

User classification	Baseline	Proposed method
Socially well-known user	A-CLB & B-CLB	HITS AuthScore: High
Reliable and active user	A-NRM & B-NRM	Activity Score
Normal user	C-NRM	HITS HubScore: High
Low active user	D-SPM	HITS Auth & Hub Score: Low

The reliable user extraction method using *FFRatio* and *RTRatio* in the static behavior analysis of Twitter user and using dynamic behavior analysis on the issue are proposed. The result of calculation of precision (P@10) for top 10 extracted reliable users from each method is as follows.

Table 5. Comparative experiment for reliable user extraction P@10

User classification	Socially well-known user		Reliable and active user		Average	
Issue	A-CLB B-CLB	Auth Score	B-NRM C-NRM	Activity Score	Baseline	Proposed method
1. Canon & Nikon	**0.5**	0.4	0.2	**0.4**	0.35	0.40
2. iPad & Galaxy tab	0.7	**0.8**	**0.9**	0.8	**0.80**	**0.80**
3. Park Ji-Sung	**0.7**	0.6	0.6	**1.0**	0.65	**0.80**
4. Kim Yu-Na	0.5	**0.9**	0.6	**0.9**	0.55	**0.90**
5. Apple	0.5	0.5	0.5	**0.7**	0.50	0.60
6. Samsung	0.5	**1.0**	0.3	**0.9**	0.40	**0.95**
7. Earthquake	0.4	**0.7**	0.2	**0.8**	0.30	0.75
8. Chonanham	0.8	**0.9**	0.5	**1.0**	0.65	**0.95**
Average	0.57	**0.72**	0.47	**0.81**	0.52	**0.76**

In the result of experiment, the proposed method showed better performance than baseline. The method *AuthScore* and *Activity Score* achieved 72% and 81% respectively. Since an experiment carried out by targeting users who are interested in the issue, a user who is socially popular about each issue, a user who has a reliable and direct correlation with an event could be extracted.

The distribution of reliable users' tweet about the earthquake issue is shown in Figure 4.

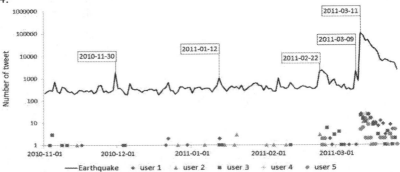

Fig. 4. The number of tweets containing the *"Earthquake"* issue word and distribution of reliable users.

Table 6. Top 5 reliable users about earthquake issue

User ID	User screen name	Description
User 1	KoreanRedCross	Official Twitter account of the Korean Red Cross
User 2	parknews9	KBS Nine O'clock News anchor
User 3	kbsnewstweet	Official Twitter account of KBS News
User 4	Russa	Blogger
User 5	mofatkr	Ministry of Foreign Affairs

In the Figure 4, the higher picks with date label represent the occurrence of earthquakes. From here, it can be seen that when the number of daily tweet frequency become higher, all reliable users wrote about an issue on the same day. Reliable and highly active top 5 users related to earthquake issue are shown in Table 6. In the case of user 1, tweets are not only about earthquake occurrence but also about donation and help. In other hand, when an earthquake occurs, user 5 posts tweet about guidelines for South Korean citizens who live in abroad. A user 2 and user 3 write the earthquake news.

5　Conclusion

In this paper, a Twitter user classification method through graph-based reliability measurement metrics and user activity metrics using timeline analysis were proposed in a network of users who are interested in the issue. Reliable user can be used for other use such as recommendation system, as well as event extraction. Eight of social issues were experimented in Twitter data to verify validity of the proposed method. The top 10 results of experiments achieved 76.8% of performance in precision (P@10). Discovering methods for better user behavior analysis and less dependent on the number of tweets are future works.

Acknowledgements. This research was supported by Basic Science Research Program through the National Research Foundation of Korea (NRF) funded by the Ministry of Education, Science and Technology (2012R1A1A2044811).

References

1. Yang, Z., Guo, J., Cai, K., Tang, J., Li, J., Zhang, L., Su, Z.: Understanding retweeting behaviors in social networks. In Proceedings of the 19th ACM international conference on Information and knowledge management, pp. 1633-1636, ACM (2010)
2. Boyd, D., Golder, S., & Lotan, G.: Tweet, tweet, retweet: Conversational aspects of retweeting on Twitter. In System Sciences (HICSS), pp. 1-10. IEEE (2010)
3. Mendoza, M., Poblete, B., Castillo, C.: Twitter Under Crisis: Can we trust what we RT?. In Proceedings of the first workshop on social media analytics, pp. 71-79, ACM (2010).
4. Tinati, R., Carr, L., Hall, W., Bentwood, J.: Identifying communicator roles in Twitter. In Proceedings of the 21st international conference companion on World Wide Web, pp. 1161-1168, ACM (2012)

5. Kwak, H., Lee, C., Park, H., Moon, S.: What is Twitter, a social network or a news media?. In Proceedings of the 19th international conference on World Wide Web, pp. 591-600, ACM. (2010)
6. Kleinberg J. M.: Authoritative Sources in a Hyperlinked Environment, Journal of the ACM, 46(5) pp. 604-632, (1999)

A Modified Artificial Bee Colony Optimization for Functional Link Neural Network Training

Yana Mazwin Mohmad Hassim and Rozaida Ghazali

Faculty of Computer Science and Information Technology
Universiti Tun Hussein Onn Malaysia (UTHM), Johor, Malaysia

{yana,rozaida}@uthm.edu.my

Abstract. Functional Link Neural Network (FLNN) has becoming as an important tool for solving non-linear classification problem. This is due to its modest architecture which required less tunable weights for learning as compared to the standard multilayer feed forward network. The most common learning scheme for tuning the weight in FLNN is a Backpropagation (BP-learning) algorithm. However, the learning method by BP-learning algorithm tends to easily get trapped in local minima which affect the performance of FLNN. This paper discussed the implementation of modified Artificial Bee Colony (mABC) as a learning scheme for training the FLNN network in overcoming the drawback of BP-learning scheme. The aim is to introduce an alternative learning scheme that can provide a better solution for training the FLNN network.

Keywords: Functional Link Neural Network, Modified Artificial Bee Colony, Learning scheme, Training.

1 Introduction

Artificial Neural Networks (ANNs) have been known to become a powerful tool applied to variety of real world tasks such as classification, prediction and clustering [1, 2]. One of the best known types of ANNs is the Multilayer Perceptron (MLP). The MLP structure consists of multiple layers of nodes which give the network the ability to solve problems that are not linearly separable. However, MLP usually requires a fairly large amount of available measures in order to achieve good classification ability. Difficulties in fixing appropriate number of neurons in layers and a challenging work on determining number of hidden layers has make the MLP architecture becomes not that easy to train. The increase number of hidden layers and neurons also make the MLP architecture become complex and resulting in slower operation. An alternative approach of avoiding this problem is by removing the hidden layers from the architecture which prompted to an alternative network architecture named Functional Link Neural Network (FLNN) [3]. The FLNN is a flat network (without hidden layers) where it reduced the neural architectural complexity while at the same time possesses the ability to solve non-linear separable problems.

T. Herawan et al. (eds.), *Proceedings of the First International Conference on Advanced Data and Information Engineering (DaEng-2013)*, Lecture Notes in Electrical Engineering 285, DOI: 10.1007/978-981-4585-18-7_8,
© Springer Science+Business Media Singapore 2014

FLNN network is usually trained by adjusting the weight of connection between neurons. The most common method for tuning the weight in FLNN is using a Backpropagation (BP) learning algorithm. However one of the crucial problems with the standard BP-learning algorithm is that it can easily get trapped in local minima especially for the non-linear problems [4], thus effect the performance of FLNN network. To overcome this, the modified Artificial Bee Colony (mABC) optimization algorithm is used to optimize the weights of FLNN instead of the BP-learning algorithm [5]. The original standard ABC was proposed by Karaboga [6] for solving numerical optimization problem. Several studies on the comparison of ABC with other well-known optimization algorithm particularly on the PSO, ACO, GA, and DE done by [6, 7] resolved that ABC algorithm turn out to be very simple, flexible and robust as compared to the existing population-based optimization algorithms in solving numerical optimization problem. In this study, we adopt the ABC algorithm with some modification in the employed bee's foraging phase to provide a thorough local search behavior instead of original random behavior to search the optimal weights set for the FLNN. Our experimental results showed that the FLNN performance trained with mABC learning scheme gives better accuracy results with less processing time.

2   Related Works

This section presents an overview on Functional Link Neural Network, related work on FLNN learning scheme and Standard Artificial Bee Colony Optimization Algorithm that will be utilized in this work.

2.1. An Overview of Functional Link Neural Network

Functional Link Neural Network (FLNN) is a class of Higher Order Neural Networks (HONNs) that utilize higher combination of its inputs created by Pao [3, 8]. The FLNN is much more modest than MLP as it has a single-layer of trainable weights whilst able to handle non-linear separable problems. The flat architecture of FLNN has also make the learning algorithm in the network less complicated [9]. In order to capture non-linear input-output mapping, the input vector of FLNN is extended with a suitable enhanced representation of the input nodes which artificially increase the dimension of input space [3, 8].

Fig. 1 and Fig. 2 show the network structure of MLP and FLNN both with 2 input nodes. The network structure of FLNN presented in Fig. 2 was enhanced up to 2nd order (the highest order) make it employed only 4 trainable parameters (3 weights + 1 bias) in it structure. As compared to Fig. 1, the MLP with the same number of input nodes (2 inputs) and with a single hidden layer of 2 nodes (the least numbers of hidden nodes and layers) formed 9 trainable parameters (6 weights + 3 biases) in it structure. Of both this network, FLNN need less trainable weights as compared to MLP for training and this make the learning scheme for FLNN less complicated.

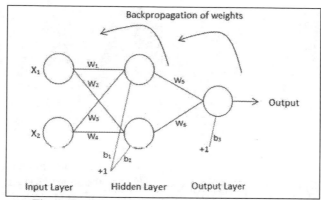

Fig. 1. Single layer MLP structure with 2 input nodes

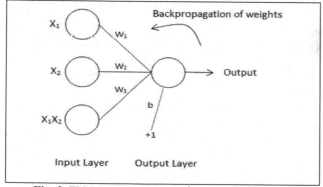

Fig. 2. FLNN structure up to 2nd order of 2 input nodes

Our focused in this work is on the FLNN with generic basis architecture which uses a tensor representation. Fig. 2 illustrates the FLNN structure of 2 inputs up to second order input enhancement. The first order consists of the original inputs which are x_1 and x_2. To provide the network with non-linear mapping capability the inputs is enhanced with additional higher order unit by extending them based on the product unit. The product unit of this network is x_1x_2 as depicted is Fig. 2 and is known as the 2nd order input enhancement for the network.

2.2. FLNN learning Scheme

Most previous learning algorithm used in the training of FLNN, is the BP-learning algorithm [4, 9-15]. As shows as in Fig. 2, the weight values between enhanced input nodes and output node are randomly initialized. The output node, $\hat{y}$ of FLNN would correspond to the input pattern x and the number of input patterns, n. For tensor representation with single output node, enhance input can be noted as n+n(n-1)/2. Let the enhanced input node of tensor x be represented as $x_t = \langle x_1, x_2, \dots x_n, x_1x_2, \dots x_{n-1}x_n \rangle$. Let f denotes the output node's activation function which in this work we applied a logistic sigmoid activation function:

$$f(s) = \frac{1}{1+\exp(-s)} \tag{1}$$

$$where \quad s = wx_t + b$$

The output value of the FLNN is obtained by:

$$\hat{y} = f(s) \tag{2}$$

where $\hat{y}$ is the output while f denotes the output node activation function and b is the bias. In Eq. (1), wx_t is the aggregate value which is the inner product of w and x_t. The square error E, between the target output and the actual output will be minimized as:

$$E = \frac{1}{2}\sum_{i=1}^{n}(y_i - \hat{y}_i)^2 \tag{3}$$

where y_i is the target output and $\hat{y}_i$ is the actual output of the ith input training pattern, while n is the number of training pattern. During the training phase, the BP-learning algorithm will continue to update w and b until the maximum epoch or the convergent condition is reached.

Although BP-learning is the mostly used algorithm in the training of FLNN, the algorithm however has several limitations which affect the performance of FLNN-BP. FLNN-BP tends to easily gets trapped in local minima especially for those non-linearly separable classification problems. This is an inherent problem that exists in the BP-learning algorithm. Employing BP-learning algorithm as learning scheme has made FLNN-BP model strictly depends on the shape of the error surface and since a common error surface may have many local minima and multimodal, this has typically makes the algorithm prone to stuck in some local minima when moving along the error surface during the training phase. In addition, FLNN-BP model also very dependent on the choices of initial values of the weights as well as the parameters in the algorithm such as the learning rate and momentum [4] which make it not very easy to meet the desired convergence criterion during the training. Therefore, further investigation to improve learning algorithm in FLNN are still desired.

2.3. Artificial Bee Colony Optimization

The Artificial Bee Colony (ABC) algorithm is an optimization tool, which simulates the intelligent foraging behavior of a honey bee swarm for solving multidimensional and multimodal optimization problem [6]. In this model, three groups of bees which are employed, onlooker and scout bees determined the objects of problems by sharing information to one another. The employed bee uses random multidirectional search space in the Food Source area (FS). They carry the profitability information (nectar quantity) of the FS and share this information with the onlookers. Onlooker bees evaluate the nectar quantity obtained by the employed and bees and choose FS depending on the probability value base on the fitness. If the nectar amount of FS is higher than that of the previous one in their memory, they memorize the new position and forget the previous one [6]. The employed bee whose food source has been abandoned becomes a scout and starts to search for finding a new food source randomly. The following is the standard ABC pseudo code:

1. Initialization population of scout bee with random solution $x_{i,j}$ $i=1,2...FS$
2. Evaluate fitness of the population
3. Cycle = 1:MCN
4. form new population $v_{i,j}$ for the employed bees using:

$$v_{i,j} = x_{i,j} + \Psi_{i,j}(x_{i,j} - x_{k,j})$$ (4)

   where k is a random selected solution in the neighborhood of i, Φ is a random number in the range [-1,1] while j is a random selected dimension vector in i and evaluate them
5. Apply greedy selection between $v_{i,j}$ and $x_{i,j}$
6. Calculate the probability values p_i for the solutions x_i using:

$$p_i = \frac{fit_i}{\sum_{i=1}^{FS} fit_i} \quad where \quad fit_i = \begin{cases} \frac{1}{1+f_i}, & if\ f_i > 0 \\ 1 + abs(f_i), & if\ f_i < 0 \end{cases}$$ (5)

7. Produce the new solutions u_i for the onlookers from the solutions x_i selected depending on p_i and evaluate them
8. Apply the greedy selection process for onlookers
9. Determine the abandoned solution for the scout, if exists, and replace it with a new randomly produced solution x_i using:

$$x_i^j = x_{min}^j + rand(0,1)(x_{max}^j - x_{min}^j)$$ (6)

10. Memorize the best solution
11. cycle=cycle+1
12. Stop when cycle = Maximum cycle number (MCN).

3 Proposed Learning Scheme

Inspired by the robustness and flexibility offered by The Artificial Bee Colony optimization algorithm, we adopted the ABC optimization algorithm as the learning scheme with some modification on employed bee's foraging phase to overcome the disadvantages caused by gradient descent BP-learning in the FLNN training. The training scheme flowchart is presented in Fig. 3. In the initial process, the FLNN architecture (weight and bias) is transformed into objective function along with the training dataset. This objective function then is fed to the modified ABC (mABC) algorithm in order to search for the optimal weight parameters. The weight changes are then tuned by the mABC algorithm based on the error calculation (difference between actual and expected results). The optimal weights set obtained from the training phase are then fed to the FLNN objective function for performance evaluation upon an unseen data (test set) in the testing phase.

In this study, a modified Artificial Bee Colony (mABC) is introduced as a learning scheme for training the FLNN. The modification is done on the part of employed bees' foraging phase so that they would exploit all weights and biases in the FLNN architecture in order to improve the network ability on searching the optimal weights set. In standard ABC algorithm, the position of a food source (FS) represents a possible solution to the optimization problem, and the nectar amount of a food source corresponds to the profitability (fitness) of the associated solution. In the case of training the FLNN with ABC, the weight, w and bias, b of the network are treated as optimization parameters to the optimization problem (finding minimum Error, E) as presented in Eq. (3).

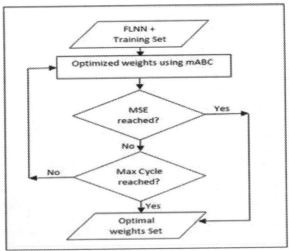

Fig. 3. The proposed Learning scheme

The FLNN optimization parameters are represented as D-dimensional vector for the solution $x_{i,j}$ where $(i = 1, 2, ..., FS)$ and $(j = 1, 2, ..., D)$ and each vector i is exploited by only one employed bee. In order to produce a candidate food source $v_{i,j}$ from the old one $x_{i,j}$ in memory, the ABC uses Eq. (4) where $k \in \{1,2, ..., FS\}$ and both k and j are a randomly chosen indexes. The food source of $x_{i,j}$ can be represented in a form of $X = FS \times D$ matrix.

$$X = \begin{bmatrix} x_{1,1} & \cdots & x_{1,D} \\ \vdots & \ddots & \vdots \\ x_{FS,1} & \cdots & x_{FS,D} \end{bmatrix} \qquad (7)$$

As can be seen from Eq. (4) and matrix representation from Eq. (7), for each row of FS only one element from D will be chosen randomly and exploited by the employed bee by using:

$$j = fix(rand * D) + 1; \qquad (8)$$

However in the case of FLNN mainly for classification tasks which are always deal with large number of optimization parameters (weights + bias), exploiting one element in each solution vector x_i will cause longer foraging cycle in finding the optimal solution [5]. Random selection of elements in each vector x_i during employed bee phase also leads to a poor ability for FLNN network in finding the optimal weights set which result to a low classification accuracy on unseen data [16]. To overcome this, we eliminate the random employed bee behavior in selecting the elements in vector dimension as in Eq. (8). In the other hand, we direct the employed bee to visit all elements in D to exploit them before evaluating the vector x_i. The modified ABC is performed as shown in pseudo code below, where the box indicates the improvement made to the standard ABC:

```
1) Cycle = 0
2) Initialize FLNN optimization parameters, D
3) Initialize population of scout bee with random solution
   xᵢ, i-1,2…F3
4) Evaluate fitness of the population
5) Cycle = 1:MCN
6) Form new population (vᵢ) for employed bees
   i.    select solution, k in the neighbourhood of i,
         randomly
   ii.   For j = 1:D
   iii.  Direct employed bee to exploit nectar value of j in
         population (vᵢ,ⱼ) using Eq. (4)
             where j = 1,2,…,D is a dimension vector in i
   iv.   j= j+1;
   v.    exit loop when j = D;
7) evaluate the new population (vᵢ)
8) Apply greedy selection between vᵢ and xᵢ
9) Calculate the probability values pᵢ for the solutions xᵢ
   using Eq. (4)
10) Produce the new solutions uᵢ for the onlookers from the
    solutions xᵢ selected depending on pᵢ and evaluate them
11) Apply the greedy selection process for onlookers
12) Determine the abandoned solution for the scout, if exists,
    and replace it with a new randomly produced solution xᵢ
    using Eq. (5)
13) Memorize the best solution
14) cycle=cycle+1
15) Stop when cycle = Maximum cycle number (MCN).
```

4 Experimentation

In order to evaluate the performance of the proposed learning scheme FLNN-ABC for classification problem, simulation experiments were carried out on a 2.30 GHz Core i5-2410M Intel CPU with 8.0 GB RAM in a 64-bit Operating System. The comparison of standard BP training and ABC algorithms are discussed based on the simulation results implemented in Matlab 2010b. In this work we have considered 3 benchmark problems, XOR, 3-bit Parity and 4-bit Encoder-decoder problem. During the experiment, simulations were performed on the training of MLP with standard BP-learning algorithm (MLP-BP), FLNN with BP-learning algorithm (FLNN-BP) and FLNN with modified ABC algorithm (FLNN-mABC). The best training accuracy for every benchmark problems were noted from these simulations. The activation function used for the network output for both MLP and FLNN is Logistic sigmoid function. Ten trials were performed on each simulations with the best network performance result is noted from these 10 trials. Table 1 below summarized the parameters considered in this simulation.

Table 1. Parameters considered for FLNN-BP, FLNN-ABC and FLNN-mABC simulation

Parameters	MLP-BP	FLNN-BP	FLNN-mABC
Learning rate	0.3	0.3	-
Momentum	0.7	0.7	-
Maximum epoch/cycle	10000	10000	10000
Minimum error	0.001	0.001	0.001

The reason of conduction a simulation on MLP and FLNN architectures is to provide a comparison on neural complexity. The neural complexity refers to the numbers of trainable parameters (weight and bias) needed to perform good approximation. The least numbers of parameters indicate that the networks is less complex as it required a small numbers of weight and bias to be updated at every epoch or cycle and thus led to a faster training time.

5 Result and Discussion

Ten trials were performed on each simulation of the MLP-BP, FLNN-BP and FLNN-mABC with the best training accuracy result is taken out from these 10 trials. Table 2 below, presents the simulation result of MLP-BP, FLNN-BP and FLNN-mABC architectures.

Table 2. The simulation result of MLP-BP, FLNN-BP and FLNN-mABC architectures

Data sets	Learning scheme	Network structure	Trainable weights	CPU time (s)	MSE	Accuracy (%)	Rank
XOR	MLP-BP	2-3-1	13	1.9344	0.0010	95.56	3
	FLNN-BP	2nd order	4	1.2979	0.0009	95.58	2
	FLNN-mABC	2nd order	4	0.2527	3.52E-06	99.93	1
3-bit	MLP-BP	3-3-1	16	3.4211	0.0010	95.63	2
	FLNN-BP	3rd order	8	3.4773	0.0016	94.35	3
	FLNN-mABC	3rd order	8	2.8660	0.0002	99.21	1
4-bit enc/ dec	MLP-BP	4-4-4	40	3.2916	0.0010	95.78	2
	FLNN-BP	2nd order	44	1.4508	0.0009	95.63	3
	FLNN-mABC	2nd order	44	0.8206	8.7E-05	99.85	1

From the simulations conducted, the best MLP structure for training the XOR dataset was 2-3-1 which employed 13 trainable weights and the best FLNN structure for training XOR dataset was 2nd order input enhancement which employed 4 trainable weights. Incorporating mABC as learning scheme for the FLNN network on training the XOR dataset has given better result in terms of MSE (3.52E-06), CPU time (0.2527s) and accuracy rate (99.93%) which outperformed both FLNN and MLP trained with standard BP-learning.

Simulation results on 3-bit Parity and 4-bit encoder/decoder datasets from table 2 also shows that our FLNN-mABC learning scheme had also provide better performance on both datasets in terms of MSE, CPU time and accuracy rate. It is also shown that, the implementation of mABC can be considered as another alternative training scheme that could provide better performance with less computational time and less neural complexity.

6 Conclusion

In this work, the experiment has demonstrated that FLNN-mABC performs the classification task quite well. The simulation results also shows that the proposed mABC Algorithm can successfully train the FLNN for solving non-linear problems with better accuracy rate and less processing time. This research work is carried out to introduce mABC as an alternative learning scheme for training the Functional Link Neural Network that can give a promising result. In future work, we will conduct an experiment on high dimensional classification benchmark data in order to explore and to proof the feasibility of the proposed training scheme.

Acknowledgement

The authors wish to thank the Ministry of Higher Education Malaysia and Universiti Tun Hussein Onn Malaysia for the scholarship given in conducting these research activities.

References

1. Zhang, G.P., Neural networks for classification: a survey. Systems, Man, and Cybernetics, Part C: Applications and Reviews, IEEE Transactions on, 2000. **30**(4): p. 451-462.

2. Liao, S.-H. and Wen, C.-H., Artificial neural networks classification and clustering of methodologies and applications – literature analysis from 1995 to 2005. Expert Systems with Applications, 2007. **32**(1): p. 1-11.

3. Pao, Y.H. and Takefuji, Y., Functional-link net computing: theory, system architecture, and functionalities. Computer, 1992. **25**(5): p. 76-79.

4. Dehuri, S. and Cho, S.-B., A comprehensive survey on functional link neural networks and an adaptive PSO–BP learning for CFLNN. Neural Computing & Applications 2010. **19**(2): p. 187-205.

5. Mohmad Hassim, Y.M. and Ghazali, R., Using Artificial Bee Colony to Improve Functional Link Neural Network Training. Applied Mechanics and Materials, 2013. **263**: p. 2102-2108.

6.   Karaboga, D., An Idea Based on Honey Bee Swarm for Numerical Optimization. 2005, Erciyes University, Engineering Faculty, Computer Science Department, Kayseri/Turkiye.

7.   Karaboga, D. and Basturk, B., On the performance of artificial bee colony (ABC) algorithm. Elsevier Applied Soft Computing, 2007. **8**: p. 687-697.

8.   Pao, Y.-H., Adaptive pattern recognition and neural networks. 1989: Addison-Wesley Longman Publishing Co., Inc. 309.

9.   Misra, B.B. and Dehuri, S., Functional Link Artificial Neural Network for Classification Task in Data Mining. Journal of Computer Science, 2007. **3**(12): p. 948-955.

10. Haring, S. and Kok, J. Finding functional links for neural networks by evolutionary computation. in In: Van de Merckt Tet al (eds) BENELEARN1995, proceedings of the fifth Belgian–Dutch conference on machine learning. 1995. Brussels, Belgium: pp 71–78.

11. Haring, S., Kok, J., and Van Wesel, M., Feature selection for neural networks through functional links found by evolutionary computation. In: ILiu X et al (eds) Adavnces in intelligent data analysis (IDA-97). LNCS 1280, 1997: p. 199–210.

12. Abu-Mahfouz, I.-A., A comparative study of three artificial neural networks for the detection and classification of gear faults International Journal of General Systems 2005. **34**(3): p. 261-277.

13. Sierra, A., Macias, J.A., and Corbacho, F., Evolution of functional link networks. Evolutionary Computation, IEEE Transactions on, 2001. **5**(1): p. 54-65.

14. Dehuri, S., Mishra, B.B., and Cho, S.-B., Genetic Feature Selection for Optimal Functional Link Artificial Neural Network in Classification, in Proceedings of the 9th International Conference on Intelligent Data Engineering and Automated Learning. 2008, Springer-Verlag: Daejeon, South Korea. p. 156-163.

15. Ghazali, R., Hussain, A.J., and Liatsis, P., Dynamic Ridge Polynomial Neural Network: Forecasting the univariate non-stationary and stationary trading signals. Expert Systems with Applications, 2011. **38**(4): p. 3765-3776.

16. Mohmad Hassim, Y.M. and Ghazali, R., Training a Functional Link Neural Network Using an Artificial Bee Colony for Solving a Classification Problems. Journal of Computing Press, NY, USA, 2012. **4**(9): p. 110-115.

A Review on Bloom Filter Based Approaches for RFID Data Cleaning

Hairulnizam Mahdin

FSKTM, Universiti Tun Hussein Onn Malaysia,
Batu Pahat, Johor, Malaysia
hairuln@uthm.edu.my

Abstract. This paper provides comprehensive review about data cleaning for RFID data streams. It serves the purpose to understand the current undertakings to ensure data quality in RFID. It focused on three major RFID data issues which are noise readings, duplicate readings and missed readings. It includes in-depth analysis on existing approaches specifying on Bloom filter based approaches. This literature can be used by researcher to understand the background of RFID data filtering, the challenges and expectation in the future.

Keywords: RFID, data filtering, noise reading, missed reading, duplicate reading, Bloom filter

1 Introduction

RFID identification works by reader reading ID on tag and send it to the middleware for processing. Before readings can be transformed into information, it needs to be filtered. The RFID data stream contains unreliable data such as noise reading and duplicates. Noise reads will produce incorrect information such as incorrect stock quantity. Duplicate readings need to be removes because it over-represents the data and does not contain new information for the system. It needs to be removed to avoid unnecessary processing being performed.

The RFID data stream is different from the common relational and data warehousing because of (i) large size of data (ii) simple tuple structure (iii) inaccuracy and (iv) temporal and spatial information that make it require new data management approach. To illustrate the RFID data size, consider small store that have 10,000 items. 10,000 readings generated. If the readings are repeated for every 10 minutes, in eight hours there will be already 480,000 tuples. This is why it needs really efficient data structure based approach to cater the large size of data. The filtering process must be carried out to ensure only correct readings are used to produce information. This paper presents and analyzes the existing filtering approaches which are based on Bloom filter.

T. Herawan et al. (eds.), *Proceedings of the First International Conference on Advanced Data and Information Engineering (DaEng-2013)*, Lecture Notes in Electrical Engineering 285, DOI: 10.1007/978-981-4585-18-7_9,
© Springer Science+Business Media Singapore 2014

1.1 RFID System Architecture

A typical RFID system consists of a transponder (i.e., tag), which is attached to the objects to be identified, an interrogator (i.e., reader) that creates an RF field for detecting radio waves, and a backend database system for maintaining expanded information on the objects and other associated objects. Figure 1 shows RFID-enabled a generic system of interest. Generally RFID system architecture is made of four layers: tags, readers, middleware and database and enterprise applications.

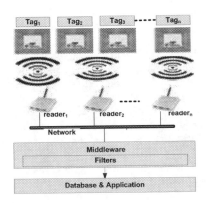

Fig. 1. An RFID-enabled system architecture

At the first layer, objects to be tracked and monitored are attached with an RFID tag. A tag contains memory to store the identifying information of the object to which it is attached and an antenna that communicates the information via radio waves. Tags can be classified based on their power sources: passive, active and semi-active tags. Active and semi-active tags have their own battery on-boards while passive tags not. Passive tag is the most commonly used tags in the market and has an indefinite operational life compare to other tags. Relative to both active and semi-active tags, passive tags are very cheap and they are widely used in very large quantities in many applications such as supply chain management.

Generally, passive tags are most used tags in many applications because of its cheap price than the others [1]. They were use massively mostly in the supply chain application. However, due to the low-powered hardware and the massive number of the tags, it raises many issues in data management and security. Most of the researches discussed in this thesis are based on the passive tags.

At the second layer, the RFID system is assumed to contain multiple networked RFID readers deployed to collaboratively collect data from tagged objects in their vicinity. The reading distance ranges from a few centimeters to more than 300 feet depending on factors such as interference from other RF devices [2]. The RFID readers query tags to obtain data and forward the resulting readings to the middleware. The middleware processes these data and then send it to the backend applications or database servers. The applications then respond to these events and orchestrate corresponding actions such as ordering additional products, sending theft alerts, raising alarms regarding harmful chemicals or replacing fragile components before failure.

1.2 RFID Reading Classes

The type of readings generate by reader could be generally classified into four classes: true positive, false positive, false negative and duplicate readings. Only true positive reading is acceptable in the RFID system. The other types of reading are the three major anomalies that need to be removed or smoothed from the data stream [3]. Each anomaly has different impact to the RFID system. The unreliable data are being listed as one of the major hindrance in achieving widespread adoption of RFID technology [4].

1.2.1   True positive

True positive is correct readings made by the reader. It returns the actual ID on the tag to the reader. The structure of tag ID according to the EPC Tag Data Standard consists of four parts: Header, EPC manager, Object Class and Serial Number. The example of tag ID is *01.0000E78.00019A.000198CFE*. Header identifies the length, type and structure of EPC, EPC Manager identifies the manufacturer, Object class identifies the type of product and Serial Number is the unique identifier for the item.

1.2.2 False positive

The second reading class in RFID readings is false positive, also known as noise reading. The reading returns tag ID that does not exist in the system. The corrupted reading can be caused by: (i) low power signal, (ii) signal interference, (iii) signal collision, and (iv) infrastructure obstacle [5]. Low power signals occur when the tag is located at the end of reader's vicinity or the reader trying to read too many tags at the same time [6]. The tag will not be having sufficient power to reply the signal back to the reader successfully. The radio signal also can be weaken by interference from metal and water [6].

The third cause of noise reading is the signal collision which is common in RFID. Signal collision occurs when two or more tags responding to the reader at the same time [7]. It also occurs when the tag is being read by more than one reader [8]. The signal collision can change the content of the signal which creates new ID that did not exist in the system. Another source of noise readings is the infrastructure obstacle such as the orientation of the objects and obstacles from the surrounding environment. When numbers of pallets are coming together, some of the tags can be buried deep in the pallet arrangement, which made them hard to be read. The amount of power coming to them is not enough for the tag to responds to the reader correctly, which results in noise readings.

The effect of false positive or noise readings is it reports an existence of an object that is not exists, causing the system to take wrong decision opposing the reality. For example a noise read at the security check-out in a shopping mall will trigger alarm indicating that a customer did not pay for an item. The noise read also can indicate inaccurate number of items available in the store which causing delays in ordering new stock. The noise readings need to be removed from the data stream to avoid such confusion in those examples. A pro-active step that can be taken to reduce the noise readings problem is to ensure that the right tag is used for the application. For example it is better to use high frequency (HF) tag to read object that is build up from metal compare to ultra high frequency (UHF) tag because HF tag have good performance with metal.

1.2.3 False negative

The third type of reading is the false negative or missed reading. False negative is a reading that supposedly performed by the reader but is being left from entering the data stream. The source of problem can be the same as noise reading. In this case the signal did not reach the reader at all to transmit the data. False negative also can occur due to the filtering process itself. Some of the filtering process put a threshold for a reading to be counted as correct reading within a specific time period. When the object only resides in the vicinity for a short time period, the number of readings made on it is less than the threshold. Therefore it is being removed from the data stream and left undetected. The effect of false negative to the system is opposed to false positive errors. While false positive add the quantity ups from the reality, the false negative reduces the real quantity. There problem with the incorrect quantity is like business loss because some of the items shipped to customer is not being detected. Simple ways to increase the chances of the tag being read is by increasing the number of read cycle or use more than one reader to read the same area [9].

1.2.4 Duplicate reading

The next reading class that considered as anomaly is duplicate reading. Duplicate reading is common problem in RFID. It is exists because RFID reader has the ability to read the same tag number of times as long the tag's is still the reader's vicinity. Duplicate reading need to be removed because it over represent the data, unnecessarily occupying the memory space and require unnecessary processing. We are only interested on the data that indicate the occurrence of events. For example on smart shelf application, the readings that are most important are when the item is put on the shelf and when it is being pickup by customer. That can indicate the current number of items that are currently available on the shelf. Another source of duplicate readings is because more than one reader is covering the same vicinity to increase the reliability of the readings. In some cases, the reader's vicinity was overlapped with each other that causing the redundancy.

2   Basic of RFID Data Filtering

Fig. 2 depict the basic pattern in RFID data stream that need to be filtered. It contains true positive, noise, missed and duplicates reading. At time 100 the reader read the tag correctly which it generate the ID 100E. However at time 200 it produced noise reading where it generate tag ID 300E, which is not exists in the reader interrogation area. At time 300 it misses to read the tag but at time 400 and 500 it read the tag correctly which create duplicate readings in the data stream.

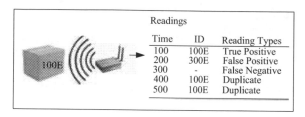

Readings		
Time	ID	Reading Types
100	100E	True Positive
200	300E	False Positive
300	-	False Negative
400	100E	Duplicate
500	100E	Duplicate

Fig. 2. Types of reading possibly generated when reader reads tag 100E

Basically, the occurrence of true positive is higher than false positive [9]. The false positive can be filtered by removing readings that have low readings. The problem now is how to set the threshold and how to perform the filtering efficiently.

The second anomaly, the false negative can be solved by adding more reading cycles, so that tag have more chances to be read. By filtering the duplicate reading too, the missed reading can be recovered. For example, it has gone missing at time 300, but at time 400 and 500 it is being read again. That's mean the tag exist in the reader interrogation area from time 100 to 500. Based on this, the problem of false negative has been solved. The problem now when we increase the reading cycle, there will be higher probability on false positive and duplicate readings occurrence. Our research will be focusing on filtering these anomalies which in the same time recover the missed readings.

3   Filtering Approaches

In this section we discuss on RFID data filtering especially on noise and duplicate reading based on Bloom filter.

3.1 Bloom-Filter based Approach

Bloom filter [10] can represent a data in its bit array of size m that is done through a number of hash functions. Each hash function that runs on the data will return a number which referring to the counter position in Bloom filter. The counter's value is turn to one from zero when it is being hashed. To test whether a data has been stored in the filter, the data need to be run through the same hash functions. If the value of all counters from the hashing process is positive, that's mean the data already have been inserted in the filter. If there is more than one counter that is zero, its mean that the data is new and have not been inserted in the filter. Bloom filter achieve efficiency by having constant operation in checking the existence of the data through the hashing process by allowing some false positive. The operation of Bloom filter is shown in Fig. 3. In Fig. 3(a) the word "Apple" is inserted into Bloom filter using 3 hash functions. In Fig. 3(b) the "Orange" is checked whether it has been inserted to the filter using the same number of hash functions. The first and third hash return counter that is zero which means "Orange" has never been inserted in the filter.

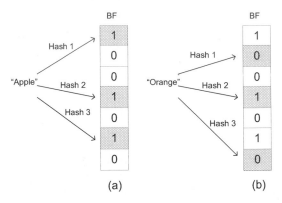

(a)   (b)

Fig. 3. (a) Insertion of "Apple" in the filter and (b) checking whether "Orange" already has been in the filter

Bloom filter has been used to filter duplicate data in its original and modified form such as in [11-13]. In [11], Bloom filter is used to filter duplicate data from being sent to the coordinator site. The readings that come to remote site will be inserted in the Bloom filter. Then the copy of the Bloom filter is sent to the coordinator sites. The coordinator site then updates its filter with the new reading and sent it to the other remote site. If the same reading being produced again at any of the remote sites, it will be ignored because it has been inserted in the filter. However, this approach is not feasible to be implemented because the coordinator needs to send the updates to each remote site whenever there is an update to its Bloom filter. To reduce this problem, they proposed the Lazy approach where the coordinator needs to send the update only to the sites that send new readings. By this the overhead processing can be reduced significantly. However this approach is still does not suitable for RFID application because RFID generate too many readings. Even if the reading cycle is set with bigger interval, when the reader performs readings, they can read all the objects in their vicinity repetitively in a short time. There will be a latency to update the coordinator and remote sites each times new reading coming in.

One weakness of Bloom filter is that it does not allow deletion. This is because a single counter in Bloom filter can be hashed number of times by different data. Turning the counter to zero will affect other data that is not involved in the deletion. To solve this problem, Counting Bloom filter (CBF) [14] introduces the use of integer array instead of bit array. It allows the counters in the filter to store 8 bits data rather than 1 bit as in Bloom filter. When a CBF counter is hashed, the counter value will be increase by 1. When deletion is made, the respective counter value will be decreased by 1. Fig. 4 shows the deletion process in CBF. Fig. 4(a) shows that word "Apple" is hashed and inserted into CBF at counter 0, 3 and 5. In Fig. 4(b) word "Orange" is hashed and inserted at counter 1, 3 and 6. Counter 3 which also has been hashed by

"Apple" is increase to 2 in CBF. At Fig. 4(c), the word "Apple" is deleted from CBF, where all the respective counters are decrease by 1.

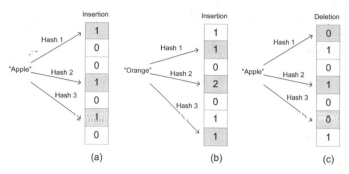

Fig. 4. (a) Insertion of "Apple" in CBF (b) Insertion of "Orange" in CBF (c) Deletion of "Apple" in CBF

From Fig. 4, we can see that to delete element from CBF, the element must be known, otherwise CBF does not have method to identify the original element in its filter. In RFID application, reading that is no longer coming in, can indicate that the object has leaved the reader vicinity. Therefore the reading can be deleted from memory. However, if we use CBF to filter RFID reading, it cannot remove the old readings by itself, because it cannot identify which counters represent this reading. In [13], they introduce Decaying Bloom filter (DBF) to solve this problem. DBF can delete old readings from its filter automatically by incorporate sliding windows movement in its filter.

The counters in DBF will start with the sliding windows size, and its value will be decreased automatically by one when new reading is coming. If the same reading occurs, all the respective counters will be set back to the sliding window size. If there is no new reading on the same tags, the reading will be evicted from the filter gracefully resulting from decrement process.

The purpose of Bloom filter is to achieve space-efficient with some allowable false positive. The false positive occurs when all the hashed counters is positive for an element that have not been inserted in the filter. At one level, all the hashed elements will be identified as false positive when the Bloom filter has become 'full'. It is when almost or all the counters have been used to represent the hashed elements. To avoid this, the size of Bloom filter must be big enough to cover the number of elements that going to be inserted in the filter. Old elements must be removed whenever is possible to reduce the probability of false positive.

4 Conclusion

RFID technology is the key to modern supply chain. It offers automation to object identification which surely minimizes businesses cost. However, due to its nature radio frequency can be distorted by external materials and thus create noise and missed readings. This faulty readings need to be filtered to ensured system provide correct information to the businesses. RFID also has too many duplicate data because of the repeated readings from its reading. In order to filter those readings, one of the methods that can be used is Bloom filter which consume very minimum space with efficient process. It can be used to filter noise and duplicate readings in large RFID data stream. However the challenge is the RFID readings are unpredictable thus it is hard to determine when to clear old readings in the filter. This can lead to the condition where the filter can become 'full' and produce false positive results. In future research we planned to overcome this problem to exploit the efficiency of Bloom filter in filtering massive data successfully.

Acknowledgments. This work is sponsored by Ministry of Education Malaysia and Universiti Tun Hussein Onn Malaysia.

References

1. Chawla, V., Ha, D.S.: An overview of passive RFID. IEEE Communications Magazine, vol. 45, pp.11-17 (2007) [1]
2. Clampitt, H.G.: RFID Certification Textbook. 3rd edition, American RFID Solution, LLC, (2007)
3. Darcy, P., Stantic, B., Sattar, A.: A fusion of data analysis and non-monotonic reasoning to restore missed RFID readings. In Proceedings of the 5th International Conference on Intelligent Sensors, Sensor Networks and Information Processing, pp.313-318, Melbourne, Victoria, (2009)
4. Nogee, A.: RFID tags and chips: Opportunities in the second generation. In-Stat/MDR Reports (2005)
5. Derakhshan, R., Orlowska, M., Li, X.: RFID data management: challenges and opportunities. In Proceedings of the IEEE International Conference on RFID, pp. 175-182, Grapevine, Texas (2007)
6. Jeffery, S. R., Alonso, G., Franklin, M.J., Hong, W., Widom, J.: A pipelined framework for online cleaning of sensor data streams. In Proceedings of the 22nd International Conference on Data Engineering, p. 140, Atlanta, Georgia, (2006)
7. Leong, K. S., Ng, M. L., Grasso, A. R., Cole, P. H.: Synchronization of RFID readers for dense RFID reader environments. In Proceedings of the International Symposium on Applications and the Internet Workshops, pp. 48-51, Phoenix, Arizona (2006)
8. Azambuja, M.C.d., Jung, C.F., Caten, C.S., Hessel, F.P.: RFID-Env: Methods and software simulation for RFID environments. Business Process Management Journal, 16, 1014 – 1038 (2010)
9. Bai, Y., Wang, F., Liu, P.: Efficiently filtering RFID data streams. In Proceedings of the CleanDB Workshop, pp. 50-57, Seoul, South Korea (2006)
10. Bloom, B.: Space/Time tradeoffs in hash coding with allowable errors. Commun. ACM, 13, 422–426 (1970)
11. Dddd
12. Wang, X., Zhang, Q., Jia, Y.: Efficiently filtering duplicates over distributed data streams. In Proceedings of the International Conference on Computer Science and Software Engineering, Wuhan, Hubei, 12-14 Dec. 2008; pp. 631 – 634.
13. Mahdin, H., Abawajy, J.: An Approach for Removing Redundant Data from RFID Data Streams. Sensors, 11, 9863-9877 (2011).
14. Shen H., Zhang Y.: Improved approximate detection of duplicates for data streams over sliding windows. Journal of Computer Science and Technology, 23, 973–987 (2008).
15. Fan L., Cao P., Almeida J., Broder A.Z.: Summary cache: A Scalable Wide-Area Web Cache Sharing Protocol. IEEE/ACM Trans. Networking, 8, 281-293 (2000)

An unsupervised, fast correlation-based filter for feature selection for data clustering

Part Pramokchon and Punpiti Piamsa-nga

Department of Computer Engineering, Faculty of Engineering, Kasetsart University, Jatujak, Bangkok, 10900, THAILAND
{g4785036, pp}@ku.ac.th

Abstract. Feature selection is an important method to provide both efficiency and effectiveness for high-dimension data clustering. However, most feature selection methods require prior knowledge such as class-label information to train the clustering module, where its performance depends on training data and types of learning machine. This paper presents a feature selection algorithm that does not require supervised feature assessment. We analyze relevance and redundancy among features and effectiveness to each target class to build a correlation-based filter. Compared to feature sets selected by existing methods, the experimental results show that performance of a feature set selected by the proposed method is comparably equal and better when it is tested on the RCV1v2 corpus and Isolet data set, respectively. However, our technique is simpler and faster and it is independent to types of learning machine.

Keywords: feature selection, unsupervised learning, clustering, filter-based method, correlation, similarity, redundancy

1   Introduction

Data clustering is an automatic process for grouping unlabeled data into a number of similar groups, called clusters, by assessment of their contents. It is an unsupervised process by nature [1-2]. Most data clustering methods typically employ feature vector models for representing data instants [3-5]. However, this representation will suffer when applying with high-dimension and sparse data, known as the curse of dimensionality. For example in text document clustering, a document is represented by a feature vector known as Bags of words (BOW) that generally uses all words in data collection as features [1,6-8]. Most words in documents are usually redundant and irrelevant to the clustering. Hence, determining clusters in this data space is not only computationally expensive but it also degrades the learning performance [2-3, 8-10].

In order to eliminate irrelevancy and redundancy, we have to select a smaller number of most relevant features to the targeted clusters [2, 11-13]. Feature selection methods can be classified as filter-based approaches, wrapper-based and hybrid approach [2, 12, 14]. Filter-based methods are to select features by "usefulness" of each individual feature for clustering. The usefulness is determined by assessment of feature's characteristics using certain statistical criteria, without any learning process [1,

T. Herawan et al. (eds.), *Proceedings of the First International Conference on Advanced Data and Information Engineering (DaEng-2013)*, Lecture Notes in Electrical Engineering 285, DOI: 10.1007/978-981-4585-18-7_10,
© Springer Science+Business Media Singapore 2014

3, 7-8, 11, 15]. On the other hand, wrapper-based methods employ a chosen learning algorithm to determine an optimal subset of features. Even though they outperform filter-based methods, they are in general more computationally expensive. The selected feature subset is usually overfit to a chosen learning algorithm [2, 11-12, 14]. Several authors have proposed hybrid methods, which take advantages of both filter and wrapper methods [1-2, 7, 12]. However, the wrapper and hybrid approach is usually not suitable when availability of time or computing resources is a constraint. Thus, the filter model is still preferable in terms of computational time [2, 7, 11-12, 14].

There are two types of filter-based feature selections, supervised and unsupervised method. For supervised methods, class label, which information to identify class of a data entity, must be provided. Some supervised feature selection methods have been successfully used in text classification [3, 16]. However, if class label is not available, many research projects introduce some unsupervised feature selection such as Document Frequency (DF) [3, 15-16], Term Contribution (TC) [1, 6, 15, 17], Term Variance (TV) [6-7, 17] and Mean Median (MM) [7]. Some researchers proposed to reduce redundancy and irrelevance for the feature selection algorithm, such as [14]; however, class label is required and it cannot be used for data clustering directly.

In this article, we propose an unsupervised filter-base feature selection for data clustering. The method is unsupervised method and it is independent to type of learning machines. Furthermore, we integrate concept of evaluating feature redundancies into the proposed algorithm. The redundancy assessment is a feature similarity measurement based on the geometric view of features. [7] Unlike classical filter-based methods, which have to predefine a threshold by expertise or empirical experiments to pick up top-ranked features, our proposed method uses a coefficient of confident to select relevant features, rather than finding a new threshold for every new set of data. The performance of the algorithm is evaluated on a corpus dataset for high-dimension data analysis, namely the RCV1v2 dataset [18] and Isolet dataset [19]. Compared by F1, Average Accuracy and Rand statistics, the experiment on RCV1v2 dataset shows that clustering accuracy of the proposed algorithm significantly is comparably equal to the baseline filter-based method. The result also shows that proposed method outperforms the baseline on the Isolet dataset.

The rest of the paper is organized as follows. Background of feature selection is summarized in Section 2. Section 3 presents our proposed feature selection algorithm and describes its characteristics. Experiments are explained and results are discussed in Section 4. Section 5 concludes this work.

2   Features and Feature Redundancy

Notations used in this paper are as follows: $F = \{f_1, \dots, f_{|F|}\}$ be a set of original distinct features; $D = \{d_1, \dots, d_{|D|}\}$ be a training dataset; A training sample d_j is represented as a feature weight vector, $\vec{d_j} = [w(f_1, d_j), \dots, w(f_{|F|}, d_j)]$. This feature weight $w(f_i, d_j)$ quantifies the importance of the feature f_i for describing semantic content of d_j. For example, text clustering prefers feature weight such as the Term Frequency or the Term Frequency Inverse Document Frequency [1-2, 6, 8-9, 15, 18].

Feature redundancy can be represented in terms of feature correlation. It is widely accepted that the features are redundant to each other if their values are completely correlated [7, 11, 14]. The traditional filter-based feature selection is incapable of removing redundant features because redundant features likely have similar rankings. As long as features are deemed relevant to the class, they will all be selected even though many of them are highly correlated to each other. For high-dimensional data which may contain a large number of redundant features, this approach may produce results far from optimal. Many research projects address some similarity/correlation/redundancy measures that have been used for feature selection, such as correlation coefficient [7, 14, 20], symmetrical uncertainty [7, 14] and absolute cosine [7]. In [7] argue that using angles to measure similarity is better suitable for high-dimensional sparse data. Therefore, absolute cosine is used as geometric view in our proposed algorithm.

3   Proposed method

The proposed method composed of two approaches: measuring feature relevance by feature score to keep highly relevance features; and measuring feature redundancy by feature correlation to identify redundant features. Then, we define a policy to eliminate less important features. The algorithm is listed in Algorithm 1 (Unsupervised fast correlation-based filter algorithm.). In Algorithm 1, scores of each feature are computed and then used it for sorting (lines 2 and 3); each feature will be compared with others in order to find relevancy (lines 4-12); feature redundancy is measured by computing feature similarity [7] (line 8); each feature is then considered to be removed from the output (lines 13-17); and after looping for every feature, the result, which is a feature set that each has high relevance and low similarity among themselves, is returned (line 20).,

Algorithm 1 Unsupervised Fast Correlation-Based Filter Algorithm

Input : Original feature set (F)
    Training data (D)
    Threshold parameter (α)
Output : Optimal feature subset (Opt)
1:   for each feature f_i in F do   // Compute score all feature
2:      $s(f_i) \leftarrow Score(f_i, D)$
3:   $ST \leftarrow SortedFeatureByScoreDes(s(f_i), F)$ // set of sorting feature in F by score descending
4:   $f_j \leftarrow GetFirstElement(ST)$
5:   do begin
6:      $F' \leftarrow GetAllNextFeature(f_j, ST)$
7:      for each feature f_i in F'

8:        $sim(f_i, f_j) \leftarrow ComputeSimilarity(f_i, f_j)$
9:      $thres_{redundant}(f_j) \leftarrow ComputeRedudantThreshold(\alpha)$
10:     $thres_{remove}(f_j) \leftarrow ComputeRemoveThreshold(\alpha)$
11:     $ST' \leftarrow SortFeatureByScoreAs(s(f_i), F')$ //set of sorting feature in F' by score ascending
12:     $cr \leftarrow CumulativeRelevance(ST')$
13:     for each feature f_i in ST' // redundant identify
and remove
14:       if $(sim(f_i, f_j) > thres_{redundant})$
15:           if $(cr - s(f_i) > thres_{remove})$
16:              remove f_i from ST
17:     $f_j \leftarrow GetNextElement(f_j, ST)$
18: end until $(f_j = NULL)$
19: $Opt = ST$
20: **return** Opt

In practical, many feature selection methods suffer the problem of selecting appropriate thresholds for both redundant feature identification and redundancy elimination. Thus, we proposed the statistics based method to compute threshold to identify redundant feature of feature fj, is defined as

$$Thres_{redundant} = \overline{sim(f_j)} + Z_{1-\frac{\alpha}{2}} \cdot \sigma_{f_j} \tag{1}$$

where α is a confidence coefficient, $Z_{1-\frac{\alpha}{2}}$ is the $1 - \frac{\alpha}{2}$ quantize of the $N(0,1)$, $\overline{sim(f_j)}$ and σ_{f_i} are average and standard deviation, respectively, of similarity between features f_i and f_j, $f_i \in ST'$. The feature f_i, that has similarity value $\left(sim(f_i, f_j)\right)$ more than $thres_{redundant}$, is identified as redundant feature of feature f_j. Next, we introduce a criterion in strategy for redundant feature elimination. The decision to remove a feature depends on a cumulative relevance (cr) measure [7].

$$cr = \sum_{i=1}^{|ST'|} s(f_i) \tag{2}$$

The cr value is used to calculate summation of relevance score of feature in subset ST'. Then, we propose removing the features f_i where $cr - s(f_i)$ is more than threshold, $Thres_{remove} = \left(1 - \frac{\alpha}{2}\right) \cdot cr$. This means that f_i is redundant and removing f_i affects cr value a little. Thus, f_i can be removed from the feature set. On the other hand, some features f_i have $s(f_i)$ that make $cr - s(f_i) < thres_{remove}$, it means that the feature is redundant but it is influent to the cumulative relevance of feature set; thus, it should be kept in the selected feature set. Finally, we can select highly-relevant features and remove highly-redundant feature for data clustering.

4 Experiments

In the experiment, we use the RCV1v2 dataset [18] and choose the data samples with the highest four topic codes (CCAT, ECAT, GCAT, MCAT) in the "Topic Codes" hierarchy which contains 19,806 training documents and 16,942 testing documents. Furthermore, we also use Isolet dataset [19]. There are 617 real features with 7797 instances and 26 classes. We generate a vector model for training data without using class label based on our selected features. Then, we use K-Means clustering onto the vector model. The result, which is a set of clusters, will be used as a new model for clustering onto the testing data (also without using class label.) The labels assigned by clustering testing data are used to compare with class labels given from corpora. In order to assess clustering performance under different feature selection method, three qualitative measures are selected. For Average Accuracy (AA) [6, 21] and RS Rand Statistics (RS) [6, 22], we count number of documents, which have the same topics, in the same cluster and number of documents, which have different topics, in different clusters. In our clusters and in the corpus, both documents are placed in the same clusters: ss. In our clusters both documents are placed in the same clusters but in corpus they are in different clusters: sd. In our clusters documents are placed in different clusters but in the corpus they are in the same clusters: ds. In our clusters and in the corpus both documents are placed in different clusters: dd. Then, AA and RS are defined as follows:

$$AA = \frac{1}{2} \times \left(\frac{ss}{ss+ds} + \frac{dd}{sd+dd} \right) \tag{3}$$

$$RS = \frac{(ss+dd)}{ss+sd+ds+dd} \tag{4}$$

Another measure for evaluating clustering is the macro F1-measure (F1) [10, 21] that is evaluated as

$$F1 = \Sigma_i \frac{n_i}{N} \left[max_{j \in C} \left\{ \frac{2 \times Recall(i,j) \times Precision(i,j)}{Recall(i,j)+Precision(i,j)} \right\} \right] \tag{5}$$

which $Recall(i,j) = \frac{n_{ij}}{n_i}, Precision(i,j) = \frac{n_{ij}}{n_j}$, where n_{ij} is the number of instances belonging to class i in corpus that falls in cluster j, and n_i, n_j are the cardinalities of class i cluster j respectively.

Our proposed selection algorithm is evaluated by comparing the effectiveness of our optimal feature subset with other subsets selected by a baseline algorithm. We use ranking wrapper-based feature selection, which is the most preferable filter-based method to determine the best number of highly relevance feature as the baseline [6, 9, 13, 23]. We applied four unsupervised feature scoring schemes, DF, TC, TV and MM, to determine feature subset on training documents of RCV1v2 dataset with different cut-off number of features ranging from 500 to 4000. At each cut-off number, the performance of the K-Means clustering for the selected feature subset is estimated by 10-fold 10-time cross-validation. The results show that the optimal number of features for DF, TC, TV and MM are 1300, 500, 500 and 500, respectively.

We then used the proposed outlier-based feature selection in Algorithm 1 to select a feature subset with these four feature scores. We set merely a parameter α, to identify the optimal threshold. From preliminary parameter setting, numbers of features selected by the proposed algorithm for DF, TC, TV, and MM are 486, 483, 537, and 470, respectively. The result shows that the proposed algorithm almost selects a smaller feature subset when compared with the feature subset selected using the baseline algorithm. The selected feature subsets are used to train the clustering; then the clustering performance is evaluated on testing documents. Table 1 shows the performance of features selected by each algorithm for DF, IC, IV, and MM. The performance values in each table are 10-run average values. We compute Student's independent two-tailed t-test in order to evaluate the statistical significance of the difference between the two averaged values: the one from the proposed method and the one from baseline method. The p-Val is the probability associated with an independent two-tailed t-Test. The "compare" means that the proposed method is statistically significant (at the 0.05 level) win or loss over the baseline methods and equal means no statistically significant difference. The experimental result shows that the proposed method with four feature scores get comparably equal clustering performance to the baseline method.

Table 1. Comparing three performance measures between feature subsets selected by the UFCBF method and the baseline method for DF, TC, TV and MM on RCV1v2 dataset.

Feature Score	Method	#feature	F1	AA	RS
DF	baseline	1300	0.688	0.553	0.626
	UFCBF	486	0.693	0.554	0.632
	p-Value		0.857	0.969	0.722
	compare		equal	equal	equal
TC	baseline	500	0.667	0.525	0.600
	UFCBF	483	0.684	0.557	0.633
	p-Value		0.588	0.278	0.234
	compare		equal	equal	equal
TV	baseline	500	0.683	0.552	0.626
	UFCBF	537	0.645	0.555	0.629
	p-Value		0.204	0.837	0.841
	compare		equal	equal	equal
MM	baseline	500	0.665	0.539	0.613
	UFCBF	470	0.702	0.563	0.638
	p-Value		0.149	0.143	0.127
	compare		equal	equal	equal

Moreover, we also evaluate performance of the proposed method on the Isolet with of baseline method presented in [10]. The Table 2 shows comparison of validation. The number of selected features from the proposed method is lower than number from the baseline method. Table 2 shows that proposed algorithm achieve higher F1 than the baseline and RS value of both methods are equal.

Table 2. A comparison of the performance on Isolet dataset

Method	#feature	F1	Rand	AA
baseline [10]	617	0.365	-	-
	274	0.336	0.94	-
	275	0.344	0.94	-
Purposed	263	0.530	0.94	0.62
Compare		Win	equal	-

5   Conclusions

In this paper, we proposed an effective and computationally efficient algorithm that dramatically reduces size of feature set in high dimensional datasets. The proposed algorithm eliminates a large number of irrelevant and redundant features and selects a subset of informative features that provide more discriminating power for unsupervised learning model. Our algorithm is developed and tested it on the RCV1v2 and Isolet data corpus. Our experimental results confirm that the proposed algorithm can greatly reduce the size of feature sets while maintaining the clustering performance of learning algorithms. The algorithm uses a simple statistic based threshold determination to develop a novel filter-based feature selection technique. Our approach does not require iterative empirical processing or prior knowledge. Compared to traditional hybrid feature selection the optimal subset from our proposed algorithm is significantly comparable or even better than the baseline algorithm. Experiments showed that proposed method works well and can be used with the unsupervised learning algorithm which class information is unavailable and the dimension of data is extremely high. Our proposed method can not only reduce the computation cost of text document analysis but can also be applied to other textual analysis applications.

References

1. Almeida, L.P., Vasconcelos, A.R., Maia, M.G.: A Simple and Fast Term Selection Procedure for Text Clustering. In: Nedjah, N., Macedo Mourelle, L., Kacprzyk, J., França, F.G., De Souza, A. (eds.) Intelligent Text Categorization and Clustering, vol. 164, pp. 47-64. Springer Berlin Heidelberg (2009)
2. Alelyani, S., Tang, J., Liu, H.: Feature Selection for Clustering: A Review. In: Aggarwal, C., Reddy, C. (eds.) Data Clustering: Algorithms and Applications. CRC Press (2013)

3. Sebastiani, F.: Machine learning in automated text categorization. ACM Comput. Surv. 34, 1-47 (2002)
4. Ferreira, A.J., Figueiredo, M.A.T.: An unsupervised approach to feature discretization and selection. Pattern Recognition 45, 3048-3060 (2012)
5. Shamsinejadbabki, P., Saraee, M.: A new unsupervised feature selection method for text clustering based on genetic algorithms. J Intell Inf Syst 38, 669-684 (2012)
6. Luying, L., Jianchu, K., Jing, Y., Zhongliang, W.: A comparative study on unsupervised feature selection methods for text clustering. In: Natural Language Processing and Knowledge Engineering, 2005. IEEE NLP-KE '05. Proceedings of 2005 IEEE International Conference on, pp. 597-601. (Year)
7. Ferreira, A.J., Figueiredo, M.A.T.: Efficient feature selection filters for high-dimensional data. Pattern Recognition Letters 33, 1794-1804 (2012)
8. Ferreira, A., Figueiredo, M.: Efficient unsupervised feature selection for sparse data. In: EUROCON - International Conference on Computer as a Tool (EUROCON), 2011 IEEE, pp. 1-4. (Year)
9. Yanjun, L., Congnan, L., Chung, S.M.: Text Clustering with Feature Selection by Using Statistical Data. Knowledge and Data Engineering, IEEE Transactions on 20, 641-652 (2008)
10. Mitra, S., Kundu, P.P., Pedrycz, W.: Feature selection using structural similarity. Information Sciences 198, 48-61 (2012)
11. Guyon, I., Andr, #233, Elisseeff: An introduction to variable and feature selection. J. Mach. Learn. Res. 3, 1157-1182 (2003)
12. Liu, H., Yu, L.: Toward integrating feature selection algorithms for classification and clustering. IEEE Transactions on Knowledge and Data Engineering 17, 491-502 (2005)
13. Somol, P., Novovicova, J., Pudil, P.: Efficient Feature Subset Selection and Subset Size Optimization. Pattern Recognition Recent Advances 75-97 (2010)
14. Yu, L., Liu, H.: Efficient Feature Selection via Analysis of Relevance and Redundancy. J. Mach. Learn. Res. 5, 1205-1224 (2004)
15. Liu, T., Liu, S., Chen, Z.: An Evaluation on Feature Selection for Text Clustering. In: In ICML, pp. 488-495. (Year)
16. Yang, Y., Pedersen, J.O.: A Comparative Study on Feature Selection in Text Categorization. In: 14th International Conference on Machine Learning, pp. 412-420. Morgan Kaufmann Publishers Inc., 657137 (Year)
17. Zonghu, W., Zhijing, L., Donghui, C., Kai, T.: A new partitioning based algorithm for document clustering. In: Fuzzy Systems and Knowledge Discovery (FSKD), 2011 Eighth International Conference on, pp. 1741-1745. (Year)
18. Lewis, D.D., Yang, Y., Rose, T., Li, F.: RCV1: A New Benchmark Collection for Text Categorization Research. Journal of Machine Learning Research 5, 361-397 (2004)
19. Bache, K., Lichman, M.: UCI Machine Learning Repository. University of California, Irvine, School of Information and Computer Sciences, Irvine, CA (2013)
20. Mitra, P., Murthy, C.A., Pal, S.K.: Unsupervised feature selection using feature similarity. Pattern Analysis and Machine Intelligence, IEEE Transactions on 24, 301-312 (2002)
21. Shamsinejadbabki, P., Saraee, M.: A new unsupervised feature selection method for text clustering based on genetic algorithms. J Intell Inf Syst 1-16 (2011)
22. Achtert, E., Goldhofer, S., Kriegel, H.-P., Schubert, E., Zimek, A.: Evaluation of Clusterings - Metrics and Visual Support. In: ICDE'12, pp. 1285-1288. (2012)
23. Ruiz, R., Riquelme, J., Aguilar-Ruiz, J.: Heuristic Search over a Ranking for Feature Selection. In: Cabestany, J., Prieto, A., Sandoval, F. (eds.) Computational Intelligence and Bioinspired Systems, vol. 3512, pp. 498-503. Springer Berlin / Heidelberg (2005)

Comparison of Feature Dimension Reduction Approach for Writer Verification

Rimashadira Ramlee, Azah Kamilah Muda, Nurul Akmar Emran

Faculty of Information and Communication Technology
Universiti Teknikal Malaysia Melaka, Melaka, Malaysia
rimashadira@gmail.com; azah@utem.edu.my;
nurulakmar@utem.edu.my;

Abstract. Dimension reduction is useful approach in data analysis application. In this paper, research is done to test whether the concept of Dimension reduction can be applied to improve writer verification process results. Two approaches have been chosen to be compared which are Features Selection and Feature Transformation, where the comparison is on the way of reducing the dimension of writer handwritten data. Both approaches have slightly difference results in reducing the data and classification accuracy. The objective of this paper is to observe the differences between both approaches according to the classification accuracy results, by using some classification techniques.

Keywords: Dimension Reduction, Writer Verification, Features Transformation, Features Selection

1 Introduction

Dimension reduction (DR) is a useful approach to solve a problem in data analysis application. Usually DR can be beneficial not only for reasons of computational efficiency but also because it can improve the accuracy of the analysis [1]. Reduction of the data dimension will help the process of identifying the most important features in handwritten data. According to [2], the process is either by transforming the existing features to a new reduced set of features or by selecting a subset of the existing features. In the data analysis, not all the features can yield important information that represents unique individualities of the writer, because maybe there is a lot of data redundancy which is not very useable in the analysis. In these issues, dimension reduction is useful to in order to improve that quality of the data used in analysis of data.

The purpose of this paper is to observe the comparison between features selection and features transformation approach in acquiring the most significant features among handwritten data via DR concept. The Comparison will be conducted by examining the classification accuracy and number of features data has been effectively reduced using both methods above. Features selection will select the feature directly from the original features and wish to keep the original meaning of the features, where Feature Transformation will allow the modification of the feature to a new feature space and

T. Herawan et al. (eds.), *Proceedings of the First International Conference on Advanced Data and Information Engineering (DaEng-2013)*, Lecture Notes in Electrical Engineering 285, DOI: 10.1007/978-981-4585-18-7_11,

wish to determine which of those important features [3]. This paper is organized as follows in section 2 the detail explanation of Writer Verification is provided. In section 3 the description about Dimension Reduction approach and the way of their performing will be showed. Section 4 will describe about the Experiment setup of the process. The experiment illustration and result explanation will be in Section 5. Finally, conclusion will be in section 6.

2   Writer Verification

In theory, Writer Identification and Writer Verification belong to the group of behavioral methods in biometrics. Both methods will come to a conclusion of identifying the unknown writer, but the difference is according to the task of their performance. Writer Verification task is determined whether two samples of handwriting is written by the same writer or not [4].

Most of the recent research focuses on signature verification especially in field of on-line writer verification, where the verification process is used to perform the matching of two sample signature from one writer. To solve the problem of forged handwriting, dynamic information such as velocity, acceleration, and force exerted on the pen are utilized [5]. In this research, the verification process is chosen to be performed in text verification, because this task consists in matching the unknown writer with each of those in the selected subset.

Fig. 1. Example of Verification Process

Based on Fig 1, Writer verification also can be defined as one-to-one comparison process to make a decision for determine the real writer of handwritten document [5]. According to Srihari, Arora and lee, the individuality of handwriting rests on the hypothesis that each individual has consistent handwriting that is distinct from the handwriting of another individual.

3   Dimension Reduction Method

3.1   Feature Selection

Feature Selection is a popular approach used in Handwriting Analysis research field. In general, feature selection techniques do not alter the original representation of the variables, but merely select a subset of them. Feature selection techniques can be organized into three categories as showed in Table 1, depending on how they combine the feature selection search with the construction of the classification model which are filter methods, wrapper methods, and embedded methods [6].

In this work, we focus on filter model in our experimental framework due to its computational efficiency and usually chosen when the number of features becomes very large [7]. Different from wrapper method and embedded method, the classifier chosen has to embed in the features selection approaches, where in this experiment features selection and classification process are separated. In further review, filter model can be categorized into two groups namely, Feature Weighting Algorithm (FWA) and Subset Search Algorithm (SSA).

Table 1. Type of Feature Selection Methods

FS space ◯ ⟶ Classifier	*Filter methods:* assess the relevance of features by looking only at the intrinsic properties of the data.
FS space Classifier	*Wrapper methods:* embed the model hypothesis search within the feature subset search.
FS U Hypothesis space Classifier	*Embedded methods*: the search for an optimal subset of feature is built into the classification construction

Additionally, the filter model relies on general characteristic of the training data to select some features without involving any learning algorithm. This will bring us to the objective of features selection to obtain the most important features by avoiding over fitting and improve model performance beside of provide faster and more cost-effective models, and gain a deeper insight into the underlying processes that generated the data [7]. There are three methods from filter model that have been selected:

- **Correlation-based Feature Subset Selection (CFS)**

CFS is a fully automatic algorithm, this method does not required user to specify any thresholds or the number of features to be selected, although both are simple to incorporate if desired [8]. The features subset evaluation function is:

$$M_{zc} = \frac{k\overline{r_{zi}}}{\sqrt{K+K(K-1)\overline{r_{ii}}}}$$ (1)

Where M_{zc} is the correlation between the summed features and the outside variable, K is the number of variables, $\overline{r_{zi}}$ is the average of the correlations between the components and the outside variable, and $\overline{r_{ii}}$ is the average inter-correlation between features.

- **Relief**

Relief is a feature weighting algorithm that is sensitive to feature interactions. A key idea of the original Relief Algorithm is introduced by Kira and Rendell [9]. Relief will search for its two nearest neighbors, one is from the same class called *nearest hit*, and the other is from different class called *nearest miss*. Below is a probability for the weight of a feature:

$$W_{feature} = P(X|Y\ of\ different\ class) - P(X|Y\ of\ same\ class)$$ (2)

Where $W_{feature}$ is a weight, X is feature's difference value, and Y is a nearest instance.

- **Fast Correlation-based Filter (FCBF)**

In this algorithm, Symmetrical Uncertainty calculates dependency of features and finds best subset using backward selection technique with sequential search strategy [10]. FCBF can remove a large number of features that are redundant peers with predominant feature in the current iteration. Concept of entropy is used in this algorithm, for example, the entropy of a variable X is defined as:

$$H(X) = -\sum_i P(x_i) log2(P(x_i))$$ (3)

And the entropy of X after observing values of another variable Y is defined as:

$$H(X|Y) = -\sum_j P(y_j) \sum_i P(x_i|y_j) \log_2(P(x_i|y_j))$$ (4)

Symmetrical uncertainty (SU) compensates for information gain's bias toward attributes with more values and normalizes its value to the range [0, 1]. Below is the equation for calculating symmetrical uncertainty coefficient:

$$SU(X,Y) = 2 \left[\frac{IG(X|Y)}{H(X)+H(Y)} \right]$$ (5)

An SU value of 1 indicates that using one feature compared to other feature's value can be totally predictable and value 0 indicates two features that are totally independent [13].

3.2 Feature Transformation using Principal Component analysis

Feature transformation refers to a family of data pre-processing techniques that transforms the original features of a data set to an alternative, a more compact set of dimensions, while retaining as much information as possible [3]. This techniques aim to reduce the dimensionality of data to a small number of dimensions which are linear or non-linear combinations of the vector coordinates in the original dimensions [11]. Principal Component Analysis (PCA) which is one of the unsupervised feature transformation techniques. Unsupervised technique does not take class labels into account so that the process became easier. Since PCA is a powerful tool for analyzing and identifying a valuable pattern in the data, we propose this technique in one of pattern recognition field which is handwriting analysis. Once the pattern of the data is found this technique will reduce the dimension without losing many features components from the original data like stated in [12].

Typically in this work, the objective of PCA is to transform the data into another set of feature f', for example x_i transformed into x'_i in k dimensions shows:

$$x'_i = W x_i$$ (6)

The transformation of PCA is by reducing the space that captures most of the variance in the data. The whole idea of PCA is rest on the covariance matrix of the data as:

$$C = \frac{1}{n-1} XX^T$$ (7)

C Captures the variance in the individual features and the off-diagonal terms quantify the covariance between the corresponding pairs of features. C can produce C_{PCA}, when the data is transformed by $Y = PX$ where the rows of P are the eigenvector of XX^T, then

$$C_{PCA} = \frac{1}{n-1} YY^T \tag{8}$$

$$C_{PCA} = \frac{1}{n-1} (PX)(PX)^T \tag{9}$$

C_{PCA}, is the quantifier of variance of the data in the direction of the corresponding principal component.

4 Experiment Setup

In this experiment, three representative feature selections are chosen in comparison with PCA like stated above.

4.1 Dataset

Handwriting sample dataset is taken from IAM Handwriting Database [13] is chosen to be used, where there are 657 classes available, however only 60 classes are used for this research. From these 60 classes, 4400 instances are collected, and are randomly divided into five different datasets to form training and testing dataset. First of all, the form of handwriting text will be extracted by using United Moment Invariance (UMI) [14]. We use both undiscretized and discretized data in this experiment in order to see the influence of verification and to improve the classification accuracy. Discretization was done by employed Equal Width Binning (EWB), and the main goal of this process is to minimize the number of intervals without significant loss of class-attribute mutual dependence [15]. Besides that, discretization process is important in order to obtain the detachment of writer's individuality and produce better data representation. Table 2 shows the example of data after UMI process:

Table 2. Example of Handwriting Data

Word	F1	F2	F3	F4	F 5	F 6	F 7	F8
the	0.75	0.38	0.44	0.19	0.67	4.62	0.50	4.59
shshine	0.71	0.37	0.49	0.23	0.73	4.66	0.63	4.13

4.2 Framework Design

We design our work following the traditional of pattern recognition task for writer verification process as shown in Fig 4 below:

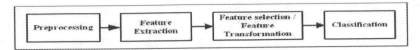

Fig. 2. Writer Verification Framework

This process begins, with preprocessing task, which is to process the data before extracting the real word features. UMI is applied in feature extraction part where all the handwriting text is changed to the word features representation. After discretization process the data becomes more clean and easy to determine the unique feature of the writer's data before we proceed to feature selection and classification task. We develop this experiment by using Waikato Environment for Knowledge Analysis (WEKA) 3.7.5 to measure the performance of feature selection, features transformation and classification method.

4.3   Verification Process

Verification will classify the data according to the same writer, where classification process will be used to verify the writer of document in sample data based on their class. There are several techniques of classification are used to perform the task which are Bayes Network, K-Nearest Neighbor (KNN), Random Forest and 1-Rule,

5   Result and Discussion

Two categories values are used to measure the performance of feature selection and feature transformation algorithm in writer verification process. The result as shown in Table 3:

Table 3. Result of the Experiment

5.1   Comparison by Classification Accuracy

The results have shown that when the data is undiscretized almost all the methods yield the same accuracy, which the average of all accuracies is less than 50%,

Sample Data	Category	Undiscretized				Discretized			
		CFS	Relief	FCBF	PCA	CFS	Relief	FCBF	PCA
Sample 1	Classification Accuracy	45.42%	45.99%	46.55%	45.54%	98.31%	98.08%	98.31%	66.33%
	Number of Selected features	1	8	8	5	6	8	8	6
Sample 2	Classification Accuracy	48.53%	45.99%	48.65%	49.24%	98.24%	98.24%	98.00%	65.22%
	Number of Selected features	1	8	8	5	6	8	8	6
Sample 3	Classification Accuracy	39.91%	45.99%	40.36%	40.02%	97.49%	97.38%	97.72%	62.14%
	Number of Selected features	1	8	8	5	6	8	8	6
Sample 4	Classification Accuracy	29.77%	45.99%	29.99%	30.43%	98.24%	97.91%	97.91%	55.57%
	Number of Selected features	1	8	8	5	5	8	8	6
Sample 5	Classification Accuracy	39.12%	45.99%	40.25%	39.46%	98.07%	97.96%	98.19%	61.90%
	Number of Selected features	1	8	8	5	6	8	8	6
AVERAGE		40.55%	45.99%	41.16%	40.94%	98.07%	97.91%	98.02%	62.23%

but the values are still slightly different as compared to the others. In comparison when the data has been discretized, the result plots better accuracy. The result of Relief method is slightly different among other feature selection methods, as compared to the chosen feature transformation method named PCA. Based on the observation, Relief method has gained higher average of classification accuracy when using undiscretized data, followed by FCBF, PCA and CFS. Even though the accuracy of relief method is decrease when the data is discretized but the different is not huge compared to others method. From the average of classification accuracy as shown in Table 3, Relief method is the best feature selection approach to verify the writer of sample data. It is caused by the process of feature selection that keeps the original meaning of every feature. Besides that, it is more helpful in verification process, rather than to modify the feature and transform it into a new feature space, So that the results will be more accurate when using this process.

5.2   Comparison by Selected Features

Second comparison is done by comparing the number of feature that can be reduced by each method from both approaches. Based on the result in TABLE III, The best method in reducing the features is CFS and PCA followed by Relief and FCBF. This because CFS has reduced 7 features and PCA has reduced 3 features using undiscretized data. Otherwise, only 2 features are reduced by both methods when the data is discretized, this because cleanliness of data can affect relation between the features. CFS and PCA can reduce the some feature according to their correlation of features between each other. CFS calculates the correlations and then searches the feature subset space which is important to represent the original. PCA also reduce the features but different concept from CFS which is less importance to represent the original data would reduced after transforming the feature by using all the original feature in data set. Relief and FCBF are not reducing any feature in both types of data. According to the Relief and FCBF concept, the relevant features are chosen rely on the dependences between each other. So that, this both methods will estimate that the entire feature are interact to each other in representing the original data then the reduction process cannot be done. Here, we able to prove that feature selection and features transformation approach successfully can reduce the dimension of data by only select the most signification feature that can verify the actual writer in verification process.

6   Conclusion

As a conclusion, the experimental result has proved that Dimension Reduction process can be used in verification process especially in processing data activities. Dimension reduction is more concern in eliminating the redundant data, so that this characteristic can improve the performance of the process. Redundancy will increase the relation among the feature and will cause the feature strongly depend on each other. CFS and PCA have different concept than Relief and FBCF as stated above. Among them CFS is the best method because, this method reduces the feature to the lowest number and accurately verify the writer of sample data.

Acknowledgement

This work is funded by the Faculty of Information and Communication Technology, Universiti Teknikal Malaysia Melaka (UTeM).

References

1. P.Cunningham, "Dimension Reduction", Technical Report UCD-CSI-2007-7, University College Dublin, 2007.
2. Ch. Aswani Kumar, "Analysis of unsupervised dimensionality reduction techniques", Computer Science and Information Systems, 6(2):217-227, doi: 10.2298/csis0902217K, 2009.
3. M.Masaeli, G.Fung, Jennifer G.Dy, "From Transformation-Based Dimensionality Reduction to Feature Selection", in Mahdokht Masaeli, G.F.J.G.D., ed. Proc IEEE, 27th International Conference on Machine Learning. Boston, USA, 2010.
4. S.Srihari, H.Arora, S.Lee, "Individuality of Handwriting", Journal of Forensic Sciences, 47(4), 2002, pp.1-17.
5. F.Leclerc, R.Plamondon: "Automatic signature verification and writer identification: The state of the art - 1989-1993", Int. J. of Pattern Recogn. Artif. Intell. (IJPRAI), 8, 3, pp.643-660(1994).
6. Pratama, S.F., Muda, A.K., Choo, Y.-H.: Feature Selection Methods for Writer Identification: A Comparative Study. In: Proceedings of 2010 Intl. Conference on Computer and Computational Intelligence, pp. 234--239. IEEE Press, Washington (2010).
7. Saeys, Y., Inza, I., & Larranaga, P., "A Review of Feature Selection Techniques in Bioinformatics". Journal of Bioinformatics, 2507-2517, 2007.
8. M. A. Hall, "Correlation-based Feature Subset Selection for Machine Learning", 1998, Hamilton, New Zealand.
9. M.Robnik-Sikonja, I.Kononenko, "Theoretical and Empirical Analysis of ReliefF and RReliefF", Machine Learning Journal, 23-69, 2003.
10. L.Yu, H.Liu, "Feature Selection for High-Dimensional Data: A Fast Correlation-Based Filter Solution", Proc of the Twentieth International Conference on Machine Learning, pp. 856-863, 2003.
11. T. Jollife, "Principal Component Analysis", Springer-Verlag, New York. 1986.
12. L. I. Smith, " A tutorial on Principal Components Analysis", PP. 13-27, February 26, 2002.
13. U.Marti, H. Bunke, "The IAM-database: an English Sentence Database for Off-line Handwriting Recognition", International Journal on Document Analysis and Recognition, Volume 5, 39-46, 2002.
14. S.Yinan, L. Weijun, W. Yuechao, "United Moment Invariants for Shape Discrimination", International Conference on Robotics, Intelligent Systems and Signal Processing, pp. 88-93, Changsha: IEEE, 2003.
15. Kotsiantis, S,K.&.D., "Discretization Techniques: A Recent Survey", GESTS International Transactions On Computer Science and Engineering, 32, pp.47-58, 2006.

Countering the problem of oscillations in Bat-BP gradient trajectory by using momentum

Nazri Mohd. Nawi[1], M. Z. Rehman[1], Abdullah Khan[1]

[1]Software and Multimedia Centre, Faculty of Computer Science and Information Technology,
Universiti Tun Hussein Onn Malaysia (UTHM).
P.O. Box 101, 86400 Parit Raja, Batu Pahat, Johor Darul Takzim, Malaysia.

nazri@uthm.edu.my, zrehman862060@gmail.com, hi100010@siswa.uthm.edu.my

Abstract. Metaheuristic techniques have been recently used to counter the problems like slow convergence to global minima and network stagnancy in back-propagation neural network (BPNN) algorithm. Previously, a meta-heuristic search algorithm called Bat was proposed to train BPNN to achieve fast convergence in the neural network. Although, Bat-BP algorithm achieved fast convergence but it had a problem of oscillations in the gradient path, which can lead to sub-optimal solutions. In-order to remove oscillations in the BAT-BP algorithm, this paper proposed the addition of momentum coefficient to the weights update in the Bat-BP algorithm. The performance of the modified Bat-BP algorithm is compared with simple Bat-BP algorithm on XOR and OR datasets. The simulation results show that the convergence rate to global minimum in modified Bat-BP is highly enhanced and the oscillations are greatly reduced in the gradient path.

Keywords: metaheuristics, slow convergence, global minima, bat algorithm, back-propagation neural network algorithm, Bat-BP algorithm, momentum.

1   Introduction

Back-propagation Neural Network (BPNN) is a very old optimization technique applied on the Artificial Neural Networks (ANN) to speed up the network convergence to global minima during training process [1-3]. BPNN follows the basic principles of ANN which mimics the learning ability of a human brain. Similar to ANN architecture, BPNN consists of an input layer, one or more hidden layers and an output layer of neurons. In BPNN, every node in a layer is connected to every other node in the adjacent layer. Unlike normal ANN architecture, BPNN learns by calculating the errors of the output layer to find the errors in the hidden layers [4]. This qualitative ability makes it highly suitable to be applied on problems in which no relationship is found between the output and the inputs. Due to its high rate of plasticity and learning capabilities, it has been successfully implemented in wide range of applications [5].

T. Herawan et al. (eds.), *Proceedings of the First International Conference on Advanced Data and Information Engineering (DaEng-2013)*, Lecture Notes in Electrical Engineering 285, DOI: 10.1007/978-981-4585-18-7_12,
© Springer Science+Business Media Singapore 2014

Despite providing successful solutions BPNN has some limitations. Since, it uses gradient descent learning which requires careful selection of parameters such as network topology, initial weights and biases, learning rate, activation function, and value for the gain in the activation function. An improper use of these parameters can lead to slow network convergence or even network stagnancy [5-6]. Previous researchers have suggested some modifications to improve the training time of the network. Some of the variations suggested are the use of learning rate and momentum to stop network stagnancy and to speed-up the network convergence to global minima. These two parameters are frequently used in the control of weight adjustments along the steepest descent and for controlling oscillations [7].

Besides setting network parameters in BPNN, evolutionary computation is also used to train the weights to avoid local minima. To overcome the weaknesses of gradient-based techniques, many new algorithms have been proposed recently. These algorithms include global search techniques such as hybrid PSO-BP [8], artificial bee colony back-propagation (ABC-BP) algorithm [9-10], evolutionary artificial neural networks algorithm (EA) [11], genetic algorithms (GA) [12] and Bat based back-propagation (Bat-BP) algorithm [13] etc. Unlike other algorithms, Bat-BP algorithm [13] shows high accuracy and avoids local minima. But, still some oscillations are detected in the gradient trajectory. In-order to avoid extreme changes in the gradient due to local anomalies [4-6], momentum coefficient is introduced in Bat-BP algorithm [13].

In this paper, Bat-BP with momentum is compared with simple Bat-BP algorithm and validated on XOR and OR datasets. We find that by using an appropriate metaheuristic technique such as BAT with BPNN enhanced with momentum coefficient can answer many limitations of gradient descent efficiently. The next two sections provide a brief discussion on BAT algorithm, followed by the proposed BAT-BP with momentum algorithm, simulation results, and conclusions.

2   The Bat Algorithm

Bat is a metaheuristic optimization algorithm developed by Xin-She Yang in 2010 [14]. Bat algorithm is based on the echolocation behavior of microbats with varying pulse rates of emission and loudness. Yang [14] has idealized the following rules to model Bat algorithm;

1) All bats use echolocation to sense distance, and they also "know" the difference between food/prey and back-ground barriers in some magical way.

2) A bat fly randomly with velocity (v_i) at position (x_i) with a fixed frequency (f_{min}), varying wavelength λ and loudness A_0 to search for prey. They can automatically adjust the wavelength (or frequency) of their emitted pulses and adjust the rate of pulse emission $r \in [0,1]$, depending on the proximity of their target.

3) Although the loudness can vary in many ways, Yang [14] assume that the loudness varies from a large (positive) A0 to a minimum constant value $Amin$.

Firstly, the initial position xi, velocity vi and frequency fi are initialized for each bat bi. For each time step t, the movement of the virtual bats is given by updating their velocity and position using Equations 2 and 3, as follows:

$$f_i = f_{min} + (f_{max} + f_{min})\beta \qquad (1)$$

$$v_i^t = v_i^{t-1} + (x_i^t + x_*)f_i \qquad (2)$$

$$x_i^t = x_i^{t-1} + v_i^t \qquad (3)$$

Where β denotes a randomly generated number within the interval [0,1]. Recall that x_i^t denotes the value of decision variable j for bat i at time step t. The result of fi in Equation 1 is used to control the pace and range of the movement of the bats. The variable $x*$ represents the current global best location (solution) which is located after comparing all the solutions among all the n bats. In order to improve the variability of the possible solutions, Yang [12] has employed random walks. Primarily, one solution is selected among the current best solutions for local search and then the random walk is applied in order to generate a new solution for each bat;

$$x_{new} = x_{old} + \in A^t \qquad (4)$$

Where, A^t stands for the average loudness of all the bats at time t, and $\in \in [-1,1]$ is a random number. For each iteration of the algorithm, the loudness A_i and the emission pulse rate r_i are updated, as follows:

$$A_i^{t+1} = \propto A_i^t \qquad (5)$$

$$r_i^{t+1} = r_i^0[1 - exp(-\gamma t)] \qquad (6)$$

Where α and γ are constants. At the first step of the algorithm, the emission rate, r_i^0 and the loudness, A_i^0 are often randomly chosen. Generally, $A_i^0 \in [1,2]$ and $r_i^0 \in [0,1][12]$.

3   The Proposed BAT-BP Algorithm

BAT is a population based optimization algorithm, and like other meta-heuristic algorithms, it starts with a random initial population. In Bat algorithm, each virtual bat flies randomly with a velocity v_i at some position x_i, with a varying frequency f_i and loudness A_i, as explained in the Section 2. As, it searches and finds its prey, it changes frequency, loudness and pulse emission rate r_i. Search is intensified by a local random walk. Selection of the best continues until stopping criterion are met. To control the dynamic behavior of a swarm of bats, Bat algorithm uses a frequency-tuning technique and the searching and usage is controlled by changing the algorithm-dependent parameters [14].

In the proposed BAT-BP algorithm [13], each position represents a possible solution (i.e., the weight space and the corresponding biases for BPNN optimization in this paper). The weight optimization problem and the position of a food source represent the quality of the solution. In the first epoch, the best weights and biases are initialized with BAT and then those weights are passed on to the BPNN where momentum coefficient, α is appended. The weights in BPNN are calculated and compared in the reverse cycle. In the next cycle BAT will again update the weights with the best possible solution and BAT will continue searching the best weights until the last cycle/ epoch of the network is reached or either the MSE is achieved.

The pseudo code of the proposed Bat-BP algorithm is shown in the Figure 1:

Step 1: BAT initializes and passes the best weights to BPNN
Step 2: Load the training data
Step 3: While MSE < Stopping Criteria
Step 4: Initialize all BAT Population
Step 5: Bat Population finds the best weight in Equation 4 and pass it on to the network, the weights, w_{ij} with momentum, α and biases, b_i in BPNN are then adjusted using the following formulae;

$$w_{ij}(k+1) = (w_{ij}k + \mu\partial_j y_i)\alpha_{ij}$$
$$b_i(k+1) = b_i k + \mu\partial_j$$

Step 6: Feed forward neural network runs using the weights initialized with BAT
Step 7: Calculate the backward error
Step 8: Bat keeps on calculating the best possible weight at each epoch until the Network is converged.
End While

Fig. 1. Pseudo code of the proposed Bat-BP algorithm

4   Results and Discussions

The simulations are carried-out on workstation equipped with a 2.33GHz Core-i5 processor, 4-GB of RAM, Microsoft Windows 7 and MATLAB 2010. Three datasets such as 2-bit XOR, 3-Bit XOR and 4-bit OR were used. The performance of the Simple BAT BP [13] algorithm is analyzed and compared with the modified BAT-BP algorithm. Three layer back-propagation neural networks is used for testing of the models, the hidden layer is kept fixed to 10-nodes while output and input layers nodes vary according to the datasets given. Log-sigmoid activation function is used as the transfer function from input layer to hidden layer and from hidden layer to the output layer. Momentum coefficient of 0.03 is found to be optimal for the weight updating in modified Bat-BP algorithm. A total of 20 trials, each trial consisting of 1000 epochs are run for each dataset. CPU time, average accuracy, and Mean Square Error (MSE) are recorded for each independent trials on XOR and OR datasets and stored in a separate file.

4.1   2-Bit XOR Dataset

The first test problem is the 2 bit XOR Boolean function consisting of two binary inputs and a single binary output. In simulations, we used 2-10-1 network architecture for two bit XOR. For the simple Bat-BP and modified Bat-BP, Table 1, shows the CPU time, number of epochs and the MSE for the 2 bit XOR test problem. Figure 2 shows the 'MSE performance vs. Epochs' of simple BAT-BP and modified Bat-BP for the 2-10-1 network architecture.

Table 1. CPU Time, Epochs and MSE for 2-bit XOR dataset with **2-10-1** ANN Architecture

Algorithms	Simple BAT-BP	Modified Bat-BP
CPUTIME	2.39	0.44
EPOCHS	23.25	7.9
MSE	0	0
Accuracy (%)	100	100

The modified Bat-BP algorithm avoids the local minima and converges on the provided network architecture successfully within 100 epochs as seen in the Table 1. The average CPU time is also reduced to a mere 0.44 from 2.39 CPU cycles in modified Bat-BP in Table 1. In Figure 2, the modified BAT-BP algorithm can be seen to converge within 5 epochs and shows a smooth gradient while simple BAT-BP converges within 10 epochs and shows a lot of oscillations in the trajectory path.

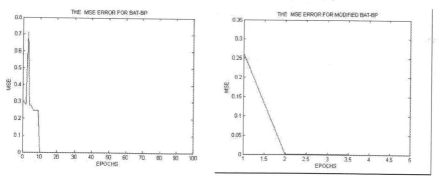

Fig. 2. (From Left to Right) Simple Bat-BP and modified Bat-BP convergence performance on 2-bit XOR with 2-10-1 ANN Architecture

4.2   3-Bit XOR Dataset

In the second phase, we used 3 bit XOR dataset consisting of three inputs and a single binary output. For the three bit input we apply 3-10-1, network architecture. The parameter range is same as used for two bit XOR problem, for the 3-10-1 the network it has forty connection weights and eleven biases. For the simple Bat-BP, and modified

Bat-BP, Table 2 shows the CPU time, number of epochs and the MSE for the 2 bit XOR test problem.

Table 2. CPU Time, Epochs and MSE for 3-bit XOR dataset with **2-10-1** ANN Architecture

Algorithms	Simple BAT-BP	Modified Bat-BP
CPUTIME	4.05	1.68
EPOCHS	23	25
MSE	0.0625	0.05
Accuracy (%)	93.69	94.99

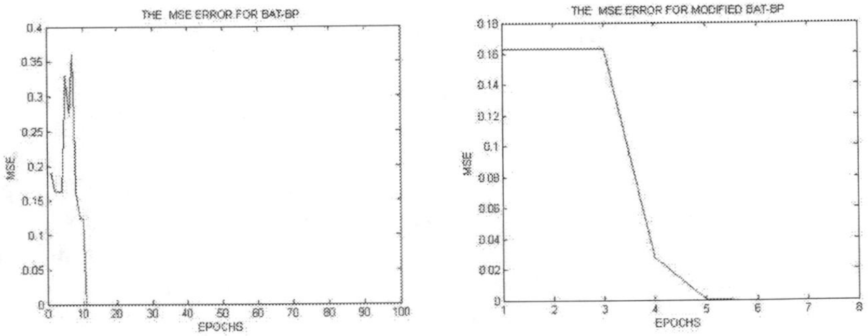

Fig. 3. (From Left to Right) Simple Bat-BP and modified Bat-BP convergence performance on 3-bit XOR with 2-10-1 ANN Architecture

In Table 2, modified BAT-BP algorithm can be seen converging with superior 1.68 CPU cycles. While 1 percent improvement in average accuracy is recorded but no improvement in MSE is detected in modified Bat-BP, as seen in the Table 2. In Figure 3, we can see the simulation results 'MSE vs. Epochs' convergence performance for 3-bit XOR dataset on simple and modified Bat-BP algorithms. Oscillations in modified Bat-BP can be seen controlled in the Figure 3.

4.3   4-Bit OR Dataset

The third dataset is based on the logical operator OR which indicates whether either operand is true. If one of the operand has a nonzero value, the result has the value 1. Otherwise, the result has the value 0. The network architecture used here is 4-10-1 in which the network has fifty connection weights and eleven biases. Table 3, illustrates the CPU time, epochs, and MSE performance of the simple Bat-BP, and modified Bat-BP, algorithms respectively. Figure 4, shows the 'MSE performance vs. Epochs' for the 4-10-1 network architecture of the proposed Bat-BP algorithm. In Figure 4, we can see that modified Bat-BP is converging within 9 epochs which is much better than the 22 epochs offered by simple Bat-BP. Also, it can be noted from the Table 3 that Bat-

BP comes with less CPU overheads and converges within 0.48 CPU cycles. Accuracy and the MSE was same for both algorithms, as shown in the Table 3.

Table 3. CPU Time, Epochs and MSE for 4-bit OR dataset with **2-10-1** ANN Architecture

Algorithms	BAT-BP	Modified Bat-BP
CPUTIME	2.88	0.48
EPOCHS	46.8	10.25
MSE	0	0
Accuracy (%)	100	100

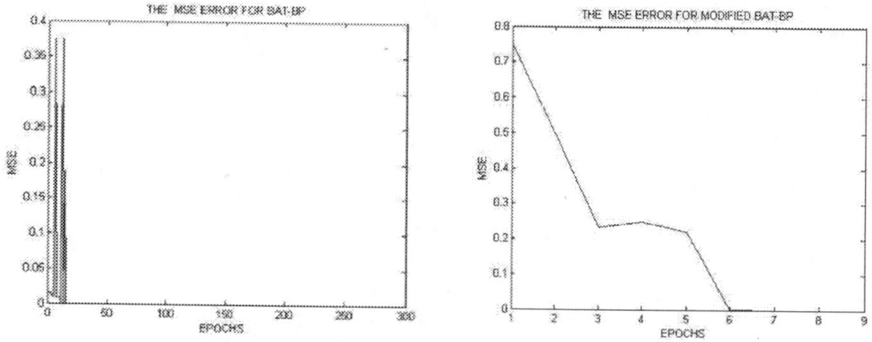

Fig. 4. (From Left to Right) Simple Bat-BP and modified Bat-BP convergence performance on 4-bit OR with 2-10-1 ANN Architecture

ACKNOWLEDGEMENTS

The Authors would like to Thank Office of Research, Innovation, Commercialization and Consultancy Office (ORICC), Universiti Tun Hussein Onn Malaysia (UTHM) and Ministry of Higher Education (MOHE) Malaysia for financially supporting this Research under Fundamental Research Grant Scheme (FRGS) vote no. 1236.

5 Conclusions

BPNN algorithm is one of the most widely used procedure to train Artificial Neural Networks (ANN). But BPNN algorithm has some drawbacks, such as getting stuck in local minima and slow speed of convergence. Nature inspired meta-heuristic algorithms provide derivative-free solution to optimize complex problems. Previously, a meta-heuristic search algorithm, called Bat was proposed to train BPNN to achieve fast convergence rate in the neural network. Although, Bat-BP [13] algorithm achieved fast convergence but it had a problem of oscillations in the gradient path, which can lead to sub-optimal solutions. In-order to remove oscillations in the BAT-BP algorithm [13] this paper, proposed the addition of momentum coefficient to the weights update in the Bat-BP algorithm. The performance of the modified Bat-BP is compared with the simple Bat-BP [13] algorithm by means of simulations on 2-bit, 3-bit XOR and 4-bit OR datasets. The simulation results show that the modified Bat-BP algorithm converges to the global minimum successfully showing a 0 MSE, less CPU cycles and 100 percent accuracy with almost no oscillations in the gradient descent path.

References

1. Deng, W. J., Chen, W. C., and Pei, W.: Back-propagation neural network based importance-performance analysis for determining critical service attributes, J. Expert Systems with Applications, vol. 34 (2). 2008
2. Kosko, B.: Neural Network and Fuzzy Systems, 1st Edition, Prentice Hall, India. (1992)
3. Rumelhart, D. E., Hinton, G. E., and Williams, R. J.: Learning Internal Representations by error Propagation, J. Parallel Distributed Processing: Explorations in the Microstructure of Cognition. (1986)
4. Rehman, M.Z., Nawi, N. M., and Ghazali, R.: Studying the effect of adaptive momentum in improving the accuracy of gradient descent back propagation algorithm on classification problems, J. International Journal of Modern Physics (IJMPCS), vol.1 (1). (2012)
5. Nawi, N. M., Ransing, M. R., and Ransing, R. S.: An improved Conjugate Gradient based learning algorithm for back propagation neural networks, J. Computational Intelligence, vol. 4. (2007)
6. Nawi, N. M., Rehman, M. Z., and Ghazali, M. I.: Noise-Induced Hearing Loss Prediction in Malaysian Industrial Workers using Gradient Descent with Adaptive Momentum Algorithm, J.International Review on Computers and Software (IRECOS), vol. 6 (5). (2011)
7. Lee, K., Booth, D., and Alam, P. A.: Comparison of Supervised and Unsupervised Neural Networks in Predicting Bankruptcy of Korean Firms, J. Expert Systems with Applications, vol. 29. (2005)
8. Mendes, R., Cortez, P., Rocha, M., and Neves, J.: Particle swarm for feed forward neural network training. In: Proceedings of the International Joint Conference on Neural Networks, vol. 2, pp. 1895--1899. (2002)
9. Nandy, S., Sarkar, P. P., and Das, A.: Training a Feed-forward Neural Network with Artificial Bee Colony Based Backpropagation Method, J. International Journal of Computer Science & Information Technology (IJCSIT), vol. 4 (4), pp. 33--46. (2012)
10. Karaboga, D., Akay, B., and Ozturk, C.: Artificial Bee Colony (ABC) Optimization Algorithm for Training Feed-Forward Neural Networks, In: 4th International Conference on Modeling Decisions for Artificial Intelligence (MDAI 2007), Kitakyushu, Japan, August 16-18. (2007)
11. Yao, X.: Evolutionary artificial neural networks, J. International Journal of Neural Systems, vol. 4(3), pp. 203--222. (1993)
12. Montana, D. J., & Davis, L.: Training feedforward neural networks using genetic algorithms, In: Proceedings of the eleventh international joint conference on artificial Intelligence, vol. 1, pp. 762--767. (1989)
13. Nawi, N. M., Rehman, M. Z., and Khan, A.: A New Bat-Based Back-propagation (BAT-BP) Algorithm, In: ICSS 2013, Wrocław, Poland, September 10-12. (2013)
14. Yang, X. S.: A new metaheuristic bat-inspired algorithm, In: Nature Inspired Cooperative Strategies for Optimization (NICSO 2010), pp. 65--74. (2010)

CSBPRNN: A New Hybridization Technique Using Cuckoo Search to Train Back Propagation Recurrent Neural Network

Nazri Mohd. Nawi[1], Abdullah khan[1], M. Z. Rehman[1]

[1]Software and Multimedia Centre, Faculty of Computer Science and Information Technology,
Universiti Tun Hussein Onn Malaysia (UTHM).
P.O. Box 101, 86400 Parit Raja, BatuPahat, Johor Darul Takzim, Malaysia.

nazri@uthm.edu.my, hi100010@siswa.uthm.edu.my,
zrehman862060@gmail.com

Abstract. Nature inspired meta-heuristic algorithms provide derivative-free solution to optimize complex problems. Cuckoo Search (CS) algorithm is one of the most modern addition to the group of nature inspired optimization meta-heuristics. The Simple Recurrent Networks (SRN) were initially trained by Elman with the standard back propagation (SBP) learning algorithm which is less capable and often takes enormous amount of time to train a network of even a moderate size. And the complex error surface of the SBP makes many training algorithms are prone to being trapped in local minima. This paper proposed a new meta-heuristic based Cuckoo Search Back Propagation Recurrent Neural Network (CSBPRNN) algorithm. The CSBPRNN is based on Cuckoo Search to train BPRNN in order to achieve fast convergence rate and to avoid local minima problem. The performance of the proposed CSBPRNN is compared with Artificial Bee Colony using BP algorithm, and other hybrid variants. Specifically OR and XOR datasets are used. The simulation results show that the computational efficiency of BP training process is highly enhanced when coupled with the proposed hybrid method.

Keywords: back propagation, cuckoo search, local minima, and artificial bee colony algorithm, meta-heuristic optimization, recurrent neural network.

1   Introduction

Recurrent neural networks have been an important focus of research and development since1990's [1].There has been large number of research on dynamic system modelling with recurrent neural networks (RNN) [2-4]. They are calculated to learn sequential or time-varying patterns. All recurrent neural networks have feedback (closed loop) connections and are categorized as; BAM, Hopfield, Boltzmann machine, and recurrent back propagation nets [5]. RNN techniques have been applied to a wide variety of problems. Such as forecasting of financial data [6], electric power demand [7] and track water quality and minimize the additives needed for filtering water [8]. The main step

T. Herawan et al. (eds.), *Proceedings of the First International Conference on Advanced Data and Information Engineering (DaEng-2013)*, Lecture Notes in Electrical Engineering 285, DOI: 10.1007/978-981-4585-18-7_13,
© Springer Science+Business Media Singapore 2014

of the RNN-based impedance extraction method is to train the neural network in such a way that it learns the dynamic performance of the test system. There are a number of training algorithms available for neural networks [9-12]. The Simple Recurrent Networks were initially trained by Elman with the standard back propagation (BP) learning algorithm, in which errors are computed and weights are updated at each time step. The BP is not as effective as the back propagation through time (BPTT) learning algorithm, in which error signal is propagated back through time [13]. However, the BP algorithm that was introduced by Rumelhart [14] suffers from two major drawbacks: low convergence rate and instability. They are caused by a possibility of being trapped in a local minimum and prospect of overshooting the minimum of the error surface [15-18]. Also, RNN generate complex error surfaces with multiple local minima, the BP fall into local minima in place of a global minimum [15, 19-21]. Many methods have been proposed to speed up the back propagation based training algorithms. A recent development in evolutionary computation technique has enabled the application of various population based search algorithm in the training of neural network. Many researchers have focused on using genetic algorithms to change the weight parameters of neural networks [12, 15]. As a relatively new stochastic algorithm the particle swarm optimization (PSO) method has gained more and more attention. A preliminary study of PSO trained feed forward networks was presented in [11]. Test results based on the training of some simple problems showed that the performance of PSO is not much better than other methods. However, the authors argued that PSO is still promising in cases where a high number of local minima are known to exist [6].

In-order to improve convergence rate in gradient descent, this papers proposed a new meta-heuristic algorithm called Cuckoo search (CS) [22] integrated with BPRNN algorithm. The proposed Cuckoo Search Back-propagation Recurrent Neural Network (CSBPRNN) convergence behavior and performance will be analyzed on XOR and OR datasets. The results are compared with artificial bee colony using BPNN algorithm, and similar hybrid variants. The main goal is to decrease the computational cost and to accelerate the learning process using a hybridization method.

The remaining paper is organized as follows: Section II gives literature review of learning algorithm. Section III described the proposed CSBPRNN. Result and discussion are explained in Section IV. Finally, the paper is concluded in the Section V.

2   Cuckoo Search (CS) Algorithm

Cuckoo Search (CS) algorithm is a novel meta-heuristic technique proposed by Xin-She Yang [22-28]. This algorithm was stimulated by the obligate brood parasitism of some cuckoo species by laying their eggs in the nests of other host birds. Some host nest can keep direct difference. If an egg is discovered by the host bird as not its own, it will either throw the unknown egg away or simply abandon its nest and build a new nest elsewhere. The CS algorithm follows three idealized rules:

a.   Each cuckoo lays one egg at a time, and put its egg in randomly chosen nest;
b.   The best nests with high quality of eggs will carry over to the next generations;

c. The number of available host nests is fixed, and the egg laid by a cuckoo is discovered by the host bird with a probability $pa \in [0, 1]$.

In this case, the host bird can either throw the egg away or abandon the nest, and build a completely new nest. The rule-c defined above can be approximated by the fraction $pa \in [0, 1]$ of the n nests that are replaced by new nests (with new random solutions).

2.1   Back Propagation Recurrent Neural Network

A Simple Recurrent Network (SRN) in its simplest form is a three layer feed forward (FF) network, but in SRN the hidden layer is self-recurrent which have short-term memory [1]. It is a special case of a general Recurrent Neural Network (RNN) and trained with full back propagation through time (BPTT) and its variants [29]. In this papers we used three layers network with one input layer, one hidden or 'state' layer, and one 'output' layer. Each layer will have its own index variable: k for output nodes, j and l for hidden, and i for input nodes. In a feed forward network, the input vector, x is propagated through a weight layer, V;

$$net_j(t) = \sum_i^n x_i(t)v_{ji} + b_j \tag{1}$$

Where, n the *number of inputs*, b_j is a bias, and f is an output function. In a SRN the input vector is similarly propagated through a weight layer, but also combined with the previous state activation through an additional recurrent weight layer, U;

$$net_j(t) = \sum_i^n x_i(t)v_{ji} + \sum_i^m y_l(t-1)u_{jl} + b_j \tag{2}$$

$$y_j(t) = f(net_j(t)) \tag{3}$$

Where, m is the number of 'state' nodes. The output of the network is in both cases determined by the state and a set of output weights, W;

$$net_k(t) = \sum_j^m y_j(t)w_{kj} + b_k \tag{4}$$

$$y_k(t) = g(net_k(t)) \tag{5}$$

Where, g is an output function. Finally the output is read off from y_k, and the error E against a target vector T is computed:

$$E = \frac{1}{2}\sum_p^n \sum_k^m (T_{pk} - y_{pk})^2 \tag{6}$$

Where, T is the desired output, n is the total number of available training samples and m is the total number of output nodes. And p, adds to the cost, overall output units k. According to gradient descent, each weight change in the network should be proportional to the negative gradient of the cost with respect to the specific weight.

$$\delta_{pk} = -\eta \frac{\delta E}{\delta y_{pk}} \tag{7}$$

Thus, the error for output nodes is;

$$\delta_{pk} = -\frac{\delta E}{\delta y_{pk}}\frac{\delta y_{pk}}{\delta net_{pk}} = (T_{pk} - y_{pk})g'(y_{pk}) \tag{8}$$

$$\delta_{pk} = (T_{pk} - y_{pk})y_{pk}(1 - y_{pk}) \tag{9}$$

And for the hidden nodes;

$$\delta_{pj} = -\left(\sum_k^m \frac{\delta E}{\delta y_{pk}}\frac{\delta y_{pk}}{\delta net_{pk}}\frac{\delta net_{pk}}{y_{pj}}\right)\frac{\delta y_{pj}}{\delta net_{pj}} \tag{10}$$

$$\delta_{pj} = \sum_k^m \delta_{pk}w_{kj}f'(y_{pj}) \tag{11}$$

Thus, the output weights are calculated as;

$$\Delta w_{kj} = \eta \sum_p^n \delta_{pk}y_{pj} \tag{12}$$

And for the input weights;

$$\Delta v_{ji} = \eta \sum_p^n \delta_{pj}x_{pi} \tag{13}$$

Adding a time subscript, the recurrent weights can be modified according to Equation 12;

$$\Delta u_{jh} = \eta \sum_p^n \delta_{pj}(t)y_{ph}(t - 1) \tag{14}$$

3   The Proposed CSBPRNN Algorithm

Step 1:CS and BPRNN are initialized
Step 2: Load the training data
Step 3: Initialize all cuckoo nests
Step 4: Pass the cuckoo nests as weights to network
*Step 5: **While MSE<STOPPING CRITERIA***
Step 6: Feed forward network runs using the weights initialized with CS
Step 7: Calculate the error
Step8: CS keeps on calculating the best possible weight at each epoch until the network is converged.
    End While

In the proposed CSBPRNN algorithm, each best nest represents a possible solution (i.e., the weight space and the corresponding biases for BPRNN optimization in this paper). The weight optimization problem and the size of the solution represent the quality of the solution. In the first epoch, the best weights and biases are initialized with CS and then those weights are passed on to the BPRNN. The weights in BPRNN are calculated. In the next cycle CS will updated the weights with the best possible solution, and CS will continue searching the best weights until the last cycle/ epoch of the network is reached or either the MSE is achieved. The pseudo code of the proposed CSBPRNN algorithm is:

4   Results and Discussion

The simulation experiments are performed on an AMD E-450, 1.66 GHz CPU with 2-GB RAM. The software used for simulation process is MATLAB 2009b. For performing simulations the following algorithms are trained on the 2-bit XOR and 4-bit OR datasets:

a. Back-Propagation Neural Network (BPNN) [14]
b. Artificial Bee Colony Back-Propagation (ABCBP) [30],
c. Artificial Bee Colony Levenberg Marquardt (ABCLM), [31]
d. Artificial Bee Colony Neural Network (ABCNN),[32] and
e. Proposed Cuckoo Search Back propagation Recurrent Neural Network (CSBPRNN).

Three layer neural networks is used for testing of the models, the hidden layer is kept fixed to 5-nodes while output and input layers nodes vary according to the data set given. Log-sigmoid activation function is used as the transfer function from input layer to hidden layer and from hidden layer to the output layer. And for CSBPRNN there is a feedback connection from hidden to input layers. For each problem, trial is limited to 1000 epochs. A total of 20 trials are run for each case. The network results are stored in the result file for each trial. Mean, standard deviation (SD) and accuracy are recorded for each independent trial on 2-bit XOR, and 4-bit OR dataset.

4.1   The 2-Bit XOR Dataset

The first test problem is the 2-bit XOR Boolean function of two binary inputs and a single binary output. In the simulations, we used 2-5-1, CSBPRNN network for two bit XOR dataset. Table 1, shows the CPU time, number of epochs, MSE and the Standard Deviation (SD) for the 2 bit XOR tested on CSBPRNN, ABCLM, ABCBP, ABCNN, and BPNN algorithms. Table 1 shows the MSE performance comparison of CSBPRNN, ABCBP, ABCLM, ABCNN, and BPNN for the 2-bit XOR. Table. 1 shows that the proposed CSBPRNN algorithm successfully avoided the local minima and converge the network within 327 epochs. From this, we can easily see that the proposed CSBPRNN can converge successfully for almost every kind of network structure. Also CSBPRNN has less MSE with high accuracy when compared with other algorithms.

Table 1. CPU Time, Epochs, MSE Accuracy, and Standard deviation for 2-5-1 ANN Structure

Algorithm	BPNN	ABCBP	ABCNN	ABCLM	CSBPRNN
CPUTIME	42.64	172.33	121.74	123.95	54.65
EPOCH	1000	1000	1000	1000	327
MSE	0.22066	2.39E-04	0.12500	0.1250	0
SD	0.01058	6.70E-05	1.05E-06	1.51E-06	0
Accuracy (%)	54.6137	96.4723	74.1759	71.6904	100

4.2   The 4-Bit OR dataset

The second dataset selected for training the proposed CSBPRNN is 4-bit OR dataset. In 4-bit OR, if the number of inputs all is 0, the output is 0, otherwise the output is 1. Again for the four bit input we applied 4-5-1, recurrent neural network structure. Table 2 illustrates the CPU time, epochs, and MSE performance accuracy and SD of the proposed CSBPRNN, ABCBP, ABCLM, ABCNN and BPNN algorithms respectively. From Table 2, we can see that the proposed CSBPRNN algorithm outperforms other algorithms in-terms of CPU time, epochs, MSE, accuracy, and SD. For the 4-5-1 network structure CSBPRNN again has 0 MSE, 100 percent accuracy and 0 SD of the MSE which is better than the BPNN, ABCNN, ABCBP, ABCLM algorithms.

Table 2. CPU Time, Epochs, MSE Accuracy, and Standard deviation for 4-5-1 ANN Structure

Algorithm	BPNN	ABCBP	ABCNN	ABCLM	CSBPRNN
CPUTIME	63.28	162.49	116.31	118.73	10.122
EPOCH	1000	1000	1000	1000	63
MSE	0.0528	1.9E-10	1.8E-10	1.8E-10	0
SD	0.008	1.5E-10	2.3E-11	2.1E-11	0
Accuracy (%)	89.83	99.97	99.99	99.99	100

5   Conclusion

Nature inspired meta-heuristic algorithms provide derivative-free solution to optimize complex problems. A new meta-heuristic search algorithm, called cuckoo search (CS) is used to train BPRNN to achieve faster convergence rate, minimize the training error, and to increase the convergence accuracy. The proposed CSBPRNN algorithm is used to train network on the 2-bit XOR, and 4-Bit OR benchmark dataset. The results show that the proposed CSBPRNN is simple and generic for optimization problems and has better convergence rate, Standard Deviation (SD), and high accuracy than the ABCLM, ABCBP, ABCNN, and BPNN algorithms. In future, the proposed algorithm CSBPRNN will be trained and tested on different benchmarks data classification tasks.

Acknowledgements

The Authors would like to thank Office of Research, Innovation, Commercialization and Consultancy Office (ORICC), Universiti Tun Hussein Onn Malaysia (UTHM) and Ministry of Higher Education (MOHE) Malaysia for financially supporting this Research under Fundamental Research Grant Scheme (FRGS) vote no. 1236.

References:

1. Elman, J. L.: Finding structure in time, J. Cognitive Science, Vol.14 (2), pp. 179--211, (1990)

2. Barbounis, T.G., Theocharis, J.B., et al., Long-term wind speed and power forecasting using local recurrent neural network models, J. IEEE Transactions on Energy Conversion, vol. 21,(1,) pp. 273--284, (2006)

3. Goedtel, A., Dasilva, I.N., and Amaral Semi, P.J., Recurrent Neural Network for Induction Motor Speed Estimation in Industry Application, In: IEEE MELECON, pp. 1134--1137, August (2006)

4. Xiao, P., Venayagamoorthy, G. K., and Corzine, K. A.: Combined Training of Recurrent Neural Networks with Particle Swarm Optimization and Back propagation Algorithms for Impedance Identification, In: Proceedings of the IEEE Swarm Intelligence Symposium (2007)

5. Hecht-Nielsen, R.: Neurocomputing, Addison-Wesley, Reading, PA, (1990)

6. Giles, C. L., Lawrence, S., and Tsoi, A. C.: Rule inference for financial prediction using recurrent neural networks, In: IEEE Conference on Computational Intelligence for Financial Engineering, IEEE Press. (1997)

7. Li, S., Wunsch II, D. C., O'Hair, E., and Giesselmann, M. G.: Wind turbine power estimation by neural networks with Kalman filter training on a SIMD parallel machine, In: International Joint Conference on Neural Networks, (1999)

8. Coulibay, P., Anctil, F., and Rousselle, J.: Real-time short-term water inflows forecasting using recurrent neural networks, In: International Joint Conference on Neural Networks, (1999)

9. Gudise, V.G. and Venayagamoorthy, G.K.: Comparison of particle swarm optimization and backpropagation as training algorithms for neural networks, In: IEEE Swarm Intelligence Symposium, pp. 110--117. April (2003)

10. Janson, D.J. and Frenzel, J.F.: Training product unit neural networks with genetic algorithms, J. IEEE Intelligent Systems and Their Applications, vol. 8 (5), pp. 26--33, Oct. (1993)

11. Salerno, J., Using the particle swarm optimization technique to train a recurrent neural model, In: Ninth IEEE International Conference on Tools with Artificial Intelligence, pp. 45--49, Nov. (1997)

12. Werbos, P.: Back propagation through time: what it does and how to do it, In: Proceedings of the IEEE, vol. 78 (10), pp. 1550--1560. (1990)

13. Temurtas, F., Yumusak, N., Gunturkun, R., Temurtas, H., and Cerezci, O.: Elman's Recurrent Neural Networks Using Resilient Back Propagation for Harmonic Detection, In: 8th Pacific Rim International Conference on Artificial Intelligent Proceedings, vol. 3157, pp. 422--428. (2004)

14. Rumelhart D.E. Hinton G.E., and Williams R.J.: Learning Representations by Back– Propagating Errors Nature, vol. 323, pp. 533-536. (1986)

15. Nawi, N. M., Rehman, M. Z., and Khan, A.: A New Bat-Based Back-propagation (BAT-BP) Algorithm, In: ICSS 2013, Wrocław, Poland, September 10-12. (2013)

16. Nawi, N. M., Khan, A., and Rehman, M. Z.: A New Back-propagation Neural Network optimized with Cuckoo Search Algorithm, In: ICCSA-2013, pp. 413--426. (2013)

17. Nawi, N. M., Khan, A., and Rehman, M. Z.: A New Cuckoo Search based Levenberg-Marquardt (CSLM) Algorithm, In: ICCSA-2013, pp. 438--451. (2013)

18. Nawi, N. M., Khan, A., and Rehman, M. Z.: A New Levenberg-Marquardt based Back-propagation Algorithm trained with Cuckoo Search, In: ICEEI-2013, pp.18--24. (2013)

19. Lahmiri, S.: A comparative study of backpropagation algorithms in financial prediction. J. International Journal of Computer Science, Engineering and Applications (IJCSEA), vol.1 (4). (2011)

20. Nawi, N. M., Ransing, R. S., Salleh, M.N.M., Ghazali,R., and Hamid, N.A, An improved back propagation neural network algorithm on classification problems, J. Communications in Computer and Information Science vol. 118, pp. 177--188. (2010)
21. Nawi, N.M., Ghazali, R., Salleh, M.N.M. The development of improved back-propagation neural networks algorithm for predicting patients with heart disease, J. LNCS, vol. 6377 (4), pp. 317--324. (2010)
22. Yang. XS., and Deb. S.: Cuckoo search via Lévy flights, In: World Congress on Nature & Biologically Inspired Computing, India, pp. 210--214. (2009)
23. Yang, X. S., and Deb, S.: Engineering Optimisation by Cuckoo Search, Int. J. of Mathematical Modelling and Numerical Optimisation, vol. 1 (4), pp. 330-- 343. (2010)
24. Tuba, M., Subotic, M., Stanarevic, N.: Modified cuckoo search algorithm for unconstrained optimization problems, In: European Computing Conference, pp. 263--268. (2011)
25. Tuba, M., Subotic, M., Stanarevic, N.: Performance of a Modified Cuckoo Search Algorithm for Unconstrained Optimization Problems, J. Faculty of Computer Science, vol. 11 (2), pp. 62--74. (2012)
26. Chaowanawate, K., and Heednacram, A.: Implementation of Cuckoo Search in RBF Neural Network for Flood Forecasting, In: Fourth International Conference on Computational Intelligence, Communication Systems and Networks, pp. 22--26. (2012)
27. Pavlyukevich, I.: Levy flights, non-local search and simulated annealing, J. Journal of Computational Physics, vol. 226 (2), pp. 1830--1844. (2007)
28. Walton, S., Hassan, O., Morgan, K., and Brown, M.: Modified cuckoo search: A new gradient free optimisation algorithm. J. Chaos, Solitons& Fractals, vol. 44 (9), pp. 710--718. (2011)
29. Williams, R. J. and Peng, J.: An efficient gradient-based algorithm for on-line training of-recurrent network trajectories, J. Neural Computation, vol.2, pp. 490--501. (1990)
30. Nandy, S., Sarkar, P. P., and Das, A.: Training a Feed-forward Neural Network with Artificial Bee Colony Based Backpropagation Method, J. International Journal of Computer Science & Information Technology (IJCSIT), vol. 4 (4), pp. 33--46. (2012)
31. Ozturk, C., and Karaboga, D.: Hybrid Artificial Bee Colony algorithm for neural network training, In: IEEE Congress of Evolutionary Computation (CEC), pp. 84--88. (2011)
32. Karaboga, D., and Ozturk, C..: Neural networks training by artificial bee colony algorithm on pattern classification, J. Neural Network World, vol. 19, no. 10, pp. 279--292, (2009).

Data Mining of Protein Sequences with Amino Acid Position-Based Feature Encoding Technique

Muhammad Javed Iqbal[1], Ibrahima Faye[2], Abas Md Said[1], Brahim Belhaouari Samir[3]

[1] Computer and Information Science Department
[2] Fundamental and Applied Science Departments
[1,2] Universiti Teknologi PETRONAS, Malaysia
[3] Colleges of Sciences, Alfaisal University, Riyadh, KSA
Javed1797@hotmail.com, ibrahima_faye@petronas.com.my, abass@petronas.com.my, sbelhaouari@alfaisal.edu

Abstract. Biological data mining has been emerging as a new area of research by incorporating artificial intelligence and biology techniques for automatic analysis of biological sequence data. The size of the biological data collected under the Human Genome Project is growing exponentially. The available data is comprised of DNA, RNA and protein sequences. Automatic classification of protein sequences into different groups might be utilized to infer the structure, function and evolutionary information of an unknown protein sequence. The accurate classification of protein sequences into family /superfamily based on the primary sequence is a very complex and open problem. In this paper, an amino acid position-based feature encoding technique is proposed to represent a protein sequence using a fixed length numeric feature vector. The classification results indicate that the proposed encoding technique with a decision tree classification algorithm has achieved 85.9% classification accuracy over the Yeast protein sequence dataset.

Keywords- Data Mining, Feature Vector, Superfamily, Protein Classification, Biological data, Feature Encoding.

1 Introduction

Biological data mining has emerged as a new area of research for the development of highly sophisticated automated tools to examine the biological data for knowledge extraction and making some predictions. The ultimate objectives of the genome sequencing projects were to discover the biological data and store it in publicly accessible databases. The sequenced data largely comprised of Deoxyribonucleic acid (DNA), Ribonucleic acid (RNA) and protein and their volume is increasing every day. Among these macromolecules, proteins are the essential building blocks of living organisms [4]. A protein sequence is consisted of 20 amino acids and these amino acids may combine in any order to form a specific protein. The essential attributes associated with a protein are: sequence, structure and function. Each protein sequence has a designated three dimensional structure and it also performs some function in the cell. Presently, UniProt Knowledge Base (UniProtKB) database includes 33,646,106

T. Herawan et al. (eds.), *Proceedings of the First International Conference on Advanced Data and Information Engineering (DaEng-2013)*, Lecture Notes in Electrical Engineering 285, DOI: 10.1007/978-981-4585-18-7_14,
© Springer Science+Business Media Singapore 2014

protein sequences from a variety of species and the sequences are increasing every-day. The experimentally determined structures in the Protein Data Bank (PDB) are approximately 85000. Mining or classification of protein sequences based on the pri-mary sequence to determine the structure and function of an unknown protein is a critical and challenging problem in Bioinformatics and computational biology.

A number of local and global alignment methods were introduced by Needleman - Wunsch, Dayhoff et al. and Smith-Waterman to find similarity between the protein and DNA sequences. Several other techniques such as: BLAST, FASTA and Hidden Markov Models (HMM) were also developed and are still being used for the biologi-cal analysis of the available protein sequences. A brief review of some most popular computational approaches using machine learning techniques from the literature are described below.

Jong et al. [1] proposed a feature extraction technique using Position Specific Scor-ing Matrix (PSSM) for classification of protein sequences into different families. Experiments were performed using four different classifiers on the yeast protein se-quence data of three families and the classification accuracy achieved was 72.5%. Bandyopadhyay et al. [2] used a 1-gram encoding method to extract features from a protein sequence. For the classification purpose, a variable length fuzzy genetic clus-tering algorithm was employed to find a number of prototypes for each superfamily. Three superfamilies namely Globin, Ras and trypsin were experimented and the over-all accuracy observed was 81.3 %. Mansoori et al. [4] proposed a feature encoding method which involved 2-gram pairs of amino acids. Fuzzy rules were created for the classification of protein sequences into superfamilies. In [7], Angadi et al. proposed unsupervised techniques to classify protein sequences to SCOP superfamilies. The authors created a database and similarity matrix of P-values generated through a BLAST. An ART2 unsupervised classification algorithm was employed in the train-ing and testing. Patrick et al. [8] proposed an alignment independent algorithm called an unaligned protein sequence classifier (UPSEC) to detect patterns or motifs in a sequence for effectively classification of protein sequences into families. Swati et al. [9] proposed an adoptive multi objective genetic algorithm (AMOGA) to optimize the structure of a radial basis function network (RFBN). A 2-gram encoding method was used to represent a protein sequence.

From the literature, it is concluded that various feature encoding or extraction techniques were developed to represent a protein sequence with a feature vector. The-se feature vectors are then analyzed by the classification and clustering algorithms for grouping of various protein sequences into their respective superfamilies. The most popular feature extraction among them is an n-gram encoding technique. This tech-nique is although very straightforward; however the machine learning experts are still working on for the development of new encoding techniques which can be effectively used to classify unknown protein sequences with maximum classification accuracy in minimum time.

In this paper, our objective is to propose an amino acid position-based feature en-coding technique to represent a primary sequence of a protein. The proposed tech-nique finds all the occurrence positions of each amino acid in a sequence. The per-formance of the encoding technique was validated by using different classification algorithms. The introduced encoding technique has successfully classified homologous sequences into different superfamilies with significantly improved classification accuracy, specificity, sensitivity and a low misclassification rate.

The rest of the paper is arranged as follows. Section 2 presents the proposed methodology. Section 3 illustrates the experimental results, performance measures metrics and the comparison of the proposed encoding technique with the previously available best classification accuracy on the same dataset. The results are discussed in section 4 and the conclusion is presented in the last section.

2   Methodology

The proposed methodology involves different phases for an accurate data mining of protein sequences into different superfamilies. This section introduces the necessary phases and provides a brief description of each of them. Following are the different phases of the proposed methodology.

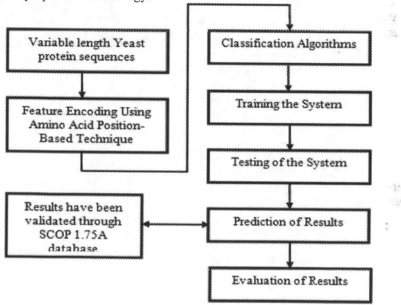

Fig. 1. Phases of the Proposed Methodology

Figure 1 shows the entire process of a protein sequence classification system. The data of each superfamily used in the experiments was taken from the UniProtKB database. The sequence's length is varied from few amino acids to hundred amino acids. The arbitrary chosen data is divided into two parts: training and testing. Both the training and testing sequence data has been encoded using amino acid position-based feature encoding method. The sequence encoding technique has been applied independently on each protein sequence of each family.

Let i be one of the amino acid:

$$i = A,C,D,E,F,G,H,I,K,L,M,N,P,Q,R,S,T,V,W,Y$$

Muhammad Javed Iqbal et al.

The proposed amino acid position-based feature encoding technique finds all the occurrence positions of each amino acid in a sequence. For example: p_1^i shows the first occurrence position of amino acid i in a sequence, p_2^i is the second occurrence position and $p_{n_i}^i$ is the last occurrence position of amino acid i in a sequence, where n_i is the maximum number of times the specific amino acid symbol i occurs in a sequence. The vector of positions for an amino acid i in the given sequence is obtained as follows:

$$p^i = (p_1^i, p_2^i, \dots, p_{n_i}^i) \tag{1}$$

From the above vector of positions, two features are calculated:

the mean: $\mu_p^i = \frac{1}{n_i} \sum_{j=1}^{n_i} p_j^i,$ (2)

the variance: $var_p^i = \frac{1}{n_i} \sum_{j=1}^{n_i} (p_j^i - \mu_p^i)^2,$ (3)

By using the proposed feature encoding technique, 2 features namely μ_p^i and var_p^i has been obtained for each amino acid in a sequence. Subsequently, the features for all 20 amino acid symbols have been calculated. The total number of features calculated for each sequence will be 40. The example sequence "ACAADADYDKVLIIFACYVCVY" explains the sequence encoding process of amino acid symbol A in detail.

Amino acid positions for a symbol A: $p_1^A = 1, p_2^A = 3, p_3^A = 4, p_4^A = 6, p_5^A = 16$

Amino acid positions for a symbol C: $p_1^C = 2, p_2^C = 17, p_3^C = 20$

Amino acid positions for a symbol D: $p_1^A = 5, p_2^A = 7, p_3^A = 9$

Similarly, we have calculated $p_1, p_2, \dots p_{n_i}$ for each symbol in a sequence. To extract a feature vector from the above calculated position vectors for each symbol and for each sequence, the mean and the variance is taken of the above positions for each symbol respectively: $mean(p_1^A, p_2^A, p_3^A, p_4^A, p_5^A)$ and $var(p_1^A, p_2^A, p_3^A, p_4^A, p_5^A)$. The extracted feature vector used for symbol A, C and D from the sample sequence is as shown in Figure 2:

μ_p^A	var_p^A	μ_p^C	var_p^C	μ_p^D	var_p^D
6	34.50	13	93	7	4

Fig. 2. Feature Vector for Amino Acid symbol A, C and D

Family Name	μ_p^A	Var_p^A	μ_p^C	Var_p^C	...	μ_p^W	Var_p^W	μ_p^Y	Var_p^Y
	168.6	8552.1	185.9	12465.9	...	27.3	555.7	169.3	801.1
	165.9	10883.8	146.3	6080.6	...	27.1	482.5	213.8	506.6
Metabolism	165.9	10883.8	146.3	6080.6	...	27.1	482.5	213.8	506.6
	...	...	...	...	...	...	...	...	...
	244.2	30067.8	188.0	10666.8	...	21.2	623.3	420.1	72.0
	169.1	11198.2	182.4	28126.3	...	164.0	46526.0	297.0	307520.0
	113.9	5612.0	243.0	19431.5	...	44.5	924.5	266.7	1521.3
Transcription	75.9	2125.5	235.0	50.0	...	196.5	64440.5	158.3	5461.2
	...	...	...	...	...	...	...	...	...
	199.9	12706.2	89.2	2628.2	...	208.3	48971.6	381.3	44445.1
	319.5	35566.3	191.2	17042.5	...	252.5	15068.7	261.1	223.9
	270.2	38241.2	315.8	15253.3	...	213.9	29757.0	315.0	383.9
CellTransport	348.7	43231.7	321.9	22025.4	...	174.0	33597.5	361.6	377.4
	...	...	...	...	...	...	...	...	...
	230.4	28181.4	527.2	27797.8	...	366.5	43540.4	314.0	319.5

Fig. 3. Sample Amino Acid Position-Based Feature Vector Space

Figure 3 shows the sample feature vector space of the given experimental data of three different families: metabolism, transcription and cell transport. The column titles represent the means and variance of each amino acid symbol for each sequence. Any unknown sequence can be encoded using the proposed encoding technique in a straightforward way with a minimum number of features. Because, there are 20 amino acid symbols, as a result the total numbers of features become 40. The dimensions of the data will be $No.\,of\,sequences \times 40$. For instance, if we have 20 sequences, the feature vector space dimension will be 20×40. In the next section, the experimental results in detail have been presented.

3 Experiments and Results

In this paper, we have extracted Yeast protein sequences from UniProtKB database (http://www.uniprot.org/help/uniprotkb). Three functional families: metabolism, transcription and cellular transport, transport facilities were used in the experiments. The sequences of the three families were selected randomly. The detail of the dataset involved in the experiments is shown in Table 1.

Table 1. Details of dataset used in the experiments

Family Name	Number of Sequences
Metabolism	752
Transcription	520
Cellular Transport, Transport Facilities	565
Total Sequences	1837

For the evaluation of results, a tenfold cross validation method was employed. In each fold, 70% sequences were used in training and remaining 30% for testing. Popular classification algorithms: naïve Bayes, decision tree, neural network, random forest and multilayer perceptron were used to classify protein sequences. The performance measure metrics such as accuracy, true positive rate (TPR), false positive rate (FPR), specificity, sensitivity, recall, F-measure and Mathews Correlation Coefficient (MCC) were used for performance evaluation [1], [5], [6].

The confusion matrices in Table 2, 3, 4, 5 and 6 show classification results obtained upon each classifier using amino acid position-based feature encoding technique. The rows of a confusion matrix show actual number of sequences and the columns represent the predicted number of sequences.

Table 2. Confusion Matrix obtained with Naive Bayes Classifier

Data\results	Metabolism	Transcription	Cellular Transport
Metabolism	541	141	70
Transcription	125	319	76
Cellular Transport	203	110	252

Muhammad Javed Iqbal et al.

Table 3. Confusion Matrix obtained with Decision Tree Classifier

Data\results	Metabolism	Transcription	Cellular Transport
Metabolism	668	30	54
Transcription	32	461	27
Cellular Transport	74	38	453

Table 4. Confusion Matrix obtained with Neural Network Classifier

Data\results	Metabolism	Transcription	Cellular Transport
Metabolism	618	85	49
Transcription	198	290	32
Cellular Transport	303	82	380

Table 5. Confusion Matrix obtained with Random Forest Classifier

Data\results	Metabolism	Transcription	Cellular Transport
Metabolism	496	126	130
Transcription	99	336	85
Cellular Transport	194	74	297

Table 6. Confusion Matrix obtained with Multilayer Perceptron Classifier

Data\results	Metabolism	Transcription	Cellular Transport
Metabolism	531	105	116
Transcription	230	229	61
Cellular Transport	133	96	336

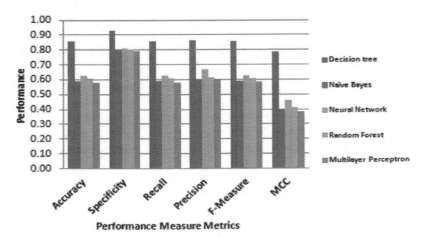

Fig. 4. Performance Measure Metrics' Values Based on the Confusion Matrices in Table 2, 3,4,5 and 6

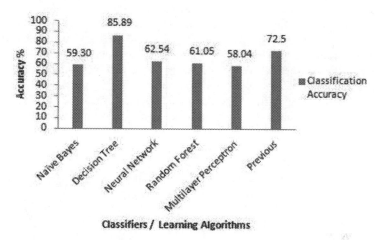

Fig. 5. Comparison of the Proposed and Previous Classification Accuracy Results

4   Discussion

The objective of the proposed work was to classify protein sequences into their respective families based on the primary sequence information solely. Moreover, this classification would be helpful in predicting the structure, function and evolutionary information of any unknown or newly discovered protein. In this paper, we have investigated variable length sequence data of three Yeast protein functional families using the distance-based encoding technique. The values of performance measure metrics obtained with different classifiers have been shown in Figure-4. The results demonstrate that the amino acid position-based method with decision tree classifier outperforms in terms of classification accuracy, specificity, recall, f-measure and MCC. The value of MCC varies from +1 (best) to -1 (worst).The average accuracy obtained with decision tree was 85.9% which shows significant improvement from the previous best accuracy result 72.5% from the literature [1]. Figure-5 shows graphical illustration of the proposed and previous classification accuracy results obtained using different classifiers. The association of the sequence with a superfamily is validated using SCOP 1.75A database. It is observed from the experiments that the amino acid position-based encoding technique selects the minimum and most informative features from a protein sequence.

Conclusion

The exploitation of data mining and machine learning techniques are becoming very popular in developing effective computational tools to analyze and generate useful information from the biological data. In this paper, our goal was to propose an amino acid position-based feature encoding technique to numerically represent a primary sequence of a protein. The fixed length feature vector space was computed from the means and variance of the position vector of each amino acid. The experiments were carried out on the numeric protein data obtained after feature encoding using five different classification algorithms. The protein sequences were successfully classified into different superfamilies with highest classification accuracy. In our experiments, the decision tree classification algorithm has shown significant improvement in the classification accuracy, specificity, precision, recall and F-measure. The results indicate that the proposed technique could be successfully used to extract different characteristics of a protein sequence like structure, function or evolutionary information from the primary sequence. The proposed technique is very simple, robust, reliable and highly accurate. In future, the amino acid position-based encoding technique can be further extended by using different level of decompositions.

Acknowledgement

The authors would like to thank UNIVERSITI TEKNOLOGI PETRONAS for supporting this work.

References

1. Jeong, J. C., Lin, X. and Chen, X. W.: "On position-specific scoring matrix for protein function prediction," IEEE/ACM Transactions on Computational Biology and Bioinformatics, vol. 8, pp. 308-315, 2011
2. Bandyopadhyay, S.: "An efficient technique for superfamily classification of amino acid sequences: Feature extraction, fuzzy clustering and prototype selection," ELSEVIER Jounal of Fuzzy Sets and Systems, vol. 152, pp. 5-16, 2005
3. Vipsita, S.,Shee, B. K. and Rath, S. K.: "An efficient technique for protein classification using feature extraction by artificial neural networks IEEE India Conference: Green Energy, Computing and Communication, INDICON 2010
4. Mansoori, E. G.,Zolghadri, M. J. and Katebi, S. D.:"Protein superfamily classification using fuzzy rule-based classifier," IEEE Transactions on Nanobioscience, vol. 8, pp. 92-99, 2009
5. Rossi, A. L. D., & De Oliveira Camargo-Brunetto, M. A.: "Protein classification using artificial neural networks with different protein encoding methods International Conference on Intelligent Systems Design and Applications, ISDA'07, pp. 169-174
6. Wang, J. T. L.,Ma, Q.,Shasha, D. and Wu, C. H.: "New techniques for extracting features from protein sequences," IBM Systems Journal, vol. 40, pp. 426-441, 2001
7. Angadi, U. B. & Venkatesulu, M.: Structural SCOP superfamily level classification using unsupervised machine learning. IEEE/ACM Transactions on Computational Biology and Bioinformatics 9, 601-608, doi:10.1109/tcbb.2011.114
8. Ma, P. C. H. & Chan, K. C. C. UPSEC: An algorithm for classifying unaligned protein sequences into functional families. J. Comput. Biol. 15, 431-443, doi:10.1089/cmb.2007.0113 (2008)
9. Vipsita, S. & Rath, S. K.: Protein superfamily classification using adaptive evolutionary radial basis function network. International Journal of Computational Intelligence and Applications 11, doi:10.1142/s1469026812500265 (2012)

Detecting Definite Least Association Rule in Medical Database

Zailani Abdullah[1], Tutut Herawan[2] and Mustafa Mat Deris[3]

[1]Department of Computer Science, Universiti Malaysia Terengganu
[2]Faculty of Computer Science & Information Technology, Universiti Malaya
[2]Universitas Teknologi Yogyakarta, Yogyakarta, Indonesia
[3]Faculty of Science Computer & Information Technology, Universiti Tun Hussein Onn Malaysia

zailania@umt.edu.my, tutut@um.edu.my, mmustafa@uthm.edu.my

Abstract. Least association rule refers to the rule that only rarely occur in database but they might reveal some interesting knowledge in certain domain applications. In certain medical datasets, finding these rules is very important and required further analysis. In this paper we applied our novel measure known as Definite Factor (DF) with SLP-Growth algorithm to mining the Definite Least Association Rule (DELAR) from a benchmarked medical datasets. DELAR is also highly correlated and evaluated based on standard Lift measure. The result shows that DF can be used as alternative measure in capturing the interesting rules and thus verify its scalability.

Keywords: Definite, Least association rules, Medical data.

1 Introduction

Association rule is one of the popular methods in data mining to analyze and understand the hidden patterns from database. It was first introduced by Agrawal *et al.* [1] and still received a great attention from knowledge discovery community. In term of the strength of association rule, it is typically measured by support and confidence. Usually, specific threshold values from both measures are employed to determine the classification of frequent association rule.

In certain domain applications, uncommon rule or least association rule is more interesting and valuables. As a result, several works have been conducted towards mining least association rule [2-15]. Generally, least association rule is a contradiction of frequent association rule and typically required more scalable techniques and measures. This rule is very important in discovering rarely occurring events but significantly important in various applications and one of them is in medical domains [16]. In fact, lease association rule can facilitate in determining and representing the real world of medical knowledge due to their simplicity, uniformity, transparency, and ease of inference [32]. Typically, many series of association rule mining

T. Herawan et al. (eds.), *Proceedings of the First International Conference on Advanced Data and Information Engineering (DaEng-2013)*, Lecture Notes in Electrical Engineering 285, DOI: 10.1007/978-981-4585-18-7_15,

algorithms are using the minimum supports-confidence framework to limit the number of rules. As a result, by increasing or decreasing the minimum support or confidence values, the interesting rules might be missing out or untraceable. Since the complexity of study, difficulties in algorithms [17] and it may require excessive computational cost, there are very limited attentions have been paid to discover highly correlated least association rule. Highly correlated of least association rule is referred to the itemsets that its frequency does not satisfy a minimum support but are closely correlated. Association rule is classified as highly definite and correlated if it has positive correlation and high certainty. Recently, statistical correlation technique has been widely applied in the transaction databases [18], which to find relationship among pairs of items whether they are highly positive or negative correlated. In reality, it is not absolutely true that only the frequent items have a positive correlation rather than the least items. In this paper, we address the problem of mining least association rule with the objectives of discovering least association rule with highly correlated and certainty. An algorithm named Significant Least Pattern Growth (SLP-Growth) with DF measure [2,13] is employed to extract the association rule from the give dataset. The algorithm imposes interval support to capture all least itemsets family first before continuing to construct a significant least pattern tree (SLP-Tree). The measure named Lift [26] to find the degree of correlation between itemset is also embedded in the algorithm. Furthermore, DF measure is used to fine-tune and finally produce Definite Least Association Rule (DELAR). In summary, the main contributions of this paper are as follows:

- We enhanced the previous SLP-Growth algorithm by embedding with the scalable DF measure
- We evaluate the performance of the enhanced algorithm against a benchmarked SPECT Heart dataset [32]
- We show that the number of DELAR being produced is relatively very small as compared to the typical measures

The reminder of this paper is organized as follows. Section 2 describes the related work. Section 3 explains the basic concepts and terminology of association rules mining. Section 4 discusses the proposed method. This is followed by performance analysis in section 5. Finally, conclusion is reported in section 6.

2   Related Work

The research on mining least association rule gaining more attentions due to its contribution in certain domain applications. Thus, there have been quite a number of works published in an attempt to improve the current limitations in term of algorithms and measures.   Hoque *et al.* [19] proposed Multi-objective Genetic Algorithm (MOGA) method and NBD-Apriori-MOFR algorithm to generate rare rules. However, one of the disadvantages of these algorithms is a high computational cost. Lavergne *et al.* [20] introduced TRARM-RelSup by combines the efficiency of targeted association mining querying with the capabilities of mining rare rule. The main limitation is, it increases the complexity in mining the targeted rules. Rawat *et al.* [21] suggested Probability Apriori Multiple Minimum Support (PAMMS) algorithm to discover rare association rule in grid environment. However, the generation of candidate itemset is uncontrollable and thus requires excessive of memory consumption. Zhou *et al.* [22] proposed an approach to mine the association rules by considering only infrequent itemset. The limitation is, Matrix-based Scheme (MBS) and Hash-based scheme (HBS) algorithms are facing the expensive cost of hash collision. Ding [23] introduced Transactional Co-occurrence Matrix (TCOM for mining association rule among rare items. However, the implementation of this algorithm is too costly. Yun *et al.* [17] suggested the Relative Support Apriori Algorithm (RSAA) to generate rare itemsets. The challenge is if the minimum allowable relative support is set close to zero, it takes similar time taken as performed by Apriori. Koh *et al.* [24] proposed Apriori Inverse algorithm to mine infrequent itemsets without generating any frequent rules. The main constraints are it suffers from too many candidate generations and time consumptions during generating the rare association rule. Liu *et al.* [25] suggested Multiple Support Apriori (MSApriori) algorithm to extract the rare association rules. In actual implementation, this algorithm is still suffered from the "rare item problem". Most of the proposed approaches [17,19-25] are using the percentage-based approach in order to improve the performance of existing single minimum support based approaches. Brin *et al.* [26] presented objective measure called lift and chi-square as correlation measure for association rules. Lift compares the frequency of pattern against a baseline frequency computed under statistical independence assumption. Instead of lift, there are quite a number interesting measures have been proposed for association rules. Omiecinski [27] introduces two interesting measures based on downward closure property called all confidence and bond. Lee *et al.* [28] proposes two algorithms for mining all confidence and bond correlation patterns by extending the frequent pattern-growth methodology. Han *et al.* [29] proposed FP-Growth algorithm which break the two bottlenecks of Apriori series algorithms. Currently, FP-Growth is one of the fastest approach and most popular algorithms for frequent itemsets mining. This algorithm is based on a prefix tree representation of database transactions (called an FP-tree).

3   Basic Concept and Terminology

4.1 Association Rule

Association rule was first introduced to study customer purchasing patterns in retail stores [1]. Nowadays, it has been applied in various types of disciplines [30,31]. Let I is a non-empty set such that $I = \{i_1, i_2, \cdots, i_n\}$, and D is a database of transactions where each T is a set of items such that $T \subset I$. An association rule is a form of $A \Rightarrow B$, where $A, B \subset I$ such that $A \neq \phi$, $B \neq \phi$ and $A \cap B = \phi$. The set A is called antecedent of the rule and the set B is called consequent of the rule. An item is a set of items. A k-itemset is an itemset that contains k items. An itemset is said to be frequent if the support count satisfies a minimum support count (minsupp). The set of frequent itemsets is denoted as L_k. The support of the association rules is the ratio of transaction in D that contain both A and B (or $A \cup B$). The support is also can be considered as probability $P(A \cup B)$. The confidence of the association rules is the ratio of transactions in D contains A that also contains B. The confidence also can be considered as conditional probability $P(B|A)$. Association rules that satisfy the minimum support and confidence thresholds are said to be strong.

3.2 Definite Factor

Definite Factor is a formulation of exploiting the support difference between itemsets with the frequency of an itemset against a baseline frequency. The baseline frequency of itemset is presumed as statistically independence. The Definite Factor denoted as DF and

$$DF(I) = |P(A) - P(B)| \times \frac{P(A \cup B)}{P(A)P(B)} \tag{1}$$

It also can be expressed as

$$DF(I) = |\text{supp}(A) - \text{supp}(B)| \times \left(\frac{\text{supp}(A \Rightarrow B)}{\text{supp}(A) \times \text{supp}(B)} \right) \tag{2}$$

DF is a new measurement to indicate the degree of certainty of association rules. The advantages of this measurement are, it takes into account the support of right-hand side (consequence) and always in the range $[0,1)$. The higher value of DF means the affinity between itemsets is more definite and convinced. In fact, the range of DF is more realistic as compared to wide-ranging of correlation values in lift measurement.

4 Methodology

4.1 Algorithm Development

Determine Interval Support for least Itemset. The Interval Support is a form of ISupp (ISMin, ISMax) where ISMin is a minimum and ISMax is a maximum values respectively, such that $\text{ISMin} \geq \phi$, $\text{ISMax} > \phi$ and $\text{ISMin} \leq \text{ISMax}$. Itemsets are said to be significant least if they satisfy two conditions. First, support counts for all items in the itemset must greater ISMin. Second, those itemset must consist at least one of the least items. In brevity, the significant least itemset is a union between least items and frequent items, and the existence of intersection between them.

Construct Significant Least Pattern Tree. A Significant Least Pattern Tree (SLP-Tree) is a compressed representation of significant least itemsets. This trie data structure is constructed by scanning the dataset of single transaction at a time and then mapping onto path in the SLP-Tree. In the SLP-Tree construction, the algorithm constructs a SLP-Tree from the database. The SLP-Tree is built only with the items that satisfy the ISupp.

Generate Significant Least Pattern Growth (SLP-Growth). SLP-Growth is an algorithm that generates significant least itemsets from the SLP-Tree by exploring the tree based on a bottom-up strategy. 'Divide and conquer' method is used to decompose task into a smaller unit for mining the desired patterns in conditional databases, which can optimize the searching space. The algorithm extracts the prefix path sub-trees that ending with any least item. The complete SLP-Growth algorithm is shown in Fig. 1.

```
1:   Read dataset, D
2:   Set Interval Support (ISMin, ISMax)
3:   for items, I in transaction, T do
4:       Determine support count, ItemSupp
5:   end for loop
6:   Sort ItemSupp in descending order, ItemSuppDesc
7:   for ItemSuppDesc do
8:       Generate List of frequent items, FItems > ISMax
9:   end for loop
10:  for ItemSuppDesc do
11:      Generate List of least items, ISMin <= LItems < ISMax
12:  end for loop
13:  Construct Frequent and Least Items, FLItems = FItems U LItems
14:  for all transactions,T do
15:      if (LItems  1 in T > 0) then
16:          if (Items in T = FLItems) then
17:              Construct items in transaction in descending order,TItemsDesc
18:          end if
19:      end if
20:  end for loop
21:  for TItemsDesc do
22:      Construct SLP-Tree
23:  end for loop
24:  for all prefix SLP-Tree do
25:      Construct Conditional Items, CondItems
26:  end for loop
27:  for all CondItems do
28:      Construct Conditional SLP-Tree
29:  end for loop
30:  for all Conditional SLP-Tree do
31:      Construct Association Rules, AR
32:  end for loop
33:  for all AR do
34:      Calculate Support and Confidence
35:      Apply Correlation
36:      Apply Definite Factor
36:  end for loop
```

Fig. 1. SLP-Growth Algorithm

4.2 Value Assignment

Apply Correlation and Definite Factor Measures. Besides the others typical measures, the values of association rule are also derived based on Lift [26] and DF [2,13] measures. The processes of generating the values of association rule are taken place after all patterns and association rules are completely produced.

Discover Definite Least Association Rule. From the list of valued association rules, the algorithm begins to scan all of them. However, only those valued association rule with the correlation value that more than one and with a certain DF value are captured and considered as DELAR. For association rules that occur less than threshold values will be pruned out and classified as low correlation.

5   Experiment Test

The performance analysis was made by comparing the total number of association rules being produced by different measures. The algorithm determines which association rules that are highly correlated and certainty based on the specified threshold values of Lift and DF measures, respectively. The interpretations are made based on the results obtained.

The experiment was conducted on Cardiac Single Proton Emission Computed Tomography (SPECT) dataset or also known as SPECT Heart dataset [32]. SPECT is a nuclear medicine technique that uses radiopharmaceuticals to produce images representing slices through the body in different planes. By embedding FP-Growth algorithm with ISupp, 31,710 association rules are produced. Association rules are formed by applying the relationship of an item or many items to an item (cardinality: many-to-one). Here, the maximum number of items appears in each association rule is set to 6. Fig. 2 depicted the correlation's classification of least association rules. The rule is categorized as DELAR if it has positive correlation and DF should be at least 0.5. Fig. 3 illustrates the summarization of correlation analysis with different ISupp.

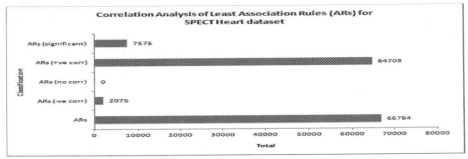

Fig. 2. Classification of association rules using correlation analysis. Only 2.82% from the total of 4,082 association rules are classified as significant least association rules.

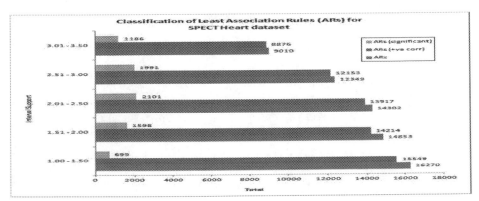

Fig. 3. Correlation analysis of least association rules using variety ISupp. The total numbers of overall association rules are decreased when the predefined ISupp thresholds are increased.

6 Conclusion

Mining least association rule is very important in revealing new information for certain domain applications. One of the potential fields is in medical domains. However from the literature, only few attentions have been paid in mining this rule as compared to mining frequent rules. In this paper we embedded SLP-Growth algorithm with Definite Factor (DF) measure to generate Definite Least Association Rules (DELAR) from a benchmarked medical dataset. The result show that DELAR is highly correlated and certainty. Moreover, it also reveals that SLP-Growth and DF measure can discover the interesting rules and thus verify its scalability.

References

1. Agrawal, R., Imielinski, T., and Swami, A.: Database Mining: A Performance Perspective. IEEE Transactions on Knowledge and Data Engineering, 5 (6), 914–925 (1993)
2. Abdullah, Z., Herawan, T. and Deris, M.M.: Mining Significant Least Association Rules using Fast SLP-Growth Algorithm. In T.H. Kim and H. Adeli (Eds.): AST/UCMA/ISA/ACN 2010, LNCS, vol. 6059, pp. 324–336 (2010)
3. Z. Abdullah, T. Herawan and M.M. Deris. (2010). Scalable Model for Mining Critical Least Association Rules. In Rongbo Zhu et al. ICICA 2010, LNCS, vol. 6377, pp. 509–516. Springer Heidelberg (2010)
4. Abdullah, Z., Herawan, T., Noraziah, A., and Deris, M.M.: Extracting Highly Positive Association Rules from Students' Enrolment Data. Procedia Social and Behavioral Sciences, 28, 107–111 (2011)
5. Abdullah, Z., Herawan, T., Noraziah, A., and Deris, M.M.: Mining Significant Association Rules from Educational Data using Critical Relative Support Approach. Procedia Social and Behavioral Sciences, 28, 97–101 (2011)
6. Herawan, T., Vitasari, P., and Abdullah, Z.: Mining Interesting Association Rules of Student Suffering Mathematics Anxiety. In J.M. Zain et al. (Eds.): ICSECS 2011, CCIS, vol. 188, part II, pp. 495–508. Springer Heidelberg (2011)
7. Herawan, T., Yanto, I.T.R., and Deris, M.M.: SMARViz: Soft maximal association rules visualization. In H. Badioze Zaman et al. (Eds.): IVIC 2009, LNCS, vol. 5857, pp. 664–674. Springer Heidelberg (2009)
8. Abdullah, Z., Herawan, T., and Deris, M.M.: Visualizing the Construction of Incremental Disorder Trie Itemset Data Structure (DOSTrieIT) for Frequent Pattern Tree (FP-Tree). In H.B. Zaman et al. (Eds.): IVIC 2011, LNCS, vol. 7066, pp. 183–195. Springer Heidelberg (2011)
9. Ahmad, N., Abdullah, Z., Herawan, T., and Deris, M.M.: WLAR-Viz: Weighted Least Association Rules Visualization. In B. Liu et al. (Eds): ICICA 2012, LNCS, vol.7473, pp.592–599. Springer Heidelberg (2012)
10. Herawan, T., Yanto, I.T.R., and Deris, M.M.: Soft Set Approach for Maximal Association Rules Mining. In D. Ślęzak et al. (Eds.): DTA 2009, CCIS, vol. 64, pp. 163–170 (2009)
11. Herawan, T., and Deris, M.M.: A Soft Set Approach for Association Rules Mining. Knowledge Based Systems, 24 (1), 186–195 (2011)
12. Abdullah, Z., Herawan, T., and Deris, M.M.: An Alternative Measure for Mining Weighted Least Association Rule and Its Framework. In J.M. Zain et al. (Eds.): ICSECS 2011, CCIS, vol. 188, part II, pp. 475–485. Springer Heidelberg (2011)

13. Abdullah, Z., Herawan, T., and Deris, M.M.: Efficient and Scalable Model for Mining Critical Least Association Rules. In a special issue from AST/UCMA/ISA/ACN 2010, Journal of The Chinese Institute of Engineer, 35 (4), 547–554 (2012)
14. Abdullah, Z., Herawan, T., Ahmad, N., and Deris, M.M.: DFP-Growth: An Efficient Algorithm for Mining Pattern in Dynamic Database. In B. Liu et al. (Eds): ICICA 2012, LNCS, vol. 7473, pp. 51–58. Springer Heidelberg (2012)
15. Ahmad, N., Abdullah, Z., Herawan, T., and Deris, M.M.: Scalable Technique to Discover Items Support from Trie Data Structure. In B. Liu et al. (Eds): ICICA 2012, LNCS, vol. 7473, pp. 500–507. Springer Heidelberg (2012)
16. Szathmary, L., Valtchev, P., and Napoli, A.: Generating Rare Association Rules Using the Minimal Rare Itemsets Family. International Journal on Software Informatics, 4, 219-238 (2010)
17. Yun, H., Ha, D., Hwang, B., Ryu, K.H.: Mining Association Rules on Significant Rare Data Using Relative Support. The Journal of Systems and Software, 67(3), 181–19 (2003)
18. Xiong, H., Shekhar, S., Tan, P-N., and Kumar, V.: Exploiting A Support-Based Upper Bond Pearson's Correlation Coefficient for Efficiently Identifying Strongly Correlated Pairs. In The Proceeding of ACM SIGKDD'04, pp. 334-343 (2004)
19. Hoque, N., Nath, B., and Bhattacharyya, D.K.: A New Approach on Rare Association Rule Mining. International Journal of Computer Applications (0975 - 8887), vol. 53, no. 3, September 2012, 1-6 (2012)
20. Lavergne, J., Benton, R., and Raghavan, V.V.: TRARM-RelSup: Targeted Rare Association Rule Mining Using Itemset Trees and the Relative Support Measure. In L. Chen et al (Eds.): ISMIS 2012, LNAI, vol. 7661, pp. 61-70 (2012)
21. Rawat, S.S., and Rajamani, L.: Discovering Rare Association Rules using Probability Apriori in Grid Environments. In K.Deep et al. (Eds). Proceedings of the International Conference on Soft Computing for Problem Solving (SocProS 2011), AISC 131, pp. 527-539 (2012)
22. Zhou, L., and Yau, S.: Association Rule and Quantitative Association Rule Mining among Infrequent Items. In The Proceeding of ACM SIGKDD'07, pp. 15-32 (2007)
23. Ding, J.: Efficient Association Rule Mining among Infrequent Items. Ph.D Thesis, University of Illinois at Chicago, (2005)
24. Koh, Y.S., and Rountree, N.: Finding Sporadic Rules using Apriori-Inverse. LNCS, vol. 3518, pp. 97–106 (2005)
25. Liu, B., Hsu, W., and Ma, Y.: Mining Association Rules with Multiple Minimum Supports, SIGKDD Explorations, pp. 337 – 341 (1999)
26. Brin, S., Motwani, R., and Silverstein, C.: Beyond Market Baskets: Generalizing ARs to Correlations. Special Interest Group on Management of Data (SIGMOD'97), pp. 265–276 (1997)
27. Omniecinski, E.: Alternative Interest Measures for Mining Associations. IEEE Trans. Knowledge and Data Engineering, 15, 57–69 (2003)
28. Lee, Y.K., Kim, W.Y., Cai, Y.D., and Han, J.: CoMine: Efficient Mining of Correlated Patterns. The Proceeding of ICDM'03, pp.581-584 (2003)
29. Han, J., Pei, H., and Yin, Y.: Mining Frequent Patterns without Candidate Generation. The Proceeding of SIGMOD'00, ACM Press, pp 1-12 (2000)
30. Han, J., and Kamber, M.: Data Mining: Concepts and Techniques. Morgan Kaufmann, 2nd edition (2006)
31. Au, W.H., and Chan, K.C.C.: Mining Fuzzy ARs in a Bank-Account Database. IEEE Transactions on Fuzzy Systems 11 (2), 238–248 (2003)
32. Mangat, V.: Swarm Intelligence Based Technique for Rule Mining in the Medical Domain. International Journal of Computer Applications 4(1), 19-24 (2010)
33. Frequent Itemset Mining Dataset Repository, http://fimi.cs.helsinki.fi/data/

Discovering Interesting Association Rules from Student Admission Dataset

Zailani Abdullah[1] Tutut Herawan[2], Mustafa Mat Deris[3]

[1]Department of Computer Science, Universiti Malaysia Terengganu
[2]Faculty of Computer Science & Information Technology, University Malaya
[2]Universitas Teknologi Yogyakarta, Yogyakarta, Indonesia
[3]Faculty of Computer Science & Information Technology, University Tun Hussein Onn
Malaysia

zailania@umt.edu.my, tutut@um.edu.my, mmustafa@uthm.edu.my

Abstract. Finding the interesting rules from data repository is quite challenging weather for public or private sectors practitioners. Therefore, the purpose of this study is to apply an enhanced association rules mining method, so called SLP-Growth (Significant Least Pattern Growth) proposed by [11,36] to mining the interesting association rules based on the student admission dataset. The dataset contains the records of preferred programs being selected by post-matriculation or post-STPM students of Malaysia via Electronic Management of Admission System (e-MAS) for the year 2008/2009. The results of this study will provide useful information for educators and higher university authority personnel in the university to understand the programs' patterns being selected by them.

Keywords: Association rule mining, significant least patterns, students.

1   Introduction

Generally, public universities are among the ultimate directions for almost post-matriculation or post-STPM students in Malaysia. After the students obtaining the actual result of the examination, they have to choose their preferred programs at Malaysian public universities via Electronic Management of Admission System (e-MAS). The issue is, for the average and the lower grades students, they might not be offered to their preferred programs. Based on this situation, many studies [1-3] have been carried out to ensure prolong of the students at university. Currently, there is an increasing interest in data mining and educational systems, making educational data mining as a new growing research community [4]. One of the popular data mining methods is Association Rules Mining (ARM) [5]. It aims at discovering the interesting correlations, frequent patterns, associations or casual structures among sets of items in the data repositories. The problem of association rules mining was first introduced by Agrawal for market-basket analysis [6,7,8]. After the introduction of Apriori [6], many studies [13-26] have been done pertinent to Association Rules (ARs). Generally, an item is said to be frequent if it appears more than a minimum support threshold. Least

T. Herawan et al. (eds.), *Proceedings of the First International Conference on Advanced Data and Information Engineering (DaEng-2013)*, Lecture Notes in Electrical Engineering 285, DOI: 10.1007/978-981-4585-18-7_16,
© Springer Science+Business Media Singapore 2014

item is an itemset whose rarely found in the database but it may produce interesting and useful ARs. In this paper, we employ SLP-Growth algorithm and Critical Relative Support (CRS) measure [11,36] to capture interesting rules from student admission dataset. The dataset was taken from Division of Academic, Universiti Malaysia Terengganu for 2008/2009 intake students in computer science program. The results of this study will provide useful information for educators or higher university personnel authority to offer more relevant programs to the potential students rather than by chance or unguided technique.

The reminder of this paper is organized as follows. Section 2 describes the related work. Section 3 describes the essential rudiments. Section 4 describes the employed method, SLP-Growth algorithm. This is followed by performance analysis through student admission dataset in section 5. Finally, conclusion of this work is reported in section 6.

2   Related Works

For the past decades, there are several efforts has been made to discover the interesting ARs. Zhou et al. [27] suggested a method to mine the ARs by considering only infrequent itemset. Ding [28] proposed Transactional Co-occurrence Matrix (TCOM for mining association rule among rare items. Yun et al. [9] introduced the Relative Support Apriori Algorithm (RSAA) to generate rare itemsets. Koh et al. [29] suggested Apriori-Inverse algorithm to mine infrequent itemsets without generating any frequent rules. Liu et al. [30] proposed Multiple Support Apriori (MSApriori) algorithm to extract the rare ARs. From the proposed approaches [9,28–30], many of them are using the percentage-based approach to improve the performance as faced by the single minimum support based approaches. In term of measures, Brin et al. [31] introduced objective measure called lift and chi-square as correlation measure for ARs. Lift compares the frequency of pattern against a baseline frequency computed under statistical independence assumption. Omiecinski [32] proposed two interesting measures based on downward closure property called all confidence and bond. Lee et al. [33] suggested two algorithms for mining all confidence and bond correlation patterns by extending the pattern-growth methodology Han et al. [34]. In term of mining algorithms, Agrawal et al. [6,7] proposed the first ARs mining algorithm called Apriori. Han et al. [35] suggested FP-Growth algorithm which amazingly can break the two limitations as faced by Apriori series algorithms. Recently, Educational Data Mining (EDM) has emerged as an important research area in order to resolve educational research issues [37]. Kumar et al. [38] enhanced the quality of students' performances at post graduation level via association rule mining. Garcial et al. [39] described a collaborative educational data mining tool based on association rule mining for the ongoing improvement of e-learning courses. Tair et al. [40] used association rule mining technique to analyze to improve graduate students' performance, and overcome the problem of low grades of graduate students. Chandra et al. [41], applied the association rule mining technique to identifies the students' failure patterns in order to improve the low capacity students' performances.

3 Essential Rudiments

3.1 Association Rules (ARs)

Throughout this section the set $I - \{i_1, i_2, \cdots, i_{|A|}\}$, for $|A| > 0$ refers to the set of literals called set of items and the set $D = \{t_1, t_2, \cdots, t_{|U|}\}$, for $|U| > 0$ refers to the data set of transactions, where each transaction $t \in D$ is a list of distinct items $t = \{i_1, i_2, \cdots, i_{|M|}\}$, $1 \leq |M| \leq |A|$ and each transaction can be identified by a distinct identifier TID.

Definition 1. *A set $X \subseteq I$ is called an itemset. An itemset with k-items is called a k-itemset.*

Definition 2. *The support of an itemset $X \subseteq I$, denoted $\mathrm{supp}(X)$ is defined as a number of transactions contain X.*

Definition 3. *Let $X, Y \subseteq I$ be itemset. An association rule between sets X and Y is an implication of the form $X \Rightarrow Y$, where $X \cap Y = \phi$. The sets X and Y are called antecedent and consequent, respectively.*

Definition 4. *The support for an association rule $X \Rightarrow Y$, denoted $\mathrm{supp}(X \Rightarrow Y)$, is defined as a number of transactions in D contain $X \cup Y$.*

Definition 5. *The confidence for an association rule $X \Rightarrow Y$, denoted $\mathrm{conf}(X \Rightarrow Y)$ is defined as a ratio of the numbers of transactions in D contain $X \cup Y$ to the number of transactions in D contain X. Thus*

$$conf(X \Rightarrow Y) = \frac{\sup p(X \Rightarrow Y)}{\sup p(X)}.$$

ARs that satisfy the minimum support and confidence thresholds are said to be strong.

4 Methodology

4.1 Algorithm Development

Determine Interval Support for least Itemset. An itemset is said to be least if the support count satisfies in a range of threshold values called Interval Support (ISupp).

The Interval Support is a form of ISupp (ISMin, ISMax) where ISMin is a minimum and ISMax is a maximum values respectively, such that $ISMin \geq \phi$, $ISMax > \phi$ and $ISMin \leq ISMax$.

Construct Significant Least Pattern Tree. A Significant Least Pattern Tree (SLP-Tree) is a compressed representation of significant least itemsets. There are three steps involved in constructing SLP-Tree. In the first step, the algorithm scans all transactions to determine a list of least items, LItems and frequent items, FItems (least frequent item, LFItems). In the second step, all transactions are sorted in descending order and mapping against the LFItems. It is a must in the transactions to consist at least one of the least items. In the final step, a transaction is transformed into a new path or mapped into the existing path.

Generate Significant Least Pattern Growth (SLP-Growth). SLP-Growth is an algorithm that generates significant least itemsets from the SLP-Tree by exploring the tree based on a bottom-up strategy. The algorithm will extract the prefix path sub-trees ending with any least item. In each of prefix path sub-tree, the algorithm will recursively execute to extract all frequent itemsets and finally built a conditional SLP-Tree.

4.2 Value Assignment

Apply Correlation and Critical Relative Support Measures. The values of association rule are derived based on Lift [26] and CRS [12, 36] measures. The processes of generating the values of association rule are taken place after all patterns and association rules are completely produced.

Discover Interesting Association Rule. From the list of valued association rules, the algorithm will begin to scan all of them. However, only those valued association rule with the correlation value that more than one and with the certain CRS value are captured and considered as Interesting ARs.

5   Result and Discussion

In order to capture the interesting ARs, the experiment employed SLP-Growth method and conducted on Intel® Core™ 2 Quad CPU at 2.33GHz speed with 4GB main memory, running on Microsoft Windows Vista. The algorithm has been developed using C# as a programming language.

The data was obtained from Division of Academic, Universiti Malaysia Terengganu in a text file and Microsoft excel format. There were in total of 822 bachelors programs offered in Malaysian public universities for July 2008/2009 students' intake. From this figure, 342 bachelor programs were selected by our 160 students and it can be generalized into 47 unique general fields. In addition, SLP-Growth algorithm with lift measurement to determine the degree of correlation of association rules was employed. There are 3,768 ARs are successfully extracted from the dataset. ARs are formed by applying the relationship of an item or many items to an item (cardinality: many-to-one). Fig. 1 depicts the correlation's classification of interesting ARs. The rule is categorized as significant and interesting if it has positive correlation, confidence is 100% and CRS value is equal to 1.0. Fig. 2 depicts the correlation of interesting ARs based on several ISupp. The result indicates that CRS successfully in producing the less number of ARs as compared to the others measures. The typical support or confidence measure alone is not a suitable measure to be employed to discover the interesting ARs. Although, correlation measure can be used to capture the interesting ARs, it ratio is still nearly 12 times larger than CRS measure. Therefore, CRS is proven to be more efficient and outperformed the benchmarked measures for discovering the interesting ARs from the dataset. Generally, the total numbers of ARs are kept decreased when the predefined Interval Supports thresholds are increased.

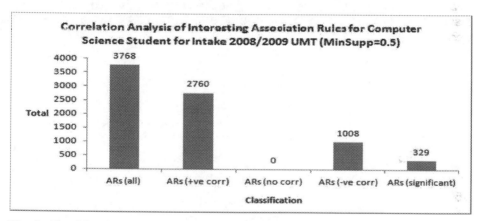

Fig. 1. Classification of ARs using correlation analysis. Only 8.73% from the total of 3,768 ARs are classified as interesting ARs

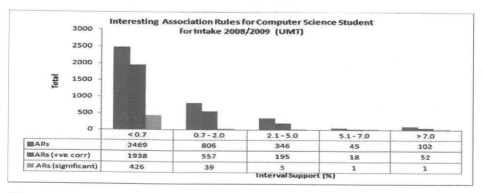

Fig. 2. Correlation analysis of interesting ARs using variety Interval Supports

6   Conclusion

Recently, there is an increasing interest in data mining and educational systems, making educational data mining as a new growing research community [4]. In this paper, we applied Significant Least Pattern Growth algorithm (SLP-Growth) and Critical Relative Support measure (CRS) proposed by [11,36] to mining the interesting association rules from student enrollment admission dataset. The dataset was taken from Division of Academic, Universiti Malaysia Terengganu (UMT) for the intake student 2008/2009. The results show that the interesting ARs can be extracted with the lesser number as compared to the common measures. Moreover, the results also can be analyzed by educators or the university' higher authority personnel in offering more appropriate programs to the prospect students rather than by chance.

Acknowledgment. The authors would like to thanks Universiti Malaysia Terengganu for supporting this work. The work of Tutut Herawan is supported by Excellent Research Grant Scheme no vote O7/UTY-R/SK/0/X/2013 from Universitas Teknologi Yogyakarta, Indonesia.

References

1. Yukselture, E.,and Inan, F.A.: Examining the Factors Affecting Student Dropout in an Online Certificate Program. Turkish Online Journal of Distance Education, 7(3), July 2006, 76-88 (2006)
2. Mohammad, S., and El-Masri, A.Z.: Factors Affecting Dropouts Students in Arab Open University - Bahrain Branch. International Journal of Science and Technology, 2(7), July 2012, 435-442 (2012)
3. Stearns, E., and Glennie, E.J.: When and Why Dropouts Leave High School. Youth Society. SAGE Journal, vol. 38, no. 1, September 2006, 29-57 (2006)
4. Romero, C., and Ventura, S.: Educational Data Mining: A Survey from 1995 to 2005. Expert Systems with Applications, 33, 135–146 (2007)
5. Ceglar,   A., Roddick, J.F.: Association Mining. ACM Computing Surveys, 38(2), 1–42 (2006)
6. Agrawal, R., Imielinski, T., and Swami, A.: Database Mining: A Performance Perspective. IEEE Transactions on Knowledge and Data Engineering, 5 (6), 914–925 (1993)
7. Agrawal, R., Imielinski, T., and Swami, A.: Mining Association Rules between Sets of Items in Large Databases. In Proceedings of the ACM SIGMOD '93 Intl. Conf. on the Management of Data, pp. 207–216 (1993)
8. Agrawal, R., and Srikant, R.: Fast Algorithms for Mining Association Rules. In Proceedings of the 20th Intl. Conf. on Very Large Data Bases (VLDB) 1994, pp.487–499 (1994)
9. Yun, H., Ha, D., Hwang, B., and Ryu, K.H.: Mining Association Rules on Significant Rare Data Using Relative Support. The Journal of Systems and Software, 67 (3), 181-19 (2003)
10. Xiong, H., Shekhar, S., Tan, P.N., and Kumar, V.: Exploiting A Support-Based Upper Bond Pearson's Correlation Coefficient for Efficiently Identifying Strongly Correlated Pairs. In The Proceeding of ACM SIGKDD 2004, pp. 334-343 (2004)
11. Abdullah, Z., Herawan, T., and Deris, M.M.: Mining Significant Least Association Rules using Fast SLP-Growth Algorithm. In T.H. Kim and H. Adeli (Eds.): AST/UCMA/ISA/ACN

2010, LNCS, 6059, pp. 324–336. Springer Heidelberg (2010)
12. Abdullah, Z., Herawan, T., and Deris, M.M.: Scalable Model for Mining Critical Least Association Rules. In Rongbo Zhu et al. ICICA 2010, LNCS, 6377, pp. 509-516. Springer Heidelberg (2010)
13. Abdullah, Z., Herawan, T., Noraziah, A., and Deris, M.M.: Extracting Highly Positive Association Rules from Students' Enrollment Data. Procedia Social and Behavioral Sciences, 28, 107–111 (2011)
14. Abdullah, Z., Herawan, T., Noraziah, A., and Deris, M.M.: Mining Significant Association Rules from Educational Data using Critical Relative Support Approach. Procedia Social and Behavioral Sciences, 28, 97–101 (2011)
15. Abdullah, Z., Herawan, T., and Deris, M.M.: An Alternative Measure for Mining Weighted Least Association Rule and Its Framework. In J.M. Zain et al. (Eds.): ICSECS 2011, CCIS, vol. 188, II, pp. 475–485. Springer Heidelberg (2011)
16. Abdullah, Z., Herawan, T., and Deris, M.M.: Visualizing the Construction of Incremental Disorder Trie Itemset Data Structure (DOSTrieIT) for Frequent Pattern Tree (FP-Tree). In H.B. Zaman et al. (Eds.): IVIC 2011, LNCS, 7066, pp. 183–195. Springer Heidelberg (2011)
17. Herawan, T., Vitasari, P., and Abdullah, Z.: Mining Interesting Association Rules of Student Suffering Mathematics Anxiety. In J.M. Zain et al. (Eds.): ICSECS 2011, CCIS, 188, II, pp. 495–508. Springer Heidelberg (2011)
18. Abdullah, Z., Herawan, T., and Deris, M.M.: Efficient and Scalable Model for Mining Critical Least Association Rules. In a special issue from AST/UCMA/ISA/ACN 2010, Journal of The Chinese Institute of Engineer, Taylor and Francis, 35, No. 4, 27 June 2012, 547–554 (2012)
19. Herawan, T., Abdullah, Z., Noraziah, A., Deris, M.M., and Abawajy, J.H.: IPMA: Indirect Patterns Mining Algorithm. In N.T. Nguyen et al. (Eds.): ICCCI 2012, AMCCISCI, vol. 457, pp. 187–196. Springer Heidelberg (2012).
20. Herawan, T., Abdullah, Z., Noraziah, A., Deris, M.M., and Abawajy, J.H.: EFP-M2: Efficient Model for Mining Frequent Patterns in Transactional Database. In N.T. Nguyen et al. (Eds.): ICCCI 2012, LNCS, pp. 7654, 29–38. Springer Heidelberg (2012).
21. N. Ahmad, Z. Abdullah, T. Herawan, and M.M. Deris. Scalable Technique to Discover Items Support from Trie Data Structure. In B. Liu et al. (Eds.): ICICA 2012, LNCS, vol. 7473, pp. 500–507. Springer Heidelberg (2012)
22. N. Ahmad, Z. Abdullah, T. Herawan, and M.M. Deris.: WLAR-Viz: Weighted Least Association Rules Visualization. In B. Liu et al. (Eds.): ICICA 2012, LNCS, 7473, pp. 592–600. Springer Heidelberg (2012)
23. Herawan, T., and Abdullah, Z.: CNAR-M: A Model for Mining Critical Negative Association Rules. In Zhihua Cai et al. (Eds): ISICA 2012, CCIS, 316, pp. 170–179. Springer Heidelberg (2012)
24. Abdullah, Z., Herawan, T., Ahmad, N., and Deris, M.M.: DFP-Growth: An Efficient Algorithm for Mining Pattern in Dynamic Database. In B. Liu et al. (Eds.): ICICA 2012, LNCS, vol. 7473, pp. 51–59. Springer Heidelberg (2012)
25. Abdullah, Z., Herawan, T., and Deris, M.M.: Detecting Critical Least Association Rules in Medical Databasess. International Journal of Modern Physics: Conference Series, World Scientific, 9, 464–479 (2012)
26. Herawan, T., Abdullah, Z., Noraziah, A., Deris, M.M., and Abawajy, J.H.: EFP-M2: Efficient Model for Mining Frequent Patterns in Transactional Database. In N.T. Nguyen et al. (Eds.): ICCCI 2012, LNCS, 7654, pp. 29–38. Springer Heidelberg (2012)
27. Zhou, L., and Yau, S.: Association Rule and Quantitative Association Rule Mining Among Infrequent Items. Rare Association Rule Mining and Knowledge Discovery, pp.15-32. IGI-Global (2010)

28. J. Ding.: Efficient Association Rule Mining among Infrequent Items. Ph.D Thesis, University of Illinois at Chicago. (2005)
29. Y.S. Koh and N. Rountree.: Finding Sporadic Rules using Apriori-Inverse. LNCS, vol. 3518, pp. 97–106. Springer Heidelberg (2005)
30. Liu, B., Hsu, W., and Ma, Y.: Mining Association Rules with Multiple Minimum Supports, SIGKDD Explorations, pp. 337 – 341 (1999)
31. Brin, S., Motwani, R., and Silverstein, C.: Beyond Market Baskets: Generalizing ARs to Correlations. Special Interest Group on Management of Data (SIGMOD'97), pp. 265–276 (1997)
32. Omniecinski, E.: Alternative Interest Measures for Mining Associations. IEEE Trans. Knowledge and Data Engineering, 15, 57–69 (2003)
33. Y.-K. Lee, W.-Y. Kim, Y.D. Cai, J. Han. CoMine: Efficient Mining of Correlated Patterns. The Proceeding of ICDM'03 (2003)
34. Han, J., Pei, J., Yin, Y., and Mao, R.: Mining Frequent Patterns without Candidate Generation: A Frequent-Pattern Tree Approach*. Data Mining and Knowledge Discovery, 8, 53–87 (2004)
35. Han, J., and Kamber, M.: Data Mining: Concepts and Techniques. Morgan Kaufmann, 2nd ed., (2006)
36. Abdullah, Z., Herawan, T., and Deris, M.M.: Tracing Significant Information using Critical Least Association Rules Model. International Journal of Innovative Computing and Applications, Inderscience, 5, 3-17 (2013)
37. Baker, R., and Yacef, K.: The State of Educational Data mining in 2009: A Review and Future Visions. Journal of Educational Data Mining, 1(1), 3–17 (2010)
38. Kumar, V., and Chadha, A.: Mining Association Rules in Student's Assessment Data. International Journal of Computer Science Issues, 9(5), 3, September 2012, 211-216 (2012)
39. Garcia, E., Romero, C., Ventura, S., and Castro, C.: An Architecture for Making Recommendations to Courseware Authors using Association Rule Mining and Collaborative Filtering. User Modeling and User-Adapted Interaction: The Journal of Personalization Research, 19, 99–132 (2009)
40. Tair, M.A.A., and El-Halees, AM.:. Mining Educational Data to Improve Students' Performance: A Case Study. International Journal of Information and Communication Technology Research, 2(2), 140-146 (2012)
41. Chandra, E., and Nandhini, K.: Knowledge Mining from Student Data. European Journal of Scientific Research, 47(1), 156-163 (2010)

Follower Classification Based on
User Behavior for Issue Clusters

Kwang-Yong Jeong*, Jae-Wook Seol*, and Kyung Soon Lee**

Dept. of Computer Science & Engineering, CAIIT, Chonbuk National University,
{kyjeong0520, wodnr754}@naver.com, selfsolee@chonbuk.ac.kr

Abstract. Recently, Social Network Service has made a meteoric rise as means
of communication to sharing important information. peoples discuss about
social issues, especially in Twitter. Besides, unlike any other social network
service, Twitter users can *follow* without the agreement of the other party, for
this reason, the users has followers with various intentions exist. To measure
followers's agree about a followee's opinion, our method builds issue clusters
by defining trust period about extracting an issue. In this paper, we propose two
methods for follower classification that are based on extraction of Influential
supporters and issues cluster that is reflected on a target user's opinion. To
evaluate the effectiveness of the proposed method, we examine behaviors of
followers of politicians from Twitter data. As a result of the experiment, the
proposed approach effectively classifies the follower based on issues reflected
opinions of the target user and Influential supporters.

Keywords: follower classification, issue cluster, social opinion, user behavior

1 Introduction

Social Network Service(SNS) recently has made a meteoric rise as means of
communication to sharing important information. Since SNS users can not only
exchange the information with strangers, but also spread the information faster than
internet cafe and blog, the study on Twitter is actively being progressed [1-2].

Twitter users can *follow* without the agreement of the other party, for this reason, the
users has followers with various intentions exist. Since these followers can favorably
follow, blindly follow or unfavorably follow, it is needed to classify by the feelings
about the user. Also these followers do not express all sympathy on *a target user's*
issues. Thus, our method classifies these followers according to each issue. In here,
the target users are as authority users who has numerous followers, they are a new
type of opinion leader who exert a big influence on the setting of the agenda or
formation of public opinion. Various opinions on an issue are formed by mutual
discussion of numerous followers of this authority user.

In this paper, we propose a method to classify feelings of followers about a target
user, namely support, non-support and neutrality. These followers are classified by
using a characteristic that the followers follow one target user or follow two or more

---

* Co-equally contributed.
** Corresponding author.

T. Herawan et al. (eds.), *Proceedings of the First International Conference
on Advanced Data and Information Engineering (DaEng-2013)*, Lecture Notes
in Electrical Engineering 285, DOI: 10.1007/978-981-4585-18-7_17,
© Springer Science+Business Media Singapore 2014

of target users at the same time in network the target users. In here, a follower who follows two or more of target users at the same time is defined as a *co-follower*.

Through the observation of followers's behavior, we find these three characteristics as follows. *i*) The followers who sympathize their opinion and actively supports exist for authority users, *ii*) Authority users want to diffuse social issues through tweets, and these tweets strongly influences on a way of behavior of followers, however, the influential followers tend to sympathize sometimes, and not sympathize sometimes, and *iii*) Co-follower following two or more of target users. The feeling about each target user is relative.

Our method classify followers of a target user as in the following based on observations, mentioned the authority users's social issues and the co-followers, *i*) A system classify the followers by extracting *influential supporters* on a target user. There are the followers who sympathize the target user's opinion and actively support, for the target user. We define the influential supporters are considered as an expanded concept of the target user. *ii*) The system classifies followers by extracting issues that the target user's opinion is reflected. The followers do not express sympathy to all issues about supporting the target user. Since they could sympathize or not sympathize on each issue, the system classifies the followers based on issue cluster, and *iii*) Since co-followers who following two or more of users have relative feeling about each target user, using *Bias Ratio* to classify them. *iv*) The system classifies followers with sentiment analysis about Support Vector Machine(SVM) and mention before *RT* of feeling about target users.

To evaluate the effectiveness of the proposed method, experiments are conducted on followers of five authority users on Korean Twitter test collection. The proposed methodology is language independent, which can be applied to other languages.

2   Related Work

Recently, there have been studies which analyze the users who use social network services such as Twitter, Orkut, MySpace, Flickr, and YouTube. Since Social network service is being used by diverse people around the world, it has various tendencies along with various behavior patterns. They analyzed these user behaviors through the number of following, followers, retweet and mentions etc on Twitter[3,4]. Also, there are studies that analyze identification and characteristics of user behaviors on Orkut, MySpace, Flickr, and YouTube [5,6]. Different from those researches, by analyzing user's behavior for a target user, our method classify they are positive or negative.

Park et.al. [8] introduced the method to automatically construct Twitter data set, and a proposed method that classifies affirmation and negation using constructed data set. Our method is trained through SVM of two groups, supporter and non-supporter by using tweet data of politician.

3   Follower Classification Based on User Behavior for Issue Clusters

In this section, we first describe a process for follower classification. The process involves four stages: *i*) Extracting influential supporters of a target user, *ii*) Extracting issues that reflected the target user's opinion among the target users's tweets and influential supporters's tweets, *iii*) Classifying co-followers using Bias detection, and *iv*) Classifying followers using sentiment analysis.

3.1   Extraction of Influential Supporters

Some of a target user's followers sympathize his opinion and actively support for the target user. Also, they seldom change the opinion of supporting the target user. These followers define *influential supporters* of the target user. In our method, influential supporters are considered as an expanded concept of the target user. When a follower retweets the tweet that the target user is appeared on between the target user's tweets and influential supporters's tweets, it is classified as supporting the follower. We apply modified HITS algorithm to extract influential supporters among the target user's followers.

The HITS[9] is very well known link analysis algorithm. Similarly, it is more authority user when the user gets more retweet in Twitter. Therefore, HITS is applied to extract influential supporter of a target user. We assume that more they have the retweet of tweet that includes the name of the target user; it is the influential supporter who supports the target user.

In order to apply modified HITS algorithm, each edge on the graph is created as follower: edge between vertexes with following condition. When user v_j retweeted the tweet that a target user's name is appeared on between tweets of user v_i, then edge becomes e_{ij}. The number that v_j retweeted v_i becomes w_{ij}.

The weight of each edge is set to set as follows:

$$Auth^{t+1}(v_i) = \sum_{j:e_{ji} \in E} w_{ji} \times Hub^t(v_j) \tag{1}$$

$$Hub^{t+1}(v_i) = \sum_{j:e_{ij} \in E} w_{ij} \times Auth^t(v_j) \tag{2}$$

The authority and hub score of each node iteratively updates the scores until convergence according to the modified HITS algorithm. The method selects followers whose authority score are 0.01 or more as an influential supporter. This is because a threshold is set to 0.01 empirically.

The initial authority score of a user is set as follows:

$$Auth^0(v_i) = avgRT(v_i) \cdot mention(v_i) \tag{3}$$

where $avgRT(v_i)$ represents the average of retweets for a user v_i, and $mention(v_i)$ is the ratio of tweet that a target user's name is appeared on between tweets of user v_i. The more a tweet is retweeted the higher initial authority score it has. Also the initial scores of the hub score are set as follows: $Hub^0(v_i) = 1$.

The extracted influential supporters show that are strongly supportive of the target user. They play important role in propagation of a target user's opinion as a medium. Using the above characteristics, the proposed method extracts issues by using the influential supporters.

3.2   Expansion of User Relationship based on Issue Clusters

Followers may disagree with entire issues that a target user mentioned. For example, follower u is a target user A's follower. But u does not express all the sympathy on the A's issues. Therefore, it is need to classify by analyzing the followers's opinion on each issue. The method classifies a follower in following order: i) Extracting issues that a target user's opinion is reflected, ii) Building an issue cluster by defining trust period about extracted an issue, and iii) Classifying them as a supporting follower when they retweet the concerned tweet in an issue cluster.

Extraction of Issue Keywords. In order to select an issue among the words extracted through *tf·idf,* give more weight when it appears more in influential supporters's tweets, give less weight when it appears less.

$$w(t) = tf(t,d) \cdot idf(t,N) \cdot sf(t) \tag{4}$$

where $w(t)$ reflects the weight of influential supporters's opinion when t appears in a day d. Here, $tf(t,d)=log(1+tf(t,d))$, $idf(t)= log(N/df)$ and $sf(t) = IS_t / IS_{all}$, where $tf(t,d)$ represents the frequency of t appeared on the target user's tweet in d. df represents the number of user that t appeared. IS_{all} is the total number of influential supporters, IS_t is number of influential supporters that t appeared in d. High weight of tf·idf does not simply means all t that the target user mentioned is an issue. Hence, extracts social issues by applying a number of influential supporters that t is appeared on as a weight. Therefore, high IS_t / IS_{all} implies the high possibility of t being social issues.

Construction of Issue Clusters. In today's society, opinions about an issue are rapidly changing, thus the opinions about an issue may be different now than it was a month ago. Hence, we define *trust period* on each issue. Our method considers the tweets about the issue within a trust period. There are three stages to build issue cluster using the trust period: i) Extracting the time of a first mentioned tweet that appeared the issue among target's tweets as a starting point, ii) Extracting the time of last retweeted tweet that appeared the issue among target's tweets and influential supporters's tweet as an ending point, and iii) When a follower retweets tweet that appeared the issue among target user's tweet and Influential supporter's tweet from starting point to ending point, classifying the follower agree with target user's tweet.

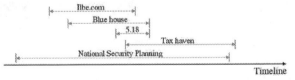

Fig. 1. Moon's issue cluster based on the timeline using retweet relation.

Fig.1 shows trust period about each issue. The figure shows the distribution of followers's retweet by the time. "Blue house" is distributed from 4th of May to 21st of May. Define this period as the trust period about "blue house".

If a follower retweets the target user's tweets and influential supporters's tweets that include the issue within trust period, they are classified as a supporting follower. The method uses an issue cluster to classify followers for each issue.

3.3   Classification of Co-followers Using Bias Ratio

A co-follower is a follower who following two or more target users at the same time. The sentiment about each target user is relative. Therefore, our method classifies follower according to preference about target users. The system calculates a number of tweet with which a target user's name is being retweeted, issues reflected which the target user's opinion is being retweeted. The system classifies co-followers with a number of retweets for each target user by using a bias ratio. Bias ratio is used to classify co-followers. When a follower u following two target users A and B, bias ratio equation is as follows:

$$\text{BiasRatio}(u, A) = \frac{Retweet(u, A)}{Retweet(u, A) + Retweet(u, B)} \tag{5}$$

$$\text{BiasRatio}(u, B) = \frac{Retweet(u, B)}{Retweet(u, A) + Retweet(u, B)} \tag{6}$$

where BiasRatio(u,X) is a bias ratio that a follower u prefer to agree with a opinion of a target user X. Retweet(u,X) is a number that the follower u retweets tweet that mentioned X among X's tweets and influential supporters's tweets, and tweet that X's opinion is reflected issues.

Table 1. Follower Classification Based on BiasRatio.

Co-follower Classification Algorithm (a follow u)
Input: A follower u's tweets , Target user pair=$\{A,B\}$ Output: u's bias result

If the follower u is co-follower of Target user A and B then // Bias Detection
    classify u to *neutral*   if   BiasRatio$(u,A)==0$ and BiasRatio$(u,B)==0$
    classify u to *support to both A and B*   if $0<|$ BiasRatio(u,A) - BiasRatio(u,B) $|<\theta$
    classify u to *support to A*   if   BiasRatio$(u,A)>$ BiasRatio(u,B)
    classify u to *support to B*   if   BiasRatio$(u,A)<$ BiasRatio(u,B)
    end

Table 1 is an algorithm that classifies co-follower by using bias ratio. If BiasRatio(u,A) > BiasRatio(u,B), the follower is detected as biased toward A's a supporting follower; otherwise, biased toward B's a supporting follower. If the difference of the BiasRatio(u,A) and the BiasRatio(u,B) is below a threshold θ, it is classified neutral (θ is set to 0.2 by training in our experiments).

The proposed method classify a co-follower by using the bias ratio of relative value on each target user when retweet a tweet with a target user's name or retweet the tweet that includes an issue within a trust period.

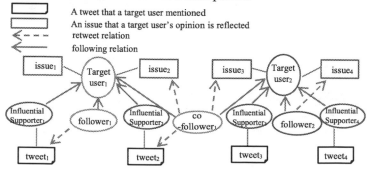

Fig. 2. Follower Classification using Influential Supporters and Issues.

Fig. 2. shows that follower classification using influential supporters and issues. Because follower₁ retweets an influential supporter's tweets, the follower₁ is classified as a Target user₁'s supporting follower. Likewise, because follower₂ retweets a tweet of issues that target user₂'s opinion is reflected, the follower₂ is classified as a Target user₂'s supporting follower. A co-follower is classified as a Target user₁'s supporting follower by followers who isn't classified bias ratio.

3.4   Tweet Sentiment Analysis

There is a case that follower classification with issues that reflects the opinion of a target user and influential supporter is not available. In this case, the system classifies follower using sentiment analysis in tweet document and SVM(Support Vector Machine) based on similarity. We use SVM light[10].

To train target user's tweets, the method set supporting tweet(+1) and non-supporting tweet(-1) of the target user. The system calculates the weight of the term appeared on tweet document using tf·idf, a formula that calculates term weight, and represent one tweet document as one vector. The system takes all tweet documents of each user as a test set then it classifies them into supporting, neutral, non-supporting follower of the target user. In the test, +1 for positive word, -1 for negative word by using tweet pattern that uses sentiment word with the target user. Also, the system classifies a follower by analyzing sentiment about the target user based on sentiment analysis of mention before the RT. A threshold is set to 0.6 empirically. When positive tweets are 60% of entire tweet, it is classified as a supporting follower.

4   Experiments

4.1   Experimental Set-up

To see the effectiveness of the proposed follower classification method, we have performed an evaluation using Twitter test collection in Korean. Target users are selected among popular politicians in Twitter, followers's tweets of five target users are retrieved by using Twitter Search API[11,12] from 15th, March, 2013 to 15th, June. It is randomly collected 10 thousand from followers of each target user. Influential supporters are extracted among 10 thousand by using HITS algorithm. Test sets of each target user are four hundred followers who are selected among 10 thousand. Collected Twitter data group is as shown in the Table 2. The answer sets for the test collection are judged by two human assessors.

Table 2.   Twitter Test Collection.

Target User	The Number of Followers	The Number of Tweets of	
		10,000 Followers	400 Followers
Moon J.I	450,286	3,718,803	134,548
Ahn C.S	263,138	3,714,305	98,124
Park G.H	329,234	2,159,034	68,255
You S.M.	610,714	3,284,739	80,351
Pyo C.W	129,090	3,618,626	56,884
Total	1,782,462	16,495,507	438,162

To verify validity of follower behavior classification using the proposed issue cluster, comparatively experimented SVM classification, classification using influential user and the proposed method.

- **SVM**: The follower classification method using the method described in 3.4.
- **The partial method**: The follower classification method using influential supporters and sentiment analysis.
- **The proposed method**: The follower classification method using issues that reflects the opinion of a target user, influential supporters and sentiment analysis.

4.2 Experimental Results and Analysis

Table 3 is the result of comparative experiments of follower classification.

Table 3. Twitter Test Collection.

Evaluation Measure	SVM	Partial method	Proposed Method
Accuracy	0.565	0.631 (+11.6%)	0.680 (+20.3%)
Precision	0.640	0.681 (+6.40%)	0.705 (+10.1%)
Recall	0.548	0.620 (+13.1%)	0.649 (+18.3%)
F_1 measure	0.590	0.648 (+9.83%)	0.675 (+14.4%)

Performance of SVM was poor. Due to the characteristic of tweet, tf·idf weight of terms that composing tweet is short within 140 letters. Hence, weight was not applied on important words, and there were not many tweets including sentiment that could know the sentiment about a target user. The method using influential supporter showed 11.6% better performance in accuracy than SVM method. Since this result expanded the target user by using influential supporter of the target user, followers who could certainly classify increased. The proposed method showed the best performance compare to SVM and influential supporter method. It showed 20.3% better performance in accuracy than SVM method. Therefore, the proposed classification method using issue cluster is meaningful.

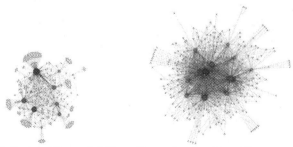

Fig. 3. The Partial Method and Expanded Users Network using Issue Clusters.

The network graphs in Fig.3 show that the result of classification for 400 of the followers of a target user "*moonriver365*". Edges of Blue color is the target user and influential supporters. The left graph shows experimental results of partial method. The right graph shows experimental results of proposed method. The right graph shows more number of edge than the left graph in network. It shows that classifiable followers increase compare to the left graph. Through this figure, we verified that method with issue cluster is efficient.

5   Conclusion

In this paper, we propose a model that follower classification based on issue clusters. A follower is classified to supporting, non-supporting and neutrality according to their sentiment about a target user. The proposed method was classified follower based on an issue cluster that reflects the opinion of the target user and influential supporters extracted by HITS algorithm, and tweet sentiment analysis. As a result of the experiment, it showed 20.3% higher performance in accuracy than tweet contents based SVM classification method. This result indicates that the proposed method to classify follower of each target user to supporting follower, non-supporting follower and neutrality is effective.

For the future work, we will research what kind of tweet do followers retweet, and how the retweet distributed based on user behavior analysis.

Acknowledgements. This research was supported by Basic Science Research Program through the National Research Foundation of Korea(NRF) funded by the Ministry of Education, Science and Technology (2012R1A1A2044811).

References

1. Xu, Z., Zhang, Y., Wu, Y., Yang, Q.: Modeling User Posting Behavior on Social Media. In: 35th international ACM SIGIR conference on Research and de development in information retrieval, pp. 545--554 (2012)
2. Kwak, H., Lee, C., Park, H., Moon, S,.: What is Twitter, a Social Network or a News Media?. In: 19th international conference on World wide web, pp. 591--600 (2010)
3. De Choudhyry, M., Diakopoulos, N., Naaman, M.: Unfolding the Event Landscape on Twitter. In: Classification and Exploration of User Categories. CSCW' 12 Proceedings of the ACM 2012 conference on Computer Supported Cooperative Work, PP. 241--244 (2012)
4. Cha, M., Haddadi, H., Benevenuto, F., Gummadi, K.P.: Measuring User Influence in Twitter: The Million Follower Fallacy. In: ICWSM' 10 Proceedings of the 4th International AAAI Conference on Weblogs and Social Media, pp. 10--17 (2010)
5. Benevenuto, F., Rodrigues, T., Cha, M., Almeida, V.: Characterizing User Behavior in Online Social Networks. In: IMC '09 Proceedings of the 9th ACM SIGCOMM conference on Internet Measurement conference, pp.49--62 (2009)
6. Maia, M., Almeida, J., Almeida, V.: Identifying User Behavior in Online Social Networks. In: SocialNets'09 Proceeding of the 1st workshop on Social Network Systems, pp.1--6 (2008)
7. Lai P .:Extracting Strong Sentiment Trends from Twitter. nlp.stanford.edu (2010)
8. Park, Alexander, and Patrick Paroubek. "Twitter as a Corpus for Sentiment Analysis and Opinion Mining." *LREC* (2010)
9. Kleinberg, J.: Authoritative sources in a hyperlinked environment. In: Journal of the ACM. 46(5), pp. 604--632 (1999)
10. SVM light, http://svmlight.joachims.org/
11. Twitter Developers, https://dev.twitter.com/
12. Twitter4J, http://twitter4j.org/

Implementation of Modified Cuckoo Search Algorithm on Functional Link Neural Network for Temperature and Relative Humidity Prediction

Siti Zulaikha Bt Abu Bakar*, Rozaida Bt Ghazali, Lokman Hakim Bin Ismail

Universiti Tun Hussein Onn Malaysia, Batu Pahat, Johor, 86400, Malaysia

GI110020@siswa.uthm.edu.my

Abstract. The impact of temperature and relative humidity changes bringing a sharp warming climate. These changes can cause extreme consequences such as floods, hurricanes, heat waves and droughts. Therefore, prediction of temperature and relative humidity is an important factor to measure environmental changes. Neural network, especially the Multilayer Perceptron (MLP) which uses Back Propagation algorithm (BP) as a supervised learning method, has been successfully applied in various problem for meteorological prediction tasks. However, this architecture still facing problem where the convergence rate is very low due to the multilayering topology of the network. Thus, this study proposes an implementation of Functional Link Neural Network (FLNN) which composes of a single layer of tunable weight trained with the Modified Cuckoo Search algorithm (MCS). The FLNN is used to predict the daily temperatures and relative humidity of Batu Pahat region. Extensive simulation results have been compared with standard MLP trained with the BP, and FLNN with that of BP. Promising results have shown that FLNN when trained with the MCS has successfully outperformed other network models with reduced prediction error and fast convergence rate.

Keywords: Neural Network, Multi-layer perceptron , Higher order neural networks, Functional Link Neural Network, Back Propagation algorithm, Modified Cuckoo Search"

1 Introduction

The increase in temperature of the earth at present clearly shows a sharp warming climate. The increase in global temperature scan also lead to other changes such as rising sea levels, changes in the amounts and forms of deposition. These changes can cause changes that occur in extreme weather such as floods, hurricanes, heat waves and droughts. Clearly an increasing of temperature is the most accurate parameter for the problem of global warming.

T. Herawan et al. (eds.), *Proceedings of the First International Conference on Advanced Data and Information Engineering (DaEng-2013)*, Lecture Notes in Electrical Engineering 285, DOI: 10.1007/978-981-4585-18-7_18,
© Springer Science+Business Media Singapore 2014

According to Malaysian Meteorological department (MMD)[4] scientific report regional climatic trends are in line with the in- crease in average temperature observed for Malaysia. The impact of climate changes is the temperature rise in the highlands like Cameron Highlands. Furthermore, unusual weather causing flooding or drought, haze, rising sea levels and depleting water resources.

Before this conventional techniques has been used for temperature and relative humidity prediction, and it's shows the technique really sophisticated and extremely complex. But the models through of data-driven techniques are more computationally fast and require less input parameters than process-based models [5]. Therefore data driven techniques provide an effective alternative to conventional process-based modelling. Chief among them are Neural Network (NN).

The characterized of neural network can be describe through the architecture of net- work topology and pattern of linked between the nodes. Multi-layer perceptrons (MLPs), which found the most widely used network architecture, are composed of a hierarchy of processing units organized in a series of two or more mutually exclusive sets of neurons or layers. The information flow in the network is restricted to a flow, layer by layer, from the input to the output, hence also called feed-forward network. However, in temporal problems, measurements from physical systems are no longer an independent set of input samples, but functions of time. Furthermore, when the number of inputs to the model and the number of training examples becomes extreme- ly large, the training procedure for ordinary neural network architectures becomes slow and unduly tedious. At the same time this model has high computational cost structure like polynomial perceptron due to the availability of hidden layers [13].

Since the BP learning algorithm its gradient descent local optimization technique which involves backward error correction of network weights, its still have several major problems needed to be solved. Hence that, the convergence rate of BP getting slow and becomes unsuitable to solve a large problems. To overcome such time-consuming operations, this research work focuses on using FLNN [6], a network of type Higher Order Neural Networks which has a single layer of learnable weights, therefore reducing the networks' complexity, lower error rate and higher convergence rate. However, convergence behaviour of BP algorithm highly depends on the choice of initial value of connection weights and other parameters used in the algorithm, produce unstable program to solving bigger problem. Therefore The MCS has been proposed to replace the BP function to and solving the problem and produce convergence rate faster than training in BP algorithm.

2   Functional Link Neural Network (FLNN)

In spite of the development of various types of ANNs, higher order neural networks (HONs) [7] have recently found that powerful methods to handle some of problem that occurred in non-HONs. Since FLNN it's one of the subclass under HONs root, so it is also very important to introduce basic of HONs in this paper.

HONs were first introduced by Giles et al [6] and further analyzed by Pao and claimed them as 'tensor networks' and retarded them as a special case of his functional link models [8]. In pattern recognition and associative recall, HONs has already proved some success in this task [7], but so far little research has been done to times series prediction applications. In HONs, the number of layers was decrease compared to MLPs that means that the problems of over-fitting and local optima can't be migrated to large degree were solved and is faster to train and executes [8].

Even though the single layer network much easier to train that multilayer networks, but its only solve for the problem of linearly separable only. So, to handle this problem, FLNN can be used [8]. Since one of the abilities of FLNN can work as a nonlinear separation as drawn from Figure 1. The FLNN works by introducing nonlinear terms into a neural network. The network can be reduced to a single layer, which also increases the speed and ease of training

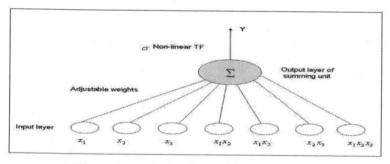

Figure 1 : Functional Link Neural Network

FLNN was introduced by Pao [8] and has been successfully used in many applications such as system identification [14], channel equalization [3], classification [16], pattern recognition [7-8] and prediction [3,15]. FLNN is much more advanced than MLP since it has a flat network compared to the MLP but still is able to handle a non-linear task. The FLNN architecture is basically a flat network without any hidden layer which has make the learning algorithm used in the network less complicated

3   Cuckoo Search (CS)

The CS is a metaheuristic search algorithm that was inspired by the cuckoo bird breeding behavior which has been proposed by Yang and Deb through laying their progenies in the nests of other species host birds. Female cuckoo bird also capable to mimic the colors and patterns of the eggs of a few chosen host species. This reduces the probability of the eggs being abandoned and therefore will enhanced their reproductive highly [9].

The main task of cuckoo egg is to replace the best solution in the host nest. Each eggs of cuckoo bird will carried as a new solution in the existing nest. Three rules of CS were declared based on:

- Cuckoo bird's will lays one egg for each nest and the egg will carry either as a set of solution at a time or also can be abandoned.
- The new generations will carry the best solution that is refer to the cuckoo egg
- The probability of nest is fixed either strange egg were discover by host will caused they abandon the egg or nest and create another new nest

4   Modified Cuckoo Search (MCS)

Given enough computation, the CS will always find the optimum has said by Yang and Deb [10] but, as the search based on whole area on random walks; it's not guaranteed to be as fast convergence as usual. To avoid these issues, two changes have been made where the main goal is still to increase the convergence rate. The modification was more applicable for a wider range of application without losing the attractive features of original method [11].

The first modification was made to the size of Levy flight step size α, which in CS α was constant and the value of $\alpha = 1$ is employed [12]. But for MCS method, If the value of α decrease, it will increase the number of generations. This is proven for the same reasons that the inertia constant is reduced in the PSO, which is to encourage more localized searching as the individuals, or the eggs, get closer to the solution.

The second modification for MCS method is adding information exchange between the eggs with the same goals that to speed up conver- gence as minimum as can be. While in CS, the process of information exchange be- tween the eggs didn't happened and the performed of searches was by their self.The full algorithm for this swarm intelligence method can be referring from Modi- fied cuckoo search: A new gradient free optimization algorithm [11].

5   Data Collection

Data collection is one of the most important part in the success of neural network problems. The dataset must be totally verify in the angle of reliability, quality and relevancy to make sure the data provided can be able to develop and run the program from critical to its success. For the implementation of temperature and relative humidity forecasting, historical data for five years of daily temperature and relative humidity measurement in Batu Pahat, ranging from 1/1/2005 to 31/12/2009 were collected from Central Forecast Office, Malaysian Meteorological Department (MMD) [4].

Based on the previous records, the maximum, minimum and average temperature and relative humidity measurements are tabulated in Table 1. The idea behind this selection was to examine the applicability of FLNN-MCS on the prediction of temperature and relative humidity measurement.

Table 1. The Maximum, Minimum and Average Temperature and Relative Humidity Measurement at Batu Pahat

Datasets	Size	Average	Max	Min
Temperature ($^{\circ}$C)	1826	26.75	29.5	23.7
Relative Humidity (%)	6392	85.9031	98.1	69.5

Regarding table 1, the datasets was divide into three section which is; 60% for the training, 20% for the testing and other 20% for the validation section.

6   Experimental Design

We have implemented the models using MATLAB 7.10.0 (R2008a) on Pentium® Core™2 Quad CPU. In this paper, the results of comparisons between MLP-BP, FLNN-BP and FLNN-MCS that have been conducted on the same training, testing and validation sets are shown on table 3 and 4. The evaluation of the prediction of temperature and relative humidity has been done on the basis of standard measuring criteria: Mean Squared Error (MSE) and Normalised Mean Squared Error (NMSE). From the table 2, For the MLP-BP and FLNN-BP the main reason to initialize with small values between (0,1) is to avoid saturation for all patterns and the insensitivity to the training process. While for the FLNN-BP, the initialize value set with (0.25,0.75) since its give the best result for the minimize the error for the training process.

In MLP-BP and FLNN-BP, if the initial weight set too small, training will tend to start very slow. On the other hand, early stopping is use as the stopping crite- ria. If the validation error continuously increased several times, the network training will be terminated. The weights at the minimum validation error are stored for network generalization purpose. In the testing phase, this set of weights from the lowest validation error that was monitored during training phase is used. The target error is set to 0.0001 and the maximum epochs to 3000. This technique is using for the pre- diction for the next day.

7   Results

Table 3 and table 4 shows the MSE results for the MLP-BP, FLNN-BP and FLNN-MCS respectively. Moreover for Table 4, its displays the result for NMSE results for the same algorithms.

Table 2. Parameters

Datasets	Initial Setting Weights Values	Learning Rate/Probability	Epoch/Cycle	Number of input nodes
MLP-BP	(0,1)	0.05	3000	5
FLNN-BP	(0,1)	0.05	3000	5
FLNN-MCS	(0.25,0.75)	0.7	3000	5

Table 3. MSE for Temperature and Relative Humidity

Datasets	Temperature			Relative Humidity		
Algorithm	3	4	5	3	4	5
MLP-BP	0.004788	0.004776	0.004771	0.000069	0.000073	0.000077
FLNN-BP	0.004794	0.004701	0.004811	0.000112	0.000981	0.000921
FLNN-MCS	**0.001905**	**0.001929**	**0.005109**	**0.000061**	**0.000052**	**0.000073**

Table 4. NMSE for Temperature and Relative Humidity

Datasets	Temperatu			Relative Humidity		
Algorithm	3	4	5	3	4	5
MLP- BP	0.63258	0.630973	0.630414	0.99954	1.002797	0.99924
FLNN- BP	0.630791	0.630712	0.630398	1.021211	1.024701	1.02409
FLNN- MCS	**0.000585**	**0.000948**	**0.000085**	**0.006485**	**0.007320**	**0.00018**

Regarding from the table 3 and 4 results, it shows that, the effectiveness of using combination of FLNN-MCS between than other algorithms. It can be seen from the MSE results of FLNN-MCS algorithm for temperature datasets produced **0.001905** in networks order 3 while for relative humidity shows **0.000052** in networks order 4 which is getting the lowest error from any other networks. For NMSE results, FLNN-MCS also prove that their algorithm its slightly better than other network through result given from temperature dataset; **0.000085** and for relative humidity datasets; **0.000183 in** the networks order. For the conclusion, it's prove that the network gives the best results in evaluation of MSE Testing and NMSE an out of sample data.

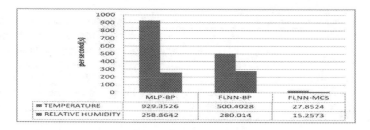

Figure 2 : CPU Times Result

Based on the Figure 2, the results shows that the FLNN-MCS network produced less time for training section compare to other algorithm like MLP-BP and FLNN-BP. The final result show that its only **27.8524s** for temperature while, **15.2573s** for relative humidity datasets while the value given by MLP-BP and FLNN-BP is incredibly higher than FLNN-MCS. From that, we can conclude the algorithm of FLNN-MCS can provide the best convergence time other than algorithms.

8   Conclusion and Future Works

Three ANN models, MLP-BP, FLNN-BP and FLNN-MCS were used and simulated for temperature and relative humidity prediction. The performance of the models is justified on the MSE, NMSE and CPU time. From the results obtained in the work, it is concluded that FLNN-MCS is computationally provided better prediction and convergence time compared to other algorithms. Hence, it can be concluded that FLNN-MCS are capable of modeling prediction. As for future works, we are considering to try the model with different data and different parameters and to expand the robustness of the network model.

Acknowledgements

The author would like to thank Ministry of Higher Education for the Economical support under grant number vote 0882.

References

1. Hayati,M.,Mohebi,Z. 2007.Temperature Forecasting Based on Neural Network Approach in World Applied Sciences Journal 2 (6):613-620,2007.IBSN:1818-4952.IDOSI Publications 2007
2. Kazemijad,M.,Deghan,M.,Motamadinejad,M.B.,Rastegar,H.,2006. A New Short Term Load Forecasting Using Multilayer Perceptron in ICIA.
3. Husaini,N.A.,Ghazali,R., Ismail,L.H.(2011). An Application of Pi-Sigma Neural Network for The Prediction of Flood Disaster.University Tun Hussein Onn Malaysia.
4. Climate change scenarios for Malaysia: 2001-2099. Malaysian Meteorological department scientific report, January 2009.
5. Paras,Mathur,s.,Kumar,A. & Chandra,M., 2007. A Feature Based Neural Network Model for Weather Forecas F32ting in WorldAcademy of Science Engineering and Technology.
6. Giles, C. L. and Maxwell, T. (1987) Learning, invariance and generalization in high-order neural networks. *In Applied Optics*, vol. 26, no. 23, Optical Society of America, Washington D. C., pp.4972-4978.
7. W. A.C. Schmidt & J.P. Davis. " Pattern Recognition Properties of Various Feature Spaces for Higher Order Neural Networks." In IEEE Transactions on Pattern Analysis and Machine Intelligence, Vol. 15, No.8, August 1993
8. Y.Pao. " Adaptive Patten Recognition and Neural Networks." Addison-Wesley, USA, 1989. ISBN: 0 2010125846
9. Payne, Robert B. (2005). The Cuckoos. OxfordUniversity Press. ISBN 0-19-850213-3. Retrieved 2007-12-19.
10. Yang X-S, Deb S. Engineering optimisation by cuckoo search. International Journal of Mathematical Modelling and Numerical Optimisation 2010;1:330–43.
11. S. Walton , O. Hassan, K. Morgan, M.R. Brown (2011). Modified cuckoo search: A new gradient free optimisation algorithm. Chaos, Solitons & Fractals 44,p.710–71
12. X. S. Yang and S. Deb, "Cuckoo search via Lévy flights," in World Congress on Nature & Biologically Inspired Computing, Coimbatore, India, p. 210–214, 2009.
13. Z. Xiang, G. Bi, and T. Le-Ngoc, "Polynomial perceptronsand their applications to fading channel equalization and cochannel interference suppression," *IEEE Transactions on Signal Processing*, vol. 42, no. 9, pp. 2470–2479, 1994.
14. J. C. Patra and C. Bornand, "Nonlinear dynamic system identification using Legendre neural network," in Neural Networks(IJCNN), The 2010 International Joint Conference on, 2010, pp. 1-7.
15. Ghazali, R., Hussain, A., and El-Dereby, W. (2006). "Application of Ridge Polynomial Neural Networks to Financial Time Series Prediction." *International Joint Conference on Neural Networks (IJCNN '06).* 913-920.
16. P. P. Raghu, et al., "A combined neural network approach for texture classification," Neural Networks, vol. 8, pp. 975-987, 1995.

Mining Indirect Least Association Rule

Zailani Abdullah[1], Tutut Herawan[2] and Mustafa Mat Deris[3]

[1]Department of Computer Science, Universiti Malaysia Terengganu
[2]Faculty of Computer Science & Information Technology, Universiti Malaya
[2]Universitas Teknologi Yogyakarta, Yogyakarta, Indonesia
[3]Faculty of Science Computer & Information Technology, Universiti Tun Hussein Onn
Malaysia

zailania@umt.edu.my, tutut@um.edu.my, mmustafa@uthm.edu.my

Abstract. Indirect pattern can be considered as one of the interesting information that is hiding in transactional database. It corresponds to the property of high dependencies between two items that are rarely appeared together but indirectly occurred through another items. Therefore, we propose an algorithm for Mining Indirect Least Association Rule (MILAR) from the real dataset. MILAR is embedded with a scalable least measure called Critical Relative Support (CRS). The experimental results indicate that MILAR is capable in generating the indirect least association rules from the given dataset.

Keywords: Mining, indirect, least, association rule.

1   Introduction

Data mining is about making analysis convenient, scaling analysis algorithms to large databases and providing data owners with easy to use tools in helping the user to navigate, visualize, summarize and model the data [1]. In summary, the ultimate goal of data mining is more towards knowledge discovery. One of the important models and extensively studies in data mining is known as association rule mining.

Since it was first introduced by Agrawal *et al.* [2] in 1993, association rule mining has been extensively studied by many researchers [3-11]. The general aim of ARM is at discovering interesting relationship among a set of items that frequently occurred together in transactional database [12]. However, under this concept, infrequent or least items are automatically considered as not important and pruned out during rules generation. In certain domain applications, least items may also provide useful insight about the data such as competitive product analysis [13], text mining [14], web recommendation [15], biomedical analysis [16], etc. Indirect association rule [17] refers to a pair of items that are rarely occurred together but their existences are highly depending on the presence of mediator itemsets. It was first proposed by Tan *et al.* [13] for interpreting the value of infrequent patterns and effectively pruning out the uninteresting infrequent patterns. Recently, the problem of indirect association mining has become more and more important because of its various domain applications [17-21]. Generally, the studies on indirect association mining can be divided into two categories, either focusing on proposing more efficient mining algorithms [14,17,21] or extending the definition of indirect association for different

T. Herawan et al. (eds.), *Proceedings of the First International Conference on Advanced Data and Information Engineering (DaEng-2013)*, Lecture Notes in Electrical Engineering 285, DOI: 10.1007/978-981-4585-18-7_19,
© Springer Science+Business Media Singapore 2014

domain applications [5,17,18]. The process of discovering indirect association rule is a nontrivial and usually relies more on the existing interesting measures that has been discussed in [13]. However, most of the measures are not properly evaluated in term of the least association rule. Therefore, in this paper we propose Mining Indirect Least Association Rule (MILAR) algorithm by utilizing the strength of Least Pattern Tree (LP-Tree) data structure [11]. In addition, Critical Relative Support (CRS) measure [22,27-35] is also embedded in the algorithm to mine the indirect least association rules among the least rules.

The rest of the paper is organized as follows. Section 2 describes the related work. Section 3 explains the proposed method. This is followed by performance analysis through two experiment tests in section 4. Finally, conclusion is reported in section 5.

2   Related Work

Indirect association is closely related to negative association. It deals with itemsets that do not have a sufficiently highest support. The negative associations' rule was first pointed out by Brin *et al.* [23]. The focused on mining negative associations is better that on finding the itemsets that have a very low probability of occurring together. Indirect associations provide an effective way to detect interesting negative associations by discovering only frequent itempairs that are highly expected to be frequent.

Until this recent, the important of indirect association between items has been discussed in many literatures. Tan *et al.* [13] proposed INDIRECT algorithm to extract indirect association between itempairs using the famous Apriori technique. Wan *et al.* [14] introduced HI-Mine algorithm to mine a complete set of indirect associations. HI-Mine generates indirect itempair set (IIS) and mediator support set (MSS), by recursively building the HI-struct from database. IS measure [24] is used as a dependence measure. Lin *et al.* [25] suggested GIAMS as an algorithm to mine indirect associations over data streams rather than static database environment. GIAMS contains two concurrent processes called PA-Monitoring and IA-Generation. In term of dependence measure, IS measure [24] is again adopted in the algorithm. Chen *et al.* [17] proposed an indirect association algorithm that was similar to HI-mine, namely MG-Growth. In this algorithm, temporal support and temporal dependence are used in this algorithm. Kazienko [15] presented IDARM* algorithm to extracts complete indirect associations rules. The main idea of IDARM* is to capture the transitive page from user-session as part of web recommendation system. A simple measure called Confidence [2] is employed as dependence measure. Lin *et al.* [36] presented EMIA-LM algorithm for mining indirect association rules over web data stream. The preliminary experiments also showed that EMIA-LM is better than HI-mine* for static data in term of computational speed and memory consumption. Liu *et al.* [37] suggested FIARM (Filtering-Based Indirect Association Rule Mining) algorithm to analyze gene microarray data. It is Apriori-based algorithm. In the analysis, Gene Ontology is employed to verify the accuracy of the relationships.

3 The Proposed Method

3.1 Association Rule

Throughout this section the set $I = \{i_1, i_2, \cdots, i_{|A|}\}$, for $|A| > 0$ refers to the set of literals called set of items and the set $D = \{t_1, t_2, \cdots, t_{|U|}\}$, for $|U| > 0$ refers to the data set of transactions, where each transaction $t \in D$ is a list of distinct items $t = \{i_1, i_2, \cdots, i_{|M|}\}$, $1 \le |M| \le |A|$ and each transaction can be identified by a distinct identifier TID.

Definition 1. *A set $X \subseteq I$ is called an itemset. An itemset with k-items is called a k-itemset.*

Definition 2. *The support of an itemset $X \subseteq I$, denoted $\mathrm{supp}(X)$ is defined as a number of transactions contain X.*

Definition 3. *Let $X, Y \subseteq I$ be itemset. An association rule between sets X and Y is an implication of the form $X \Rightarrow Y$, where $X \cap Y = \phi$. The sets X and Y are called antecedent and consequent, respectively.*

Definition 4. *The support for an association rule $X \Rightarrow Y$, denoted $\mathrm{supp}(X \Rightarrow Y)$, is defined as a number of transactions in D contain $X \cup Y$.*

Definition 5. *The confidence for an association rule $X \Rightarrow Y$, denoted $\mathrm{conf}(X \Rightarrow Y)$ is defined as a ratio of the numbers of transactions in D contain $X \cup Y$ to the number of transactions in D contain X. Thus*

$$\mathrm{conf}(X \Rightarrow Y) = \frac{\mathrm{supp}(X \Rightarrow Y)}{\mathrm{supp}(X)}$$

Definition 6. (Least Items). *An itemset X is called least item if $\mathrm{supp}(X) < \alpha$, where α is the minimum support (minsupp)*

The set of least item will be denoted as Least Items and

$$\text{Least Items} = \{X \subset I \mid \mathrm{supp}(X) < \alpha\}$$

Definition 7. (Frequent Items). *An itemset X is called frequent item if $\mathrm{supp}(X) \ge \alpha$, where α is the minimum support.*

The set of frequent item will be denoted as Frequent Items and

$$\text{Frequent Items} = \{X \subset I \mid \mathrm{supp}(X) \ge \alpha\}$$

3.2   Indirect Association Rule

Definition 8. *An itempair* $\{X\,Y\}$ *is indirectly associated via a mediator M, if the following conditions are fulfilled:*
1. $\text{supp}(\{X,Y\}) < t_s$ (itempair support condition)
2.   There exists a non-empty set M such that:
     a. $\text{supp}(\{X\}\cup M) \geq t_m$ and $\text{supp}(\{Y\}\cup M) \geq t_m$ (mediator support condition)
     b. $\text{dep}(\{X\}, M) \geq t_d$ and $\text{dep}(\{Y\}, M) \geq t_d$, where $\text{dep}(A, M)$ is a measure of dependence between itemset A and M (mediator dependence measure)

The user-defined thresholds above are known as itempair support threshold (t_s), mediator support threshold (t_m) and mediator dependence threshold (t_d), respectively. The itempair support threshold is equivalent to *minsupp* (α). Normally, the mediator support condition is set to equal or more than the itempair support condition $(t_m \geq t_s)$

The first condition is to ensure that (X, Y) is rarely occurred together and also known as least or infrequent items. In the second condition, the first-sub-condition is to capture the mediator M and for the second-sub-condition is to make sure that X and Y are highly dependence to form a set of mediator.

Definition 9. (Critical Relative Support). *A Critical Relative Support (CRS) is a formulation of maximizing relative frequency between itemset and their Jaccard similarity coefficient.*

The value of Critical Relative Support denoted as CRS and

$$\text{CRS}(A, B) = \max\left(\left(\frac{\text{supp}(A)}{\text{supp}(B)}\right), \left(\frac{\text{supp}(B)}{\text{supp}(A)}\right)\right) \times \left(\frac{\text{supp}(A \Rightarrow B)}{\text{supp}(A) + \text{supp}(B) - \text{supp}(A \Rightarrow B)}\right)$$

CRS value is between 0 and 1, and is determined by multiplying the highest value either supports of antecedent divide by consequence or in another way around with their Jaccard similarity coefficient. It is a measurement to show the level of CRS between combination of the both Least Items and Frequent Items either as antecedent or consequence, respectively. Here, Critical Relative Support (CRS) is employed as a dependence measure for 2(a) in order to mine the desired Indirect Association Rule.

3.2   Algorithm Development

Determine Minimum Support. Let I is a non-empty set such that $I = \{i_1, i_2, \cdots, i_n\}$, and D is a database of transactions where each T is a set of items such that $T \subset I$. An itemset is a set of item. A k-itemset is an itemset that contains k items. From Definition 6, an itemset is said to be least (infrequent) if it has a support count less than α.
Construct LP-Tree. A Least Pattern Tree (LP-Tree) is a compressed representation of the least itemset. It is constructed by scanning the dataset of single transaction at a

time and then mapping onto a new or existing path in the LP-Tree. Items that satisfy the α (Definition 6 and 7) are only captured and used in constructing the LP-Tree.

Mining LP-Tree. After the LP-Tree is fully constructed, the mining process will begin by implementing the bottom-up strategy. Hybrid 'Divide and conquer' method is employed to decompose the tasks of mining desired pattern. LP-Tree utilizes the strength of hash-based method during constructing itemset in support descending order.

Construct Indirect Patterns. The pattern is classified as indirect association pattern if it fulfilled with the two conditions. The first condition is elaborated in Definition 8 where it contains three sub-conditions. One of them is mediator dependence measure. CRS from Definition 9 is employed as mediator dependence measure between itemset in discovering the indirect patterns. Fig. 1 shows a complete graphical representation of Mining Indirect Least Association Rule Algorithm (MILAR).

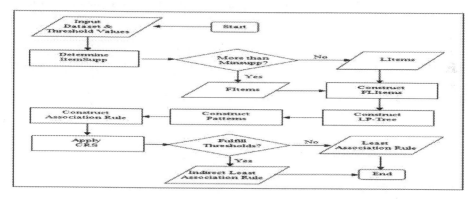

Fig. 1. A Complete Graphical Representation of MILAR Algorithm

4 Experiment Test

In this section, the analysis is made by comparing the total number of association rules being extracted based on the predefined thresholds using our proposed algorithm, MILAR. Here, three items are involved in forming a complete association rule; two items as an antecedent and one item as a consequence. The mediator is appeared as a part of antecedent. The experiment has been performed on Intel® Core™ 2 Quad CPU at 2.33GHz speed with 4GB main memory, running on Microsoft Windows Vista. All algorithms have been developed using C# as a programming language.

The dataset used in the experiment is a language anxiety dataset. It was taken from a survey on exploring language anxiety among engineering students at Universiti Malaysia Pahang (UMP) [3]. The respondents were 770 students, consisting of 394 males and 376 females. They are undergraduate students from five engineering based faculties, i.e., 216 students from Faculty of Chemical and Natural Resources Engineering (FCNRE), 105 students from Faculty of Electrical and Electronic Engineering (FEEE), 226 students from Faculty of Mechanical Engineering (FME),

178 students from Faculty of Civil Engineering and Earth Resources (FCEER), and 45 students from Faculty of Manufacturing Engineering and Technology Management (FMETM). To this, we have a dataset comprises the number of transactions (student) is 770 and the number of items (attributes) is 5.

Different Interval Supports were employed in the experiment. Fig. 2 shows the performance analysis against the dataset. Minimum Support (minsupp or α) and Mediator Support Threshold (t_m) are set to 30% and 10%, respectively. Varieties of minimum CRS (min-CRS) were employed in the experiment. During the performance analysis, 286 least association rules and 152 indirect least association rules were produced, respectively. The general trend was, the total number of indirect least association rules were kept reducing when the values of min-CRS were kept increasing. However, there are no changes in term of total least association rules and indirect least association rules when the min-CRS values were in the range of 0.15 until 0.20.

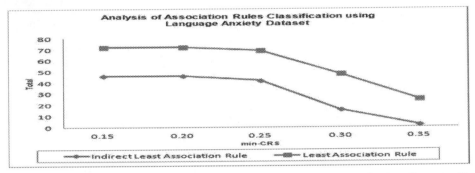

Fig. 2. Analysis of the Generated Association Rules against Language Anxiety Dataset with variation of min-CRS

5   Conclusion

Mining indirect least association rules from data repository is a very useful and nontrivial study in dealing with the rarity cases. In fact, it may contribute into discovering of a new knowledge which cannot be obtained through typical association rules approach. Therefore, we proposed Mining Indirect Least Association Rule (MILAR) algorithm to extract the hidden indirect least association rules from the data repository. MILAR algorithm embeds with a scalable measure called Critical Relative Support (CRS) rather than the common interesting measures in data mining. We conducted the experiment based on a real dataset. The result shows that MILAR algorithm can successfully generate the different number of indirect least association rules based on the variety of threshold values. It is also expected that the obtained information can provide a new insight for subject-matter experts to do further investigation.

References

1. Fayyad, U., Piatetsky-Shapiro, G., & Smyth, P.: From Data Mining to Knowledge Discovery in Databases. Advances in Knowledge Discovery and Data Mining. American Association for Artificial Intelligence, Menlo Park, USA (1996)
2. Agrawal, R., & Srikant, R.: Fast Algorithms for Mining Association Rules in Large Databases.: In Proceedings of the 20th International Conference on Very Large Data Bases, pp. 487-499 (1994)
3. Mannila, H., Toivonen, H., and Verkamo, A.I.: Discovery of Frequent Episodes in Event Sequences. Data Mining and Knowledge Discovery, 1, 259–289 (1997)
4. Park, J.S., Chen, M.S., and Yu, P.S.: An Effective Hash-based Algorithm for Mining Association Rules. In: Proceedings of the ACM-SIGMOD Intl. Conf. Management of Data (SIGMOD'95), pp. 175–186. ACM Press (1995)
5. Savasere, A., Omiecinski, E., and Navathe, S.: An Efficient Algorithm for Mining Association Rules in Large Databases. In: Proceedings of the 21st Intl. Conf. on Very Large Data Bases (VLDB'95), pp. 432–443. ACM Press (1995)
6. Fayyad, U., Patesesky-Shapiro, G., Smyth, P. and Uthurusamy, R.: Advances in Knowledge Discovery and Data Mining. MIT Press, MA (1996)
7. Bayardo, R.J.: Efficiently Mining Long Patterns from Databases. In: Proceedings of the ACM-SIGMOD International Conference on Management of Data (SIGMOD'98), pp. 85–93. ACM Press (1998)
8. Zaki, M.J. and Hsiao, C.J.: CHARM: An Efficient Algorithm for Closed Itemset Mining. In: Proceedings of the 2002 SIAM Intl. Conf. Data Mining, pp. 457–473. SIAM (2002).
9. Agarwal, R., Aggarwal, C., and Prasad, V.V.V.:A Tree Projection Algorithm for generation of Frequent Itemsets. Journal of Parallel and Distributed Computing 61, 350–371 (2001)
10. Liu, B., Hsu, W. and Ma, Y.: Mining Association Rules with Multiple Minimum Support. In: Proceedings of the 5th ACM SIGKDD International Conference on Knowledge Discovery and Data Mining, pp.337 – 341. ACM Press (1999)
11. Abdullah, Z., Herawan, T. and Deris, M.M.: Scalable Model for Mining Critical Least Association Rules. In Rongbo Zhu et al. ICICA 2010, LNCS 6377, pp. 509-516. Springer Heidelberg (2010)
12. Leung, C.W., Chan S.C and Chung, F.: An Empirical Study of a Cross-level Association Rule Mining Approach to Cold-start Recommendations. Knowledge-Based Systems, 21(7), October 2008, 515–529 (2008)
13. Tan, P.N., Kumar, V. and Srivastava, J.: Indirect Association: Mining Higher Order Dependences in Data. In: Proceedings of the 4th European Conference on Principles and Practice of Knowledge Discovery in Databases, pp. 632–637. Springer Heidelberg (2000)
14. Wan, Q. and An, A.: An Efficient Approach to Mining Indirect Associations. Journal Intelligent Information Systems, 27(2), 135–158 (2006)
15. Kazienko, P.: Mining Indirect Association Rules for Web Recommendation. International Journal of Applied Mathematics and Computer Science, 19(1), 165–186 (2009)
16. Tsuruoka, Y., Miwa, M., Hamamoto, K., Tsujii, J. and Ananiadou, S.: Discovering and Visualizing Indirect Associations between Biomedical Concepts. Bioinformatics, 27(13), 111-119 (2011).
17. Chen, L., Bhowmick, S.S. and Li, J.: Mining Temporal Indirect Associations. In: PAKDD2006, LNAI 3918, pp. 425-434. Springer Heidelberg (2006)
18. Cornelis, C., Yan, P., Zhang, X. and Chen, G.: Mining Positive and Negative Association from Large Databases. In: Proceedings of the 2006 IEEE International Conference on Cybernatics and Intelligent systems, pp. 1-6. IEEE (2006)
19. Kazienko, P. and Kuzminska, K.: The Influence of Indirect Association Rules on Recommendation Ranking Lists. In: Proceeding of the 5th Intl. Conf. on Intelligent Systems Design and Applications, pp. 482-487 (2005)

20. Tseng, V.s, Liu, Y.C and Shin J.W.: Mining Gene Expression Data with Indirect Association Rules. In Proceeding of the 2007 National Computer Symposium (2007)
21. Wu, X., Zhang, C. and Zhang, S. Efficient Mining of Positive and Negative Association Rules. ACM Transaction on Information Systems, 22(3), 381-405 (2004)
22. Abdullah, Z., Herawan, T., Noraziah, A. and Deris, M.M.: Mining Significant Association Rules from Educational Data using Critical Relative Support Approach. Procedia Social and Behavioral Sciences, 2011, 28, 97 – 191. Elsevier (2011)
23. Brin, S., Motwani, R., Ullman, J., and Tsur, S.: Dynamic Itemset Counting and Implication Rules for Market Basket Data. In: Proceedings of the International ACM SIGMOD Conference, pp. 255–264. ACM Press (1997)
24. Tan, P., Kumar, V., and Srivastava, J.: Selecting the Right Interestingness Measure for Association Patterns. In: Proceedings of the 8th Intl. Conf. on Knowledge Discovery and Data Mining, pp.32-41. (2002)
25. Lin, W-Y., Wei, Y-E., and Chen, C-H. Generic Approach for Mining Indirect Association Rules in Data Streams. LNCS 6703, pp. 95-104. Springer Heidelberg. (2011).
26. UCI Machine Learning Repository, http:// archive.ics.uci.edu/ml/datasets/
27. Herawan, T., Vitasari, P., and Abdullah, Z.: Mining Interesting Association Rules of Students Suffering Study Anxieties Using SLP-Growth Algorithm. International Journal of Knowledge and Systems Sciences, 3(2), 24-41 (2012)
28. Abdullah, Z., Herawan, T. Noraziah, A. and Deris, M.M.: Detecting Critical Least Association Rules in Medical Databasess. International Journal of Modern Physics: Conference Series, World Scientific, 9, 464–479 (2012)
29. Herawan, T. and Abdullah, Z.: CNAR-M: A Model for Mining Critical Negative Association Rules. In Zhihua Cai et al. (Eds): ISICA 2012, CCIS, 316, pp. 170–179. Springer Heidelberg (2012)
30. Abdullah, Z., Herawan, T. Noraziah, A. and Deris, M.M and Abawajy, J.H..: IPMA: Indirect Patterns Mining Algorithm. In N.T. Nguyen et al. (Eds.): ICCCI 2012, AMCCISC, 457, pp. 187–196. Springer Heidelberg (2012)
31. Herawan, T., P. Vitasari, and Abdullah, Z.: Mining Interesting Association Rules of Student Suffering Mathematics Anxiety. In J.M. Zain et al. (Eds.): ICSECS 2011, CCIS, vol. 188, II, pp. 495–508. Springer Heidelberg (2011)
32. Abdullah, Z., Herawan, T. and Deris, M.M.: Efficient and Scalable Model for Mining Critical Least Association Rules. In a special issue from AST/UCMA/ISA/ACN 2010, Journal of The Chinese Institute of Engineer, Taylor and Francis, 35, No. 4, 27 June 2012, 547–554 (2012)
33. Abdullah, Z., Herawan, T. Noraziah, A. and Deris, M.M.: Extracting Highly Positive Association Rules from Students' Enrollment Data. Procedia Social and Behavioral Sciences, 28, 107–111 (2011)
34. Abdullah, Z., Herawan, T. Noraziah, A. and Deris, M.M.: Mining Significant Association Rules from Educational Data using Critical Relative Support Approach. Procedia Social and Behavioral Sciences, 28, 97–101 (2011)
35. Abdullah, Z., Herawan, T. and Deris, M.M.: An Alternative Measure for Mining Weighted Least Association Rule and Its Framework. In J.M. Zain et al. (Eds.): ICSECS 2011, CCIS, vol. 188, II, pp. 475–485. Springer Heidelberg (2011)
36. Lin, W-Y., and Chen, Y-C.: A Mediator Exploiting Approach for Mining Indirect Associations from Web Data Streams. IEEEXplore in the 2nd Intl. Conf. on Innovations in Bio-inspired Computing and Applications (IBICA) 2011, pp. 183-186 (2011)
37. Liu, Y-C., Shin, J.W., and Tseng, V.S.: Discovering Indirect Gene Associations by Filtering-Based Indirect Association Rule Mining, International Journal of Innovative Computing, Information and Control, 7(10), 6041–6053 (2011)

Negative Selection Algorithm: A Survey on the Epistemology of Generating Detectors

Ayodele Lasisi[1], Rozaida Ghazali[1], and Tutut Herawan[2]

[1] Faculty of Computer Science and Information Technology
Universiti Tun Hussein Onn Malaysia
86400, Parit Raja, Batu Pahat, Johor, Malaysia
lasisiayodele@yahoo.com, rozaida@uthm.edu.my
[2] Faculty of Computer Science and Information Technology
University of Malaya, 50603, Kuala Lumpur, Malaysia
tutut@um.edu.my

Abstract. Within the Artificial Immune System community, the most widely implemented algorithm is the Negative Selection Algorithm. Its performance rest solely on the interaction between the detector generation algorithm and matching technique adopted for use. Relying on the type of data representation, either for strings or real-valued, the proper detection algorithm must be assigned. Thus, the detectors are allowed to efficaciously cover the non-self space with small number of detectors. In this paper, the different categories of detection generation algorithm and matching rule have been presented. Briefly, the biologial and artificial immune system, as well as the theory of negative selection algorithm were introduced. The exhaustive detector generation algorithm used in the original Negative Selection Algorithm laid the foundation at proferring other algorithmic methods based on set of rules in generating valid detectors for revealing anomalies.

Keywords: negative selection algorithm, data representation, detector generation algorithm, matching rule

1 Introduction

Negative Selection Algorithm (NSA), a significant set of rules within Artificial Immune System (AIS), announces its presence in the computer security domain. It is especially directed for use in anomaly detection and change detection. The process imitates T-cells detection execution of foreign invaders in the human body. In other to attain its target of recognition and elimination accordingly, the detector generation algorithm and matching technique plays a crucial role. Depending on the data representation, strings (binary) or real-valued representation, the appropriate detector generation algorithm and matching rule will be utilized. Forrest et al. [1] used the exhaustive detector generation algorithm based on r-contiguous matching rule. The drawback is that it is a costly computation in terms of time and space. Some other improved detector generation

T. Herawan et al. (eds.), *Proceedings of the First International Conference on Advanced Data and Information Engineering (DaEng-2013)*, Lecture Notes in Electrical Engineering 285, DOI: 10.1007/978-981-4585-18-7_20,
© Springer Science+Business Media Singapore 2014

algorithms were proposed, these are linear-time, binary, greedy, and Negative
Selection using Mutation (NSMutation) detector generation algorithms. Also
different matching algorithms have been introduced as well. These are related
to binary string representation. The detector generation schemes are different
for real-valued representation. This goes for the matching techniques too. More
information shall be presented in the latter section of this paper.

This paper reports and focus on the detector generation methods of negative selection algorithm. The different matching algorithms were discussed. It
examines in detail the process of generating detectors from the original method
to the various improvements carried out by researchers. These are geared at
reducing the number of detectors to its minimum moving a step higher than
the previous methods. It is believed that this review will further enhance our
understanding and knowledge behind the detector generation process of NSA.
The arrangement of the paper is organized as follows: Section 2 introduces biological immune system. Section 3 briefly describes artificial immune system. The
concept of negative selection algorithm is mentioned in section 4. Afterwards,
data representation and matching techniques occupies section 5, while detector
generation algorithms used in NSA are elaborated in section 6. Conclusion sums
up the paper in section 7.

2   Biological Immune System

The Biological Immune System (BIS), an integral part of the vertebrate immunity over centuries, is a dynamic, powerful, intelligent, and interconnection of
different components of the body, working in totality to fight, defend, and prevent pathogenic organisms' entrance into the body. The postulation of immunological concept of the body mechanisms defending against pathogens through
immunoglobulin called antibodies gave birth to immunity [2]. Two major functions attributed to BIS are: protection from foriegn invaders, and maintaining
homeostasis [3, 4]. However, there are still inquiries by immunologist about the
precise function of the immune system because of its sophisticated nature. As
such, it goes to show that the immune system has inherent capabilities which
surpasses what is being obtained now (i.e innate and adaptive system; and humoral and cellular processes) in guarding the body from pathogen invasion [5].
Boukerche et al. [6] gave the properties of immune systems as detection, diversity, learning, and tolerance. The immune system utilizes two lines of defenses
known as the innate immune system and adaptive immune system [7, 8]. Innate
immunity is the first line defense and its non-specific. Non-specific in that, it
does not concentrate on a particular kind of pathogen. Adaptive immunity on
the other hand handles such invasions that bypass the innate immunity line of
defense. It is specific as it targets, matches a particular pathogen, and stores in
memory the structure of that pathogen for faster detection and elimination if
encountered again.

3   Artificial Immune System

Artificial Immune System (AIS) is a computational paradigm that has evolved over the past two decades with algorithms developments mimicking the immune system processes. It connects and fosters the immunology, computer science, and engineering disciplines [9, 10]. Theoretical immunology models the immune system for in-depth knowledge of its behaviour [11]. Coupling the theoretical immunology with observed immune functions, principles and models introduces AIS as being able to cope effectively with changing situations and also suitable for problem solving [12]. Such problem solving tasks include but not limited to pattern recognition, learning, memory acquisition, distributed detection, and optimization [13]. Among AIS researchers, three (3) definitions have gained popularity, and only one of these is widely accepted as defined in [12].

Undoubtedly, the works of Bersini and Varela [14], pioneer in the use of metaphor for immune network theory, and Forrest et al. [1] announces the pathway from immunology to computing. The immune network theory was the focus of Hugues Bersini and Francisco Varela, abstracted from the way the natural immune network memorizes and functions leading to models and algorithms. Collaboration with researchers of the same field of interest popularized the concept [14, 15]. Self-non-self discrimination as it applies to computer security was the intention of Stephanie Forrest and her colleagues [1]. They were inspired at how the immune system recognizes self (normal) from non-self (abnormal) and a Negative Selection Algorithm (NSA) was proposed, thus becoming the pioneers of AIS algorithm development. It sets in motion the modeling and development of immune functions and properties of a number of AIS algorithms. A detailed review on theories and algorithms of artificial immune system can be found in [10].

4   Negative Selection Algorithm

In the biological immune system, there exist cells responsible for battling and annihilating intoxicated foreign molecules which are harmful to the body. T-cells, a special kind of white blood cell called lymphocytes, falls under the umbrella of the protecting cells. The receptors of T-cells are generated in a pseudo-random manner which undergoes a censoring process in the thymus called negative selection. In the thymus, the T-cells reacting to self cells are terminated while those not reacting leave the thymus into maturation stage. At this stage, they are equipped with the full functionality of protecting the body. Based on the negative selection principle, Forrest et al. [1] proposed and developed the Negative Selection Algorithm (NSA) for detection applications. Two principal stages of the NSA are the generation stage and the detection stage. The production of detector set is carried out at the generation stage, and these detector set are now ultimately used for change detection. Steps in NSA execution is summarized as follows [16]:

Given a universe U which contains all unique bit-strings of length l, self set

$S \subset U$ and non-self set $N \subset U$, where
$$U = S \subset U \qquad \text{and} \qquad S \cap N = \emptyset$$
1. Define self as a set S of bit-strings of length l in U.
2. Generate a set D of detectors, such that each fails to match any bit-string in S.
3. Monitor S for changes by continually matching the detectors in D against S.

Clearly, it can be deduced from the algorithmic steps that a kind of matching rule is required, reflected in both stages of the algorithm. This matching rule is hinged to a data representation method, invariably working in togetherness for performing the change detection task. Thus, in the next section, we shall elaborate on the data representation and matching techniques used by NSA for generating detectors.

5 Data Representation and Matching Techniques

The success of the detector generation algorithms depends solely on how the data is being represented and the matching technique adopted. For negative selection algorithm, strings (or binary) representation and real-valued representation has been widely used. Also, there is hybrid of both data representations which consist of different data types such as integer, real value, categorical information, boolean value, text information, etc. String representation has proved advantageous owing to the fact that: 1) it can be eventually represented in binary form; 2) anaylzed easily; and 3) beneficial for either textual or categorical information [17]. However, it suffers from space complexity and scability issues [18]. As a result, real-valued representation emerged in dealing with real value data types and also being suitable for real world applications. While data represented in strings can be used with a variety of matching techniques, euclidean distance is the primary matching technique used for real-valued representation [19]. Other utilized techniques for representing real-valued data and those of string representation are discussed below.

5.1 Matching Rule for Strings Representation

R-Contiguous Matching Rule. The interaction between antigens and antibodies needs a proper representation and there must be an affinity funtion. The r-contiguous matching rule was proposed by Percus et al. [20] in mapping antibodies to antigens, and matching process is defined as follow:

Suppose we denote antigens as set of binary strings $x = x_1, x_2, \ldots, x_n$ and antibodies denoted by a detector $d = d_1, d_2, \ldots, d_n$. This notation shall be used for the rest of this paper. The antibody matches the antigen if (1) holds:

$$\exists i \leq n - r + 1 \; \Big| \; x_j = d_j, \forall \, j = i, \ldots, i + r - 1 \tag{1}$$

where $|$ denotes such that, and $\forall$ is for-all or for any.

The original NSA in [1] made use of the r-contiguous matching rule. With a pre-defined window size r, two binary strings are set to match if identical.

R-Chunk Matching Rule. The r-contiguous matching rule described above, and the matching rules for classifier systems in [21] inspired r-chunk matching rule by Balthrop et al. [22]. The r-chunk rule encapsulate r-contiguous rule in that all the r-bits in the window must be matched with the binary strings. Therefore, any r-contiguous detectors can equally be represented as a set of r-chunk detectors. It is defined as follow:

Given detector $d = d_1, d_2, \ldots, d_m$ and binary strings $x = x_1, x_2, \ldots, x_n$ with $m \leq n$ and $i \leq n - m + 1$. The detector matches the binary strings if and only if (2) is satisfied:

$$x_j = d_j \ \forall \ j = i, \ldots, i + m - 1 \tag{2}$$

However, the distinguishing factor between r-contiguous and r-chunk matching rule is that full length r-contiguous bits develops crossover holes as well as length-limit holes, while r-chunk matching rule is devoid of this by eradicating the problem posed by length-limit holes.

Hamming Distance. Jerne [23] proposed a computational model based on idiotypic network theory which uses binary representation for the antibodies and antigens. Hamming distance is the matching rule implemented for this model. It is defined as follow:

Given detector $d = d_1, d_2, \ldots, d_n$ and binary strings $x = x_1, x_2, \ldots, x_n$, the detector matches binary strings if (3) is satisfied:

$$\sum_i \overline{x_1 \oplus d_i} \geq r \tag{3}$$

where $\oplus$ is the exclusive-or (XOR) operator, the line over $\overline{x_1 \oplus d_i}$ is the NOT operator, and $0 \leq r \leq n$ is a threshold value.

Additionally, variation of the Hamming distance known as Rogers and Tanimoto (R&T) matching rule was compared with several binary matching techniques and results shows it stands out to be the best [24]. This hamming distance has computational issues because it requires a huge number of steps in its execution. Thus, limits its application area.

5.2   Matching Rule for Real-Valued Representation

Euclidean Distance. This method of matching rule is widely incorporated for real-valued representation [19]. Inspite of its suitability for real valued cordinates, it yield undesirable results under large real-valued cases. Therefore, the best performance is achieved with limited real-valued cases [25]. Given the cordinates of detector $d = d_1, d_2, \ldots, d_n$ and binary strings cordinates as $x = x_1, x_2, \ldots, x_n$, the distance D existing between the detectors and binaries is shown in (4):

$$D = \sqrt{\sum_{i=1}^{n}(d_i - x_i)^2} \qquad (4)$$

Manhattan Distance. This is an alternative distance measure to euclidean distance, also used for real-valued representation. It executes based on sum of the absolute value of the detectors and binary strings as against the square of the sum in euclidean distance. Given the cordinates of detector $d = d_1, d_2, \ldots, d_n$ and binary strings cordinates as $x = x_1, x_2, \ldots, x_n$, the distance D existing between them is shown in (5):

$$D = \sqrt{\sum_{i=1}^{n}|d_i - x_i|} \qquad (5)$$

Minkwoski Distance. Minkwoski distance is an abstraction of the Euclidean distance and Manhattan distance [26], used by Dasgupta et al. [27] for aircraft fault detection. The distance D of the Minkwoski distance is defined as in (6):

$$D = \sqrt[h]{\sum_{i=1}^{n}|d_i - x_i|^h} \qquad (6)$$

where h is real number such that $h \geq 1$. When $h = 1$, it represents the Manhattan distance; while for $h = 2$, Euclidean distance is being represented. Hamaker and Boggess [28] presented several other matching techniques which are useful for real-valued representation.

6   NSA Detector Generation Algorithms

Insight into the various detector generation algorithms with respect to the above matching mechanism for strings representation and real-valued representation shall be discussed. For string (or binary) representation, the Exhaustive Detector Generating Algorithm (EDGA) using the r-contiguous bits [20] was incorporated in the original work by Forrest et al. [1]. It imitates the T-cells generation processes of the BIS by random generation of detectors, and matching with self strings for creating a database of legitimate detectors to be used in detection purpose. Time complexity and space complexity need to be considered greatly in determining the degree at which detectors exert their authority. In other to ascertain the computational complexities of the original NSA, a mathematical expression was derived by D'Haeseleer et al. [29]. This, coupled with the experiments carried out by Ayara et al. [30] proves that it is computationally expensive as most randomly generated detectors are discarded.

Furthermore, a modified version to EDGA using somatic hypermutation was proposed in [12] called NSMutation. The proposition of other improved detector generation algorithm, the linear-time detector generating algorithm and greedy detector generating algorithm were reported in [31]. They are more deterministic as against the randomized method of exhaustive detector generating algorithm. The former has higher space complexity than EDGA, whereas for the latter, a higher time complexity is observed but demonstrate to have increased coverage area with limited number of detectors leading to higher detection rate [17]. Still on the deterministic approach, Wierzchon [32] introduced a binary template detector generating algorithm with increased efficiency which produces less number of detector to cover the search space [30] went on to compare all the above detector generating algorithm and results shows that NSMutation is more extensible. Table 1 below shows the time complexity and space complexity of the above detector generation algorithms [29]. The terms used in the table denotes the following: l is the length of strings; r is the matching threshold; m, matching size; N_S is the population of self data; N_R, population of competent detectors.

Table 1. Complexities for detector generating algorithms based on strings (or binary) representation

Algorithm	Time	Space
Exhaustive	$O(m^l \cdot N_S)$	$O(l \cdot N_S)$
Linear	$O((l - r + 1) \cdot N_S m^r) + O((l - r + 1) \cdot m^r) + O(l \cdot N_R)$	$O((l - r + 1)^2 \cdot m^r)$
Greedy	$O((l - r + 1) \cdot N_S m^r) + O((l - r + 1) \cdot m^r \cdot N_R)$	$O((l - r + 1)^2 \cdot m^r)$
Binary Template	$O(m^r \cdot N_S) + O((l - r + 1) \cdot m^r \cdot N_R)$	$O((l - r + 1) \cdot m^r) + O(N_R)$
NSMutation	$O(m^l \cdot N_S) + O(N_R \cdot m^r) + O(N_R)$	$O(l(N_S + N_R))$

Moreover, for real-valued representation, euclidean distance has been the predominant matching rule used by detector generation algorithm [19,33]. The detector generation scheme can be hyper-rectangle [34], hyper-sphere [35], multi-shape [36], and convex hull [18] based on research work at rightly representing the detectors. They gained attention due to problems that surfaced from binary representation and are deemed fit for solving real world problems. Its first representation was in the characterization of self and nonself space using genetic algorithm in evolving the detectors. These resulted into hyper-rectangles and called detector rules [34]. Thereafter, Real Valued Negative Selection (RNS) that uses a detector generation algorithm resting on the idea of heuristic was proposed [35]. As with the desired goal of other detectors, they tend to maximize the coverage of the non-self space. The matching technique used was fuzzy membership function and thus, the detectors were hyper-spheres.

Also, Gonzalez et al. [37] put forward a RandomizedRNS (RRNS) by replacing the heuristic method with Monte Carlo simulation method. To further proliferate the distribution of detectors in the non-self space, simulated annealing was employed. In multi-shaped detector generation scheme, a structured genetic algorithm was merged with various shapes of detectors. Monte Carlo estimation method evolves the detectors which adequately cover non-self space [36]. A variable detector method called V-detector used euclidean distance in matching and generating detectors. It shows to be efficient in terms of the number of detectors generated [19]. Taking advantage of pseudo-random detector generation algorithmic method, [18] proposed a Convex-Hull NSA (CH-NSA) using a matching algorithm specifying if a point is within the convex hull. It supports dissimilar anatomy of shapes, thus no special preference for any. Number of detectors generated are significantly less, yielding good performance with regards to coverage area.

Generally, the aim of researchers are directed towards the creation of small number of detectors that can competently cover the non-self space. Ma et al. [38] gave a guided rule in effectively generating detectors stated as: (1) generating detector covering the area of shape space, and (2) generating detectors that will be in the surrounding of the inhabitant within the shape space. Strictly adhering to this rule will increase accuracy and performance.

7   Conclusion

The overview of the detector generation algorithm as applied within NSA has been outlined and given light in this paper. While the algorithms have provided researchers with varying options based on data represenation, continual investigation at improving the existing ones marches on. The matching mechanism, integrated with the detector generation algorithm, signifies both as the major components for negative selection algorithm. The r-contiguous bit matching rule and euclidean distance have established themselves as the dominant forces for both string and real-valued representation respectively. Instructions leading to generating less number of detectors was provided, and altogether producing an increased performance. This will spur computer scientist at trying to have the least minimum detectors as possible. While proper recognition has been duly accorded due to the success rate of the detector generation mechanisms of NSA, further experimental investigations are needed at collapsing the detectors with minimum overlap so as to optimize its overall process.

References

1. Forrest, S., Perelson, A.S., Allen, L., Cherukuri, R.: Self-nonself discrimination in a computer. In: Research in Security and Privacy, 1994. Proceedings., 1994 IEEE Computer Society Symposium on, IEEE (1994) 202–212
2. Silverstein, A.M.: Paul ehrlich, archives and the history of immunology. Nature immunology 6(7) (2005) 639–639
3. Immune, A.: Artificial immune systems. (2006) 107–118
4. Greensmith, J., Whitbrook, A., Aickelin, U.: Artificial immune systems. In: Handbook of Metaheuristics. Springer (2010) 421–448
5. Aickelin, U., Dasgupta, D.: Artificial immune systems. In: Search Methodologies. Springer (2005) 375–399

6. Boukerche, A., Jucá, K.R.L., Sobral, J.B., Notare, M.S.M.A.: An artificial immune based intrusion detection model for computer and telecommunication systems. Parallel Computing **30**(5) (2004) 629–646

7. Janeway Jr, C.A.: How the immune system recognizes invaders. life, death and the immune system. Scientific American **269**(3) (1993) 72

8. Ou, C.M.: Multiagent-based computer virus detection systems: abstraction from dendritic cell algorithm with danger theory. Telecommunication Systems (2011) 1–11

9. Dasgupta, D.: An overview of artificial immune systems. In Dasgupta, D (Ed.), Artificial Immune Systems and Their Applications (1998) 3–19

10. Dasgupta, D., Yu, S., Nino, F.: Recent advances in artificial immune systems: models and applications. Applied Soft Computing **11**(2) (2011) 1574–1587

11. De Castro, L.N., Timmis, J.: Artificial immune systems: a novel approach to pattern recognition. (2002) 67–84

12. de Castro, L.N., Timmis, J.: Artificial immune systems: a new computational intelligence approach. Springer Verlag (2002)

13. De Castro, L.N., Von Zuben, F.J.: Artificial immune systems: Part i– basic theory and applications. Technical Report - RT DCA 01/99, School of Computing and Electrical Enginnering. State University of Campinas, Brazil (1999)

14. Bersini, H., Varela, F.J.: Hints for adaptive problem solving gleaned from immune networks. In: Parallel problem solving from nature. Springer (1991) 343–354

15. Bersini, H., Varela, F.: The immune learning mechanisms: reinforcement, recruitment and their applications. Computing with Biological Metaphors **1**(2) (1994) 166–192

16. Stibor, T., Timmis, J., Eckert, C.: The link between r-contiguous detectors and k-cnf satisfiability. In: Evolutionary Computation, 2006. CEC 2006. IEEE Congress on, IEEE (2006) 491–498

17. Ji, Z., Dasgupta, D.: Revisiting negative selection algorithms. Evolutionary Computation **15**(2) (2007) 223–251

18. Majd, Mahshid, A.H., Hashemi, S.: A polymorphic convex hull scheme for negative selection algorithms. International Journal of Innovative Computing, Information and Control **8**(5A) (2012) 2953–2964

19. Ji, Z., Dasgupta, D.: Real-valued negative selection algorithm with variable-sized detectors. In: Genetic and Evolutionary Computation–GECCO 2004, Springer (2004) 287–298

20. Percus, J.K., Percus, O.E., Perelson, A.S.: Predicting the size of the t-cell receptor and antibody combining region from consideration of efficient self-nonself discrimination. Proceedings of the National Academy of Sciences **90**(5) (1993) 1691–1695

21. Holland, J.H., Holyoak, K.J., Nisbett, R.E., Thagard, P.R.: Induction: Processes of inference, learning, and discovery. computational models of cognition and perception (1986)

22. Balthrop, J., Esponda, F., Forrest, S., Glickman, M.: Coverage and generalization in an artificial immune system. In: Proceedings of the Genetic and Evolutionary Computation Conference, Citeseer (2002) 3–10

23. Jerne, N.K.: Towards the network theory of the immune system. Ann. Immunol.(Inst. Pasteur) **125C** (1974) 373–389

24. Harmer, P.K., Williams, P.D., Gunsch, G.H., Lamont, G.B.: An artificial immune system architecture for computer security applications. Evolutionary computation, IEEE transactions on **6**(3) (2002) 252–280

25. Chen, J., Yang, D., Naofumi, M.: A study of detector generation algorithms based on artificial immune in intrusion detection system. WSEAS TRANSACTIONS on BIOLOGY and BIOMEDICINE **4**(3) (2007) 29–35
26. Han, J., Kamber, M., Pei, J.: Data Mining: Concepts and Techniques. Morgan Kaufmann (2011)
27. Dasgupta, D., KrishnaKumar, K., Wong, D., Berry, M.: Negative selection algorithm for aircraft fault detection. In: Artificial Immune Systems. Springer (2004) 1–13
28. Hamaker, J.S., Boggess, L.: Non-euclidean distance measures in airs, an artificial immune classification system. In: Evolutionary Computation, 2004. CEC2004. Congress on. Volume 1., IEEE (2004) 1067–1073
29. D'Haeseleer, P., Forrest, S., et al.: An immunological approach to change detection. In: Proc. of IEEE Symposium on Research in Security and Privacy, Oakland, CA. (1996)
30. Ayara, M., Timmis, J., de Lemos, R., de Castro, L.N., Duncan, R.: Negative selection: How to generate detectors. In: Proceedings of the 1st International Conference on Artificial Immune Systems (ICARIS). Volume 1., Canterbury, UK:[sn] (2002) 89–98
31. D'Haeseleer, P., Forrest, S., Helman, P.: An immunological approach to change detection: Algorithms, analysis and implications. In: Security and Privacy, 1996. Proceedings., 1996 IEEE Symposium on, IEEE (1996) 110–119
32. Wierzchon, S.T.: Discriminative power of the receptors activated by k-contiguous bits rule. Journal of Computer Science & Technology **1**(3) (2000) 1–13
33. Yu, S., Adviser-Dasgupta, D.: Exploration of sense of self and humoral immunity for artificial immune systems: algorithms and applications. 361
34. Dasgupta, D., González, F.: An immunity-based technique to characterize intrusions in computer networks. Evolutionary Computation, IEEE Transactions on **6**(3) (2002) 281–291
35. González, F.A., Dasgupta, D.: Anomaly detection using real-valued negative selection. Genetic Programming and Evolvable Machines **4**(4) (2003) 383–403
36. Balachandran, S., Dasgupta, D., Nino, F., Garrett, D.: A framework for evolving multi-shaped detectors in negative selection. In: Foundations of Computational Intelligence, 2007. FOCI 2007. IEEE Symposium on, IEEE (2007) 401–408
37. Gonzalez, F., Dasgupta, D., Niño, L.F.: A randomized real-valued negative selection algorithm. In: Artificial Immune Systems. Springer (2003) 261–272
38. Ma, W., Tran, D., Sharma, D.: A practical study on shape space and its occupancy in negative selection. In: Evolutionary Computation (CEC), 2010 IEEE Congress on, IEEE (2010) 1–7

Nonparametric Orthogonal NMF and Its Application in Cancer Clustering

Andri Mirzal

Faculty of Computing, N28-439-03
Universiti Teknologi Malaysia
81310 UTM Johor Bahru, Malaysia
andrimirzal@utm.my

Abstract. Orthogonal nonnegative matrix factorization (NMF) is an NMF objective function that enforces orthogonality constraint on its factor. There are two challenges in optimizing this objective function: the first is how to design an algorithm that has convergence guarantee, and the second is how to automatically choose the regularization parameter. In our previous work, we have been able to develop a convergent algorithm for this objective function. However, the second challenge remains unsolved. In this paper, we provide an attempt to answer the second challenge. The proposed method is based on the L-curve approach and has a simple form which is preferable since it introduces only a small additional computational cost. This method transforms the algorithm into nonparametric, and is also extendable to other NMF objective functions as long as the functions are differentiable with respect to the corresponding regularization parameters. Numerical results are then provided to evaluate the feasibility of the method in choosing the appropriate regularization parameter values by utilizing it in cancer clustering tasks.

Keywords: cancer clustering, gene expression, nonnegative matrix factorization, nonparametric learning, orthogonality constraint

1 Introduction

The nonnegative matrix factorization (NMF) is a recent development in matrix decomposition and factor analysis. The NMF was first introduced by Paatero & Anttila [1, 2] and made popular by Lee & Seung [3, 4] in which the latter authors proposed a simple NMF algorithm and showed its uses in image and document analysis. The NMF has been successfully applied in many application domains including document clustering [5, 6], spectral analysis [7, 8], image processing [9–13], blind source separation [14–18], and cancer clustering and classification [19–30].

Orthogonal NMF was introduced by Ding et al. [31] to improve clustering capability of the NMF. There are two challenges in optimizing this objective function that have not been addressed by the authors: (1) how to design a convergent algorithm, and (2) how to develop a method to automatically choose

T. Herawan et al. (eds.), *Proceedings of the First International Conference on Advanced Data and Information Engineering (DaEng-2013)*, Lecture Notes in Electrical Engineering 285, DOI: 10.1007/978-981-4585-18-7_21,
© Springer Science+Business Media Singapore 2014

the regularization parameter. In our previous work [32], we have been able to answer the first challenge. In this paper, we provide an attempt to answer the second challenge. The proposed method is based on the L-curve approach and has a simple form which is preferable since it introduces only a small additional computational cost. Numerical results are then provided to evaluate the feasibility of the method in choosing the appropriate regularization parameter values by utilizing it in cancer clustering tasks.

2   Orthogonal NMF

Since there are two factors produced by the NMF, i.e., the basis matrix $\mathbf{B}$ and the coefficient matrix $\mathbf{C}$, orthogonality constraints can be imposed on rows or columns of $\mathbf{B}$ and/or $\mathbf{C}$. Here we only discuss orthogonality constraint on rows of $\mathbf{C}$, similar results can be obtained for other cases by following the same procedure. The objective function of orthogonal NMF thus can defined with the following:

$$\min_{\mathbf{B},\mathbf{C}} J(\mathbf{B},\mathbf{C}) = \frac{1}{2}\|\mathbf{A} - \mathbf{BC}\|_F^2 + \frac{\alpha}{2}\|\mathbf{CC}^T - \mathbf{I}\|_F^2 \tag{1}$$
$$\text{s.t. } \mathbf{B} \geq 0, \mathbf{C} \geq 0,$$

where the first component of the right hand side part denotes the approximation error, the second component denotes the orthogonality constraint, and α denotes the regularization parameter.

An algorithm based on the multiplicative update rules (MUR) [3, 4] for minimizing eq. 1 can be derived by utilizing the Karush-Kuhn-Tucker (KKT) optimality conditions. The KKT function of the objective can be defined with:

$$L(\mathbf{B},\mathbf{C}) = J(\mathbf{B},\mathbf{C}) - \text{tr}\left(\boldsymbol{\Gamma}_\mathbf{B}\mathbf{B}^T\right) - \text{tr}\left(\boldsymbol{\Gamma}_\mathbf{C}\mathbf{C}\right),$$

where $\boldsymbol{\Gamma}_\mathbf{B} \in \mathbb{R}_+^{M \times R}$ and $\boldsymbol{\Gamma}_\mathbf{C} \in \mathbb{R}_+^{N \times R}$ denote the KKT multipliers, and $\text{tr}\left(\mathbf{X}\right)$ denotes trace of $\mathbf{X}$. Partial derivatives of L with respect to $\mathbf{B}$ and $\mathbf{C}$ are:

$$\nabla_\mathbf{B} L(\mathbf{B}) = \nabla_\mathbf{B} J(\mathbf{B}) - \boldsymbol{\Gamma}_\mathbf{B}, \text{ and}$$
$$\nabla_\mathbf{C} L(\mathbf{C}) = \nabla_\mathbf{C} J(\mathbf{C}) - \boldsymbol{\Gamma}_\mathbf{C}^T,$$

with

$$\nabla_\mathbf{B} J(\mathbf{B}) = \mathbf{BCC}^T - \mathbf{AC}^T, \text{ and}$$
$$\nabla_\mathbf{C} J(\mathbf{C}) = \mathbf{B}^T\mathbf{BC} - \mathbf{B}^T\mathbf{A} + \alpha\mathbf{CC}^T\mathbf{C} - \alpha\mathbf{C}.$$

Then, the KKT optimality conditions can be written with the following:

$$\mathbf{B}^* \geq \mathbf{0}, \qquad\qquad\qquad \mathbf{C}^* \geq \mathbf{0},$$
$$\nabla_\mathbf{B} J(\mathbf{B}^*) = \boldsymbol{\Gamma}_\mathbf{B} \geq \mathbf{0}, \qquad\qquad \nabla_\mathbf{C} J(\mathbf{C}^*) = \boldsymbol{\Gamma}_\mathbf{C}^T \geq \mathbf{0},$$
$$\nabla_\mathbf{B} J(\mathbf{B}^*) \odot \mathbf{B}^* = \mathbf{0}, \qquad\qquad \nabla_\mathbf{C} J(\mathbf{C}^*) \odot \mathbf{C}^* = \mathbf{0}, \tag{2}$$

where $(\mathbf{B}^*, \mathbf{C}^*)$ is a point that satisfies the KKT optimality conditions (the stationary point), and $\odot$ denotes the Hadamard product. Eq. 2 is also known as the complementary slackness.

A MUR algorithm is derived by utilizing the complementary slackness:

$$b_{mr}^{(k+1)} \longleftarrow b_{mr}^{(k)} \frac{\left(\mathbf{A}\mathbf{C}^T\right)_{mr}}{\left(\mathbf{B}\mathbf{C}\mathbf{C}^T\right)_{mr}} \quad \forall m, r$$

$$c_{rn}^{(k+1)} \longleftarrow c_{rn}^{(k)} \frac{\left(\mathbf{B}^T \mathbf{A} + \alpha \mathbf{C}\right)_{rn}}{\left(\mathbf{B}^T \mathbf{B}\mathbf{C} + \alpha \mathbf{C}\mathbf{C}^T \mathbf{C}\right)_{rn}} \quad \forall r, n,$$

where $b_{mr}^{(k)}$ denotes (m, r) entry of $\mathbf{B}$ at k-th iteration. As shown, in order to update $\mathbf{B}$ and $\mathbf{C}$, α value needs to be chosen. The common approach is to use a fixed value from the start (this is the approach used both in the original work [31] and our work [32]). Thus, it is difficult to say that the chosen value is optimal in some sense.

3 An L-curve based method for choosing α

The L-curve is a log-log plot of the approximation error versus the solution size as the regularization parameter varies. This technique is often employed to search for an optimum solution to the Tikhonov regularized least square problem [33]. The solution is proposed to be located at L-corner, a point where both the approximation error and the solution size are at their minimum values. As the objective of the Tikhonov regularized least square is to minimize both the approximation error and the solution size, it is intuitive to use L-corner to find the optimum solution.

The L-curve concept can be directly extended to the orthogonal NMF problem by substituting the solution size with the orthogonality constraint. The problem now is how to find a corner where both the approximation error $\frac{1}{2}\|\mathbf{A} - \mathbf{B}\mathbf{C}\|_F^2$ and the orthogonality constraint $\frac{1}{2}\|\mathbf{C}\mathbf{C}^T - \mathbf{I}\|_F^2$ are at their minimum values. Let $J_1 = \frac{1}{2}\|\mathbf{A} - \mathbf{B}\mathbf{C}\|_F^2$ and $J_2 = \frac{1}{2}\|\mathbf{C}\mathbf{C}^T - \mathbf{I}\|_F^2$. By using the L-curve approach, α that corresponds to the L-corner can be found by solving the following problem:

$$\alpha = \arg\min_{\alpha} \left(J_1 \cdot J_2\right).$$

Since this is a line search, an optimum α can be found by:

$$\frac{\partial\left(J_1 \cdot J_2\right)}{\partial \alpha} = \frac{\partial J_1}{\partial \alpha} J_2 + \frac{\partial J_2}{\partial \alpha} J_1 = 0,$$

where

$$\frac{\partial J_1}{\partial \alpha} = \frac{\partial J_1}{\partial \mathbf{C}} \frac{\partial \mathbf{C}^T}{\partial \alpha} = \frac{1}{2}\left(\mathbf{B}^T \mathbf{B}\mathbf{C} - \mathbf{B}^T \mathbf{A}\right) \frac{\partial \mathbf{C}^T}{\partial \alpha},$$

Algorithm 1 A MUR algorithm for orthogonal NMF.

Initialization: $\mathbf{B}^{(0)} > \mathbf{0}$, $\mathbf{C}^{(0)} > \mathbf{0}$, and $\alpha = 0$.
for $k = 0, \ldots, K$ **do**

$$b_{mr}^{(k+1)} \longleftarrow b_{mr}^{(k)} \frac{\left(\mathbf{A}\mathbf{C}^{(k)T}\right)_{mr}}{\left(\mathbf{B}^{(k)}\mathbf{C}^{(k)}\mathbf{C}^{(k)T}\right)_{mr} + \delta} \quad \forall m, r$$

$$c_{rn}^{(k+1)} \longleftarrow c_{rn}^{(k)} \frac{\left(\mathbf{B}^{(k+1)T}\mathbf{A} + \alpha\mathbf{C}^{(k)}\right)_{rn}}{\left(\mathbf{B}^{(k+1)T}\mathbf{B}^{(k+1)}\mathbf{C}^{(k)} + \alpha\mathbf{C}^{(k)}\mathbf{C}^{(k)T}\mathbf{C}^{(k)}\right)_{rn} + \delta} \quad \forall r, n$$

$$\alpha \longleftarrow \frac{\|\mathbf{A} - \mathbf{B}^{(k+1)}\mathbf{C}^{(k+1)}\|_F^2}{M \times N \|\mathbf{C}^{(k+1)}\mathbf{C}^{(k+1)T} - \mathbf{I}\|_F^2}$$

end for
---

and

$$\frac{\partial J_2}{\partial \alpha} = \frac{\partial J_2}{\partial \mathbf{C}}\frac{\partial \mathbf{C}^T}{\partial \alpha} = \frac{1}{2}\left(\mathbf{C}\mathbf{C}^T\mathbf{C} - \mathbf{C}\right)\frac{\partial \mathbf{C}^T}{\partial \alpha}.$$

Thus $\frac{\partial(J_1 \cdot J_2)}{\partial \alpha} = 0$ leads to:

$$\left(\mathbf{B}^T\mathbf{B}\mathbf{C} - \mathbf{B}^T\mathbf{A}\right)\|\mathbf{C}\mathbf{C}^T - \mathbf{I}\|_F^2 + \left(\mathbf{C}\mathbf{C}^T\mathbf{C} - \mathbf{C}\right)\|\mathbf{A} - \mathbf{B}\mathbf{C}\|_F^2 = 0. \qquad (3)$$

By setting $\nabla_\mathbf{C} J(\mathbf{C})$ to zero, we get:

$$\mathbf{B}^T\mathbf{B}\mathbf{C} - \mathbf{B}^T\mathbf{A} = \alpha\mathbf{C} - \alpha\mathbf{C}\mathbf{C}^T\mathbf{C}.$$

Substituting this into eq. 3 leads to:

$$\alpha = \frac{\|\mathbf{A} - \mathbf{B}\mathbf{C}\|_F^2}{\|\mathbf{C}\mathbf{C}^T - \mathbf{I}\|_F^2}.$$

And because α value is an accumulation of $M \times N$ elements, the normalized value is used instead:

$$\alpha = \frac{\|\mathbf{A} - \mathbf{B}\mathbf{C}\|_F^2}{M \times N \|\mathbf{C}\mathbf{C}^T - \mathbf{I}\|_F^2}. \qquad (4)$$

Algorithm 1 gives our proposed algorithm with α is updated in each iteration using eq. 4, and δ denotes a small positive number to avoid division by zero. Note that $\mathbf{B}$ and $\mathbf{C}$ must be initialized using positive matrices to prevent zero locking from the start. More discussions on zero locking phenomenon can be found in e.g., [34, 35].

4   Cancer clustering using the NMF

Let $\mathbf{A} \in \mathbb{R}^{M \times N}$ denotes gene-by-sample cancer expression dataset, the standard prosedur in clustering using the NMF is to compute the coefficient matrix $\mathbf{C}$, and then cluster assignment vector $\mathbf{x}$ can be determined by $x_n =$

Algorithm 2 Clustering procedure using the NMF.

---

1. Input: gene-by-sample matrix $\mathbf{A} \in \mathbb{R}^{M \times N}$, and #cluster K.
2. If $\mathbf{A}$ contains negative entries, shift all entries to the nonnegative orthant.
3. Normalize each column of $\mathbf{A}$, i.e., $a_{mn} \leftarrow \frac{a_{mn}}{\sum_m a_{mn}}$ for $\forall n$.
4. Compute $\mathbf{C}$ by setting $R = K$ using any NMF algorithm.
5. Apply k-means clustering on columns of $\mathbf{C}$ to obtain K clusters of the samples.

$\arg_{k \in [1,K]} \max \mathbf{c}_n$ $\forall n \in [1, N]$, where x_n denotes n-th entry of $\mathbf{x}$, $\mathbf{c}_n$ denotes n-th column of $\mathbf{C}$, and K denotes the number of clusters. Another way to infer $\mathbf{x}$ from $\mathbf{C}$ is to apply kmeans clustering on columns of $\mathbf{C}$. We used the latter because we found that it produced better results in all cases. Algorithm 2 outlines the clustering procedure using the NMF. To evaluate clustering quality, we used two metrics: Adjusted Rand Index (ARI) [36–38] and Accuracy [20].

5   Experimental results

To evaluate the clustering capability of the proposed algorithm, we used six publicly available cancer datasets from the work of Souto et al. [39]. Tables 1 summarizes the information of the datasets. As shown the datasets are quite representative as the number of classes varied from 2 to 10, the number of samples varied from tens to hundreds, and we also have one dataset, Su-2001, that contains multiple type of cancers.

The clustering results of the proposed algorithm (ONMF) are compared to the results of the standard NMF algorithm by Lee & Seung [3, 4] (NMFLS). Table 2 shows the results, and figure 1 plots them. As shown, ONMF outperformed NMFLS in all cases. Performance improvements are more visible in four datasets, i.e., Armstrong, Tomlins, Pomeroy, and Su. Only in two datasets, Nutt and Yeoh, ONMF and NMFLS have comparable results.

6   Conclusion

We have presented a method to automatically choose the regularization parameter in orthogonal NMF. Because the proposed method was designed based on

Table 1. Cancer datasets.

Dataset	Tissue	#Samples	#Genes	#Classes
Nutt-2003-v2	Brain	28	1070	2
Armstrong-2002-v2	Blood	72	2194	3
Tomlins-2006-v2	Prostate	92	1288	4
Pomeroy-2002-v2	Brain	42	1379	5
Yeoh-2002-v2	Bone	248	2526	6
Su-2001	Multi	174	1571	10

Table 2. Average ARI and Accuracy over 1000 trials.

Dataset	NMFLS		ONMF	
	ARI	Acc	ARI	Acc
Nutt-2003-v2	0.00203	0.571	0.0107	0.571
Armstrong-2002-v2	0.520	0.737	0.670	0.852
Tomlins-2006-v2	0.0553	0.445	0.155	0.543
Pomeroy-2002-v2	0.301	0.607	0.519	0.759
Yeoh-2002-v2	0.0424	0.393	0.0432	0.393
Su-2001	0.465	0.428	0.586	0.457

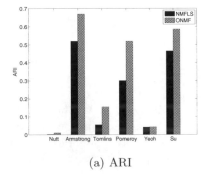
(a) ARI

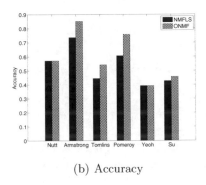
(b) Accuracy

Fig. 1. Performance comparison between NMFLS and ONMF.

a rigorous approach, optimality of the method can be guaranteed. The method also allows the proposed algorithm to be used in a fully unsupervised manner because the regularization parameter can be learned directly from the dataset. In addition, the formulation of the method is simple and does not change the computational complexity of the algorithm: $O(M \times N \times K)$ where $M \times N$ denotes the size of the input matrix and K denotes the number of iterations which is the same as the computational complexity of the standard NMF algorithm, NMFLS. The performances of the proposed algorithm were then evaluated in cancer clustering tasks in which as shown in table 1 and figure 1, it can confidently outperformed the standard NMF algorithm.

Acknowledgement

The author would like to thank the reviewers for useful comments. This research was supported by Ministry of Higher Education of Malaysia and Universiti Teknologi Malaysia under Exploratory Research Grant Scheme R.J130000.7828. 4L095.

References

1. Paatero, P., Tapper, U.: Positive matrix factorization: A non-negative factor model with optimal utilization of error estimates of data values. Environmetrics 5, 111–126 (1994)
2. Anttila, P., et al.: Source identification of bulk wet deposition in Finland by positive matrix factorization. Atmospheric Environment 29(14), 1705–1718 (1995)
3. Lee, D., Seung, H.: Learning the parts of objects by non-negative matrix factorization. Nature 401(6755), 788–791 (1999)
4. Lee, D., Seung, H.: Algorithms for non-negative matrix factorization. In Proc. Advances in Neural Processing Information Systems, 556–562 (2000)
5. Xu, W., et al.: Document clustering based on non-negative matrix factorization. In Proc. ACM SIGIR, 267–273 (2003)
6. Shahnaz, F., et al.: Document clustering using nonnegative matrix factorization. Information Processing & Management 42(2), 373–386 (2006)
7. Pauca, V.P., et al.: Nonnegative matrix factorization for spectral data analysis. Linear Algebra and Its Applications 416(1), 29–47 (2006)
8. Jia, S., Qian, Y.: Constrained nonnegative matrix factorization for hyperspectral unmixing. IEEE Transactions on Geoscience and Remote Sensing 47(1), 161–173 (2009)
9. Li, S.Z., et al.: Learning spatially localized, parts-based representation. In Proc. IEEE Comp. Soc. Conf. on Computer Vision and Pattern Recognition, 207–212 (2001)
10. Hoyer, P.O.: Non-negative matrix factorization with sparseness constraints. The Journal of Machine Learning Research 5, 1457–1469 (2004)
11. Wang, D., Lu, H.: On-line learning parts-based representation via incremental orthogonal projective non-negative matrix factorization. Signal Processing 93(6), 1608–1623 (2013)
12. Pascual-Montano, A., et al.: Nonsmooth nonnegative matrix factorization. IEEE Transactions on Pattern Analysis and Machine Intelligence 28(3), 403–415 (2006)
13. Gillis, N., Glineur, F.: A multilevel approach for nonnegative matrix factorization. J. Computational and Applied Mathematics 236(7), 1708–1723 (2012)
14. Cichocki, A., et al.: Extended SMART algorithms for non-negative matrix factorization. LNCS 4029, 548–562, Springer (2006)
15. Zhou, G., et al.: Online blind source separation using incremental nonnegative matrix factorization with volume constraint. IEEE Transactions on Neural Networks 22(4), 550–560 (2011)
16. Bertin, N., et al.: Enforcing harmonicity and smoothness in bayesian non-negative matrix factorization applied to polyphonic music transcription. IEEE Transactions on Audio, Speech, and Language Processing 18(3), 538–549 (2010)
17. Bertrand, A., Moonen, M.: Blind separation of non-negative source signals using multiplicative updates and subspace projection. Signal Processing 90(10), 2877–2890 (2010)
18. Virtanen, T., et al.: Bayesian extensions to non-negative matrix factorisation for audio signal modelling. In IEEE Int'l Conf. on Acoustics, Speech and Signal Processing, 1825–1828 (2008)
19. Brunet, J.P., et al.: Metagenes and molecular pattern discovery using matrix factorization. Proc. Natl Acad. Sci. USA 101(12), 4164–4169 (2003)
20. Gao, Y., Church, G.: Improving molecular cancer class discovery through sparse non-negative matrix factorization. Bioinformatics 21(21), 3970–3975 (2005)

21. Kim, H., Park, H.: Sparse non-negative matrix factorizations via alternating non-negativity constrained least squares for microarray data analysis. Bioinformatics 23(12), 1495–1502 (2007)
22. Devarajan, K.: Nonnegative matrix factorization: an analytical and interpretive tool in computational biology. PLoS Computational Biology 4(7), e1000029 (2008)
23. Kim, H., Park, H.: Nonnegative matrix factorization based on alternating non-negativity constrained least squares and active set method. SIAM J. Matrix Anal. Appl. 30(2), 713–730 (2008)
24. Carmona-Saez, et al.: Biclustering of gene expression data by non-smooth non-negative matrix factorization. BMC Bioinformatics 7(78) (2006)
25. Inamura, K., et al.: Two subclasses of lung squamous cell carcinoma with different gene expression profiles and prognosis identified by hierarchical clustering and non-negative matrix factorization. Oncogene (24), 7105–7113 (2005)
26. Fogel, P., et al.: Inferential, robust non-negative matrix factorization analysis of microarray data. Bioinformatics 23(1), 44–49 (2007)
27. Zheng, C.H., et al.: Tumor clustering using nonnegative matrix factorization with gene selection. IEEE Transactions on Information Technology in Biomedicine 13(4), 599–607 (2009)
28. Wang, G., et al.: LS-NMF: A modified non-negative matrix factorization algorithm utilizing uncertainty estimates. BMC Bioinformatics 7(175) (2006)
29. Wang, J.J.Y., et al.: Non-negative matrix factorization by maximizing correntropy for cancer clustering. BMC Bioinformatics 14(107) (2013)
30. Yuvaraj, N., Vivekanandan, P.: An efficient SVM based tumor classification with symmetry non-negative matrix factorization using gene expression data. In Int'l Conf. on Information Communication and Embedded Systems, 761–768 (2013)
31. Ding, C., et al.: Orthogonal nonnegative matrix t-factorizations for clustering. In 12th ACM SIGKDD Int'l Conf. on Knowledge Discovery and Data Mining, 126–135 (2006)
32. Mirzal, A.: A convergent algorithm for orthogonal nonnegative matrix factorization. To appear in J. Computational and Applied Mathematics.
33. Hansen, P.C.: Analysis of discrete ill-posed problems by means of the L-curve. SIAM Review 34(4), 561–580 (1992)
34. Lin, C.J.: On the convergence of multiplicative update algorithms for nonnegative matrix factorization. IEEE Transactions on Neural Networks 18(6), 1589–1596 (2007)
35. Lin, C.J.: Projected gradient methods for non-negative matrix factorization. Technical Report ISSTECH-95-013, Department of CS, National Taiwan University (2005)
36. Rand, W.M.: Objective criteria for the evaluation of clustering methods. Journal of the American Statistical Association 66(336), 846–850 (1971)
37. Hubert, L., Arabie, P.: Comparing partitions. Journal of Classification 2(1), 193–218 (1985)
38. Vinh, N.X., et al.: Information theoretic measures for clustering comparison: Is a correction for chance necessary? In 26th Annual Int'l Conf. on Machine Learning, pp. 1073–1080 (2009)
39. Souto, M.C.P., et al.: Clustering cancer gene expression data: a comparative study. BMC Bioinformatics 9(497) (2008)

Optimal Parameters Estimation in AWJ Machining Process using Active Set Method

Andri Mirzal

Faculty of Computing, N28-439-03
Universiti Teknologi Malaysia
81310 UTM Johor Bahru, Malaysia
andrimirzal@utm.my

Abstract. Parameters estimation in abrasive waterjet (AWJ) machining process is a task of searching a set of values that lead to the optimal machining performance. This task is usually conducted by using soft computing techniques like genetic algorithms (GA) and simulated annealing (SA). However, we found that the objective function of the problem is a simple quadratic formula with box constraints. Accordingly many established optimization methods from quadratic programming study can be employed to search for the optimal parameter values. In this paper, we demonstrate the use of the active set method for solving the problem and show that this method can confidently outperform GA and SA both in the machining performance and computational times. These results suggest that this kind of problems should be addressed by using established gradient based optimization algorithms first before using soft computing techniques because the formers have better convergence property and also usually are faster than the latters.

Keywords: abrasive waterjet machining, active set method, genetic algorithms, simulated annealing, quadratic programming.

1 Introduction

Abrasive waterjet (AWJ) is a cutting machine that uses a highly concentrated stream of water mixed with abrasive particles to cut through a wide variety of materials. AWJ machining technology is often used in industry for materials processing and machine parts fabrication. And since this method generates no thermal distortion [1], it is the preferable technology when the materials are sensitive to high temperature. In addition, it also has advantages of high machining versatility, flexibility, and small cutting forces [1] and thus can be used for processing both hard materials like titanium, steel, brass, aluminium, stone, inconel, glass, and any kind of composites [2] and soft materials like food, rubber and foam [3].

The performance of AWJ machining process is determined by the quality of the processed products which usually is measured by using *surface roughness,*

T. Herawan et al. (eds.), *Proceedings of the First International Conference on Advanced Data and Information Engineering (DaEng-2013),* Lecture Notes in Electrical Engineering 285, DOI: 10.1007/978-981-4585-18-7_22,
© Springer Science+Business Media Singapore 2014

a metric that describes geometry and surface textures of the products [4]. This metric is defined by the following equation [5]:

$$R_a = \frac{1}{L} \int_0^L |y(x)||dx|, \tag{1}$$

where R_a denotes surface roughness (average value), L denotes sampling length, y denotes profile curve, and x denotes profile direction. There are five dominant parameters that drive the performance of the machining process: traverse speed (V), waterjet pressure (P), standoff distance (h), abrasive grit size (d), and abrasive flow rate (m) [6]. The task of machining optimization often involves adjusting these parameters to achieve minimum R_a value.

In the literature, there are two directions in the research of AWJ machining process optimization. The first is how to model the machining process so that errors between the model and the corresponding experimental results can be minimized. The model is usually estimated by using artificial neural networks (ANN) approach [5–14] or regression analysis [6]. And the second is how to minimize the surface roughness using the estimated model by searching for optimal parameter values [15–17]. The usual approach is to use soft computing techniques like genetic algorithms (GA) and simulated annealing (SA). However, we found that the objective function of the model is a simple quadratic formula with box constraints so that many established methods from quadratic programming study can be used to search for the optimal solutions.

In this paper, we demonstrate the use of the active set method for solving the problem and show that this method can confidently outperform GA and SA both in the machining performance and computational times. These results suggest that this kind of problems should be addressed by using established gradient based optimization algorithms first before using soft computing techniques because the formers have better convergence property and also usually are faster.

2   AWJ machining process model

The task of modeling the AWJ machining process is the task of estimating transfer function $\mathcal{F}$ that takes V, P, h, d, and m as the inputs and produces R_a as the output so that the differences between the approximations and the corresponding experimental R_a values can be minimized. Certainly to obtain an accurate model, the experiments should be conducted for the entire input parameters space. However, since the cost of conducting experiments is high and only a fraction of those experiments are necessary for modeling purpose, Taguchi's orthogonal array method is often employed to reduce the number of experiments [6, 18, 19].

In the original work by Caydas & Hascalik [6], the authors have been able to reduce the number of required experiments from 3^5 (three-value levels for each parameter, shown in table 1) to only 27 experiments by utilizing the Taguchi's method. Table 2 outlines the R_a values for these 27 experiments. The authors

Table 1. Process parameters and the value levels.

Process parameters	Units	Level		
		1	2	3
Traverse speed (V)	mm/min	50	100	150
Waterjet pressure (P)	MPa	125	175	250
Standoff distance (h)	Mm	1	2.5	4
Abrasive grit size (d)	μm	60	90	120
Abrasive flow rate (m)	g/s	0.5	2	3.5

then used these results to estimate the transfer function $\mathcal{F}$ using two methods: second order polynomial regression and ANN, and reported that the ANN performed better than the regression method in predicting the R_a values. The following is the regression model used by Caydas & Hascalik [6] to estimate $\mathcal{F}$:

$$\begin{aligned}
\hat{R}_a = \mathcal{F}_{reg} = &- 5.07976 + 0.08169V + 0.07912P - 0.34221h \\
&- 0.08661d - 0.34866m - 0.00031V^2 - 0.00012P^2 \\
&+ 0.10575h^2 + 0.00041d^2 + 0.07590m^2 - 0.00008Vm \\
&- 0.00009Pm + 0.03089hm + 0.00513dm, \quad\quad (2)
\end{aligned}$$

where $\hat{R}_a$ and $\mathcal{F}_{reg}$ respectively denote the approximate R_a and the estimate transfer function using the regression model.

In ref. [15], Zain et al. took eq. 2 as a problem of finding optimal values for V, P, h, d, and m that minimize $\hat{R}_a$ given that each parameter is bounded by its corresponding levels (see table 1):

$$\begin{aligned}
50 &\leq V \leq 150 \\
125 &\leq P \leq 250 \\
1 &\leq h \leq 4 \\
60 &\leq d \leq 120 \\
0.5 &\leq m \leq 3.5.
\end{aligned}$$

As shown the problem turns into a quadratic programming problem with box constraints. The author then proposed to use two soft computing methods: GA and SA to compute the optimal V, P, h, d, and m values, and showed that SA performed better than GA in minimizing $\hat{R}_a$.

There is actually one fundamental problem related to the work of Zain et al. [15]: the authors used eq. 2 given the constraints defined above as the objective function to minimize $\hat{R}_a$ even though eq. 2 is only an estimate to the unknown AWJ machining process. So, $\hat{R}_a$ value corresponds to the optimal parameters computed by minimizing eq. 2 is unlikely equal to R_a value when those optimal values are used as the inputs to the AWJ machine. Thus, comparing $\hat{R}_a$ with R_a is a misleading effort since the equation is an estimate model to the real process (the correct effort should be: the authors propose the using of soft computing

Table 2. Experimental data [6].

No.	Machining settings used in the experiments					R_a (μm)
	V (mm/min)	P (MPa)	h (mm)	d (μm)	m (g/s)	
1	50	125	1	60	0.5	2.124
2	50	125	1	60	2	2.753
3	50	125	1	60	3.5	3.352
4	50	175	2.5	90	0.5	4.311
5	50	175	2.5	90	2	4.541
6	50	175	2.5	90	3.5	5.123
7	50	250	4	120	0.5	6.789
8	50	250	4	120	2	7.524
9	50	250	4	120	3.5	9.123
10	100	125	2.5	120	0.5	3.575
11	100	125	2.5	120	2	4.457
12	100	125	2.5	120	3.5	5.628
13	100	175	4	60	0.5	7.01
14	100	175	4	60	2	7.535
15	100	175	4	60	3.5	7.893
16	100	250	1	90	0.5	8.121
17	100	250	1	90	2	8.312
18	100	250	1	90	3.5	9.163
19	150	125	4	90	0.5	4.328
20	150	125	4	90	2	5.12
21	150	125	4	90	3.5	5.852
22	150	175	1	120	0.5	6.143
23	150	175	1	120	2	6.721
24	150	175	1	120	3.5	7.78
25	150	250	2.5	60	0.5	8.89
26	150	250	2.5	60	2	9.12
27	150	250	2.5	60	3.5	10.035
R_a (minimum)						2.124

techniques to also estimate the machining process and compare the accuracies of their models to the accuracies of the models proposed in the original work). Moreover, the authors also do not give any justification for the using of the constraints (it is possible that the real physical constraints of the AWJ machine can be different since the above values were the machining settings for certain experiments conducted in the original work). In addition, there is no guarantee that the optimal values are physically implementable.

In this paper, we treat this quadratic programming as a mathematical task regardless of the problems stated above. But instead of using soft computing techniques, we used the active set method, an established gradient based optimization method for quadratic programming. To use the active set method, first we rewrite the problem into the standard quadratic programming form:

$$\min_{\mathbf{x}} \mathcal{F}(\mathbf{x}) = \mathbf{x}^T \mathbf{A} \mathbf{x} + \mathbf{b}^T \mathbf{x} + c \tag{3}$$

$$\text{s.t. } \mathbf{l} \le \mathbf{x} \le \mathbf{u},$$

where vector $\mathbf{x}$ contains the parameters $(V, P, h, d,$ and $m)$ in its entries, so that lowerbound $\mathbf{l}$, upperbound $\mathbf{u}$, matrix $\mathbf{A}$, vector $\mathbf{b}$, and constant c can be written as:

$$\mathbf{l} = \begin{bmatrix} 50, & 125, & 1, & 60, & 0.5 \end{bmatrix},$$

$$\mathbf{u} = \begin{bmatrix} 150, & 250, & 4, & 120, & 3.5 \end{bmatrix},$$

$$\mathbf{A} = \begin{bmatrix} -0.00031 & 0 & 0 & 0 & -0.00004 \\ 0 & -0.00012 & 0 & 0 & -0.000045 \\ 0 & 0 & 0.10575 & 0 & 0.015445 \\ 0 & 0 & 0 & 0.00041 & 0.002565 \\ -0.00004 & -0.000045 & 0.015445 & 0.002565 & 0.0759 \end{bmatrix},$$

$$\mathbf{b} = \begin{bmatrix} 0.08169, & 0.07912, & -0.34221, & -0.08661, & -0.34866 \end{bmatrix}, \text{ and}$$

$$c = -5.07976.$$

As shown the problem transforms into the standard quadratic programming, and consequently any suitable gradient algorithm can now be used to search for the optimal solution.

3 Experimental results

In this section we present the result of using the active set method to find the minimum $\hat{R}_a$ value and the corresponding parameter values for the quadratic programming eq. 3. The result is compared to the results of using GA and SA

Table 3. Summary of the results.

Method	$\hat{R}_a$ (μm)	Optimal parameters				
		V (mm/min)	P (MPa)	h (mm)	d (μm)	m (g/s)
GA [15]	1.5549	50.024	125.018	1.636	94.973	0.525
SA [15]	1.5355	50.003	125.029	1.486	107.737	0.5
Active set	1.5223	50	125	1.545	102.494	0.5

Table 4. Performance comparison for 100 cases.

	GA	SA	Active set
Comp. times (average)	0.0982	3.4159	0.0074
#times performed the best	0	8	92

reported in ref. [15]. We used the built-in `quadprog` function in MATLAB that can be run by selecting one out of three available algorithms (trust region, active set, and interior point). The algorithms can be chosen by setting the `options` argument in the function. Concretely, the following lines were entered into the MATLAB command prompt to get the results:

```
options = optimset ('Algorithm', 'active−set');
[x, Ra] = quadprog(2 A,b,[],[],[],[],l,u,[], options);
Ra = Ra + c;
```

where A, b, l, u, and c correspond to the **A**, **b**, **l**, **u**, and c defined above, x denotes the vector **x**, and Ra denotes the $\hat{R}_a$. The last line was used to shift the $\hat{R}_a$ value because we removed the constant c in line two to compute $\hat{R}_a$. Table 3 summarizes the experimental results by using the active set method and also displays the results reported by Zain et al. [15] for GA and SA. As shown, the active set method outperformed both GA and SA in searching for the optimal parameters.

To evaluate and compare the computational costs of the methods, using only the problem eq. 2 certainly is insufficient and inconclusive. Therefore we created 100 equivalent problems of the same size as the problem eq. 2 randomly, and computed the average computational times (in second). Because it is also interesting to investigate which method performed the best in each of these 100 cases, we counted the number of times the methods performed the best in these 100 cases. Note that for a fair treatment, both GA and SA were also implemented using the built-in `ga` and `simulannealbnd` functions provided by MATLAB. Table 4 summarizes the results. As shown, the active set method was not only much faster than GA and SA (about 13 times faster than GA and 462 times faster than SA), but also performed much better as it outperformed the other methods in 92 cases out of 100 cases. These results imply that the active set method (or possibly other gradient based optimization algorithms) should be used first to solve this kind of problems before proceeding further with soft computing techniques.

4 Conclusion

The task of estimating AWJ machining process is the task of how to build a model that can predict surface roughness value for each set of process parameters as accurate as possible. The model is only an estimate to the real machining process, thus using the model to find optimal parameters that lead to the minimum surface roughness certainly is a misleading effort because the real surface roughness is unlikely equal to the surface roughness computed using the model. This is the fundamental problem in the work of Zain et al. [15].

Regardless of this problem, we found that the problem solved by the authors is a simple quadratic programming with box constraints. Accordingly, established gradient based optimization methods should be employed first before using soft computing techniques because the formers have better convergence property and also usually are faster than the latters. In particular, we showed that the active set method not only performed better, but also was much faster than GA and SA. This implies that the active set method (or possibly other gradient based algorithms) should be employed first to solve this kind of problems.

Acknowledgement

The author would like to thank the reviewers for useful comments. This research was supported by Ministry of Higher Education of Malaysia and Universiti Teknologi Malaysia under Exploratory Research Grant Scheme R.J130000.7828. 4L095.

References

1. Hascalık, A., et al.: Effect of traverse speed on abrasive waterjet machining of Ti-6Al-4V alloy. Mater. Des. 28, 1953–1957 (2007)
2. Akkurt, A., et al.: Effect of feed rate on surface roughness in abrasive waterjet cutting applications. J. Mater. Process. Technol. 147, 389–396 (2004)
3. About Waterjets: Basic Information, http://waterjets.org (2013)
4. Nalbant, M., et al.: Application of Taguchi method in the optimization of cutting parameters for surface roughness in turning. Mater. Des. 28, 1379–1385 (2007)
5. Ozcelik, B., et al.: Optimum surface roughness in end milling Inconel 718 by coupling neural network model and genetic algorithm. Int. J. Adv. Manuf. Technol. 27, 234–241 (2005)
6. Caydas, U., Hascalik, A.: A study on surface roughness in abrasive waterjet machining process using artificial neural networks and regression analysis method. J. Mater. Process. Technol. 202, 574–582 (2008)
7. Chien, W.T., Chou, C.Y.: The predictive model for machinability of 304 stainless steel. J. Mater. Process. Technol. 118, 442–447 (2001)

8. Erzurumlu, T., Oktem, H.: Comparison of response surface model with neural network in determining the surface quality of moulded parts. Mater. Des. 28, 459–465 (2007)
9. Kumar, S., Choudhury, S.K.: Prediction of wear and surface roughness in electrodischarge diamond grinding. J. Mater. Process. Technol. 191, 206–209 (2006)
10. Lee, B.Y., et al.: Modeling and optimization of drilling process. J. Mater. Process. Technol. 74, 149–157 (1998)
11. Nabil, B.F., Ridha, A.: Ground surface roughness prediction based upon experimental design and neural network models. Int. J. Adv. Manuf. Technol. 31, 24–36 (2006)
12. Oktem, H., et al.: Prediction of minimum surface roughness in end milling mold parts using neural network and genetic algorithm. Mater. Des. 27, 735–744 (2006)
13. Risbood, K.A., et al.: Prediction of surface roughness and dimensional deviations by measuring cutting forces and vibrations in turning process. J. Mater. Process. Technol. 132, 203–214 (2003)
14. Spedding, T.A., Wang, Z.Q.: Study on modeling of wire EDM process. J. Mater. Process. Technol. 69, 18–28 (1997)
15. Zain, A.M., et al.: Genetic algorithm and simulated annealing to estimate optimal process parameters of the abrasive waterjet machining. Eng. Comput. 27, 251–259 (2011)
16. Zain, A.M., et al.: Application of GA to optimize cutting conditions for minimizing surface roughness in end milling machining process. Expert Syst. Appl. 37, 4650–4659 (2010)
17. Zain, A.M., et al.: Simulated annealing to estimate the optimal cutting conditions for minimizing surface roughness in end milling Ti-6Al-4V. Mach. Sci. Technol. 14, 43–62 (2010)
18. Ko, D.C., et al.: Application of neural network and Taguchi method to perform design in metal forming considering workability. Int. J. Mach. Tool Manuf. 39, 771–785 (1999)
19. Jegaraj, J.J.R., Babu, N.R.: A soft approach for controlling the quality of cut with abrasive waterjet cutting system experiencing orifice and focusing tube wear. J. Mater. Process. Technol. 185, 217–227 (2007)

Orthogonal Wavelet Support Vector Machine for Predicting Crude Oil Prices

Haruna Chiroma[1], Sameem Abdul-Kareem[1], Adamau Abubakar[2] Akram M.
Zeki[2] and Mohammed Joda Usman[3]

[1]Department of Artificial Intelligence, University of Malaya, Kuala Lumpur, Malaysia. freedonchi@yahoo.com, sameem@siswa.um.edu.my
[2]Department of Computer Science, Faculty of Information and Com-munication Technology, International Islamic University Malaysia, Gombak, Kuala Lumpur, Malaysia.100adamu@gmail.com,
[3]School of Electronic and Information Engineering, Liaoning University of Technology, Jinzhou, China. usmanjodah1@yahoo.com

Abstract. Previous studies mainly used radial basis, sigmoid, polynomial, linear, and hyperbolic functions as the kernel function for computation in the neurons of conventional support vector machine (CSVM) whereas orthogonal wavelet requires less number of iterations to converge than these listed kernel functions. We proposed an orthogonal wavelet support vector machine (OSVM) model for predicting the monthly prices of West Texas Intermediate crude oil prices. For evaluation purposes, we compared the performance of our results with that of the CSVM, and multilayer perceptron neural network (MLPNN). It was found to perform better than the CSVM, and the MLPNN. Moreover, the number of iterations, and time computational complexity of the OSVM model is less than that of CSVM, and MLPNN. Experimental results suggest that the OSVM is effective, robust, and can efficiently be used for crude oil price prediction. Our proposal has the potentials of advancing the prediction accuracy of crude oil prices, which makes it suitable for building intelligent decision support systems.

Keywords: Support vector machine, orthogonal wavelet, crude oil prices, Kernel function

1   Introduction

The fluctuation of crude oil prices could be attributed to macroeconomic theory of oil demand and supply factors, probably caused by economic development and behavior of oil producing countries. Uncertainty events related to crude oil such as war, revolution, extreme weather, among others; significantly contribute to oil price volatility. Volatile nature of oil prices has direct effects on individuals because prices of petrol (fuel) as well as goods and services also increase as the price of crude oil increases [1]. Factors that interact to have accumulated effects on crude oil prices are economic,

T. Herawan et al. (eds.), *Proceedings of the First International Conference on Advanced Data and Information Engineering (DaEng-2013)*, Lecture Notes in Electrical Engineering 285, DOI: 10.1007/978-981-4585-18-7_23,
© Springer Science+Business Media Singapore 2014

political, military action, natural disaster, speculations, market demand, and supply
[2].
    Crude oil price volatility prompted academicians, researchers, and institutions in
search for a reliable crude oil price prediction system. As a result of searching for a
reliable system of predictions, many researches were conducted in the area of
artificial intelligence using several soft computing techniques in an attempt to come
up with a lasting solution. For example, research in [3] applied wavelet neural
network to forecast crude oil prices. The network uses orthogonal or continuous
wavelets as activation function instead of the sigmoid function because it requires less
iteration to converge to the optimal solution. In [4] conventional support vector
machine (CSVM) model was built to predict crude oil prices since its immune of
being stuck in a local minima. The model uses a radial basis function as the kernel
function whereas the orthogonal wavelet function requires less number of iterations to
converge to the optimal solution. Moreover, the wavelet function constructively
initializes the parameters of the CSVM unlike the case of the commonly use radial
basis function. Investigation of the crude oil prices is of interest as the orthogonal
support vector machine has not been applied to the crude oil price prediction problem,
and our research covers the existing gap.
    Other sections of the paper comprised of section 2 which covers the
description of our proposed algorithm. Section 3 gives the detail description of the
application. Section 4 present results and discussion before concluding remarks in
section 5.

2   The proposed algorithm

In CSVM, the unknown function is approximated by mapping x to higher dimensional space through a function ϕ. The CSVM determine linear maximum – margin
hyper plan [5]:

$$f(x) = \omega' \phi(x) + b = 0 \tag{1}$$

$$\frac{1}{2} \|\omega\|^2 + C \sum_{i=1}^{m} L_\varepsilon(y_i) \tag{2}$$

Eq. 1 is the hyper plane of CSVM, where ω and b are estimated by minimizing
Eq.2.
$\|\omega\|$ = Regularize term

$$\sum_{i=1}^{m} L_\varepsilon(y_i) \tag{3}$$

Eq. 3 is empirical error and C = regularization constant and $C > 0$
The CSVM penalize if it deviaa from y_i by means of an ε-insensitive loss function

$$L_\varepsilon(y_i) = \begin{cases} 0 & if\, |f(x_i)-y_i<\varepsilon| \\ |f(x_i)-y|-\varepsilon & otherwise \end{cases} \tag{4}$$

$$y_i - (\omega'\phi(X_i)+b) \le \varepsilon + \xi_i^- \quad \forall_i \tag{5}$$

$$(\omega'\phi(X_i)+b) - y_i \le \varepsilon + \xi_i^+ \tag{6}$$

$$\tag{7}$$

$$\xi_i^- \ge 0, \xi_i^+ \ge 0$$

Introducing ξ_i^- and ξ_i^+ as slack variables for minimizing Eq. 2

$$\min_{\omega,b,\xi_i^-,\xi_i^+} \frac{1}{2}\|\omega\| + C\sum_{i=1}^n (\xi_i^- + \xi_i^+) \tag{8}$$

Subject to Eqs. 5 – 7
Eq. 9 is a solution to the minimization problem

$$f(x) = \sum_{i=1}^m (\lambda_i - \lambda_i^*)K(X_i,X)+b \tag{9}$$

$$y_i - (\omega'\phi(X_i)+b) \le \varepsilon + \xi_i^- \tag{10}$$

$$(\omega'\phi(X_i)+b) - y_i) \le \varepsilon + \xi_i^+ \tag{11}$$

λ_i and λ_i^* are Lagrange multipliers related to constrains Eqs. 10 – 11 [6].
Eqs. 12 – 15 are widely use kernel functions in literature for computation in the hidden layer of the CSVM but Eq. 12 is the popular choice.

Radial basis function

$$K(X_i,X_j) = \ell^{(-\gamma\|X_i-X_j\|^2)} , \gamma > 0 \tag{12}$$

Sigmoid function

$$K(X_i,X_j) = \tanh(\gamma X_i'X_j + r) \tag{13}$$

Polynomial function

$$K(X_i, X_j) = (\gamma X_i' X_j + r)^d, \ \gamma > 0 \tag{14}$$

Linear function

$$K(X_i, X_j) = X_i' X_j \tag{15}$$

In this paper, wavelet analysis is proposed. Wavelet is a function that is used for localizing a function in a position and scaling. A wavelet has to satisfy Eqs. (16), (17), and (18).

$$\psi(\bullet) = \int_{-\alpha}^{\alpha} \psi(v)dv = 0 \tag{16}$$

$$\psi(\bullet) = \int_{-\alpha}^{\alpha} \psi^2(v)dv = 1 \tag{17}$$

$$k_\psi \equiv \int_0^\alpha \frac{|\Psi(f)|^2}{x} dx \quad 0 < k_\psi < \alpha \tag{18}$$

Where $\psi(\bullet)$, $(-\alpha, \alpha)$, k_ψ, Ψ and $\psi(v)$ are wavelet function, limits, adminissibility, cruscal and line, respectively.
The kernel function in the hidden layer of the CSVM in this research is subtituted with orthogonal wavelet which is now referred to as orthogonal support vector machine (OSVM). Consider the decomposition of signal $x(t)$ at $t \in [0, T]$ into a finit number of scales as given in Eq. (19)

$$x(t) = w_0 + \sum_{j=0}^{\alpha} \sum^{u-1} w_{2l+k} \psi(2^i t - KT) \tag{19}$$

$$w_o = \int_0^T x(t)dt \tag{20}$$

$$w_{2+k}^j = \int_0^T x(t)\psi(2_t^j - KT)dt \tag{21}$$

Where w_0 and w_{2+k}^j are the orthogonal wavelet for computation in OSVM [8]: where Eq. (22) satisfied Eq. (23) and (24)

$$\psi(2^j t - k) \tag{22}$$

$$\int_0^T (2^r t - KT)dt = 0 \tag{23}$$

$$\int_0^T (2^s t - KT)\psi(2^s t - lT)dt = \begin{cases} 1 & when \quad r=s,-k=l \\ 0 & otherwise \end{cases} \tag{24}$$

Finally, the CSVM constrain is expressed as:

$$f(x) = \sum_{i=1}^m (\lambda_i - \lambda_i^*)\int_0^T KT(\psi^{sr} + \psi K)d\psi(X_i, X) + b$$

The OSVM is train with Kernel-Adatron learning algorithm as its performance is superior to orther learning algorithms [8].

3 Application

3.1 Data

The experimental data use monthly West Texas Intermediate crude oil prices. The data were collected from US Department of Energy starting from January 1986 to December 2011. In this research, 208 data points were used as training data, 44 as validation dataset and 44 as independent dataset for determining the generalization capability of the proposed OSVM model. In other words, the data were partition into 70%: 15%: 15% according to the convention in [9]. The data used in this research was not normalized because normalization has no effect on neural network performance as argued in [10]. Monthly frequency data were used due to non- availability of variables such as organization for economic co – operation and development (OECD) inventory on daily or weekly frequency [3].

3.2 Orthogonal Support Vector Machine Modeling

The OSVM used in this research has 10 inputs, 1 output neuron, and 12 hidden neurons which the OSVM was trained using the data described in the preceding section. Insensitivity loss function was set to 0.000234, and C was 97. The validation was set to terminate after 6 epochs without performance improvement in validation MSE. The training was conducted using Kernel-Adatron learning algorithm with an epoch of 1000. The initial training parameters were realized through preliminary experimentation since the ideal framework for determining such parameters are scarce in the literature. The Kernel-Adatron algorithm maps the crude oil prices into higher dimensional feature space and then separates the set of data that share complex boundaries. Computation in the neurons was performed using orthogonal wavelet as initially proposed. The parameter Center of the width was selected as 1 which was adopted in [11]. The experiment is as well performed with CSVM in order to obtained different results and compared with our proposal results for evaluation purposes. The spread of radial basis function used is 0.6. The same dependent and independent variables as well as other initial training parameters used in modeling OSVM were also used in CSVM modeling for fair comparison. Several experiments were conducted with different kernels including polynomial, radial basis function kernel, Gaussian, and sigmoid functions. Results indicated that the radial basis kernel function outperformed others, and subsequently, was selected as the optimum. Multilayer perceptron neural network (MLPNN) with 10 input neurons, 10 hidden neurons, 1 hidden layer, the sigmoid activation function in the hidden layer neurons, 1 output neuron, linear activation function in the output neuron was trained with Levenberg–Marquardt algorithm to build a model for predicting crude oil prices. The parameters were realized after preliminary experimental trials.

4   Results and Discussion

Several experiments were conducted in search for optimal OSVM and CSVM models, and the optimal among the results are reported in this section. The training and validation MSE obtained for OSVM are 0.0007234 and 0.0001281, respectively. In contrast, MSE for CSVM after training and validation were found to be 0.001241 and 0.001193, respectively. For an MLPNN the MSE for training and validation are 0.07843 and 0.05462, respectively. The validation MSE indicated that there is an improvement over training MSE for the OSVM, the CSVM and the MLPNN. Validation in all the situations was instructed to be terminated when validation MSE stop improving after 6 epochs as shown in Fig. 1. The MSE improvement of CSVM, and MLPNN was a little compared to the OSVM validation MSE improvement. In summary, observations of these MSE clearly indicated that the training performance of OSVM outperforms both CSVM, and MLPNN. In-sample performance of OSVM give an insight that it can perform better in out-sample. Although, is not always true as argued in [12] that a model has the possibility of injecting false optima, whereas it performs excellently on the training dataset.

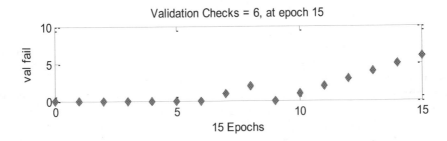

Fig. 1. Validation checks during training

Performance of OSVM, MLPNN, and CSVM were evaluated on the independent test dataset and results are depicted in Fig. 2, Fig. 3 and Table 1. A close observation of Fig. 2, and Fig. 3, shows the OSVM is closer to the actual crude oil price than its counterpart. The pattern of the graph reveals the superior performance accuracy of our proposal. However, the CSVM, and the MLPNN do not perform poorly because they are competing with the OSVM with good results.

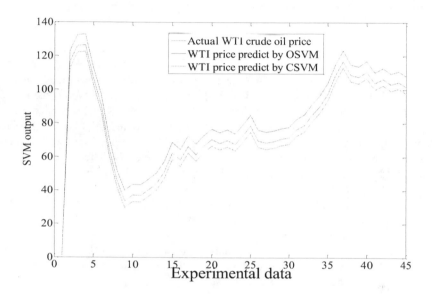

Fig. 2. Plots of predicted and actual crude oil prices

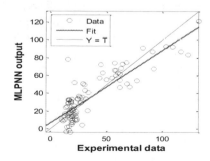

Fig. 3 Regression plot for MLPNN

Table 1. Comparing performances of OSVM, CSVM, and MLPNN

	MSE	NMSE	MAE	R	Num. Itr.	TCC
OSVM	0.000368	0.00003467	0.000356	0.967234	15	8.34E+07 ns
CSVM	0.001896	0.000098763	0.002135	0.911571	18	1.0000 sec.
MLPNN	0.00761	N/A	N/A	0.85849	21	2.0000 sec.

Number of iteration (Num. Itr.), Time computational complexity (TCC), Seconds (sec.), Nanoseconds (ns), Mean Square Error (MSE), Normalized Mean Square Error (NMSE), Mean Absolute Error (MAE), Regression (R), Not Applicable (N/A).

Table 1 indicated that OSVM clearly outperformed CSVM, and MLPNN in terms of the performance evaluation metrics shown in section 3.1. The number of iterations and time required by OSVM to converge to optimal solution is less than that of CSVM, and MLPNN. Although, CSVM, and MLPNN also performs well considering the values of the performance metrics. The CSVM and MLPNN can as well predict crude oil prices with reasonable performance accuracy. As noted in [13], an effective and robust algorithm is required to converge to the best solution with a minimum number of iterations, and a less time computational complexity. There are some applications, in which the time computational complexity of the algorithm is equally important as the accuracy of the prediction. Therefore, suggestions from our experimental findings shows that OSVM performs better than the CSVM and MLPNN. The results were expected as already observed in previous literature that the orthogonal wavelet is faster than the conventional kernel functions. The probable cause of this improve accuracy is most likely due to the harmonic nature of the orthogonal wavelet. Another probable caused is the ability of the wavelet analysis to project data into time scale domain and conduct multi scale analysis.

5 Conclusions

This study presents OSVM for prediction of crude oil prices. The proposed OSVM used orthogonal wavelet instead of the commonly use sigmoid kernel function, radial basis function, linear function, polynomial function, and hyperbolic function. The results obtained were found to be impressive when compared to the CSVM, and MLPNN. Time computational complexity, number of iterations required to converge to minimum MSE, RMSE, MAE and maximum R for OSVM was better than CSVM, and MLPNN which makes our proposed approach ideal for predicting crude oil prices with satisfactory performances. With little additional information, our proposal has the potential of being transformed into intelligent decision support systems for helping government, private businesses, investors, and risk managers in decision making processes.

References

1. Haidar, I., Kulkarni, S., Pan, H.: Forecasting model for crude oil prices based on artificial neural networks. In: The Proceeding of fourth International Conference on Intelligent Sensors, Sensor Networks and Information Processing, pp. 103 - 108. Sydney (2008)
2. Yu, L., Wang, S., Lai, K. K.: A rough set refined text mining approach for crude oil market tendency forecasting. Int. J. Knowl. Syst. Sci. 2, 1, 33 - 46 (2005)
3. Pang, Y., Xu, W., Ma, J., Lai, K. K., Wang, S. Xu, S.: Forecasting crude oil spot price by wavelet neural networks using OECD petroleum inventory levels. New Math. Nat. Comput. 2, 28 - 297 (2011)
4. Khashman, A., Nwulu, I.N.: Intelligent prediction of crude oil using support vecto machines. In: 9th IEEE International Symposium on Applied Machine Intelligence Informatics, 65 -169. Smolenice (2011)
5. Cortes, C., Vapnik, V.: Support vector networks. Mach. Learn. 20, 273 - 297 (1995)
6. Hamel, L.: knowledge discovery with support vector machines. John Wiley & Sons, Inc. New Jersey (2009)
7. Mallat, S.G.: Multiresolutionapproximationsandwaveletorthonormalbases of L2(R). Transactions of American Mathematical Society 315, 69 - 87 (1989)
8. Frie, T., Cristianini, N., Campbell C.: The Kernel-Adatron Algorithm: a Fast and Simple Learning Procedure for Support Vector Machines http://citeseerx.ist.psu.edu/viewdoc/download?doi=10.1.1.42.2060&rep=rep1&type=pdf (2006). Sons, Inc. New Jersey (2009)
9. Beale, M.H., Hagan, M.T., Demuth, H.B.: Neural network toolbox^{TM} user's guide. The MathWorks, Inc., Natick (2013)
10. Shanker, M., Hu, M.Y., Hung, M.S.: Effect of data standardization on neural network training. Omega 24, 385 - 397 (1996)
11. Dunis, C. L., Rosillo, R., Fuente, D., Pino, R.: Forecasting IBEX-35 moves using support vector machines. Neural Comput. Appl. DOI 10.1007/s00521-012-0821-9, (2012)
12. Jin, Y.: A comprehensive survey of fitness approximation in evolutionary computation. Soft. Comput. 9, 3 - 12, (2005)
13. Alpaydin, E.: Introduction to machine learning. MIT press, Cambridge (2004)

Performance Comparison of Two Fuzzy Based Models in Predicting Carbon Dioxide Emissions

Herrini Mohd Pauzi[1] and Lazim Abdullah[2*]

[1&2]Department of Mathematics, Faculty of Science and Technology, University Malaysia Terengganu
21030, Kuala Terengganu, Terengganu
[1]mherrini@yahoo.com, [2*]lazim_abdullah@umt.edu.my

Abstract. Many studies have been carried out worldwide to predict carbon dioxide (CO_2) emissions using various methods. Most of the methods are not sufficiently able to provide good forecasting performances due to the problems with non-linearity of the data. Fuzzy inference system (FIS) and adaptive neuro fuzzy inference system (ANFIS) are two of the well-known methods with its ability to handle the problems of non-linearity. However, the performances of these two fuzzy based models in predicting CO_2 emissions are not immediately known. This paper offers the performance comparison of the two fuzzy based models in prediction of CO_2 emissions in Malaysia. The inputs for the models were simulated using the Malaysian data for the period of 1980 to 2009. The prediction performances were measured using root means square error, mean absolute error and mean absolute percentage error. The performances of the two models against the CO_2 emission clearly show that the ANFIS outperforms the FIS model.

Keywords: Artificial intelligence, CO_2 emissions, Fuzzy inference systems, Prediction.

1 Introduction

Air pollution is one of the major environmental issues concerned worldwide. This environmental problem is mainly caused by the high concentrations of greenhouse gases (GHGs) such as carbon dioxide (CO_2), sulphur dioxide (SO_2), nitrogen oxide (NO_x), carbon monoxide (CO) and others in the atmosphere. The pollution has been recognized as a threat to environment's ecosystem and human health. Climate change, global warming and health effect of air pollution have been studied in many different parts of the world. The anthropogenic driver of global warming is the increasing concentration of GHG in the atmosphere, and among the GHGs, CO_2 is the largest contributing gas to the greenhouse effects. The greenhouse gas has a leading role to air pollution where its emissions into the atmosphere naturally through the carbon cycle and human activities associated primarily with the combustion of fossil fuels. Emissions of CO_2 are responsi-

T. Herawan et al. (eds.), *Proceedings of the First International Conference on Advanced Data and Information Engineering (DaEng-2013)*, Lecture Notes in Electrical Engineering 285, DOI: 10.1007/978-981-4585-18-7_24,
© Springer Science+Business Media Singapore 2014

ble for the enhanced GHG effect, resulting from the increasing concentrations of CO_2 [1].

Many studies have been conducted especially to predict CO_2 emissions in order to overcome its effect towards environment. However, many researchers face difficulties in prediction of air quality as it involves non-linear relationships and uncertainty of variables. The emissions that caused by some natural and anthropogenic sources, for example, forest fires, industry, road and air traffic are often uncertainty and not normally available. On the other hand, it is known that official emission data cannot be considered as completely reliable [2]. Thus, it is a necessary to have a reliable method to predict CO_2emissions due to its uncertainty and non-linearity data. Accurate and robust prediction of the air quality would be very significant in order to overcome the environmental problems.

Environmental data are typically very complex to model due to the underlying correlation among several variables of different type which yields intricate mesh relationships [3]. The development of tools for predicting air quality has been drawing the attention of many scientists. The prediction is done on the basis of information including, past historical data, meteorological parameters such as temperature, humidity, speed and wind direction and pressure, demographic and anthropogenic parameters, and also environmental pollutant concentrations such as particulate matters (PM_{10}), ozone (O_3), CO and CO_2. Apart from that, there are numerous methods have been applied in predicting the CO_2 emissions. From the literature review, most of the methods to the prediction of CO_2emissions are meant to increase the accuracy of the results. Among the popular methods are grey model [4-7], neural network [8-12] and computer based simulation model [13-14]. However, all these methods are lack of reasoning process.

In recent years, artificial intelligence (AI) based techniques have been proposed as alternatives to traditional statistical ones in many scientific disciplines. In order to handle uncertainty of variables in prediction, many researchers propose the application of fuzzy logic in this area. Fuzzy logic (FL) is another modelling method that is classified as a black box model. Based on the principle that variables are often imprecise or uncertain, FL provides effective solutions for non-linear and partially known systems [15]. Implementation of reasoning process (rule) to FL yields fuzzy inference system (FIS) which has been applied by many researchers in various areas. For example, Peche and Rodriguez [16] applied a new approach based on FL to assess the environmental impact related to the execution of activities and projects, in which the information related to the impacts was limited, inaccuracy and uncertain. Huang et al., [17] developed and found out that fuzzy-based simulation method was useful for studying hydrological processes within a system containing multiple factors with uncertainties and providing support for identifying proper water resources management strategies. Additionally, many researchers conducted researches based on FL to do forecasting. Among them, there are a few who had predicted and modelled concentration of many type of gases in atmosphere such surface ozone concentration and also to predict air quality in different areas of case studies [15], [18], [19]. Nevertheless, there is still no paper which discusses about the model in predicting CO_2 emissions.

Apart from that, one of the most popular AI models is neural networks (NN). It has been discussed by many researchers regarding the prediction of air quality. The NN is considered to be simplified mathematical models which imitate brain-like systems. It is able to determine nonlinear relationship between variables in input datasets and output datasets. Antanasijević et al., [20] described development of artificial neural network (ANN) model for PM_{10} emission forecasting. Yu et al., [21] evaluated the levels of manufacturers developing low-carbon using ANN. Lim et al., [22] used ANN model prediction of ammonia emission form field-applied manure.

Instead of good performances of the both methods mentioned, the models still have their own drawbacks. Major drawbacks of FL, it only can handle linguistic variables and not numerical values. Meanwhile, the NN depend too much on trial and error method in determining number of hidden layers and nodes hence it is unable to converge to global minimum even by complex back propagation learning [23]. As a result, many researchers proposed an enhancement on FIS which lead to a new method well known as adaptive neuro fuzzy inference system (ANFIS). The model is a hybridization of the both methods mentioned before. It complements each disadvantages possessed by each method. A few researchers [23-25] had successfully applied ANFIS for modelling and prediction air pollutant concentration levels in different area. However, it has been seen from literature survey that there are not many studies using this method to forecast CO_2 emissions levels.

Furthermore, a few comparison studies have been carried by using these aforementioned methods. For example, Campbell [26] presented a comparative study of soft computing models which included multilayer perceptron networks, partial recurrent neural networks, radial basis function network, FIS and hybrid fuzzy NN for the hourly electricity demand forecast in Northern Ireland. The author found out that the hybrid fuzzy NN and radial basis function networks were the best models for the analysis and forecasting of electricity demand. Badri et al., [27] investigated the application of ANN and FL model as forecasting tools on predicting the load demand in short term category. ANN represents the more accurate results in comparison to FL especially for short term procedure. Lohani et al., [28] compared the ANFIS model with ANN and autoregressive techniques in hydrological time series modelling. They found that in all cases that had been tested ANFIS gives more accurate forecast than the AR and ANN models. Chen et al., [29] also found out that that the ANFIS model has better forecasting performance than the fuzzy time series model, grey forecasting model and Markov residual modified model. From the brief review of these comparative studies, it can be clearly seen that most of the researchers are interested to study the comparison between fuzzy model with its hybrid model (ANFIS) or with NN model alone. This might be due to the mutual relationship between FIS and NN to produce a hybrid model of ANFIS that is believe to be more precise and robust especially in forecasting area. Hence, this present study aims to investigate performance comparison between these two fuzzy based models, FIS and ANFIS which indirectly will involve the NN model.

This paper is organized as following. Section 2 describes the variables for simulations. Performance evaluation criteria are presented in Section 3. Results and discussion are provided in Section 4 and finally Section 5 concludes the paper.

2   Simulation Data

This paper considers socio-economic and demographic parameters as the variables. These variables are also well known as anthropogenic factors. The experimental data were drawn from the website of World Bank's World Development Indicators [30]. Models were generated from the data collected. Inputs will include annual historical data for Malaysia from 1980 to 2009. These parameters are considered as the most influential variables towards the CO_2 emissions. The data obtained were gross domestic product per capita (GDP) with the unit constant 2000 US$, energy use in kg of oil equivalent per capita, population density (people per sq. km of land area), combustible renewable and waste (% of total energy), CO_2 intensity (kg per kg of oil equivalent energy use) were selected as the input variables for CO_2 emission model. The ANFIS model will use the NNs algorithm to train the input data where, upon the satisfactory training NN should be able to provide output for previously "unseen" inputs [31]. That means the data should be at least divided into two sets which are training dataset and testing dataset. As for the present study, 80% of data has been using for training data which include the data from 1980 to 2004. Meanwhile the other 20% of the data are provided for testing data which include the data from 2005 to 2009.

3   Performance Evaluation Criteria

Three statistical evaluation criteria are used to assess the performances of the applied models. These criteria are root mean square error (RMSE), mean absolute error (MAE), and mean absolute percentage error (MAPE). All these measures are expressed as follows:

$$RMSE = \sqrt{\sum_{i=1}^{n} \frac{(t_1 - z_1)^2}{n}}$$

$$MAE = \sum_{l=1}^{n} \frac{|t_l - z_l|}{n}$$

$$MAPE = \frac{100}{n} * \sum_{l=1}^{n} |\frac{t_l - z_l}{z_l}|$$

Where t = actual value, z= predicted value of t, $\bar{t}$= mean of all the t values, n = total number of points.

Small values of RMSE, MAE and MAPE indicate the accuracy of the models. Lewis (1983) interprets the MAPE as a way to judge the accuracy of forecast, where less than 10% is a highly accurate forecast; 10%-20% is a good forecast; 20%-50% is a reasonable forecast; and more than 50% is an inaccurate forecast.

4.   Results and Discussion

This study employs fuzzy based models approach to predict CO_2 emissions in Malaysia. The employed approach includes two models namely FIS and ANFIS. Table 1 depicts the performances of the constructed model for training datasets and testing datasets. The performance of the models regarding to accuracy are compared.

Table 1 Prediction Performances of FIS and ANFIS

Models	Training Set	Testing Set
FIS		
RMSE	0.0010	0.0492
MAE	0.0079	0.0335
MAPE(%)	0.2881	1.2807
ANFIS		
RMSE	8.1379e-004	0.0427
MAE	7.2628e-004	0.0331
MAPE(%)	0.0281	1.1936

As it is shown in Table 1, an observation can be made which is that all evaluation criteria indicate that training set as well as testing set of ANFIS possesses the smallest values compare to FIS. For the training dataset period from 1980 to 2003, the MAPE value of ANFIS is 0.0281% which is very small when comparing to FIS value which is 0.2881%. Furthermore, for the testing dataset period from 2004 to 2009, the MAPE value of ANFIS is smaller than FIS value. As mentioned before, when the value of MAPE is less than 10%, it indicates that the prediction is highly accurate. Both models show MAPE values less than 10%, however ANFIS shows the smallest value.

Fig. 1 shows comparison between actual values and predicted values of CO_2 emissions for the testing dataset. Based on the figure, the results from FIS and ANFIS for the period between 2004 and 2009 are approaching the same values as the actual values of CO_2 emissions. However, ANFIS yields better results with smaller differences between the actual values and its' predicted values compared to FIS values. Furthermore, it is also can be clearly seen that the results from FIS prediction differ substantially from actual values. Therefore, these scenarios indicate that ANFIS outperforms FIS for the case of predicting CO_2 emissions.

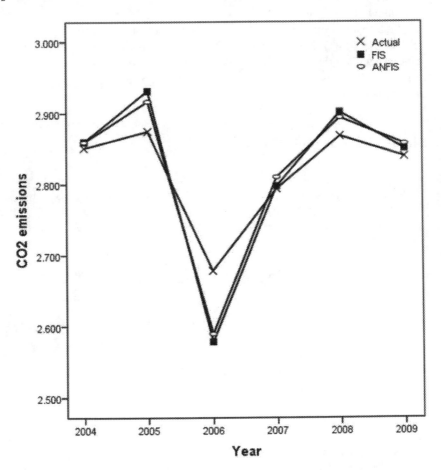

Fig. 1 Comparison between actual values and predicted values of CO_2 emissions

5 Conclusions

Air pollution in many developing countries is currently an issue of a great concern due to its economic activities. Public awareness of this problem has arisen among its citizens due to the available information on the harmful air pollution on human health and environmental sustainability.

Most of the surveyed literatures focused on applying NNs to predict CO_2 emissions. However, this paper attempted to compare predictability of two fuzzy based models; Takagi-Sugeno FIS and ANFIS in CO_2 emissions prediction. Subtractive clustering was used to initialize the FIS and ANFIS optimization. The proposed methods were applied on data taken from the country of Malaysia ranging between the years 1980 with 2009. Five anthropogenic factors were considered. The ANFIS is an improved model of FIS by using NN learning algorithm. Therefore, a better result is expected from the ANFIS rather than FIS.

Results proved that ANFIS outperforms FIS regarding to accuracy of prediction. The model recorded smaller values for all of the three evaluation criteria that were used in this study (RMSE, MAE and MAPE) when compared to the values from the FIS model. The model performed with RMSE, MAE and MAPE values that are differed about 1%-10% from the outcomes of the FIS model respectively.

The improvement of prediction accuracy and adaptation of other environmental quality indicators should be carry out in future. Improvement of the other computational intelligence tools for prediction area such as interval type-2 fuzzy logic system can be explored in future research.

Prediction of CO_2 emissions can be very essential especially towards policy makers in order to reduce its emissions. Besides that, in the real world, CO_2 emissions are influenced by many other factors such as technological advances, fuel consumption types, agricultural growth and political initiatives. Hence, the improvement of prediction accuracy and adaptation of other environmental quality indicators should be carry out in future research. Additionally, improvement of the other computational intelligence tools for prediction area such as interval type-2 fuzzy logic system can be explored in future research.

Acknowledgments

The authors are grateful to the Malaysian Ministry of Higher Education and University Malaysia Terengganu for financial support under the FRGS grant number 59243.

References

1. Intergovernmental Panel Climate Change (IPCC). Synthesis Report, Geneva, Switzerland (2007).
2. Pokrovsky, O.M., Kwok, Roger H.F., Ng, C.N. : Fuzzy logic approach for description of meteorological impacts on urban air pollution species.. A Hong Kong case study. Computers & Geosciences, 28, 119-127 (2002).
3. Marino, D., Morabito, F. C., Ricca, B.:Management of uncertainty in environmental problems: an assessment of technical aspects and policies. Handbook of Uncertainty, J. Gil Aluja, Ed., Kluwer Academic Publisher (2001).
4. Pao, H.T., Fu, H.C., Tseng, C.L.: Forecasting of C_{O2} emissions, energy consumption and economic growth in China using an improved grey model.Energy.40(1),400-409 (2012).
5. Pao, H.T., Tsai, C.M. : Modeling and forecasting the CO_2 emissions, energy consumption, and economic growth in Brazil. Energy. 36(5), 2450-2458 (2011).
6. Lin, C.S., Liou, F.M.,Huang, C.P.: Grey forecasting model for CO_2 emissions: A Taiwan study. Applied Energy. 88, 3816-3820 (2011).
7. Lu, I.J., Lewis, C., Lin, S.J.: The forecast of motor vehicle, energy demand and CO_2 emission from Taiwan's road transportation sector, Energy Policy 37(8), 2952-2961 (2009).
8. Radojević, D., Pocajt, V., Popović, I., Perić-Grujić, A., Ristić, M. Forecasting of Greenhouse Gas Emission in Serbia Using Artificial Neural Networks, Energy Sources, Part A: Recovery, Utilization, and Environmental Effects. 35(8), 733-740 (2013).
9. Liu, P., Zhang, G., Zhang, X., Cheng, S. : Carbon Emissions Modeling of China Using Neural Network. Computational Sciences and Optimization (CSO), Fifth International Joint Conference, pp.679-682 (2012).
10. Yap, W.K, Karri, V. Emissions predictive modelling by investigating various neural network models.Expert Syst. Appl. 39(3), 2421-2426 (2012).
11. Li, S., Zhou, R., Ma, X. : The forecast of CO_2 emissions in China based on RBF neural networks. Industrial and Information Systems (IIS), 2nd International Conference, pp.319-322 (2010).
12. Sözen, A., Gülseven, Z., Arcaklioğlu, E. : Forecasting based on sectoral energy consumption of GHGs in Turkey and mitigation policies. Energy Policy. 35(12), 6491-6505 (2007).
13. Feng, Y.Y., Chen, S.Q., Zhang, L.X. :System dynamics modeling for urban energy consumption and CO_2 emissions: A case study of Beijing, China, Ecol. Model. 252,44-52 (2013).
14. Zhao, J., Zhang, J., Jia, S., Li, Q., Zhu, Y. A: MapReduce framework for on-road mobile fossil fuel combustion CO_2 emission estimation.Geoinformatics, 19th International Conference, pp.1-4 (2011).
15. Mintz, R., Young, B. R.,Svrcek, W. Y.: Fuzzy logic modeling of surface ozone concentrations.Computers and Chemical Engineering.29, 2049-2059 (2005).
16. Peche, R., Rodríguez, E.: Environmental impact assessment procedure: A new approach based on fuzzy logic. Environmental Impact Assessment Review.29, 275–283 (2009).
17. Huang, Y., Chen, X. , Li, Y. P., Huang, G.H., Liu, T. : A fuzzy-based simulation method for modelling hydrological processes under uncertainty.Hydrol.Process.24, 3718–3732 (2010).
18. Carbajal-Hernández, J. J., Sánchez-Fernández, L.P., Carrasco-Ochoa, J. A., & Martínez-Trinidad, J. F.: Assessment and prediction of air quality using fuzzy logic and autoregressiveModels. Atmospheric Environment, 60, 37-50. (2012).

19. Yetilmezsoy, K., & Abdul-Wahab, S.A. : A prognostic approach based on fuzzy-logic methodology to forecast PM_{10} levels in Khaldiya residential area, Kuwait. Aerosol and Air Quality Research,12,1217-1236 (2012).
20. Antanasijević, D.Z., Pocajt, V.V., Povrenović, D.S., Ristić, M.Đ., Perić-Grujić, A.A., : PM^{10} emission forecasting using artificial neural networks and genetic algorithm input variable optimization. Sci. Total Environ 443,511-519 (2012).
21. Yu, Z., Liangsheng, L.,& Changhai, S.: Evaluation of the Levels of Manufacturers Developing Low-Carbon on BP Neural Network.*E-Product E-Service and E-Entertainment (ICEEE), 2010 International Conference,*pp.1-4 (2010).
22. Lim, Y., Moon, Y.S.,& Kim, T.W. :Artificial neural network approach for prediction of ammonia emission from field-applied manure and relative significance assessment of ammonia emission factors.Europ. J. Agronomy. 26, 425–434 (2007).
23. Jain, S., & Khare.M.: Adaptive neuro-fuzzy modeling for prediction of ambient CO concentration at urban intersections and roadways. Air Quality, Atmosphere & Health,3, 203-212 (2010).
24. Morabito, F.C., & Versaci, M.: Fuzzy Neural Identification and forecasting techniques to process experimental urban air pollution data. , *Neural Network.*16, 493-506 (2003).
25. Noori, R., Hoshyaripour, G., Ashrafi, K., & Araabi, B.N.: Uncertainty analysis of developed ANN and ANFIS models in prediction of carbon monoxide daily concentration. Atmospheric Environment 44(4), 476–482 (2010).
26. Campbell, P. R J. : Comparison of fuzzy modelling techniques for load forecasting,. Fuzzy Systems Conference, 2007.FUZZ-IEEE 2007. IEEE International , pp.1,5. (2007).
27. Badri, A., Ameli, Z., & Birjandi, A. M.: Application of artificial neural networks and fuzzy logic methods for short term load forecasting, Energy Procedia , 1883-1888 (2012).
28. Lohani, A.K., Kumar, R., & Singh, R.D. :Hydrological time series modeling: A comparison between adaptive neuro-fuzzy, neural network and autoregressive techniques. Journal of Hydrology, 442–443 ,23–35(2012).
29. Chen, M.S., Ying, L.C., & Pan, M.C.: Forecasting tourist arrivals by using the adaptive network-based fuzzy inference system.Expert Systems with Applications. 37,1185–1191 (2010).
30. World Bank 2011. World Development Indicators [Online].<http://data.worldbank.org/>, (2011).
31. Palani, S.,Liong, S.Y., & Tkalich, P.: An ANN application for water quality forecasting. Mar. Pollut. Bull., 56, 1586-1597 (2008).

Survey on Product Review Sentiment Classification and Analysis Challenges

Mubarak Himmat, Naomie Salim

Faculty of Computing
Universiti Teknologi Malaysia
81310 UTM Skudai-Johor, MALAYSIA
barakamub@yahoo.com, naomie@utm.com

Abstract. There is no doubt that the process of using the internet to post comments and to get others' comments has become a common daily practice on the Web. Nowadays, a huge amount of information is available on the internet .The data which is posted by users and customers who visit these websites every day contain significant information. Some companies ask their customers about a product or services, for feedback analysis and to evaluate the satisfaction ratio of their products and services. The reviews by customers of products are rapidly growing. This paper provides ground knowledge and covers the most important scholarly papers and research that have been done in the area of sentiment analysis and the classification of opinion. This work presents opinion definitions and more detailed opinion classifications, and explains the related topics. This review will provide an introduction to the most common and significant information related to sentiment analysis, and it will answer many questions that have been asked in opinion mining, analysis, classifications and challenges.

Keywords: Opinion mining, Text mining, Sentiment classification, Customer reviews.

1   Introduction

The development of the internet and its applications and the wide use of it technologies has created a mania for using the internet. Nowadays, most users and customers are becoming familiar and comfortable with the internet's applications and these increases with each day. The development that occurred in the generation and the fourth generation of mobile phone technology has helped and increased the wide use of the internet by people all over the world. People now share information, opinions and reviews rapidly, within few moments. Now, when you want to know information about any specific items, you do not need to ask others as in the past to get their opinion. You needn't follow the advertisements of companies, for these contain many fakes, and they contains information that also more over the truth to gain and attract

T. Herawan et al. (eds.), *Proceedings of the First International Conference on Advanced Data and Information Engineering (DaEng-2013)*, Lecture Notes in Electrical Engineering 285, DOI: 10.1007/978-981-4585-18-7_25,
© Springer Science+Business Media Singapore 2014

customers. If want to get true information that is based on experience, you must get the information from real users who are using or used the product or service before. You just seek the reviews of people of the specific target and you will find huge information and reviews that spread and are available on the online selling products' websites, firm's websites, forums, social media and so forth. The growth of opinion has generated a good ground for opinion mining and sentiment analysis. The products manufacturers need to find out the customers' viewpoints about their products for feedback and enhancement reasons. There are many works which focus on opinion mining analysis [1 6] generating a summary of the product reviews [3, 7-13] and opinion analysis that have been attracting many researchers recently.

In this paper, we present a discussion and examine the works that have been done for opinion mining of products and general opinion classification. We also provide the series development, what and how to support the knowledge discovery of internet text mining. What is opinion and how is it classified, how can it be extracted and how is it analysed and classified? The rest of this paper will be classified as follows. Firstly, an introduction, then defining opinion and sentiment analysis, and then sentiment classification and discussing the sentiment analysis level, followed by a conclusion.

2   Opinion Mining

Recently, opinion mining has become one of the rich data mining fields. The review of the customers has significant information as we mentioned above. This lets many researchers try to work in this area using data mining techniques and methods and by applying new techniques within Natural Language Processing (NLP) tools, and modified tools of text analysis and summarisation.

The term sentiment analysis was used for the first time in 2003 with [14] when they evolved a Sentiment Analyser (SA) that extracts sentiment (or opinion) about a subject from online text documents, but really much research had been done on sentiment and topic detection before that date[1, 15, 16].

3   Sentiment Analysis and Classification

Opinion definition: opinion mining and sentiment analysis can be described as the process of automatically extracting and analysing the opinions, sentiments, thoughts and feelings of opinion writers on a specific target. This target could be a product, some issue like politics, economics, events, phenomena, services, etc.).[16] define sentiment to be a personal positive or negative feeling.[2, 4, 5, 10-12, 15, 17-19]have works on sentiment analysis and they focus on classifying each customer review as positive, negative and neutral.

2.1   3.1. Opinion Definitions

Opinion has two definitions. The first one is provided by [14]and the second is modified and redefined by [6]. Here we will examine the two definitions and the development of that definition that happened.

2.2   3.1.1 Defining Opinion as Quadruple

This definition was given by [3]. They divided the opinion into four major basic components (g_i, so_{ijkl}, h_i, t_l),
   Where

- g_i is a target
- so_{ijkl} is the sentiment value of the opinion from opinion holder h_i of target g_i at time t_l. So_{ijl} is positive, negative or neutral, or a rating score.
- h_i is an opinion holder.
- t_l is the time when the opinion is expressed.

   Opinion Definition: An opinion is a quadruple, (g, so, h, t), where g is the opinion (or sentiment) target, s is the sentiment about the target, h is the opinion holder and t is the time when the opinion was expressed. Let us look at the following example.

   E.g. *"(1) I bought a Sony Mobile phone two weeks ago. (2) I really like It. (3) I have captured a clear nice image. (4) The sound voice is good. (5) My friends said it needs not a lot of money to buy it."*

   The above opinion example describes the opinion components and demonstrates that opinion analysis is not an easy way in the real practice of online reviews of products, services, for the complexity target description, because of the implicit and indirect expression. For example, in sentence (3), the opinion target meant here is "picture quality of Sony mobile", but the sentence does not mention that directly. It indicates it by the following sentence: *"a clear nice image"* and this case will usually be found. Case 1 actually leads to new definitions of opinion by it the entities of the target. To know clearly which aspect of the product is the opinion about and what is the sentiment of the opinion holder on specific aspects.

2.3   3.1.2 Defining Opinion as Quintuple

An opinion is defined as a quintuple component [6](e_i, a_{ij}, s_{ijkl}, h_k, t_l) where
- e_i is the name of an entity.
- a_{ij} is an aspect of e_i.
- s_{ijkl} is the sentiment on aspect a_{ij} of entity ei.
- h_k is the opinion holder.
- t_l is the time when the opinion is expressed by h_k.

   The sentiment s_{ijkl} is positive, negative, neutral, or expressed with different strength/intensity levels, e.g. 1 to 5 stars, as used by most review sites on the Web. When an opinion is on the entity itself as a whole, the special aspect GENERAL is used to denote it. Here, e_i and a_{ij} together represent the opinion target. This definition can be considered more accurate, especially in product and services opinions because any product has different component or aspects and it is not fair to judge the product at all by the overall opinion. Sometimes the opinion holder's sentiment is varied from one aspect to another.

3.2 Opinions' Sentences General Classification.

Just as general the opinion sentences can be classified into to two main classes, a fact which expresses factual information from sentences, and subjective that expresses the opinion holder's attitude and sentiment. Each class is derived to subclasses. Figure 1 illustrates the General Classifications of opinions' Sentences.

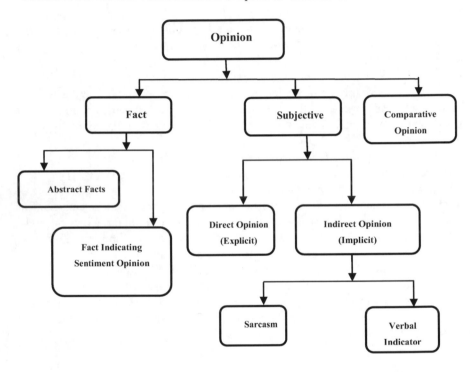

Fig .1 General Classifications of opinions' Sentences.

Within the below opinion example, we will demonstrate and discuss each class:

E.g. *"Proton Saga cars are made in Malaysia (1), and it has a very strong engine (2), its design is very beautiful (3) You do not need a lot of money to buy it (4)"*.

3.3.1 Abstract Facts

To define the abstract fact let us look at the following example.

> E.g. *Proton Saga cars are made in Malaysia (1), and it has a very strong engine (2), its design is very beautiful (3). You do not need a lot of money to buy it (4).*

The first sentence, "*Proton Saga cars are made in Malaysia*", this sentence is not telling us the sentiment of the opinion holder, and it tells us about truth and facts "that the proton saga cars are made in Malaysia". When the sentence is not telling us about what the opinion holder really feels about the target, but just tells us some information and facts on that target we call this sentence a fact sentence.

3.3.2 The facts indicating sentiment

Sometimes facts can contain implicitly some sentiment. Let us examine the following opinion sentence: "*this kid is two years old, he has the ability to read and write*". This seems like a fact, because it talks about truth and fact, but it implicitly indicates that the kid is intelligent, for there are hidden sentiments in facts sometimes.

3.4 Subjective Opinions

The subjective opinion is the sentence that tells us what the opinion holder really feels about the product or target. Let us take a look at the second, third and fourth sentences."*It has a very strong engine* ","*its design is very beautiful", you need not a lot of money to buy it.*" These sentences are telling us about the sentiment of this opinion holder because he/she looks at the engine as being strong, and at the design as being beautiful, and this can be considered as his own opinion or sentiment, maybe someone else has a different opinion. The subjective opinion analysis has been more effective in sentiment analysis. For that reason,[20]identified subjective classification as a task that investigates whether a paragraph presents the opinion of its author or reports facts. Many researchers have shown that there is a very tight relation between subjectivity classification and sentiment classification[10, 11, 13, 17, 21, 22].and all these approaches are used to automatically assign subjective opinion.

3.4.1 Explicit opinion (Direct)

This type of opinion we can call explicit if a feature or any of its synonyms appears in a sentence. The feature could be identified as explicit or direct opinion. The explicit features are features which appear directly in a review, such as phone speed in the following sentence: "*The speed of the phone is slow*". The opinion that directly points to the target feature is called a direct opinion, e.g."It has a very strong engine"," its design is very beautiful" - these two sentences are direct opinions. The first sentence indicted to the engine feature, and the second sentence indicated directly to the design feature. This kind of sentiment is called a direct opinion[10, 12, 13, 18][10, 12, 13, 18] and these kinds of opinions provide good and easy analysis compared with others opinion classes.

3.4.2 Implicit opinions (Indirect)

The opinion has been called implicit if the feature or any of its synonyms do not appear in a sentence. The feature can be identified as explicit or an indirect opinion, such as "*my friend said that you lost your money by purchasing this phone*". Here the sentence is free of any feature or synonym of features, but the sentence indicates that it means the phone is bad.

Sometimes, the opinion holder does write and express his/her sentiment explicitly without writing the feature name, let's look at the last sentence of another example: "*you need not a lot of money to buy it.*" This sentence indicates that the price of the car is cheap. But it does not mention the feature (price) directly, he assigned it explicitly. This kind of sentence is found in many reviews and it can be classified into two classes:

1. **Verbal indicator.**

The opinion holder writes and expresses his sentiment explicitly without writing the feature's name. Let us look at the last sentence of the four in the previous mentioned examples: "*you do not need a lot of money to buy it.*" This sentence indicated that the price of the car is cheap, but he does not mention the price feature directly; he assigns it explicitly. This kind of sentence is quite difficult to understand using natural language processing and it has been expressed in many commented opinions.

2. **Sarcasm definition**

Sarcasm is defined as the use of irony to mock or convey contempt (http://oxforddictionaries.com/). It is the activity of saying or writing the opposite of what you mean, or of speaking in a way intended to make someone else feel stupid or show them that you are angry, as defined in (http://www.macmillandictionary.com/)or the use of unpleasant remarks intended to hurt a person's feelings, as defined in (the free English dictionary).

The sarcastic type of opinion is very popular and many opinion holders use it when they are not satisfied with the target. They describe their opinion by using joke or sarcastic expression [23]. Sarcasm (also known as verbal irony) is a sophisticated form of speech act in which the speakers provide enhancement of the semi-supervised sarcasm identification algorithm (S A S I).Their algorithm employs two modules, semi-supervised pattern acquisition for identifying, sarcastic patterns that serve as features for a classifier, and a classification stage that classifies each sentence to a sarcastic class. Convey their message in an implicit way. This type of opinion is very difficult to be explored by using the available natural language processing tools because there is no standard for the way of writing sarcasm and jokes. Below are two examples of sarcastic opinion sentences. Some work has been done in the area of sarcasm detection. This has investigated the impact of lexical and pragmatic factors on machine are investigate the impact of lexical and pragmatic factors on machine learning effectiveness for identifying sarcastic utterances.

E.g.1 *"I bought a new novel, it was expensive, it isn't interesting, but it's so good for the kindergarten"*.

E.g.2.*"It's great for a fun read or if you are looking for some funny material for gags or jokes"*.

In the first example, the opinion holder posts an opinion about a new story book. The first sentence and second sentences are direct opinions, the book is expensive and not interesting. The third sentence at the first glance seems like a positive opinion, but really, when you take a deep look, you will find that the opinion holder here meant that the book is not good and it is better to be scaled for kids, and they are sarcastic that the contents of the book do not convince him.

In the second example, the opinion is telling us that the book is great for fun and for funny material, and this means the book's content is not important and not serious. But he does not indicate this directly. He uses a sarcastic expression.

3. Comparative Opinion

Opinion holders sometimes prefer to compare products with each other. The comparison opinion expresses the differences or similarities of the product or specific aspects or features by stating the preferences. E.g.: "Samsung galaxy II camera is better than the IPhone camera." [7] discussed different types of comparisons.

4   Sentiment Analysis Level

The process of opinion mining analysis can be classified into three general levels. Many works have done with researches. Here, we will try to describe each level.

4.1 Document level

At the document level, which is to obtain an overall opinion value for the whole document, the task of analysis at this level is classifying the opinion as the overall meaning, what is really the conclusion of sentiment that has been expressed, is it positive or negative sentiment. For example, if we have a product review, the system will determine the opinion in general whether the review as general evaluated positive, negative, or neutral sentiment. This kind of analysis is known as document-level sentiment classification.

Many works have been done at this level. [22]studied the effect of dynamic adjectives, semantically oriented adjectives, and gradable adjectives on a simple subjectivity and how they can be classified, and they proposed a trainable method for statistically combining two indicators of gradable. They proposed a system called Opinion Finder that automatically identifies the appearance of opinions, sentiments, speculations, and other private states in a text, via the subjective analysis, and applied machine learning techniques to address the sentiment classification problem for movie review data, and show how it can become effective. They there algorithms: Naive Bayes, classification, maximum entropy classification, and support vector machines. Work to address the rating inference problem, i.e. how to summarise a user review to a virtual rating from 0 to 5, or binary value (thumb up/down). They presented a graph based on a semi-supervised learning algorithm to resolve the rating inference problem. It inferred numerical ratings from unlabelled documents based on the sentiment expressed in the text. Concretely, they achieved and solved an optimisation problem to obtain good rating function over the whole graph that was created on the reviewer's data. They proved that, when limited labelled data is available, their method can achieve significantly better predictive accuracy over other methods. Das and Hu et al. (2007)studied sentiment classification for financial documents.

However, all the above works discovered the sentiment to represent the reviewer's overall opinion results, but did not find which features the reviewer actually preferred and liked and the one that they disliked. And in all the mentioned approaches, an overall negative sentiment on an object will not indicate that the reviewer dislikes every feature of the object.

4.2 Sentence level

In this kind of analysis level, the analysis will be concerned with the sentences and determining whether each sentence polarity (positive, negative, or neutral).Neutral means no opinion on the sentence. The sentence level of analysis is closely related to the subjectivity classification, which distinguishes sentences that express subjective views and objective sentences that express factual information from sentences. Many works, like[21], do subjective analysis. Work[3]identifies opinion sentences in each review and decides whether each opinion sentence is positive or negative[4, 5, 17, 19, 24].

4.3 Aspect or Feature-based level

As mentioned above, the usage of document level and sentence level analyses is not exactly determined and does not explore what exactly people like and dislike in a product's features. Aspect (feature) performs a perfect analysis in some cases. The feature based level [3], or aspect-based level as it has been called recently is directly looking at the opinion itself instead of an aspect of the target that is mentioned in each sentence, for opinion holders comment on different target features or aspects and it is not fair to judge an opinion at overall. For example, the sentence "*the Samsung mobile phone camera is fantastic (1), and it has a very nice design (2), the memory is not enough for many applications*". In this example we can…"We cannot say that the voice or the sentiment of the customers here is positive, because the first two sentences are positive opinions about the camera and design, but the last sentence is a negative opinion about the memory. In many comments on specific targets, opinions are described by target aspects or entities. Thus, the aim of the aspect level is to discover sentiments on each entity and/or their aspects that are mentioned throughout opinion sentences. The aspect-level is quite difficult. It really faces some challenges.[24] presented a how regular opinion expresses a sentiment only on a particular entity or an aspect of the entity, e.g. *"mobile battery is bad ,"* which expresses a negative sentiment on the aspect battery of a mobile. [9] proposed a graphical model to extract and visualise comparative relations between products from customer reviews, with the interdependencies among relations taken into consideration[7]. They adapted the CRFs (Conditional Random Fields) model for accomplishing the feature-level of web opinion mining tasks for analysing online consumer reviews.

5   Conclusion

In this literature survey paper, we have illustrated a sentiment analysis approach and discussed sentiment analysis and classification in more detail. This work discusses many related problems, sentiment analysis and classification, subjectivity classification, and we found that the major challenges and the problems faced that need to be solved are the following challenges: the opinion holder has written and expressed their opinion without regard for the grammatical syntax and also there are many comment opinions which could include difficult sentences. Some comments also contain sarcastic expressions and this type of opinion is difficult to detect because there are no standards for sarcasm; it depends on human recognition. Also comparison sentence are used by many researchers, as we mentioned, but we also need more works. Slang and privation words are difficult to detect. Using comparison sentences also needs more work. But we think that the future of opinion mining will deal with the outstanding problems identified above. The work can be further extended to emerging areas like sarcasm and slang and privation words and comparisons analysis and detection using natural language processing tools and improving new tools.

Acknowledgement

We would like to thank the Ministry of Higher Education (MOHE) and Research Management Centre (RMC) at the Universiti Teknologi Malaysia (UTM) under Research University Grant Category (VOT Q.J130000.2528.02H99) for supporting this work."

References

1. Turney, P.D. Thumbs up or thumbs down?: semantic orientation applied to unsupervised classification of reviews. in Proceedings of the 40th Annual Meeting on Association for Computational Linguistics. 2002: Association for Computational Linguistics.
2. Yi, J., et al. Sentiment analyzer: Extracting sentiments about a given topic using natural language processing techniques. in Data Mining, 2003. ICDM 2003. Third IEEE International Conference on. 2003: IEEE.
3. Hu, M. and B. Liu. Mining and summarizing customer reviews. in Proceedings of the tenth ACM SIGKDD international conference on Knowledge discovery and data mining. 2004: ACM.

4. Kim, S.-M. and E. Hovy. Determining the sentiment of opinions. in Proceedings of the 20th international conference on Computational Linguistics. 2004: Association for Computational Linguistics.
5. Liu, B., Sentiment analysis: A multi-faceted problem. the IEEE Intelligent Systems, 2010.
6. Liu, B., Sentiment analysis and opinion mining. Synthesis Lectures on Human Language Technologies, 2012. 5(1): p. 1-167.
7. Chen, L., L. Qi, and F. Wang, Comparison of feature-level learning methods for mining online consumer reviews. Expert Systems with Applications, 2012.
8. Aguwa, C.C., L. Monplaisir, and O. Turgut, Voice of the customer: Customer satisfaction ratio based analysis. Expert Systems with Applications, 2012.
9. Xu, K., et al., Mining comparative opinions from customer reviews for Competitive Intelligence. Decision Support Systems, 2011. 50(4): p. 743-754.
10. Eirinaki, M., S. Pisal, and S. Japinder, Feature-based opinion mining and ranking. Journal of Computer and System Sciences, 2011.
11. Aurangzeb, K., B. Baharum, and K. Khairullah, Sentiment Classification from Online Customer Reviews Using Lexical Contextual Sentence Structure. 2011.
12. Lu, Y., et al., Exploring the sentiment strength of user reviews, in Web-Age Information Management. 2010, Springer. p. 471-482(2010).
13. Hu, M. and B. Liu. Mining opinion features in customer reviews. in Proceedings of the National Conference on Artificial Intelligence. 2004: Menlo Park, CA; Cambridge, MA; London; AAAI Press; MIT Press; 1999.
14. Nasukawa, Tetsuya, and Jeonghee Yi. "Sentiment analysis: Capturing favorability using natural language processing." Proceedings of the 2nd international conference on Knowledge capture. ACM(2003)
15. Rajaraman, K. and A.-H. Tan, Topic detection, tracking, and trend analysis using self-organizing neural networks, in Advances in Knowledge Discovery and Data Mining. 2001, Springer. p. 102-107.
16. Pang, B., L. Lee, and S. Vaithyanathan. Thumbs up?: sentiment classification using machine learning techniques. in Proceedings of the ACL-02 conference on Empirical methods in natural language processing-Volume 10. 2002: Association for Computational Linguistics.
17. Pang, B. and L. Lee, Opinion mining and sentiment analysis. 2008: Now Pub.
18. Zhan, J., H.T. Loh, and Y. Liu, Gather customer concerns from online product reviews–A text summarization approach. Expert Systems with Applications, 2009. 36(2): p. 2107-2115.
19. Liu, B., Opinion mining and sentiment analysis. Web Data Mining, 2011: p. 459-526.
20. Tang, H., S. Tan, and X. Cheng, A survey on sentiment detection of reviews. Expert Systems with Applications, 2009. 36(7): p. 10760-10773.
21. Wiebe, J.M., R.F. Bruce, and T.P. O'Hara. Development and use of a gold-standard data set for subjectivity classifications. in Proceedings of the 37th annual meeting of the Association for Computational Linguistics on Computational Linguistics(1999).
22. Hatzivassiloglou, V. and J.M. Wiebe. Effects of adjective orientation and gradability on sentence subjectivity. in Proceedings of the 18th conference on Computational linguistics-Volume 1.(2000)s.
23. Davidov, D., O. Tsur, and A. Rappoport. Semi-supervised recognition of sarcastic sentences in twitter and amazon. in Proceedings of the Fourteenth Conference on Computational Natural Language Learning. 2010: Association for Computational Linguistics.
24. Jindal, N. and B. Liu. Mining comparative sentences and relations. in Proceedings of the National Conference on Artificial Intelligence. 2006: Menlo Park, CA; Cambridge, MA; London; AAAI Press; MIT Press; 1999.

SVD based Gene Selection Algorithm

Andri Mirzal

Faculty of Computing, N28-439-03
Universiti Teknologi Malaysia
81310 UTM Johor Bahru, Malaysia
andrimirzal@utm.my

Abstract. This paper proposes an unsupervised gene selection algorithm based on the singular value decomposition (SVD) to determine the most informative genes from a cancer gene expression dataset. These genes are important for many tasks including cancer clustering and classification, data compression, and samples characterization. The proposed algorithm is designed by making use of the SVD's clustering capability to find the natural groupings of the genes. The most informative genes are then determined by selecting the closest genes to the corresponding cluster's centers. These genes are then used to construct a new (pruned) dataset of the same samples but with less dimensionality. The experimental results using some standard datasets in cancer research show that the proposed algorithm can reliably improve performances of the SVD and kmeans algorithm in cancer clustering tasks.

Keywords: cancer clustering, DNA microarray datasets, gene selection algorithm, kmeans, singular value decomposition

1 Introduction

Cancer clustering using microarray gene expression datasets is one of the most important researches in medical community [1–9]. Cancer clustering is an unsupervised task of grouping tissue samples from patients with cancers so that samples with the same cancer type can be clustered in the same group. It is worth to mention that this task is complement but different to cancer classification [10–23], a supervised task where the classifiers are trained first using training datasets before being used to classify the samples.

Gene expression datasets have one particular characteristic: they usually consist of only a few samples (hundreds at most), but each sample is represented by thousands of gene expressions. This characteristic makes the clustering tasks challenging because usually clustering algorithms perform poorly when the number of samples is small. Additionally, the huge dimensionality of the samples implies that the datasets contain many irrelevant and potentially misleading gene expressions. Thus, gene selection is a prerequisite to prepare the datasets for further analysis.

In this paper, we propose an unsupervised gene selection algorithm based on the SVD to determine the most informative genes from a gene expression

T. Herawan et al. (eds.), *Proceedings of the First International Conference
on Advanced Data and Information Engineering (DaEng-2013)*, Lecture Notes
in Electrical Engineering 285, DOI: 10.1007/978-981-4585-18-7_26,
© Springer Science+Business Media Singapore 2014

dataset. The algorithm is designed by making use of clustering capability of the SVD to find the natural groupings of the genes. The most informative genes are then chosen by selecting the closest genes to the corresponding cluster's centers. These genes are then used to construct a new (pruned) dataset of the same samples but with less dimensionality.

2   The SVD

The SVD is a matrix decomposition technique that factorizes a rectangular real or complex matrix into its left singular vectors, right singular vectors, and singular values. Some applications of the SVD include clustering [24, 25], matrix approximation [26], pseudoinverse computation [27], and determining the rank, range, and null space of a matrix [28].

The SVD of matrix $\mathbf{A} \in \mathbb{C}^{M \times N}$ with $\text{rank}(\mathbf{A}) = r$ is defined with $\mathbf{A} = \mathbf{U}\mathbf{\Sigma}\mathbf{V}^T$, where $\mathbf{U} \in \mathbb{C}^{M \times M} = [\mathbf{u}_1, \ldots, \mathbf{u}_M]$ denotes a unitary matrix that contains the left singular vectors of $\mathbf{A}$, $\mathbf{V} \in \mathbb{C}^{N \times N} = [\mathbf{v}_1, \ldots, \mathbf{v}_N]$ denotes a unitary matrix that contains the right singular vectors of $\mathbf{A}$, and $\mathbf{\Sigma} \in \mathbb{R}_+^{M \times N}$ denotes a matrix that contains the singular values of $\mathbf{A}$ along its diagonal.

Rank-K SVD approximation of $\mathbf{A}$ is defined with $\mathbf{A} \approx \mathbf{A}_K = \mathbf{U}_K\mathbf{\Sigma}_K\mathbf{V}_K^T$, where $K < r$, $\mathbf{U}_K$ and $\mathbf{V}_K$ contain the first K columns of $\mathbf{U}$ and $\mathbf{V}$ respectively, and $\mathbf{\Sigma}_K$ denotes a $K \times K$ principal submatrix of $\mathbf{\Sigma}$. $\mathbf{A}_K$ is also known as the truncated SVD of $\mathbf{A}$, and according to the Eckart-Young theorem, $\mathbf{A}_K$ is the closest rank-K approximation of $\mathbf{A}$ in Frobenius norm criterion [26, 27].

3   The proposed algorithm

Algorithm 1 outlines the proposed gene selection algorithm. The output of the algorithm, $\hat{\mathbf{A}}$, is then inputted to the SVD or kmeans for clustering purpose. Note that the samples in $\hat{\mathbf{A}}$ is the same as the samples in $\mathbf{A}$ but its columns consist of only the top genes selected by algorithm 1. Algorithm 2 describes the standard clustering procedure using the SVD. Clustering procedure using kmeans is conducted by simply applying kmeans clustering on rows of $\hat{\mathbf{A}}$. When gene selection is not used, then the clustering procedures will be the same, but the input is $\mathbf{A}$ instead. As both the gene selection procedure and the clustering algorithms are unsupervised, this strategy can be implemented in a fully unsupervised fashion.

4   Experimental results

We now use of the proposed algorithm to improve clustering performances of the SVD and kmeans algorithm using four standard datasets in cancer clustering research. The datasets were downloaded from http://algorithmics.molgen.mpg.de/ Static/Supplements/CompCancer/datasets.htm [29]. To evaluate clustering quality, we used two metrics; Adjusted Rand Index (ARI) [30–32] and Accuracy [1].

Algorithm 1 SVD based gene selection algorithm.

---

1. Input: sample-by-gene matrix $\mathbf{A} \in \mathbb{R}_+^{M \times N}$, and #cluster K.
2. Normalize each column of $\mathbf{A}$, i.e., $a_{mn} \leftarrow \frac{a_{mn}}{\sum_m a_{mn}}$ for $\forall n$.
3. Compute $\mathbf{V}_K = [\mathbf{v}_1, \ldots, \mathbf{v}_K]$ of $\mathbf{A}$.
4. Apply k-means clustering on rows of $\mathbf{V}_K$ to obtain K clusters of the genes.
5. Compute cluster's centers by averaging all gene vectors $\mathbf{a}_n$ in the same cluster.
6. Compute Euclidean distances of all genes from the corrensponding centers.
7. Sort the genes according the the distances (in ascending order), and select $G < N$ top genes from the list.
8. Form the pruned matrix $\hat{\mathbf{A}} \in \mathbb{R}_+^{M \times G}$ where the colums are the top genes selected from step 7.

Algorithm 2 Clustering procedure using the SVD.

---

1. Input: sample-by-gene matrix $\mathbf{A} \in \mathbb{R}_+^{M \times N}$ (or $\hat{\mathbf{A}} \in \mathbb{R}_+^{M \times G}$), and #cluster K.
2. Normalize each column of $\mathbf{A}$ (or $\hat{\mathbf{A}}$).
3. Compute $\mathbf{U}_K = [\mathbf{u}_1, \ldots, \mathbf{u}_K]$ of $\mathbf{A}$ (or $\hat{\mathbf{A}}$).
4. Apply k-means clustering on rows of $\mathbf{U}_K$ to obtain K clusters of the samples.

There is one parameter needs to chosen in the proposed algorithm: the number of top genes G. After several attempts, G was set to 50, 2000, 1000, and 1000 for Nutt-2003-v2, Armstrong-2002-v2, Tomlins-2006-v2, and Pomeroy-2002-v2 respectively.

Table 2 and 3 give clustering performances of the SVD and kmeans without the gene selection, and table 4 and 5 show the results when the gene selection was employed. Because the proposed algorithm does not have uniqueness property (due to the use of kmeans to obtain clustering assignments), the experiments were repeated 1000 times for each case, and the values are displayed in format average $\pm$ standard deviation values.

Figure 1 plots average values in table 2–5. It is interesting to see that without the gene selection, the performances of the SVD and kmeans are quite comparable (first row of figure 1). Then when the procedure was employed, the SVD seems to benefit more (second row of figure 1).

The performance improvements of the SVD and kmeans due to the gene selection procedure are summarized in first row and second row of figure 2 re-

Table 1. Cancer datasets.

Dataset name	Tissue	#Samples	#Genes	#Classes
Nutt-2003-v2	Brain	28	1070	2
Armstrong-2002-v2	Blood	72	2194	3
Tomlins-2006-v2	Prostate	92	1288	4
Pomeroy-2002-v2	Brain	42	1379	5

Table 2. Results of the SVD without gene selection.

Dataset name	ARI	Accuracy
Nutt-2003-v2	0.0107 ± 0.0	0.571 ± 0.0
Armstrong-2002-v2	0.518 ± 0.0399	0.720 ± 0.0247
Tomlins-2006-v2	0.0828 ± 0.0244	0.489 ± 0.0350
Pomeroy-2002-v2	0.287 ± 0.125	0.598 ± 0.0901

Table 3. Results of kmeans without gene selection.

Dataset name	ARI	Accuracy
Nutt-2003-v2	0.0185 ± 0.0191	0.583 ± 0.0272
Armstrong-2002-v2	0.491 ± 0.0997	0.729 ± 0.0454
Tomlins-2006-v2	0.164 ± 0.0439	0.559 ± 0.0405
Pomeroy-2002-v2	0.181 ± 0.0939	0.481 ± 0.0755

Table 4. Results of the SVD with gene selection.

Dataset name	ARI	Accuracy
Nutt-2003-v2	0.101 ± 0.101	0.644 ± 0.0961
Armstrong-2002-v2	0.637 ± 0.119	0.825 ± 0.0876
Tomlins-2006-v2	0.143 ± 0.0336	0.535 ± 0.0336
Pomeroy-2002-v2	0.455 ± 0.110	0.717 ± 0.0762

Table 5. Results of kmeans with gene selection.

Dataset name	ARI	Accuracy
Nutt-2003-v2	0.0624 ± 0.0767	0.621 ± 0.0743
Armstrong-2002-v2	0.601 ± 0.149	0.785 ± 0.0954
Tomlins-2006-v2	0.171 ± 0.0481	0.564 ± 0.0415
Pomeroy-2002-v2	0.377 ± 0.115	0.663 ± 0.0810

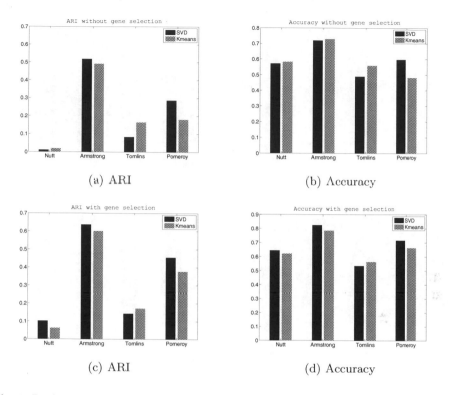

Fig. 1. Performance comparison between the SVD and kmeans (first row: without gene selection, second row: with gene selection).

spectively. Table 6 and 7 give more quantitative results of the improvements. As shown, the gene selection procedure improved the clustering performances for both the SVD and kmeans in all cases with the better improvements were observed in the SVD cases.

5 Conclusion

We have presented an unsupervised gene selection algorithm based on the SVD. The proposed algorithm was designed by making use of clustering capability of the SVD to select the most informative genes from a gene expression dataset. The experimental results showed that the proposed algorithm improved clustering performances of the SVD and kmeans algorithm with more visible improvements were observed in the SVD cases. In addition to improving the clustering performances, the gene selection procedure also has a benefit in reducing the sizes of the datasets. These results suggest that some mechanism of gene selection should be employed to remove irrelevant and misleading gene expressions before analyzing the datasets further.

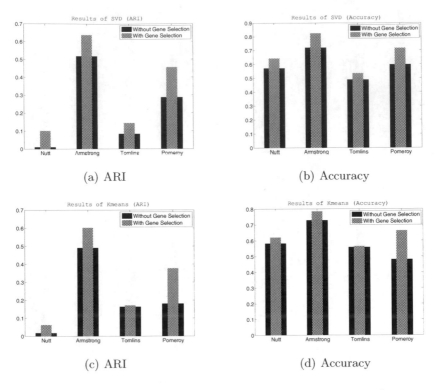

(a) ARI   (b) Accuracy

(c) ARI   (d) Accuracy

Fig. 2. Improvements gained by employing the gene selection algorithm (first row corresponds to the SVD and second row to kmeans).

Table 6. Percentages of improvements (SVD).

Dataset name	ARI	Accuracy
Nutt-2003-v2	846.3	12.62
Armstrong-2002-v2	22.94	14.55
Tomlins-2006-v2	72.88	9.417
Pomeroy-2002-v2	58.63	19.99
Average	250.2	14.14

Table 7. Percentages of improvements (kmeans).

Dataset name	ARI	Accuracy
Nutt-2003-v2	237.3	6.489
Armstrong-2002-v2	22.47	7.765
Tomlins-2006-v2	4.093	0.9220
Pomeroy-2002-v2	108.5	37.80
Average	93.08	13.24

Acknowledgement

The author would like to thank the reviewers for useful comments. This research was supported by Ministry of Higher Education of Malaysia and Universiti Teknologi Malaysia under Exploratory Research Grant Scheme R..J130000.7828. 4L095.

References

1. Gao, Y., Church, G.: Improving molecular cancer class discovery through sparse non-negative matrix factorization. Bioinformatics 21(21), 3970–3975 (2005)
2. Dueck, D., et al.: Multi-way clustering of microarray data using probabilistic sparse matrix factorization. Bioinformatics 21(1), 145–151 (2005)
3. Brunet, J.P., et al.: Metagenes and molecular pattern discovery using matrix factorization. Proc. Natl Acad. Sci. USA 101(12), 4164–4169 (2003)
4. Kim, H., Park, H.: Sparse non-negative matrix factorizations via alternating non-negativity constrained least squares for microarray data analysis. Bioinformatics 23(12), 1495–1502 (2007)
5. Carmona-Saez, et al.: Biclustering of gene expression data by non-smooth non-negative matrix factorization. BMC Bioinformatics 7(78) (2006)
6. Inamura, K., et al.: Two subclasses of lung squamous cell carcinoma with different gene expression profiles and prognosis identified by hierarchical clustering and non-negative matrix factorization. Oncogene (24), 7105–7113 (2005)
7. Fogel, P., et al.: Inferential, robust non-negative matrix factorization analysis of microarray data. Bioinformatics 23(1), 44–49 (2007)
8. Zheng, C.H., et al.: Tumor clustering using nonnegative matrix factorization with gene selection. IEEE Transactions on Information Technology in Biomedicine 13(4), 599–607 (2009)
9. Wang, J.J.Y., et al.: Non-negative matrix factorization by maximizing correntropy for cancer clustering. BMC Bioinformatics 14(107) (2013)
10. Golub, T.R., et al.: Molecular classification of cancer: class discovery and class prediction by gene expression monitoring. Science 286(5439), 531–537 (1999)
11. Guyon, I., et al.: Gene selection for cancer classification using support vector machines. Machine Learning 46(1-3), 389–422 (2002)
12. Yuvaraj, N., Vivekanandan, P.: An efficient SVM based tumor classification with symmetry non-negative matrix factorization using gene expression data. In Int'l Conf. on Information Communication and Embedded Systems, pp. 761–768 (2013)

13. Pirooznia, M., *et al.*: A comparative study of different machine learning methods on microarray gene expression data. BMC Genomics 9(Suppl 1), S13 (2008)

14. Liu, X., *et al.*: An entropy-based gene selection method for cancer classification using microarray data. BMC Bioinformatics 6 (2005)

15. Wang, L., *et al.*: Accurate cancer classification using expressions of very few genes. IEEE/ACM Transactions on Computational Biology and Bioinformatics 4(1), 40–53 (2007)

16. Chuang, L.Y., *et al.*: Improved binary PSO for feature selection using gene expression data. Computational Biology and Chemistry 32(1), 29–37 (2008)

17. Mitra, P., Majumder, D.D.: Feature selection and gene clustering from gene expression data. In 17th Int'l Conf. on Pattern Recognition, pp.343–346 (2004)

18. Furey, T.S., *et al.*: Support vector machine classification and validation of cancer tissue samples using microarray expression data. Bioinformatics 16(10), 906–914 (2000)

19. Moon, S., Qi, H.: Hybrid dimensionality reduction method based on support vector machine and independent component analysis. IEEE Transactions on Neural Networks and Learning Systems 23(5), 749–761 (2012)

20. Lee, Y., Lee, C.K.: Classification of multiple cancer types by multicategory support vector machines using gene expression data. Bioinformatics 19(9), 1132–1139 (2003)

21. Zhang, X., *et al.*: Recursive SVM feature selection and sample classification for mass-spectrometry and microarray data. BMC Bioinformatics 7(197) (2006)

22. Lu, Y., Han, J.: Cancer classification using gene expression data. Information Systems 28(4), 243–268 (2003)

23. Zhang, H.H., *et al.*: Gene selection using support vector machines with non-convex penalty. Bioinformatics 22(1), 88–95 (2006)

24. Dhillon, I.S.: Co-clustering documents and words using bipartite spectral graph partitioning. In 7th ACM SIGKDD Int'l Conference on Knowledge Discovery and Data Mining, pp. 269–274 (2001)

25. Drineas, *et al.*: Clustering large graphs via the singular value decomposition. Machine Learning 56(1-3), 9–33 (2004)

26. Eckart, C., Young, G.: The approximation of one matrix by another of lower rank. Psychometrika 1, 211–218 (1936)

27. Golub, G.H., Kahan, W.: Calculating the singular values and pseudo-inverse of a matrix. J. SIAM Numerical Analysis 2(2), 205–224 (1965)

28. Golub, G.H., van Loan, C.F.: Matrix computations 3rd edition. Johns Hopkins University Press (1996)

29. Souto, M.C.P., *et al.*: Clustering cancer gene expression data: a comparative study. BMC Bioinformatics 9(497) (2008)

30. Rand, W.M.: Objective criteria for the evaluation of clustering methods. Journal of the American Statistical Association 66(336), 846–850 (1971)

31. Hubert, L., Arabie, P.: Comparing partitions. Journal of Classification 2(1), 193–218 (1985)

32. Vinh, N.X., *et al.*: Information theoretic measures for clustering comparison: Is a correction for chance necessary? In 26th Annual Int'l Conf. on Machine Learning, pp. 1073–1080 (2009)

The social network role in improving recommendation performance of collaborative filtering

Waleed Reafee[1, 2] and Naomie Salim[1]

[1] Faculty of Computing,
Universiti Teknologi Malaysia, 81310, Skudai, Johor, Malaysia
[2] Faculty of Pure and Applied Science,
International University of Africa, 2469, Khartoum, Sudan
rcafce@gmail.com, naomie@utm.my

Abstract. Recently a recommender system has been applied to solve several different problems that face the users. Collaborative filtering is the most commonly used and successfully deployed recommendation technique. Despite everything, the traditional collaborative filtering (TCF) operates only in the two-dimensional user-item space. The explosive growth of online social networks in recent times has presented a powerful source of information to be utilised as an extra source for assisting in the recommendation process. The purpose of this paper is to give an overview of collaborative filtering (CF) and existing methods used social network information to incorporate in collaborative filtering recommender systems to improve performance and accuracy. We classify CF-based social network information into two categories: TCF-based trust relation approaches and TCF-based friendship relation approaches. For each category, we review the fundamental concept of methods that can be used to improve recommendation performance.

Keywords: Recommender Systems, Collaborative Filtering, Social Networks.

1. Introduction

We live today in the era of technology and rapid developments especially online and the web's applications have grown significantly, thus providing a huge source of information to users. The amount of information and resources available on the Internet is increasing every moment, and users face difficulty in accessing, searching and retrieving the information they need. More recently, the online social networking phenomenon has enabled access to users' profiles and preferences. This has allowed several databases available on the Internet to be collected and used in experiments by researchers. One way to overcome this situation is to employ the most successful approaches for increasing the level of relevant content over "noise" that continually grows as more and more content becomes available online and lies in recommender systems [1]. The basic idea of recommender systems is to generate recommendations that depend on users' preferences [2], and today's recommender systems are being applied to solve several different problems in web applications in order to overcome

T. Herawan et al. (eds.), *Proceedings of the First International Conference on Advanced Data and Information Engineering (DaEng-2013)*, Lecture Notes in Electrical Engineering 285, DOI: 10.1007/978-981-4585-18-7_27,
© Springer Science+Business Media Singapore 2014

problems such as information overload. In recent years recommender systems have become more popular and important and they are employed in a large number of applications, as several web applications depend on them in recommending operations [3]. The most popular recommender system techniques recommend that users depend on the preferences of other similar users through collaborative filtering (CF). Collaborative filtering (CF) collects information about a user by asking them to rate items and makes recommendations based on highly rated items by users with similar tastes [4]. The current generation of collaborative filtering operates in the two-dimensional user-item space based on users' attributes (profile), item attributes and a user-item rating matrix. That allows their recommendations to be based only on the user and their item information and does not take into the consideration additional information that may be very important for recommendation [5]. The role of the information available on social networks has become very important as it can improve and enhance the recommendation process. Nowadays, some recommender systems allow users to build their social networks and use these networks as additional information to suggest items to users [6] such as [7], [8], [9].

The rest of the paper is organised as follows: We present the recommender system collaborative filtering approach in section 2. We then introduce social network relations as an additional CF source in section 3. Finally, section 4 presents our conclusion of the review.

2. Traditional Collaborative Filtering (TCF)

Recommender systems identify exactly what the user's need, which means providing support and assistance to users and helping them to make decisions on products. Collaborative filtering (CF) is the most popular recommender system technique predicts items for a particular user depending on the items previously rated by other users. In other words, CF utilises choices of similar neighbours (users) to generate recommendations for users. CF is very successful in the real world, and is easy to understand and implement [10]. There are two main types of CF algorithms: memory-based and model-based.

2.1 Memory-Based CF

This operates on the entire user-item rating matrix to make predictions. The memory-based approaches are the most popular prediction methods and work based on three steps:

1. *Calculate the similarity between active users and other users.*
2. *Select the neighbours (similar users).*
3. *Generate recommendations.*

Similarity measures: Similarity measures are one of the basic elements of the work of the collaborative filtering technique. The more similar two users are, the more likely it is that a new item liked by one of these users is going to be liked by the other.

Different similarity measures can be applied to the problem. The most common similarity measures are Cosine distance and Pearson's correlation coefficient. The latter is used to calculate the similarity between items in recommender systems and can be given by their correlation, which measures the linear relationship between objects [11]. Cosine similarity is another approach to consider the items as document vectors of n-dimensional and compute their similarity as the cosine of the angle.

The memory-based approaches are widely adopted in many commercial domains such as [12] Amazon, who use three approaches: search-based methods, classical collaborative filtering, and cluster models to solve the recommendation problems. In [13] a memory-based approach is also used on a GroupLens system for NetNews, to help users find articles they will need. There are two types of memory-based (CF): user-based and item-based approaches. User-based approaches use the ratings of similar users to predict the ratings of active users, and [14] presented a framework for collaborative filtering, whereby the work of [15] presents an optimisation method to automatically calculate the weights based on ratings from training users. Item-based approaches predict the ratings of active users depending on the calculated information on items similar to those chosen by the active user. In [16], development concerned model-based recommendation to address the scalability problems, and the authors [12] used three approaches: search-based methods, classical CF and cluster models, to solve the recommendation problems. The work of [17] explored the item-based approach to address increases the number of participants in the system, and was based on identifying the relationship between items in the recommender system.

2.2 Model-Based CF

The model-based approaches, using training datasets, are used to train a predefined model and then use the model to predict recommendation [9]. The obvious difference between memory-based and model-based algorithms is that the latter estimate the ratings by utilising machine learning techniques and statistics to learn about a model from the basic data, while the former uses some heuristic rules to predict the ratings [18]. Many models are used in model-based approaches, such as Bayesian models and clustering models. The authors in [19] described a new model-based algorithm and proposed an algorithm based on a generalisation of probabilistic latent semantic analysis to continuous-valued response variables. In [20] statistical model higher accuracy and we constant time prediction worked to standard memory-based methods. The work of [21] presented CF as a flexible mixture model (FMM). The matrix factorisation methods have recently been proposed for collaborative filtering [22] for direct gradient-based optimisation method. [23] used Bayesian for Probabilistic Matrix Factorisation (PMF), [24] presented the Probabilistic Matrix Factorisation model, which scales linearly with the number of observations and, more importantly, performs very well on the large and sparse. In [25] the authors provided simple and efficient algorithms for solving weighted low-rank approximation problems and applying the method to collaborative filtering.

3. Using Social Network Information in Collaborative Filtering

With the explosion of Web 2.0 applications such as forums, blogs and microblogging, social networks, social bookmarking and several other types of social media [9], users' online activities have changed [4]. Recently, social networks have become more significant for the information society. The basic idea of a social network is that it has a set of actors, i.e. a group of users or organisations, which are the nodes of the network, and relations are linked between the nodes by one or more relations [26], [27]. Social networks are everywhere, exponentially increasing in volume, and changing everything about our lives: the way we do business, how we understand ourselves and the world around us. They have become very important for many research areas because they offer very good information and can be incorporated in different methods. Today some recommender systems allow users to build their social networks and use these networks to suggest items to users [6]. Existing social networks allow discovery and connections between users and social referrals [28], and thereby offer many opportunities for recommendations. Collaborative filtering algorithms are based on everyday life, because they are based on recommendations from other people [29]. In a previous study, the prediction methodologies that depend on incorporating social network information may be more accurate than those based on just mathematical algorithms [30]. The importance of incorporating social network elements such as social relations, in terms of "friendship and trust relation" and "social contents", into traditional collaborative filtering systems (TCF) have appeared significantly in several studies, for example [31] and [32].

In collaborative filtering (CF) some existing methods use social network information to improve performance and accuracy. We classify CF-based social network into two categories in Table 1: TCF-based trust relations and TCF-based friendship relations approaches.

Table 1. Overview of classified CF based on social network information

Social Network & Dataset	Recommendation Domain	Data used for Recommendations		Evaluation Methods	Ref
		CF data	Social network Information		
Collect data from a Cyworld SN web site	Skin items (wallpapers, background, music, and virtual appliances)	Rating by users	Friendship relation	MAE	[30]
Collect data from the Facebook SN web site	Movie.	Rating by users	Users publish Users vote Friendship relation	MAE, Precision, Recall, MAP and R-Precision	[31]
Last.fm dataset Delicious datasets	URL bookmarks. Music artists.	Rating by users	Friendship relation	Precision & Recall	[9]

Douban dataset	Movies, books, music.	Rating by users	Friendship relation	MAE & RMSE	[33]
Epinions dataset	Products (movies, books, music, etc.).				
Movielens dataset	Movie.	Rating by users	Tags	MAE & RMSE	[34]
Epinion dataset	Products (movies, books, music, etc.).		Trust relation		
Epinion dataset	Products (movies, books, music, etc.).	Rating by users	Trust relation	MAE	[7]
Film trust dataset.	Movies.				
Flixster dataset.	Movies.	Rating by users	Trust relation	MAE	[35]
Epinions dataset	Products (movies, books, music, etc.).				
Movielens dataset	Movies.				
LibimSeTi dataset	Users.	Rating by users	Trust relation	MAE & RMSE	[32]
Epinions dataset	Products (movies, books, music, etc.).				

3.1 TCF-Based Social Network Trust Relations Approaches

The trust between users of social networks is one of the active elements for improving the work of the collaborative filtering approach. A trust relationship exists between two users when one user has an opinion about the other's trustworthiness and a recommendation is a communicated opinion about the trustworthiness of a third party [36]. Many previous studies have suggested various methods to incorporate trust into collaborative filtering in reference to the recommendation process; for example, [35] and [32]. The work of [32] presented an empirical analysis of several of the techniques used to incorporate trust into collaborative filtering through exploiting the method proposed by [37] to infer trust relations between actor pairs in a social network, based only on the structural information of a bipartite graph. For the purpose of comparison three datasets are used: "Epinions, LibimSeTi and MovieLens". MAE and RMSE are utilised to measure the performance of each of the recommendation algorithms. The results of incorporating trust into collaborative filtering algorithms can increase both the accuracy and the coverage of personalised recommendations.

In [35], a method named "Merge", incorporate social trust information into collaborative filtering. The authors merge the ratings of an active user's directly trusted neighbours by using Eq(1) according to how much the neighbours are trusted by the active user and then incorporating this with CF.

$$\tilde{r}_{u,i} = \frac{\sum_{v \in TN_u} t_{u,v} r_{v,i}}{\sum_{v \in TN_u} t_{u,v}} \tag{1}$$

Where $\tilde{r}_{u,i}$ is the merged rating for the active user u on item i, according to the ratings of her trusted neighbourhood TN_u. For finding a set of similar users for the active user u, the Pearson Correlation Coefficient (PCC) was used in Eq(2). Finally, Eq(3) predicts the rating for the active user.

$$S_{u,v} = \frac{\sum_{i \in I_{u,v}} \left(\bar{r}_{u,i} - \bar{r}_u \right) \left(r_{v,i} - \bar{r}_v \right)}{\sqrt{\sum_{i \in I_{u,v}} \left(\bar{r}_{u,i} - \bar{r}_u \right)^2} + \sqrt{\sum_{i \in I_{u,v}} \left(\bar{r}_{v,i} - \bar{r}_v \right)^2}} \tag{2}$$

$$\hat{r}_{u,j} = \frac{\sum_{v \in N \, Nu} S_{u,v} r_{v,j} + \sum_{v \in T \, Nu} t_{u,v} r_{v,j}}{\sum_{v \in N \, Nu} S_{u,v} + \sum_{v \in T \, Nu} t_{u,v}} \tag{3}$$

Experimental results of the method is based on three real data sets, "Film Trust, Flixster and Epinions", and shows that the proposed method is more effective in relation to the accuracy and coverage of recommendations than other approaches based on the MAE value.

Moreover, there is also previous work on exploiting the trust relation in CF recommender systems such as [7], [8] and [38]. In paper [7], the SNSCF method was proposed and employed four types of ties: friend, fan, follower and member. This identification of trustworthy users in a social network can improve the performance of traditional collaborative filtering systems. The researchers in [8] used the Film Trust website and studied the utility of the trust level in social networks. They obtained subjective trust values about a specific user from each rater, and then the raters with the highest trust values were selected. In [38], similar users are found by using trust metrics, and user preferences are predicted based on propagated trust metrics. The result shows that trust information works most effectively in terms of accuracy.

3.2 TCF-Based Social Network Friendship Relation Approaches

There are several relationships in social networks, but friendship is the most crucial one seen between users whereby links are continually created over time and the 'friend' relationship represents how the concerned users indulge in mutual interaction on social networks [7]. Traditional CF methods cannot distinguish between friends or strangers with similar interests [39]. All of the work of CF-based methods can be applied to nearest neighbours, who may be friends or strangers, but recently many researchers have used the friendships from social network as additional information to address this problem; for example, [9] and [30].

Authors of [30] proposed an approach to increase recommendation effectiveness by incorporating social networking information into CF. They are using the friendship relation "close friend" and user ratings to collect data from Cyworld, a social networking site, by distributing a survey to users to obtain data about their preferences and their social networking. In this approach they create four experiments: the first one used traditional collaborative filtering (TCF); in the second experiment utilised friends instead of nearest neighbours; in the third experiment, they combined nearest neighbours and friends in a new neighbourhood group, and in the fourth they incorporated levels of interaction to emphasise the influence of friends among neighbours. All experiments used a Pearson correlation coefficient (PCC) Eq(4) to calculate the similarity between two users and used Eq(5) to predict the ratings of the active user.

$$\omega(a,i) = \frac{\sum_j (v_{aj} - \bar{v}_a)(v_{ij} - \bar{v}_i)}{\sqrt{\sum_j (v_{aj} - \bar{v}_a)^2 \sum_j (v_{ij} - \bar{v}_i)^2}} \qquad (4)$$

$$P_{i,j} = \bar{v}_i + \frac{\sum_{a \in Raters}(r_{a,j} - \bar{v}_a) w_{(a,i)}}{\sum_{a \in Raters}|w_{(a,i)}|} \qquad (5)$$

The results indicate that utilising social network information is effective in CF for enhancing recommendation performance.

In [9], the researchers proposed an approach, namely community-based CF, to recommend users based on the preferences of friends or communities, an approach that consists of two steps. In the first step, they extract communities of users using the structure of two algorithms – attribute-clustering SAC1 and SAC2 proposed by [40] to discover the communities based on social link "friendship". In the second step they use Eq(6) to predict the rating of an active user u based on the users in the same community that exist alongside u.

$$\hat{r}_{u,i} = \bar{r}_u + \frac{\sum_{v \in C(u)} w(u,v) \cdot (r_{v,i} - \bar{r}_v)}{\sum_v |w(u,v)|} \qquad (6)$$

Where $\bar{r}_u$ Is the average rating of user u, $w(u,v)$ is the attribute similarity between u and v, $C(u)$ is the set of users in $u's$ community, and $r_{v,i}$ is the rating of user v on item i. The authors used two datasets from last.fm for recommendations of music artist and a delicious dataset for the recommendations of URL bookmark. They then used precision and recall respectively to evaluate the recommendation quality; the results showed that community-based methods yield better performance than a traditional CF approach that makes use only of the ratings. [31] explored the possibility of utilising user preferences and used explicit rating data and data from Facebook for the recommendation process. In order to understand users' preferences they use the content that users publish and voting on in their Facebook pages, and then using the cross domain concept [31] to achieve improvements in recommendation performance by using the cross domain data, which is integrated as input to the collaborative filtering. The authors collected two data sets from Facebook: explicit user ratings on 170 popular movies and Crawled Facebook accounts to obtain data from user profiles about the user preferences of "music items, TV series and movies" and also about the users' first and second degree friends (160 friends), who play a very important role in the selection of users' preferences. For the purposes of the experiment, different types of cross-domain data from Facebook was used, incorporating various examined methods to improve recommendations. To evaluate the results, the authors applied measures such as MAE, Precision, Recall, MAP and R-Precision. The results show that, when cross domain data is available, the recommendations based only on Facebook data are similar to those based on ratings, specifically when no data is available.

Some researchers found that the utilised friendship with a matrix factorisation framework can improve the recommendation process. For example, [33] proposed two social recommendation algorithms, SR1 and SR2, which impose social regularisation terms to constrain matrix factorisation. The results show that users' social friendship information can help to improve the prediction accuracy of recommender systems.

4. Conclusion

The purpose of this paper is to review and describe the state-of-the-art of CF and the role of social relationships in SN used to improve recommender systems. The social network environment includes helpful information such as social relations (friendship and trust between users) and social network contents (where users publish, vote, watch, etc.). It can be used as additional input for the collaborative filtering recommendation process. We discussed the different techniques used for recommendations based on two categories: TCF-based trust relation approaches and TCF-based friendship relation approaches. In recent years, many researchers have investigated how the elements of social networks can play an effective role in recommender systems and this has become a common trend in the research area.

Acknowledgment. This work is supported by the Ministry of Higher Education (MOHE) and Research Management Centre (RMC) at the Universiti Teknologi Malaysia (UTM) under Research University Grant Category (VOT Q.J130000.2528.02H99).

References

1. Marinho, L.B., et al., Social tagging recommender systems, in Recommender systems handbook. 2011, Springer. pp. 615-644.
2. Lops, P., M. de Gemmis, and G. Semeraro, Content-based recommender systems: State of the art and trends, in Recommender Systems Handbook. 2011, Springer. pp. 73-105.
3. Lynch, C. Personalization and recommender systems in the larger context: New directions and research questions. in Second DELOS Network of Excellence Workshop on Personalisation and Recommender Systems in Digital Libraries. 2001.
4. Zhou, X., et al., The state-of-the-art in personalized recommender systems for social networking. Artificial Intelligence Review, 2012. 37(2): pp. 119-132.
5. Adomavicius, G., et al., Incorporating contextual information in recommender systems using a multidimensional approach. ACM Transactions on Information Systems (TOIS), 2005. 23(1): pp. 103-145.
6. Pham, M.C., et al., A clustering approach for collaborative filtering recommendation using social network analysis. Journal of Universal Computer Science, 2011. 17(4): pp. 583-604.
7. Kim, S.-C., J.-W. Ko, and S.K. Kim, Collaborative Filtering Recommender System Based on Social Network, in IT Convergence and Services. 2011, Springer. pp. 503-510.
8. Golbeck, J., Generating predictive movie recommendations from trust in social networks. 2006: Springer.

9. Viennet, E. Collaborative filtering in social networks: A community-based approach. in Computing, Management and Telecommunications (ComManTel), 2013 International Conference on. 2013: IEEE.

10. Adomavicius, G. and A. Tuzhilin, Toward the next generation of recommender systems: A survey of the state-of-the-art and possible extensions. Knowledge and Data Engineering, IEEE Transactions on, 2005. 17(6): pp. 734-749.

11. Amatriain, X., et al., Data mining methods for recommender systems, in Recommender Systems Handbook. 2011, Springer. pp. 39-71.

12. Linden, G., B. Smith, and J. York, Amazon. com recommendations: Item-to-item collaborative filtering. Internet Computing, IEEE, 2003. 7(1): pp. 76-80.

13. Resnick, P., et al. GroupLens: an open architecture for collaborative filtering of netnews. in Proceedings of the 1994 ACM conference on Computer supported cooperative work. 1994: ACM.

14. Herlocker, J.L., et al. An algorithmic framework for performing collaborative filtering. in Proceedings of the 22nd annual international ACM SIGIR conference on Research and development in information retrieval. 1999: ACM.

15. Jin, R., J.Y. Chai, and L. Si. An automatic weighting scheme for collaborative filtering. in Proceedings of the 27th annual international ACM SIGIR conference on Research and development in information retrieval. 2004: ACM.

16. Deshpande, M. and G. Karypis, Item-based top-n recommendation algorithms. ACM Transactions on Information Systems (TOIS), 2004. 22(1): pp. 143-177.

17. Sarwar, B., et al. Item-based collaborative filtering recommendation algorithms. in Proceedings of the 10th international conference on World Wide Web. 2001: ACM.

18. Al Falahi, K., N. Mavridis, and Y. Atif, Social Networks and Recommender Systems: A World of Current and Future Synergies, in Computational Social Networks. 2012, Springer. pp. 445-465.

19. Hofmann, T. Collaborative filtering via gaussian probabilistic latent semantic analysis. in Proceedings of the 26th annual international ACM SIGIR conference on Research and development in information retrieval. 2003: ACM.

20. Hofmann, T., Latent semantic models for collaborative filtering. ACM Transactions on Information Systems (TOIS), 2004. 22(1): pp. 89-115.

21. Si, L. and R. Jin. Flexible mixture model for collaborative filtering. in Machine Learning - International Workshop then Conference. 2003.

22. Rennie, J.D. and N. Srebro. Fast maximum margin matrix factorization for collaborative prediction. in Proceedings of the 22nd international conference on Machine learning. 2005: ACM.

23. Salakhutdinov, R. and A. Mnih. Bayesian probabilistic matrix factorization using Markov chain Monte Carlo. in Proceedings of the 25th international conference on Machine learning. 2008: ACM.

24. Salakhutdinov, R. and A. Mnih, Probabilistic matrix factorization. Advances in neural information processing systems, 2008. 20: pp. 1257-1264.

25. Srebro, N. and T. Jaakkola. Weighted low-rank approximations. in Machine Learning - International Workshop then Conference. 2003.

26. Golbeck, J. and J. Hendler, Accuracy of metrics for inferring trust and reputation in semantic web-based social networks, in Engineering knowledge in the age of the semantic web. 2004, Springer. pp. 116-131.

27. Adamic, L.A. and E. Adar, Friends and neighbors on the web. Social networks, 2003. 25(3): pp. 211-230.

28. Kautz, H., B. Selman, and M. Shah, Referral Web: combining social networks and collaborative filtering. Communications of the ACM, 1997. 40(3): pp. 63-65.

29. Resnick, P. and H.R. Varian, Recommender systems. Communications of the ACM, 1997. 40(3): p. 56-58.

30. Liu, F. and H.J. Lee, Use of social network information to enhance collaborative filtering performance. Expert Systems with Applications, 2010. 37(7): pp. 4772-4778.

31. Shapira, B., L. Rokach, and S. Freilikhman, Facebook single and cross domain data for recommendation systems. User Modeling and User-Adapted Interaction, 2013. 23(2-3): pp. 211-247.

32. O'Doherty, D., S. Jouili, and P. Van Roy. Trust-based recommendation: an empirical analysis. in Submitted to: Proceedings of the Sixth ACM SIGKDD Workshop on Social Network Mining and Analysis SNA-KDD, Beijing, China, ACM. 2012.

33. Tavakolifard, M. and K.C. Almeroth, Social computing: an intersection of recommender systems, trust/reputation systems, and social networks. Network, IEEE, 2012. 26(4): pp. 53-58.

34. Guo, G., J. Zhang, and D. Thalmann, A simple but effective method to incorporate trusted neighbors in recommender systems, in User Modeling, Adaptation, and Personalization. 2012, Springer. pp. 114-125.

35. O'Doherty, D., S. Jouili, and P. Van Roy. Towards trust inference from bipartite social networks. in Proceedings of the 2nd ACM SIGMOD Workshop on Databases and Social Networks. 2012: ACM.

36. Massa, P. and P. Avesani. Trust-aware recommender systems. in Proceedings of the 2007 ACM conference on Recommender systems. 2007: ACM.

37. Loni, B., Toward the More Personalized Web by Extracting Knowledges From Social Networks.

38. Viennet, E. Community Detection based on Structural and Attribute Similarities. in ICDS 2012, The Sixth International Conference on Digital Society. 2012.

39. Ma, H., et al., Improving recommender systems by incorporating social contextual information. ACM Transactions on Information Systems (TOIS), 2011. 29(2): p. 9.

40. Ma, H., et al. Recommender systems with social regularization. in Proceedings of the fourth ACM international conference on Web search and data mining. 2011: ACM.

Time Series Clustering: A Superior Alternative for Market Basket Analysis

Swee Chuan Tan, Jess Pei San Lau

SIM University, School of Business
535A Clementi Road, Singapore
{jamestansc, pslau002}@unisim.edu.sg

Abstract. Market Basket Analysis often involves applying the de facto association rule mining method on massive sales transaction data. In this paper, we argue that association rule mining is not always the most suitable method for analysing big market-basket data. This is because the data matrix to be used for association rule mining is usually large and sparse, resulting in sluggish generation of many trivial rules with little insight. To address this problem, we summarise a real-world sales transaction data set into time series format. We then use time series clustering to discover commonly purchased items that are useful for pricing or formulating cross-selling strategies. We show that this approach uses a data set that is substantially smaller than the data to be used for association analysis. In addition, it reveals significant patterns and insights that are otherwise hard to uncover when using association analysis.

1 Introduction

Association analysis is sometimes known as affinity analysis or more specifically, association rule mining [1]. It is a method commonly used for Market Basket Analysis. In retail, Market Basket Analysis helps the retailer to find commonly purchased products so as to identify cross-selling opportunities, optimise store layout, and manage inventory [4, 8].

While association rule mining is currently the standard method for Market Basket Analysis, we argue that it is not always the most suitable method for analysing big market-basket data. When there is a large volume of sales transactions with high number of products, the data matrix to be used for association rule mining usually ends up large and sparse, resulting in longer time to process data as well as generation of trivial rules with little insight.

One possible solution to the above problems is to apply clustering on sales transaction data formatted as time series. Clustering of time series sales data may be a superior alternative (to association analysis) for Market Basket Analysis because of the following reasons:

- A data matrix required by association analysis becomes very large when there are many sales transactions and products. The data matrix also

T. Herawan et al. (eds.), *Proceedings of the First International Conference on Advanced Data and Information Engineering (DaEng-2013)*, Lecture Notes in Electrical Engineering 285, DOI: 10.1007/978-981-4585-18-7_28,
© Springer Science+Business Media Singapore 2014

becomes sparse when each transaction involves the sales of only a few products.

- On the other hand, time series clustering of sales transactions requires data to be summarised as time series, which can result in a substantially smaller data set that requires less time to process.

- Time series clustering can be used to identify products that are commonly purchased across a certain time period. Such patterns are otherwise hard to discover using association rule mining, which analyses transactions without temporal consideration.

So, why should clustering work? Our intuition is that the sales quantities of any two commonly purchased products should be positively correlated. When the sales quantities of these two products are observed over different time points, the time series of the sales quantities should be similar in their upwards and downwards temporal patterns. When clustering is applied, these two time series should be assigned to the same cluster. Conceptually, each cluster contains a set of items analogous to an itemset in association rule mining.

In this paper, we present a practical case of applying Market Basket Analysis on a real-world sales transaction data set using time series clustering, rather than using association rule mining. We find that time series clustering requires a very small data set that is a fraction of the size of data matrix required by association analysis. Yet, the clustering process helps to discover many sets of complementary parts, where each set of parts are used to make the same product. Such information is useful for cross-selling and pricing. Furthermore, the time series patterns can also be analysed to provide useful information for inventory control.

The rest of this paper is organised as follow. Section 2 reviews related work. Section 3 discusses a real-world case where Market Basket Analysis can be performed using time series clustering of sales data. Section 4 presents the results and discusses the insights gained. Finally, Section 5 concludes this paper and discusses implications of our work on future research directions.

2 Related Work

Most existing applications of Market Basket Analysis involve applying association rule mining on sales transaction records. The most popular story (or fable) about such applications is the unexpected discovery that diapers being commonly purchased with beer in the retail sector [8]. In the following, we review some existing applications and issues about applying association rule mining for Market Basket Analysis.

Three Example Applications of Market Basket Analysis: The first application is on analysing library circulation data [7]. The idea is to detect subject-classification categories that co-occur in loan records of books. The second application is in the domain of Bioinformatics where association rule patterns are discovered from gene expression data [6]. The third example is on extending the application of Market Basket Analysis form a single-store to a multiple-store environment [5]; while this extension from single to multiple stores helps identifying new insights, the number of transactions to be processed also increases considerably. This will increase data processing time when association rule mining algorithm is applied. In the following, we discuss some related issues.

Issues in using Association Rule Mining for Market Basket Analysis: The first issue is the generation of too many redundant rules. While there have been attempts in the data mining community to address this problem (e.g., see [2], [11]), research is still ongoing. The second issue concerns the "interestingness" of rules. Association rule mining tends to generate many trivial rules that are already known to the user. These rules often distract users from identifying rules that are really interesting and useful. The task of finding interesting association rules is an important topic in data mining research and has received a lot of attention [13]. The third issue is the long computation time required to discover large item-set patterns. To overcome this problem, some attempts have been made to develop more efficient algorithms (e.g., see [9]) or using sampling technique [3, 12] to reduce the amount of data to be processed.

As mentioned earlier, this work considers the application of cluster analysis instead of association rule mining to perform Market Basket Analysis. In the following, we review some topics related to time series clustering.

Cluster analysis: While association analysis aims at identifying groups of attributes (i.e., products, as in the context of Market Basket Analysis), cluster analysis focuses on identifying groups of similar records (i.e., sales transactions). One of the most commonly used clustering methods is K-Means clustering [10].

K-Means is a simple partitional clustering algorithm that clusters observations into K groups, where K is a user-specified input parameter for the number of clusters. K-Means assigns each observation to its nearest cluster. When the value of K is not known in advance, it is necessary to generate different clustering solutions using different values of K. The clustering solution that best describes the actual clustering pattern in the data can be found using cluster quality measures. One of the commonly used measures is the Silhouette coefficient. This is a good indicator of cluster quality because it gives an objective measure of the cohesion and separation of clusters in the clustering solution.

Most sales data sets are represented as a table recording sales transactions sequenced by time. To apply cluster analysis on such data, the format has to be converted into time series format. A time series is a sequence of observations ordered by time points [16]. Figure 1 shows a good example of time series, which is a sequence of monthly sales quantity of a part (D23) ordered by a company.

When clustering time series, the method can be applied directly on the raw time series data vectors; but if the data spans across many time points, then global features (such as skewness, kurtosis, seasonality, etc) be extracted for clustering in order to achieve scalability [16].

In the next section, we present a real-world case where Market Basket Analysis was performed using time series clustering of sales data.

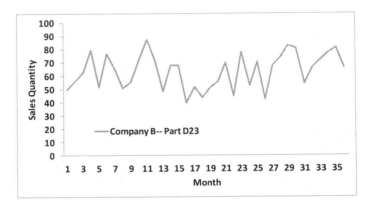

Figure 1: An example of time series showing the sales quantity of Part D23 being purchased by Company B over 36 monthly time points.

3   A Real-world Case of Market Basket Analysis using Clustering

Z Company[1] is a global supplier of raw parts to many manufacturing companies. According to the annual financial report by the Headquarter of Z Company, the outlook is not rosy due to fierce competition. The company sees the urgent need to be more rigorous in managing its productions, stock control and distributions.

In response to the situation, one of the business objectives is to better understand customer needs. While customer questionnaire survey may be used to achieve this goal, the approach is expensive, time consuming, and requires careful execution to get reliable and useful results. One alternative is to mine the existing data to discover useful customer information. Towards this end, this project aims to analyse the sales transaction data to discover customer purchasing behaviours.

Data Preparation: Z Company has about 56 thousand sales records of more than 700 products over the past three years. The sales transaction data set also contains more than ten transaction-related variables. For the purpose of time series clustering, we only require the Customer Number, Item Part Number, Sales Date and Quantity Ordered of each transaction as the data set for analysis.

---

1   The actual identity of this company cannot be disclosed due to confidentiality reasons.

During the data preparation stage, a small number of invalid records (mainly canceled orders) were identified and removed before further analysis. The data was sorted by Customer Number and then Item Part Number and Sales Date. The records are then aggregated by month and the total quantity ordered in each month was then computed. This resulted in a table with Customer Number, Item Part Number, Month Identifier (an integer than ranges from 1 to 36), and Monthly Sales Quantity. This table is then reformatted to have each Month Identifier being represented as a column capturing the sales quantity of the month. The resulting table ends up with about two thousand records over a period of 36 monthly time-points. Each record captures 36 monthly sales quantities of a part sold to a particular customer over the three-year period. Figure 2 gives an example of the data conversion from sales transaction format to time series format. Since the data spans over only 36 time points, the data size is small and we can apply clustering directly on the time series data.

Customer Num	Item Part Num	Month ID	Sales Qty
Company B	D23	1	73
Company B	D23	2	59
Company B	D23	...	...
Company B	D23	36	67
Company B	D24	1	15
Company B	D24	2	22
Company B	D24	...	...
Company B	D24	36	23
...	...	...	...

		Month ID			
Customer Num	Item Part Num	1	2	...	36
Company B	D23	73	59	...	67
Company B	D24	15	22	...	23
...	...	...	...	...	...

Figure 2: Data conversion from sales transaction table format to time series format.

One characteristic of the original sales data is that the ordered quantity of the same part may vary greatly. For example, Customer A may purchase 5000 pieces of one part, while Customer B may purchase only 500 pieces of the same part. When performing cluster analysis on such data, we want the algorithm to cluster time series records with similar temporal patterns without being concerned with the actual quantities being purchased. This is particularly important for clustering algorithms that make use of distance function to compute similarity between any two time series. In this project, we normalize each time series data record to a standard range, which results in assigning an equal weight to every time series when performing clustering.

4 Results and Discussions

We use K-Means clustering algorithm to generate a series of clustering solutions with different number of clusters, namely K = 35, K = 40, K = 45 and K = 50. The solution with 45 clusters gives a good score based on Silhouette coefficient; so this clustering solution is being selected for further analysis.

In the 45 clusters generated, we found many interesting sets of complementary parts that are used to make different types of products. These sets of complementary parts are previously not known to us, and the new information allows sales staff to derive useful strategies for pricing, sales and marketing.

Since most of patterns found are similar, we present a typical example pattern to highlight the key characteristics. Deriving from one of the clusters, Figure 3 shows a high correlation between the sales of Part D-12 and Part D-14 supplied to D Company. This company is an existing customer of Z Company and the company specialises in making components used in production of white goods. Further clarification with D Company confirms that these two parts are used to manufacture the same product. Hence, Parts D-12 and D-14 are complementary parts.

This finding is useful in two ways. Firstly, basic economics principle tells us that the demands of complementary products are positively correlated [4]. This implies that price reduction of Part D-12 may lead to an increase of its demand, as well as the demand of Part D-14. Hence, this piece of information can help Z Company in deriving appropriate pricing strategy of commonly purchased parts discovered in the clustering exercise.

Secondly, the information may be useful for stock control because Figure 3 shows that when one part has zero demand, it tends to lead the other part to have zero demand as well. For example, when there is no sales transaction of part D-14 at time-points 4, 16, and 28, there is also no sales transaction of D-12 at time-points 5, 17 and 29 respectively. This suggests that Z Company should be prudent in stocking up D-12 for D Company in the upcoming month when there is no sales transaction of D-14 in the current month. Note that this type of insight is not easy to discover when applying association rule mining.

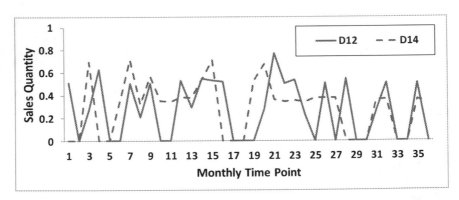

Figure 2: The sales quantities of Parts D-12 and D-14 are highly correlated in demand.

5 Concluding Remarks

This paper shows that time series clustering of sales data can help reveal interesting information about customers, which contributes towards competing on customer intelligence. For example, the discovery of complementary products can lead to new cross-selling opportunities and help more careful inventory control. The information can also be used to improve pricing.

This study also suggests that applying time series clustering on sales data could complement and reduce the cost of traditional customer surveys. We have shown that mining customer sales data has helped Z Company to obtain useful information (e.g., discovery of previously unknown complementary products) that is otherwise hard to obtain using questionnaire surveys.

One interesting point to note is the conversion of the original sales data to the time series format. The conversion leads to a significant reduction in the size of data to be mined. If the original sales data were to be plainly analysed using association rule mining, it would have been a 56000-by-700 data matrix (which is based 56000 transactions and 700 products); but when converted to time series, the data matrix size is reduced to about 2000 by 36. The total number of data cells needed is reduced by 544 times (i.e., 56000*700/(2000*36))! This reduction in data size has helped our analysis tremendously. Moreover, the problem of having sparse matrix is also resolved as a result of this conversion.

Our work shows that certain market basket data can be analysed more easily using time series clustering instead of association analysis. Since Market Basket Analysis has wide applicability in many domains, we believe some problems previously solved using association analysis can now be better tackled using time series clustering. Furthermore, more advanced clustering methods (e.g., [14, 15]) can be used to enhance the analysis.

References

1. Agrawal, R., Srikant, R.: Fast algorithms for mining association rules. Proceedings of the International Conference on Very Large Data Bases. pp. 487–499. (1994)

2. Baralis, E., Cagliero, L., Cerquitelli, T., Garza, P.: Generalized association rule mining with constraints. Information Sciences. Vol. (194). pp. 68-84. (2012)

3. Basel, A.M., Amer F.A., and Mohammed Z. Z.: A new sampling technique for association rule mining. Journal of Information Science. Vol. 35. pp. 358–376. (2009)

4. Blattberg, R.C., Kim, B-D., Neslin, S.A.: Database Marketing, Analyzing and Managing Customers. Series: International Series in Quantitative Marketing. Vol. 18. (2008)

5. Chen, Y.L., Tang, K., Shen, R.J., Hu, Y.H.: Market basket analysis in a multiple store environment, Decision Support Systems. Vol. 40(2). pp. 339–354. (2005)

6. Creighton, C., Hanash S.: Mining gene expression databases for association rules. Bioinformatics. Vol. 19 (1), pp.79–86. (2003)

7. Cunningham, S.J., Frank, E.: Market basket analysis of library circulation data. Proceedings of 6th International Conference on Neural Information Processing. pp.825–830. (1999)

8. Gutierrez, N.: Demystifying Market Basket Analysis. DM Review Special Report. (2006)

9. Luís C.: A scalable algorithm for the market basket analysis. Journal of Retailing and Consumer Services. Vol. 14(6). pp. 400–407. (2007)

10. MacQueen, J. B.: Some Methods for classification and Analysis of Multivariate Observations. Proceedings of 5th Berkeley Symposium on Mathematical Statistics and Probability 1. University of California Press. pp. 281–297. (1967)

11. Mafruz, Z.A., David, T., Kate, S.: Redundant association rules reduction techniques. International Journal Business Intelligent Data Mining. Vol. 2 (1). pp. 29–63. (2007)

12. Matteo, A. R., Eli, A.U.: Efficient Discovery of Association Rules and Frequent Itemsets through Sampling with Tight Performance Guarantees. Proceedings of European Conference on Machine Learning and Principles and Practice of Knowledge Discovery in Databases (ECML PKDD) 2012. pp. 25–41. (2012)

13. Stéphane, L., Olivier, T., Elie, P.: Association rule interestingness: measure and statistical validation. Quality measures in data mining. Springer. (2006)

14. Tan, S.C.: Simplifying and improving swarm-based clustering. In Proceedings of IEEE Congress on Evolutionary Computation. pp. 1–8. (2012)

15. Tan, S.C., Ting, K.M., Teng, S.W.: A general stochastic clustering method for automatic cluster discovery. Pattern Recognition. Vol. 44 (10). pp. 2786–2799. (2011)

16. Xiaozhe, W., Kate A.S., Rob, H., Damminda, A.:A Scalable Method for Time Series Clustering. Technical Report. Monash University. (2004)

To Boost Graph Clustering Based on Power Iteration by Removing Outliers

Amin Azmoodeh *, 1 , Sattar Hashemi *,2

*Department of Computer Science and Engineering, Shiraz University,Shiraz,Iran

1 azmoodeh@cse.shirazu.ac.ir, 2 s_hashemi@shirazu.ac.ir

Abstract. Power Iteration Clustering (PIC) is an applicable and scalable graph clustering algorithm using Power Iteration (PI) for embedding the graph into the low-dimensional *eigenspace* what makes graph clustering tractable.

This paper investigates the negative impacts of outliers on power iteration clustering and based on this understanding we present a novel approach to remove outlier nodes in a given graph what in turn brings significant benefits to graph clustering paradigms. In the original PIC algorithm, outliers have a high potential to be mis-clustered, and hence, they impress the output of PI embedding. As a result, the embedded space offered by PIC couldn't be deemed as a reasonable illustration of the graph in question. The statistical outlier detection method applied in this paper detects and removes outlier nodes in an iterative manner to enhance the embedded space for clustering.

Experiments on several datasets across different domain show the advantages of the proposed method compared to the rival approaches.

Keywords: Clustering, Graph Clustering, Outlier Detection, Spectral Clustering, Power Iteration Method, Eigenspace, Spectrum

1 Introduction

Clustering is a fundamental task of machine learning aims to organize the data in homogeneous groups. Graphs are complex data structure for both supervised and unsupervised learning .In the unsupervised context, there are many algorithms for clustering the graphs. Common objective of graph clustering algorithms is grouping the nodes that maximize the similarity, or links, intra the cluster(s) and in the same way, minimize inter the cluster connections .Ideal algorithm belongs to NP class. Global and local algorithms are two main trends of graph clustering methods to approximate admissible clustering result. In a global clustering, each vertex of the input graph is assigned a cluster in the output of the method, whereas in a local clustering, the cluster assignments are only done for a certain subset of vertices, commonly only one vertex. Graph clustering has a wide range of applications in social network analysis, machine vision, bioinformatics, VLSI design etc. [2].

Power Iteration Clustering (PIC) is an accurate and scalable algorithm that embeds a given graph into very low-dimensional space [1]. Each point in the embedded space

T. Herawan et al. (eds.), *Proceedings of the First International Conference on Advanced Data and Information Engineering (DaEng-2013)*, Lecture Notes in Electrical Engineering 285, DOI: 10.1007/978-981-4585-18-7_29,
© Springer Science+Business Media Singapore 2014

representing a node in the graph. After embedding, algorithm runs a general clustering algorithm, e.g. *k-means*, on the embedded space. But our PIC evaluations on different datasets show that outlier data affected the PIC accuracy. We present an improved version of PIC by detection and then rejection the outlier nodes of the graph. Because detection of outlier nodes on the graph structure is a challenge and could be confront from different views, we prefer to detect them in the embedded space .Experiments on datasets demonstrate the improvements of our approach.

This Work is presented as follows: in the Sec.2. a brief history of related works would described. We explain the PIC with more details in Sec. 3 and some popular methods for outlier detection presents in Sec. 4. Next, we propose our algorithm in Sec. 5 and show the experiments results in Sec. 6. Finally, we conclude about why our improved algorithm generates better result in more datasets and suggest some future work in Sec.7.

2   Related Work

Spectral Clustering [3-5] is a branch of global methods that using the *spectrum* of graph [6] for clustering the node instead the native space that data represented. A wide range of algorithms proposed based on spectral clustering [7, 8] that try to optimize a criteria on graph and find its solution in graph *eigenspace* .PIC and spectral clustering are similar in that both embed the data points in low-dimensional eigenspace and different in what is this embedding and how it is derived [1].

Basis spectral clustering algorithms based on the relaxation of some graph partitioning problems and uses the benefits of linear algebra. Shi and Malik [7] relaxed the *Ncut* objective and solution of the problem is the set of *k* first *eigenvectors* of graph adjacency matrix. They applied the *k-means* algorithm to cluster the rows of the joined *k* first *eigenvectors*, in which each row represents an object from a dataset. A k-way *Ncut* spectral clustering algorithm proposed by Ng et al [8], differs from [7] in the sense that the *eigenvectors* to be clustered are obtained from the graph Laplacian matrix. A complete history of spectral graph clustering algorithms could be found on [4, 16].

Despite the ability and scalability of PIC, there are few activity based on it. It might relate to the simple way of PI embedding .One of the few works is by Anh Pham et al. [17] to propose an improved version of PIC based on deflation technique for computing multiple orthogonal pseudo-eigenvectors of graph to overcome the inter-cluster collision problem. Encountering to noise is a challenge of clustering and some researches focus on this problem [18, 19].

3   Power Iteration Clustering (PIC) and Outlier Detection

3.1 Power Iteration

Power Iteration (PI) is a fast, simple and scalable method for calculating the dominant *eigenvector* of the graph. It embeds the graph into a very low-dimensional space. And this embedding is very useful domain for graph clustering .PIC and other spectral methods are similar in that both use *eigenspace* to cluster the nodes and different in the embedding space and the way of embedding.

Algorithm I   The original PIC algorithm [1]

Input: A row-normalized affinity matrix W and the number of clusters k

Pick an initial arbitrary and non-zero vector v^0

Repeat

$$\text{Set } v^{t+1} \leftarrow \frac{Wv^t}{\|Wv^t\|} \text{ and } \delta^{t+1} \leftarrow |v^{t+1} - v^t|$$

Increment t

Until $|\delta^t - \delta^{t-1}| \neq 0$

Use *k-means* to cluster points on v^t

Output: Clusters $C_1, \dots, C_k$

Shi and Malik [7] proved that *eigenspace* could be a very useful alternative space for clustering the graphs and in the *eigenspace, Normalized Cut* equation maximizes approximately .In a graph including k cluster, we have an *eigengap* between the kth and $(k+1)^{th}$ eigenvalue [8]. Experiments in [1] shows that a combination of *eigenvectors* with respect to their *eigenvalues* is an appropriate embedding space for graph clustering instead the eigenvector(s). Given an affinity matrix W, PI embeds it to an *eigenvalue-weighted* combination of eigenvector(s).

$$
\begin{aligned}
v^t &= cWv^{t-1} = \cdots = c'W^t v^0 \\
&= c_1 W^t e_1 + c_2 W^t e_2 + \cdots + c_n W^t e_n \\
&= c_1 \lambda_1{}^t e_1 + c_2 \lambda_2{}^t e_2 + \cdots + c_n \lambda_n{}^t e_n \quad\quad (1)\\
&\text{assumption}: v^0 = c_1 e_1 + c_2 e_2 + \cdots + c_n e_n
\end{aligned}
$$

Where v^t is an arbitrary non-zero vector at the beginning of iterations ($t = 0$) that it would converge to our slightly low-dimensional embedding space. c is a normalizing constant to keep v^t from getting too large. (Here $c = \frac{1}{\|Wv^t\|}$). λ_i Is the ith eigenvalue of the graph and e_i is its corresponding *eigenvector*.PI dose not converge to a significant vector finally. But observation shows that an intermediate state of v^t is useful for clustering. So we should define stopping criteria for the PI [1].

3.2   Outlier Detection

Outliers are data that do not perform expected normal behavior [9]. These data have various and also significant effect on machine learning algorithms and their outputs. Outlier detection has been found to be directly applicable in a large number of domains. In some application, e.g. medical and public health, fraud detection and network instruction detection. The objective of these problems is finding the outliers as an important pattern [10, 11]. But in some other machine learning algorithms outliers or anomalies are pattern that disturb the correctness of method.

Some probable causes of the outliers are malicious activity, instrumentation error, change in the environment and human error [9]. In our problem domain noise on spatial datasets and also out of common structure nodes in the inherently graph datasets could be detect as outlier. There are many approaches proposed for outlier detection like supervised [12], semi-supervised and unsupervised [13], nearest neighbor based, statistical based, information theory based and spectral decomposition based [9].

The underlying principle of any statistical approach for outlier detection arises from the following definition - *An outlier is an observation which is suspected of being partially or wholly irrelevant because it is not generated by the stochastic model assumed* [14]. Statistical outlier detection methods are model-based techniques. They estimate a statistical model and then evaluate the data instances with respect to how well they fit the model. If the probability of a data instance to be generated by this model is very low, the instance is deemed as an outlier. Two main categories of Statistical approaches for the training phase are:

i.   **Parametric**: techniques assume that the data is generated by a known distribution .Gaussian model, regression model and mixture of models [9] are some parametric statistical outlier detection approaches.

ii. **Non-Parametric**: in this category do not make any assumptions about the statistical distribution of the data .he most popular approaches for outlier detection have been histograms and finite state automata (FSA) [9].

4   Proposed Algorithm

Analyzing the outputs of PI shows that it suffers from the effect of outliers. In a Graph without outliers including k cluster, we have an *eigengap* between the kth and $(k+1)^{th}$ eigenvalue and it's the strength of the PI regarding to equation (1) because the output contains the k main *eigenvector*(s) respect to their *eigenvalue* value and effect of unimportant *eigenvectors* debilitate by their weak coefficients . But in presence of outlier(s), they allocate some *eigenvector*(s) to themselves that these *eigenvector*(s) are at the important side of *eigengap*.

Observations for a graph including outlier(s) shows: Outliers affects the embedding space by a big deviation from the general behavior of PI output for valid data (Figure 1). This behavior is due to fact that outliers dedicate some important eigenvector(s) to themselves and PI couldn't generate precise embedding for valid nodes. Output of PI for valid nodes can be model with a normal distribution approximately. Outliers have low *probability* in this distribution (Figure 2).

Results as diagrams shows after removing the outliers, if exists in the dataset, embedding space is more useful for clustering (Fig. 3, 4.).After the removing outliers,

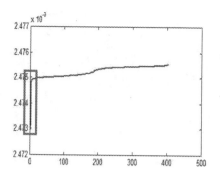

Fig. 1. Power Iteration output of UMBGBlog

(X-axis is data indices and Y-axis is PI Output)

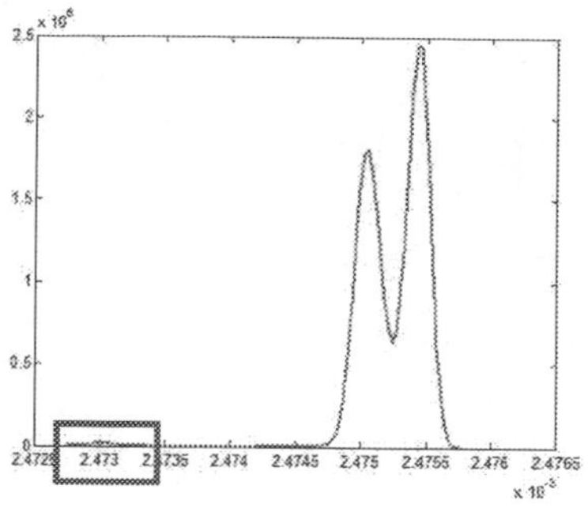

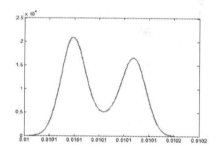

Fig. 2. Probability Density Function (PDF) for PI output of UMBGBlog Dataset.

embedding space is more separable, in other words *k-means* in the last step of algorithm could generate better clusters, because embedded value of nodes are more separable.

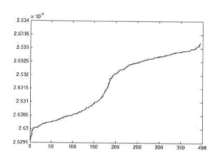

Fig. 3. PI Output UMBGBlog Dataset **after** removing Outliers

Fig. 4. PDF for UMBGBlog Dataset **after** removing Outliers

So in Boosted algorithm we apply a statistical parametric approach for outlier detection. By an iterative process we embed the dataset to low-dimensional space, detect the outliers (Algorithm II) and then delete them from dataset for next iteration.

Algorithm II   The Outlier Detection algorithm

Input: A vector $v_{n \times 1}$ and a Probability threshold p_0

Set $\mu = mean(v), \sigma = Statndard\ Deviation(v)$

Repeat

   if $NormalDistributionProbability(v_i, \mu, \sigma) < p_0$

         Add *i* to Outlier indices

For *i* =1 ... n

Output: Outlier indices

Algorithm III Boosted PIC algorithm

Input: A row-normalized affinity matrix W, number of clusters k and a Probability threshold p_0

Repeat

      Pick an initial arbitrary and non-zero vector v^0

      Repeat

            Set $v^{t+1} \leftarrow \dfrac{Wv^t}{\|Wv^t\|}$ and $\delta^{t+1} \leftarrow |v^{t+1} - v^t|$

            Increment t

      Until $|\delta^t - \delta^{t-1}| \neq 0$

      Set *OutlierIndices* $\leftarrow$ OutlierDetector (v^t, p_0)

      Delete the *OutlierIndices* from W

Until size (*OutlierIndices*) $\neq 0$

Use *k-means* to cluster points on v^t

Output: Clusters $C_1, \dots, C_k$

---

5 Experiments

We show the improvement of our algorithm (AlgorithmIII) compared to the PIC and other state-of-the-art spectral algorithms on a set of real datasets. They are labeled data and use for both classification and clustering tasks [1].

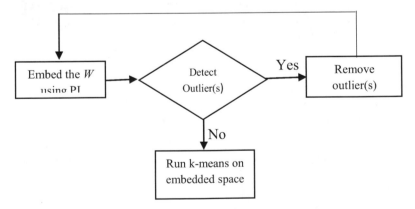

Fig 5. Flow diagram of Boosted algorithm (Algorithm III)

Table 1. Purity measurement of Boosted algorithm, PIC and spectral clustering algorithms on several real datasets. Bold numbers are the highest in its row if results are different.

Dataset	Cluster	Boosted Algorithm	Original PIC	NCut	NJW
Iris	3	**0.98**	**0.98**	0.67	0.76
PenDigits17	2	**0.76**	0.75	0.75	0.75
PolBooks	3	**0.87**	0.86	0.81	0.82
UMBGBlog	2	**0.95**	0.94	0.95	0.95
AGBlog	2	**0.96**	0.95	0.52	0.52
20ngA	2	**0.96**	**0.96**	**0.96**	**0.96**
20ngB	2	**0.95**	0.87	0.5	0.52
20ngC	3	0.69	0.69	0.61	0.63
20ngD	4	0.51	**0.54**	0.47	0.51
Average		**0.84**	0.83	0.69	0.71

Table 2. NMI measurement of Boosted algorithm, PIC and spectral clustering algorithms on several real datasets. Bold numbers are the highest in its row if results are different.

Dataset	Cluster	Boosted Algorithm	Original PIC	NCut	NJW
Iris	3	**0.93**	**0.93**	0.72	0.60
PenDigits17	2	**0.22**	0.20	0.20	0.20
PolBooks	3	**0.66**	0.62	0.57	0.51
UMBGBlog	2	**0.90**	0.71	0.74	0.73
AGBlog	2	**0.76**	0.74	0.01	0.01
20ngA	2	**0.92**	0.92	0.92	0.92
20ngB	2	**0.74**	0.55	0.01	0.08
20ngC	3	**0.49**	0.44	0.32	0.31
20ngD	4	**0.29**	**0.29**	0.23	0.29
Average		**0.65**	0.60	0.41	0.40

For the network datasets (Polbook's, UMBGBlog, AGBlog) the Affinity Matrix is simply $A_{ij} = 1$ if node i has a link to j. And for other datasets adjacency matrix calculated by cosine similarity between feature vectors: $\frac{x_i x_j}{\|x\|_i \|x\|_j}$. One of the main advantages of this similarity measurement is avoiding from parameter tuning like σ^2. In all experiments probability threshold is $p_0 = 0.02$.

Clustering results are evaluated by three measurements: *Purity, Normalized Mutual Information (NMI)* and *Rand Index (RI)* [15].We compare the results with original PIC,

Table 3. RandIndex measurement of Boosted algorithm, PIC and spectral clustering algorithms on several real datasets. Bold numbers are the highest in its row if results are different.

Dataset	Cluster	Boosted Algorithm	Original PIC	NCut	NJW
Iris	3	**0.97**	**0.97**	0.77	0.79
PenDigits17	2	**0.65**	0.63	0.63	0.63
PolBooks	3	**0.88**	0.86	0.84	0.83
UMBGBlog	2	0.91	0.90	**0.91**	**0.91**
AGBlog	2	**0.92**	0.91	0.50	0.50
20ngA	2	0.76	0.76	0.76	0.76
20ngB	2	**0.92**	0.78	0.50	0.50
20ngC	3	0.69	**0.73**	0.67	0.68
20ngD	4	0.58	0.65	0.63	**0.68**
Average		**0.80**	0.79	0.69	0.69

Normalized Cut (*NCut*) [7] and the Ng-Jordan-Weiss (NJW) algorithm [8] (Table1-3).

$$Purity(\Omega, C) = \frac{1}{N}\sum_k \max|\omega_k \cap c_j|$$

Where $\Omega = \{\omega_1, \omega_2, ..., \omega_k\}$ is the set of clusters and $C = \{c_1, c_2, ..., c_J\}$ is the set of classes.

$$NMI(\Omega, C) = \frac{I(\Omega, C)}{\sqrt{H(\Omega) + H(C)}}$$

Where $I(\Omega, C)$ is mutual information and $H(\Omega)$ is entropy of Ω [15].

$$Rand\ Index = \frac{TP + TN}{TP + FP + FN + TN}$$

For *iris* and *20ngA* dataset algorithm didn't find any outlier data and so results are equal to original PIC. Also for *20ngD* Boosted algorithm generate lower results than PIC. Regarding to its other method's low-results seems that this dataset hasn't proper data structure for generating good graph clustering result. For other datasets experiments show better results.

Rejected data points in all the datasets are less than 1% of dataset size. For example only one sample rejected in *20ngB* dataset but a big improvement achieved. But deciding about the cluster label of rejected nodes could be an improvement for increasing the measurements.

Analyzing the PI output for network datasets, e.g. *UMBGBlog* compared to other non-network datasets that their affinity matrix obtained by distance metrics, shows that outliers in these datasets play a bigger role to deviate the output of PI. Unusual and abnormal behavior in these datasets is more distinct and isolated from the valid nodes in affinity matrix and therefore *eigenvalue(s)* of these node get a bigger value .But in non-network datasets although outliers affect the correctness of PI from generating *eigenvalue-weighted* combination of our desired *eigenvectors*, their small link with valid nodes avoid big deviation from ideal state.

6 Conclusion and Future work

In the preceding sections we proposed an improved version of Power Iteration Clustering (PIC). By analyzing the behavior of PIC we found that outliers affect the result of it. Their two main effects on the result are described in the following:

i.   Decreasing the accuracy of embedding

Regarding to equation1 in presence of Outliers some important terms $\lambda_k{}^t e_k$ of result should dedicated to the *eigenvectors* that described the behavior of Outliers. Unfortunately their corresponding *eigenvalues* are at the important side of *eigengap* and PI should contain them in the embedded space with high weight.In other words, in presence of Outliers PI is not precise and careful for embedding of valid nodes .So embedding has less detail and some nodes would finally miss-cluster.

ii.   Decreasing the accuracy of clustering on embedding

Because in many cases the structure of outliers is very different from the structure of valid nodes of graph, their *eigenvalue(s)* have extra weight and so regarding to equation1, we have two quite different part in embedding .one part described the outliers in the new space and another part dedicate to the valid nodes of graph .So in last step of PIC, *k-means* couldn't cluster the nodes carefully and some of centers would converge to the outliers.

In this paper we empirically analyze the effects of outliers and present an improved version of PIC to overcome this undesirable state. Analyzing the effect of outliers mathematically and combine it with PI mathematics could generate a significant framework for graph clustering. Investigating the results of other outlier detection algorithms could be another approach. Also derive a mathematical criteria for p_0 could help to make proposed algorithm more transparent.

On the other, rejected data in our algorithm might be important and useful. So proposing an algorithm to decide about their proper cluster label after rejecting them from dataset could avoid us from losing useful information.

References

1.   Frank Lin,William.W.Cohen : "Power Iteration Clustering" ,International Conference on Machine Learning.pp.655-662,2010

2. Satu Elisa Schaeffer : "Graph clustering, survey " ,Computer Science Review,2007, pp. 27-64
3. Benjamin Auffarth, "Spectral Graph Clustering ", Tech.Rep. Polytechnic University of Catalunya.2007
4. Ulrike von Luxburg. "A tutorial on spectral clustering". Max Planck Institute for Biological Cybernetics.Tech.Rep.2007
5. D. Verma, M. Meila, "A comparison of spectral clustering algorithms", Tech.Rep, Department of CSE University of Washington Seattle, WA, 98195–2350, 2005
6. Chung, F. R. K. "Spectral graph theory". American Mathematical Society.1996
7. Shi, J. and Malik, J. "Normalized cuts and image segmentation". IEEE Transactions on Pattern Analysis and Machine Intelligence, 2000, 888 – 905.
8. Ng, A., Jordan, M., and Weiss.Y. "On spectral clustering: analysis and an algorithm". Advances in Neural Information Processing Systems 2002 , pp. 849 - 856
9. V. Chandola, A. Banerjee, and V. Kumar. "Anomaly Detection: A Survey". ACM Computing Surveys, Volume 41 Issue 3, July 2009
10. Suzuki, E.; Watanabe, T.; Yokoi, H.; Takabayashi, K., "Detecting interesting exceptions from medical test data with visual summarization". IEEE International Conference on Data Mining, Nov 2003, pp.315-322
11. Richard J. Bolton and David J. Hand. "Unsupervised profiling methods for fraud detection ". In Proceedings of the Conference on Credit Scoring and Credit Control VII, Edinburgh, UK,2001
12. Sun, P. and Chawla, S. "On local spatial outliers ". In Proceedings of 4th IEEE International Conference on Data Mining. 2004, pp.209-216.
13. Vinueza, A. and Grudic, G. "Unsupervised outlier detection and semi-supervised learning ".Tech. Rep. Univ. of Colorado at Boulder. May 2004
14. Anscombe, F. J. and Guttman, I. "Rejection of outliers". Technimetrics. Vol. 2, No. 2.pp. 123-147. May 1960
15. Manning, Christopher D., Raghavan, Prabhakar, and Schtze, Hinrich. "Introduction to Information Retrieval ". Cambridge University Press, 2008.
16. Mariá C.V. Nascimento and André C.P.L.F. de Carvalho. "Spectral methods for graph clustering – A survey". European Journal of Operational Research,Vol.211.No.2,pp. 221–231,June 2011
17. Anh Pham et al. "Deflation based power iteration clustering ". Applied Intelligence .February 2013 [Online].Available : http://link.springer.com/article/10.1007/s10489-012-0418-0 [Accessed Aug. 11, 2013]
18. Zhenguo Li; Jianzhuang Liu; Shifeng Chen; Xiaoou Tang, "Noise Robust Spectral Clustering," Computer Vision, 2007. ICCV 2007. IEEE 11th International Conference on , vol., no., pp.1,8, 14-21 Oct. 2007
19. Sivaraman Balakrishnan , Min Xu , Akshay Krishnamurthy , Aarti Singh , "Noise thresholds for spectral clustering," Advances in Neural Information Processing Systems 2011 , pp. 954 - 962

Weight-Based Firefly Algorithm for Document Clustering

Athraa Jasim Mohammed[1], Yuhanis Yusof[2], and Husniza Husni[2],

School of Computing, College of Arts and Sciences, Universiti Utara Malaysia,
06010 Sintok, Kedah, Malaysia
[1]autoathraa@yahoo.com
[2]{yuhanis, husniza}@uum.edu.my

Abstract. Existing clustering techniques have many drawbacks and this includes being trapped in a local optima. In this paper, we introduce the utilization of a new meta-heuristics algorithm, namely the Firefly algorithm (FA) to increase solution diversity. FA is a nature-inspired algorithm that is used in many optimization problems. The FA is realized in document clustering by executing it on Reuters-21578 database. The algorithm identifies documents that has the highest light intensity in a search space and represents it as a centroid. This is followed by recognizing similar documents using the cosine similarity function. Documents that are similar to the centroid are located into one cluster and dissimilar in the other. Experiments performed on the chosen dataset produce high values of Purity and F-measure. Hence, suggesting that the proposed Firefly algorithm is a possible approach in document clustering.

Keywords: Firefly algorithm, partitional clustering, hierarchical clustering, text clustering.

1 Introduction

Clustering is a process of grouping documents into a cluster. Similar documents are grouped in the same cluster and dissimilar documents in another cluster [1]. Furthermore, it is an unsupervised learning that does not require pre-defined classes for the intended documents. In general clustering algorithms can be classified into two categories; hierarchical clustering and partition clustering algorithms [2, 3].

Hierarchical clustering algorithm is a technique to build a hierarchy of clusters. There are two approaches of this technique [1]. The first approach is agglomerative hierarchical clustering which started by working from bottom to top, meaning that every object in a single cluster is merged based on similarity between clusters [4]. The second approach is the divisive hierarchical clustering which operates from top to bottom. In this approach, objects in cluster are separated using one of the partition clustering techniques. In undertaking a hierarchical clustering, one does not require to determine the number of output clusters [5]. On the other hand, Partition clustering returns an unstructured set of clusters. Partition techniques have some drawbacks; they require a pre-defined number of cluster and an initial cluster centers. This paper

T. Herawan et al. (eds.), *Proceedings of the First International Conference on Advanced Data and Information Engineering (DaEng-2013)*, Lecture Notes in Electrical Engineering 285, DOI: 10.1007/978-981-4585-18-7_30,
© Springer Science+Business Media Singapore 2014

uses Weight-based Firefly Algorithm (WFA) in a divisive hierarchical clustering to overcome such problems. The proposed WFA attempts to overcome the problem of trapping in local optima (of identifying the k number of clusters) by using Firefly Algorithm (FA). The total weight of document is assigned as the initial light intensity of a firefly. Furthermore, the attractiveness between documents is later undertaken based on Euclidean distance. The proposed approach then finds the center of cluster that produces the highest brightness.

The remainder of the paper is structured as below: In section 2, we provide related work. Section 3, present the proposed Weight-based Firefly Algorithm (WFA) approach. The evaluation is presented in section 4, followed by the conclusion in section 5.

2 Related Works

Text clustering is a useful technique for organizing text documents as clusters, the similar text is in one group and dissimilar text is in another group [6]. One of the most famous clustering techniques is the K-means which is classified as a type of partitioning clustering. Another well-known algorithm is the Principal Direction Divisive Partitioning (PDDP) for divisive hierarchical clustering [7]. Divisive clustering is one type of hierarchical clustering that it is used to construct a hierarchy of clusters. One drawback of divisive clustering is its low performance, hence, leading to the combination with Intelligent Swarm methods. Particle Swarm Optimization algorithm is one type of Intelligent Swarm methods that it is integrated with divisive clustering approach [8]. The experiment result indicates high performance and robustness with lower running time.

The K-means algorithm has been widely utilized in the domain of clustering. Nevertheless, due to its random initial centroids, work has been reported to fall into local optima. Such situation has led researchers to integrate K-means with optimization techniques such as Particle Swarm Optimization algorithm (PSO) [9]. However, the result of the proposed sequential approach was no better than having PSO as an individual clustering method. Despite such result, it was learned that for the Wine and Iris [10] datasets, the combination of K-means and PSO generate better results.

Another type of Intelligent Swarm methods is the Firefly Algorithm (FA). Firefly algorithm was developed by Xin-She Yang in 2007 at Cambridge University. It has two important issues, the light intensity and the attractiveness. For optimization problems, the light intensity, I, of a Firefly at a particular location, x, can be determined by $I(x) \propto f(x)$. The attractiveness β is relative. It changes depends on the distance between two fireflies [11]. Firefly algorithm is utilized in many optimization problems such as image processing which it is used to search for multiple thresholds [12]. Furthermore, the performance of Firefly algorithm was also studied in numerical clustering [13]. The objective function of such work was to minimize the distance between a center and the documents. The result was efficient, robust, and reliable that generates optimal cluster centers compared with two nature inspired algorithms;

Particle Swarm Optimization (PSO) and Artificial Bee Colony (ABC). The proposed FA solved the local optima problem but the number of k cluster was pre-defined.

There has also been effort in integrating FA and K-means for data clustering. The FA was employed to find the center of clusters while the K-means was utilized in finding its clusters [14]. The proposed hybrid KFA minimized the intra-cluster distance and reduced the clustering error compared to the K-means, PSO and KPSO. However, this hybrid approach was implemented only on numerical data sets. In addition, a hybrid of Firefly Algorithm and K-harmonic Means algorithm was also undertaken on numerical data sets [15]. In this paper, we propose a clustering technique utilizes FA to identify the initial cluster center using an objective function that is formulated based on the term frequency of a document. The FA will also identify documents that will be grouped into the identified centroid.

3 Method

Theproposed Weight-based Firefly Algorithm consists of the following steps:

3.1 Data Preprocessing

The data preprocessing is an important phase in web mining as it extract various information from websites and represent it as a database. This phase includes seven steps and they are follows. First, extraction of important text is done, the tags that contains title and body. Second, the selected text is cleaned from digit and special characters. Each text from documents that was cleaned is split to words. All of the words in each document are analyzed for its length. If the words length is less than two then it will be removed because it is useless in search process otherwise remain. Fifth, the stop words is removed from the list of words and this includes words like the, on, in, etc. The sixth step is on utilizing a stemmer algorithm that transforms a word into its root. Lastly, the word frequency is calculated based on its occurrence in the document [17].

3.2 Development of Vector Space Model

Vector space model VSM has been widely used to represent data in document clustering. Each document is represented as a vector in the vector space [6]. The VSM includes several steps: In the first step, a term-frequency (TF) database is created. The rows include terms (words) and the columns represent the documents. The intersection between row and column contain the occurrence of each terms (term frequency) [16]. Secondly, a normalized matrix is created that where the occurrence of terms is normalized between (0, 1) through calculate the length of each document by using equation 1 [17]:

$$Length = \sqrt{\sum_{i=1}^{m} V_i(d)^2} \tag{1}$$

Where, m is the number of term in a collection, V is the term frequency, d is the document. Then, normalized term frequency is divided on document length by using equation 2:

$$EN = \frac{TF}{Length} \tag{2}$$

Where, TF is term frequency. Then the weighting matrix tf-idf is created to find the weight of each term in the document [17]. In order to do so, we need to calculate the inverse documents frequency idf by using equation 3:

$$idf = \log N/dft \tag{3}$$

Where, N is the total number of documents in the collection, dft is the number of documents in the collection that contain a term. Following to that, we determine the weight of a term by using equation 4 [17]:

$$tf - idf_{t,d} = tf_{t,d} * idf_t \tag{4}$$

The total weighted of each document is obtained using equation 5:

$$total\ weight_{d_j} = \sum_{i=1}^{m} tf - idf_{t_i,d_j} \tag{5}$$

Where, j is the number of documents, i is the number of the terms.

3.3 Text Clustering

Firefly algorithm has two important features, (a): the light intensity and (b): the attractiveness. In an optimization problems, the light intensity I of a firefly at a particular location x can be determined using objective function f(x). The attractiveness β is relative. It changes depending on the distance between two fireflies. The attractiveness β formula is shown in equation 6 [11]:

$$\beta = \beta_0 exp^{(-\gamma r_{ij}^2)} \tag{6}$$

The movement of one document i to another document j is determined based on equation 7:

$$X^i = X^i + \beta * (X^j - X^i) + \alpha \qquad (7)$$

Where, Xi is the position of first document; Xj is the position of second document in training data set.

In this paper, we propose that each document is represented by a single firefly and the total weight of the document is the initial brightness I of the firefly. The highest brightness is indicated by the highest total weight value and represents the best point. The best point hence indicates the center of a cluster. The distance between two documents is then calculated using Euclidean distance function [14] as shown in equation 8:

$$Euclidean\ distance(X_i, X_j) = \sqrt[2]{(X_i - X_j)^2} \qquad (8)$$

The WFA is used to determine document in the collection that has the highest brightness and is later used as centroid of cluster. Once this is done, we then find the similar documents for the centroid using cosine similarity matrix. Documents having high similarity value is located in the first cluster while the ones with lower values in a second cluster. Such an approach requires threshold. The cosine intra-similarity is defined in equation 9 [5]:

$$Intra - sim(C_i) = \sum_{j=1}^{m}(X_j * V_j) \qquad (9)$$

Where, Ci is the output cluster, j is the number of terms in the collection, Xj is the documents in cluster C, Vj is the center of cluster.

The second cluster that contains documents with low similarity value against the centroid will again enter the text clustering phase (weight-based firefly algorithm to find new centroids). The process of finding a centroid and its cluster continues until it reaches the last document. The proposed Weight-based Firefly Algorithm is shown in Figure 1.

```
Generate Initial population of firefly randomly x^i where i=1, 2, .., n,
n=number of fireflies (documents).
Initial Light Intensity, I=total weight of document.
Define light absorption coefficient γ, initial γ=1
Define the randomization parameter α, α=0.7
Define initial attractiveness β_0 = 1.0
While t < N
For i=1 to N
For j=1 to N
If (total weight I_i < total weight I_j) {
Calculate distance between i, j using equation 8.
Calculate attractiveness using equation 6.
Move document i to j using equation 7.
Update light intensity I^i = I^i + β
End For j
End For i
Loop
Rank to find best document.
```

Fig.1. The proposed Weight-based Firefly Algorithm (WFA)

4 Evaluation

The proposed WFA is tested on a standard text classification dataset which is the Reuters-21578 [18]. In the undertaken experiment, data is divided into three sets; two parts are used for training and one part is used for testing. The training data is used to build the clusters of documents and the test data is used to measure the clustering quality of the proposed algorithm. Two data collection from Reuters-21578 was chosen which are the RE0 and RE1. Description of the chosen documents is presented in Table I [19].

Table I. Description of Data

Data Set	No. of Documents	Classes	No. of Training data	No. of Testing data	No. of Terms
RE0	201	13	134	67	2149
RE1	192	25	128	64	2156

As for the evaluation, the Classification Error Percentage (CEP) [12], Purity and F-measure [19] are used as performance measurement. CEP is calculated by counting the number of documents that is wrongly classified. It is shown in equation 10 [13]:

$$CEP = \frac{number\ of\ wrong\ classified\ documents}{total\ number\ of\ documents\ in\ test\ data\ set} * 100 \tag{10}$$

Purity on the other hand is a measure of clustering quality [19]. The purity depends on the maximum number of documents in class Ωk and in cluster Cj respectively. The equation is in 11 [19]:

$$P(\Omega_k, C_j) = \text{Max}_k \ |\Omega_k \cap C_j| \tag{11}$$

The cluster purity calculated as in equation 12[19]:

$$Purity = \sum_{\Omega_k \in \{\Omega_1, \dots, \Omega_c\}} \frac{P(\Omega_k, C_j)}{N} \tag{12}$$

To measure the accuracy, the F-measure [19] is employed and it depends on the recall and precision values [20]. The total F-measure is the summation average of F-measure for all class. The equation to collect maximum value of F-measure is in equation 13 [19]:

$$F(\Omega_k) = \frac{\text{max}}{C_j \in \{C_1, \dots, C_k\}} \left(\frac{2 * R(\Omega_k, C_j) * P(\Omega_k, C_j)}{R(\Omega_k, C_j) + P(\Omega_k, C_j)} \right) \tag{13}$$

Where: $R(\Omega_k, C_j)$ is recall measure and $P(\Omega_k, C_j)$ is precision measure. The equation for total F-measure is in equation 14 [19]:

$$Total\ F - measure = -\sum_{k=1}^{C} \frac{|\Omega_k|}{N} * \max(F(\Omega_k)) \tag{14}$$

The CEP, Purity and F-measure results are shown in Table II. Form the table, it is noted that the first dataset, RE0, has CEP of 23.9, purity of 0.7089 and F-measure of 0.5535. As for the RE1, the CEP is equal to 21.9, while its purity and F-measure are 0.7890 and 0.5768 respectively. Based on literature, it is learned that a good clustering is when the CEP value is low and the F-measure and Purity values are high [19]. Hence, the obtained result of CEP which is less than 30, and Purity and F-measure values higher than 0.5 indicates that the Firefly algorithm produces good clusters.

Table II. Result of WFA

Data Sets	CEP	Purity	F-measure
RE0	23.880	0.7089	0.5535
RE1	21.875	0.7890	0.5768

5 Conclusion

A new approach for document clustering is presented using a meta-heuristics algorithm which is the Firefly. In this paper, we propose that each document is represented by a single firefly and the total weight of a document is the initial brightness of the firefly. The point (document) with the highest brightness is later identified as the centroid. Such an operation is assumed to be a new approach in utilizing Firefly in document clustering. Such an approach operates by defining that the significance of total weight of documents is equal to the light intensity in firefly. In theory, the firefly which has the highest light will attract other fireflies. Hence, in this work, the proposed WFA uses the highest total weight of document to represent the centroid and attract other documents based on similarity between centroids and documents. The performance of the proposed WFA is tested on a standard text classification dataset which is the Reuters-21578 and is evaluated using three performance measurements which are the Classification Error Percentage (CEP), Purity and F-measure. The obtained results indicated that the proposed Weight-based Firefly Algorithm would become a competitor in the area of data clustering.

Additionally, the proposed WFA can be operationalized in the form of a search engine. It could be used to optimize the organization of index file structures into clusters. Hence, may lead to a better precision and recall of a search engine and reduces its computational time.

References

1. Das, S., Abraham, A., Konar, A.: Metaheuristic Clustering, Springer, Heidelberg (2009).
2. AnithaElavarasi, S., Akilandeswari, J., Sathiyabma, B.: A survay on Partition Clustering Algorithms. In: International journal of Enterprise Computing and Business Systems, vol. 1, issue 1, (2011).
3. Ye, N., Gauch, S., Wang, Q., Luong, H.:An Adaptive Ontology based Hierarchical Browsing System for CiteSeerX. In: Second International Conference on Knowledge and Systems Engineering (KSE), pp. 203–208, IEEE, (2010).
4. Wilson, II., Doots, B., Millward, A. A.: A Comparison of Hierarchical and Partitional Clustering Techniques for Multispectral Image Classification. vol.3, pp. 1624-1626, (2002).
5. Xu, Y.: Hybrid clustering with application to web mining. In: Proceedings of the International Conference on Active Media Technology (AMT 2005), pp. 574–578,IEEE, (2005).
6. Aliguliyev, R. M.: Clustering of Document Collection- A Weighted Approach. In: Expert Systems with Applications,vol. 36, issue 4, pp. 7904–7916,Elsevier, (2009).
7. Boley, D.: Principal Direction Divisive Partitioning. In: Data Mining and Knowledge Discovery, vol. 2, issue. 4, pp. 325 – 344, ACM, (1998).
8. Feng, L., Qiu, M.H., Wang, Y.X., Xiang, Q.L., Yang, Y.F., Liu, K. A.: Fast Divisive Clustering Algorithm Using an Improved Discrete Particle Swarm Optimizer. In:Pattern Recognition Letters, vol. 31, issue. 11, pp. 1216-1225,Elsevier, (2010).
9. Rana, S., Jasola,S., Kumar,R.: A Hybrid Sequential Approach for Data Clustering using K-means and Particle Swarm Optimization Algorithm. In: International Journal of Engineering, Science and Technology,vol. 2, No. 6, pp. 167-176, (2010).
10. Bache, K., Lichman, M.: UCI Machine Learning Repository [http://archive.ics.uci.edu/ml]. Irvine, CA: University of California, School of Information and Computer Science,(2013).
11. Yang,X. S.: Nature-inspired Metaheuristic Algorithms, 2nd ed., Luniver press, United Kingdom, (2011).
12. Horng, M. H., Jiang, T. W.: Multilevel Image Thresholding Selection based on theFirefly Algorithm. In: 7th International Conference on Ubiquitous Intelligence & Computing and 7th International Conference on Autonomic & Trusted Computing (UIC/ATC), pp. 58 – 63, IEEE, (2010).
13. Senthilnath, J., Omkar, S. N., Mani, V.: Clustering Using Firefly Algorithm: Performance Study. In: Swarm and Evolutionary Computation, vol. 1, issue. 3, pp. 164-171, Elsevier, (2011).
14. Hassanzadeh, T., Meybodi, M. R.:A New Hybrid Approach for Data Clustering Using Firefly Algorithm and K-means. In: 16thIEEECSI International Symposium on Artificial Intelligence and Signal Processing (AISP), pp. 007 – 011, (2012).
15. Abshouri, A. A., Bakhtiary,A.: A New Clustering Method Based on Firefly and KHM. In: Journal of Communication and Computer, vol. 9, pp. 387-391, (2012).
16. Xu, G., Zhang,Y., Li, L.: Web mining and social networking, Techniques and application, New York, Springer, (2011).
17. Manning, C. D., Raghavan,P., Schütze,H.: Introduction to Information Retrieval, 1 ed., Cambridge University Press, (2008).
18. Lewis,D.: The reuters-21578 text categorizationtest collection, 1999.[Online].Available:http://kdd.ics.uci.edu/database/reuters21578/reuters21578.html.
19. Murugesan, K, Zhang,J.: Hybrid Bisect K-means Clustering Algorthim. In: IEEE International Conference on Business Computing and Global Informatization (BCGIN), pp. 216 – 219, IEEE, (2011).
20. Meghabghab, G., Kandel, A.: Search Engines ,Link Analysis ,and User's Web Behaviour, Berlin Heidelberg: Springer-Verlag, (2008).

Part III
Image and Video Processing

A Comparative Study of Tree-based Structure Methods for Handwriting Identification

Nooraziera Akmal Binti Sukor, Azah Kamilah Muda, Noor Azilah Muda, Choo Yun Huoy

Faculty of Information and Communication
UniversitiTeknikal Malaysia Melaka, Melaka, Malaysia
mai_sukor@yahoo.com; azah@utem.edu.my; azilah@utem.edu.my;
huoy@utem.edu.my

Abstract. Handwriting Identification is a process to determine the author of the writing and it involves some of process. Classification process is a final stage of Handwriting Identification process where it will analyze the classification accuracy and based on the number of features selected. In this study, classification process was conducted using various tree-based structure methods. Tree-based structure method is one of the classification methods where it is able to generate a compact subset of non-redundant features and hence improves interpretability and generalization. However its focus is still limited especially in Writer Identification domain. Several of tree-based structure selected and performed using image dataset from IAM Handwriting Database. The results also analyze and compared of each methods of Writer Identification. Random Forest Tree classifier gives the best result with the highest percentage of accuracy followed by J48, Random Tree, REP Tree and Decision Stump.

Keywords: feature selection; writer identification; tree-based structure; comparative study.

1   Introduction

Feature selection is a common technique used in the field of pattern recognition, machine learning and data mining and it has become the focus of researchers for a long time. Feature selection aims to obtain the minimal sized subset of features, increasing the classification accuracy, reduces computational complexity and help finding easier algorithmic models [1] [2]. Besides, it also removes redundant and noisy features from data sets and selects the most significant features for creating robust learning models [3].

Handwriting is a behavioral biometric [11] [12] [13] characteristic, and is considered unique for each of individual. Hence, based on the individuality of the handwriting, the authors of the handwriting can be identified. The process of determining the author of sample handwriting comes from a set of writers also known as Writer Identification (WI) [14]. This process based on the handwriting, ignores the meaning of the words. Feature selection process that takes place in WI domain was conducted by

T. Herawan et al. (eds.), *Proceedings of the First International Conference on Advanced Data and Information Engineering (DaEng-2013)*, Lecture Notes in Electrical Engineering 285, DOI: 10.1007/978-981-4585-18-7_31,

[15] [16] and [17]. The main issue in WI is how to acquire the features that reflect the authors of handwriting [11-13] and [18-30].

This paper proposes to evaluate the performance of various tree-based structures for writer identification on small-sized data sets, where the number of features is less than 20 features. The remainder of this paper is organized into six sections. In Section 2, explains about WI domain details. Section 3 explains each selected feature selection algorithms. In Section 4 and 5, experimental setup is presented and the results are shown. In the last section, conclusion and future works are proposed.

2 Handwriting Identification

Handwriting is a behavioral biometric [11] [12] [13] characteristic, and consider being unique for each of individual. Handwriting identification problems was introduced long time ago and various studies were conducted using image processing and pattern recognition technique [11] [15] [18] [40] where the result is to identify or verify the author's handwritten document.

Handwriting Analysis divided into Handwriting Identification and Handwriting Recognition. Handwriting Identification tries to identify and verify the writer of the given handwritten document while Handwriting Recognition deals with the content and meaning of handwritten text. In addition, there are two models in handwriting identification; identification and verification.

Identification model [16] and [17] is the process to determine who are the writer among the candidates for the given handwriting samples while verification [24] deals with a given a few samples of handwriting and then determine whether this samples belong to same writer or a different writer among the candidates.

The main issue in Writer Identification is how to acquire the features that reflects the authors of handwriting. Extracted features may consist of excessive features. Some of the features are useless in classification, does not provide useful information for the task in Writer Identification and this will interrupt the performance of the classifier. However, the process of selecting the most significant features in handwriting identification has not been entirely researched.

Therefore this paper proposes several techniques of tree-based structure of classification process in Handwriting Identification. This paper is contribution from [34] presenting comparison of tree-based structures for feature selection. Classification is the final stage in handwriting identification after Preprocessing and Feature Extraction. The accuracy of classifier depends on quality of the features [16] selected passesd during Feature Extraction stage.

3 Tree-based Structure of Feature Selection

The objective of feature selection are to obtain the most minimal sized subset of features, increasing the classification accuracy, reduces computational complexity and help find simpler algorithmic models [1] [2]. Besides, it also removes redundant and noisy features from data sets and selects the most significant features for creating robust learning models [3].

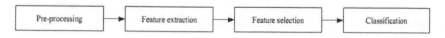

Fig 1. Framework of the experiment

Figure 1 showed the framework of the experiment that has been used in [34]. It is found that feature selection process takes place between feature extraction and classification process. Meaning that the extracted features from Feature Extraction stage will be processed in Feature Selection stage whereas; only the relevant and significant features were allowed to classification stage. This is to ensure that the performance of the classifier is increased and also to avoid the learning process to slowdown and the risk of the learned classifier over-fitting the training data [4].

In this paper, five tree-based structures of classification methods was selected and retrieved from various literatures by using Correlation-based Feature Selection (CFS) as the feature selection method. The output of this research will compare and analyze the classification accuracy and based on the number of features selected.

3.1   Decision Stump Tree

The decision stump has only one level of decision tree. It is showing that one internal node (root node) of decision tree connected to the terminal nodes (leaves node). A decision stump makes a prediction based on the value of just a single input feature. Sometimes they are also called 1-rules. Other than that, decision stumps are often used as components (called "weak learners" or "base learners") in machine learning ensemble techniques such as bagging and boosting.

3.2   J48 Tree

In classification, J48 starts with create a decision tree based on the attribute values of the available training data. The internal nodes of a decision tree denote the different attributes; the branches between the nodes tell us the possible values that these attributes can have in the observed samples, while the terminal nodes tell us the final value (classification) of the dependent variable. In J48, the splitting process terminates if all the attributes in a subset belong to the same class. Then the leaf node is created in a decision tree telling to choose that class.

3.3   Random Forest Tree

A random forest is an ensemble of random decision trees with a randomized selection of features at each split and classification is based on majority voting of the trees. Each tree in the forest has been trained using a bootstrap sample of individuals from the data, and each split attribute in the tree is chosen from among a random subset of attributes. Classification of individuals is based on aggregate voting over all trees in the forest. Repetition of this algorithm yields a forest of trees, each of which have been trained on bootstrap samples of individuals. Thus, for a given tree, certain individuals will be left out during training. Prediction error and attribute importance is estimated from these "out-of-bag" individuals. The out-of-bag (unseen) individuals are used to estimate the importance of particular attributes.

3.4   Random Tree

A random tree is a collection of tree predictors that is called forest further. During classification, the random trees classifier takes the input feature vector, classifies it with every tree in the forest, and outputs the class label that received the majority of "votes". In case of a regression, the classifier response is the average of the responses over all the trees in the forest. In random trees there is no need for any accuracy estimation procedures, such as cross-validation or bootstrap, or a separate test set to get an estimate of the training error. The error is estimated internally during the training.

3.5   REP Tree

REP Tree (Reduced Error Pruning Tree) generates a quick efficient decision tree using information gain and also variance to identify the class information. Pruning should reduce the size of a learning tree without reducing predictive accuracy as measured by a test set or using cross-validation. Starting at the leaves, each node is replaced with its most popular class. If the prediction accuracy is not affected then the change is kept. While somewhat naive, reduced error pruning has the advantage of simplicity and speed. It uses fast pruning algorithm to increase accurate detection rate. The pruned tree provided enables to reduce the complexity of the process working by finding the best sub-tree of the initially grown tree with the smallest error for test set.

4   Experiment

This experiment was carried out using; Mark Hall, Eibe Frank, Geoffrey Holmes, Bernhard Pfahringer, Peter Reutemann, Ian H. Witten (2009); The WEKA Data Mining Software: An Update; SIGKDD Explorations, Volume 11, Issue 1.

The experiment was executed using dataset from IAM Handwriting Database, which is contains sample of handwritten English word.This word image was extracted using United Moment Invariants (UMI) to generate feature vector and here discretization data was produced. UMI was proposed to use in extraction feature because of its ability to generate features that are unique in each word.

The samples from this dataset were chosen randomly. Only 60 classes from 657 classes were used to implement in this experiment. From these classes, 4400 instances are collected. According to figure 2, the 4400 instances were divided randomly into five groups to form training and testing data in the classification task. The quality of feature subset produce by each method can be explore by testing against classification using five tree-based structure whereby CfsSubsetEval use as feature selection and the result will show the comparison of performance among them.

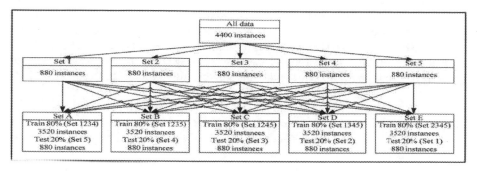

Fig 2. Data collection for training and testing

5   Experimental Result and Discussion

Table 1. Experimental Result

Tree-based Structure	Dataset	Calssification Accuracy (%)			
		Undisc.	No. Features Selected	Disc.	No. Features Selected
DecisionStump	Set A	3.8418	1	5.1977	6
	Set B	3.8778	1	5.4054	6
	Set C	3.9909	1	5.2452	6
	Set D	4.0794	1	5.6229	6
	Set E	4.0816	1	5.5556	6
	Average	3.9743	1	5.40536	6
J48	Set A	33.2203	1	98.6441	6
	Set B	40.5405	1	98.7074	6
	Set C	36.488	1	99.0878	6
	Set D	23.7045	1	98.5667	6
	Set E	35.6009	1	98.8662	6
	Average	33.91084	1	98.77444	6
RandomForest	Set A	46.4407	1	98.0791	6
	Set B	48.1786	1	98.7074	6
	Set C	40.4789	1	98.2896	6
	Set D	31.2018	1	97.7949	6
	Set E	40.0227	1	98.6395	6
	Average	41.26454	1	98.3021	6
RandomTree	Set A	46.4407	1	96.2712	6
	Set B	48.6486	1	96.7098	6
	Set C	40.821	1	97.1494	6
	Set D	30.2095	1	95.9206	6
	Set E	39.229	1	97.619	6
	Average	41.06976	1	96.734	6
REPTree	Set A	22.7119	1	97.0621	6
	Set B	25.3819	1	98.0024	6
	Set C	23.4892	1	96.9213	6
	Set D	19.1841	1	97.9052	6
	Set E	22.449	1	97.7324	6
	Average	22.64322	1	97.52468	6

Table 1 show the result of this experiment. The dataset used in this study are divided into two categories which are discretize and undiscretize dataset. Found that in any given set of dataset for the undiscretize dataset, single features are selected whereby discretize dataset are 6 features selected. This shows that categorizing dataset has substantial impact for the feature selection process and hence classification process.

Although the number of features selected remains same on all dataset for each data category, the classification's accuracy was different. Forundiscretize dataset, Random Forest (RF) gives the highest accuracy; 41.26454%, followed by Random Tree (RT); 41.06976%, J48; 33.91084%, REP Tree; 22.64322% and Decision Stump (DS); 3.9743%. Surprisingly, the result show that none of methods in undiscretize dataset gives the classification accuracy more than 50%. This is due to the number of feature selected. Single feature may not have discriminatory power to differentiate each class. Another reason is may be because of the nature of data itself due to the various style of words represent a writer. However the classification accuracy improved in discretize dataset. Out of 5 methods, J48 gives the highest classification accuracy which is 98.77445 followed by RF; 98.3021%, REP Tree; 97.52468%, RT;96.734% and DS; 5.40536%. In this study, although J48 give the highest classification accuracy, Random Forest is considered as the best method for classification process in handwriting identification. This is because RF tree give the best result compare with other methods especially in handling both undiscretize and discretize data type. J48 gives the best performance in classification accuracy in handling discretize data but is weaker in undiscretize data.

The good performance of RF as a classifier in handwriting identification is based on how it's works along the classification process. RF consists of ensemble of randomized decision trees and the output is depends on plurality vote of all these decision trees. During classification, each tree casts votes, assigning a class to each sample. The result class refers to the class that received the most votes. Each decision-tree of RF is constructed using a bootstrapped set which is only about two-third of the original training data and another is left out as out-of-bag (OOB) cases. During the development of tree, the selection of features is used to determine the split of each node.

6   Conclusion and Future Work

The comparative study on tree-based structure for handwriting identification has been presented. This study compared five methods of tree-based structure; Decision Stump, J48, Random Forest, RandomTree and REPTree. Based on the result in Table 1, Random Forest Tree classifier gives the best result in both category (undiscretize, discretize) of dataset with average accuracy (41.26%, 98.30) followed by J48 (33.91%, 98.77%), Random Tree (41.07%, 96.73%), REP Tree (22.64%, 97.52%) and Decision Stump (3.97%, 5.41%).

As Random Forest gives the best result, this method will be carrying on the next study.For the future work, the Random Forest Tree decided to use as a feature selection method for handwriting identification process, with the same dataset.

Acknowledgement

This work is funded by the Faculty of Information and Communication Technology, UniversitiTeknikal Malaysia Melaka (UTeM).

References

1. Xie,J., Wu,J. , Qian,Q. Feature Selection Algorithm Based on   Association Rules Mining Method. Eigth IEEE/ACIS International Conference on Computer and Information Science, 2009.
2. Dash, M., & Liu, H. Feature Selection for Classification. Journal of Intelligent Data Analysis, pp.131-156, 1997.
3. Lewis P M. The characteristic selection problem in recognition system. IRE Transaction on Information Theory, 1962, 8, pp.171-178.
4. L. Yu, H. Liu. Efficiently handling feature redundancy in high-dimensional data, in: Proceedings of The Ninth ACM SIGKDD International Conference on Knowledge Discovery and Data Mining (KDD-03), Washington, DC, August, 2003, pp. 685-690.
5. Zexuan Zhu, Yew-Soon Ong, Manoranjan Dash. Wrapper-filter feature selection algorithm using a memetic framework. IEEE transactions on systems, man, and cybernetics. Part B, Cybernetics: a publication of the IEEE Systems, Man, and Cybernetics Society 2007; 37(1): 70-6.
6. R. Kohavi and G. H. John. Wrapper for Feature Subset Selection. Artificial Intelligence, vol. 97, no. 1-2, pp.273-324, 1997.
7. Saeys, Y., Inza, I., &Larranaga, P. A Review of Feature Selection Techniques in Bioinformatics. Journal of Bioinformatics, 2507-2517, 2007.
8. Hall, M. A. (1999). Correlation-based Feature Subset Selection for Machine Learning. Hamilton: University of Waikato.
9. Gadat, S., &Younes, L. A Stochastic Algorithm for Feature Selection in Pattern Recognition. Journal of Machine Learning Research, 509-547, 2007.
10. Portinale, L., &Saitta, L. Feature Selection: State of the Art. In L. Portinale, & L. Saitta, Feature Selection, pp. 1-22, 2002. Alessandria: UniversitadelPiemonte Orientale.
11. S.N. Srihari; Sung-Hyuk Cha and Sangjik Lee, Establishing handwriting individuality using pattern recognition techniques, Document Analysis and Recognition, 2001. Proceedings. Sixth International Conference on, 10-13 Sept. 2001, Pages: 1195 – 1204.
12. S. N. Srihari; Cha, S.-H.; Arora, H.; and Lee, Individuality of Handwriting, Journal of Forensic Sciences, 47(4), , pp. 1-17, July 2002.
13. Bin Zhang and Srihari, S. N., Analysis of Handwriting Individuality Using Word Features, Document Analysis and Recognition. Proceedings. Seventh International Conference Page(s):1142 – 1146, 2003.
14. R. Plamondon and G. Lorette. Automatic signature verification and writer identification – the state of the art. In PatternRecognition, volume 22, pages 107–131, 1989.
15. Schlapbach, A., Bunke, H.: Off-line Handwriting Identification Using HMM Based Recognizers. In: Proc. 17th Int. Conf. on Pattern Recognition, pp. 654–658. IEEE Press, Washington (2004)
16. Bensefia, A., Nosary, A., Paquet, T., Heutte, L.: Writer Identification by Writer's Invariants. In: Eighth Intl. Workshop on Frontiers in Handwriting Recognition, pp. 274–279. IEEE Press, Washington (2002)
17. B. Zhang, S. Srihari, and S. Lee, Individuality of handwritten characters, in: International Conference on Document Analysis and Recognition, (Edinburgh, Scotland), pp. 1086–1090, August 3-6 2003.

18. S. Srihari, S. Cha, H. Arora, and S. Lee, Individuality of handwriting: a validation study, in International Conference on Document Analysis and Recognition, pp. 106–109, 2001.
19. G. Leedham and S.Chachra, Writer identification using innovative binarised features of handwritten numerals, in International Conference on Document Analysis and Recognition, 2003.
20. X. Wang Ding, H. Liu, Writer identification using directional element features and linear transform, in: Proceedings of the 7th International Conference on Document Analysis and Recognition, pp. 942–945, 2003
21. C TomaiI., B. Zhang and S. N. Srihari, Discriminatory power of handwritten words for writer recognition, in International Conference on Pattern Recognition, vol. 2, (Cambridge, UK), pp. 638–641, 2004.
22. E. Zois and V. Anastassopoulos, Morphological waveform coding for writer identification, Pattern Recognition, vol. 33, pp. 385–398, March 2000
23. H. Said, G. Peake, T. Tan, and K. Baker, Writer Identification from Non-Uniformly Skewed Handwriting Images, Proc. Ninth British Machine Vision Conf., pp. 478-487, 1998.
24. A. Bensefia, T. Paquet, L. Heutte, Information retrieval based writer identification, in: Proceedings of the 7th International Conference on Document Analysis and Recognition, 2003, pp. 946–950.
25. S.K.Chan, C. Viard-Gaudin, Y.H. Tay, Online writer identification using character prototypes distributions, in: Proceedings of SPIE—The International Society for Optical Engineering, 2008.
26. G.X. Tan, Automatic writer identification framework for online handwritten document susing character prototypes, Pattern Recognition(2009), Elsevier Ltd.
27. [40] Y. Zhu; Tieniu Tan and Yunhong Wang, Biometric Personal Identification Based on Handwriting, Pattern Recognition, 2000. Proceedings. 15th International Conference on Volume 2, 3-7 Sept 2000 Page(s):797 - 800 vol.2. Recognition, Cambridge, August 23-26, 2004, pp. 654–658 (2004)
28. M. Tapiador, J.A. Sigüenza, Writer identification method based on forensic knowledge, Biometric Authentication: First International Conference, ICBA 2004, Hong Kong, China, July 2004.
29. Chapran, Biometric writer identification: feature analysis and classification, In J. Pattern Recognition Artif. Intell. 20 (4) pp. 483–503, 2006.
30. AzahKamilahMuda&SitiMariyamShamsuddin, A Framework of Artificial Immune System in Writer Identification, Proceeding of International Symposium on Bio-Inspired Computing, 5-7th September, Johor Bahru, 2005
31. AzahKamilahMuda, SitiMariyamHj. Shamsuddin, MaslinaDaru, Embedded Scale United Moment Invariant for Identification of Handwriting Individuality, ICCSA (1) 2007: 385-396
32. F.P. Satrya, Muda A.K, Choo Y.H, MudaN.Computationally Inexpensive Sequential Forward Floating Selection for Acquiring Significant Features for Authorship Invarianceness in Writer Identification, International Journal on New Computer Architectures and Their Applications (IJNCAA) 1(3): 581-598 The Society of Digital Information and Wireless Communications, 2011 (ISSN: 2220-9085).
33. F.P. Satrya, Muda A.K, Choo Y.H, Feature Selection Methods for Writer Identification: A Comparative Study, 2010 International Conference on Computer and Computational Intelligence (ICCCI 2010).

An Audio Steganography Technique to Maximize Data Hiding Capacity along with Least Modification of Host

Lukman Bin Ab. Rahim[1], Shiladitya Bhattacharjee[2], Izzatdin B A Aziz[3]

Universiti Teknologi PETRONAS,
Bandar Seri Iskandar, Malaysia

{[1]lukmanrahim, [3]izzatdin}@petronas.com.my,

[2]shiladityaju@gmail.com

Abstract. Steganography is the concept of concealing information within a cover media. In this paper, we present a technique to hide the text information within an audio file using a higher Least Significant Bit (LSB) layer. To hide text information within the audio file, we propose an approach that uses two LSB layers of the host file. The data hiding is done by minimum modification to the host file, which implies that changes of host audio file do not impact the Human Auditory System (HAS). Hiding of dual bit in the 4th and 1st LSB layer improves the hiding capacity of the host file. The proposed approach is also robust with respect to the errors, occurred during embedding of text in the audio files, as minimum modifications are performed on the original host audio.

Keywords: Higher LSB layer; HAS; text hiding; host audio; robust; embedding of text.

1   Introduction

Steganography is a new approach of providing information security by incorporating messages within a different media such as text, audio or video. The use of such media as a mean to cover and hide the secure information increases the proficiency of masked communication [1]. Sometimes the authenticity of such media can be provided by embedding copyright information. In several occasions, steganography is used along with cryptographic mechanism to safeguard the message from illegal persons. Generally, in the cryptographic mechanism [1, 6], the unwanted third party intends to detect, modify, or distorted information during transmission; and most of the time they succeed to do that. But in steganography, the information is embedded within the cover media in such a way that human auditory system or visual system cannot recognize them [2].

These days, most commercial organizations use the steganography approach by mean of communication with respect to transmission of secure information such as transaction information, business dealing, etc. [3]. But sometimes, these information hiding schemes are not enough to provide security for the aforementioned confidential

T. Herawan et al. (eds.), *Proceedings of the First International Conference
on Advanced Data and Information Engineering (DaEng-2013)*, Lecture Notes
in Electrical Engineering 285, DOI: 10.1007/978-981-4585-18-7_32,
© Springer Science+Business Media Singapore 2014

information. The enormous use of electronic communication and huge availability of different free data hiding software makes the situation more critical [4, 5]. The correctness of information sometimes suffers by the data hiding scheme as the information loss its originality during different types of transformations [5].

To solve the above issues, the use of audio steganography is quite successful up to a certain extend as it provides better security of information and robustness [4, 5]. Audio steganography is advantageous [7] over the other steganography approaches because it provides better security and privacy with respect to information leaking as the stego media sounds the same as the original cover media. That is why the use audio steganography is becoming more popular.

1.1 Current Problems

As we are concerned with audio steganography, the problems [6, 7] of audio steganography are:

- Large embedded file is needed to conceal large information.
- Original information cannot be extracted from the embedded file due to erroneous data hiding scheme.

1.2 Objectives

To solve the problem regarding data hiding capacity and disturbance of cover media during the hiding of original large data within it, we should develop such a system which can:

- Hide large text data [7] within a small audio cover media so that we can enhance the space utilization of cover media.
- Hide the original information without losing its originality and the changes cannot be detected by Human Auditory System (HAS).

2 Background Study

A list of nomenclature is shown in this section that will be used throughout the whole paper,

Nomenclature	
Σ	Set of all characters, digits, numbers and symbols
$\langle \rangle$	Used for representing element sequence of a set
$\emptyset$	Denotes the null sequence

Data hiding or steganography is narrated in the following sections by dividing it into a few sub modules as follows:

2.1 Types of Steganography

Depending upon the cover media, the steganography approach can be classified into several categories, i.e. the text based, image, video, Steganography in OSI Network Model, and the audio based steganography.

Audio steganography can be categorized into three category i.e. Temporal Domain [9], Transform Domain [13] and the Encoding Domain [17]. In the proposed technique, we have used audio as the covering media. A brief description has been drawn in the following section about the above categories of audio steganography.

2.1.1 Temporal Domain Audio Steganography

In temporal domain steganography [9], several well-known approaches so far have been applied to hide actual information with in the audio file. Low Bit Encoding and Echo Hiding are the suitable example of such approaches.

The incorporation of target message into the cover media is quite an easy task, as well as the capacity of hiding is also measurably good in this method. But it is not very efficient to prevent the different network attacks i.e. IP attack, application layer attack, TCP attack, etc. and it provides low robustness against the error. Sometime the filtering of noise can destroy the original message. As the decoding is a very easy task, the attacker can easily retrieve the original message from it.

2.1.2 Transform Domain Audio Steganography

Transform domain [13] is further classified into frequency domain and the wavelet domain. Several approaches of hiding information into audio cover media for both domains are discussed in this section.

Frequency domain [11] steganography includes several techniques i.e. tone insertion, phase coding, amplitude modulation, Cepstral domain and spread spectrum. The tone insertion method is based on addition of low power inaudible tone to hide information in presence of the apparently higher tones. The masking is used in the host audio file that cannot be detected by the HAS. The low data insertion capacity and easy recovery of the inserted data are the main disadvantages of this system.

In the Discrete Wavelet Transform (DWT) [17] method, the actual data is embedded at the LSB of the wavelet coefficient of cover media to provide higher hiding capacity. To ensure the inaudibility of modified signal, a threshold value is settled to incorporate the original data. The probability of occurring error in this method is very high as it provides higher embedding rate.

2.1.3 Encoding Domain Audio Steganography

In this method different types of sub-band amplitude modulation techniques are used to hide the data within the speech or audio file. The linear prediction analysis is used to retrieve hidden information from the cover medium [17]. Fourier transform of cover speech is done to hide data in the LSB position. A linear predictive coding filter is used to reduce the flickering LSB noise. Some threshold value has to be set for a few acoustic parameters to make inaudible changes of the cover medium.

2.2 Important features of Audio Steganography

- Incorporation of information into host should be such in a way that, it should create minimum impact to cover media.
- Hiding capacity of original information with in the host medium should be large.
- Confidentiality and originality of actual data should be maintained.
- Steganography process should be flexible with respect to the host audio file format.

3 Proposed Methodology

Having no perceptually changes, incorporation of text within the host audio is a challenging job. We have tried to achieve this by incorporating the information at the 4th and 1st LSB position of some selected samples of cover audio. The process flow of the proposed hiding technique has been shown in Fig.1.

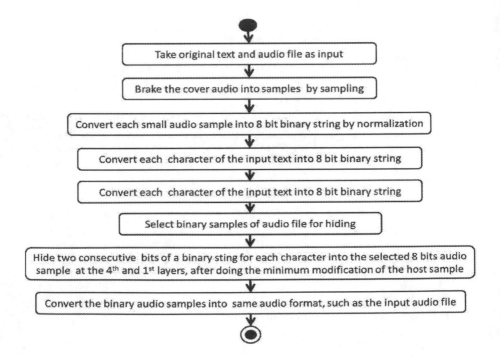

Fig. 1. Incorporation of text into Host Audio

3.1   Binarization of input files

To convert input audio (.wav) files into binary strings, we first break the audio file into several small samples using sampling theorem. The sampling theorem fixes a threshold value depending upon its amplitude value. Then normalization is done with these samples and to conversion is performed on these samples' value in the range of -255 to 254.Tthe normalized values are multiplied with 10. These values are then converted into 8 bit binary strings. While the ASCII values are calculated of each character in the input text and then they are also converted into 8 bit binary strings.

3.2   Selection of Sample

To embed text within the cover audio with minimal changes in the cover audio, we select some particular sample strings. This selection can be done using the following method:

1. Select the sample from input audio samples $\langle A_n \rangle$, whose values are prime numbers and arranged in a sequence i.e. $\langle A'_p \rangle$ where n is total number of samples, $n \in \Sigma$, $p < n$ and $A' \subseteq A$.
2. Find position value of each A'_p and put them in $\langle POS_p \rangle$. Add each A'_p and POS_p to find the actual sample numbers to which character strings are to be embedded as,

$$A''p = A'p + POSp \tag{1}$$

3.3   Embedding Mechanism

Find all 8 bit samples S_p, according to position value A''_p from A_n (from equation (1)). Bit sequences of all S_p and total numbers of character string in the input text are denoted by $\langle S_{pm} \rangle$ and T_q, where $\langle T_{qm} \rangle$ are the 8 bit sequences of each T_q. Here q is the total number of characters, $q \in \Sigma$ and $0 \le m \le 7$. To embed the character string into sample string performs the following:

1. If the 4th bit of the sample string S_{p3} is modified during incorporation from 0 to 1 then perform,

```
a1) If S_p2 = 1 and S_p4 = 1 then,    b1) If S_p2 = 1 and S_p4 = 0 then,
a2)    For i=0 to 2 do S_pi = 0;      b2)    For i=0 to 2 do S_pi = 0;
a3)  End                              b3)  End
a4) End                              b4)  End
```

Line a1 to a4 and b1 to b4 depict that 1st, 2nd and the 3rd bit are to be converted into 0 when either the 3rd (i.e. S_{p2}) and 5th (i.e. S_{p4}) bit of the cover sample string are 1 or 3rd bit is 1 and 5th bit is 0.

```
c1) If S_{p2} = 0 and S_{p4} = 1 then,    d1) If S_{p2} = 0 and S_{p4} = 0 then,
c2)    S_{p4} = 0;                          d2)    For i=0 to 2 do S_{pi} = 1;
c3)    For i=0 to 2 do                      d3)    End
c4)        S_{pi} = 1;                      d4)    For j=4 to 7 do
c5)    End                                  d5)        If S_{pj} = 1 then S_{pj} = 0;
c6) End                                     d6)            Break;
                                            d7)        End
                                            d8)        Else S_{pj} = 1;
                                            d9)    End
                                            d10)   End
                                            d11) End
```

Line c1 to c6 depicts that when the 3rd bit (i.e. S_{p2}) is 0 and the 5th bit (i.e. S_{p4}) is 1 of the cover sample string, then convert 5th bit into 0 while convert 1st, 2nd and 3rd bit of the cover sample string into 1.

Line number d1 imposes the condition when 3rd bit (i.e. S_{p2}) and 5th bit (i.e. S_{p4}) of the cover sample string are 0 then convert 1st, 2nd and 3rd bit of the cover sample string into 1, shown at line d2 and d3 at the same time invert the value of 5th to 8th bit which are described at d4 to d11.

2. If we have to modify the 4th bit of S_p, i.e. S_{p3} from 1 to 0 in the time of data hiding then perform,

```
e1) If S_{p2} = 0 and S_{p4} = 0 then,    f1) If S_{p2} = 0 and S_{p4} = 1 then,
e2)    For i=0 to 2 do S_{pi} = 1;         f2)    For i=0 to 2 do S_{pi} = 1;
e3)    End                                  f3)    End
e4) End                                     f4) End
```

Line e1 to e4 and f1 to f4 depict that 1st, 2nd and the 3rd bit are to be converted into 1 when either the 3rd (i.e. S_{p2}) and 5th (i.e. S_{p4}) bit of the cover sample string are 1 or 3rd bit is 0 and 5th bit is 1.

So we can see that the information is embedded at dual positions of the each se-lected cover sample string which signifies the fact that our proposed technique can hide large information which satisfies our 1st objective, while from the above algo-rithm we can see that, we have done insignificant modification of actual value of each selected cover sample string during incorporation of actual information by without disturbing its originality, which can resolve our second objective.

```
g1)If S_{p2} = 1 and S_{p4} = 0 then,     h1)  If S_{p2} = 1 and S_{p4} = 1 then,
g2)     S_{p4} = 1;                        h2)      For i=0 to 2 do S_{pi} = 0;
g3)     For i=0 to 2 do                    h3)      End
g4)         S_{pi} = 0;                    h4)      For j=4 to 7do
g5)     End                               h5)          If S_{pj} = 0 then S_{pj} = 1;
g6) End                                   h6)              Break;
                                          h7)          End
                                          h8)          Else S_{pj} = 0;
                                          h9)          End
                                          h10)     End
                                          h11) End
```

Line g1 to g6 depicts that when the 3rd bit (i.e. S_{p2}) is 1 and the 5th bit (i.e. S_{p4}) is 0 of the cover sample string, then convert 5th bit into 1 while convert 1st, 2nd and 3rd bit of the cover sample string into `0.

Line number h1 imposes the condition when 3rd bit (i.e. S_{p2}) and 5th bit (i.e. S_{p4}) of the cover sample string are 1 then convert 1st, 2nd and 3rd bit of the cover sample string into 1, shown at line h2 and h3 at the same time invert the value of 5th to 8th bit which are described at h4 to h11.

3. If the 4th LSB i.e. S_{p3} remains same during the incorporation as the original, then there no modification is needed in the 8 bit sample string S_p.
4. When we incorporate LSB of each T_q group into the 1st LSB of sample string i.e. S_{p0}, just change the value of each S_{p0} according to the LSB value of each T_q group.

3.4   Reconstruction of Stego Audio file

After incorporation of charter strings (T_q) into sample strings (S_p), we have reform the stego audio file from audio sample stings A_n, where $S_p \subseteq A_n$. First we convert all A_n strings into decimal number (in between -255 to 254). Then divide them by 10. The denormalization is done, depending upon the predetermined threshold value to get the sample value of the stego audio file. Finally combine them into a single file to get the stego audio file.

A property of a good steganography approach is that there should be no perceptually differences between the original cover file and the stego file. We have also maintained this property in our proposed approach. We have verified with different audio files of different sizes (up to 200 GB big file size) to hide texts of different sizes (depending upon the cover size). To demonstrate big audio clips is not possible due to limitation of space, so we have shown a small audio clip in the following example (i.e. Fig.2 and Fig.3) to hide a small text through which we are trying to justify our claim that stego audio file generated by of our proposed approach is perceptually same as the original input audio file.

In following Fig. 2 we have shown a small cover audio clip test1.wav to hide the small text "SHILADITYA

Fig. 2. Sample Cover Audio Clip test1.wav

The following Fig. 3 stego audio clip test1_stego.wav is shown after hiding the above text within the audio clip test1.wav.

Fig. 3. Stego Audio Clip test1_stego.wav

3.5   Retrieval of Target Text from the Stego Audio File

After receiving the stego audio file at the receiver end, the receiver tries to extract the text from the received file. This process has three phases. In the 1st phase the receiver initially calculate the length of the received file. Based on this length, the stego audio file is sampled and each sample file is then converted into 8 bit binary strings by the above described methods. In the 2nd phase the target samples strings are selected by the above described series of position values to extract character string of the intended text file.

In the third phase the selected sample strings are taken as input and then from the each selected sample string, the 4th and 1st LSB are taken. By accumulating such four consecutive selected sample strings, a single character is formed. After forming all the character they have placed in the set O_q, where initially $\langle O_q \rangle = \emptyset$ to form actual text.

4 Result Analysis

To demonstrate properly, we have divided the outcome into two sections. The first section describes about how we have achieved our first objective, i.e. the efficiency of our proposed approach with respect to time and space where as in the other section, we discuss about the achievement of our second objective i.e. efficiency of noise reduction during incorporation with in the cover audio file.

4.1 Efficiency calculation of the proposed approach with respect to time and space

In the experimental setup, we have used different sizes (up to 200 GB big file) of audio (.wav) file to hide the different sizes of text files or .txt files (depending upon the size of the cover file shown in the following Table-1), which is also known as the *Hiding Capacity* of the cover files. The time efficiency of the proposed algorithm i.e. how long time it will take to hide a certain size text within the cover can be determined by calculating its *Throughput*. The *Hiding Capacity* is measured in terms of percentage of total cover audio size by the following formula,

$$HidingCapacity = \left(\frac{InputTextSize}{CoverAudioSize} \times 100\right)\%$$ (2)

To calculate *Throughput*, the encoding time is taken in Milliseconds whereas the encoded file size is converted into Kilobytes. *Throughput* of the proposed hiding approach can be formulated as,

$$Throughput = \left(\frac{SizeofEncodedData}{Encoding\ Time}\right)$$ (3)

Some cover audio files are taken as input and their *Hiding Capacity* and *Throughput* are calculated and tabulated in the following Table-1. In Table-1, we can see that the *Hiding Capacity* of the cover audio file is quite high and it is increasing with respect to their size. In our experiment, we vary input cover file size form 3MB to 200 GB (We consider 200 GB is as big input cover file size) and their *Hiding Capacity* varies from 2.87% to 4.76% of cover file size.

From experiment we can also see that the *Throughput* of the proposed approach is also very high and it is increasing with respect to increment of the file size too. Depending upon the input cover file size we can see from the Table-1 that, *Throughput* of the proposed algorithm varies from the 0.86 to 2.14.

Table 1. Hiding capacity and Throughput calculation of cover audio

Name of Cover Audio	Size of Cover Audio	Hiding Capacity w.r.t. Cover Size (%)	Throughput
test1.wav	3 MB	2.87	0.86
test2.wav	45 MB	3.08	1.05
test3.wav	500 MB	3.24	1.36
test4.wav	1.2 GB	3.33	1.69
test5.wav	20 GB	4.14	1.85
test6.wav	80 GB	4.39	1.96
test7.wav	200 GB	4.76	2.14

If we look at the Table-1, the *Hiding Capacity* of the proposed algorithm is quite high; it signifies that small audio file can hide considerably large data depending upon its size. So our proposed technique can fulfill our first objective as it enhances the hiding space. The *Throughput* of proposed algorithm is also very high, which implies that the efficiency of hiding big data with respect to the time is very good, i.e. it can hide big text file in a short time.

4.2 Efficiency of noise reduction during incorporation of text within the cover audio file

The other aspect of measuring the efficiency of our proposed algorithm is to measure the quantity of errors occurred during incorporation of input text within the cover audio file. If the bigger percentage of error is presented in the stego file, then this will make sense in HAS and it can easily differ between original audio and the stego audio file and the original text can be hampered or distorted. So, the main aim is to reduce the noise that cannot differ between the original and stego audio file, and the hidden text should not loss its originality by any distortion. The error during hiding information can be measured in term of *Signal to Noise Ratio or SNR*. The *SNR* is formulated as,

$$SNR = 10 \times Log_{10}^{\{\Sigma_n x^2(n) / \Sigma_n[x^2(n) - y^2(n)]\}} \tag{4}$$

In the equation (4), $x(n)$ is the strength of original sample sequence, whereas $y(n)$ is the strength of stego sample sequence of different input audio files. Some sample audio files and their stego versions of different sizes have been taken as input and their respective *SNR* values have been plotted in the following Fig.4.

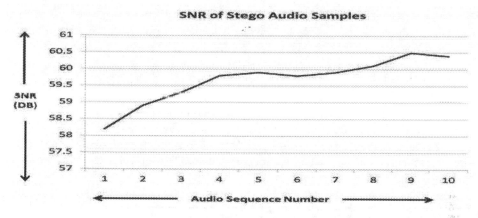

Fig. 4. SNR Values of Different stego audio samples

From Fig.4, we can see that *SNR* of input files are quite high and consistent i.e. the range is between 58.2 dB to 60.5 dB. *SNR* of stego files are high and consistent signify that the error occurred during the incorporation of text files within the cover audio files are very low. As the result of it, the distortion of the hidden information will be negligible. So the probability of losing originality of the hidden information will also be very less. Thus we can claim that we have achieved our second objective through our proposed algorithm.

If the amplitude differences of the cover audio files and the stego audio files is very high then it can impact on the HAS, and they can be easily distinguished. So to show the amplitude differences between the stego audio samples and its original samples, Fig.5 has been plotted in the following,

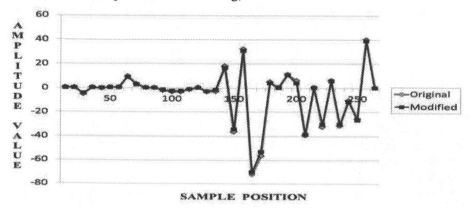

Fig. 5. Comparison of stego audio over original audio amplitudes

In Fig.5, we can see that the amplitude differences between the stego and its original audio sample are quite negligible, which signifies that HAS cannot differentiate between the original audio files and its stego version. So from the above discussion we have proved that our proposed algorithm can fulfill all of the above discussed objectives.

From the above discussion, we have established that our proposed algorithm has excellent *Hiding Capacity* and high imperceptibility, i.e. there are very negligible differences between the stego files and its original version, which cannot be detected for the HAS. As the *Throughput* of the proposed algorithm is also high, this signifies that the encoding time is low. So we can also claim that the Real Time Suitability is also high. As *SNR* of the stego file is high, so the possibilities of error occurrence during incorporation of text within the cover audio are very negligible. So the proposed approach is robust against errors. A comparison has been drawn among the proposed technique and the existing approaches in the following Table-2.

Table 2. Comparison of proposed approach with some existing technique

Methods	Payload Capacity	Imperceptibility	Real Time Suitability	Robustness
Existing LSB Method	High	Medium	High	Low
Parity Coding	Medium	Medium	Low: Delay	Low
Phase Coding	Low	High	Medium	High
Spread Spectrum	High	Low	Low: Bandwidth	High
Echo Hiding	High	Low	Medium	Medium
Proposed Technique	High	High	High	High

From the above table we can see that our proposed is better in every aspect with comparison of existing steganographic approaches as it has high Payload Capacity, Imperceptibility, Real Time Suitability, and Robustness.

5   Conclusion

In this paper the authors have presented an LSB based audio steganographic approach, which is robust with respect to noise and error, capable of embedding big data, embedding time is also low and the proposed approach reduces the perceptually differences between the original audio files and the stego audio files. Our proposed approach is also compared with other existing techniques of audio steganography. It is also experimentally tested with the different size of cover audio files. From the comparison and experiment, we can establish the fact that our proposed technique can fulfill our aforesaid objectives. It can be also proved that our proposed technique is better than the different existing techniques of audio steganography. But still there are many areas to improve our proposed algorithm with respect to hiding capacity and robustness.

6 References

1. Djebbar, B. Ayad, K. Meraim, H. Hamam, Comparative study of digital audio steganography techniques, EURASIP Journal on Audio, Speech, and Music Processing, Springer International Publishing AG, Volume 2012, October 2012, Pages 1-16.
2. K. Bhowal, D. Bhattacharyya, A. J. Pal, T. H. Kim, A GA based audio steganography with enhanced security, Telecommunication Systems, Springer US, Volume 52, April 2013, Pages 2197-2204.
3. N. Samphaiboon, Steganography via running short text messages, Multimedia Tools and Applications, Springer US, Volume 52, April 2011, Pages 569-596.
4. Li Z, Qin, X Zhang, X Wang, Auditory Cryptography Security Algorithm With Audio Shelters, Procedia Engineering, Volume 15, 2011, Pages 2695-2699.
5. S. Xu, P. Zhang, P. Wang, H. Yang, Performance Analysis of Data Hiding in MPEG-4 AAC Audio, Tsinghua Science & Technology, Volume 14, February 2009, Pages 55-61.
6. K. Sakthisudhan, P. Prabhu, P. Thangaraj, C. M. Marimuthu, Dual Steganography Approach for Secure Data Communication, Procedia Engineering, Volume 38, 2012, Pages 412-417.
7. R. Petrovic, D. T. Yang, Audio watermarking in compressed domain, Telecommunication in Modern Satellite, Cable, and Broadcasting Services, TELSIKS '09. 9th International Conference on, October 2009, Pages 395-401.
8. K. Ren, H. Li, Large capacity digital audio watermarking algorithm based on DWT and DCT, Mechatronic Science, Electric Engineering and Computer (MEC), 2011 International Conference on, August 2011, Pages 1765-1768.
9. M. B. Begum, Y. Venkataramani, LSB Based Audio Steganography Based On Text Compression, Procedia Engineering, Volume 30, 2012, Pages 703-710.
10. Yang, X. Wang, T. Ma, A robust digital audio watermarking using higher-order statistics, AEU - International Journal of Electronics and Communications, Volume 65, June 2011, Pages 560-568.
11. Yan, R. Wang, X. Yu, J. Zhu, Steganography for MP3 audio by exploiting the rule of window switching, Computers & Security, Volume 31, July 2012, Pages 704-716.
12. Satir, H. Isik, A compression-based text steganography method, Journal of Systems and Software, Volume 85, October 2012, Pages 2385-2394.
13. M. K. Khan, Research advances in data hiding for multimedia security, Multimedia Tools and Applications, Springer US, Volume 52, April 2011, Pages 257-261.
14. W. Zeng, R. Hu, H. Ai, Audio steganalysis of spread spectrum information hiding based on statistical moment and distance metric, Multimedia Tools and Applications, Springer US, Volume-55, December 2012, Pages 525-556.
15. X.Y. Wang, P. P. Niu, H. Y. Yang, A robust digital audio watermarking based on statistics characteristics, Pattern Recognition, Volume 42, November 2009, Pages 3057-3064
16. A. Cheddad, J. Condell, K. Curran, P. M. Kevitt, Digital image steganography: Survey and analysis of current methods, Signal Processing, Volume 90, March 2010, Pages 727-752.
17. O. Cetin, A. T. Ozcerit, A new steganography algorithm based on color histograms for data embedding into raw video streams, Computers & Security, Volume 28, October 2009, Pages 670-682.

Carving linearly JPEG images using Unique Hex Patterns(UHP)

Nurul Azma Abdullah[1], Rosziati Ibrahim[1,] Kamaruddin Malik Mohamad[1] ,
Norhamreeza Abdul Hamid[1]

[1]Faculty of Computer Science and Information Technology
Universiti Tun Hussein Onn Malaysia, Johor, Malaysia

{azma,rosziati,malik}@uthm.edu.my,
gi090007@siswa.uthm.edu.my

Abstract. Many studies have been conducted in addressing problem of fragmented JPEG. However, there are many scenarios in fragmentation yet to be solved. This paper is discussing of using pattern matching to identify single linear fragmented JPEG images. The main contribution of this paper is introducing Unique Hex Patterns (UHP) to carve single linear fragmented JPEG images.

Keywords: JPEG image, pattern matching, file carving, DFRWS 2006/07

1 Introduction

A method to recover files from the disk collected from the crime scene which is known as cyber evidences is called file carving [1, 2, 3]. Year by year, with the increasing number of computers and other digital devices usages, file carving technique also evolve drastically. In carving a JPEG images, the focus is in fragmentation file carving. It is an important issue in file carving since handling fragmentation can significantly improve the number of potential images successfully obtained from the crime scene. Statistic presented in [3] shows the fact that fragmentation in today's file system is relatively infrequent. However, the capability of carving fragmented files which is not extensively explored is important for computer forensic because the possibility of files that interest forensic investigation to be fragmented is relatively high [3].

This paper discusses a technique in handling fragmentation especially fragmentation occurs in among JPEG files. In order to enhance the functionality of file carver is handling fragmentation efficiently. A file can be fragmented with another whether same types, different types or random data. Fragmentation can occur either linearly or nonlinear. According to [4], linear fragmentation occurs when a file has been fragmented and split into multiple parts with all parts are present in the dataset in their original order while nonlinear fragmentation is when the parts not in their original order or in reverse order. Joachim Metz, Bas Kloet and Robert-Jan Mora have developed Revit07 to handle linearly fragmented files including JPEG. However, they handle thumbnails same as handling original. This may result

T. Herawan et al. (eds.), *Proceedings of the First International Conference on Advanced Data and Information Engineering (DaEng-2013)*, Lecture Notes in Electrical Engineering 285, DOI: 10.1007/978-981-4585-18-7_33,
© Springer Science+Business Media Singapore 2014

thumbnail is assumed as a fragmentation point which may lead to falsely detect fragmentation point.

In this paper, with consideration of thumbnail, we proposed new Unique Hex Patterns (UHP) to carve JPEG files including linearly fragmented JPEG files. The datasets used are from DFRWS 2006 and DFRWS2007 (DFRWS).

2 Related Works

2.1. An Overview of JPEG Standard

Computer forensics is to recover evidences resides on a computer, by mean to solve pornography cases [1], [7], [8]. This involves image files obtained from the perpetrator in certain format like Bitmap and JPEG but most common format is JPEG. JPEG is popular because of its compressed file that can reduce the size required to allocate an image. Joint Photographic Experts Group (JPEG) was formed by International Telegraph and Telephone Consultative Committee in 1986 inspired by an effort of International Organization of Standard (ISO) to find ways to use high resolution graphics and pictures in computers [2]. JPEG introduced compression standard for both grayscale and color continuous-tone images. The details of JPEG compressed data formats can be found in [9]. There are two types of JPEG that are mostly used today, JPEG File Interchange Format (JFIF) and JPEG Exchangeable Image File Format (Exif) [10]. JFIF is popular for internet file while EXIF is the popular image file format used for digital camera [13].

JFIF (JPEG File Interchange Format) was introduced by Eric Hamilton in 1991. It is a minimal file format that allows JPEG bitstreams to be exchanged in multiplatform environment and wide variety range of applications. According to work done by [9], the APPO marker is included into JFIF structure as an additional requirement while maintaining the compatibility of JFIF with the standard JPEG format interchange. A JPEG file that apply JFIF standard contains a head signature that starts with SOI followed by hexadecimal string *0xFF E0 XX XX 4A 46 49 46 00* and ends with EOI [12].

In 1996 Exif (Exchangeable image file format) is introduced by JEITA standard who aimed to provide standard image format for digital cameras and other related equipments. JPEG *Exif* standard was created to stipulates the method of recording image data in files which specifies the structure of image data files, tags used by this standard and definition and management of format versions. Exif can be recognized by SOI and "Exif\0" identifier.

2.2. Fragmentation Point

Less than 10% fragmentation occurred in a typical disk. However, Garfinkel [3] also found that most of fragmented files are potentially useful for forensics investigation. There are limitations of carving tools available today. Many concentrate on carving contiguous data file and do not provide with validator which resulting with many false positive files [3].

A fragmented file is a condition where a file is split into two or multiple parts which can be in different locations in a dataset [4]. Pal, Sencar & Memon [11] explained that fragmented files are when a file is stored in non contiguous clusters. This will result in recovering process using traditional file carving technique will be failed and when using earlier automate file carving tool, incorrect files will be carved.

A signature in header and footer has been used to carve in straightforward carving. This is a simple technique and has been proved successfully carve a contiguous files with an assumption that files clusters remain in order [5]. However, if files are fragmented, files can be disconnected and becomes unordered which cause the straightforward carving fail. In order to solve the fragmentation, detecting fragmentation point is required before carving those files. However, if files are fragmented, files can be disconnected and becomes unordered which cause the straightforward carving fail. Mohamad et al. in [6] pointed out the importance of focusing on fragmentation problem especially within DHT (Define Huffman Table) area because any damaged in DHT can cause image distortion or worse, corruption. Therefore DFRWS initiate efforts to encourage research within fragmented files carving by preparing data set contains some contiguous files while others were fragmented.

Fragmentation point exists only when a file is fragmented into more than two parts. There are three approaches that have been discussed in [11] which are syntactical tests, statistical tests and basic sequential validation. Alternative technique to identify fragmentation is by using cluster information. In a computer file system, a cluster or in DOS 4.0 known as an allocation unit or in UNIX System as block, is the smallest logical set that is created to perform actual erasure for files and directories [16], [17]. A cluster may contain the whole file or portion of files but a cluster only stores data for one file [18], [19]. Therefore, it is important to determine the cluster's size to determine the start of file. This information is useful for both steganography and file carving.

In file carving, the whole content of the hard disk used for evidence need to be imaged (will be referred to as image file) in preparation for forensics investigation. The image file contains thousands of hex code representing all files in the hard disk used for evidence. It is impossible to read line by line manually in order to extract all files into their original form. Information about the start of file can be used to identify each file that resides in any dataset. PredClus as proposed in [15] is developed to automatically display the predicted cluster size according to probabilistic percentage. It concentrates on JPEG images. The information can be used to distinguish original with thumbnail/embedded JPEG files which can help to determine the real fragmentation point. However, this technique has limitation. For some rare cases where thumbnail is pushed to the start of cluster, it can be falsely identified as an original JPEG image. Hence, PattrecCarv is introduced where thumbnails and embedded JPEG files are carved using Unique Hex Patterns (UHP) as described in [20]. It carves more thumbnails compared to PredCLus.

3   Frag_Carv Model

This section discusses on the experiment designed for Frag_Carv model (as illustrated in Fig. 1) is adapted from myKarve [14]. Two algorithms, PattrecCarv [20] and Pattern_temp are installed into the model. PattrecCarv accepts data sources from either image from physical memory, removable storage, hard disk or any JPEG file. However, for this particular experiment, we use dataset from DFRWS 2006 and 2007. This algorithm also searches for JPEG files location and store in ADB (Address Database). The other algorithm, Pattern_temp matches the header and footer with predefined patterns for original, thumbnails and embedded JPEG files. PattrecCarv is developed using C++ language while UHP_template is implemented in Matlab environment to simplify the pattern matching process.

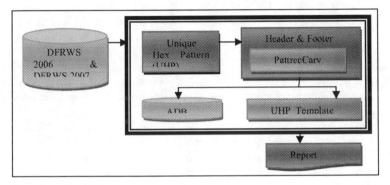

Fig. 1. Frag_Carv model

The following are some of the assumptions for this experiment:
- Baseline JPEG is being used for the experiment because of its popularity and simple file structure.
- Only JFIF and Exif format can be accepted by this model. JPEG 2000 is not compatible with this model.
- All the file headers and footers are in sequential order and not corrupted.

3.1. PattrecCarv

PattrecCarv is inserted in Frag_Carv to identify thumbnails and embedded JPEG files. The algorithm of PattrecCarv[25] is introduced (illustrated in Fig.. 2) to read raw data from DFRWS 2006 and 2007 dataset and mark thumbnails and embedded JPEG files with a different marker from original marker.

```
1.    Read data image
2.    Initialize hex_values
3.    Initialize thumb_markers
4.    Initialize embedded_markers
5.    Find JPEG header
6.    If found
7.           Jump to 9ᵗʰ hex_values
8.           If hex_values== embedded_markers
9.                  Read CurrentOffset
10.                 Save the CurrentOffset
11.          Else
12.                 Read next hex_values
13.                 If hex_values==thumb_markers
14.                        Go to the most recent SOI
15.                        Save the CurrentOffset (location of SOI)
16.                 Endif
17.          Endif
18.   If not end of data image, repeat step 5
19.   Generate report
```

Fig. 2. Algorithm used in PattrecCarv for carving thumbnails and embedded JPEG files

First, data from dataset is read. These data are in hex values. The hex values then matched with the standard JPEG header. However, in this experiment, additional markers are also used instead of standard JPEG headers and footers, 0xFFD8 alone. The additional markers which are 0xFFE0, 0xFFE1, 0xFFE2, 0xFFC4 and 0xFFDB and standard headers/footers are known as validated markers. The validated headers are used to reduce false detection of JPEG files. When matched, the offset for each markers matched is retrieved. Using UHP as described in [Abdullah], all thumbnails and embedded JPEG files are identified and standard headers for the files are renamed with a special name to indicate the type, either thumbnail (TN_FFD8) or embedded JPEG file (EF_FFD8). This is to differentiate them from original header (FFD8).

After all thumbnails and embedded files are determined, all locations for JPEG standard headers/footers are then stored in ADB (Address database). The markers and locations will be the input to Pattern_temp.

3.2. UHP_Template

UHP_Template is a database contains UHP for different scenarios of JPEG as shown in Fig. 4 and Fig. 5. First, it reads ADB for headers and matches the header with predefined patterns. So, from the output, we can view different patterns of JPEG files, either JPEG file without thumbnail, JPEG file with one thumbnail or JPEG file with two thumbnails. If we found a pattern where two headers (original JPEG file) found consecutively, this indicate fragmentation scenario. Below is the algorithm for Pattern_temp (as illustrated in Fig..3) which is according to patterns template shown in Fig. 4 and Fig. 5.

1. Read ADB
2. If match 'pattern 1'
        a.   Write to pattern 1 column in output file.
3. Endif
        .
        .
        .
4. Else if match 'pattern *n*'
        a.   Write to pattern *n* column in output file
5. Endif
6. If not end of data image, repeat step 1
7. Exit

Fig. 3. Algorithm used in Pattern_temp.

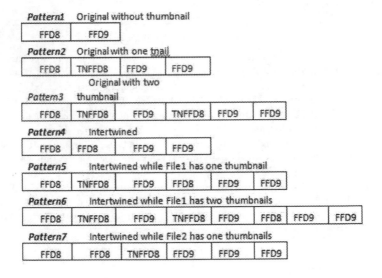

Fig. 4. Pattern 1-7 for different scenarios of JPEG files

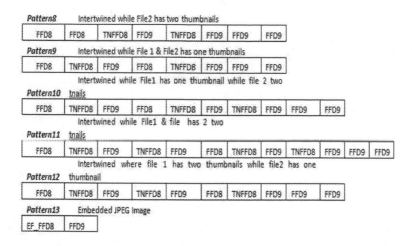

Fig. 5. Pattern 8-13 for different scenarios of JPEG files

There are 13 patterns simulating different JPEG scenarios. Pattern 1 to Pattern 3 describes scenarios for original image with or without thumbnail while Pattern 4 to Pattern 12 describes scenarios for fragmented JPEG images and Pattern 13 is for embedded JPEG files. These patterns help in sorting JPEG images found in the dataset into certain categories which will ease the process of identify the fragmented JPEG images.

4 Experimentation

There are three major steps involved in Frag_Carv model to complete this experiment namely pre-processing, pattern matching and JPEG image file carving. During the pre-processing, a dataset is read and searched for JPEG headers and footers. The real concern is to automatically determine the correct header-footer pairing to match the predefined pattern. In this model, a header-footer is paired if the validated header-footer is strictly matched one of the introduced UHP. This is to validate the read data are belongs to JPEG files. Only then, with the addition of special markers to reduce false detection, these headers and footer along with their locations are populated in ADB. Next, the ADB is ready for pattern matching process.

During pattern matching process, all headers and footers are read from ADB. These headers and footers then matched with the predefined patterns namely pattern 1, pattern 2, and pattern 3 until pattern 13. Pattern 1 is a pattern for JPEG image without thumbnail, pattern 2, for JPEG image with one thumbnail, Pattern 3, for JPEG image with two thumbnails, Pattern 4 is for two JPEG images without thumbnail intertwined with each other, Pattern 5 and 7 for intertwined JPEG images where one JPEG image has one thumbnail, while Pattern 6 and 8 are same as Pattern 5 and 7 but this time, the file contains two thumbnails. The same goes with Pattern 9-12, they are differ with each other depending on which file contain one or two thumbnails.

Finally, the last step in this model is carving JPEG images according to patterns described above. Once, a pattern is determined, the headers and footers are stored in an output file along with their locations. For this experiment, the output file is in Excel format to simplify the analysis process later. All patterns are organized in different columns so that it is easy to view and read.

5   Result and Discussion

Although there are 13 patterns discussed above, by implementing datasets from DFRWS 2006 and 2007, only few patterns are carved. The screenshot of the output can be clearly examined in Fig. 6, Fig. 7 and Fig. 8. When we differentiate thumbnail's header from the standard JPEG header, we can easily confident that Pattern 2 is capturing fragmented JPEG images, where a non complete JPEG image is followed by another JPEG image. This scenario will cause distortion when viewed in image viewer. Now, the investigator can concentrate on these fragmented image instead wasting time investigate a complete image that is mistaken to be fragmented JPEG images.

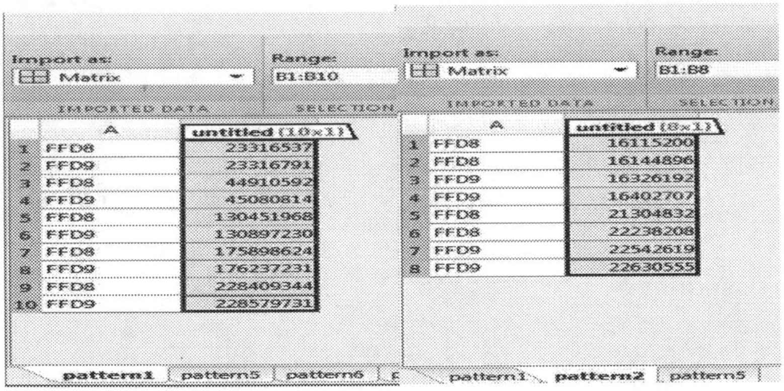

Fig. 6. Example of output for pattern 1 and 2.

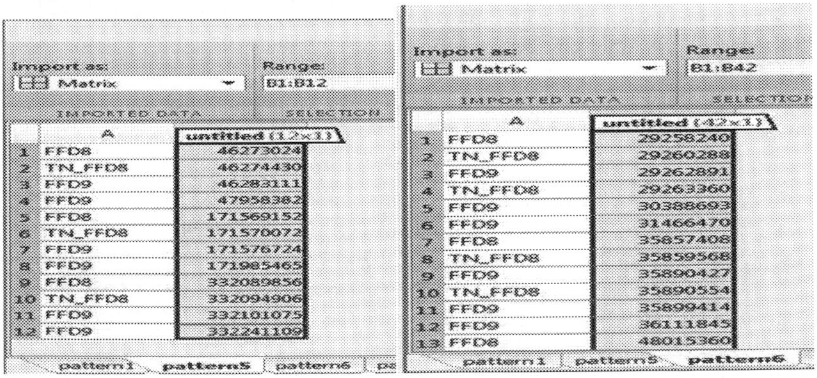

Fig. 7. Example of output for pattern 5 and pattern 6

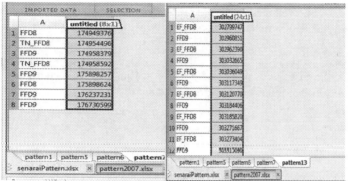

Fig. 8. Example of output for pattern 5 and pattern 6

Furthermore, in Fig. 6, we can see clearly the different of carving a contiguous images and fragmented images. For Pattern 1, the output shows that there is a consecutive order of headers and footers found. This indicates a complete JPEG image can be carved using this pattern. In other hand, for Pattern 2, there is a second header found before the footer. This indicates, there is a fragmentation occurs at that point. From the output, we can easily assume that there is a complete JPEG image in the middle of another JPEG image. Hence, we can carve these files as two separate images.

6   Conclusion

A JPEG image can contains none, single or two thumbnails in the image itself. A thumbnail which is a reduced version of an image carried similar feature as the original. This thumbnail is always mistaken as another JPEG image. Therefore, knowledge of thumbnail's existence helps investigator to separate JPEG files with thumbnail/s and concentrates to investigate correct point where the real fragmentation occurs. They can then identify which JPEG image is fragmented with another JPEG images. With this way, they can ascertain that those fragments are belongs to another JPEG file, not file/s (thumbnail) within a JPEG file. This is important because during the reassembling process, if a thumbnail is mistakenly identified as another JPEG files, the original file may corrupt because of missing fragments and also wasting investigator's time in looking for fragmentation while the fragmentation does not exist. Subsequently, this is also accelerating the reassembling process by allowing investigators to concentrate on real fragmentation situation. In conclusion, by recognizing thumbnail, false fragmentation detection can be reduced significantly.

7   Future Work

There is a limitation of this technique in carving fragmented JPEG images. From the experiment, we found that, Pattern 2 is also valid for a scenario where two JPEG files intertwined with each other where each file is split into two parts and not in contiguous location. Using only pattern matching will result distorted images carved. Further investigation using different technique is required to solve this scenario. Nevertheless, using pattern matching technique, we can save time for investigating fragmentation point which require more times because we have to check upon the files by its contents. By applying this technique, we can separate the complete carved image from the images with distortion where it can be further investigated.

Acknowledgement

The authors would like to thank Ministry of Science, Technology and Innovation (MOSTI), for granting Science Fund (Vote s019) to support this research.

References

1. S. L. Garfinkel, "Digital Forensic Research :The next 10 years," Digital Investigation, vol. 7(1), 2010, pp. S64-S73.
2. M. I. Cohen, "Advanced Carving Techniques," Digital Investigation, vol. 4(1-4), 2007, pp. 119-128.
3. S. L. Garfinkel, "Carving Contiguous and Fragmented Files with Fast Object Validation," Digital Investigation, vol. 4(1), 2007, pp. S2-S12.
4. S. J. J. Kloet, Measuring and Improving the Quality of File Carving Methods, Master Thesis, Endhoven University of Technology, 2007.
5. C. J. Veenman, Statistical Disk Cluster Classification for FIle Carving. Proc. of the Third International Symposium on Information Assurance and Security, Machester, 2007.
6. K. M. Mohamad, M. Mat Deris, Fragmentation Point Detection of JPEG Images at DHT Using Validator. Proc. of the 2009 FGIT, 2009, pp.173-180.
7. A., Pal, & N. Memon, "Automated reassembly of the file fragmented images using greedy algorithms," IEEE Trans. Image Processing, vol. 15(2), pp. 385-393, 2003.
8. M. Karresand, & N. Shahmehri, Reassembly of fragmented jpeg images containing restart markers. in 2008 European conference on computer network defense, 2008.
9. The International Telegraph and Telephone Consultative Committee (CCITT). Information technology—digital compression and coding of continuous-tone still images–requirements and guideline (ITU-T T.81), 1992. Retrieved Sept. 5, 2012, from World Wide Web Consortium (W3C): http://www.w3.org/Graphics/JPEG/itu-t81.pdf
10. S. Bettelli, " The structure of JPEG pictures," 2006. Retrieved Sept. 5, 2012 from : http://search.cpan.org/dist/Image-MetaData-JPEG/lib/Image/MetaData/JPEG/Structures.pod
11. Pal, J. T. Sencar, & N. Memon, "Detecting File Fragmentation Point Using Sequential Hypothesis Testing," Digital Investigation,vol. 5, 2008, pp. S2-S13 2008.
12. K. M. Mohamad, & M. Mat Deris, Visualization of JPEG metadata. in: Proceeding of the 2009 first International Visual Informatics Conference on Visual Informatics,2009.
13. P. Alvarez, "Using extended file information (exif) file headers in digital evidence analysis," International Journal of Digital Evidence. Vol. 2(3), 2004.
14. K.M. Mohamad, A. Patel, T. Herawan, & M. Mat Deris, "myKarve: JPEG image and thumbnail carver," Journal of Digital Forensic Practice, vol. 3, 2011, pp. 74-97.
15. N. A. Abdullah, R. Ibrahim, & K. M. Mohamad, Cluster size determination using JPEG files. in Proceedings of the 12th international conference on Computational Science and Its Applications, 2012.
16. S. W. Ng, "Advances in Disk Technology: Performance Issues," Computer, vol. 31,1998, pp. 75-81.
17. File Allocation Table, http://en.wikipedia.org/wiki/File_Allocation_Table#Boot_Sector
18. R. P. Jemigan & S. D. Quinn, Two-Pass Defragmentation of Compressed Hard Disk Data with a Single Data Rewrite. U.S Patent 5574907 .
19. Cluster Size for NTFS, FAT, and ExFAT, http://support.microsoft.com/kb/140365
20. N. A. Abdullah, R. Ibrahim & K. M. Mohamad, " Carving Thumbnail/s and Embedded JPEG Files Using Image Pattern Matching, " JSEA, vol. 6, 2013, pp. 62-66.

Distinguishing Twins by Gait via Jackknife-Like Validation in Classification Analysis

Wan-Noorshahida Mohd-Isa[1], Junaidi Abdullah[1], Chikkanan Eswaran[1],

[1] Faculty of Computing and Informatics, Multimedia University
63100 Cyberjaya, Malaysia
{noorsha, junaidi, eswaran}@mmu.edu.my

Abstract. This paper is about analysing the uniqueness of twins by gait biometric. The motivation arises due to twins, having facial similarity may lay difficulties to a video-based recognition system employing face biometric. Gait, a biometric based on the way a person walk, can perhaps be a useful descriptor. Due to the small size data set, classification via leave-one-out cross validation may not be sufficient to test gait's viability as a descriptor for twins. Thus, this paper proposes a jackknife-like validation in a matched-pair classification. Comparing between the results of both validation approaches, results of the proposed method have shown to be promising. The results perhaps may point to the uniqueness of each individual twin by gait biometric.

Keywords: Twins biometric, gait recognition, jackknife validation

1 Introduction

Research in face biometric for recognition within twins has begun to take the pace [1] [2] [3] [4] [5] [6]. Still, for a security system that employs face biometric, the issue of facial similarity poses a risk and becomes a problem when such cases occur. Gait is a biometric based on the way human walks. It is believed to be unique to every person [7] [8].

As opposed to face, a gait biometric recognition system looks into human walking motion that comes naturally to a person, where this can be captured with common closed-circuit television (CCTV) cameras. It can overcome the drawbacks of face recognition systems, which requires details with high resolution camera setup. In such settings, gait may be an alternative biometric to face, or may be adopted in parallel to complement the shortcomings of face biometric.

Since there has not been a study on the gait of identical twins, this paper investigates the potential of gait as a biometric to discriminate within twins. This paper proposes to adopt a matched-pair classification with a novel jackknife-like validation to discriminate twins.

T. Herawan et al. (eds.), *Proceedings of the First International Conference on Advanced Data and Information Engineering (DaEng-2013)*, Lecture Notes in Electrical Engineering 285, DOI: 10.1007/978-981-4585-18-7_34,
© Springer Science+Business Media Singapore 2014

2 Methodology

2.1 Video Data Collection

A gait recognition system is a typical video-based recognition system with cameras as the sensors and the processing of inputs are done on the computing machines. The inputs are video recordings, which are later sliced into image-frames. At each image-frame, raw data is extracted. After that the raw data are further processed as a suitable gait descriptor before classification is performed on the developed descriptor. A descriptor is a signature of gait for each person that gives out the identity of the person by gait.

In this work, videos of 24 twins have been recorded. The twins are healthy individuals between the age of 16-40 years old with no known walking problems. These twins have been recorded walking their usual walking pace, sideways from left to right and right to left. After slicing the videos into image-frames, the motion of walking is tracked for one gait cycle. A gait cycle as defined by Murray [7] contains two walking steps that start from a heel-strike of one leg to the next heel-strike of the same leg.

It has been shown in research on motion perception that human observers perceive gait by relying on features from lower limb motion dynamics [9]. Mimicking that on computers, the joint angular trajectory of a person is employed as a descriptor. Thus, the coordinates of the tracked leg are gathered where using these coordinates, angular measurements are calculated, as in Fig.1. This follows other successful literatures on gait [10] [11] .

2.2 Angular Gait Descriptors Extraction

In this paper, the angular variable, α and β represent the angle measured between the hip to knee and knee to ankle, respectively. The measurements of α and β values are calculated using the standard triangle trigonometry. The resultant α and β trajectories are as shown in Fig.2 and Fig.3.

The collection of α and β trajectories for each person i can then become a feature data set (feature set) consisting of,

$$\boldsymbol{\alpha}_i = \alpha\left(t_{[n]}\right) \quad \boldsymbol{\alpha}_i \in [0, 2\pi]^{L_i} \tag{1}$$

$$\boldsymbol{\beta}_i = \beta\left(t_{[n]}\right) \quad \boldsymbol{\beta}_i \in [0, 2\pi]^{L_i} \tag{2}$$

with $t_{[n]}$ refers to the image-frame number and L_i refers to the length of the gait cycle for subject i . Due to the different speeds of walking for each person, the feature set across all i has unequal length n . For a robust classification analysis, the feature set has to be standardized, hence resampling of the trajectories is essential.

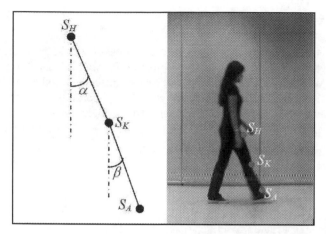

Fig. 1. The coordinates of the tracked leg; S_H is the hip location, S_K is the knee location, and S_A is the ankle location with the angular measurements, denoted by α and β, which are calculated based on the coordinates S_H, S_K, and S_A of the tracked leg.

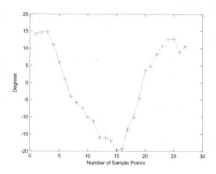

Fig. 2. A right-to-left gait cycle as defined by Murray [7] of thigh displacement trajectory, α for Twin 1 Pair 7.

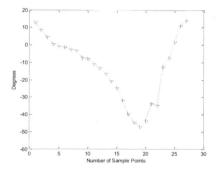

Fig. 3. A right-to-left gait cycle as defined by Murray [7] of lower leg displacement trajectory, β for Twin 1 Pair 7.

2.3 Feature Vector Resampling

Via the SVR framework, the trajectories are regressed and interpolated. A cubic spline kernel is chosen for use as an approximating function. This choice is based on Murray's [7] work that has concluded gait to be continuous and periodic for every gait cycle. The cubic spline function guarantees continuous first and second derivatives across all polynomials' segments, which makes it smooth and attractive for interpolating and regressing gait motion data. The SVR formulation for β is as follows (formulation for α works the same way).

Given discrete data of n points,

$$\left\{\left(t_{[1]}, \beta_{[1]}\right), \left(t_{[2]}, \beta_{[2]}\right), ..., \left(t_{[n]}, \beta_{[n]}\right)\right\} \qquad t, \beta \in \Re \tag{3}$$

The SVR finds an estimated function,

$$\hat{\beta}(t) = \sum_{d=1}^{m} \gamma_d K\left(t_{[d]}, t\right) + \delta \tag{4}$$

where $\gamma_{[d]} \in \Re$ are the support vectors, $K\left(t_{[d]}, t\right)$ is a kernel function, δ is a bias term and a resampling value of $m = 30$; chosen based on the average number of image-frames across all gait cycles. These γ and δ are found by minimizing the regularized risk, R,

$$R = \sum_{d=1}^{m} Loss\left(\beta(t), \hat{\beta}(t)\right) + \zeta \|\gamma\|^2 \tag{5}$$

where ζ is a regularization parameter, and

$$K\left(t_{[n]}, t\right) = 1 + t_{[n]} \cdot t + t_{[n]} \min\left(t_{[n]}, t\right) - \frac{\left(t_{[n]} + t\right)}{2}\left(\min\left(t_{[n]}, t\right)\right)^2 + \frac{1}{3}\left(\min\left(t_{[n]}, t\right)\right)^3 \tag{6}$$

is the cubic spline kernel. The resultant feature vector for each subject i video j becomes,

$$\mathbf{w}_{ij} = \left[\boldsymbol{\alpha}_{ij}^L \mid \boldsymbol{\beta}_{ij}^L\right] \tag{7}$$

By combining all $\mathbf{w}_{ij}$ as a matrix, $\mathbf{W}$ is formed,

$$\mathbf{W} = \left[\mathbf{w}_{1,1} \cdots \mathbf{w}_{1,4} \quad \mathbf{w}_{2,1} \cdots \mathbf{w}_{2,4} \quad \cdots \quad \mathbf{w}_{24,1} \quad \cdots \quad \mathbf{w}_{24,4}\right]^T \tag{8}$$

3 Matched-Pair Classification with Jackknife-Like Validation

To investigate the uniqueness of the extracted gait descriptors, pattern recognition method of classification has been employed. There is one data set, **W** but within **W** there are two sets that are hypothesized as to be dependent due to being twins. The term matched-pair in statistics implies samples of a data set to be related (or correlated) in some way. In psychology, the matched-pair validation has been used in a situation when husbands and wives are matched with each other. These are two separate groups of supposedly random subjects but they are believed to have selected each other because of some similarities. The same idea has been applied to experiments on twins, where each twin serves as a control group for the other. Such validation is useful when only a small number of people is available for research; similarly with this paper.

In statistics, the mechanism for comparing two dependent groups is a paired-test. To paired-test is to investigate the within-pair differences of two dependent groups. In the context of pattern recognition, this means to look into dissimilarity and uniqueness of each twin. Commonalities shared by each twin pair will be removed by subtracting one from the other. This concept of deleting or omitting samples from data set is known as jackknifing.

The original idea of jackknife was invented by Quenouille [12]. It was for a limited purpose of correcting possible bias in an estimator for a small size data set. Also it can be viewed as a type of supervised classification's cross-validation strategy, where it is employed to estimate the accuracy of classification across all samples especially for a small data set.

In this paper, for measuring the correct classification rate (CCR), the classifiers of linear discriminant analysis (LDA) and k-nearest neighbour (k-NN) are applied to the matrix **W**. These classifiers are standard classifiers and have been performing well for small sample problem in gait biometric literatures [10] [11]. LDA gives out a discriminant function that indicates the maximum class separability information based on the eigenvectors of matrix **W**. Similarity distance from a test sample to the discriminant function produces the class information of the test sample. Similarity distance for the k-NN is based on measurements made from the test sample to its neighbours, which are samples located around it. The standard similarity measures of Euclidean and City-Block Distance are used in k-NN.

In any supervised classification, a data set is divided into training and test sets. A cross-validation performs permutation of training and test sets across samples in the data set. Some number of samples (either fixed or random) are selected to become the training set and classification is performed on them. After that the remaining samples (test set) are classified into defined classes based on the class information given out by the classification outcome done on the training set. The classification runs some number of times in permutative order. The performance estimate is obtained as the average of the classification runs.

Similar to cross validation, the jackknife method performs validation by permuting training and test sets across samples in the data set. However the jackknife omits some samples during permutations from both the training and test sets. Thus the training set is said to be trained without some samples but these samples are also not in the test set. When applied in a matched-pair classification, the absence of samples from the other twin allows uniqueness validation of the test set to the training set with the deleted samples.

In this paper, a jackknife-like procedure is proposed within the matched-pair classification. By jackknifing other samples of a selected twin's class, the matched-pair analysis lets the test sample be classified only with its other twin's samples as the training samples representing its own class. The rational for doing this is to check if the test sample can be classified into its other twin's class. The assumption in a matched-pair classification here is that the training and test sets are dependent data that would have a shared matching characteristic. Therefore, this is an analysis to find out if there occurs any similarities between the twins.

3.1 Jackknife-like Procedure Formulation

Given a data set $X = \{P_1, P_2, ..., P_n\}$ with n number of classes and $P_i = \{B_1, B_2\}$ representing paired siblings, B_j for P_i of class i and twin j. Also each $B_j = \{s_1, s_2, ..., s_K\}$, denoted as s_k with a total of K number of samples. Let $C_n(X) = C_n(P_1, P_2, ..., P_n)$ be the class label for the data set X. The jackknife-like procedure defines a kth pseudoclass for $C_n(X)$ to be,

$$ps_{i,j}^k(X) = C_n(P_1, P_2, ..., P_n) - C_n\left((P_1, P_2, ..., P_n)_{[i,j]}^{[k]}\right) + C_n\left((P_1, P_2, ..., P_n)_{[i,\sim j]}^{[k]}\right) \qquad (9)$$

The $P_{[i,j]}^{[k]}$ is the sample $B_j = \{s_1, s_2, s_3, s_4\} \in P_i = \{B_1, B_2\} \in X = \{P_1, P_2, ..., P_n\}$ with i^{th} class, j^{th} twin, k^{th} sample, $s_k \in B_j \in P_i$ retained in the data set, but other $K-1$ samples of ith class and j^{th} twin deleted from the data set, so that X is a data set of size $2Kn - K + 1$ at each classification run.

In a matched-pair classification, the hypothesis is that the pseudoclasses, $ps_{i,1} = ps_{i,2}$. In classification term, if truly gait is unique for each twin, the test sample should be unable to be classified correctly in its other twin's class, hence will result in low average correctly classified estimate. The proposed strategy can be used to measure and validate the uniqueness of a twin.

4   Results and Discussion

Table 1 summarizes the average CCR for two types of validation across the twins dataset. They are the proposed jackknife-like validation and the standard leave-one-out cross validation (LOO). The standard LOO is chosen among others because it gives very accurate estimate of performance due to the reduce bias via the large number of folds. It is a preferential choice in order for the classifier to train on as many samples as possible especially with a small data set like ours. The LOO takes the test set across all samples' combinations for validation at each classification analysis. Then the average CCR is calculated. The higher the percentage value, the better the estimation.

From Table 1, the best LOO results are 76% via the 1-NN with both Euclidean and City-Block distance. The LOO tests the ability of the classifiers to identify the individual twin as a unique individual in a population of 24 number of classes. The percentage would not be considered high because previous literatures on non-twin individuals have achieved results as high as 98% [11]. This value of 76% may mean that 14% from the population of twins in the data set have been incorrectly recognized perhaps due to the similarity in their gait.

In contrast to the results of LOO, the worst average CCR is the value to pay attention to in Table 1. The values in bracket are the complemented percentage in order to compare them with the results of LOO. The worst average CCR is 4.17% (95.83%) by the City-Block distance k-NN classifier. Even with the Euclidean distance, the average CCR is 5.21% (94.79%), which is not far from the City-Block distance. With the jackknife-like validation (JLV), at each time the classification runs, all other samples of the current test class is removed. That is, the test set is classified only with its twin's samples as the training samples of its class. In pattern recognition terms, the resultant low average CCR from the JLV approach means that the test set has been *unable* to be classified into any of the trained set classes.

5 Conclusion

This paper has looked into distinguishing twins by gait. Its aim is to find out the existence or non-existence of similarity within a pair of twins. This is perhaps useful since the faces of twins have high similarity where surveillance system employing face recognition may fail.

A supervised classification analysis has been considered. A data set of 24 twins is gathered and angular trajectories are set as its gait descriptors. Due to the small size data set, two validation techniques under the supervised classification analysis have been applied. One is the proposed JLV by matched-pair classification. Another is the standard LOO cross validation. The JLV seems to be better at pointing to the uniqueness of each individual twin than the LOO.

Future work may be to re-look on the JLV approach via k-NN. Since there seems to be albeit a very small percentage of similarity, which indicates that the test data may have been matched to one of its siblings in one of the JLV classification runs. A k-NN classifier is a naïve classifier, perhaps a more sophisticated classifier may give a better estimation. Thus the results of this paper may confirm and perhaps is consistent with previous literatures that have stated gait is believed to be unique for each individual [7].

Table 1. Average CCR in a Matched-Pair Classification Analysis by the Proposed Jackknife-
Like Validation and Leave-One-Out Cross-Validation.

Classifiers \ Validation	Average CCR (%)	
	Jackknife-Like Validation (JLV)* *The lower the better	Leave-One-Out Cross-Validation (LOO)# #The higher the better
LDA	10.42 (89.58)	43.75
5-NN via Euclidean	5.21 (94.79)	45.83
3-NN via Euclidean	5.21 (94.79)	65.63
1-NN via Euclidean	5.21 (94.79)	**76.00**
5-NN via CityBlock	**4.17 (95.83)**	58.33
3-NN via CityBlock	**4.17 (95.83)**	65.63
1-NN via CityBlock	**4.17 (95.83)**	**76.00**

References

1. Park, U., & Jain, A. K.: Face matching and retrieval using soft biometrics. Information Forensics and Security, IEEE Transactions on, 5(3), pp. 406-415 (2010)
2. Phillips, P. J., Flynn, P. J., Bowyer, K. W., Bruegge, R. V., Grother, P. J., Quinn, G. W., & Pruitt, M.: Distinguishing identical twins by face recognition. In Automatic Face & Gesture Recognition and Workshops (FG 2011), 2011 IEEE International Conference on, pp. 185—192 (2011)
3. Jain, A. K., Klare, B., & Park, U.: Face matching and retrieval in forensics applications. IEEE MultiMedia, 19(1), 20 (2012)
4. Li, H., Huang, D., Chen, L., Wang, Y., & Morvan, J. M.: A group of facial normal descriptors for recognizing 3D identical twins. InBiometrics: Theory, Applications and Systems (BTAS), 2012 IEEE Fifth International Conference on, pp. 271-277 (2012)
5. Srinivas, N., Aggarwal, G., Flynn, P., & Vorder Bruegge, R. W.: Analysis of Facial Marks to Distinguish Between Identical Twins. IEEE Transactions on Information Forensics and Security 7(5), pp. 1536—1550 (2012)
6. Dinakardas, C. N., Sankar, S. P., & George, N.: MultiModal Identification System in Monozygotic Twins. International Journal of Image Processing (IJIP), 7(1), 72 (2013)
7. Murray, M.P.: Gait as a total pattern of movement. American Journal of Physical Medicine 46 (1), pp. 290-332 (1967)
8. Stevenage, S.S., Nixon M. S., and Vince K.: Visual analysis of gait as a cue to identity. Applied Cognitive Psychology 13, pp. 513–526 (1999)
9. Todd, J. T.: Perception of gait. Journal of Experimental Psychology: Human Perception and Performance, 9(1), pp. 31--42 (1983)
10. Yam, C., Nixon, M. S., & Carter, J. N.: Automated person recognition by walking and running via model-based approaches. Pattern Recognition, 37(5), pp. 1057--1072 (2004)
11. Mohd-Isa, W. N. (2005, October). Analysis on spatial and temporal features of gait kinematics. In Automatic Identification Advanced Technologies, 2005. Fourth IEEE Workshop on, pp. 130-133 (2005)
12. Quenouille, M. H.: A large-sample test for the goodness of fit of autoregressive schemes. Journal of the Royal Statistical Society, 110(2), 123--129 (1947)

Fingertips Tracking Based Active Contour for General HCI Application

Kittasil Silanon[1], Nikom Suvonvorn[1]

[1] Department of Computer Engineering, Faculty of Engineering, Prince of Songkla University, Hat Yai, Songkhla, Thailand 90112
{kittasil.silanon, nikom.suvonvorn}@gmail.com

Abstract. This paper presents a real time estimation method for 3D trajectory of fingertips. Our approach is based on depth vision, with Kinect depth sensor. The hand is extracted using hand detector and depth image from sensor. The fingertips are located by the analysis of the curvature of hand contour. The fingertips detector is implemented using concept of active contour which combine the energy of continuity, curvature, direction, depth and distance. The trajectory of fingertips is filtered to reduce the tracking error. The experiment is evaluated on the fingers movement sequences. Besides, the capabilities of the method are demonstrated on the real-time Human-Computer Interaction (HCI) application.

Keywords: We Fingertips Detection and Tracking, Hand Posture Estimation, Human-Computer Interaction (HCI).

1 Introduction

Hand gesture recognition has been a popular research in recent year. It provides more natural human-computer interaction. Many researches in this field related to real-time hand gesture recognition are proposed. Most of these system use trajectories of hand motion to recognize the commands [1,2,3,4,5]. However, the most important aspect of hand gesture recognition is to recognize commands accurately can be done with the accurate position of fingertips. Thus, some works have introduced on contours-based method for 2D fingertips tracking [6,7,8,9]. However, this approach cannot track fingertips robustly and usually design to track only stretched fingertips. Other systems for tracking the fingertips are using finger kinematic model [10,11,12] which search for the special form of the fingertips. These systems can work robustly. However, the computation cost is still too high. Using stereo vision has proposed to analyze the 3D fingertip positions [13,14,15,16]. The most problems in 3D fingertips are failure tracking when the fingertips are bending into the palm or overlapped each other. Therefore, this paper presented system to deal with such situations by using the depth image from Kinect. The hand region is segmented from the depth image and initial hand features are detected. The 3D fingertips are tracked using concept of active contour which occurs from internal and external energy that use features of

T. Herawan et al. (eds.), *Proceedings of the First International Conference on Advanced Data and Information Engineering (DaEng-2013)*, Lecture Notes in Electrical Engineering 285, DOI: 10.1007/978-981-4585-18-7_35,
© Springer Science+Business Media Singapore 2014

continuity, curvature, depth and distance from candidate fingertips that are detected at hand region in each frame. The energy represents the possible of candidate fingertips to be the fingertip in next frame. The tracking experiments are tested on basic finger movement. In addition, we develop an HCI application based on the fingertips tracking result. The rest of paper is organized in four sections by the following: initial hand segmentation, finger detection and tracking, experimentation results and conclusion respectively.

2 Initial Hand Segmentation

2.1 Hand Segmentation

From our previous work [5], we proposed hand detector by using object detection method which provides accurate result (Fig.1.a). The hand detector will be used to search hand's region in image. From experimentation, we found that hand detector failed when other parts of body such as face, arm move close to hand's region (Fig.1.b).

(a)   (b)

Fig. 1. Hand segmentation: (a) hand detector (b) wrong segmentation.

Therefore, depth image (Fig.2.c) is used in the system. The depth is a 3D position vector (x,y,z) which obtained from the Kinect camera, where the x and y are the rows and columns in an image and z is the depth readings that are stored in the pixels, for detecting how pixel far away from camera.

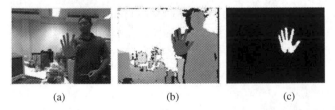

(a)   (b)   (c)

Fig. 2. Depth information: (a) complex background (b) depth image (c) hand's region.

The depth can be solved the problem of complex background (Fig.2.a) by setting initial depth value from hand detector to remove any object behind the hand. Thus, our system takes depth information to extract only hand's region in image. After extraction hand's region, the initial hand's features will be estimated. For example, hand center, fingertips position, palm size etc. All initial hand features are used to be reference value to compare changing of hand gesture in command recognition system.

2.2 Hand Center Point

We obtained the center point of hand's region that can be easily computed from the moments of pixels in hand's region, which is defined as:

$$X_c = \frac{M_{10}}{M_{00}}, \; Y_c = \frac{M_{01}}{M_{00}} \quad (M_{ij} = \sum_x \sum_y x^i y^j I(x, y)) \tag{1}$$

In the above equations, $I(x,y)$ is the pixel value at the position (x,y) of the image, x and y are range over the hand's region. The center point is calculated as Xc *and* Yc (Fig.3.a). The palm size is defined as the distance between the center point and the closest pixel on hand contour (Fig.3.b).

(a)   (b)

Fig. 3. Initial hand feature: (a) hand center (b) palm size.

2.3 Fingertips Position

Since the user is required to initialize the system by producing the pose of an "open" right hand facing the camera. Therefore, it is simple to locate fingertips from curvature of boundary point of hand's region. However, it may not be necessary to consider all the boundary points of hand's region. Thus, we use the polygon approximation method to extract key point [18] and is stored in a new series of key point P1,...,Pn (Fig.4.a). Each key point Pi has two parameters, the angle (θ) and slope. The angle can be estimated by using k-curvature [6] which calculates the angle of key point by two vectors $[P(i-k)P(i)]$ and $[P(i)P(i+k)]$ with the same range (k) (Fig.4.b). The key point, with curvature value is in the threshold and slope is positive, is an initial fingertip (Fig.4.c).

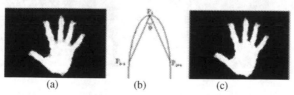

(a)   (b)   (c)

Fig. 4. Fingertips position: (a) key point (b) curvature calculation (c) fingertip points.

The initial positions of the five fingertips are detected. Each of them will be given a label that corresponds to the thumb, index, middle, ring and pinky fingers (f_j). As we control initial open hand posture to frontal view. We can simply label fingers based on sorting the five points by clockwise arranging around palm center. Nevertheless, it can only detect fingertip in open hand. Thus, tracking method is used for tracking fingertips which can be changed position at all time in hand gesture sequence.

3   Finger Detection and Tracking

3.1   Finger Location

Most movements in hand gesture are finger movements (Fig.5.a). Therefore, we defined two conditions to segment finger locations (Fig.5.b). For a stretching finger, use distance condition and a bending finger, use depth conditions follow the rules:

$$D(p, C_0) > k.R \quad \| \quad |Z(p) - Z(C_0)| < \tau_D \tag{2}$$

Where k is a scaling factor, D (p, C_0) is distance between considering pixel and hand center, Z is depth value and τ_D is a pre-defined depth threshold.

(a)   (b)

Fig. 5. Finger locations: (a) origin hand posture (b) extraction finger location.

3.2   Candidate Fingertips

We define searching area to locate the candidate fingertip positions (Cf_i). We assume that fingertip positions should be points on hand contour. Therefore, it may not be necessary to consider all points on hand contour. Hence, the polygon approximation algorithm is used again to find candidate fingertips. Finger.6.b shows points on contour for stretching finger. In addition, candidate fingertip positions can be found by assuming that they are closest to the camera in each finger region. Thus, we use depth to find point which has minimum depth to be candidate fingertips. Fig. 6.b shows these points inside contour.

(a)   (b)

Fig. 6. Candidate fingertips: (a) origin hand posture (b) candidate fingertips.

3.3   Fingertip Tracking

The tracking of the fingertip position between successive frames is built by concept of active contour [18,19]. The possible candidate fingertips of each fingertip will be assigned energy and then the maximum energy point is chosen to be the fingertip in the next frame. The discrete formulation of energy function which can be written as:

$$E(Cf_i) = E_{internal}(Cf_i) + E_{external}(Cf_i) \qquad (3)$$

The energy for each candidate fingertips can be decomposed into two basic energy term. $E_{internal}$ represents the internal energy of the candidate due to bending or stretching of finger and $E_{external}$ is the external constraint introduced by user. The internal Energy of candidate fingertips is defined as:

$$E_{internal} = E_{continuity} + E_{curvature} + E_{depth} \qquad (4)$$

The $E_{continuity}$, It forces the candidate fingertip points to be continuous, because the fingertip should not change much from the current point to the next one. Therefore, this term tries to keep point which has appropriate distance between the average distance ($\overline{d}$) and candidate fingertip point. The form for $E_{continuity}$ is the following:

$$E_{continuity} = \overline{d} - \left\| f_j - Cf_i \right\|^2 \qquad (5)$$

The $E_{curvature}$, this term will find smoothness of the candidate fingertips by considering contour curvature between the two vectors, $A = (x_i - x_{i-k}, y_i - y_{i-k})$ and $B = (x_{i+k} - x_i, y_{i+k} - y_i)$, k is constraint. If the candidate fingertips have the curvature value that fall under a threshold, these points will be kept to be possible candidate fingertips. The formula for $E_{curvature}$ is given by:

$$E_{curvature} = \cos^{-1} \frac{\mathbf{A}.\mathbf{B}}{\|\mathbf{A}\| \|\mathbf{B}\|} \qquad (6)$$

Because we have use property of depth, the E_{depth} has been established. The E_{depth} is distance between candidate fingertips to camera that represent in 16-bit depth data units of millimeter. As we have assumed, the fingertips should be found at the closest point to the camera. Therefore, the E_{depth} becomes maximum value when the candidate fingertips get close to a camera. The closest point will be given more priority than the rest points in order of depth. As we mentioned previously, The $E_{external}$ is the external constraint. Thus, in our system, the distances from all fingertips to considering candidate fingertip in image are used to describe about the $E_{external}$. For instance, if we are considering the movement of index fingertip, the distances from other fingertips to considering candidate fingertip point are equivalent to the external energy which will be used to estimate suitability for choosing the considering candidate fingertip to be the index fingertips. The distance from index fingertip to considering candidate fingertip should be shorter than other fingertip. On the other hand, if the distances from other fingertips are shorter than index fingertip distance, the considering candidate fingertip should be assigned to another fingertip. The energy function represents the importance of candidate fingertips relative to each fingertip. The candidate fingertip with maximum energy is selected to be new location of fingertip in next frame. In order to reduce the tracking error due to losing depth value, a simple low-pass filter is applied for the smoothness trajectory of fingertip tracking. The average point between selected point and current fingertip point will be estimated.

4 Experimentation Result

4.1 Fingertips Tracking Precision

We evaluate the fingertip tracking precision on basic finger movements between the tracked fingertips and the ground truth. We have defined the ground truth using the end point contour of each finger. The basic finger movements have included five sequences (Fig.7), bending finger (seq.1), moving finger into the palm (Seq.2), for crossing finger (Seq.3), hand movement (up, down, left and right) (Seq.4). 45^0 hand rotation (counterclockwise and clockwise) (Seq.5). Each sequence is tested at 10 rounds. Table1 shows the precision in terms of the Euclidean distance.

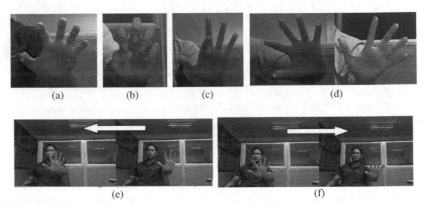

Fig. 7. Basic fingertip movements: (a) bending finger (b) moving finger into the palm (c) crossing finger (d) 45^0 hand rotation (e-f) hand movement.

Table 1. Fingertips Tacking Precision

Seq.	Tracking Precision(Pixel)					Avg.
	Thumb	Index	Middle	Ring	Pinky	
1	7.94	7.65	8.50	9.25	8.51	8.37
2	25.35	14.28	14.96	23.12	15.69	18.68
3	7.47	6.06	5.12	5.60	5.10	5.87
4	5.60	4.16	3.06	5.05	5.77	4.72
5	4.17	3.44	5.80	3.76	7.19	4.87

From experimentation result, we found that the tracking cases for crossing, rotation and movement (Seq.3, Seq.4 and Seq.5) give a good precision ($\approx$ 4-6 pixels error) because the candidate fingertips obtained from polygon approximation algorithm are quite well located to the fingertip ground truth. But the tracking cases for bending and moving into palm (Seq.1 and Seq.2) give less precision ($\approx$ 9-20 pixels error) due to the imperfect depth data received from Kinect that is not suitable for near mode function. Hence, depth data of finger may lose in some frames. From these factors, there might produce some errors in the fingertips tracking process.

4.2 Human-Computer Interaction Application

We carried out experiments on a general HCI Application that used as the commands to interface with Window Media Center on Window 7 (Fig.8.). For example, pressing up, down, left, right button. You can see video showing the real-time interactive at: http://www.youtube.com/watch?v=OnQra4We-4o

Fig. 8. HCI Application for controlling Window Media Center.

5 Conclusion

Fingertips and palm positions are significant features for hand gesture recognition. The most of previous works cannot track 3D fingertip positions because the complexity of finger movement. In this paper, we present the method to deal with these issues by using depth data feature for correctly hand segmentation and apply concept of active contour to track fingertips over finger movement. Our method shows good performance in term of real-time and also has capability to expansion to Human-Computer Interaction application. However, our method still has some limitation in tracking fingertips. The fingertips tracking procedure fails if the neighborhood candidate fingertips are lost, which is the case of the finger movements are too fast. For the possible applications, our system can be combined with other Human-Computer Interaction applications, such as finger-spelling process, robot controlling, visual input device and etc.

Acknowledgments. We would like to thank the National Research University Project of Thailand's Office of the Higher Education Commission for financial support.

References

1. Deng, J.W., Tsui, H.T.: An HMM-based approach for gesture segmentation and recognition. In: 15th International Conference on Pattern Recognition Proceedings, vol. 3, pp. 679-682 (2000)

2. Ho-Sub Yoon, Jung Soh, Younglae J. Bae, Hyun Seung Yang: Hand gesture recognition using combined features of location, angle and velocity. In: Pattern Recognition, vol. 34, pp. 1491–1501 (2001)
3. Feng-Sheng Chen, Chih-Ming Fu, Chung-Lin Huang: Hand gesture recognition using a real-time tracking method and hidden Markov models. In: Image and Vision Computing, vol. 21, pp. 745–758 (2003)
4. Elmezain M., Al-Hamadi: Gesture Recognition for Alphabets from Hand Motion Trajectory Using Hidden Markov Model. In: IEEE International Symposium on Signal Processing and information Technology, pp.1192-1197 (2007)
5. Kittasil Silanon, Nikom Suvonvorn: Hand Motion Analysis for Thai Alphabet Recognition using IIMM. In: International Journal of Information and Electronics Engineering vol. 1, pp. 65-71 (2011)
6. Wei Du, Hua Li: Vision based gesture recognition system with single camera. In: 5th International Conference on Signal Processing Proceedings WCCC-ICSP, vol. 2, pp.1351-1357 (2000)
7. Antonis A. Argyros, Manolis I. A. Lourakis: Vision-based interpretation of hand gestures for remote control of a computer mouse. In: Computer Vision in Human-Computer Interaction, pp. 40-51 (2006)
8. Ko-Jen Hsiao, Tse-Wei Chen, Shao-Yi Chien: Fast fingertip positioning by combining particle filtering with particle random diffusion. In: IEEE International Conference on Multimedia and Expo, pp. 977-980 (2008)
9. J. Ravikiran, Mahesh Kavi, Mahishi Suhas, R. Dheeraj, S. Sudheender, Pujari Nitin V.: Finger Detection for Sign Language Recognition In: International MultiConference of Engineers & Computer Scientists, pp. 489 (2009)
10. Lee, J., Kunii, T.L.: Model-based analysis of hand posture. In: IEEE Computer Graphics and Applications, vol.15, no.5, pp.77-86 (1995)
11. Cheng-Chang Lien, Chung-Lin Huang: Model-based articulated hand motion tracking for gesture recognition. In: Image and Vision Computing, vol. 16, Issue 2, pp. 121-134, (February 1998)
12. Lathuiliere, F., Herve, J. Y.: Visual tracking of hand posture with occlusion handling. In: 15th International Conference on Pattern Recognition Proceedings, vol.3, pp.1129-1133 (2000)
13. Dung Duc Nguyen, Thien Cong Pham, Jae Wook Jeon: Fingertip detection with morphology and geometric calculation. In: IEEE/RSJ International Conference on Intelligent Robots and Systems, pp. 1460-1465 (2009)
14. M. Do, T. Asfour, R. Dillman: Partical filter-based fingertips tracking with circular hough transform feature. In: Proceedings of the 12th IAPR Conference on Machine Vision Application (2011)
15. Raheja, J.L., Chaudhary, A., Singal, K.: Tracking of Fingertips and Centers of Palm Using KINECT Computational Intelligence. Third International Conference on Modeling and Simulation (CIMSiM), pp. 248-252 (2011)
16. Hui Liang, Junsong Yuan, Daniel Thalmann: 3D fingertip and palm tracking in depth image sequences. In: Proceedings of the 20th ACM international conference on Multimedia (MM '12). ACM, New York, NY, USA, pp.785-788 (2012)
17. Ramer-Douglas-Peucker algorithm, http://en.wikipedia.org/wiki/Ramer-Douglas-Peucker_algorithm
18. Michael Kass, Andrew Witkin, Demetri Terzopoulos: Snakes: Active contour models. In: International journal of computer vision, vol. 1, no. 4, pp. 321-331 (1988)
19. Donna J. Williams, Mubarak Shah: A Fast algorithm for active contours and curvature estimation. In: CVGIP: Image Understanding, vol. 55, no. 1, pp. 14-26 (January 1992)

Parallel Full HD Video Decoding for Multicore architecture

S.Sankaraiah* Lam Hai Shuan, C. Eswaran, and Junaidi Abdullah

Centre for Visual Computing, Multimedia University, Cyberjaya, Malaysia
{*sankar2510@ieee.org}{hslam,eswaran,junaidi}@mmu.edu.my

Abstract. Nowadays, the multi-core architecture is adopted everywhere in the design of contemporary processors in order to boost up the performance of multitasking applications. This paper mainly exploits the multi-core capability for full HD video decoding speedup to meet realtime display. Hantro 6100 H.264 decoder is chosen as the reference decoder. The serial decoding algorithm in the Hantro 6100 H.264 decoder is replaced with a parallel decoding algorithm. . In this research work, macroblock level parallelism is implemented using the enhanced version of macroblock region partitioning (MBRP) is implemented for the parallel video decoding of H.264 video. The results show that the workloads are well-balanced among the processor cores. It is observed that the maximum speedup values are attained when the decoder is running with 4 threads on a 4 core system and 8 logical core system configuration. Moreover, it is also observed that there is no degradation of visual quality throughout the decoding process.

Keywords: Multicore processor, speed-up, macroblock, parallel, load balancing, Hantro decoder.

1   Introduction

In general multi-core architecture platforms, without parallelism implementation, it overloads all the work on one processor and keeps other processors idle. This process does not utilize the available resources and works like a single processor. But, with parallel implementation, the multi-core architecture provides a platform for speeding up the application by performing thread-level and data-level parallelism rather than by increasing the operating frequency of the processor. In the proposed research work, Hantro 6100 H.264 decoder is chosen, since it is fully open-source and implemented in the Android mobile operating system [1]. The decoder is designed to support only the baseline profile of H.264 video decoding and it primarily targets for the mobile devices. The source code of the decoder is fully written in ANSI C standard, so it can be easily ported to different platforms. The decoder is implemented for the mobile platform and it can support a maximum of CIF resolution (352X288). In order to see the real full HD quality, the Hantro decoder implementation is modified for Windows platform using advanced data structures and some windows supported libraries. However, the decoder is still unable to decode and render full HD video smoothly, this is because only one processor is running and the frame rate of video playback is perceived at approximately 10 FPS on average. This value is far from the actual frame rate for real-time video playback, which is at least 30 FPS on average. From this it is observed that, the Hantro H.26 decoder is implemented in serial manner and one processor cannot handle the full-HD decoding task, since the H.264 decoding process is computation-intensive and time-consuming [2]. From the analysis, this research work is motivated by the need

---

* Corresponding author: S.Sankaraiah, email:sankar2510@ieee.org

T. Herawan et al. (eds.), *Proceedings of the First International Conference on Advanced Data and Information Engineering (DaEng-2013)*, Lecture Notes in Electrical Engineering 285, DOI: 10.1007/978-981-4585-18-7_36,
© Springer Science+Business Media Singapore 2014

318 S. Sankaraiah et al.

for parallel decoding of H.264 and non-uniform distribution of work among the processor cores. Hence, the Hantro H.264 video decoding process can be sped up using the data-level parallelism. As per Amdahls law, theoretically the speed-up of the decoding process will be increased by a factor equal to the number of cores in the processor [3]. In this research work, the macro-block level parallelism is implemented on the video decoder and made the workload equally among the processor cores.

The remaining part of the paper is organized as follows: Section 2 gives the literature review on the concepts of H.264 video decoding process and techniques of H.264 parallelization. Section 3 provides the detailed explanation on the design and implementation. Section 4 presents the results of the proposed implementation on the performance (speed), visual quality, thread utilization and cache misses are analyzed and discussed. Section 5 concludes the research work with the recommendation of future work.

2   Literature Review

The H.264 decoding process starts with the extraction of a compressed video frame from H.264 bitstream. The compressed video frame undergoes the process of entropy decoding to obtain the transformed residual image and its associated information of prediction. The transformed residual image then undergoes an inverse quantization process and followed by inverse Discrete Cosine Transform (IDCT) to obtain the original residual image. The reconstructed image undergoes deblocking filter before storing it into the frame buffer as the reference for next frame or before sending it out for display. The different stages of H.264 decoding process with serial implementation takes long time, which is not possible for real-time decoding. Thus, H.264 process requires parallelization to speed up the decoding process. According to C. Meenderinck et. Al ., H.264 parallelization can be performed at either task-level or data-level [4]. In task-level parallelism, the H.264 decoding task is decomposed into smaller entities or sub-tasks. Each sub-task is then assigned to different processors or cores for parallelism. In data-level parallelism, different portions of data are handled by different processors or cores for parallelism. Data-level parallelism has several advantages over task-level parallelism. Data-level parallelism has a higher scalability than task-level parallelism. This is due to the huge amount of data that stored in the video frame for processing [5]. Besides, parallelization of H.264 decoder using data-level parallelism, the load-balancing among the processors can be more easily achieved when compared to task-level parallelism since the time taken for processing each data is relatively the same.

According to C. Meenderinck et. al., data-level parallelism can be exploited at different levels of H.264 structures, such as GOP level, frame level, slice level and macroblock level[4]. In GOP-level parallelism, each Group Of Picture (GOP) is assigned to different processors or cores, so each processor and cores will decode its assigned GOP independently [6]. According to A. Gurhanli et. al. and S. Sankaraiah et. al., the size of GOP affects the speedup and memory usage [6]-[9]. From their results, by using small size of GOP, it yields a significant speedup due to lower memory usage. The lower memory usage will eventually cause less cache pollution, which is one of the factors that affect the performance. Frame-level parallelism suffers the problem of scalability due to the limited amount of B-frames that can exist within a video sequence . According to Y. Chen et. al., the performance of frame-level parallelism can be sped up by accelerating the P-frame decoding in order to unlock the more B-frames available for decoding to exploit more parallelism [10] , [11]. The advantage of slice-level parallelism is that there is no communication overhead and synchronization between processors and cores as the slice are independent of each other. Besides, the implementation of slice-level parallelism, it also consumes less memory resources when compared to GOP-level parallelism and the effect of the cache pollution can be minimized.The disadvantage of slice-level parallelism is that it suffers the problem of scalability because there can be no more than 8 slices in a frame for H.264 video [12]. Hence, slice-level parallelism may not be suitable to be implemented on decoder due to limited parallelism. Macroblock-level parallelism has the advantage of low data communication overhead. Besides, it also consumes lesser memory resources when compared to other levels of parallelism [13] , [14]. In macroblock-level parallelism, several macroblocks are assigned to different processors for concurrent decoding.

3   DESIGN AND IMPLEMENTATION

In this research work, the source code of Hantro 6100 H.264 decoder is modified in two aspects. The original source code of Hantro 6100 H.264 decoder implementation is only for the Android mobile platform. This is not suitable to analyze for full HD resolution applications. In order to exploit the capability of the multicore processor and to render full HD video, as a first task Hantro 6100 H.264 decoder is modified for Windows platform. The second aspect of modification is to replace the serial decoding algorithm with a parallel decoding algorithm to improve the speedup and render real-time full HD resolution video. This paper focuses on H.264 baseline profile. The main contribution of this research work is to implement decoding of full HD video resolution at 1920X1080 without any loss of quality. The proposed macroblock-level parallelism is implemented using enhanced macroblock-region partitioning (MBRP) to exploit data-level parallelism of H.264 video [13] . The programming model that used in this research work is based on fork-join model as specified by OpenMP standard [3]. In the parallel region, several threads are created and each thread handles the decoding task in one macroblock-region only. The width of each macroblock region is determined by dividing the width of the frame with the number of cores or processor. Each thread identifies starting and ending position of its region based on its own thread ID. Each thread decodes the macroblocks within its own region after all the data dependencies are resolved. However, each thread will perform entropy decoding if there are no macroblocks available for decoding. Whenever there is unresolved inter-thread dependency, each thread will continue its execution on entropy decoding instead of polling for the dependency to be resolved. So that, each thread will not fall into idle state. This idea is to effectively balance the workload among the threads, so that no threads will fall into an idle state. Hence threads do not have to wait for the availability of macroblocks to decode within its region. The detailed step by step procedure in the proposed algorithm is as follows:

1. Obtain the first macroblock index of the region.
2. Check whether current macroblock is entropy decoded, if it is not entropy decoded performs entropy decoding of one macroblock, otherwise proceed to the next step.
3. Check whether the data dependencies of current macroblock are resolved, if the dependencies are not resolved, perform entropy decoding of one macroblock, otherwise proceed to the next step.
4. Decode the current macroblock and then followed by performing deblocking filter.
5. Check whether it reaches to the end of the region, if it is the end of the region, proceed to step 7, otherwise proceed to the next step.
6. Obtain the next macroblock index of the region and proceed to step 2.
7. All the macroblocks within the region are completely decoded and wait for all threads to reach the barrier.

The video rendering is implemented with the use of the Simple DirectMedia Layer (SDL) library API [15]. SDL is a fully open-source multimedia development API, which can be used in different types of platforms. The proposed research work invokes the overlay functions defined in SDL library for rendering the decoded video frames. The tools for evaluating the visual quality (PSNR) and elapsed time are incorporatedine the source code as a data loggerr.

4   RESULTS AND DISCUSSION

4.1   Testing Environment and Platform

In this research work, Intel Core i7-930 multicore processor operates with windows 7, 64-bit operating system is used and has a clock speed of 2.8 GHz turbo up to 3.06GHz, with 32KB D-Cache (L1), 32KB I-Cache (L1), 256KB cache (L2), 8MB L3 cache and 8GB RAM for the profiling of the parallel Hantro 6100 H.264 decoder. Before performing the profiling on the decoder, some of the additional precautions have taken to isolate the unnecessary interferences, which are as follows:

1. Disabled the LAN card
2. Changed the theme from Windows Aero to Windows Basic
3. Closed all background processes running unnecessarily
4. Disabled the real-time protection of Anti Virus software
5. Disabled the disk defragmentation task

All these steps are considered to minimize the processor consumption before performing the profiling on the decoder. The samples of both PSNR and elapsed time with respect to the number of threads are recorded by using a data logger embedded inside the decoder. The profiles of both thread workload and thread concurrency are obtained by using the Intel Parallel Studios Parallel Amplifier. The compiler used is Intel C++ compiler. The Big Buck Bunny is the H.264 video sequence is used throughout the whole profiling process.

4.2   Performance metrics

The measurement of the decoders elapsed time is intended for benchmarking the performance of decoder with respect to the number of threads created. The measurement of PSNR is the visual quality representation of the video with respect to the number of threads created. The performance of the decoder is evaluated based on the elapsed time of decoding. The elapsed time is defined as the difference between the start time and end time of the decoding process. The shorter is the elapsed time, the higher is the performance of the decoder and the higher the speed up. The measurement of both PSNR and elapsed time are embedded inside the source code of the decoder as a data logger. The data collected when the system is running with Hyper-threading (HT) technology enabled and with HT technology disabled. Figures 1and 2 shows the data collected before the thread affinity is implemented, whereas Figure 3 shows the comparison of data collected before and after the thread affinity is implemented. The results of the data collection process are performed with video rendering disabled. For each sample of elapsed time, the speedup is calculated as the ratio of the elapsed time during serial execution to the parallel execution according to the Amdahls Law [3].

4.2.1 Performance analysis without hyper threading technology is enabled

When HT technology is not enabled, only 4 out of 8 processing elements or 4 physical cores (1 thread per core) are activated in the Intel Core i7 processor. From Figure 1, the speedup is gradually increasing with an increase in the number of threads from 1 to 4. This proves that the decoding performance can be speed up by utilizing multicore architecture.However when HT technology is not enabled,the decoder is running from

4 threads onwards the speedup decreases drastically, which is shown in Figure 1. This problem arises because of the sharing of 4 physical cores among the 6 or 8 threads. When the number of threads exceeds the number of available physical cores, the operating system will perform the task of thread switching to switch each physical core from one thread to another. This will pose a serious problem to the decoder speedup. This is because the threads keep waiting since threads are functioning like a thread pool.

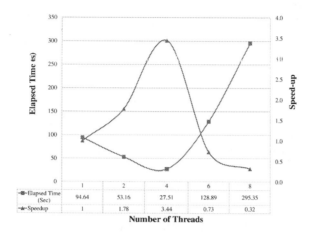

	1	2	4	6	8
Elapsed Time (Sec)	94.64	53.16	27.51	128.89	295.35
Speedup	1	1.78	3.44	0.73	0.32

Number of Threads

Fig. 1. The graph of elapsed time and speedup with respect to the number of threads (HT Technology is not enabled and no thread affinity implemented)

Discussion:

The parallel algorithm of decoder is designed in such a way that each thread is waiting for another thread to continue its execution by checking the status of reference flag. Thus, this might cause data race condition while reading the value of reference flag due to the thread switching activity. This data race condition will incur extra latency and eventually degrade the overall decoding performance. Another problem arises due to the extra waiting cycle of thread in an attempt to acquire the Mutex (Mutual exclusion) lock. Whenever there is no macroblock available for decoding, each thread will enter the shared region to perform entropy decoding [16]-[19]. In shared region, only one thread is allowed to access at one time. Before entering the shared region, each thread will try to acquire a Mutex lock to gain an access and execute the code inside the shared region. Therefore, the thread will possibly fall into waiting state due to repeated failure in acquiring the Mutex lock. This issue will also incur extra latency and eventually degrade the overall decoding performance. Therefore, this might give rise to the thread synchronization problem due to the data race condition and thread locking overhead.

4.2.2 Performance analysis with hyper threading technology is enabled

When HT technology is enabled, all the 8 processing elements or 8 logical cores (2 threads per core) are activated in the Intel Core i7 processor. From Figure 2, there is an improvement in the speedup when compared to the speedup shown in Figure 1. This is due to the advantage of hyper technology, as it provides Simultaneous Multithreading (SMT) to allow seamlessly concurrent execution of multiple threads. By using the 8 logical cores, up to 8 threads can be independently executed without sharing of cores between the threads. This will effectively minimize the effect of data race condition and extra waiting cycle in acquiring the Mutex lock. However, the speedup is not that much as expected. This might be due to the possibility of the occurrence of thread switching by operating system. In order to resolve thread switching problem, thread affinity has to be implemented.

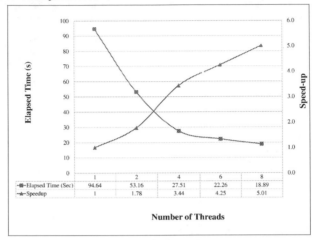

Fig. 2. Comparison of elapsed time and speedup with respect to the number of threads (HT technology is enabled and no thread affinity implemented)

4.2.3 Performance analysis with hyper threading technology is enabled and thread affinity is implemented

Thread affinity is a process of binding a single thread to a single core or processor in order to utilize the same cache resources on the same processor. As a result of implementing the thread affinity, the process may run more efficiently by minimizing the number of cache misses. After the thread affinity is implemented, each thread is permanently mapped to one logical core. In this case, the thread will not be switched to another core during the program runtime by operating system.

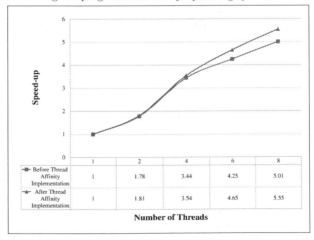

Fig. 3. The comparison of speedup before and after thread affinity is implemented

Discussion:

As shown in Figure 3 it is observed that, thread affinity implementation increases the speedup progressively when compared to the speedup of without implementing the thread affinity. It is also observed that the speedup is increased with the number of threads from 1 to 8 after implementing the thread affinity. However, when starting from 4 threads onwards, the speedup is slowly ramping up. This is due to the fact that the speedup is gradually approaching to its limit. According to the Amdahls Law, the maximum speedup is restricted by some part of the program cannot be parallelized and also the logical cores can give a maximum speedup of 33% only [?]. Therefore, the speedup of decoder is restricted by the sequential process, such as the initialization process. From the graphs shown in Figures 1, 2 and 3, it is observed that the number of threads created should be at least equal to the number of processor cores in order to obtain the maximum performance. Hence, it is advisable that the number of threads should not create more than the number of processing cores. Thus, it is justified that the implementation of thread affinity improves the stability in the performance of the decoder.

5   CONCLUSION AND RECOMMENDATION

In the proposed implementation, each thread manages its own task without the control of master thread. The results obtained from the proposed parallel algorithm (MBRP) shows that, the speedup values are marginal when it is running with 4 threads without thread affinity is implemented. This is partly due to the sharing task of entropy decoding (CAVLC) among the threads. After implementing the thread affinity, the stability in the performance of the decoder is gradually improved. From the results, it is observed that the workloads among the threads are well-balanced with the efficient design of the algorithm without any deviation among the workload of threads. Besides, the video rendered on screen shows the same visual quality (PSNR) irrespective of the number of threads created. Also, the results show that the decoder exhibited a good level of thread concurrency and CPU utilization. Furthermore, it is observed that the effect of cache misses is minimized after implementing the thread affinity. In order to improve the performance in terms of speedup, it is recommended that the entropy decoding task should be separated from the parallel region. The entropy decoding is a slow process as it has to be performed sequentially. Hence, the entropy decoding task should be assigned to another core, which has a higher processing speed. Besides, it is recommended that the MBRP should be implemented in both spatial and temporal domains. In this case, the decoding of macroblocks in next frame can start without waiting for all macroblocks in the current frame to finish. This way waiting time of each thread for macroblocks can be minimized.

References

1. Hantro   6100   H.264/AVC   Decoder,   http://www.on2.com/hantro-embedded/ software-video-codecs/6100-h.264-avc-decoder
2. I. E. Richardson.: The H.264 Advanced Video Compression Standard. 2nd ed,Wiley,UK (2010)
3. M. J. Quinn.: Parallel Programming in C with MPI and OpenMP. McGraw-Hill,USA (2003)
4. Cor Meenderinck , Arnaldo Azevedo , Ben Juurlink , Mauricio Alvarez Mesa , Alex Ramirez.: Parallel Scalability of Video Decoders. Journal of Signal Processing Systems. 57,(2), 173–194 (2009)
5. E. B. Van der Tol, E. Jasper, and R. H. Gelderblom.: Mapping of H.264 Decoding on a Multiprocessor Architecture. In: Proceeding of SPIE Conference on Image and Video Communications, pp. 707–709. SPIE Press, New York (2003)
6. A. Gurhanli, C. C. Chen, and S. Hung.: GOP-level parallelization of the H.264 decoder without a start-code scanner. In: 2nd International Conference on Signal Processing Systems (ICSPS), pp. (2010)

7. S. Sankaraiah, H. S. Lam, C. Eswaran and Junaidi Abdullah.: GOP Level Parallelism on H.264 Video Encoder for Multicore Architecture. In: IPCSIT, pp. 127–132. IACSIT Press, New York (2011)
8. S. Sankaraiah, H. S. Lam, C. Eswaran, and J. Abdullah.: Performance Optimization of Video Coding Process on Multi-Core Platform Using Gop Level Parallelism. International Journal of Parallel Programming.Springer, 1–17 (2013)
9. A. Gurhanli, C. C. Chen, and S. Hung.: Coarse Grain Parallelization of H.264 Video Decoder and Memory Bottleneck in Multi-Core Architectures. International Journal of Computer Theory and Engineering. 3,(3), 375–381 (2011)
10. Y. Chen, E. Li, X. Zhou, and S. Ge.: Implementation of H. 264 Encoder and Decoder on Personal Computers. Journal of Visual Communications and Image Representation. 17, (2006)
11. Y. Chen, X. Tian, S. Ge, and M. Girkar.: Towards Efficient MultiLevel Threading of H.264 Encoder on Intel Hyper-Threading Architectures. In: Proceedings of the 18th International Parallel and Distributed Processing Symposium, pp. 63–67. IEEE Press, New York (2004)
12. A. Azevedo, C. Meenderinck, B. Juurlink, A. Terechko, J. Hoogerbrugge, M. Alvarez, and A. Rammirez.: Parallel H.264 Decoding on an Embedded Multicore Processor. In: 4th International Conference on High Performance and Embedded Architectures and Compilers (2009)
13. S. Sun, D.Wang, and S. Chen.: A Highly Efficient Parallel Algorithm for H.264 Encoder Based on Macroblock region partition. In: 3rd international conference on High Performance Computing and Communications, pp.577–585 181–184. IEEE Press, New York (2007)
14. M. A. Mesa, A. Ramirez, A. Azevedo, C. Meenderinck, B. Juurlink, and M. Valero.: Scalability of Macroblock-level Parallelism for H.264 Decoding. In: 15th International Conference on Parallel and Distributed Systems, pp.236-243 (2009)
15. Simple DirectMedia Layer, http://www.libsdl.org
16. Y. Kim, J. Kim, S. Bae, H. Baik, and H. Song.: H.264/AVC decoder parallelization and optimization on asymmetric multicore platform using dynamic load balancing. In: IEEE International Conference on Multimedia and Expo, pp.1001-1004 (2008)
17. Cor Meenderinck , Arnaldo Azevedo , Mauricio Alvarez , Ben Juurlink , Alex Ramirez.: Parallel Scalability of H.264. In: first Workshop on Programmability Issues for Multi-Core Computers, pp.1–12. Goteborg, Sweden (2008)
18. Mauricio Alvarez Mesa, Alex Ramirez, Arnaldo Azevedo, Cor Meenderinck, Ben Juurlink, Mateo Valero.: Scalability of Macroblock-level Parallelism for H.264 Decoding. In: 15th International Conference on Parallel and Distributed Systems , pp.236–243. Shenzhen, China (2009)
19. M. Kim, J. Song, D. H. Kim, and S. Lee.: H.264 Decoder on Embedded Dual core with Dynamically Load-balanced Functional Partitioning. In: 17th IEEE International Conference on Image Processing, IEEE Press, New York (2010)

Part IV
Information Processing and Integration

A Comparison Study on Different Crowd Motion Estimation Algorithms Using Matlab

Ibrahim kajo, NidalKamel, AamirSaeed Malik

MSc in Electrical and Electronic Engineering
CISIR, Universiti Teknologi PETRONAS

Abstract. The optical flow describes the direction and time rate of pixels in a time sequence of two consequent images. A two-dimensional velocity vector, carrying information on the direction and the velocity of motion is assigned to each pixel in a given place in the picture. This article describes different motion detection methods, gives a brief illustration of the optical flow conception, and presents in details the Lucas-Kanade and Horn-Schunk algorithms for optical flow estimation and their implementation using MATLAB.

1 Introduction

Motion recognition is a very active research topic in computer vision with many important applications, including human–computer interfaces, content-based video indexing, video surveillance, and robotics, among others. Historically, visual action recognition has been divided into sub-topics such as gesture recognition [1, 2], facial expression recognition [3], and crowd behavior recognition for video surveillance [4].

Video surveillance has become a topic more and more important with the advent of technology and because of the increasing need for security. Central to the topic is automatic analysis and detection of abnormal events in public places or during public events. One particular class of community security issues is that involving a large number of people gathering together (crowding), sport competitions, demonstrations (e.g., strikes, protests), etc. Because of the high level of degeneration risk, the security of public events involving a large crowd has always been of high concern to relevant authorities. In recent years, a number of security agencies specialized in crowd management have turned out to respond to the need. Technically speaking, crowd behavior analysis can be divided into two tasks: motion information estimation and abnormal behavior modeling. The former usually extents to crowd tracking. It is a process by which we estimate the speed, direction and location of the crowd in a video sequence. Higher of crowd behavior can be used to detect anomalous events. One of the important aspects of video analytics is to recognize human behaviors. The survey performed by [5] presented many existing techniques that had been used to recognize human behaviors in distinct environments. Apart from analyzing the crowd behavior, there is a need to recognize it using video analytics. Recognizing the human behavior from a large crowd is a difficult task, especially, when the data is acquired from the live video streams [5, 6]. 3D based methods have also been reported in the literature for visual surveillance [7]. 3D methods are based on various depth

T. Herawan et al. (eds.), *Proceedings of the First International Conference on Advanced Data and Information Engineering (DaEng-2013)*, Lecture Notes in Electrical Engineering 285, DOI: 10.1007/978-981-4585-18-7_37,
© Springer Science+Business Media Singapore 2014

estimation techniques [8, 9, 10]. However, these methods are limited by the specialized hardware and/ or non-real time processing.

There is a large amount of existing work that has been employed to analyze human behavior in diverse situations using video cameras. Two types of video sequences can be analyzed [11] i.e. real time streaming and offline videos. Using these two types, the image processing strategy for recognition the human behavior involves the following general techniques: 1) - Motion Detection & Segmentation 2) - Classification 3) - Tracking of the moving objects.

The paper is organized as follows. First, we present a general overview of motion detection methods. Then, we review the definition of the conception optical flow in Section 2 and the methods Horn-Schunck and Lucas-Kanade for optical flow estimation will be described in Section 3. In conclusion the implementation of these methods and the development for Lucas-Kanade algorithm will be performed.

2 Motion Detection

Motion segmentation is essential to cut streams of motions into single motion. The task of object detection is to segment regions corresponding to the objects from the rest of an image. This process usually involves segmentation and object classification. There are five conventional approaches to moving object detection: statistical methods, temporal difference, block-based matching and optical flow.

To identify the motion of the complex background and consider the slow movement and the updates in the background, the current image frame is subtracted either by the previous frame or the next frame of the image sequences. This is termed as temporal differencing.

Statistical methods decompose the image/video into smaller regions, not linked to body parts or image coordinates. Instead, motions are recognized based on the statistics of local features from all regions. A direct advantage of those approaches is that the detection of the motion need not be performed explicitly for the computation of the space–time features. Such approaches are typically based on bottom-up strategies, which first detect interest points in the image, mainly at corner or blob like structures, and then assign each region to a set of preselected vocabulary-features. Image classification reduces them to computations on so-called bag of features (BOF), i.e. histograms that count the occurrence of the vocabulary-features within an image. On the downside, this approach is sensitive to the size or the scale, especially when it depends on geometrical features such as corner or edge features [11].

In block matching approach [12], each image frame is divided into non-overlapping blocks of equal size. Typically, the algorithm finds matching block – a block in the reference frame that matches the current block best. The best candidate block is found and the relative distances between a source block and its candidate blocks are called motion vectors. Differential motion estimation algorithms have been used commonly such as- Full Search (FS), Three-Step Search (TSS), New Three-Step Search (NTSS), Four-Step Search (FSS), Diamond Search Algorithm (DS), and Hexagon Based Search Algorithm (HEXBS).

3 Optical Flow

Assume $I(x,y,t)$ is the center pixel in a $n \times n$ neighborhood and moves by $\delta x, \delta y$ in time δt to $I(x+\delta x, y+\delta y, t+\delta t)$. Since $I(x,y,t)$ and $I(x+\delta x, y+\delta y, t+\delta t)$ are the images of the same point (and therefore the same) we have [13]:

$$I(x, y, t) = I(x + \delta x, y + \delta y, t + \delta t) \tag{1}$$

This assumption forms the basis of the 2D Motion Constraint Equation and is illustrated in Figure 1 below. The assumption is true to a first approximation (small local translations) provided $\delta x, \delta y, \delta t$ are not too big. We can perform a 1st order Taylor series expansion about $I(x,y,t)$ in (1) to obtain:

$$I(x + \delta x, y + \delta y, t + \delta t) = I(x, y, t) + \frac{\partial I}{\partial x} \delta x + \frac{\partial I}{\partial y} \delta y + \frac{\partial I}{\partial t} \delta t + H.O.T \tag{2}$$

Where H.O.T. are the Higher Order Terms, which we assume are small and can safely be ignored. Using the above two equations (1) (2) and dividing the both sides by δt, we obtain:

$$\frac{\partial I}{\partial x} Vx + \frac{\partial I}{\partial y} Vy + \frac{\partial I}{\partial t} = 0 \tag{3}$$

Here $Vx = \frac{\delta x}{\delta t}$ and $Vy = \frac{\delta y}{\delta t}$ are the x and y components of image velocity or optical flow and $\frac{\partial I}{\partial x}, \frac{\partial I}{\partial y}$ and $\frac{\partial I}{\partial t}$ are image intensity derivatives at (x,y,t). Note that the difference between (Vx, Vy) which are the x and y components of optical flow and (Ix, Iy, It) which is intensity derivatives. This equation can be rewritten more compactly as:

$$(Ix, Iy) \cdot (Vx, Vy) = -It \tag{4}$$

$$\nabla I \cdot \vec{V} = -It \tag{5}$$

Where $\nabla I = (Ix, Iy)$ is the spatial intensity gradient and $v = (Vx, Vy)$ is the image velocity or optical flow at pixel (x,y) at time t. $\nabla I \cdot v = -It$ is called the 2D Motion Constraint Equation and is one equation in two unknowns.

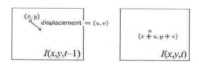

Fig. 1. The image at position (x,y,t) is the same as at (x+u,y+v,t+δt)

Differential methods belong to the most widely used techniques for optical flow computation in image sequences [14]. They can be classified into local methods such as the Lucas–Kanade technique, and into global methods such as the Horn/Schunck approach and its extensions .Often local methods are more robust under noise, while global techniques yield dense flow fields.

3.1   Lucas and Kanade method:

Lucas and Kanade implemented a weighted least-squares (LS) fit of local first-order constraints (5) to a constant model for $\vec{V}$ in each small spatial neighborhood Ω by minimizing:

$$E(Vx, Vy) = \sum_{x,y \,\in\Omega} W^2(x,y)[\nabla I(x,y,t).\vec{V} + I(x,y,t)]^2 \qquad (6)$$

Where *(x,y)* are weights allocated to individual pixels in Ω (2-D Gaussian coefficients or difference filter).The solution to (6) is given by:

$$W^2 A\,\vec{V} = W^2\vec{b} \qquad (7)$$

After multiplying the both side by A^T and dividing by$A^T W^2 A$:

$$A^T W^2 A\vec{V} = A^T W^2\vec{b} \qquad (8)$$

$$\vec{V} = [A^T W^2 A]^{-1} A^T W^2\vec{b} \qquad (9)$$

Where, for N pixels (for a n×n neighborhoods N $=n^2$), *(xi,yi)* $\in\Omega$ at a single time t:

$$A = [\nabla I(x1, y1), ...;\nabla I(xN, yN)]$$
$$W = diag[W(x1, y1) ..., W(xN, yN)]$$
$$\vec{b} = -\bigl(It(x1, y1) ... It(xN, yN)\bigr) \qquad (10)$$

The solution to equation (9) can be solved in closed form when $A^T W^2 A$ is a nonsingular matrix.

By performing an eigenvalue/eigenvector decomposition on$A^T W^2 A$, We will obtain two eigenvalues,$\lambda 2 \geq \lambda 1 \geq 0$, and their corresponding orthogonal unit eigenvectors, ˆe1 and ˆe2. If the smallest eigenvalue, $\lambda 1 \geq \tau D$, where τD is a user specified threshold (we use 1.0 typically), then the $\vec{v}$ computed by equation (9) is accepted as a reliable full velocity. If $\lambda 1 < \tau D$ but $\lambda 2 \geq \tau D$ then we can compute a least squares normal velocity by projecting $\vec{v}$ in the direction of the larger eigenvalue:

$$\vec{v}n = (\vec{v}.\,ˆe2)\,ˆe2 \qquad (11)$$

C. Liu. [18] Presented a new development for Lucas and Kanade method. The major difference is that they used conjugate gradient for solving large linear systems instead of Gauss-Seidel or successive over-relaxation (SOR).

3.2   Horn-Schunk Algorithm (HS):

This algorithm is based on a differential technique computed by using a gradient constraint (brightness constancy) with a global smoothness to obtain an estimated velocity field (Horn and Schunk 1981) [15]. The brightness of each pixel is constant along its motion trajectory in the image sequence. The relationship in continuous images sequence will be taken into account to estimate the original intensity for a gradient constraint. Let *I(x,y,t)* denote the gradient intensity (brightness) of point *(x,y)* in the images at time t. In each image sequence, *Ix*, *Iy* and *It* are computed for each pixel.

In practice [16], the image intensity or brightness measurement may be corrupted by quantization or noise. According to the equation for the rate of change of image brightness:

$$I(x,y,t) = I(x+Vx, y+Vy, t+1) \tag{12}$$
$$\varepsilon = Vx\,Ix + Vy\,Iy + It = 0 \tag{13}$$

Where Vx and Vy are the horizontal and vertical motion vectors of optical flow, respectively, one cannot expect ε to be zero. The problem is to minimize the sum of errors in the equation for the rate of change of image brightness as near as 0. So, the smoothness weight (α) is iteratively presented as:

$$Vx^{k+1} = Vx^{-k} - \frac{Ix[Ix\,Vx^{-k} + Iy\,Vy^{-k} + It]}{\alpha^2 + Ix^2 + Iy^2}$$
$$Vy^{k+1} = Vy^{-k} - \frac{Iy[Ix\,Vx^{-k} + Iy\,Vy^{-k} + It]}{\alpha^2 + Ix^2 + Iy^2}$$
$$\tag{14}$$

Where Vx^{-k} and Vy^{-k} denote horizontal and vertical neighborhood averages (Vx^k and Vy^k), which initially are set to zero and then the weighted average Kernel at neighboring points is applied for further iterations using Eq.(14) The smoothness weight (α) plays an important role where the brightness gradient is small, for which the suitable value should be determined.

Thomas Brox et al. [17] developed this method by studying energy functional for computing optical flow that combines three assumptions: a brightness constancy assumption, a gradient constancy assumption, and a discontinuity-preserving spatio-temporal smoothness constraint. In order to allow for large displacements, linearization in the two data terms is strictly avoided.

4 Implementation of Optical Flow Using Matlab

The main criteria parameters for this implementation will be: Time (elapsed time) of the optical flow calculation (speed vectors) and the Result after thresholding operation of separate images of the whole film sequence (its shape and visual quality). The performance of the proposed methods will be assessed using the PETS2009 dataset.

4.1 Testing the Horn-Schunck method:

In this method it is possible to adjust a few input parameters, the most important of which are: Smoothness factor λ, Number of iterations Size of the image Calculated values elapsed times with different values of the previous parameters are shown in table 1.

Table 1. Testing the Horn-Schunck method

Smoothness factor	Number of iteration	Size of the image	Elapsed time [s]between two frames
1	10	288x384	0.07
		576x768	0.3
	50	288x384	0.3
		576x768	1.4
	100	288x384	0.7
		576x768	2.8
0.1	10	288x384	0.08
0.01	10	288x384	0.09

Figure 2. Shows the results of calculated optical flow speed vectors with different number of iterations including the result after thresholding:

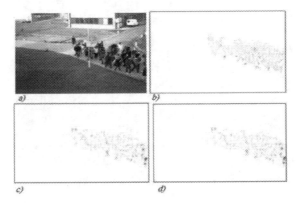

Fig. 2. (a) Frame 150 of PETS2009 S1.L1.view1 sequence. (b)   Number of iteration =10. (c) Number of iteration =50. (d) Number of iteration =100

In most cases we found that Horn-Schunck produces slightly lower error than Lucas-Kanade for the respective best parameter settings. The explicit regularization term in Horn-Schunck also provides visually superior results without the over-smoothing artifacts of the Brox algorithm. On the other hand, Horn-Schunck is somewhat sensitive to the choice of parameter settings.

4.2   Testing the Lucas-Kanade method:

Temporal gradient filter and the size of the image have the greatest effect on this method temporal gradient filter. The calculated values elapsed times are shown in the table 2 and the figure 3 shows the result after thresholding:

Table 2. Testing the Lucas and Kanade method

Temporal gradient filter	Size of the image (pixels)	Elapsed time[s] between two frames
Difference filter	116 x 154	2.6
	288 x 384	18.1
	576 x 768	76.2
Derivative of Gaussian (C.liu)	116 x 154	0.4
	288 x 384	3.3
	576 x 768	15.2
Derivative of Gaussian	576 x 768	60.3

Fig. 3. (a) Frame 150 of PETS2009 S1.L1.view1 sequence. (b)   Derivative of Gaussian. (c) Difference filter. (d) Difference filter and different colormap.

On the other hand, the implementation demonstrates that FBMAs depends strongly on motion content. For example, the TSS is good for blocks with rapid motion content. For moderate and slow motion (quasi-stationary and stationary) blocks, the FSS and DS perform better. The NTSS can effectively solve blocks with both rapid and slow motion. Furthermore, in the case of the block-matching and Lucas-Kanade algorithms, window size (w) is the key parameter that balances stability with spatial resolution. Larger windows provide a higher degree of confidence in the vector field, but do not reveal fine-scale detail.

4   Conclusion

Due to the calculated values and the depicted result it can be observed that the value of the smoothness factor in Horn-Schunck method cannot be defined exactly because the suitable value is varying upon different image sequences. The suitable iteration times also cannot be defined for the best outcome, which impacts the processing time for the best motion vector at the output. On the other hand Lucas-Kanade method is dependent on the window size which is the key parameter that balances stability with spatial resolution. On the other hand this method appears to be faster under experiment and good for the high texture and small motion.

References

1. A. Erol, G. Bebis, M. Nicolescu, R.D. Boyle, X. Twombly, Vision-based hand pose estimation: a review, Computer Vision and Image Understanding 108 (1-2) (2007) 52–73.
2. V.I. Pavlovic, R. Sharma, T.S. Huang, Visual interpretation of hand gestures for human–computer interaction: a review, Transactions on Pattern Analysis and Machine Intelligence 19 (7) (1997) 677–695.
3. W. Zhao, R. Chellappa, P.J. Phillips, A. Rosenfeld, Face recognition: a literature survey, ACM Computing Surveys 35 (4) (2003) 399–458.
4. W. Hu, T. Tan, L. Wang, S. Maybank, A survey on visual surveillance of object motion and behaviors, IEEE Transactions on Systems, Man and Cybernetics 34 (2004) 334–352.
5. J. Candamo, M. Shreve, D. B. Goldgof, D. B. Sapper, and R. Kasturi, "Understanding Transit Scenes: A Survey on Human Behavior-Recognition Algorithms," Intelligent Transportation Systems, IEEE Transactions on, vol. 11, pp. 206-224, 2010.
6. N. Gagvani, "Challenges in Video AnalyticsEmbedded Computer Vision," B. Kisačanin, et al., Eds., ed: Springer London, 2009, pp. 237-256.
7. Yasir Salih, Aamir Saeed Malik, "Comparison of Stochastic Filtering Methods for 3D Tracking", Pattern Recognition (IF: 3.402), Vol. 44, Issue 10, pp. 2711-2737, October 2011.
8. Aamir Saeed Malik, Tae-Sun Choi, "Effect of noise and source illumination on 3D shape recovery", International Journal of Pattern Recognition & Artificial Intelligence, Vol. 22, No. 5, pp. 945-958, August, 2008.
9. Muhammad Asif, Aamir Saeed Malik, Tae-Sun Choi, "3D shape recovery from image defocus using wavelet analysis", Proceedings of IEEE International Conference on Image Processing (ICIP), Genova, Italy, vol. 1, pp. 1025-1028, 11-14 September, 2005.
10. Aamir Saeed Malik, Seong-O Shim, Tae-Sun Choi, "Depth map estimation using a robust focus measure", Proceedings of IEEE International Conference on Image Processing (ICIP), San Antonio, Texas, USA, pp 564-567, 16-19 September, 2007.
11. D. Weinland et al. "survey of vision-based methods for action representation, segmentation and recognition"Computer Vision and Image Understanding 115 (2011) 224–241.
12. S. Immanuel Alex Pandian, Dr. G. Josemin Bala, Becky Alma George,"A Study on Block Matching Algorithms for Motion Estimation", International Journal on Computer Science and Engineering (IJCSE), Vol. 3, No. 1, pp. 34-44, Jan 2011.
13. Barron J. L., Thacker N. A. Tutorial: Computing 2D and 3D Optical Flow, Last updated 2005-1-20 [cit. 2010-10-19], Available www: http://www. tina-vision.net/docs/memos /2004-012.pdf.
14. Barron J. L., Fleet D. J. Performance of Optical Flow Techniques, International Journal of Computer Vision , Vol. 12, No. 1, pp 43-77, 1994.
15. Horn, B.K.P.; and Schunck, B.G. 1981. Determining optical flow. Artificial Intelligence 17(1-3): 185-203.
16. Darun Kesrarat et al. "Tutorial of Motion Estimation Based on Horn-Schunk Optical Flow Algorithm in MATLAB" IEEE AU J.T. 15(1): 8-16 (Jul. 2011).
17. Thomas Brox et al. "High Accuracy Optical Flow Estimation Based on a Theory for Warping" in Proc. 8th European Conference on Computer Vision, Springer LNCS 3024, T. Pajdla and J. Matas (Eds.), vol. 4, pp. 25-36, Prague, Czech Republic, May 2004.
18. C. Liu. Beyond Pixels: Exploring New Representations and Applications for Motion Analysis. Doctoral Thesis. Massachusetts Institute of Technology. May 2009.

A Modified LRE-TL Real-time Multiprocessor Scheduling Algorithm

Hitham Alhussian[1], Nordin Zakaria[1], Fawnizu Azmadi Hussin[2], Hussein T. Bahbouh[3]

{[1]IT Dept., [2]EE Dept.}, Universiti Technologi Petronas, Bandar Seri-Iskandar, 31750 Tronoh, Malaysia
[3]EE Dept., Damascus University, 12837, Damascus, Syria

{halhussian,htbahbouh}@gmail.com,
{nordinZakaria,Fawnizu}@petronas.com.my

Abstract. A modified version of Local Remaining Execution-TL (LRE-TL) real-time multiprocessor scheduling algorithm is presented. LRE-TL uses two events to make scheduling decisions: The Bottom (B) event and the Critical (C) event. Event B occurs when a task consumes its local utilization meaning that it has to be preempted. Event C occurs when a task's local laxity becomes zero meaning that the task should directly be scheduled for execution. Event C always results in a task migration, therefore we modified the initialization procedure of LRE-TL to select the tasks that have a higher probability of firing a C event to be scheduled for execution firstly, and thereby tasks migration will be reduced. We have conducted an independent-samples t-test to compare tasks migration using the original LRE-TL algorithm and the modified algorithm. The results obtained showed that there was a significance reduction in tasks migration when the proposed solution is applied.

Keywords: Real-time, Multiprocessor, Scheduling, Preemptions, Migrations

1 Introduction

Real-time Systems are systems in which the correctness of the system doesn't depend only on the logical results produced, but also the physical time when these results are produced [1]. Meeting the deadlines of a real-time task set in a real-time multiprocessor system, requires the use of an optimal scheduling algorithm. A scheduling algorithm is said to be optimal if it successfully schedules all tasks in the system without missing any deadline provided that a feasible schedule exists for the tasks [2-3]. The scheduling algorithm decides which processor the task will be executed on, as well as the order of the tasks execution. Although a scheduling algorithm may be optimal, but sometimes it can't be applied practically [4]. This because of the scheduling overheads, in terms of task preemptions and migrations that accompany its work.

In this paper we consider the possibility of reducing scheduling overhead incurred by tasks migration in LRE-TL algorithm by firstly sorting the tasks with Largest Local Remaining Execution First LLREF or Least Laxity First LLF, when initializing

T. Herawan et al. (eds.), *Proceedings of the First International Conference on Advanced Data and Information Engineering (DaEng-2013)*, Lecture Notes in Electrical Engineering 285, DOI: 10.1007/978-981-4585-18-7_38,
© Springer Science+Business Media Singapore 2014

the TL-plane. Then we select the first m tasks to be scheduled for execution. We have realized that when tasks are sorted with LLREF, a significance reduction in tasks migration is noticed. However, it is worth mentioning that sorting tasks will increase the complexity of the TL-plane initialization procedure. To avoid this problem, we propose an alternative solution to be discussed in section 7.

The rest of this paper is organized as follows: Section 2 describes the task model and defines the terms that will be used in this paper. Section 3 gives an overview of real-time multiprocessor scheduling and reviews related algorithms. In section 4 we show how to reduce tasks migration in LRE-TL. In section 5 we present and discuss the simulated results. Lastly we conclude in section 6.

2 Model and Terms Definition

In real-time systems, a periodic task [5] is one that is released at a constant rate. A periodic task T_i is usually described by two parameters; its execution e_i and its period p_i. The release of a periodic task is called a job. Each job of T_i is described as $T_{i,k} = (e_i, p_i)$ where k=1, 2, 3, ... The deadline of the kth job of T_i i.e $T_{i,k}$, is the arrival time of job $T_{i,(k+1)}$ i.e, at $(k + 1)p_i$.

A task's utilization is one of the important parameters and is described as $u_i = {e_i}/{p_i}$. A task's utilization is defined as the portion of time that the task needs to execute after it is released and before it reaches its deadline. The total as well as the maximum utilization of a task set T are described as U_{sum} and U_{max} respectively. A periodic task set is schedulable on m identical multiprocessor *iff* $U_{sum} \leq m$ and $U_{max} \leq 1$ [6].

3 A Review of Real-time Multiprocessor Scheduling

Scheduling on real-time multiprocessor systems can be classified into three categories: partitioning, global, and cluster scheduling. In the partitioning category, the scheduling process is divided into two steps. In the first step, tasks are allocated statically to processors and they are not allowed to migrate between processors later. The second step is the scheduling, in which each processor is scheduled using a uni-processor scheduling algorithm. The advantage of partition scheduling is that the scheduling problem is reduced from multiprocessor scheduling to a uni-processor one which has been extensively studied and are known to be well optimal. However partition scheduling suffers from two problems. Firstly, assigning tasks to processors is a bin-packing problem, which is known to be NP-hard. Secondly, there are task systems that can't be schedule unless they are not partitioned. The global scheduling category maintains a global task queue ordered according to a specific policy. The scheduler then allocates the highest priority tasks to the available processors. In the global scheduling, tasks are allowed to migrate between processors. Unfortunately uni-processor scheduling algorithms can't be used here since they produce low processor utilization. However, recently some global scheduling algorithms have been proposed that can achieve processor utilization of m such as P-fair, LLREF and LRE-

TL. Cluster scheduling, is a combination of both partitioned and global scheduling. Cluster scheduling, uses one scheduler at minimum and maximum N schedulers (where N is the number of processors). Each scheduler is assigned a subset of processors which may contain minimum 1 processor and maximum N processors (where N is the number of processors). Then each scheduler may schedules its tasks using one of the global, partitioned, or cluster scheduling algorithms [6-9].

3.1 LLREF

Largest Local Remaining Execution First, LLREF, is a real-time multiprocessor scheduling algorithm based on the fluid scheduling model, in which all tasks are executed at a constant rate. LLREF divides the schedule into Time and Local execution time planes (TL-planes), which are determined by task deadlines. The algorithm schedules tasks by creating smaller "local" jobs within each TL-plane. The only parameters considered by the algorithm during a TL-plane are the parameters of the local jobs. When a TL-plane completes, the next TL-plane is started. The duration of each TL-plane is the amount of time between consecutive deadlines [8].
For example, if we have the following task set

Table 1. Sample Task set

T	e	p
T1	3	7
T2	5	11
T3	8	17

Then, the intervals of the TL-planes will be as follows:

Table 2. TL-pane Intervals for task set in Table 1

TL-plane	Interval
TL-0	[0, 7)
TL-1	[7, 11)
TL-2	[11, 14)
TL-3	[14, 17)
TL-4	[17, 21)
TL-5	[21, 22)
.	.
.	.

Within each TL-plane, the local execution is calculated for all tasks. For example, if t_{f_0} and t_{f_1} is the starting and ending time of a TL-plane, T_i's local execution is calculated as $l_{i,0} = u_i(t_{f_1} - t_{f_0})$ i.e the local remaining execution of each task is proportional to its utilization. If task T_i starts its execution at time t_x then its local remaining execution l_i, x starts to decrease. Whenever a scheduling event occurs, LLREF selects the m highest remaining execution tasks for execution. The selected tasks will continue to execute until one of the following events occur [8, 10].

➢ *Event B*, the bottom (B) event occurs when a task completes its local remaining execution (i.e., when $l_{i,x} = 0$) [8].

➢ *Event C*, the critical (C) event occurs when a task consumes its local laxity and can't wait anymore therefore; it must be selected directly for execution; otherwise it will miss its deadline. (i.e., $l_{i,x} = t_{f_1} - t_{f_x}$) [8].
Fig. 1, below shows both events.

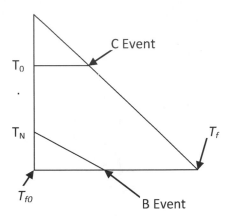

Fig. 1. B and C events

LLREF continue to execute until all tasks within the TL-plane complete their local remaining execution [8], then the next TL-plane is initialized and the process is repeated.

3.2 LRE-TL

LLREF introduces high overhead in terms of running time as well as preemptions and migrations [6]. LRE-TL (local remaining execution-TL) is a modification of LLREF. The key observation of LRE-TL is that there is no need to select tasks for execution based on largest local remaining execution time when a scheduling event occurs. In fact, any task with remaining local execution time will do. This observation greatly reduces the number of migrations within each TL-plane compared to LLREF. Moreover, LRE-TL is extended to support scheduling of sporadic tasks with implicit deadlines while achieving utilization bound of *m* [6]

LRE-TL algorithm contains four procedures. The main procedure starts by calling the TL-plane initializer procedure at each TL-plane boundary. Then it checks for each type of scheduling event and calls the respective handler when an event occurs. After that, the main procedure instructs the processors to execute their designated tasks [6].
At each TL-plane boundary, LRETL calls the TL-plane initializer. Within a TL-plane, LRE-TL processes any A, B or C events. The TL-plane initializer sets all parameters for the new TL-plane. The A event handler determines the local remaining execution of a newly arrived sporadic task, and puts the task in one of the heaps (HB or HC). The B and C event handler maintains the correctness of HB and HC [6].

LRE-TL maintains three heaps, H_D for the deadline heap, H_B for the bottom event, and H_C for the critical event. The algorithm starts by firstly initializing the TL-plane, in which the deadline heap will be populated with tasks that arrived at time T_{cur}, and then the algorithm starts adding tasks to be scheduled for execution on heap H_B until all processors are occupied. After that, all remaining tasks will be added to heap H_C. For tasks added to heap H_B and H_C, their keys are set to the time at which the task will trigger a scheduling event [6].

LRE-TL algorithm will not preempt a task unless it is absolutely necessary. When a B event occurs, the task generated the B event will be preempted and replaced by the minimum of heap H_C, the closest task to fire a C event. All task that was executing prior to the B event will continue to execute (on the same processor) after the B event is handled. On the other side, when a C event occurs, the task that fired the C event should be immediately scheduled for execution. This is done by preempting the minimum of heap H_B and replacing it with the task that fired the C event. The preempted task in turn, will be added to heap H_C [6].

4 Reducing Tasks Migration

The overheads incurred by global scheduling can potentially be very high specially when considering the hardware architecture. The fact that jobs can migrate from one processor to another can result in additional communication loads and cache misses, leading to increased worst-case execution times [9]. As we mentioned before, LRE-TL starts execution by firstly initializing the TL-plane wherein the deadline heap is updated with the deadline of tasks arrived at time T_{cur}. After that, both heaps H_B and H_C are populated with tasks selected for execution, and tasks that will remain until they consume their local laxity respectively. We realized that if we firstly sort the task with Largest Local Remaining Execution First (LLREF), before populating heaps H_B and H_C, a significant reduction of event C- which results in a task migration- is noticed. The following example clearly explains this.

Example 1
- *Without Sorting Tasks*
    The following table contains 8 tasks with their execution *e*, period *p* and local remaining execution *l* for the first TL-plane which has the interval [0, 10).

<table>
<tr><th colspan="4">Table 3. Task set for example 1</th></tr>
<tr><th>T</th><th>e</th><th>p</th><th>L_i [0, 10)</th></tr>
<tr><td>T1</td><td>8</td><td>17</td><td>4.7</td></tr>
<tr><td>T2</td><td>10</td><td>30</td><td>3.3</td></tr>
<tr><td>T3</td><td>5</td><td>11</td><td>4.5</td></tr>
<tr><td>T4</td><td>8</td><td>29</td><td>2.8</td></tr>
<tr><td>T5</td><td>1</td><td>10</td><td>1.0</td></tr>
<tr><td>T6</td><td>11</td><td>13</td><td>8.5</td></tr>
<tr><td>T7</td><td>3</td><td>26</td><td>1.2</td></tr>
<tr><td>T8</td><td>15</td><td>18</td><td>8.3</td></tr>
</table>

If we would like to schedule the tasks on a system of 4 processors, then both heaps H_B and heap H_C will be initialized as follows

Table 4. Initialization of heap H_B and heap H_C for
the first TL-plane for task set in Table 3

TL-Plane 0 [0, 10)			
Heap H_B			
T4	T2	T3	T1
2.8	3.3	4.5	4.7
Heap H_C			
T6	T8	T7	T5
1.5	1.7	8.8	9.0

It can be clearly seen that here both of T6 as well as T8 will fire C event at time T_{cur}
=1.5 and T_{cur}=1.7 respectively. This will result in the migration of both T4 and T2
which will be scheduled to execute later at time T_{cur}=8.7 and 8.4 respectively.
- *Tasks Sorted with LLREF*

On the other hand, when tasks are sorted with their local remaining execution, we
get the following order:

Table 5. Task set of Table 3 after sorting

T	e	p	L_i [0, 10)
T6	11	13	8.5
T8	15	18	8.3
T1	8	17	4.7
T3	5	11	4.5
T2	10	30	3.3
T4	8	29	2.8
T7	3	26	1.2
T5	1	10	1.0

Table 6. Initialization of heap H_B and heap H_C for
the first TL-plane for task set in Table 5

TL-Plane 0 [0, 10)			
Heap H_B			
T5	T6	T3	T7
4.5	4.7	8.3	8.5
Heap H_C			
T8	T1	T2	T4
6.7	7.2	8.8	9.0

In this case we can see that no C event will be fired since all tasks of heap H_B will
finish their execution before tasks of heap H_C consumes their local laxity. So no task
migration will occur.

4.1 Proposed Solution

We believed that, as mentioned before, sorting of tasks will increase the complexity of the TL-plane initialization procedure [6]. To overcome this problem we propose not to sort the tasks, instead we will utilize heap H_C to overcome the sorting problem. So, firstly, we populate heap H_C with all active tasks after calculating their local laxity. Since heap H_C is a minimum heap i.e. it maintains the element with the minimum key at the top; we can get the tasks back from it ordered accordingly to their least laxity first which is also equivalence to the largest local remaining first order. After heap H_C is populated, we extract the first m tasks from it, add them to heap H_B and assign them to the m processors. In this case the complexity of the TL-plane initialization procedure will remain the same and will not be affected. The algorithms of the original TL-plane initialize procedure as well as the modified one are depicted in Figure 2 below.

```
1.  Start                                  1.  Start
2.  Update the deadline heap H_D          2.  Update the deadline heap H_D with
    with tasks that arrived at                 tasks that arrived at T_cur
    T_cur
3.  Z=1                                    3.  For all active tasks
4.  For all active tasks                   4.     l = u_i(T_f - T_cur)
5.     l = u_i(T_f - T_cur)                5.     T_i.key = T_f - l
6.     If Z <= M then                      6.     H_C.insert(T_i)
7.        T_i.key = T_cur + l              7.  End for
8.        T_i.proc-id = Z                  8.  Z=1
9.        Z.task-id = T_i                  9.  While    (Z<=M    and    NOT
                                               Hc.isEmpty())
10.       H_B.insert(T_i)                  10.    T=H_C.extract-min()
11.       Z=Z+ 1                           11.    T.key= T_f - T.key + T_cur
12.    Else                                12.    T.proc-id = Z
13.       T_i.key = T_f - l               13.    Z.task-id = T
14.       H_C.insert(T_i)                  14.    H_B.insert(T)
15.    End if                              15.    Z=Z+ 1
16. End for                                16. End while
17. Z' = Z + 1                             17. Z' = Z + 1
18. While Z' <=M                           18. While Z' <=M
19.    Z'.task-id=NULL                     19.    Z'.task-id=NULL
20. End while                              20. End while
21. End                                    21. End
```

Fig. 2. LRE-TL *initialize* procedure

(a) Original procedure   (b) Modified LRE-TL *initialize* procedure

5 Result and discussions

We have conducted experimental work to verify our work. We have tested the algorithm using random task sets of 4, 8, 16, 32 and 64 that run on 2, 4, 8, 16 and 32 processors respectively. For each task set, we have generated 1000 samples. Fig. 3 below shows the difference between the total tasks migration for the first TL-plane when using the original TL-plane initialize procedure and when we apply our proposed TL-plane initialize procedure. We have also conducted an independent-samples t-test to compare tasks migration when using the original TL-plane initialize procedure, and when we use our proposed TL-plane procedure. There was a significant reduction in tasks migration when we use our proposed TL-plane initializer procedure and when the original TL-plane initializer procedure conditions. Table 7 below summarizes these results. These results suggest that the proposed TL-plane initializer procedure really does have an effect on tasks migration. Specifically, our results suggest that when the proposed TL-plane initialize procedure is used, tasks migration is reduced significantly.

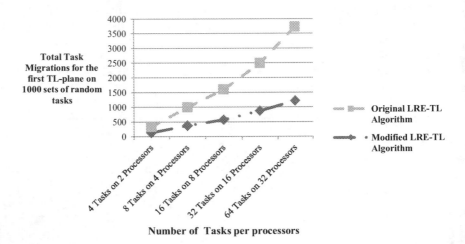

Fig. 3. Total tasks migration when using the original LRE-TL algorithm against the modified one

Table 7: T-test Results

Test Sample (1000 sets of)	Original Algorithm		Modified Algorithm		T	p	H
	M	SD	M	SD			
4 tasks on 2 Processors	0.206	0.404632831	0.132	0.33866014	4.4349	9.7124e-06	1
8 tasks on 4 Processors	0.624	0.811962612	0.373	0.698827131	7.4092	1.8652e-13	1
16 tasks on 8 Processors	1.026	1.364253	0.575	1.151391	7.9890	2.2758e-15	1
32 tasks on 16 Processors	1.617	2.32071	0.876	1.930894	7.7618	1.3263e-14	1
64 tasks on 32 Processors	2.517	4.044015958	1.212	3.254789	11.5798	4.64E-30	1

Furthermore, we have implemented the modified algorithm as well as the original one using the task set example given in Table 3 above from time t=0 until time t=29 i.e the first 10 TL-planes. The implementation has been conducted on a machine with Core I7 processor equipped with 4 cores. We have used Java^{TM} Visual VM of Oracle [11] to trace the tasks. Fig. 4 and Fig. 5 below show the result of the CPU profiler of Java^{TM} Visual VM for both algorithm. It can be clearly seen that from both figures the reduction of the number of invocations of the procedure *handleBorCEvent* that handles both scheduling events B and C, as well as the helper procedure used to manage the heaps.

Hot Spots - Method	Self time [%] ▼	Self time		Invocations
javaapplication1.Task.run ()		1040982 ms	(100%)	4
javaapplication1.LRE_TL.handleBOrCEvent ()		22.2 ms	(0%)	90
javaapplication1.TaskHeap.getMinimum ()		19.6 ms	(0%)	9694
javaapplication1.LRE_TL.TL_Plane_Initialize ()		15.7 ms	(0%)	10
javaapplication1.Task.myWakeUp ()		7.92 ms	(0%)	9965
javaapplication1.TaskHeap.isEmpty ()		7.64 ms	(0%)	9854
javaapplication1.TaskHeap.printHeap ()		4.43 ms	(0%)	20
javaapplication1.Task.<init> ()		1.75 ms	(0%)	80
javaapplication1.TaskHeap.siftDown (int)		0.729 ms	(0%)	240
javaapplication1.TaskHeap.siftUp (int)		0.577 ms	(0%)	290
javaapplication1.Task.proceed ()		0.419 ms	(0%)	60
javaapplication1.TaskHeap.insert (javaapplication1...		0.379 ms	(0%)	160
javaapplication1.TaskHeap.removeMin ()		0.371 ms	(0%)	160

Fig. 4. The result of CPU Profiler of Java Visual VM for the original algorithm

Hot Spots - Method	Self time [%] ▼	Self time		Invocations
javaapplication1.Task.run ()		946921 ms	(100%)	4
javaapplication1.LRE_TL.TL_Plane_Initialize ()		20.5 ms	(0%)	10
javaapplication1.TaskHeap.getMinimum ()		20.3 ms	(0%)	9892
javaapplication1.LRE_TL.handleBOrCEvent ()		16.5 ms	(0%)	80
javaapplication1.Task.myWakeUp ()		8.69 ms	(0%)	9925
javaapplication1.TaskHeap.isEmpty ()		7.76 ms	(0%)	10012
javaapplication1.TaskHeap.printHeap ()		4.41 ms	(0%)	20
javaapplication1.Task.<init> ()		2.75 ms	(0%)	120
javaapplication1.TaskHeap.siftDown (int)		0.456 ms	(0%)	160
javaapplication1.TaskHeap.siftUp (int)		0.403 ms	(0%)	190
javaapplication1.TaskHeap.<init> (int)		0.392 ms	(0%)	20
javaapplication1.LRE_TL$1.compare (Object, Obj...		0.345 ms	(0%)	170
javaapplication1.Task.proceed ()		0.330 ms	(0%)	80
javaapplication1.TaskHeap.insert (javaapplicatio...		0.292 ms	(0%)	120
javaapplication1.TaskHeap.removeMin ()		0.267 ms	(0%)	120

Fig. 5. The result of CPU Profiler of Java Visual VM for the Modified algorithm

6 Conclusion

Scheduling overhead in terms of tasks preemption and migration reduces the performance of a multiprocessor systems, since the processors will be more busy executing the scheduler to handle the preemption and migration than doing the actual work. In this paper, we have presented a modification of LRE-TL algorithm to reduce tasks migration. We have noticed that tasks with largest local remaining executions have always the minimum laxity, which means that not considering them for execution first will increase their probability of firing a C event, which in turn, will result in a task migration. We have modified the initialization procedure to consider such tasks for execution first. The experimental work that have been conducted showed a significant reduction of tasks migration when we apply the proposed modification. The significant of the modified algorithm has also been verified using an independent-samples t-test.

7 References

[1]   H. Kopetz, *Real-Time Systems: Design Principles for Distributed Embedded Applications*: Springer, 2011.

[2]   G. Nelissen, V. Berten, V. Nélis, J. Goossens, and D. Milojevic, "U-EDF: An unfair but optimal multiprocessor scheduling algorithm for sporadic tasks," in *Real-Time Systems (ECRTS), 2012 24th Euromicro Conference on*, 2012, pp. 13-23.

[3]   S. Funk, G. Levin, C. Sadowski, I. Pye, and S. Brandt, "DP-Fair: a unifying theory for optimal hard real-time multiprocessor scheduling," *Real-Time Systems,* vol. 47, pp. 389-429, 2011.

[4]   G. Nelissen, V. Berten, J. Goossens, and D. Milojevic, "Reducing preemptions and migrations in real-time multiprocessor scheduling algorithms by releasing the fairness," in *Embedded and Real-Time Computing Systems and Applications (RTCSA), 2011 IEEE 17th International Conference on*, 2011, pp. 15-24.

[5]   C. L. Liu and J. W. Layland, "Scheduling algorithms for multiprogramming in a hard-real-time environment," *Journal of the ACM (JACM),* vol. 20, pp. 46-61, 1973.

[6]   S. Funk, "LRE-TL: An optimal multiprocessor algorithm for sporadic task sets with unconstrained deadlines," Van Godewijckstraat 30, Dordrecht, 3311 GZ, Netherlands, 2010, pp. 332-359.

[7]   J. Carpenter, S. Funk, P. Holman, A. Srinivasan, J. Anderson, and S. Baruah, "A categorization of real-time multiprocessor scheduling problems and algorithms," *Handbook on Scheduling Algorithms, Methods, and Models, pages,* pp. 30.1-30.19, 2004.

[8]   H. Cho, B. Ravindran, and E. D. Jensen, "An optimal real-time scheduling algorithm for multiprocessors," in *Real-Time Systems Symposium, 2006. RTSS'06. 27th IEEE International*, 2006, pp. 101-110.

[9]   R. I. Davis and A. Burns, "A survey of hard real-time scheduling for multiprocessor systems," *ACM Computing Surveys (CSUR),* vol. 43, p. 35, 2011.

[10]   S. H. Funk and A. Meka, "U-LLREF: An Optimal Scheduling Algorithm for Uniform Multiprocessors," in *The 9th Workshop on Models and Algorithms for Planning and Scheduling Problems*, 2009, p. 262.

[11]   Oracle. (2013, *VisualVM - All-in-One Java Troubleshooting Tool*. Available: https://visualvm.java.net.

Border Noise Removal and Clean Up Based on Retinex Theory

Marian Wagdy[12]   Ibrahima Faye [13]   Dayang Rohaya [12]

[1] Centre of Intelligent Signal and Imaging Research (CISIR),
[2] Department of Computer and Information Sciences,
[3] Department of Fundamental and Applied Sciences
Universiti Teknologi Petronas, Malaysia

marian_wagdy_labeeb@yahoo.com,
{ibrahima_faye, roharam}@petronas.com.my

Abstract. Conversion from gray scale or color document image into binary image is the main step in most of Optical Character Recognition (OCR) systems and document analysis. After digitization, document images often suffer from poor contrast, noise, uniform lighting, and shadow. Also when a page of book is digitized using a scanner or a camera, a border noise, which is an unwanted text coming from the adjacent page, may appear. In this paper we present a simple and efficient document image clean up by border noise removal and enhancement based on retinex theory and global threshold. The proposed method produces high quality results compared to the previous works.

Keywords: Binarization; Thresholding; Border Noise; Retinex theory.

1   Introduction

Document image clean up is an interesting research topic in image processing, used in most of document image analysis and retrieval. The quality of character segmentation, analysis and recognition depends highly on the cleanup of document images. In recent years, transforming cultural and historical documents and books into digital format has become an important trend. When digitizing a page from a book, a border noise may appear. Border noise can be classified into textual noise, which is an unwanted text coming from the adjacent page and non-textual noise which can be black border, lines or speckles.

Old document images are usually subject to degradation due to uniform lighting from camera, dust and dirt on the glass of the scanner, poor storage, and bad environment [1]. Fig. 1 shows an example of poor quality document image.

The purpose of a clean up is to enhance document images suffering from degradation and border noise to make them readable by human or OCR systems. Many documents clean up techniques have been developed in the past years but it is still difficult when dealing with document images with very low quality due to variable background intensity, very low local contrast, smear, smudge and shadows [2].

Most common clean up techniques depend on binarization which convert the gray or color document image into bi-level form to solve the degradation problem. In gen-

T. Herawan et al. (eds.), *Proceedings of the First International Conference on Advanced Data and Information Engineering (DaEng-2013)*, Lecture Notes in Electrical Engineering 285, DOI: 10.1007/978-981-4585-18-7_39,
© Springer Science+Business Media Singapore 2014

eral the techniques, which deal with document image binarization, are classified into local and global. In global techniques, a single threshold value is selected for the whole document image as in [3-7].But in local techniques the thresholds are computed individually for each pixel, using the information from the local neighborhood of the pixel as in [8-15]. Global techniques are efficient to convert any grayscale image into a binary image if the document has good separation between background and foreground. They are inappropriate for degraded documents. Global binarization methods tend to produce marginal noise along the page borders. Local adaptive methods achieve better results in the case of degradation, but most of the local thresholding document image binarization techniques suffer from the following major limitations as Region size dependent, Individual image characteristics and Time consuming.

Fig. 1. Example for poor quality document image.

Another way to clean up a document image is the border noise removal. The most common approach, which removes the non-textual border noise, is to perform a document cleaning by filtering out the connected components based on their size and aspect ratio as in [16-18]. These techniques are efficient to remove the black border and isolated specks but cannot remove the textual noise.

The X-Y Cut algorithm as in [19- 21], is still widely used in border noise removal and page segmentation. The idea of this algorithm is to subdivide the document image into regions by recursively analyzing its projection profile until a stopping criterion is satisfied. Then, the horizontal and vertical projection is computed by projecting all the black pixels onto the Y-axis, and the vertical projection is obtained by projecting all the black pixels onto the X-axis, the subscript y will denote the parameters related to the horizontal direction and x will represent those for the vertical direction. This algorithm is very simple, easy to implement and also fast. However it fails when the documents have skew. Another limitation is that the document must be Manhattan layout like journals and books.

Shafait et al. [22] presented a method based on projection profile analysis. A run-length smearing step is first used for smoothing. A clean up step follows by transforming all black pixels that belong to a connected component to white if it at least one pixel is lying outside the connected component. The page frame detection is used instead of the removal of the border noise as in [23, 24].It ignores the margin noise along the page border and defines the page frame by using a geometric matching algorithm by maximizing a quality function which increases with the number of the text lines touching the boundary of the rectangular region.

In this paper, we propose an efficient clean up method which makes enhancement on degraded and poor quality document image. The proposed algorithm is based on Retinex theory, which treats the degradation problems of the document image followed by a simple method to remove the border noise.

The rest of this paper is arranged as follows. Section 2 presents the proposed scheme in details. Section 3 illustrates some examples to show the effectiveness of the proposed method. Finally the conclusion is drawn in section 4.

2   Proposed Method

Most of the previous binarization methods depend on local thresholding. They produce high quality results but consume more time. The other methods depend on global threshold. They are fast but don't perform well when the document image is degraded. Our proposed binarization method combines the advantages of local and global threshold by using Retinex theory, which can effectively enhance the degraded and poor quality document image.

The proposed document clean up method is based on two main steps. The first is an enhancement step and the second is a border noise removal. By using these steps, the quality of the result image is improved and the processing time is fast compared to other methods.

2.1   Enhancement

To solve the degradation problems, which result from converting the document into digital form, we will first use Retinex for enhancement, followed by a global threshold.

The concept of Retinex comes from the biological phenomenon of human visual system. The formation of the Retinex theory is illustrated in equation 1, where $I(x,y)$ is the intensity of the image with reflectance $R(x, y)$ and illumination L (x, y) which can approximated by using low-frequency component of the measured image [25]. If $I(x,y)$ is a degraded grayscales image, the Lightness image $L(x, y)$ is obtained by dividing it by its smooth version $M(x,y)$ as illustrated in equation 2:

$$I(x,y) = R(x, y).L(x, y) \qquad (1)$$

$$L(x,y) = {I(x,y)}/{M(x,y)} \qquad (2)$$

Most common binarization methods (as in [13]) use Gaussian filter with large kernel as a smoothed version but the result is not clear and still have some limitations. The proposed work uses the Median filter instead of Gaussian to obtain a smoother version of the image. Median filter operates over a window by selecting the median intensity in the window. Median filter is more robust to outliers, noise removal, and gives smoother result compared to Gaussian filter. Fig. 2 shows a comparison between Gaussian and Median filter in lightness document image.

A binarization is necessary to convert the document image to bi-level one. The use of Retinex overcomes the limitation of global threshold methods by removing the illumination and the degradation before performing global thresholding. The global thresholding becomes easier and accurate after Retinex. The most common global thresholding method in [8] is used to convert the lightness image into black (for text) and white (for background).

2.2   Border Noise Removal

In this step we aim to remove the border noise (textual and non-textual noise). Removal of non-textual noise could be by filtering out connected components based on their size and aspect ratio. But textual noise cannot be filtered out using this method. The proposed method is simple and efficient to detect the actual content area for all types of document structures (magazine, book, article … etc.). The proposed method to remove the border noise consists of the Foregrounds detection; the detection of the left and the right border of the page frame; the removal of the upper and lower noise. The steps will be detailed in the following sub-sections.

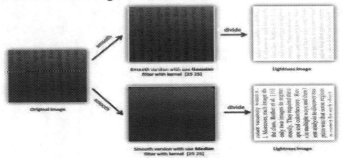

Fig. 2. Comparison between the proposed median filter and Gaussian filter in [13].

2.2.b Foreground Detection

To detect all foregrounds of the document image, Erosion Morphological operation will be used. The idea of Erosion is, let f and se (a small window) be functions representing $N \times N$ image and $(2M + 1) \times (2M + 1)$ structuring element, respectively, where M is the width of the largest character stroke. The binary Erosion of f by se is defined as

$$g(x,y) = f\theta se = \cap_{i=-M}^{M} \cap_{j=-M}^{M} se(i,j)\cup f(x+i,y+j) \qquad (3)$$

That means if all '1's in se match with the input signal then the output is '1'. In binary images, the erosion operator is to get rid of irrelevant details from the images. That means the holes within those areas become bigger.

In the proposed method we will use line as structure element with degree 270 and length 10 (fig. 3.b) to cover all foreground to detect the page border in left and right side, then use the structure element with degree 180 to remove the non-textual noise in upper and lower part.

2.2.b Left and Right Border Detection

Let $x(i,j)$ be a binary document image with width W, and height H. The document is divided into three parts; each with width $W/3$.The search is done in the right and left parts to define the page frame border as shown in fig 4.a. Since the pixel values in the binary document are 0 (for black) or 1 (for white), the sum of each column represents the number of white pixels.

In the left side, we search from the end of the left side for the first column having a sum larger than a threshold valueλH, where $\lambda \in (0,1)$ that column will be taken as the border as in equation 1. The remaining will be cropped. In the right side, similarly the search starts from the end of the right side until finding the first column with a sum larger than λH as in equation 2.

$$\text{For } j = \frac{W}{3}, \dots, 0$$

$$S(j) = \sum_{i=0}^{H} x(i,j) \quad (3)$$

$$Crop\,if\,S(j) > \lambda H$$

End

$$\text{For } j = \frac{2W}{3}, \dots, W$$

$$S(j) = \sum_{i=0}^{H} x(i,j) \quad (4)$$

$$\text{Crop if } S(j) > \lambda H$$

End

Where x(i, j) is a binary document image with width W, and height H, and S(j) represents the summation of each column. The value λ=0.9 was empirically found to be efficient.

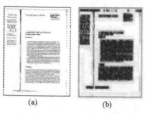

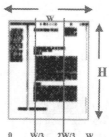

Fig. 3.(a) Example image showing document image has border noise, (b) document after detect the foregrounds.

Fig. 4. (a) Example image showing how to detect left and right border.

2.2.c Removal of Upper and Lower Noise

Most of the time, the noise in the Upper and Lower part of a document image is non-textual: black border, lines, etc. This noise can be removed by filtering out connected components based on their size and aspect ratio. The black border noise can be removed through the following steps:

(i) Edge detection.
(ii) Erosion (using line as structural element with 180 degree).
(iii) Median filter in the Upper and the Lower part only in the document image using a vertical kernel (1 × 5) to remove the lines only and do not make any effect for textual part in case of header, footer.

Fig. 5 shows the block diagram for upper and lower border noise removal. The quality of the text in the lower part is still the same after using the vertical kernel.

3   Discussion of Results

The proposed method has been implemented using MATLAB R2009a and tested on a PC with Pentium Dual Core 1.8 GHz CPU. To demonstrate its effectiveness, it is tested on a variety of degraded and noisy document images. Each document image illustrates the efficiency of the proposed method in addressing one of the clean-up challenges like low contrast, bad illumination, double side noise, and smeared documents.

Fig. 6 presents the results of the proposed binarization method on a collection of degraded document images shown in fig. 1. Some of these images are smeared handwritten document, bad illumination and others have double side noise. Based on the visual criteria, the proposed method works well and produces high quality and readable results in all cases of degradation.

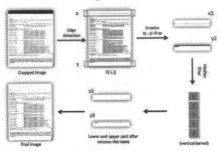
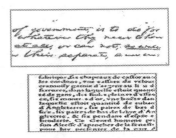

Fig. 5.block diagram for upper and lower border noise removal.

Fig. 6. Proposed binarization results of the document image in figure 2.

For more experimental results the proposed method is compared with Otsu's method [8], Niblack's method [10], Sauvola method [11], and Kim [14]. Fig. 7 illustrates the results of the proposed work and the other methods on poor quality historical handwritten document image. Based on the visual criteria our proposed method is efficient and more readable than others methods.

The evaluation of the proposed method for border noise removal algorithm was done on different poor quality historical documents as shown in fig. 8. The results show that the proposed method is efficient and performs well. In the case of upper and lower border noise, the use of a vertical kernel is efficient for removing black border and lines without affecting the text. This is particularly important in case of presence of header or footer.

In order to compare our approach with other state of-the-art approaches [26-28], as shown in fig. 9. All these methods have been proposed to remove only noisy black borders and not noisy text regions.

(a)   (b)   (c)   (d)   (e)   (f)

Fig.7. Binarization of poor quality historical handwritten document image. (a) Original image, (b) Otsu's method [8], (c) Niblack's method [10], (d) Sauvola method [11], (e) Kim method [14], and (f) our proposed method.

Fig. 8. Example image showing the result of document clean-up and border noise removal.

4 Conclusion

In this work, we proposed a fast and efficient document image clean up and border noise removal method based on Retinex theory and global thresholding. The proposed method combines the advantages of local and global thresholding by using the concept of Retinex theory, which can effectively enhance the degraded and poor quality document image. Subsequently a fast global threshold is used to convert the document image into binary form. The proposed method overcomes the limitations of the related global threshold techniques. It works well for any type of degradation and includes a simple and efficient technique for border noise removal. A projection profile technique is used to remove the noise in the left and right side; and a filtering method is used to remove the noise in the upper and the lower part. The algorithm is simple and easy to implement.

(a) (b) (c)

Fig.9. (a) Original image, (b) our proposed method, (c) Fan et al.

5 References

1. Y. Chen and G. Leedham, "Decompose Algorithm for Thresholding Degraded Historical Document Images" IEEE Proceedings on Vision, Image and Signal Processing, vol. 152 No.6, pp. 702–714, 2005.
2. G. Agam, G. Bal, G. Frieder, and O. Frieder, "Degraded Document Image Enhancement" in Document Recognition and Retrieval XIV, Proc. SPIE, vol. 6500, pp. 65000C-1 - 65000C-11, 2007.
3. J. M. White and G. D. Rohrer, "Image Thresholding for Optical Character Recognition and Other Applications Requiring Character Image Extraction" IBM Journal of Research and Development vol. 27, No. 4, pp. 400-411, 1983.
4. L. Gorman "Binarization and Multithresholding of Document Image Using Connectivity" CVGIP, Graph. Models Image Processing, vol. 56, No. 6, pp. 496-506, 1994.
5. R. Cattoni, T. Coianiz, S. Messelodi, and CM Modena, " Geometric Layout Analysis Techniques for Document Image Understanding: a Review", ITC-irst Technical Report 9703 (09), 1998.
6. P. Viola and M. J. Jones, "Robust Real-Time Face Detection," Int. Journal of Computer Vision, vol. 57, No. 2, pp. 137– 154, 2004.
7. F. Shafait, D. Keysers, and T. M. Breuel, "Performance Comparison of Six Algorithms for Page Segmentation," in 7th IAPR Workshop on Document Analysis Systems, pp. 368– 379, 2006.
8. N. Otsu, "A Threshold Selection Method FromGray-Level Histograms," IEEE Trans. Systems, Man, and Cybernetics, vol. 9, No. 1, pp. 62–66, 1979.
9. Y. Solihin, and C. G. Leedham, "Integral Ratio: A New Class of Global Thresholding Techniques for Handwriting Images", IEEE Trans. Pattern Anal. Mach. Intell., vol. 21, No. 8, pp. 761 – 768, 1999.

10. W. Niblack "An Introduction to Digital Image Processing" Prentice-Hall, Englewood Cliffs, New Jersey, 1986.
11. J. Sauvola and M. Pietikainen, "Adaptive Document Image Binarization," Proc. of Pattern Recognition, vol. 33, No. 2, pp. 225–236, 2000.
12. T.Romen "A New Local Adaptive Thresholding Technique in Binarization" IJCSI International Journal of Computer Science Issues, Vol. 8, Issue 6, No. 2, pp. 271-277 ,2011.
13. J. G. Kuk, and N. I. Cho, "Feature Based Binarization of Document Images Degraded by Uneven Light Condition" in 10th inter. Conf. On Document Analysis and Recognition (ICDAR), pp. 748-752, 2009.
14. I. K. Kim, D. W. Jung, and R. H. Park, "Document Image Binarization Based on Topographic Analysis Using a Water Fow Model" Proc. of Pattern Recognition , vol. 35, pp. 265–277, 2002.
15. Bolan Su, Shijian Lu, and Chew Lim Tan "Binarization of Historical Document Images Using the Local Maximum and Minimum" 9th IAPR International Workshop on Document Analysis Systems, pp. 159-166, 2010.
16. Baird, H.S.: Background structure in document images. In: Bunke, H. Wang, P. , B aird, H.S. (eds.) Document Image Analysis. World Scientific, Singapore, pp. 17–34 (1994).
17. Breuel, T.M.: Two geometric algorithms for layout analysis. In: Proceedings of Document Analysis Systems. Lecture Notes in Computer Science, vol. 2423, Princeton, NY, USA, pp. 188–199 (2002).
18. O'Gorman, L .: The document spectrum for page layout analysis. IEEE Trans. Pattern Anal. Mach. Intell. 15(11), 1162– 1173 (1993).
19. S. Mao and T. Kanungo, "Empirical Per formance Evaluation Methodology and Its Application to Page Segmentation A lgorithms," IEEE Trans. Pattern Analysis and M achi ne Intelligence, vol. 23, no. 3, pp. 242-256, Mar. 2001.
20. F. Shafait, D. Keysers, and T.M. Breuel, "Performance Evaluation and Benchmarking of Six Page Segmentation Algorithms," IEEE Trans. Pattern Analysis and Machine Intelligence, vol. 30, no. 6, pp. 941-954, June 2008.
21. F. Shafait, D. Keyser s, and T .M. B reuel, "Pixel-Accurate Representation and Evaluation of Page Segmentation in Document Images," Proc. 18th Int'l Conf. Pattern Recognition, pp. 872-875, Aug. 2006.
22. N. Stamatopoulos, B .Gatos , and A . K esidis, "Automatic Borders Detection of Camera DocumentImages ," Proc. Second I nt'l Workshop Camera-Based Document Analys is and Recognition, pp. 71-78, Sept. 2007.
23. F. Shafait, J. van B euseko m, D. Keysers, and T.M .Breuel, " Do cumentCleanup Using Page Frame Detectio n," Int'l J. Document Analysis and Recognition, vol. 11, no. 2, pp. 81-96, 2008.
24. F. Shafait, J. van B eusekom , D . K eysers, and T .M. B reuel, " Page Frame Detection for Marginal Noise Removal from S canned Documents," Proc. Scandinavian Conf. I mage Analys is, pp. 651-660, June 2007.
25. Edwin H. Land, "The Retinex Theory of Color Vision," Scientific American, Vol. 237, No. 6, pp. 108-128, 1977.
26. Kuo-Chin Fan, Yuan-Kai Wang, Tsann-Ran Lay, "Marginal Noise Removal of Document Images", Pattern Recognition, 35(11), 2002, pp. 2593-2611.

Chinese and Korean Cross-Lingual Issue News Detection based on Translation Knowledge of Wikipedia

Shengnan Zhao, Bayar Tsolmon, Kyung-Soon Lee[1] and Young-Seok Lee[1]

Division of Computer Science and Engineering, CAIIT, Chonbuk National University,
567 Baekje-daero, Deokjin-gu, Jeonju-si, Jeollabuk-do, 561-756 Republic of Korea
snzhao21@outlook.com, bayar_277@yahoo.com,
{ selfsolee, yslee }@chonbuk.ac.kr

Abstract. Cross-lingual issue news and analyzing the news content is an important and challenging task. The core of the cross-lingual research is the process of translation. In this paper, we focus on extracting cross-lingual issue news from the Twitter data of Chinese and Korean. We propose translation knowledge method for Wikipedia concepts as well as the Chinese and Korean cross-lingual inter-Wikipedia link relations. The relevance relations are extracted from the category and the page title of Wikipedia. The evaluation achieved a performance of 83% in average precision in the top 10 extracted issue news. The result indicates that our method is an effective for cross-lingual issue news detection.

Keywords: Issue news detection, Cross-Lingual link discovery, Wikipedia knowledge

1   Introduction

Cross-lingual link discovery is a way of automatically finding potential links between documents in different languages [1]. With the rapidly increasing amount of data on the internet lately, the research on the multilingual issue detection attracts more and more attention. Twitter's biggest feature is the anytime, anywhere access of real time information. Therefore, many people are using Twitter at all over the world. Thus, detecting cross-lingual issue news on Twitter data is challenging.

The existing machine translators are not an effective to translate peoples' names, place names, and event names. Because of this, an efficient translation method is being required.

In this paper, we propose the Chinese and Korean cross-lingual issue news detection method based on translation knowledge of Wikipedia. Firstly, we will build the Chinese and Korean translation knowledge based on the Wikipedia out-links (anchor text) as well as the Wikipedia category information on the page title. Secondly, we detect cross lingual issues from Chinese and Korean Twitter data set.

The present research work is correlated with previous reports which have been conducted by different research groups. Reliable translation is a key component of

---

[1] Corresponding author.

T. Herawan et al. (eds.), *Proceedings of the First International Conference on Advanced Data and Information Engineering (DaEng-2013)*, Lecture Notes in Electrical Engineering 285, DOI: 10.1007/978-981-4585-18-7_40,
© Springer Science+Business Media Singapore 2014

effective Cross Language Information Access systems (CLIAs). Jones et al., [2] introduces the domain-specific query translation for multilingual information access using machine translation augmented with dictionaries mined from Wikipedia. Tang et al., [1] studies in terms of how to improve information retrieval by leveraging cross-lingual links discovery, this task is focused on linking between English source articles as well as Chinese, Korean, and Japanese target articles for the Wikipedia.

The rest of the paper is organized as follows: Section 2 introduces the translation knowledge based on the Wikipedia. The proposed cross-lingual issue detection method is described in detail in Section 3. Section 4 shows the experimental results. Lastly we conclude this paper in Section 5.

2   Building Wikipedia-based Translation Knowledge

In this section, we propose how to build translation knowledge using Chinese and Korean Wikipedia. Wikipedia is a collaboratively edited, multilingual and free Internet encyclopedia supported by the non-profit Wikimedia Foundation. We build translation knowledge for three international issues by using Wikipedia page's title, anchor texts in the definition part of the page and category information of the anchor text and the certain page. The main purposes of the translation knowledge based on the Wikipedia as follows.

-   Extracting cross lingual relevant words of the issue word.
-   Determining the closest category of the issue word.
-   Translating and mapping source language word to the target language word.

Detailed information is in the following sections.

2.1   Extracting Cross-lingual Relevant Words based on Wikipedia

In the Wikipedia, the articles have many out-links (anchor text). Especially, the anchor texts in the definition part of the page have high relevance to the article. Because of the definition part summarizes the page title. Anchor texts are directly or indirectly linked to other articles via inter-wiki link. We define relevant words as a co-occurrence of anchor texts in the definition part of Chinese and Korean Wikipedia article.

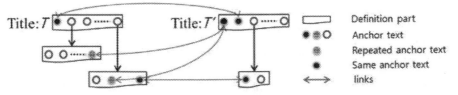

Fig. 1. The structure of Wikipedia page title and anchor text.

The source language page title T and target language page title T's anchor texts represents of inter-Wiki links are shown in Fig. 1. The relevance words of title T is obtained of anchor text (circles) $T\{A_1, A_2, A_3,..., A_N\}$ and its target language title T's anchor text $T'\{A_1, A_2, A_3,..., A_M\}$. If the certain anchor text appears on many pages, thus, its frequency becomes higher and consequently the relevance becomes higher as well. For example, Table 1 shows the relevant words of 'Common cold'. The Chinese and Korean Wikipedia page of 'Common cold' has same anchor texts like 'Sneeze' and the 'Virus' is shown in Table 1.

Table 1. An example of extracting cross lingual relevance words for 'Common cold.'

Language	Chinese		Korean	
Title	普通感冒(Common cold)		감기(Common cold)	
	Term	Frequency	Term	Frequency
Out-link (anchor text)	病毒(Virus)	4	바이러스(Virus)	3
	喷嚏(Sneeze)	3	새채기(Sneeze)	3
	疾病(Disease)	1	독감(Influenza)	1

2.2 Extracting Relevant Categories of Wikipedia Page Title Word

We extracted relevant words of the title using the frequency of co-occurrences of anchor texts is presented in section 2.1. However, each anchor text has many categories. In order to extract the closest category of the title and anchor text, we use the category of the title and the category of the anchor in the definition part of the title. The category relation between two Wikipedia pages is shown in Fig. 2.

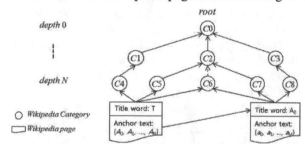

Fig. 2. Category relation between two Wikipedia pages.

In Fig. 2, the title word A_0 (anchor text of title T) is derived from definition part of title T. Title T has three categories $C4$, $C5$ and $C6$. Title A_0 has three categories $C6$, $C7$ and $C8$. The category $C6$ has a higher relevance between T and title A_0. Then the category $C4$ and $C7$ has middle relevance, $C4$ and $C8$ have lower relevance for the title T and A_0. Formula (1) calculates the closest category of title T and A_0.

$$conceptdist = 1 - \frac{length(c_i, c_j)}{depth(c_i) + depth(c_j) - depth(c_i, c_j)} \tag{1}$$

where c_i and c_j are Wikipedia category nodes of title T and A_0 respectively. The $length(c_i, c_j)$ represents the shortest number of nodes along with the shortest path between the two page nodes (e.g., the $length$ from $C5$ to $C7$ is 2 in Fig. 2.) [3], $depth(c_i)$ and $depth(c_j)$ is the number of category nodes on the path from c_j or c_i to root ($depth(C4)$ is equal to 2), $depth(c_i, c_j)$ is the number of categories on the path from the common category of node c_i and c_j to the root.

Table 2. An example of closest relevant category of title 'Common cold' and anchor text.

Wikipedia articles			Expected categories	
Title	Categories	Anchor texts	Categories	Value
Common cold	Viral diseases	Rhinovirus	Infectious diseases	1.0
	Inflammations	Cough	Inflammations and Reflexes	0.85
	Infectious diseases	Sneeze	Inflammations and Sneeze	0.56

Table 2 shows the words 'Rhinovirus' and the 'common cold' has same category 'infectious diseases'. The common cold has a contagious. Meanwhile, common cold is an infectious disease. Therefore the common cold and the infectious disease have relevance. So, we can transform to select relevant words (page title) within the same categories.

2.3   Extracting Semantic Relations between Wikipedia Page Title Words

We extracted relevant words for the given title in section 3.1. Also relevance categories of the title words are extracted by Formula 1 in section 3.2. In this section, those relevant words and relevant categories are as input in the following model. Extracting semantic relationships between relevant words and relevant categories are calculated by Probabilistic Latent Semantic Indexing (PLSI) model.

The PLSI approach is a statistical latent class model for factor analysis of count data in [5]. This approach has important theoretical advantages over standard LSI, since it is based on the likelihood principle, defines a generative data model, and directly minimizes word perplexity. It can also take advantage of statistical standard methods for model fitting, over fitting control, and model combination [4].

In this model, document d, word w and category z are as input. The word and document matrix of PLSI is shown in Fig. 3.

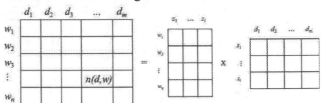

Fig. 3. Word-document matrix of PLSI.

In the Fig. 3, d represents a Wikipedia title, w is the anchor text in page d and z represents a category of d. In typical PLSI, initial input values for $P(d \mid z)$ and $P(w \mid z)$ are randomly selected. In the case of our problem, we assigned the initial values for $P(d \mid z)$ and $P(w \mid z)$ are as follows.

$P(w \mid z) = conceptdist(c_i, c_j)$

$P(d \mid z)$ = Taking the random value between 0 and 1.

In PLSI, an article word d and an anchor text w are assumed to be conditionally and independent selection of a category of z. The equation in [5] is effectively utilized in order to calculate the joint probability $P(d, w)$ as follows.

$$P(d,w) = \sum_{z \in Z} P(z)P(w|z)P(d|z) \qquad (2)$$

$P(d), P(d \mid z)$ and $P(w \mid z)$ are estimated by maximizing the log-likelihood function L, which is calculated as;

$$L = \sum_{d \in D} \sum_{w \in W} n(d, w) \log P(d, w) \tag{3}$$

where $n(d, w)$ represents the frequency of word w in article d. L is maximized using the EM algorithm, in which the E-step and M-step are given below [4][5].

E-Step:

$$P(z|d, w) = \frac{P(z)P(d|z)P(w|z)}{\sum_{z'} P(z')P(d|z')P(w|z')} \tag{4}$$

M-Step:

$$P(w|z) = \frac{\sum_d n(d, w)P(z|d, w)}{\sum_d \sum_{w'} n(d, w')P(z|d, w')} \tag{5}$$

$$P(d|z) = \frac{\sum_w n(d, w)P(z|d, w)}{\sum_{d'} \sum_w n(d', w)P(z|d', w)} \tag{6}$$

$$P(z) = \frac{\sum_d \sum_w n(d, w)P(z|d, w)}{\sum_d \sum_w n(d, w)} \tag{7}$$

The EM algorithm iterates through these steps until convergence. The example result of PLSI is shown in Table 3. As shown in Table 3, if the news contains the words earthquake and house then the category of this news is treated as a disaster. The words in the category disaster become relevant words of the news. Also if the news data contains words like earthquake and hypocenter, its category becomes earthquake.

Table 3. The probabilistic table is used Keywords extraction from the news.

Categories	Anchor Texts					
	Hypocenter	Landslide	Blizzard	Tsunami	Flood	Acid rain
Earthquake	0.18519	0.02941	0.00000	0.03448	0.00000	0.00000
Tsunami	0.07407	0.05882	0.00000	0.10345	0.00824	0.00028
Natural disasters	0.00345	0.17241	0.14706	0.06897	0.11448	0.22222

3 Cross-lingual Issue News Detection

In this section, we introduce a cross-lingual issue news detection based on Wikipedia translation knowledge. In order to detect issue news, we use Chinese and Korean Twitter data. Twitter is a popular real-time micro blogging service that allows its users to share short pieces of information known as "tweets". When any new events occur on a certain day, there might be used of certain terms and the issue related news URL increase on that day. Kwak et al. [7], classified the trending topics based on the active period and the tweets and show that the majority (over 85%) of topics are headline news or persistent news in nature. Because of this, we use the frequency of a news URL to detect issue news for each issue.

3.1   Extraction of the Issue News and Keywords

Detecting issue news. In the tweet, the most URLs of website addresses converted to the shortened URLs. The shorten URLs can convert to the source URLs. In order to gain news about the issues on the topic in the tweet, we extract all URLs for converting source URLs and calculate the source URLs frequency. Because many different shorten URLs are pointing to same source URLs. After the analyzing Twitter's issues, we have learned the hot issues starting point which is sourced news URL. Therefore, we used the frequency of source URLs to extract the issues news.
Extracting issue keywords. After detecting the issue news based on its URL frequency, we selected the higher URL frequency news as issues news. Then, in order to extract issue news keyword, firstly, we removed stop words and selected noun and verb terms by using a POS tagger[2,3]. After the pre-processing, the system calculates term frequency of the news data. If the term frequency becomes higher relevance increased respectively. The system selects top 10 words as news keywords.

3.2   Cross-lingual Issue News Detection

We extracted issue news keyword which is described in section 3.1. In order to detect cross-lingual issue news, we translate issue keyword from source language to target language. The extracted keywords are translated by Wikipedia based translation knowledge. From the keywords group via Wikipedia translation knowledge to find the words pairs of corresponds to each cross-lingual. If the translation knowledge does not have inter-Wiki links to the target language, we use an MS machine translator[4] in [8]. After translation of corresponding issue keywords ·from source language to target language, the system retrieves target language issue news URL. If the target language's tweet contains 2 or 3 source language's issue keywords with news URL, then we select it to the issue news. Following formula calculates the categorical relation of between source language news word and target language news word.

$$relatedwords(w_i, w_j) = \sum_{z \in Z} P(w_i|z)\, P(w_j|z) \tag{8}$$

where $P(w_i \mid z)$ represents the probability of category z given source language news word w_i and $P(w_j \mid z)$ is the probability of category z contain the target language word w_j. The result of the *relatedwords*(w_i, w_j) is greater than 0. 5 which are defined as the same issue news for Chinese and Korean

4   Evaluation

We have evaluated the effectiveness of the proposed method on tweet collection. Three issues are chosen and tweet documents for the issues are collected by Twitter

---

[2] Korean Stemmer (KLT) http://nlp.kookmin.ac.kr/
[3] Chinese Segmentation(ICTCLAS50) http://ictclas.org/
[4] Microsoft translate API http://microsoft.com/en-us/translator/developers.aspx

API (all issues and tweets are written in Chinese and Korean). Table 4 shows the number of tweet and number of Wikipedia pages. Each record in this tweet data set contains the actual tweet body and the time when the tweet was published. The Table 5 shows the comparison experiments of PLSI and weighted PLSI calculated by cross collection likelihood. The evaluation method is described in [9].

Table 4. Twitter and Wikipedia data set.

Data	Language		Date
	Chinese	Korean	
Twitter	2,221,805	7,716,455	2013.04.20~2013.04.25
Wikipedia	902,710	369,064	2013.04.27

Table 5. The result of comparison experiments of PLSI and weighted PLSI (P@20).

Language	Topics	PLSI	Weighted PLSI
	Earthquake	18 / 20 (0.90)	17 / 20 (0.85)
Chinese	Influenza	13 / 20 (0.65)	15 / 20 (0.75)
	Territorial dispute	7 / 20 (0.35)	9 / 20 (0.45)
	Earthquake	14 / 20 (0.70)	15 / 20 (0.75)
Korean	Influenza	13 / 20 (0.65)	17 / 20 (0.85)
	Territorial dispute	9 / 20 (0.45)	9 / 20 (0.45)
Total		0.61	0.68

The number of extracted Chinese and Korea issue news based on Wikipedia translation knowledge is shown in Table 6. First, we chose the three international issue topics then its extracted issue news and issue keyword described in section 4. Table 7 shows the detailed issue news set clustered by Wikipedia category information.

Table 6. The extracted Chinese and Korean issue news.

NO	Issue topic	# of news (Chinese)	# of news (Korean)
1	Earthquake	118	95
2	Influenza	125	80
3	Territorial dispute	83	66

Table 7. The probabilistic table is used Keywords extraction from the news.

No	Issue topic	# of news (Chinese)	# of news (Korean)	Issue news title
1	Sichuan Earthquake, China	80	45	"Sichuan earthquake, which destroyed 99 percent of houses in the village"
2	Influenza in China	65	40	From China 'bird flu' emergency measures, Seoul...... 'full strength'
3	Territorial dispute of Senkaku Island	72	23	"Imminent defeat in the Sino-Japanese"

The experimental results for each issue are shown in Table 8. The answer set is judged by two human assessors for the Chinese news set and the other two human assessors for the Korean news set. We extracted 5 keywords from the each News.

Table 8. The experimental result of the same topic issue for Chinese-Korean News.

Language	Issue 1	Issue 2	Issue 3	Accuracy
Chinese	80/80 (1.0)	65/65 (1.0)	45/72 (0.62)	0.873
Korean	45/45 (1.0)	32/40 (0.80)	13/23 (0.56)	0.786

As can be seen from the Table 8, the proposed method achieves 83.0% of the average evaluation for Chinese and Korean issue news.

5   Conclusion

Cross-lingual link discovery is a way of automatically finding potential links between documents in different languages. In this paper, we propose Chinese and Korean Cross-Lingual Issue Detection based on translation knowledge of Wikipedia. Firstly, we built the Chinese and Korean translation knowledge based on the Wikipedia out-links (anchor text) as well as the Wikipedia category information on the page title. Secondly, we detected cross lingual issues from Chinese and Korean Twitter data set. The experimental results on Chinese and Korean Twitter test collection shows 83.0% of performance. This result indicates that the proposed method based on Wikipedia translation knowledge is effective for cross lingual issue detection.

Future work includes comparing our approach with existing algorithm and adapting our work to apply in other languages as well.

Acknowledgments. This research was supported by the Basic Science Research Program through the National Research Foundation of Korea (NRF) funded by the Ministry of Education, Science and Technology (2012R1A1A2044811).

References

1. L.X.Tang, S.Geva, A.Trotman, Y.Xu, and K.Y.Itakura.: Overview of the NTCIR-9 Crosslingual Link Discovery. Proceedings of NTCIR-9, 2011
2. G.J.Jones, F.Fantino, E.Newman, and Y.Zhang.:Domain-specific query translation for Multilingual information access using machine translation augmented with dictionaries mined from Wikipedia. Proceedings of CLIA'08, 2008.
3. Leacock, C.&M.Chodorow (1998). Combining local context and WordNet similarity for word sense identification. In C. Fellbaum (Ed.), WordNet. An Electronic Lexical Database, Chp. 11, pp. 265–283. Cambridge, Mass.: MIT Press.
4. Dempster, A., Laird, N., and Rubin, D.: Maximum likelihood from incomplete data via the EM algorithm. J. Royal Statist. Soc.B39 (1977)
5. Thomas, H.: Probabilistic Latent Semantic Indexing, Proceedings of the Twenty-Second Annual International SIGIR
6. M.Strube,   S.P.Ponzetto.:WikiRelate! Computing Semantic Relatedness Using Wikipedia. Proceedings of AAAI, 2006
7. H.Kwak, C.Lee, H.Park, and S.Moon.:What is Twitter, a Social Network or a News Media?, Proceedings of WWW, 2010
8. D. Zhang, Q. Mei and C.X., Zhai.: Cross-Lingual Latent Topic Extraction, Proceedings of ACL, pp.1128-1137, 2010

Collision Avoidance Path for Pedestrian Agent Performing Tawaf

Aliyu Nuhu Shuaibu[1], Ibrahima Faye[2], Aamir Saeed Malik[1] and Mohammed Talal Simsim[3],

[1,2] Universiti Teknologi PETRONAS
[1] Department of Electrical and Electronics Engineering, [2] Department of Fundamental and Applied Sciences.
Bandar Seri Iskandar, Perak Darul Ridzuan, 31750, Malaysia
[3] Umm Al-Qura University, Makkah, Saudi Arabia

Abstract. Collision is one of the major problems affecting the flow of pedestrian in a dense environment. The case study of this research work is the ground flow of Tawaf area (Mataf) at Masjid Al-Haram, Saudi Arabia. We propose a spiral model, that simulates the movement of 1000 agents toward a unify direction while ensuring minimal collision among pedestrians during Tawaf. Based on our findings the spiral path movement is recommended for Tawaf movement. Several simulation trials were run. Outcomes such as average speed, duration and density were computed for different combination of spiral turn.

Keywords: Tawaf; Pedestrian; Simulation

1 Introduction

Crowd is often defined as large group of populous, individuals, audience, mass, aggregation etc. sharing a common goal in a given environment. Crowd movement is very important for management, planning and safety of public places especially in dense environment consisting of heterogeneous group of pedestrian agents under normal or panic situation.

Tawaf is an Arabic word which refers to the Islamic rituals performed by Muslims during Hajj or Umrah at Masjid Al-Haram, where all Muslims around the world turns towards when performing prayers. During Hajj and Umrah, Muslims are to perform circumambulatory movements around the Ka'aba seven times, in counter-clockwise. The circling is believed to demonstrate unity of believers in the worshipping of Allah, as they move in harmony together around the Ka'aba, while supplicating to Allah.

Annually, approximately around six million Muslim pilgrims perform both Hajj and Umrah [1]. Basically, Hajj has several stages and is being performed on some days thus, pilgrims move through different stages. This resulted to high pilgrim densities during the peak period. During Hajj, about 50,000 pilgrims perform Tawaf per hour within the Mataf area a place where the Tawaf ritual is being performed [1]. Fig. 1 presents the distribution of pedestrian agents performing Tawaf on Mataf area of Masjid Al-Haram.

T. Herawan et al. (eds.), *Proceedings of the First International Conference on Advanced Data and Information Engineering (DaEng-2013)*, Lecture Notes in Electrical Engineering 285, DOI: 10.1007/978-981-4585-18-7_41,
© Springer Science+Business Media Singapore 2014

Fig. 1. Pedestrian agents performing Tawaf at Mataf area of Masjid Al-Haram

Collision avoidance is a basic problem in crowd movement especially for pedestrians during Tawaf. In this paper, we address a spiral path planning for optimum flow and less collision among pilgrim agents performing Tawaf. In practice it is difficult to simulate the behavior of pilgrim agents during Tawaf due to some factors like complex motion flow, random flow of pilgrim, heterogeneous population, pilgrim clustering, touching of black stone, high density, velocity change, bilateral and unilateral exit at entrances to Mataf area [2].

Velocity obstacles proposed by [3] the approach was applied to collision avoidance of robots with moving obstacles. This method is extended for reciprocal velocity obstacles involving group of robots. Based on the findings, there is free collision scenario under some specific conditions. However, it does not guarantee a sufficient condition for collision avoidance.

Several researchers have written a comprehensive analysis on crowd modeling and simulation involving group dynamic of agents in a given setup of environment and path planning [4]. The simulation of dynamic groups of agents-based pedestrians and the effect of density threshold on group behavior was analyzed. Another study for optimum flow of pedestrian agents proposed in [5] tested a different pattern of movement based on spiral combination to determine the duration, average density and velocity of pedestrians. Comparison of Tawaf performance for both spiral pattern and circular pattern was computed. Based on the findings, spiral combination gives more convincing performance than circular pattern. However, choice of optimum spiral pattern needs validation.

There is an extensive literature on crowd simulation. Cellular Automata (CA) are some of the approaches applied for dynamic crowd simulation. Cellular Automata comprises a workspace of agents divided into discrete grid cells that can be occupied by zero or unit agent. A cellular automaton was applied to different scenario involving individual on scale of interactions and space utilization based on discrete and continuity of the crowd distribution. The concept of two-dimensional CA Moore neighborhood was presented in [6] proposed the improved method to compute the values of static flow filed for evacuation systems consisting of obstacles based on Euclidean distance. Two-dimensional CA for evacuation scenario involving obstacles is presented in [7] and the effects of the obstacles on pedestrian movement are studied. The study of bi-directional flow of pedestrian in critical density of transition phase as the result of counter flow was proposed by [8]. Another approach for evacuation scenario by means of stochastic CA, which incorporated the foraging ants and human behavior, was proposed by [9]. Concept of stochastic CA model was extended by [10] to explore the effect of friction and clogging during evacuation scenario.

Another approach to model the behavior of the crowd is the social force model [11]. Decision making capability was incorporated into social force model based on investigation ability of pedestrian agents [12]. Many researchers have attempted using the social force model in modeling the agent behavior because of it higher accuracy compared to other models. A critical review based on seven methodological approaches presented in [13] for crowd evacuation models includes the following cellular automata models, lattice gas models, agent based models, social force models, game theoretical models and animal based experimental models.

Computer vision methods can also be used with above models in cases like collision avoidance etc. There are 2D and 3D computer vision techniques [14-16] that can help in various scenarios for crowd behavior analysis. However, there are carious challenges that are yet to be resolved like illumination, real-time processing, occlusion etc. [17].

2 Problem Formulation

2.1 Problem Definition

Large crowd moving around the Mataf area of Masjid Al-Haram experience the so called "faster is slow effect". Complex motions due to pilgrim over-flowing in clusters towards attraction points: Hajarul Aswad (Black stone), Maqam Ibrahim, sometimes result to clogging and Jams. This slows down the Tawaf performance of other pilgrims and affects the flow, especially at peak period of Hajj. Crowd behavior analysis in general can be classified either as panic or normal situation, but this research work is specifically based on the normal situation by presenting a basic framework for the movements of pedestrians around the Mataf area while performing Tawaf.

Let the pedestrian agents be subjected to social forces as proposed by [11] the summation of forces that act upon the pedestrian resulted to acceleration as depicted in Eq. (2) and (3). The generalized social force model (SFM) is expressed according to the following equation of motion. The temporal change in location of pedestrian at time is given by:

$$\frac{dx_i(t)}{dt} = v_i(t) \tag{1}$$

The velocity change is given by acceleration equation:

$$m_i \frac{dv_i(t)}{dt} = f_i(t) + \xi_i(t) \tag{2}$$

Where $f_i(t)$ is the sum of all the forces influencing pedestrian i and ξ_i is the individual fluctuation. The total force acting on the pedestrian is given by:

$$m_i \frac{dv_i(t)}{dt} = m_i \frac{v_o(t) - v_i(t) + \xi_i(t)}{\tau} + \sum f_{ij} + \sum f_{ib} + \sum f_{ik} \qquad (3)$$

Where:
m_i= Mass of pedestrian i
v_o=Initial desire velocity
$v_i(t)$=Velocity of pedestrian i at time t
f_{ij}=Interaction force between pedestrian i and boundary b
f_{ik}=Interaction force between pedestrian i and group of pedestrian k

2.2 Spiral Algorithm

Suppose a virtual agents which move on spiral trajectory in Fig. 1 possesses the following parameters: n revolutions, angular velocity θ and radius r. The spiral path relations are given below:

$$x_i(n+1) = r_i(n+1)\cos(\theta_i) \qquad (4)$$

$$y_i(n+1) = r_i(n+1)\sin(\theta_i) \qquad (5)$$

The initial conditions for spiral parameters is given by: $0 \leq n \leq n_m$, $0 \leq \theta_i \leq 2\pi n$ for the iteration $i=1,2,...m$ for Tawaf $n_m=7$. x represents agent's coordinates for Tawaf trajectory along x-axis and y represents agent's coordinates for Tawaf trajectory along y-axis.

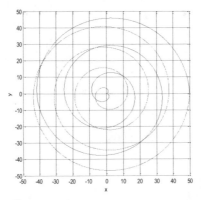

Fig. 2. Spiral path for Tawaf movement

3.0 Proposed Methodology

A spiral is a curve in the plane or in the space which runs round a centre in a special way. The model of spiral path planning uses the Logarithmic spiral which runs both clockwise and counter-clockwise depending on the nature of the movement. We generate a population of 1000 pedestrian agents with 60% male and 40% female and 5% handicap agents. The Level of Service (LoS) which is the density that determines the tendency of free flow or break flow of individual agents was assigned consistent with the observed crowd at Tawaf. The LoS-A was assigned 1agent/m^2 for free flow of individual agents and LoS-F assigned 10 agents/m^2 for shuffling movement. Surrounding temperature may vary from 30 $^\circ$C to 40 $^\circ$C. In our simulation we considered the average temperature as 35 $^\circ$C during Tawaf, individual pedestrian agents classified as kids assigned 5% of the population within the age range <1-15 years old and the remaining population assigned within 16-90 years old. Simulation is also performed by incorporating attraction points such as Maqam Ibrahim, Hajrul Asward and Hijri Isma'il.

Pedestrian behaviors during Tawaf is model based on the following design modality as presents on system flow chart in Fig. 3 as follow: Create a new project and import into the simulator which is achieved via CAD toolbox and import to 2D simulator framework, place waypoints in the architecture, the process is followed by plan building, grouping of agents based on priority (Attraction and non-attraction trajectory), scheduling agent walking path, set agent category, multiple simulation runs to optimize project settings and analyze simulation results to verify the propose concept and critical areas. All waiting points are set by initializing coordinates parameters for every pedestrian agent.

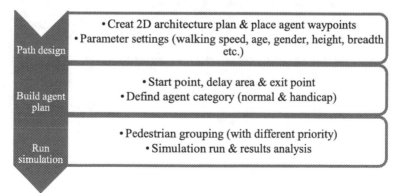

Fig. 3.   System flowchart

Simulation trials were run for the propose spiral model using all possible permutation as follows: movement pattern in order of six inward one outward (6in1out), movement pattern in order of five inward two outward (5in2out), movement pattern in order of three inward four outward (3in4out), movement pattern in order of seven inward arbitrary exit (7in-out), movement pattern in order of one inward six outward (1in6out), movement pattern in order of two inward five outward (2in5out), movement pattern in order of four inward three outward (4in3out) and

movement pattern in order of seven outward from arbitrary entrance (in-7out) in each
case Tawaf duration, density and average velocity of pedestrian agent were computed.
A statistical paired test is performed to determine the significance pattern.

4.0   Result Analysis and Discussion

This section presents the simulations outcomes of the proposed spiral model for
Tawaf movement. Simulation was carried out on each pattern for 10 iterations. The
fastest average durations is achieved with 4in_out pattern in 37 minutes. After
successful iterations, a hypothesis test is performed to evaluate the significance of the
difference between the performances of all possible spiral patterns by pairing each
pattern with 4in3out. In each case the following were computed: maximum density of
the crowd, average velocity of individual agents and Tawaf duration. The result
performances at level of significance (P=0.05) are presented in Table 1. The null
hypothesis is rejected at significance value for all pairs. This means that 4in3out
pattern gives better outcome than other patterns and the difference is statistically
significant. The result presented in Fig. 3 shows Tawaf duration of each pattern
against average density. Fig. 4 presents Tawaf duration against average speed of
pedestrian agents.
         The simulation scenario was extended by incorporating complexity
parameter for group of pedestrian X and Y. The simulation is set up by varying
population capacity and handicapped percentage. Simulation result is presented in
Table 2. In each case egress duration of each group is recorded. It can be seen that in
each simulation run, group X pedestrian exit the simulation area faster than group Y
because of attraction points effect on pedestrians egress dynamic.

Table 1. Result of t-test at significance value of 0.05.

Method	P-value	Null hypothesis (H_0)
4in3out Vs 3in4out	5.954×10^{-18}	Rejected
4in3out Vs 6in1out	2.979×10^{-24}	Rejected
4in3out Vs 1in6out	7.225×10^{-23}	Rejected
4in3out Vs 5in2out	2.051×10^{-17}	Rejected
4in3out Vs 2in5out	4.934×10^{-17}	Rejected
4in3out Vs 7in-out	6.289×10^{-18}	Rejected
4in3out Vs in-7out	5.744×10^{-20}	Rejected

Table 2. Summary of simulation performed by incorporating complexity

Experiment No.	Duration (minutes)		Population category	
	Group X	Group Y	Group X	Group Y
1	55.11	66.06	900	100
2	65.05	72.02	800	200
3	46.18	59.42	700	300
4	54.22	62.00	600	400
5	57.18	62.50	500	500

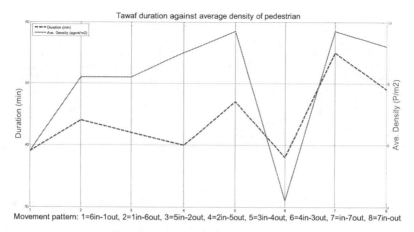

Fig. 3. Tawaf duration against average density of pedestrians

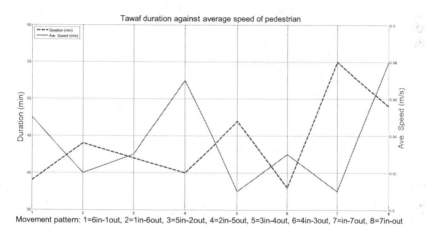

Fig. 4. Tawaf duration against average velocity of pedestrians

5.0 Conclusion

Tawaf exhibits some behaviors that are not properly modeled by existing techniques. This study has presented a simulation framework built based on social force model couple with parameterization of spiral path to model the behavior of crowd performing Tawaf. Pedestrians were categorized based on priorities, walking desires, age group, gender, handicap and normal. Some complex behaviors such as queuing at black stone and Hatim were modeled. The result outcomes such as average velocity, average completion time for single lapse of Tawaf, average density were computed for all pattern. All computations were carried out in 2D space. The simulation outcome for pattern spiraling four inward three outward (4in3out) gives an optimum Tawaf flow of pedestrian agents and the differences between the patterns are statistically significant.

References

1. Masjid Al-Haram Dataset, http://www.bdr130.net/vb/t87824.html
2. Curtis, S., Guy, S. J., Zafar, B., Manocha, D.: A Case Study in Simulating the Behavior of Dense, Heterogeneous Crowds. In: Computer Vision Workshops, IEEE International Conference, pp. 128--135, (2011)
3. Van den Berg, J., Snape, J., Guy, S., Manocha, D.: Reciprocal Collision Avoidance with Acceleration-Velocity Obstacles. In: IEEE International Conference on Robotics and Automation, pp. 3475--3482, (2011)
4. Qiu, F., Hu, X.: Modeling Dynamic Groups for Agent-Based Pedestrian Crowd Simulations. In: IEEE/WIC/ACM International Conference on Web Intelligence and Intelligent Agent Technology, pp. 461--464, (2010)
5. Nuhu, A. S., Faye, I., Simsim, M. T., Malik, A. S.: Spiral Path Simulation of Pedestrian Flow during Tawaf. In: IEEE International Conference on Signal and Image Processing Application, 13--17, Malaysia, (2013)
6. Alizadeh, R.: A Dynamic Cellular Automaton Model for Evacuation Process with Obstacles. In: Safety Science, pp. 315--323, (2011)
7. Varas, A., Cornejo, M. D., Mainemer, D., Toledo, B., Rogan, J., Muñoz, V., Valdivia, J. A.: Cellular Automaton Model for Evacuation Process with Obstacle. In Physica A: Statistical Mechanics and its Applications, pp. 382, 631--642, (2007)
8. Weifeng, F., Lizhong, Y., Weicheng, F.: Simulation of Bi-Direction Pedestrian Movement using a Cellular Automata Model. In Physica A: Statistical Mechanics and its Applications, 321, pp. 633--640, (2003)
9. Nishinari, K., Sugawara, K., Kazama, T., Schadschneider, T. A., Chowdhury, D.: Modelling of Self-Driven Particles Foraging Ants and Pedestrians. In Physica A: Statistical Mechanics and its Applications, 372, pp. 132--141, (2006)
10. Kirchner, A., Nishinari, K., Schadschneider, A.: Friction Effects and Clogging in a Cellular Automaton Model for Pedestrian Dynamics. In Physical Review E: Statistical, Nonlinear, and Soft Matter Physics, 21, pp. 056122/1--056122/10, (2003)
11. Helbing, D., Farkas, I., Vicsek, T.: Simulating Dynamical Features of Escape Panic. In: Nature, 407, pp. 487--490, (2000)
12. Zainuddin, Z. and Shuaib, M.: Incorporating Decision Making Capability into the Social Force Model in Unidirectional Flow. In: Research Journal of Applied Sciences, 5, pp. 388—393, (2010)
13. Zheng, X. P., Zhong, T. K., Liu, M. T.: Modeling Crowd Evacuation of a Building Based on Seven Methodological Approaches. In: Building and Environment, 44, pp. 437--445, (2009)
14. Asif, M., Malik, A. S., Tae-Sun, C.: 3D Shape Recovery from Image Defocus using Wavelet Analysis. In: IEEE International Conference on Image Processing, pp. 1025--1028, Genova, Italy, (2005)
15. Malik, A. S., Tae-Sun, C.: Finding Best Focused Points using Intersection of Two Lines. In: IEEE International Conference on Image Processing, pp. 1952--1955, San Diego, California, USA, (2008)
16. Malik, A. S., Shim, S., Tae-Sun, C.: Depth Map Estimation using a Robust Focus Measure. In: IEEE International Conference on Image Processing, pp 564--567, San Antonio, Texas, USA, (2007)
17. Malik, A. S., Tae-Sun, C.: Effect of Noise and Source Illumination on 3D Shape Recovery. In: International Journal of Pattern Recognition & Artificial Intelligence, Vol. 22, No. 5, pp. 945--958, (2008)

Cross-reading by leveraging a hybrid index of heterogeneous information

Shansong Yang, Weiming Lu, and Baogang Wei

Zhejiang university,
Hangzhou, Zhejiang 310000, P.R.China
{yangshansong,luwm,wbg}@zju.edu.cn

Abstract. In this paper, we present a novel application named Cross-reading, which is derived from user's reading process. Cross-reading is essentially a searching by document task from large-scale text corpus. The state-of-the-art approaches utilize similarity hashing to address this issue by modeling it as a high-dimensional data similarity search problem. However, most approaches only consider document's lexical information while ignoring documents semantic information and metadata. Moreover, searching similar hash codes from massive hash codes quickly is still a major bottleneck. To address those problems, we propose a Fast Searching By Document approach, which considers the Cross-reading from the perspective of semantic similarity and time efficiency.

Keywords: Topic-Sensitive Similarity Hash, Hybrid Index, HashCode Extension, ReRank

1 Introduction

With the rapid development of Internet and digital library, Cross-reading during user's reading process is a challenging and promising application. Cross-reading will recommend related chapters of the other books for the purpose of acquiring related knowledge. This service provides more knowledge from different perspectives when users are trying to understand some content. Conventional reading without cross-reading tends to focus on a narrow aspect of some concept, but ignores the diversity of the concept and its extended knowledge.

Cross-reading is essentially a searching by document task from large-scale text collection. Hashing methods are widely used in high-dimensional data similarity search, and have achieved good performance in many applications. SimHash [3, 5, 1] projects similar documents to similar fingerprints, which are compact binary codes and can be used for similar documents detection. However, it treats all words in the document equally, and doesn't consider the topic of the document. Intuitively, some words which reflect the topic of the document are more important than others. So this kind of information should be encoded into the fingerprints. Besides document's content, other information, e.g., metadata and user's click-through data are also useful to document similarity search. But it is difficult to integrate the heterogeneous information into a hybrid index for similarity search.

T. Herawan et al. (eds.), *Proceedings of the First International Conference on Advanced Data and Information Engineering (DaEng-2013)*, Lecture Notes in Electrical Engineering 285, DOI: 10.1007/978-981-4585-18-7_42,

2   Fast Searching by Document

In this section, we will introduce our approach for Cross-reading in detail, which can be called Fast Searching By Document.

We define our problem as follows: *Consider a corpus of documents, where each document has its category and its meta-data such as title, author, tag, etc. Given a query document and the document can also has some meta-data, the goal is to find similar documents in the corpus with the query document quickly.*

Our approach consists of three steps as shown in Figure 1. At first, we extract topic related words, document specific words and background words from each document, and encode these words into fingerprints through topic-sensitive similarity hashing. Then, a hybrid index, which can integrate the fingerprint and metadata, is constructed for fast searching. Finally, we rerank the search results by integrating user's click-through data into manifold ranking.

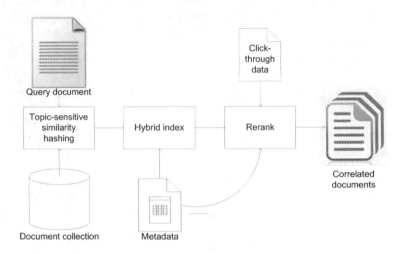

Fig. 1. The Architecture of Fast Searching By Document.

2.1   Topic-Sensitive Similarity Hashing for Document

Based on SimHash, we adopt a topic-sensitive similarity hashing for documents by treating each word in the document differently. First we compute *tfidf* of terms for each document and each class respectively. We denote tf(t,d) and t-f(t,c) as the frequency of term t occurs in document d and class c respectively, idf(t,D) and idf(t,C) as the inverse document frequency of the term t with respect to documents and classes respectively. Obviously, term t with high tfidf(t,c) is likely to be a topic related word of class c. That is, term t can reflect the semantic of class c. Moreover, term t with high tfidf(t,d) and low tfidf(t,c) is likely to be the document specific word of document d, and term t with low tfidf(t,d) is likely to be the background word. We choose words whose tfidf(t,c) greater than a threshold δ as the topic related words of class c, which can be denoted as $T(c) = \{t | tfidf(t,c) > \delta, c \in C\}$. So we can obtain document vector $\boldsymbol{d} = \{w_1(d), w_2(d), \ldots, w_{|w|}(d)\}$ and class vector $\boldsymbol{c} = \{w_1(c), w_2(c), \ldots, w_{|w|}(c)\}$, where $w_i(d) = tfidf(i,d)$, $w_i(c) = \begin{cases} tfidf(i,c), i \in T(c) \\ \delta \qquad\qquad i \notin T(c) \end{cases}$. Finally, we adjust the weight of each term according to document vector $\boldsymbol{d}$ and its class vector $\boldsymbol{c}$ as $\boldsymbol{d}' = \{w_1'(d), w_2'(d), \ldots, w_{|w|}'(d)\}$, where $w'_i(d) = w_i(d) * w_i(c)$, and the new feature vector reflects the semantic information of the document. Then the f-bit hash code of each document d with feature vector $\boldsymbol{d}'$ is computed by leveraging SimHash Algorithm[3]

2.2 A Hybrid Index for Heterogeneous information

When all documents are encoded into fingerprints, there are still two problems: (1) how to quickly discover other fingerprints similar to the query fingerprint. (2) how to integrate other information, i.e. metadata, to enhance the similarity measurement among documents.

Although hamming distance between two binary codes can be computed efficiently by using binary XOR operation, it is impractical to linearly scan the whole corpus. Two simple approaches are impractical for large-scale fingerprints collection. One is to probe the corpus with several fingerprints q' whose hamming distance from the query fingerprint q is almost k. However, the number of probes would be huge and reach to $\sum_{b=0}^{k} \binom{f}{b}$. Another is to pre-compute the similar fingerprints p' for each fingerprint p in the corpus, and index all these fingerprints. The drawback of this method is that the number of the pre-computed fingerprints will be $\sum_{b=0}^{k} \binom{f}{b}$ times of the number of the original fingerprints in the corpus, which would consume large storage. In these equations, the character f indicates the length of fingerprints.

Inspired by [4], we generate a set of substrings from each fingerprint to achieve the balance between the number of probes and the required storage. We partition the f-bit fingerprint p into m disjoint substrings $p_1, p_2, \ldots, p_m$, where the length of each substring is f/m. According to the Pigeonhole Principle, if the hamming distance between two fingerprints p and q is k, at least one of their substrings differs at most $\lfloor \frac{k}{m} \rfloor$. That is to say, if the hamming distance between all corresponding substrings is large than $\lfloor \frac{k}{m} \rfloor$, the hamming distance between two fingerprints must larger than k. If the number of the matches between two corresponding substrings is larger, the two fingerprints would be more similar.

According to the above analysis, we generate a "bag of substring" from the fingerprint to represent the document. This representation has some advantages compared with the "bag of word" representation: (1) We needn't to select keywords or key phrases from the document, since "the bag of substring" encodes all information of the document. (2) The size of the "bag of substring" is much smaller than the length of the document, which makes the inverted index suitable.

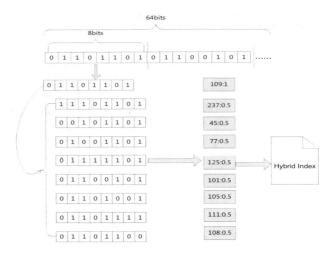

Fig. 2. The schematic diagram of "Bag of Semantic Word" generation, with the hamming distance k=8, the number of substring m=8, the length of fingerprint f=64.

In addition, the required storage of B fingerprints is

$$B * m * \sum_{b=0}^{\lfloor k/m \rfloor} \binom{f/m}{b} * \frac{f}{m} \tag{1}$$

which is much smaller than the pre-computed fingerprints requires, which is

$$B * \sum_{b=0}^{k} \binom{f}{b} * f \tag{2}$$

In order to distinguish the substrings with different positions, we convert the substring to an integer according to its position and the value of the substring:

$$I(p_i) = v(p_i) + (i - 1) * 2^{\frac{f}{m}} \tag{3}$$

where $v(p_i)$ is the value of the substring p_i. A schematic diagram is shown in Figure 2 and the detailed description is shown in Algorithm 1. These integers can be called semantic words, so finally the document is converted to a "bag of semantic word" from "bag of substring". Finally, the "bag of semantic word" and metadata can be be indexed by inverted index.

2.3 Reranking by Click-through Data

Here, we retrieve K similar documents from the hybrid index, which can be called candidate documents, and then resort to Peer Indexing [8] and Manifold Ranking [7] to rerank the candidate documents. The peer index of a document

---

Algorithm 1 "Bag of Semantic Word" generation for document.

Input: f-bit fingerprint F; The hamming distance k; The weight of substring τ;
Output: Bag of Integers S;
 1: Partition f-bit fingerprint F into m substrings $p_1, p_2, \ldots, p_m$, where the length of each string is f/m. The weight of each substring p_i is $w(p_i) = \tau$.
 2: **for** $i = 1$ to m **do**
 3:    **for** j = 1 to $\lfloor \frac{k}{m} \rfloor$ **do**
 4:       Flip every j bit in p_i, and get a new substring $p_i^{\pi_j}$, where π_j is the positions of the j bit. This weight of the substring $p_i^{\pi_j}$ is $w(p_i^{\pi_j}) = \frac{\tau}{2^j}$;
 5:    **end for**
 6: **end for**
 7: Convert substring p_i and $p_i^{\pi_j}$ to integer through

$$I(p_i) = v(p_i) + (i - 1) * 2^{\frac{f}{m}} \qquad I(p_i^{\pi_j}) = v(p_i^{\pi_j}) + (i - 1) * 2^{\frac{f}{m}}$$

 8: **return** a bag of semantic words with weight:

$$S = \left\{ I(p_i) : w(p_i), I(p_i^{\pi_j}) : w(p_i^{\pi_j}) | i \in [1, m], j \in [1, \lfloor \frac{k}{m} \rfloor] \right\}$$

---

is a list of semantically correlated documents. Through peer index, candidate documents' similarity defined from user's perspective can be computed by exploiting user's click-through data. This type of similarity is then integrated into manifold ranking through adjusting the element of affinity matrix.

Denote candidate documents as $D = \{d_1, d_2, \ldots, d_K\}$, where d_1 is the query document, and the distance between two documents d_i and d_j is $d(d_i, d_j)$. Then, we use peer index to integrate the click-through data into similarity measurement. Formally, document d_i can be represented by a set of documents $D_i = \{< d_i^1, w_1 >, < d_i^2, w_2 >, \ldots, < d_i^N, w_N >\}$, which are clicked with d_i together and the weight w_i is related to the number of clicks, then TFIDF scheme can be used to convert D_i to a vector with each dimension:

$$w_j' = w_j * (log \frac{M}{M_j} + 1) \tag{4}$$

where M is the total number of documents and M_j is the number of documents whose peer index has d_i^j in it. Therefore, the vector for peer index of document d_i is $\boldsymbol{P_i} = [w_1', w_2', \ldots, w_N']$, and the similarity between two peer indices can be computed by

$$R_{ij} = \frac{\boldsymbol{p_i} * \boldsymbol{p_j}}{\left\|\boldsymbol{p_i}\right\| * \left\|\boldsymbol{p_j}\right\|} \tag{5}$$

Let $f : D \to R$ denotes the ranking function which assigns to each document d_i a value f_i, forming a vector $f = [f_1, f_2, \ldots, f_K]^T$. An initial ranking vector $y = [y_1, y_2, \ldots, y_k]^T$, where $y_1 = 1$ and $y_i \neq 0$. Then, the detailed manifold ranking algorithm is shown in Algorithm 2.

Algorithm 2 Reranking by Manifold Ranking.
Input: candidate documents D;
 1: Calculate the affinity matrix W by

$$W_{ij} = exp\left(- \frac{d^2(d_i, d_j)}{2\sigma^2} \right) * (1 + R_{ij})$$

where $d(d_i, d_j)$ is the distance between documents, R_{ij} is the Peer Index similarity.
 2: Symmetrically normalize W by

$$S = D^{-\frac{1}{2}} W D^{-\frac{1}{2}}$$

where D is the diagonal matrix with (i,i)-element is the sum of the i-th row of W;
 3: Iterate
$$f(t + 1) = \alpha * S * f(t) + (1 - \alpha) * y$$

until convergence, where $\alpha \in [0, 1)$;
 4: Rank the candidate documents D according to the convergent result f^*;

---

3   Experimental Results

3.1   Dataset and Experiment Settings

We evaluated the proposed approach on three datasets:

20 Newsgroup Dataset: The 20 Newsgroups dataset is a collection of approximately 20,000 newsgroup documents, which are partitioned across 20 different cybergroups.

SRAA UseNet Dataset: The SRAA UseNet dataset contains 73,218 UseNet articles from four discussion groups.

CADAL Dataset: The documents for this dataset were crawled from CADAL[1] digital library. This dataset contains $704, 355$ documents, which are classified by human experts.

We randomly sampled 200 documents across all categories from each dataset as queries. The recommended document is supposed to be related to the query document if they share common category. The average precision and the normalized discounted cumulative gain(NDCG) of the top N search results were adopted to evaluate the effectiveness of our approach.

In our experiment, we implemented five baseline approaches: cosine similarity, KL-divergence, LDA-based similarity[2], DDM-based similarity[6] and SimHash.

3.2   Performance of Fast Searching By Document

Figure 3 illustrates that topic-sensitive similarity hashing can improve the precision nearly 10 percent at each hamming distance respectively. And the hamming distance between relevant document's fingerprints generated through topic-sensitive similarity hashing would be 2 bits less than that of SimHash.

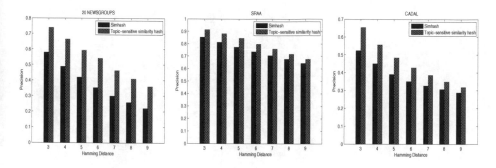

Fig. 3. Average precision at hamming distance.

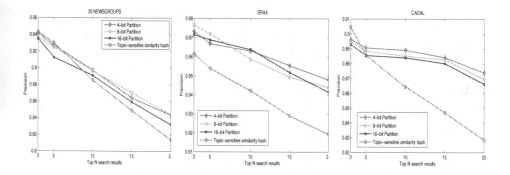

Fig. 4. Average precision of hybrid index with different length of partitions.

Figure 4, shows that the partition operation can improve the accuracy by almost 1.5 percent under SRAA and CADAL datasets. We suppose the reason for this phenomenon is the difference between inverted index's ranking mechanism and the ranking based on hamming distance. Given a great amount of fingerprints, there are plenty of fingerprints at the same distance from query fingerprint. The coarse-grained distance measurement of hamming distance leads to its underperformed ranking performance.

From Figure 5, we can see the precision of Fast Searching By document can reach to above 90 percent. We also note that SimHash is much more effective than Cosine similarity and KL-divergence, we think it's probably because the fingerprint's generative process incorporates much more discriminating information. Fast searching by document improves the searching accuracy by nearly 4 percent compared with SimHash, which is due to the identification of topic related words and document specific words. LDA-based similarity and DDM-based similarity compute document's similarity under the semantic level. But mapping the document to semantic or topic space is dimensionality reduction, which cannot but lose information.

---

[1] www.cadal.zju.edu.cn/

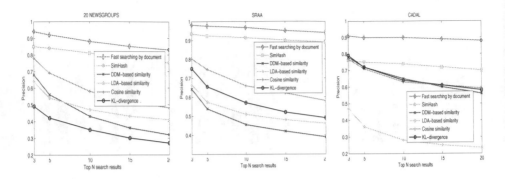

Fig. 5. Average precision of Top N search results about different methods.

Table 1. The NDCG score of CS(Cosine Similarity), KL(KL-divergence), LDA(LDA-based similarity), DDM(DDM-based similarity), SH(SimHash) and FS(Fast searching by document) on different position.

Method	20NEWSGROUPS				SRAA				CADAL			
	5	10	15	20	5	10	15	20	5	10	15	20
CS	0.968	0.959	0.954	0.950	0.971	0.961	0.955	0.952	0.966	0.957	0.949	0.947
KL	0.974	0.960	0.953	0.950	0.984	0.972	0.965	0.961	0.988	0.982	0.977	0.975
LDA	0.989	0.976	0.969	0.965	0.986	0.969	0.960	0.955	0.997	0.995	0.995	0.996
DDM	0.990	0.978	0.970	0.966	0.982	0.963	0.954	0.949	0.989	0.980	0.975	0.971
SH	0.999	0.999	0.999	0.998	0.999	0.999	0.999	0.998	1.00	1.000	0.999	0.999
FS	**0.999**	**0.999**	**0.999**	**0.999**	**0.998**	**1.000**	**0.999**	**0.999**	**1.000**	**0.999**	**0.998**	**0.998**

Table 1 shows that Fast searching by document and SimHash obtain almost the best performance compared with other methods. We also notice that the score in Table 1 is generally high. This partly because the binary value of document's relevance is coarse-grained, which cannot reflect relevant documents' difference exactly. In spite of these blemishes, different methods' performance about raking quality can be reflected by the relative degree of NDCG scores.

In our reranking experiment setting, the system returned 100 documents as the candidate results. For each query, we respectively selected 5,10 and 15 pairs of related documents to build the Peer Index. Figure 6 indicates the reranking algorithm can achieve 2.5 percent improvement compared with without reranking. It's worth noting that the manifold ranking algorithm without feedback information gets very weak performance improvement. This is perhaps because the intrinsic manifold structure among data points can only be revealed by a great amount of data. The result shows that after adding more user's click-through data, the accuracy increases gradually.

Figure 7 shows that the inverted index can significantly reduce average response time on different dataset sizes. The gradient of our approach's response time is relatively very small compared with SimHash method's. So our Fast searching by document can efficiently handle large-scale text collection.

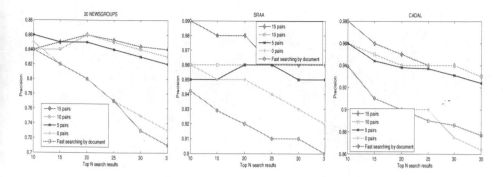

Fig. 6. Average precision of rerank with different number of related document pairs.

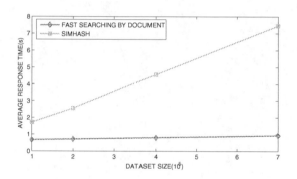

Fig. 7. Average response time on different data-set size.

References

1. C. C. Aggarwal, W. Lin, and P. S. Yu. Searching by corpus with fingerprints. In *Proceedings of the 15th International Conference on Extending Database Technology.* ACM, 2012.
2. D. Blei, A. Ng, and M. Jordan. Latent dirichlet allocation. *the Journal of machine Learning research,* 2003.
3. M. Charikar. Similarity estimation techniques from rounding algorithms. In *Proceedings of the thiry-fourth annual ACM symposium on Theory of computing.* ACM, 2002.
4. M. Norouzi, A. Punjani, and D. Fleet. Fast search in hamming space with multi-index hashing. CVPR, 2012.
5. S. Sood and D. Loguinov. Probabilistic near-duplicate detection using simhash. In *CIKM,* pages 1117–1126. ACM, 2011.
6. L. Weng, Z. Li, R. Cai, Y. Zhang, Y. Zhou, L. Yang, and L. Zhang. Query by document via a decomposition-based two-level retrieval approach. SIGIR, 2011.
7. B. Xu, J. Bu, C. Chen, D. Cai, X. He, W. Liu, and J. Luo. Efficient manifold ranking for image retrieval. In *SIGIR,* pages 525–534. ACM, 2011.
8. J. Yang, Q. Li, and Y. Zhuang. Image retrieval and relevance feedback using peer indexing. ICME, 2002.

Human Activity Recognition from Basic Actions Using Finite State Machine

Nattapon Noorit[1], and Nikom Suvonvorn[2]

[1,2]Department of Computer Engineering, Faculty of Engineering, Prince of Songkla University,
Hatyai, Songkhla, 90112
[1]nattapon.nr@gmail.com, [2]kom@coe.psu.ac.th

ABSTRACT.
High-level human activity recognition is an important method for the automatic event detection and recognition application, such as, surveillance system and patient monitoring system. In this paper, we propose a human activity recognition method based on FSM model. The basic actions with their properties for each person in the interested area are extracted and calculated. The action stream with related features (movement, referenced location) is recognized using the predefined FSM recognizer modeling based on rational activity. Our experimental result shows a good recognition accuracy (86.96% in average).

Keywords : human activity recognition, finite state machine, FSM recognizer, rational activity definition

1   Introduction

High-level human activity recognition is an important method for the automatic event detection and recognition application, such as, surveillance system, patient monitoring system, and etc. Previously, many researchers proposed many human activity representation and recognition strategies. For instance, some researches tried to learn pattern of activity within the scene [1,2], while some interesting works used hand-crafting model to recognize a particular activity [3,4]. However, in general these strategies use a similar concept that is matching an unknown sequence with the references to recognize a particular activity. Several techniques were introduced for solving the activity recognition problems, such as, Hidden Markov Model (HMM), Dynamic Time Warping (DTW), and Finite State Machine (FSM).

Hidden Markov Model is a kind of stochastic state machine. Since an activity can be represented as a sequence of actions, it can be described by the HMM representation via training process by given the observation sequences. Eventually, HMM cannot apply to recognize the activity where its action sequence is uncertain. For example, if the current activity is interrupted by another, the state become unpredictable that make HMM inapplicable. This technique has been found in many researches such as [8,9,10].

Dynamic Time Warping is a template-based dynamic programming matching technique. The advantage of DTW is that the reliable time alignment between refer-

T. Herawan et al. (eds.), *Proceedings of the First International Conference
on Advanced Data and Information Engineering (DaEng-2013)*, Lecture Notes
in Electrical Engineering 285, DOI: 10.1007/978-981-4585-18-7_43,
© Springer Science+Business Media Singapore 2014

ence and test patterns is provided. The disadvantage of using DTW is the heavy computational problem that requires determining the optimal time alignment path. DTW has been used in several activity recognitions [5,6,7].

Finite State Machine is a predefined state machine with specific transitions. Both states and transitions are used for describing the specific problem, defined by its nature mostly from observation. FSM is lightweight, human-readable and easy to parse. However, FSM may easily fail in the presence of noise. It is applied as recognizer in various methods [13,14,15,16].

In this paper, we propose a human activity recognition method based on FSM model. The basic actions (standing, walking, sitting, bending, laying) with their properties (location, movement) for each person in the interested area, as an action stream, are recognized using the predefined FSM based on rational activity. We focus to recognize four general activities that are: (1) walk through the scene (2) observation (3) rest (4) browse. We found that in some activities, such as, observation, its behavior can be described randomly by the stream of walking and standing states that finally make it unpredictable. So, the HMM is not the quite suitable technique for our test activities. Moreover, the DTW have a major weakness on the heavy computation costs that not allows the recognition in real time. Finally, FSM is chosen for its lightweight and capacity to deal with our activities.

2   System Overview

In this section, we present our proposed activity recognition process that describes a complex activity by a stream of basic actions with its properties. Figure 1 shows the system overview. The system is divided into 3 main parts: (1) Image Calibration, (2) Action extraction and features calculation, (3) Activity Recognition. The system run sequentially from (1) to (3) as the flowchart detailed as follows:

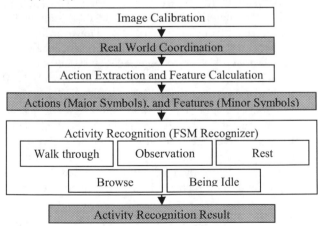

Fig. 1. System overview

3   Image Calibration

In this part, the human location detected in the image frames from video data is transformed from the pixel-based coordinate to a reference coordinate, related to the real-world location. This enables us to use data from any sensors referred to the same international unit system, such as, centimeter, that is very useful for the real-life activity definition, and for forming recognition condition, which is easily understandable from the real-life perception.

In this step, a perspective transformation is applied. The four points from the considering image with its corresponding points from the real-world coordinate are selected for calculating the transformation matrix. The Fig. 2 (a) shows the original image with our interested areas (red rectangles), corresponding the six locations associated to our recognition testing cases. Fig 2 (b) shows the transformed locations (red rectangle) in the real-world coordinate. Each blue square represents the 30x30 cm^2 square in the real-world scale.

In our experimentation, three objects (O1, O2 and O3), and three seats (C1, C2 and C3) are considered in the interested area. These objects and locations must be transformed to the referenced real-world location. Any interactions between people and objects in the ROI area using in our recognition process can compute the features in real-world scale, such as, movement distance, velocity, and acceleration.

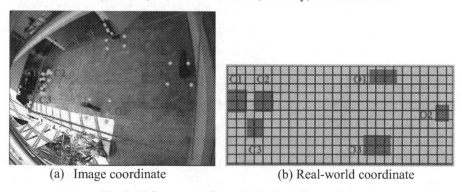

(a) Image coordinate   (b) Real-world coordinate

Fig. 2. (a) Image coordinate (b) Real-world coordinate

4   Action Extraction and Feature Calculation

This section explains how to define actions and its additional features. We represent each feature as a symbol that can be separated into two groups: major symbol and minor symbol. Firstly, the major symbol is defined from the basic actions: standing, walking, sitting, bending, and lying. For each person who does an activity will produce a stream of these actions over time: one person/action/frame. In this paper, we use the action symbol defining from the ground truth. The major symbols are described in Table 1.

Table 1. Major symbols and meaning

Symbol	Meaning
a_en	A person appear in interested area
a_st	Act standing action
a_wk	Act walking action
a_si	Act sitting action
a_bn	Act bending action
a_ly	Act laying action
a_ex	A person disappear from interested area

Major symbol consist of five basic actions and two appearing statuses (enter and exit from interested area).

Secondly, the minor symbols as the additional features are defined from the movement of person, which are divided into seven groups: (1) action time period (start with t_) (2) movement direction (start with d_) (3) direction variation (start with dv_) (4) velocity (start with v_) (5) acceleration (start with ac_) (6) object interaction (start with oi_) (7) current location (start with lo_). The minor symbols are detailed in the Table 2.

Table 2. Minor symbols

Symbols	Meaning	Symbols	Meaning	Symbols	Meaning
t_l	Take the action with short time period	dv_l	Have little movement direction variation	d_n	Move to the north
t_m	Take the action with middle time period	dv_m	Have middle movement direction variation	d_s	Move to the south
t_h	Take the action with long time period	dv_h	Have large movement direction variation	d_w	Move to the west
t_un	Unable to specify time period for first time appearing	dv_un	Unable to specify movement direction variation for first time appearing	d_un	Unable to specify movement direction for first time appearing and no movement action
ac_l	Low acceleration	v_l	Low velocity	d_nw	Move to the north-west
ac_m	Middle acceleration	v_m	Middle velocity	d_ne	Move to the north-east
ac_h	High acceleration	v_h	High velocity	d_sw	Move to the south-west
ac_un	Unable to specify acceleration for first time appearing	v_un	Unable to specify velocity for first time appearing	d_se	Move to the south-east
oi_b	Take the object out from video scene	lo_se	Person stay on seat	d_e	Move to the east
oi_l	Leave the object in video scene	lo_sp	Person stay near some special object that can browse or have interaction with it such as signboard or ATM	lo_bd	Person stay in bed
oi_no	No interaction with objects			lo_fl	Person stay on floor

The minor symbol features are computed from location changing from frame to frame in the real-word coordinate. The unit is cm/second. However, these minor symbols are defined by floating point number, such as, distance, velocity, and acceleration, which cannot apply to our state machine recognition system. From the observation, we found that its values can be represented to the Gaussian distribution. Then, the value is bounded into three values: (1) low value (lower than -1σ) (2) middle value (between -1σ and 1σ) (3) high value (higher than 1σ). The minor symbol features are detailed in the Fig. 3.

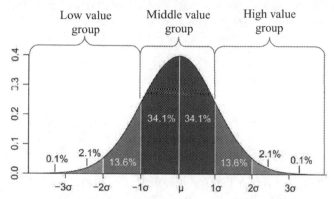

Fig. 3. Movement features clustering by Gaussian distribution

5 Action Recognition

In this section, we defined the Finite State Machine for recognizing the five activities: walk through the scene, observation, rest, browse and idle. The definition of FSM for each activity is defined from its observed behavior, which is directly related to the stream of major and of minor symbols.

5.1 Walk Through the Scene

This activity means that a person walks through the interested area without stop or having interaction with any objects in the scene. FSM recognizer for this activity consist of three major symbols, which is Enter(a_en), Walk(a_wk) and Exit(a_ex). The minor symbols used in this activity are: low direction variation(dv_l), and middle direction variation (dv_m), which means that the person must walk through the scene with low changing direction. See Fig. 4 (a) for the finite state machine.

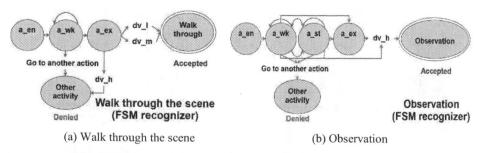

(a) Walk through the scene   (b) Observation

Fig. 4. FSM recognizer: (a) Walk through the scene, (b) Observation

5.2   Observations

This activity means that a person repeats the following action pattern, walking-standing-walking, with the large direction variation inside the interested area. So, the observation activity should have a high value of direction variation dv_h. However, the sit (a_si), lay (a_ly) and bend (a_bn) action states are not included in this FSM recognizer for preventing the undefined activities that may have much more meaning than observation activity. The Finite State Machine is shown in Fig. 4 (b).

5.3   Rest

This activity can be defined as two sub-activities: normal rest and abnormal rest.

a) Normal rest activity means that a person sits or lays on the normal rest area, such as, seat or bed.

b) Abnormal rest activity means that a person sits or lays on the unusual rest area, such as, corridor.

This activity is defined by two major symbols including sit (a_si) and lay (a_ly) with four additional current location symbols: seat (lo_se), bed (lo_bd), floor (lo_fl) and special object (lo_sp). See Fig. 5 (a) for the finite state machine.

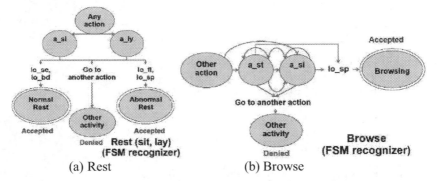

(a) Rest   (b) Browse

Fig. 5. FSM recognizer: (a) Rest, (b)Browse

5.4   Browse

This activity means that a person walks to the special object then stop near it. Browse activity is accepted when a person stand or sit near special objects that can browse or having interaction with it. This activity use the major symbols sit (a_st) and stand (s_st) with the predefined special object location (O1, O2 and O3). See Fig. 5 (b) for the finite state diagram.

5.5   Idle

This activity means that the current sequence of symbols is not matched to any activities above.

6 Experimental Result and Discussion

We test our activity recognition process with the well-known dataset from CAVIA dataset. Our experiment uses 13 video-clips including: six videos for browsing, four videos for rest, and four videos for walk. For the overall testing dataset, there are 23 activities, which are divided into 10 walking, five resting, five browsing, and three observations. The recognition result for each activity is detailed in Table 3.

Table 3. Experimental result

		True Activity				
		walk through	normal rest	browse	abnormal rest	observation
Recognized Activity	walk through	10				
	normal rest		1			
	browse			5	1	
	abnormal rest				3	
	observation					1
	idle					2
Accuracy (%)		100	100	100	75	33.33

From experimental result, our activity recognition method provides a good accuracy for walk through the scene, normal rest and browse because the sequence pattern of actions for these activities can be well defined by our FSM. However, the accuracy may depend directly on the basic action identification; if the actions are correctly classified, then the activity would be properly recognized.

For the case of observation activity, it uses the high value of direction variation feature as acceptance condition. We found that the error may occur when a person make an observation activity with short walk distance during a period of time. Short movement will ignore the direction variation value, which leads to the wrong recognition result.

For the case of abnormal rest activity, we found that when a sit occur near to a special object, there have an ambiguity in recognition process, which may consider it as a browse activity. This shows that our method lack of some crucial features, such as, face direction.

Overall, our method has ability to recognize activities with a good accuracy rate, but need more additional features in the browse and rest activities.

7 Conclusion

In this paper we proposed an activity recognition system using the FSM. We studied and defined the specific characteristic of six human activities from the stream of five basic actions and its corresponding properties, which are necessary for the recognition of complex activities using the FSM. Our proposed method can recognize activities at 86.96% in average on using the well-known dataset. As a future work, in some activities we need more additional features for better configuring the recognition model.

8   References

1. Liu, C., & Yuen, P. C. (2010). Human action recognition using boosted EigenActions. Image and Vision Computing, 28(5), 825–835.
2. Makris, D., & Ellis, T. (2005). Learning semantic scene models from observing activity in visual surveillance.IEEE Transactions on Systems, Man and Cybernetics, 35(3), 397–408.
3. González, J., Rowe, D., Varona, J., & Roca, F. X. (2009). Understanding dynamic scenes based on human sequence evaluation.Image and Vision Computing, 27(10), 1433–1444.
4. Jan, T., Piccardi, M., Hintz, T., (2002). Detection of suspicious pedestrian behavior using modified probabilistic neural network. InProceedings of Image and Vision Computing (pp. 237–241).
5. Htike, Z.Z.; Egerton, S.; Kuang Ye Chow, "Monocular viewpoint invariant human activity recognition," Robotics, Automation and Mechatronics (RAM), 2011 IEEE Conference on , vol., no., pp.18,23, 17-19 Sept. 2011.
6. Shariat, S.; Pavlovic, V., "Isotonic CCA for sequence alignment and activity recognition," Computer Vision (ICCV), 2011 IEEE International Conference on , vol., no., pp.2572,2578, 6-13 Nov. 2011.
7. Gupta, S.K.; Kumar, Y.S.; Ramakrishnan, K. R., "Learning Feature Trajectories Using Gabor Filter Bank for Human Activity Segmentation and Recognition," Computer Vision, Graphics & Image Processing, 2008. ICVGIP '08. Sixth Indian Conference on , vol., no., pp.111,118, 16-19 Dec. 2008.
8. Kang Li; Yun Fu, "ARMA-HMM: A new approach for early recognition of human activity," Pattern Recognition (ICPR), 2012 21st International Conference on , vol., no., pp.1779,1782, 11-15 Nov. 2012.
9. Ping Guo; Zhenjiang Miao, "Multi-person activity recognition through hierarchical and observation decomposed HMM," Multimedia and Expo (ICME), 2010 IEEE International Conference on , vol., no., pp.143,148, 19-23 July 2010.
10. Uddin, M.Z.; Nguyen Duc Thang; Tae-Seong Kim, "Human Activity Recognition via 3-D joint angle features and Hidden Markov models," Image Processing (ICIP), 2010 17th IEEE International Conference on , pp.713,716, 26-29 Sept. 2010.
11. Chun Zhu, Weihua Sheng, Motion- and location-based online human daily activity recognition, Pervasive and Mobile Computing, Volume 7, Issue 2, April 2011, Pages 256-269.
12. Hongqing Fang; Lei He, "BP Neural Network for Human Activity Recognition in Smart Home," Computer Science & Service System (CSSS), 2012 International Conference on , vol., no., pp.1034,1037, 11-13 Aug. 2012.
13. Antonio Fernández-Caballero, José Carlos Castillo, José María Rodríguez-Sánchez, "Human activity monitoring by local and global finite state machines," Expert Systems with Applications, Volume 39, Issue 8, 15 June 2012, Pages 6982-6993.
14. L. Rodriguez-Benitez, C. Solana-Cipres, J. Moreno-Garcia, L. Jimenez-Linares, "Approximate reasoning and finite state machines to the detection of actions in video sequences, " International Journal of Approximate Reasoning, Volume 52, Issue 4, June 2011, Pages 526-540.
15. Chun Yuan; Wei Xu, "Multi-object events recognition from video sequences using extended finite state machine," Image and Signal Processing (CISP), 2011 4th International Congress on , vol.1, no., pp.202,205, 15-17 Oct. 2011.
16. Trinh, H.; Quanfu Fan; Jiyan Pan; Gabbur, P.; Miyazawa, S.; Pankanti, S., "Detecting human activities in retail surveillance using hierarchical finite state machine," Acoustics, Speech and Signal Processing (ICASSP), 2011 IEEE International Conference, pp.1337,1340, 22-27 May 2011.

Part V
Information Retrieval and Visualization

A Comparative Study of Cancer Classification Methods Using Microarray Gene Expression Profile

Hala Alshamlan, Ghada Badr, and Yousef Alohali

King Saud University, College of Computer and Information Sciences,
Riyadh, Kingdom of Saudi Arabia
{halshamlan,yousef}@ksu.edu.sa,badrghada@hotmail.com

Abstract. Microarray based gene expression profiling has been emerged as an efficient technique for cancer classification, as well as for diagnosis, prognosis, and treatment purposes. The primary task of microarray data classification is to determine a computational model from the given microarray data that can determine the class of unknown samples. In recent times, microarray technique has gained more attraction in both scientific and in industrial fields. It is important to determine the informative genes that cause the cancer to improve early cancer diagnosis and to give effective chemotherapy treatment. Classifying cancer microarray gene expression data is a challenging task because microarray is a high dimensional-low sample dataset with lots of noisy or irrelevant genes and missing data. Therefore, finding an accurate and an effective cancer classification approach is very significant issue in medical domain. In this paper, we will make a comparative study and we will categorize the effective binary classification approaches that have been applied for cancer microarray gene expression profile. Then we conclude by identifying the most accurate classification method that has the highest classification accuracy along with the smallest number of effective genes.

Keywords: Cancer classification, Microarray, Classification methods, Gene expression.

1 Introduction

Microarrays, known as DNA chips or sometimes called gene chips, are chips that are hybridized to a labeled unknown molecular extracted from a particular tissue of interest. This makes it possible to measure simultaneously the expression level in a cell or tissue sample for each gene represented on the chip [19][22]. DNA microarrays can be used to determine which genes are being expressed in a given cell type at a particular time and under particular conditions. This allows us to compare the gene expression in two different cell types or tissue samples, where we can determine the more informative genes that are responsible for causing a specific disease or cancer [24].

T. Herawan et al. (eds.), *Proceedings of the First International Conference on Advanced Data and Information Engineering (DaEng-2013)*, Lecture Notes in Electrical Engineering 285, DOI: 10.1007/978-981-4585-18-7_44,
© Springer Science+Business Media Singapore 2014

Recently, microarray technologies have opened up many windows of opportunity to investigate cancer diseases using gene expressions. The primary task of a microarray data analysis is to determine a computational model from the given microarray data that can predict the class of the given unknown samples. The accuracy, quality, and robustness are important elements of microarray analysis. The accuracy of microarray dataset analysis depends on both the quality of the provided microarray data and the utilized analysis approach or objective. However, the curse of dimensionality, the small number of samples, and the level of irrelevant and noise genes make the classification task of a test sample more challenging [6][29]. Those irrelevant genes not only introduce some unnecessary noise to gene expression data analysis, but also increase the dimensionality of the gene expression matrix. This results in the increase of the computational complexity in various consequent research objectives such as classification and clustering [14]. Feature (gene) selection approaches eliminate those irrelevant genes and identify the informative genes [7].

As for classification, there are several methods in the literature for cancer classification approaches using microarray. However, most proposed approaches do not concerned on identifying minimum number of informative genes with high classification accuracy. Therefore, in this comparative study, we seek to determine the most efficient cancer classification method using microarray gene expression profile. Ideally, the efficiency of classifier is measured by accurate classification performance with minimum number of genes.

2 Cancer Classification Methods for Microarray Gene Expression Profile

Microarray based gene expression profiling has become an important and promising dataset for cancer classification that are used for diagnosis and prognosis purposes. The most important motivation for using microarray datasets is to classify unknown tissue samples according to their expression profiles. For example, it can be used in classifying cancerous or normal samples, or to discriminate different types or subtypes of cancer [6]. Moreover, since different subtypes of a cancer respond differently to the same therapy, it is important to diagnose the cancer type of a patient correctly, and then customize the treatment for that patient

Classification tasks are widely used in real-world applications, some of them involves only binary classifies and many of them involve more than two classes, the so-called multi-class classification problem. In this section, we investigate the most accurate and effective cancer classification method. A vast of cancer classification methods that are based on the analysis of microarray gene expression data has been proposed in literature. The main research approaches in microarray cancer classification can be divided into three main categories.

In order to investigate the accurate and more effective cancer classification method, in this section a preliminary literature review about the cancer classification method was presented. Notably, a vast of cancer classification methods based on the analysis of microarray gene expression data has been proposed in literature. The main research direction in microarray cancer classification was found to be using statistical and data mining approaches, machine learning approaches, or multiple classifiers in one of two forms that is called ensemble classifier. Therefore, in this research study we categorized the cancer classification methods proposed in literature into these three categories.

2.1 Statistical and Data Mining Classification Approaches

Many statistical and data mining microarray classification methods have been proposed for cancer classification. Statistics classification approach deals with the quantification, collection, analysis, interpretation, and drawing conclusions from data [10]. While, Data mining classification approach involves the analysis of large existing data bases such as microarray gene expression profile, in order to discover patterns and relationships in the dataset [8].

In [21], (JingJing et al., 2006) introduced an effective and novel data mining technique, named principal component accumulation (PCAcc). The proposed method (PCAcc) is based on PCA method which is an effective data mining technique and has been used for extracting information and reducing data dimensionality. Moreover, in [17] (Huang et al, 2007), an intelligent genetic algorithm (IGA) is applied to the maximum likelihood (MLHD) classification method.

Meanwhile, (Wang and Gotoh, 2009) in [31], explored the use of rule-based methods to construct cancer classification the rule-based methods are more likely to be accepted by biologists and clinicians for they are easily understood. There are many researchers in literature focus on rule-based classifiers , for instance, in paper [13] (Iwen et al., 2008) focused on association rule-based classifiers for microarray dataset. In this paper, the authors tried to extending accurate association rule-based classification methods to larger datasets by develop a scalable rule-based classifier named Boolean Structure Table Classification (BSTC). Furthermore, (He et al., 2006) in [8], presented a new Fuzzy Association Rules (FARs) mining algorithm, called FARM-DS, which is suitable for binary classification problems in the bioinformatics field.

In Table 1, and Figure 1, we summarize the classification performance of these statistical and data mining techniques for specific cancer microarray dataset along with their references from which we pick up the results. We observed that Maximum Likelihood (MLHD) method produce accurate classification results with minimum number of genes for all measurable dataset.

2.2 Machine Learning-Based Classification Approaches

Machine learning is a branch of artificial intelligence, is about the study and train of systems that can learn from data. For instance, a machine learning cancer classifier system could be trained on microarray profile to learn to distinguish between cancerous and non cancerous samples. After learning, it can then be used to classify new patient sample into cancerous and non cancerous tissue [28]. Notably, applied machine learning methods to DNA microarray gene expression

Table 1. Statistical and Data Mining-Based Classification Approaches, Accuracy (No. of Genes)

Classification Approach [Reference]	Leukemia	Lung	Prostate	DLBCL	Colon	Overian
Maximum Likelihood (MLHD) [10]	100(4)	99.45(7)	99.4(4)	90.7(5)		
Decision rules (DR) [31]		82(2)		90(2)	90(2)	
Principal component Accumulation (PCAcc) [21]	97.06(7129)	99.33(12533)		92(7129)		
Penalized logistic regression (PLR) [30]	100(256)	100(6)			99.3(16)	96.4(256)
Boolean Structure Table Classification (BSTC) [13]	82.35(866)	100(2175)	100(1554)			100(2769)
Fuzzy Association Rules Mining for Decision Support (FARM-DS) [8]	96(2)					

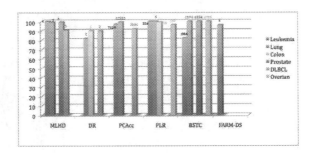

Fig. 1. Statistical and data mining Classification Methods

data allows high throughput analysis procedures when compared to histological and other approaches [99]. Moreover, there are several machine learning methods that are proposed in literature, which are used for cancer classification. These include SVMs (Support Vector Machines), k-NNs (k-Nearest Neighbors), ANNs (Artificial Neural Networks), Naive Bayes (NB), and DTs (Decision Trees).

It is worth mentioning, that the Support Vector Machines (SVM) classification methods are extensively applied, and they achieved accurate cancer diagnosis performance by using gene expression data. Because, the SVM is very suitable to works with high dimensional data, such as microarray gene expression data. Also, it is aims to find the hyperplane that is separating the feature with the largest margin (distance between itself and the closest samples from each classes). Generally, the better SVM classifier seeks to a trade-off between maximizing the margin and minimizing the number of errors. Notable, the preliminary experimental evidence currently available suggests that some SVM classification methods perform well in gene expression-based cancer diagnostic experiments

[11]. Moreover, (Lee et al, 2005) in [18] present extensive comparison study to evaluate the performances of 21 classification methods applied for seven various types of cancer microarray datasets. The contribution for this study show that the classifier based on SVM gives the best performance among the machine learning methods in most cancer datasets regardless of the gene selection.

In Table 2, and Figure 2, we compared and classified the different Machine Learning approaches to SVM-based and non-SVM-based approaches that have been proposed in literature.

Table 2. Machine Learning Classification Approaches, Accuracy (No. of Genes)

Classification Method [Reference]	Leukemia	Lung	Colon	Prostate	DLBCL	Lymphoma
Support Vector Machine (SVM) Based Classifier [1] [2] [5] [17] [23] [27]	100(3)	99.6(3)	99(3)	100(3)	100(4)	96(7)
Neural Network (NN) [16] [28] [33]	98(8)	93.43(40)	98.8(7)			95(10)
K-nearest neighbor method (KNN) [4] [9] [20] [25] [26]	100(21)	99.1(197)	87(21)	98(17)	93(27)	
Naive Bayes (NB) [20]	100(11)		90.5(11)	98.5(11)		

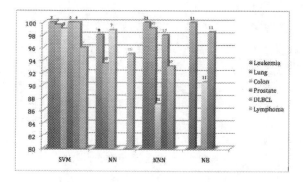

Fig. 2. Machine Learning Classification Methods

2.3 Ensemble-Based Classification Approaches

An ensemble is considers as a supervised learning classifier, because it can be trained and then used to make predictions. The trained ensemble classifiers , therefore, represents a single hypothesis. Than, Ensemble classifiers combine multiple hypotheses to form a (hopefully) better hypothesis [3]. In other words, an ensemble classifiers aim to combine multiple classifiers together as a committee in order to make more appropriate decisions for classifying microarray data instances. Thus, it offers improved accuracy and more reliability.

It is worth mentioning, that a large number of ensemble methods have been applied to biological data analysis like Microarray gene expression profile. Moreover, there are three popular ensemble methods, bagging, boosting, and random forests. These ensemble techniques have the advantage for small sample size problem by averaging multiple classification models to reduce the over fitting problem [6]. Notably, some ensemble methods such as random forests are widely used for high-dimensional datasets because increased classification accuracy can be achieved by generating multiple prediction models each with a different feature subset.

In Table 3, and Figure 3, we compared the performance of recent and more effective ensemble classification methods that have been proposed in literature.

Table 3. Ensemble Classification Approaches, Accuracy (No. of Genes)

Classification Method [Reference]	Leukemia	Lung	Colon	Prostate
Ensemble Classifier (PSO+EDA) [3]	98.6(30)	95.2(30)	100(30)	
SREC (Simple Rule-based Ensemble Classifier) [12]	100(12)	93.5(17)		
GA+SVM [32]		100(25)	99.41(10)	
CAR+SVM [15]		97.06(3)		95.97(2)	95.72(5)

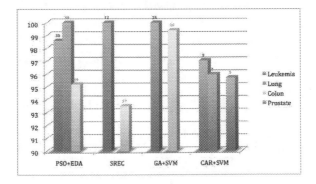

Fig. 3. Ensemble Classification Methods

3 Analysis and Discussion

There are many statistical and data mining techniques that have been used for the analysis of microarray data and for cancer classification. However, the statistical and data mining techniques are proven to be improper, because of the nature of microarray data with very high dimensional along with a limited number of patterns and very little replication. For the past few decades, extensive studies have focused on the accuracy of Microarray classification using machine learning technologies such as neural networks or support vector machine. They have achieved good results. A wide variety of research considered the use of popular machine learning techniques, such as Neural Network, Decision Trees, Bayesian Network, and Support Vector Machines in classification problems. Researches tried to get better accuracy techniques by combining different approaches. Unfortunately, ensemble classifiers that have been investigated so far, did not improve final classification performance for the following two reasons: First, samples misclassified by non-SVM classification methods are almost always a strict superset of samples misclassified by a classifier that is based on SVM algorithms. Second, stand alone SVM algorithms are fairly stable in a sense that small changes in the training data do not result in large changes in the predictive model behavior [11]. In addition, we can contribute that the superior performance of SVM-based methods compared to KNN, NB, NN and DT reacts that SVM classifiers are less sensitive to the curse of dimensionality and more robust to a small number of high dimensional gene expression samples than other non-SVM techniques.

In Table 4, and Figure 4, we compared the classification performance of statistical and data mining methods, machine learning methods, and ensemble methods. We proved that the machine learning classification method have superior performance, and the classifiers that are base on SVM are produce high classification accuracy with minimum number of selected genes. Moreover, when we compared SVM classifiers with other machine learning methods, SVM has many advantages. For many real-world applications, it is challenging to construct a linear classifier to separate the classes of the given data. SVM addresses this problem by mapping the input space into a high-dimensional feature space; then constructs linear classification decision to classify the input data with a maximum margin hyper plane. SVM was also found to be more effective and faster than other machine learning methods, such as neural networks and k -nearest neighbor classifiers [15].

4 Conclusion

DNA microarray technique has recently gained more attention in both scientific and in industrial fields. It is also important to determine the informative genes that cause the cancer to improve cancer classification and early cancer diagnosis. Classifying cancer microarray gene expression data is challenging task because microarray has a high dimensional-low sample dataset with a lots of

Table 4. Comparison between all Classification Methods Categories, Accuracy (No. of Genes)

Classification Method	Leukemia	Lung	Colon	Prostate
Statistical and Data Mining Classifiers	95(1225)	96(1438)	94(9)	99(779)
Machine Learning Classifiers				
Support Vector Machine (SVM) Based Classifier	100(3)	99.6(3)	99(3)	100(3)
Non Support Vector Machine (SVM) Based Classifier	99(13)	96(118)	91.7(13)	98(14)
Ensemble Classifier	98.7(18)	97.9(16)	95.6(19)	95.72(5)

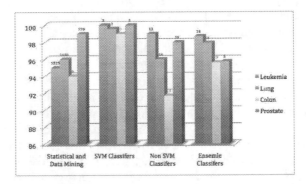

Fig. 4. Comparison between all Classification Methods Categories

noisy or irrelevant genes and missing data. Therefore, cancer classification using microarray gene expression profile is very significant issue in medical domain. In order to find accurate cancer classification methods, we conducted a comprehensive study in the area of microarray cancer classification approaches. From our study, we concluded that machine learning classification methods produce accurate result with minimum number of genes. In addition support vector machine classifiers have superior classification performance. As a future extensions: More detailed explanation about each of the presented approaches along with suggested extensions and combinations are under preparation for future extensions of the paper. In addition, a comprehensive survey about the multi-classification approaches will be analyzed and included as well.

References

1. Abderrahim, A., Talbi, E., Khaled, M.: Hybridization of genetic and quantum algorithm for gene selection and classification of microarray data. In: Parallel Distributed Processing, 2009. IPDPS 2009. IEEE International Symposium on. pp. 1–8 (2009)

2. Alba, E., Garcia-Nieto, J., Jourdan, L., Talbi, E.: Gene selection in cancer classification using pso/svm and ga/svm hybrid algorithms. In: Evolutionary Computation, 2007. CEC 2007. IEEE Congress on. pp. 284–290 (2007)

3. Chen, Y., Zhao, Y.: A novel ensemble of classifiers for microarray data classification. Applied Soft Computing 8(4), 1664 – 1669 (2008)

4. Chuang, L.Y., Yang, C.H., Wu, K.C., Yang, C.H.: A hybrid feature selection method for dna microarray data. Computers in Biology and Medicine 41(4), 228 – 237 (2011)

5. El Akadi, A., Amine, A., El Ouardighi, A., Aboutajdine, D.: A new gene selection approach based on minimum redundancy-maximum relevance (mrmr) and genetic algorithm (ga). In: Computer Systems and Applications, 2009. AICCSA 2009. IEEE/ACS International Conference on. pp. 69–75 (2009)

6. Ghorai, S., Mukherjee, A., Sengupta, S., Dutta, P.: Multicategory cancer classification from gene expression data by multiclass nppc ensemble. In: Systems in Medicine and Biology (ICSMB), 2010 International Conference on. pp. 41–48 (2010)

7. Guo, S.B., Lyu, M.R., Lok, T.M.: Gene selection based on mutual information for the classification of multi-class cancer. In: Proceedings of the 2006 international conference on Computational Intelligence and Bioinformatics - Volume Part III. pp. 454–463. ICIC'06, Springer-Verlag, Berlin, Heidelberg (2006)

8. He, Y., Tang, Y., Zhang, Y.Q., Sunderraman, R.: Mining fuzzy association rules from microarray gene expression data for leukemia classification. In: Granular Computing, 2006 IEEE International Conference on. pp. 461–464 (2006)

9. Huang, H., Li, J., Liu, J.: Gene expression data classification based on improved semi-supervised local fisher discriminant analysis. Expert Systems with Applications 39(3), 2314 – 2320 (2012)

10. Huang, H.L., Lee, C.C., Ho, S.Y.: Selecting a minimal number of relevant genes from microarray data to design accurate tissue classifiers. Biosystems 90(1), 78 – 86 (2007)

11. Huang, H.L., Lee, C.C., Ho, S.Y.: Selecting a minimal number of relevant genes from microarray data to design accurate tissue classifiers. Biosystems 90(1), 78 – 86 (2007)

12. Huerta, E.B., Duval, B., kao Hao, J.: A hybrid ga/svm approach for gene selection and classification of microarray data. In: EvoWorkshops 2006, LNCS 3907. pp. 34–44. Springer (2006)

13. Iwen, M., Lang, W., Patel, J.: Scalable rule-based gene expression data classification. In: Data Engineering, 2008. ICDE 2008. IEEE 24th International Conference on. pp. 1062–1071 (2008)

14. Javed Khan, Jun Wei, M.R.: Classification and diagnostic prediction of cancers using gene expression profiling and artificial neural networks. Nature Medicine 7(6), 673679 (2001)

15. Kianmehr, K., Alhajj, R.: CARSVM: A class association rule-based classification framework and its application to gene expression data. Artificial Intelligence in Medicine 44, 7–25 (2008)

16. Kumar, P.G., Victoire, T.A.A., Renukadevi, P., Devaraj, D.: Design of fuzzy expert system for microarray data classification using a novel genetic swarm algorithm. Expert Systems with Applications 39(2), 1811 – 1821 (2012)

17. Lee, C.P., Leu, Y.: A novel hybrid feature selection method for microarray data analysis. Appl. Soft Comput. 11(1), 208–213 (Jan 2011)

18. Lee, J.W., Lee, J.B., Park, M., Song, S.H.: An extensive comparison of recent classification tools applied to microarray data. Computational Statistics and Data Analysis 48(4), 869 – 885 (2005)
19. Linder, R., Dew, D., Sudhoff, H., Theegarten, D., Remberger, K., Poppl, S.J., Wagner, M.: The subsequent artificial neural network (sann) approach might bring more classificatory power to ann-based dna microarray analyses. Bioinformatics 20(18), 3544–3552 (2004)
20. Liu, H., Liu, L., Zhang, H.: Ensemble gene selection by grouping for microarray data classification. Journal of Biomedical Informatics 43(1), 81 – 87 (2010)
21. Liu, J., Cai, W., Shao, X.: Cancer classification based on microarray gene expression data using a principal component accumulation method. Science China Chemistry 54, 802–811 (2011)
22. Liu, K.H., Xu, C.G.: A genetic programming-based approach to the classification of multiclass microarray datasets. Bioinformatics 25(3), 331–337 (2009)
23. Mohamad, M., Omatu, S., Yoshioka, M., Deris, S.: An approach using hybrid methods to select informative genes from microarray data for cancer classification. In: Modeling Simulation, 2008. AICMS 08. Second Asia International Conference on. pp. 603–608 (2008)
24. Okun, O., Priisalu, H.: Multi-class cancer classification using ensembles of classifiers: Preliminary results. In: In Proceedings of the Workshop on Probabilistic Modeling and Machine Learning in Structural and Systems Biology. pp. 137–142 (2007)
25. Osareh, A., Shadgar, B.: Microarray data analysis for cancer classification. In: Health Informatics and Bioinformatics (HIBIT), 2010 5th International Symposium on. pp. 125–132 (2010)
26. Peng, S., Zeng, X., Li, X., Peng, X., Chen, L.: Multi-class cancer classification through gene expression profiles: microrna versus mrna. Journal of Genetics and Genomics 36(7), 409 – 416 (2009)
27. Peng, Y.: A novel ensemble machine learning for robust microarray data classification. Computers in Biology and Medicine 36(6), 553 – 573 (2006)
28. Perez, M., Rubin, D., Marwala, T., Scott, L., Featherston, J., Stevens, W.: The fuzzy gene filter: An adaptive fuzzy inference system for expression array feature selection. In: Garca-Pedrajas, N., Herrera, F., Fyfe, C., Bentez, J., Ali, M. (eds.) Trends in Applied Intelligent Systems, Lecture Notes in Computer Science, vol. 6098, pp. 62–71. Springer Berlin Heidelberg (2010)
29. Scott Pomeroy, P.T.: Prediction of central nervous system embryonal tumour outcome based on gene expression. Nature 415(6870), 436442 (2002)
30. Shen, L., Tan, E.C.: Dimension reduction-based penalized logistic regression for cancer classification using microarray data. IEEE/ACM Trans. Comput. Biol. Bioinformatics 2(2), 166–175 (Apr 2005)
31. Wang, X., Gotoh, O.: Microarray-based cancer prediction using soft computing approach. Cancer Informatics 7, 123–139 (05 2009)
32. Yu, H., Xu, S.: Simple rule-based ensemble classifiers for cancer dna microarray data classification. In: Computer Science and Service System (CSSS), 2011 International Conference on. pp. 2555–2558 (2011)
33. Zhang, R., Huang, G.B., Sundararajan, N., Saratchandran, P.: Multicategory classification using an extreme learning machine for microarray gene expression cancer diagnosis. Computational Biology and Bioinformatics, IEEE/ACM Transactions on 4(3), 485–495 (2007)

An Efficient Perceptual Color Indexing Method for Content-based Image Retrieval Using Uniform Color Space

Ahmed Talib[1,2] , Massudi Mahmuddin[1], Husniza Husni[1], Loay E. George[3]

[1] Computer Science Dept., School of Computing, University Utara Malaysia, 06010 Sintok, Kedah, Malaysia

s91707@student.uum.edu.my, {ady, husniza}@uum.edu.my

[2] IT Dept., Technical College of Management, Foundation of Technical Education, 10047 Bab Al-Muadham, Baghdad, Iraq

[3] Computer Science Dept., College of Science, Baghdad University, 10071 Al-Jadriya, Baghdad, Iraq

loayedwar57@yahoo.com

Abstract. Dominant Color Descriptor (DCD) is one of the famous descriptors in Content-based image retrieval (CBIR). Sequential search is one of the common drawbacks of most color descriptors especially in large databases. In this paper, dominant colors of an image are indexed to avoid sequential search in the database where uniform RGB color space is used to index images in LUV perceptual color space. Proposed indexing method will speed up the retrieval process where the dominant colors in query image are used to reduce the search space. Additionally, the accuracy of color descriptors is improved due to this space reduction. Experimental results show effectiveness of the proposed color indexing method in reducing search space to less than 25% without degradation the accuracy.

Keywords: Color indexing, Dominant color descriptor, LUV color space, RGB color space, Database search space.

1   Introduction

Image retrieval become one of the most famous research directions nowadays because it uses to search an image in archive, domain-specific, personal and web image databases. For retrieving images from multimedia database, low level features and especially color feature, have been widely used in this regard. This is because color represents the most distinguishable feature compared with other visual features, such as texture and shape [1-3]. From perspective of feature extraction, color-based image descriptors can be divided into two categories: (i) global descriptors that consider the whole image to obtain their features, there is no partitioning or pre-processing stage during feature extraction process and (ii) local descriptors that obtain their features from local regions or partitions of image. This can be achieved by dividing the image into either fixed-size or different-size regions. These descriptors usually have better accuracy than others but introduce more complexity of feature extraction process.

T. Herawan et al. (eds.), *Proceedings of the First International Conference on Advanced Data and Information Engineering (DaEng-2013)*, Lecture Notes in Electrical Engineering 285, DOI: 10.1007/978-981-4585-18-7_45,
© Springer Science+Business Media Singapore 2014

In this respect, MPEG-7 committee proposed many color, texture and shape descriptors to be used in image and video retrieval [4]. Human visual system as mentioned in [5, 6] firstly identifies prominent colors in the image then it processes any other details. MPEG-7's DCD provides compact and effective representations for colors in an image or region of interest [4]. Recently, compactness property of dominant colors representation becomes more attractive for many researchers to reduce size of color descriptors from several hundred bins (histogram-based methods) into few colors (8 colors as in MPEG-7 DCD) such as the works that have been achieved in [3, 7, 8].

Recently, researchers focus on either retrieval robustness (retrieval accuracy) or retrieval efficiency (retrieval speed) [9]. Basically, the searching a large images database imposed delay of image retrieval process. Therefore, this paper focuses on the retrieval efficiency that is closely related to the retrieval robustness. Color histogram and dominant color descriptors are the widely used methods in content based image retrieval [2, 3]. Histogram suffers from high-dimensional indexing problem, hence color quantization techniques are used for color reduction but fixed quantization lead to accuracy degradation because different colors may be mapped to the same quantized color bin. Therefore, dominant colors (DCs) extraction methods (a dynamic quantization approach) were proposed as the most effective solutions in this context where few colors are extracted to represent the image. DCs are proposed for image indexing in this research, but it suffers from "color approximation" problem. Another issue that need to be mentioned here, the characteristics of RGB color distribution is uniform; it is simple and straightforward color representation, however is not a perceptual color space [10]. Perceptual color spaces match human visual system and have non-uniform color distribution such as LUV, Lab and YUV. As a result of this, complicated and time consuming quantization methods must be used for these perceptual spaces. For perceptual color indexing, two methods can be used which are color clustering such in [11, 12] or utilizing uniform color spaces [13]. The shortcomings of these two methods can be summarized as follows: (i) color clustering methods (such as k-Means and adapted version of k-Means [14]) have color approximation problem and (ii) utilizing uniform quantization (such as *Octree* [15]) for perceptual (non-uniform) color spaces will not take into account the perceptual similarity between different bins. RGB and HSV color spaces do not exhibit perceptual uniformity whereas LUV, Lab and YUV color spaces have this perceptual effect. Therefore, from all above cited problems, perceptual color indexing method is proposed.

The paper is organized in the following way. Section 2 explicates the general CBIR indexing methods and specifically color-based indexing methods. Section 3 is mainly concerned with the proposed indexing method and the newly proposed color percentage filtering scheme that helps improve and speed up the retrieval process. Section 4 illustrates the quantitative results of experiments that conducted to show the effectiveness of proposed indexing method. Finally, the conclusion can be seen in Section 5.

2   Related Work

In large image database, indexing is an urgently demanded to reduce the search space of the retrieval process and in turn to speed up the process. For indexing the image features, there are two main approaches in general: multi-dimensional indexing and vector quantization techniques. Multi-dimensional indexing is divided into two approaches namely space partitioning (SP) and data-partitioning (DP). Both of them divide the space into small partitions. SP divides the whole space into disjoint partitions without consideration of the data (feature vectors). In DP methods, feature space is divided depending on features (data) distribution in the database. The advantage of SP method is it performs complete and disjoint partitions of the whole space that means there is no overlapping between these partitions. Disadvantage of SP method occurred when the query point locates at the border of partition. This will lead to degrade the retrieval performance at least in two situations: a) retrieval accuracy will decrease if the search on this point was made within that partition only and there are some similar points in neighbor partitions due to ignoring some similar points in the search space or b) computational will increase if all neighbor partitions are taken into account where the search space will increase after considering many partitions and many unrelated data will be compared.

In vector quantization techniques, there are many color clustering techniques that have been proposed including hierarchical k-Means clustering, randomized tree and self-organizing map. In these methods, no partitioning of space involves, instead grouping the data into clusters or groups. The properties of these clusters are the distance between minimizing the intra-cluster members whereas the distance between the different inter-clusters should be maximized. Each cluster is represented by a cluster's centroid, thus the query point is compared with cluster's centroid instead of original value of the members. Disadvantages of these methods are the initial number of clusters, k needs to be known prior. Additionally, majority of these methods do not preserve ordering structure of data space. This will lead to expensive online distance computation with all clusters' centroids to select the nearest one. Moreover, comparing with clusters' centroids instead of original values lead to inaccurate results because some cluster's members are far from the query points in spite of having suitable distance from cluster's centroid which known as "color approximation problem".

For 3-D color indexing, several methods have been proposed in CBIR field. High dimensional histogram indexing method, that is used by [16] for comparison, is considered as the simplest and most expensive indexing method. This method suffers from high dimensional problem, 1024-D of color histogram bins. Combination of color clustering and spatial indexing method (R-Tree) for color indexing can be seen in [11]. A method proposed by [12] on color clustering and indexing by using the mean shift algorithm for color clustering, R*-Tree for spatial indexing and perceptually uniform LUV color space. The above two methods depend on clustering that are suffered from aforementioned problems of clustering (one of VQ techniques). NeTra system proposed by [17], used binary color table for color indexing that depended on 256 colors codebook in RGB color space. However, restricting with 256 colors certainly will lead to accuracy degradation as a result of color quantization (similar to color quantization method for histogram).

In general, color-based indexing methods depend on fixed range of colors in similarity measure. Therefore, spatial indexing methods such as R-Tree and R*-Tree are not necessary and fixed indexing structure is more efficient [16]. Accordingly, Lattice structure [16] is characterized by efficient finding the nearest neighbors of given point (color) in 3-D LUV color space. Nevertheless, this efficiency depends on careful selection of radius in hexagonal lattice cell and this is not a straight forward process, hence there is no comparison (in the literature) has been made with this method. Additionally, it suffers from same problem of SP and clustering methods, as the query point may locate at the border of lattice cell, as depicted in Fig. 1. Moreover, lattice structure has better performance in uniform distribution than non-uniform distribution [18]. Thus, it is recommended for RGB color space instead of LUV color space.

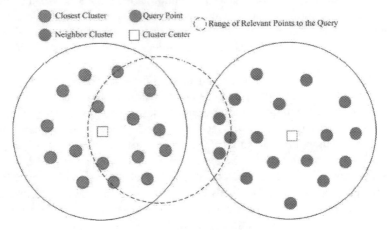

Fig. 1. "Locating query at the cluster border" problem.

Recent research is proposed by [14] to solve this problem. The significant contribution of that research was introducing two threshold values C_G and C_S that can be considered as search space parameters to improve searching process.

As conclusion, all vector quantization indexing methods, which most color indexing methods are depended on, suffered from color approximation problem. Therefore, this problem is addressed in this paper.

3   Proposed Indexing Method

In this section, DC-based indexing method is proposed to reduce search space of color-based methods (i.e. it dedicates to all color-based methods, not only DC-based methods) to speed up retrieval process as well as preserve or increase retrieval accuracy. This paper mainly working on of reducing the search space rather than doing whole database search. For that, a fixed space partitioning method is used in this research. Before building database index structure, similarity between two colors must be considered and maximum distance between these colors also needs to be determined because the index structure will depend upon them.

3.1   Indexing Structure of Maximum Distance

The key of the proposed method is a similarity among colors within fixed range. The searching can be done only in a specific range within distance, which is the maximum distance between two colors to consider them as similar colors. The Euclidian distance between two *3-D* colors to assume them similar was 10, 20 or 25 [4, 5]. Therefore, the maximum difference value (MxDV) for each color channels (Red, Green and Blue in RGB color space) is set to 25.

In DC-based methods (e.g. MPEG-7 DCD), DCs are extracted using dynamic quantization method (GLA) and most likely the image is quantized to most significant 5-bits of color channels (bit3 to bit7). Since the maximum difference between two channels is 25, the changing of the two bits (bit3 and bit4) is within this range. This because the weights of these bits are 8 and 16 respectively; their summation is 24 that approximately equals to MxDV (25). In this regard, bit7, bit6 and bit5 are out of tolerance range of colors to be similar. Thus, these three bits are the first level of color similarity; they are used to differentiate among not similar colors. The other two bits (bit3 and bit4) are used separately to be second and third level respectively of color similarity. Hence, first indexing dimension contains 512 cell (3-bits from each channel = 9-bits, no. of cells = 2^9 = 512). Second indexing dimension contains 8 cells (1-bit from each channel = 3-bits, 2^3 = 8). The third dimension has 8 cells that represent the remaining one bit (bit3), as presented in Fig. 2.

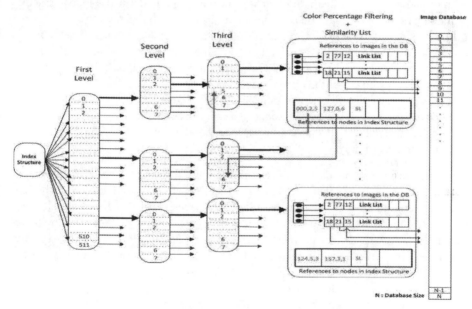

Fig. 2. Proposed indexing method for perceptual color spaces.

Color percentage plays an important role in similarity measure of certain color with its corresponding color in other images where the similar colors consider as dissimilar if their percentages have large difference such as similarity measure of different color descriptors such as MPEG-7 DCD, LBA [5], BIC [19], and Correlogram [20]. Therefore, filtering images (that have large difference in percentage) in early stage helps in reducing search space and in turn speeding up the retrieval process. Deng [16] used filtering process during query processing. This online filtering is impractical because it is time consuming. Therefore, the proposed indexing structure is extended to include partitions of color percentages. Single level B+-tree [21] is used to represent color percentages. B+-tree can be added to all leaf nodes (in the third index level), as shown in Fig. 2. To achieve the perceptual concept of LUV color space, LUV color distance is used. Each color in the 5-bits quantized space (each node in the third level of the index structure) is compared with other colors in the space and stores the perceptually similar colors in a list (similarity list) that belonging to this color as depicted in Fig. 2. The searching for similar colors does not include all color space; instead, it searches within specific range in the whole color space. This range depends on the color similarity thresholds where the similarity between each two colors must be less than or equal the certain color threshold. Color thresholds that are selected in this research are 10, 15, 20 and 25.

3.2 Searching Process

In this process, a query is required to find its similar images in the database. Searching process includes the following steps:

1. For each DC in the query image, find database images that have similar in both color and percentage; this is by reaching to the single node in the index structure and in turn to the database images that are associated with this node.
2. In each single node (of point 1) that corresponds to each DC in the query, there is a similarity list that is used to reach all similar colors in LUV color space. It contains references to all nodes (colors) in the index structure and thus to all images that contain similar colors of the query image.
3. Merging images that resulted from each DC of the query image to produce search space of the query, it represents Reduced Search Space (RSS).
4. Calculate dissimilarity distance between query and all images in the RSS, and then rank them accordingly.

In step 1, images that have same DCs of the query are reached. In step 2, all images that have perceptually similar colors to the query DCs are collected to produce RSS. It is worth mentioning, step 1 and 2 also filter the images that have different color percentages using B+-tree that exists in each node in the index structure. This helps reducing the search space that is emerged in step 3. Dissimilarity distances (of specific descriptor) are computed to all images in the RSS to obtain the most similar images to the query. The important characteristics of the proposed index structure are it uses to index RGB and LUV color spaces; it is dynamic to the database updating process (insertion and deletion). Lastly, it does not have any color approximation.

4   Proposed Indexing Method Performance

Performance of the proposed technique is evaluated based on the following:

1. *Number of Indexed colors:* Indexing of images database can be performed with different numbers of colors (8, 5, 3 or 1) to measure the effect of each one on the retrieval performance. The database is indexed firstly according to maximum DCs in the image that equal 8; then reduce the number of indexed DCs into 5, 3 and 1. It is noteworthy to mention here that DCs is sorted in descending order according to their color percentage before the indexing process.
2. *Evaluation Metrics:* Two types of metrics are used in this research:

(a) Efficiency Metrics: The main goal of indexing is to reduce search time compared to sequential search by reducing database images that can matched the query image. It eliminates unlikely (irrelevant) images from the matching process to produce RSS. The reduced time needed for searching in the RSS can be computed by the percentage RSS/WSS, WSS denoted to the Whole Search Space. The percentage RSS/WSS% can be called as Search Space Ratio (SSR) that represents ratio of images that are actually searched to the all images in the database.

(b) Accuracy Metrics: Three quantitative performance metrics are utilized to measure the accuracy of different color descriptors that are used in the proposed indexing method. These metrics are ARR, ANMRR and P(10).

3. **Evaluation Datasets:** Evaluating the proposed indexing techniques will be conducted on two datasets, newly introduced Cartoon-11K (11,120 images) and well-known Corel-10K (10,800 images). These datasets are different in terms of image content (color and variety) as well as their sizes are large enough to fit the objective of designing the indexing methods. The main dataset in this research is cartoon dataset that is used to evaluate color descriptors. This is because the characteristic of the most cartoon characters is appearing with the same colors in all or most images [22, 23].

4. **Competing Indexing Methods:** Indexing methods that are selected to compete with the proposed method are sequential search, k-Means (KM), and recent k-Means with B+-tree methods (KMB) [14]. Sequential search is a conventional method in CBIR for searching in the database. The accuracy resulted from sequential search is considered as optimal accuracy because searching in this method include whole database. Therefore, all competing indexing methods accuracies are compared with it to check the degradation that can be obtained from these methods due to the reduction of search space.

5. **Evaluation Color Descriptors:** the color descriptors that can be used to test the proposed indexing method are MPEG-7 DCD and Color Correlogram (*ColGrm*). MPEG-7 DCD is used because it contains dominant colors whereas the Correlogram is complicated color descriptor and it is general color descriptor (it does not have dominant colors). This is to prove that the proposed indexing method can be generalized for all color descriptors not just for DCDs.

Table 1. Performance metrics for *ColGrm* descriptor using sequential search, KM, KMB, and proposed LUV indexing method applied on Cartoon-11K Dataset with 158 Queries.

ColGrm	Indexed color = 8		Indexed color = 5		Indexed color = 3		Indexed color = 1	
	P(10)/ ARR/ ANMRR	% SSR	P(10) / ARR/ ANMRR	% SSR	ARR/ ANMRR/ P(10)	% SSR	P(10)/ ARR/ ANMRR	% SSR
Sequential Search			0.350/ 0.118/ 0.852			100%		
KM	0.310/ 0.100/ 0.874	45.8	0.320/ 0.102/ 0.872	40.8	0.270/ 0.089/ 0.889	24.1	0.220/ 0.076/ 0.905	14.5
KMB	0.350/ 0.115/ 0.856	76.6	0.350/ 0.116/ 0.856	71.0	0.310/ 0.104/ 0.870	39.8	0.230/ 0.080/ 0.899	26.7
LUV Indexing,	0.350/	90.5	0.360/	82.1	0.360/	69.6	0.360/	45.6

Color threshold = 25	0.118/ 0.852		0.119/ 0.851		0.120/ 0.851		0.117/ 0.855	
LUV Indexing, Color threshold = 20	0.360/ 0.120/ 0.850	78.0	0.360/ 0.121/ 0.850	68.0	0.360/ 0.121/ 0.849	54.4	0.340/ 0.114/ 0.859	31.4
LUV Indexing, Color threshold = 15	0.360/ 0.122/ 0.848	59.4	0.360/ 0.122/ 0.848	51.4	0.360/ 0.122/ 0.847	41.6	0.340/ 0.114/ 0.860	25.8
LUV Indexing, Color threshold = 10	0.360/ 0.120/ 0.849	37.1	0.360/ 0.121/ 0.849	33.0	0.360/ 0.117/ 0.854	28.2	0.330/ 0.107/ 0.867	19.8

Note: Bold values represent the accuracy values that are better than or equals to sequential search.

Table 1 shows the results of all the indexing methods using ColGrm descriptor applied on Cartoon-11K dataset with 158 queries. Accuracy of the results using LUV indexing method is better than those of sequential search, KM and KMB in most settings (in four LUV color distances 25, 20, 15, and 10 as well as in different indexed colors 8, 5, 3). This leads to retrieve all images that have similar colors (to the query DCs) where there is no color approximation at all. The unique case that has accuracy less than that of sequential search is when using 1 color for indexing.

SSR of the proposed LUV indexing method is ranged from high search space (the worst) to very low search space. The worst case occurs when LUV color distance equals to 25 (it is similar to KMB) where both of them have large SSR. The medium case occurs when distance equals to 20 (it lower than KMB and higher than KM). The lower case occurs when LUV distance equals to 15 (it similar to KM). The very low search space (the best) case (lower than KM) occurs when the distance equals to 10 with SSR equal 19% without degradation to the accuracy. According to these different search space ratios of different settings of LUV indexing method, the accuracy of it is higher than sequential search and all other indexing methods (except when indexed color equal 1) with reduction the search space to less than 20%.

Experiments on MPEG-7 DCD also show similar accuracy to the *ColGrm* descriptor. This ensures effectiveness of the proposed indexing method in different color methods. Performance of the experiment as represented in Table 2.

Table 2. Performance metrics for MPEG-7 DCD using sequential search and all indexing methods including proposed LUV indexing applied on Cartoon-11K Dataset with 158 queries.

MPEG-7 DCD	Indexed color=8 P(10)/ ARR/ ANMRR	% SSR	Indexed color=5 P(10)/ ARR/ ANMRR	% SSR	Indexed color=3 P(10)/ ARR/ ANMRR	% SSR	Indexed color=1 P(10)/ ARR/ ANMRR	% SSR
Sequential Search	0.230/ 0.060/ 0.922				100%			
KM	0.200/ 0.057/ 0.926	45.8	0.210/ 0.059/ 0.926	40.8	0.180/ 0.051/ 0.935	24.1	0.140/ 0.041/ 0.947	14.5
KMB	0.230/ 0-.059/ 0.922	76.6	0.230/ 0.059/ 0.923	71	0.210/ 0.056/ 0.927	39.8	0.150/ 0.042/ 0.945	26.7
LUV Indexing, Color threshold =25	0.230/ 0.060/ 0.922	90.5	0.240/ 0.060/ 0.922	82.1	0.240/ 0.061/ 0.922	69.6	0.240/ 0.060/ 0.922	45.6
LUV Indexing,	0.240/	78.0	0.240/	68.0	0.240/	54.4	0.230/	31.4

Color threshold =20	0.060/ 0.922		0.060/ 0.922		0.060/ 0.922		0.060/ 0.923	
LUV Indexing, Color threshold =15	0.240/ 0.060/ 0.922	59.4	0.240/ 0.060/ 0.922	51.4	0.240/ 0.060/ 0.922	41.6	0.230/ 0.059/ 0.924	25.8
LUV Indexing, Color threshold =10	0.240/ 0.061/ 0.922	37.1	0.240/ 0.061/ 0.921	33.0	0.240/ 0.059/ 0.923	28.2	0.230/ 0.057/ 0.926	19.8

5 Conclusion

In this paper, indexing methods of CBIR with advantages and disadvantages of each method are presented. Specifically, the problems of color-based indexing methods such as high-dimensional problem for histogram-based methods and color approximation problem of DC-based methods are addressed. Accordingly, DC-based indexing method is proposed where it dedicates for non-uniform LUV color space. The proposed method has similar superiority of SP methods with some enhancements. The supremacy of the proposed LUV method over SP is overcoming the problem of locating the query on the border of partitions (or cluster) and preserving the perceptual distance between colors in the color space. Additionally, existence the static representation (array) of index structure and color percentage filtering scheme (using B+-tree) help with speed up the retrieval process.

References

1. Penatti, O. A. B., Valle, E., and Torres, R. d. S., "Comparative Study of Global Color and Texture Descriptors for Web Image Retrieval," *Journal of Visual Communication and Image Representation (Elsevier),* 2012.
2. Talib, A., Mahmuddin, M., Husni, H., and George, L. E., "A weighted dominant color descriptor for content-based image retrieval," *Journal of Visual Communication and Image Representation,* vol. 24, pp. 345-360, 2013.
3. Talib, A., Mahmuddin, M., Husni, H., and George, L. E., "Efficient, Compact, and Dominant Color Correlogram Descriptors for Content-based Image Retrieval," presented at the MMEDIA 2013: Fifth International Conference on Advances in Multimedia, Venice, Italy, 22-26 April 2013, 2013.
4. Yamada, A., Pickering, M., Jeannin, S., and Jens, L. C., "MPEG-7 Visual Part of Experimentation Model Version 9.0-Part 3 Dominant Color," *ISO/IEC JTC1/SC29/WG11/N3914, Pisa,* 2001.
5. Yang, N.-C., Chang, W.-H., Kuo, C.-M., and Li, T.-H., "A fast MPEG-7 dominant color extraction with new similarity measure for image retrieval," *Journal of Visual Communication and Image Representation,* vol. 19 (2008), pp. 92–105, 2008.
6. Mojsilovic, A., Hu, J., and Soljanin, E., "Extraction of perceptually important colors and similarity measurement for image matching, retrieval, and analysis," *Transaction of Image Processing,* vol. 11 (11), pp. 1238–1248, 2002.
7. Kiranyaz, S., Birinci, M., and Gabbouj, M., *Perceptual Color Descriptors.* Foveon, Inc. / Sigma Corp., San Jose, California, USA: Boca Raton, FL, CRC Press, 2012.

8. Wong, K.-M., Po, L.-M., and Cheung, K.-W., "Dominant Color Structure Descriptor For Image Retrieval," *IEEE International Conference on Image Processing, 2007. ICIP 2007,* vol. 6, pp. 365-368, 2007.

9. Jouili, S. and Tabbone, S., "Hypergraph-based image retrieval for graph-based representation," *Pattern Recognition,* vol. 45, pp. 4054-4068, 2012.

10. Park, D.-S., Park, J.-S., Kim, T. Y., and Han, J. H., "Image indexing using weighted color histogram," in *Image Analysis and Processing, 1999. Proceedings. International Conference on,* 1999, pp. 909-914.

11. Babu, G. P., Mehtre, B. M., and Kankanhalli, M. S., "Color indexing for efficient image retrieval," *Multimedia Tools and Applications,* vol. 1 (November), pp. 327–348, 1995.

12. Sudhamani, M. and Venugopal, C., "Grouping and indexing color features for efficient image retrieval," *International. Journal of Applied Mathematics and Computer Sciences. v4 i3,* pp. 150-155, 2007.

13. Sclaroff, S., Taycher, L., and Cascia, M. L., "Image-Rover: a content-based image browser for the world wide web," *Proceedings of IEEE Workshop on Content-based Access Image and Video Libraries, Puerto Rico,* pp. 2-9, 1997.

14. Yildizer, E., Balci, A. M., Jarada, T. N., and Alhajj, R., "Integrating wavelets with clustering and indexing for effective content-based image retrieval," *Knowledge-Based Systems,* vol. 31, pp. 55-66, 2012.

15. Gervautz, M. and Purgathofer, W., "A simple method for color quantization: Octree quantization," in *New trends in computer graphics,* ed: Springer, 1988, pp. 219-231.

16. Deng, Y., Manjunath, B. S., Kenney, C., Moore, M. S., and Shin, H., " An efficient color representation for image retrieval," *IEEE Trans. Image Process,* vol. 10 (1), pp. 140–147, 2001.

17. Ma, W.-Y. and Manjunath, B. S., "Netra: A toolbox for navigating large image databases," *Multimedia systems,* vol. 7, pp. 184-198, 1999.

18. Pauleve, L., Jegou, H., and Amsaleg, L., "Locality sensitive hashing: A comparison of hash function types and querying mechanisms," *Pattern Recognition Letter,* vol. 31, pp. 1348-1358, 2010.

19. Renato, O. S., Mario, A. N., and Alexandre, X. F., "A Compact and Efficient Image Retrieval Approach Based on Border/Interior Pixel Classification," *Proceedings Information and Knowledge Management,* pp. 102-109, 2002.

20. Kunttu, I., Lepistö, L., Rauhamaa, J., and Visa, A., "Image correlogram in image database indexing and retrieval," *Proceedings of 4th European Workshop on Image Analysis for Multimedia Interactive Services, London, UK,* pp. 88-91, 2003.

21. Lightstone, S. S., Teorey, T. J., and Nadeau, T., *Physical Database Design: the database professional's guide to exploiting indexes, views, storage, and more.,* 2010.

22. Jiebo, L. and Crandall, D., "Color object detection using spatial-color joint probability functions," *IEEE Transactions on Image Processing,* vol. 15, pp. 1443-1453, 2006.

23. Khan, F. S., Rao, M. A., Weijer, J. v. d., Bagdanov, A. D., Vanrell, M., and Lopez, A., "Color Attributes for Object Detection," *Twenty-Fifth IEEE Conference on Computer Vision and Pattern Recognition (CVPR 2012),* 2012.

Holographic Projection System with 3D Spatial Interaction

Sooyeon Lim, Sangwook Kim

Department of Digital Media Art, Kyungpook National University, Korea
School of Computer Science and Engineering, Kyungpook National University, Korea
{sylim,kimsw}@knu.ac.kr

Abstract. We propose a Holographic projection system that has direct and natural interaction technology between 3D contents and users using voice and motion. For this, we had to regard user space and holographic image space as three-dimensional cube. And we use a depth camera to track user's voice and motion. The information obtained is used to manipulate the state of the holographic contents.

The proposed method allows users to interact directly with the holographic projection contents. Through this spatial interaction research that combines 3D stereoscopic images and 3D interface, the user is immersed in the content through sensory experience. And two-way communication between user and 3d images is also possible.

Keywords: 3D Spatial Interaction, Depth Camera, Holographic Projection.

1 Introduction

Lately, 3D image display technology has developed from stereoscopic method based on binocular disparity into a Hologram display over convergence of IT and Arts. By using a variety of interaction devices, it is trying to make a new way of communication with the user.

While playing video games at home, full body interaction like the actual situation, as the advent of low-cost commercial video game sensor. Some sensors Such as Microsoft's Kinect are designed to interact with the user directly.

A study on direct interaction between the user and the holographic images are still in its early stages as with Hologram application research. The most interactions in Holographic performances or exhibitions perform while maintaining some distance between the user and the contents.

We propose a Holographic projection system that has direct and natural interaction technology between 3D contents and users using voice and motion.

This method sees its surroundings of holographic contents as computing surrounding and translates the motion to spatial. For this, we use 3D (depth) camera that can recognize the depth of where the people's behavior and motion is located. Holographic contents will be controlled by spatial information that is

T. Herawan et al. (eds.), *Proceedings of the First International Conference on Advanced Data and Information Engineering (DaEng-2013)*, Lecture Notes in Electrical Engineering 285, DOI: 10.1007/978-981-4585-18-7_46,
© Springer Science+Business Media Singapore 2014

obtained from depth-camera. In addition, depth camera is used for the user's speech recognition and output.

2 Backgrounds

3D spatial interaction is human and computer interaction that included 3D UI and technology to effectively control computer-generated 3D contents.
In space with 3D interaction, users motion information is divided into 4 steps such as Navigation, Selection, Manipulation and System Control.
Navigation is for physical movement without any special motion, Selection is approaching object with intention, and Manipulation is taking an action of changing objects property. Generally, Selection and Manipulation occurs simultaneously. System Control means changing the contents condition when particular motion is sensed.

2.1 3D User Interfaces

3D user interface (3D spatial interaction) is a UI that involves human computer interaction where the users tasks are carried out in a 3D spatial context with 3D input devices or 2D input devices with direct mappings to 3D. In other words, 3D UIs involve input devices and interaction techniques for effectively controlling highly dynamic 3D computer-generated contents[7].
3D user interface with tactile interface and non-tactile interfaces, depending on the relationship of contents and user interface types can be classified. Of course, the role of the interface and type may vary, depending on the intent of the contents provider that are inherent in the contents.
In this study, it has classified the following four types of interfaces.
The first interface is to detect hand gesture using controller and joystick based on motion recognition sensor. Users enter their information by direct manipulation of touch devices such as sensor attached Gloves and Nintendo's Wii.
The second interface is to detect body movement using the video input on the camera based on vision system. In order to obtain information on the behavior of the user and the user in 3D space, video camera-based object tracking method is typically used.
The third interface is using smart devices Due to the development of the internet, smart devices have become a medium to exchange information and share the user's experience and senses through a variety of applications. For example, with one touch on the smart devices, people are able to connect and communicate beyond the temporal and spatial constraints.
In particular, user's interactions with smart devices are quite natural without reluctance because the user is familiar with smart devices.
The last method is using a combination of several interfaces. The touch screen or the user's general possession can also be used as an interface for interaction.

Motion Recognition Equipment, the physical input devices and spatial interaction techniques are needed for 3D interaction. Motion recognition is to measure the location and coordination for the movement of people in 3D space and record the types of information that can be used by computer. Of course, the information obtained from motion recognition tasks are also possible to play back the three-dimensional.

2.2 Holographic Projection System

Holographic technology allows users to see 3D stereoscopic images without special glasses. However, there are several technical problems using the hologram. The standard holographic image representation has not yet been established.
In recent performances or exhibitions, similar hologram technology has been implemented but not real. The similar hologram technology doesnt use real laser equipment. It is holographic projection technology to project 3D images on a two-dimensional transparent screen from a high-resolution projector.
This research offers reflective holographic projection display unit. The composition for the display unit is beam projector and transparent screen which is made up of Musions floating hologram which is based on principle of regeneration[11].

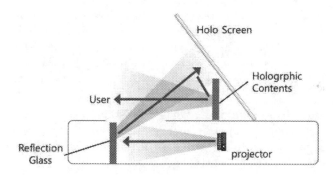

Fig. 1. Holographic projection system architecture

2.3 Related Work

Now, we will describe artworks using the spatial interaction technology on a case-by-case.

Virtual Sculptures using CavePainting. CavePainting is an application of CAVE in the field of Art and Design. [7]s 3D virtual sculptures are an artwork utilizing CavePainting system.

This system is to provide an environment that is familiar with artist using traditional desktop modeling tools. Feature of this system is designed to select the tools such as brush by artist directly for creating 3D sculpture.

Video Games using Wii Remote. The Video Games in [10] has the design of a low-cost 3D spatial interaction approach using the Wii Remote for immersive Head-Mounted Display (HMD) virtual reality.

If the user has the controller in both hands and moves, the video of the user changes real-time depending on the recognized direction

Building block tracking system using Kinect. [1] is a prototype system for interactive construction and modification of 3D physical models using building blocks. This is characterized by implementation of World-In-Miniature(WIM) technique. The WIM technique uses a miniature version of the world to allow the user to do indirect manipulation of the objects in the environment.

This system uses a depth sensing camera and a Lattice-First algorithm for acquiring and tracking the physical models. Users can interactively construct the models using their hands while the system acquires additions or deletions to the model.

Depth camera is monitoring the physical structure of the block in the pile to determine the status of the removal or additional state of blocks. Depth sensors such as Microsofts Kinect have been designed to face the user directly for interaction. But users want to be able to move freely in every possible direction.

Especially in multiple users Interaction environment, a larger space and an advanced technology for multiple users recognition is needed. Multiple Kinects are interworked in order to overcome the limits of the rear recognition.

Interaction on Mobile Devices. [8] is the system to show how a large number of users interact directly using the smart devices This system was to extend the individuals mobile screen to the touch projector interface to take advantage of the city walls as a canvas to control the video.

The significance of this research is media facades in urban space which can be manipulated by the collaborative multi-user scenarios.

Images shown in the media facade appear in each mobile screen immediately. Users interact freely with each other using desired colors by changing the LED colors shown on the wall. However, its too bad that interface does not recognize a particular user, if multiple users access at the same time.

3D interaction with fabric. General media, such as cloth can be used as 3D input device. As an input device, fabric holds potential benefits for three dimensional (3D) interaction in the domain of surface design, which includes designing objects from clothing to metalwork[2].

This system supports recognition of a point:a flexible curve, and a flexible surface. According to the direction of the fabric, Users Interact in real time with the screen in the video via physical user movement.

3   Design Considerations

For interaction between user and holographic contents, we defined the user space and the space of holographic contents as a three-dimensional cube type. We traced the occupied state of cube and the user's behavior using a depth camera. If the user approaches within the depth camera recognition range, two tasks come into action. For one thing, the occupied space of the user and the occupied space of the holographic image are activated. Another depth camera begins to track the status of the occupied space.

Fig. 2 is an overview of three-dimensional space in the cube form for obtaining 3D spatial information in this research. The size of the user space and the holographic space is designed to the maximum size of the holographic images by considering gesture recognition such as the move and rotate of the object.

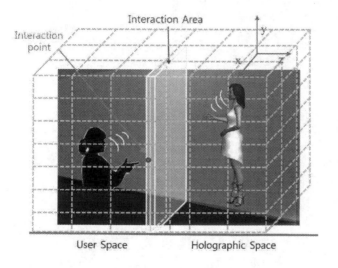

Fig. 2. Design of 3D interactive space

In the figure above, the holographic space is a virtual 3D space, 2D space that is occupied by the holographic projection images. We use virtual z-axis for manipulate the size of the image in application program.

The user can access directly in the interaction region, intersection region of the user space and the holographic space. The area, in which users and holographic content direct contact between, is called the interaction area. And we define the interaction point as the hand point of the user.

By comparison with predefined gesture operations to manipulate video information to take advantage of the user's behavior in interaction area is determined whether or not.

4   System Framework

In this research, we use Kinect for interaction with the hologram. Kinect is one of the integrated sensing devices and uses the skeleton algorithm for the motion capture.

Kinect consists of RGB camera, depth perception sensors and multi-array microphone. And using this system, the overall shape of the person's Kinect 3D motion capture, facial recognition, and voice recognition is possible.

First, obtained image from depth-camera will be used to track users fingers tip to get two interaction points space coordinate and create status graph. Tracking interaction point will be continued until user selects specific object.

Selecting object seems the user is approaching interaction point that is in the object range. Approaching interaction point shows the difference with object range which is less than a critical value. Selected object gesture is expressed in the form of gesture group and vector. To change holographic contents property with recognition gesture of users, we use Bruteforce method to search gestures/properties set. Gestures/properties set are sorted to increase efficiency of searching.

The voice recognition and output in Kinect can be specified in conjunction with the Microsoft Speech API for limit of predefined words.

If you say one word of pre-specified by the user, Kinect is able to recognize the voice and the correct the answer.

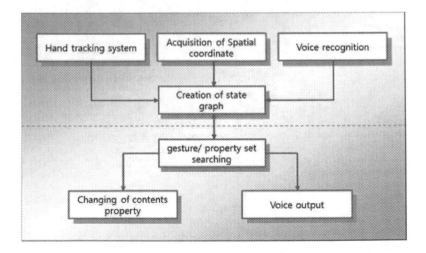

Fig. 3. 3D interactive system model

5   Conclusion

We propose a Holographic projection system that has direct and natural interaction technology between 3D contents and users using voice and motion without any special portable devices.

The proposed interactive holographic projection system allows users to interact directly with the holographic projection contents.

Through this spatial interaction research that combines 3D stereoscopic images and 3D interface, the user is immersed in the content through sensory experience. Two-way communication between user and 3d images is also possible.

Furthermore, if an abundant education about operation method (gesture) is achieved, users flexibility of input will be provided and recognize an error with users satisfaction will be increased. From now on, suggested interaction method will be applied to holographic art field.

In addition to the user's interface in conjunction with a wide variety of objects, such as the user's belongings and more familiar and easy, and you will be able to provide a natural interaction.

In order to encourage the active participation of the user, it is necessary to design of the optimized information structure and select the interface for smooth communication between the users.

Acknowledgments. This research is supported by Ministry of Culture, Sports and Tourism(MCST) and Korea Creative Content Agency(KOCCA) in the Culture Technology(CT) Research & Development Program(Immersive Game Contents CT Co-Research Center).

References

1. A. Miller, B. White, E. Charbonneau, Z. Kanzler, J. LaViola: Interactive 3D Model Acquisition and Tracking of Building Block Structures. In: IEEE Transactions on Visualization and Computer Graphics, vol. 18, no. 4, pp. 651-659 (2012)
2. A. Leal, D. A. Bowman, L. Schaefer, F. Quek, and C. Stiles: 3D sketching using interactive fabric for tangible and bimanual input. In: Proceedings of Graphics Interface 2011, pp. 49-56 (2011)
3. Barry G. Blundell: 3D Displays and Spatial Interaction: Exploring the Science, Art, Evolution and Use of 3D Technologies. From Perception to Technology, Vol. 1, Walker & Wood Limited (2011)
4. B. Williamson, C. Wingrave and J. LaViola: RealNav: Exploring Natural User Interfaces for Locomotion in Video Games. In: Proceedings of the IEEE Symposium on 3D User Interfaces, pp. 3-10 (2010)
5. B. Williamson, J. LaViola, T. Roberts and P. Garrity: Multi-Kinect Tracking for Dismounted Soldier Training. In: Proceedings of the Interservice/Industry Training, Simulation and Education Conference(I/ITSEC), pp. 1-9 (2012)
6. C. Hand: A Survey of 3D Interaction Techniques. In: Computer Graphics Forum, pp. 1-9 (1997)
7. J. LaViola, D. Keefe: 3D spatial interaction: applications for art, design, and science. In: ACM SIGGRAPH 2011 Courses, No. 1 (2011)

8. .S Boring, S. Gehring, A. Wiethoff, M. Blckner, J. Schning, A. Butz: Multi-user
   interaction on media facades through live video on mobile devices. In: Proceedings
   of the SIGCHI Conference on Human Factors in Computing Systems, pp. 2721-2724
   (2011)
9. S. Lim, S. Kim: A Study on Case Analysis of media art works using Spatial Interaction. In: Journal of Digital Design, Vol. 13, No. 2, pp. 255-264 (2013)
10. Y. Chow: 3D spatial interaction with the Wii remote for head-mounted display
   virtual reality. In: Proceedings of World Academy of Science, Engineering and Technology, pp. 377- 383 (2009)
11. MUSION, http://www.musion.co.uk/

Leveraging Post Level Quality Indicators in Online Forum Thread Retrieval

Ameer Tawfik Albaham, Naomie Salim, and Obasa Isiaka Adekunle

Faculty of Computing,
Universiti Teknologi Malaysia, Skudai, Johor, Malaysia
ameer.tawfik@gmail.com, naomie@utm.my, iaobasa@yahoo.com

Abstract. Thread retrieval is an essential tool in knowledge-based forums. However, forum content quality varies from excellent to mediocre and spam; thus, search methods should find not only relevant threads but also those with high quality content. Some studies have shown that leveraging quality indicators improves thread search. However, these studies ignored the hierarchical and the conversational structures of threads in estimating topical relevance and content quality. In that regard, this paper introduces leveraging post quality indicators in ranking threads. To achieve this, we first use summary statistics measures to convert post level quality features into thread level features. We then train a learning to rank method to combine these thread level features. Preliminary results with some features reveal that representing threads as collections of posts is superior to treating them as concatenations of their posts.

Keywords: Forum thread search, Post content quality, Summary statistics, Learning to Rank

1 Introduction

Online forums are platforms that facilitate discussion and knowledge sharing on the Web. In forums, a discussion starts when a user posts an initial post requesting help or initiating a conversation in a particular matter. Afterwards, the other users read and reply to the initial post. Each pair of an initial post and its replies forms a thread.

Online forums are rich knowledge communities for several reasons. First, the asynchronous nature and the public accessibility of forums enable communication between community members regardless of the physical and the temporal boundaries. That empowers users with various areas and levels of expertise to share and seek knowledge through in-depth discussions. Second, forums have accumulated a huge amount of content for a long time. Third, the archived content contains not only factual information but also detailed solutions and troubleshooting materials [1, 2]. In addition, the content is more comprehensive and objective than what is on the Web [2]. However, utilizing such rich content requires thread retrieval [3].

T. Herawan et al. (eds.), *Proceedings of the First International Conference on Advanced Data and Information Engineering (DaEng-2013)*, Lecture Notes in Electrical Engineering 285, DOI: 10.1007/978-981-4585-18-7_47,
© Springer Science+Business Media Singapore 2014

Nevertheless, thread retrieval is not a trivial task because of the mismatch between the retrieval and the indexing units. Thread retrieval returns a list of threads, whereas thread posts are the indexing units. Several solutions have been proposed to address this problem: the virtual document representation [4], the Voting Model [3, 5], and the multi-representation model [1, 2, 6]. However, these methods have assumed that all threads have equal quality; this assumption is inadequate within the context of knowledge-based forums for several reasons. First, when reviewing the query set of our used dataset [1], we found that all queries are question-oriented queries. In other words, a user is looking for information to adapt as a solution to his or her problem. As a result, a search algorithm should cater to such needs. The notion of adapting and using content leads to the Knowledge Adaption Model [7]. According to this model, the willingness of knowledge workers to adopt the advice that they receive in computer mediated communication (emails) is governed by the perceived information usefulness of the advise (message) . In addition, two factors influence the perceived usefulness of information: the message's content augmented quality and source credibility. Content augmented quality refers to quality dimensions associated with the content such as relevance, amount of data and value added. Source credibility refers to the source credibility aspects, such as trustworthiness and expertise. In forums, these factors have also been found to influence the perceived usefulness of content [8].

Another reason to develop quality aware search methods is that the equal quality assumption in traditional information retrieval is not adequate in forum retrieval tasks. In traditional retrieval, documents are news articles, which are professionally edited. However, in user-generated content, such as forums, content quality varies widely. In fact, quality aware retrieval methods have shown significant improvements over conventional methods in various user-generated content platforms [9, 10, 11, 12]. Some studies have addressed incorporating quality into ranking threads [13, 1, 14]. However, these studies do not consider post level quality indicators in ranking threads. In addition, several studies have shown that treating threads as collections of posts is better than treating them as concatenations of post texts (virtual documents) when ranking threads [4, 3, 5, 6]. In this paper, we argue that when leveraging quality features in ranking threads, representing threads as collections of posts is more appropriate than treating them as virtual documents for the following reasons. The first reason is that it allows incremental indexing. However, in the virtual document representation, we have to recalculate quality features whenever users create or update posts. The second reason is that the post-based approach allows filtering of results while ranking better. The third reason is that we will not be able to estimate some quality dimensions, such as relevance and value added, when treating threads as virtual documents.

The contributions of this study are: 1) we propose leveraging post quality features in ranking threads, and 2) we show how to leverage post quality indicators in ranking threads. To the best of our knowledge, no prior research has given attention to utilizing post quality indicators in ranking threads.

The rest of this paper is as follows. First, we discuss related works in Section 2. Afterwards, we detail our method of leveraging post quality indicators in thread ranking (Section 3). We then describe the quality features used in Section 4. We describe our experimental design in Section 5, then we report preliminary results in Section 6 and we discuss these results in Section 7. Lastly, Section 8 concludes this paper and outlines future work.

2 Related work

Many thread retrieval studies focused on estimating topical relevance between a user query and a thread. Generally, these studies are divided into two approaches: the thread-, and the post-oriented approaches. Thread-oriented methods assume a thread as the unit of indexing, such as the virtual document [4] and the structured document [1] representations. However, post-oriented methods, such as the Voting Model [3, 5], assume a single post as the unit of indexing. In these methods, an initial ranked list of posts is generated, then threads are ranked by aggregating their ranked posts' relevance scores. Generally, post-oriented methods are better than thread oriented studies [4, 5, 6]. However, these studies have ignored content quality in ranking threads except for [1]. Bhatia and Mitra [1] investigated leveraging three indicators of thread quality. These quality features are the number of posts, the linkage information and the authority of posts' authors. However, all of these features are thread level quality features except the authority feature. In addition, no post content quality features were investigated.

Zhang [13] and Jiao [14] investigated using learning to rank methods [15] to leverage quality-based features in ranking threads. Both studies assumed threads as virtual documents. However, we can still use the same features even though we approach them as post level quality features. For instance, the number of words in a thread, used by [13] and [14], is actually the sum of the number of words of all thread posts.

Zhang [13] and Jiao [14] also used features that are based on source credibility of post authors. Some of these features are the author's age and number of created posts. Because authors are associated with posts, [13] and [14] converted an author-based feature into a thread feature using the average of the individual post feature values. Although these features are indicators of quality, there is an inconsistency in the representations used, because when [13] and [14] generated features based on lexical content, they assumed a virtual document representation. However, when they used author-based features, they treated posts as textual units.

3 Developing quality aware thread retrieval

The focus of this paper is leveraging content quality features of posts in thread retrieval. In this aspect, we propose to use the learning to rank paradigm [15] to learn ranking methods. However, learning to incorporate post level quality features is not trivial. The issue is that any learning to rank algorithm utilizes query relevance judgments to train a ranking method. However, query relevance judgments are defined at the thread level. To tackle this issue, we use many summary statistics measures to convert post level features into thread level features.

In this method, a retrieval system first retrieves a list of threads, then it calculates the sum, the minimum, the maximum, the median and the average of all post quality features for every thread in the initial list. In other words, each post quality feature generates 5 thread quality features. The advantage of this method is that we could generate the thread quality features offline. Note that using only the average-based features is equivalent to what was used by [1, 13, 14] to convert authority features into thread features. Therefore, our approach subsumes previous work. In addition, authors in forums have various levels of activities and expertise [1]. Furthermore, the size of thread posts varies widely; it is common to find a lengthy and a two-word posts in a single thread. As a result, the average method might introduce noise as it is sensitive to outliers. In our work, we introduce more summary statistics-based features to rectify this problem.

4   Quality Features Used

In forums, there are many dimensions of content quality such as amount of data, ease of understanding, relevance, objectivity, timeliness and completeness [16]. In this study, we limit our discussion to features related to the amount of data dimension. As discussed previously, our argument in this paper is that — in leveraging content quality — representing threads as collections of posts is better than the virtual document representation. Since these features can be estimated on both representations, then results based on these features allow us to validate our argument accurately. Table 1 shows the quality features used.

Table 1. Quality Features Used

Feature	Description
nc	Number of characters in a post
nw	Number of words in a post
nuw	Number of unique words in a post
ns	Number of sentences in a post
nl	Number of web links in a post

5   Experimental Design

In evaluating the quality-based methods, we use the Ubuntu forum dataset [1]. The dataset contains 25 queries with ternary relevance judgments. In addition, the dataset was collected from the Ubuntu[1] forums, a popular open source community. A sample of queries is given in Table 2, and some statistics about the dataset are given in Table 3.

Table 2. Sample of queries

firefox no sound with flash
virtualbox keyboard problem
how to dual boot windows and ubuntu
cisco vpn client for ubuntu
how to adjust screen resolution in ubuntu
running microsoft office with wine
grub error while booting
wireless connection not working
gnome amarok alternative
which is better gnome kde
playing windows media files on ubuntu

Table 3. Ubuntu dataset statistics

No. of threads	113,277
No. of users	103,280
No. of messages	676,777
No. of evaluated threads	4,512

We indexed each post as a document and performed stemming and removed stop words. We used the Lucene[2] search library in indexing and retrieval. In particular, we leveraged the Lucene built-in analysizer[3] to preprocess text.

To evaluate our methods, we report Normalized Discounted Cumulative Gain (NDCG) at different cutoffs (1, 3, 5, 10, 20, 100 and all).

After generating the quality features, we trained a learning to rank algorithm, an instance of the Coordinate Ascent algorithm [17], to combine features. In fact, we used other learning to rank methods such as AdaRank [18] and LambdaMART [19]. However, the Coordinate Ascent method produced the best results. We used the RankLib's[4] implementation of these methods using default parameters. In addition, we also normalized features using the sum normalization method, which is also provided by RankLib. To train the learning function, we performed 5-folds cross validation using NDCG as the measure to optimize.

---

[1] ubuntuforums.org

[2] http://www.apache.org/dyn/closer.cgi/lucene/java/4.4.0

[3] the EnglishAnalyzer class

[4] http://people.cs.umass.edu/ vdang/ranklib.html

In learning to rank, the first step is to generate a list of documents. The features are then generated for those documents [15]. In this study, to generate the initial list of threads, we use the CombSUM method from the Voting Model [3]. In this method, we first retrieved the top 1,000 relevant posts to a user query, then we scored threads via the sum of their ranked posts' relevance scores. The initial list was generated using the Okapi BM25 text retrieval method with default parameters as provided by Lucene. In addition to the quality features, we used relevance scores of threads based on CombSUM as a feature as well. In our experiments, we aim to answer the following research questions:

1. Does leveraging quality features improve retrieval?
2. In generating quality features, is the post-oriented approach better than the virtual document representation?
3. Do we need to use all summary statistics measures of post quality features?

6 Result

Table 4. Performance of leveraging post content quality in ranking threads

#	Method	N@1	N@3	N@5	N@10	N@20	N @50	N@100	NDCG
1	BM25	0.2400	0.2428	0.2400	0.2185	0.2444	0.2575	0.3181	0.4722
2	nw-sum	**0.2800**	0.2445	0.2257	0.2153	0.2378	0.2522	0.3045	0.4644
3	nw-all	0.2400	**0.2610**	**0.2374**	0.2186	0.2404	0.2569	**0.3146**	**0.4665**
4	all-sum	0.2600	0.2563	0.2226	0.2160	0.2345	0.2488	0.3027	0.4627
5	all-all	**0.2800**	0.2581	0.2308	**0.2243**	**0.2412**	**0.2601**	0.3106	0.4657

Table 4 shows the performance of leveraging post quality indicators in ranking threads. The first row contains the result of the baseline method that does not leverage quality. The second and the third rows present the performance of leveraging one quality feature — the Number of Words (NW). In particular, the second row employs only the sum aggregation method to convert post level features into thread level features. This aggregation method simulates the extraction of quality features using the virtual document representation. The third row adds more summary statistics-based features. The last two rows present the performance of leveraging multiple features using single and multiple summary statistics measures, respectively.

From the table above, we can observe several patterns. First, in high precision searches (NDCG@1, NDCG@3, NDCG@10), the quality-based methods are comparable to or better than the content-based method (BM25). However, in NDCG@5, BM25 is better than all quality-based methods. It is also better in other searches as well with the exception of the all-all method in NDCG@20 and NDCG@50.

Pertaining to the performance of the quality-based methods, we can see that leveraging more summary statistics measures outperforms leveraging only the sum method. For instance, using only a single feature with multiple summary statistics methods (nw-all) outperforms using only the sum aggregation method (nw-sum) in 7 out of 8 measures. Similarly, the all-all method is always better than the all-sum method. With respect to using many features (all-sum) vs using a single feature (nw-sum), we can see that nw-sum beats all-sum in 6 out of 8 measures.

7 Discussion

The result presented above gives several insights. First, the performance of the quality-based methods indicates that leveraging quality features is good only for high precision searches. This answers the first research question introduced in Section 5.

The overall good performance of nw-all over nw-sum and all-all over all-sum indicates that the post-oriented approach is better than the thread-oriented approach; recall that nw-sum and nw-all simulate using quality features based on the virtual document representation which we are able to do using the post-oriented approach. Therefore, our proposed method is a better approach to incorporate content quality while ranking threads. This answers the second research question.

Another insight from the performance of the all-all and the all-sum methods is that, when leveraging many features, it is necessary to include many summary statistics methods. This answers the third research question. In future work, we aim to investigate the sensitivity of features to the aggregation methods used.

8 Conclusion and Future Work

In this work, we addressed exploiting post quality features in ranking threads. We proposed to use a learning to rank method to combine post quality features. In addition, we used summary statistics measures to convert post level quality features into thread level features. We also showed that our proposed method can simulate current thread quality aware methods that treat threads as virtual documents.

Although the reported initial results indicate that leveraging quality features is good only for high precision searches, the results also indicate the superiority of our proposed method to the virtual document representation. In order to make our findings more conclusive, we plan to explore more post quality dimensions. In addition, we plan to investigate the effects of adding thread level quality features such as the linkage information [1] in performance.

Bibliography

[1] Sumit Bhatia and Prasenjit Mitra. Adopting inference networks for online thread retrieval. In *Proceedings of the Twenty-Fourth AAAI Conference on Artificial Intelligence*, pages 1300–1305, Atlanta, Georgia, USA,, July 11–15 2010.

[2] Jangwon Seo, W. Bruce Croft, and David Smith. Online community search using conversational structures. *Information Retrieval*, 14:547–571, 2011. ISSN 1386-4564.

[3] Ameer Tawfik Albaham and Naomie Salim. Adapting voting techniques for online forum thread retrieval. In AboulElla Hassanien, Abdel-BadeehM. Salem, Rabie Ramadan, and Tai-hoon Kim, editors, *Advanced Machine Learning Technologies and Applications*, volume 322 of *Communications in Computer and Information Science*, pages 439–448. Springer Berlin Heidelberg, 2012. ISBN 978-3-642-35325-3.

[4] Jonathan L. Elsas and Jaime G. Carbonell. It pays to be picky: an evaluation of thread retrieval in online forums. In *Proceedings of the 32nd international ACM SIGIR conference on Research and development in information retrieval*, SIGIR '09, pages 714–715, New York, NY, USA, 2009. ACM. ISBN 978-1-60558-483-6.

[5] Ameer Tawfik Albaham and Naomie Salim. Online forum thread retrieval using pseudo cluster selection and voting techniques. In Ruay-Shiung Chang, Lakhmi C. Jain, and Sheng-Lung Peng, editors, *Advances in Intelligent Systems and Applications - Volume 1*, volume 20 of *Smart Innovation, Systems and Technologies*, pages 297–306. Springer Berlin Heidelberg, 2013. ISBN 978-3-642-35451-9.

[6] Ameer Tawfik Albaham and Naomie Salim. Meta search models for online forum thread retrieval: Research in progress. In Ali Selamat, Ngoc Thanh Nguyen, and Habibollah Haron, editors, *Intelligent Information and Database Systems*, volume 7802 of *Lecture Notes in Computer Science*. Springer Berlin Heidelberg, 2013. ISBN 978-3-642-36545-4.

[7] Stephanie Watts Sussman and Wendy Schneier Siegal. Informational influence in organizations: An integrated approach to knowledge adoption. *Info. Sys. Research*, 14(1):47–65, March 2003. ISSN 1526-5536.

[8] Xiao-Ling Jin, Christy M.K. Cheung, Matthew K.O. Lee, and Hua-Ping Chen. How to keep members using the information in a computer-supported social network. *Computers in Human Behavior*, 25(5):1172 – 1181, 2009. ISSN 0747-5632.

[9] Jiwoon Jeon, W. Bruce Croft, Joon Ho Lee, and Soyeon Park. A framework to predict the quality of answers with non-textual features. In *Proceedings of the 29th annual international ACM SIGIR conference on Research and development in information retrieval*, SIGIR '06, pages 228–235, New York, NY, USA, 2006. ACM. ISBN 1-59593-369-7.

[10] Jaeho Choi, W. Bruce Croft, and Jin Young Kim. Quality models for microblog retrieval. In *Proceedings of the 21st ACM international conference on Information and knowledge management*, CIKM '12, pages 1834–1838, New York, NY, USA, 2012. ACM. ISBN 978-1-4503-1156-4.

[11] Kamran Massoudi, Manos Tsagkias, Maarten de Rijke, and Wouter Weerkamp. Incorporating query expansion and quality indicators in searching microblog posts. In *Proceedings of the 33rd European conference on Advances in information retrieval*, ECIR'11, pages 362–367, Berlin, Heidelberg, 2011. Springer-Verlag. ISBN 978-3-642-20160-8.

[12] Chien Chin Chen and You-De Tseng. Quality evaluation of product reviews using an information quality framework. *Decision Support Systems*, 50(4): 755 – 768, 2011. ISSN 0167-9236.

[13] Xiaoyu Zhang. Effective search in online knowledge communities: A genetic algorithm approach. Master of science in computer science and applications, Faculty of the Virginia Polytechnic Institute and State University, September 2009.

[14] Jian Jiao. *A framework for finding and summarizing product defects, and ranking helpful threads from online customer forums through machine learning*. PhD thesis, Virginia Polytechnic Institute and State University, April 2013.

[15] Tie-Yan Liu. *Learning to Rank for Information Retrieval*. Springer Berlin Heidelberg, 2011. ISBN 978-3-642-14266-6.

[16] Weiguo Fan Gang Wang, Xiaomo Liu. A knowledge adaption model based framework for finding helpful user generated ccontent in online communities:. In *Thirty Second Internationl Conference on Information Systems*. AIS Electronic Library (AISeL), 2011.

[17] Donald Metzler and W. Bruce Croft. Linear feature-based models for information retrieval. *Inf. Retr.*, 10(3):257–274, June 2007. ISSN 1386-4564.

[18] Jun Xu and Hang Li. Adarank: a boosting algorithm for information retrieval. In *Proceedings of the 30th annual international ACM SIGIR conference on Research and development in information retrieval*, SIGIR '07, pages 391–398, New York, NY, USA, 2007. ACM. ISBN 978-1-59593-597-7.

[19] Qiang Wu, Christopher J. Burges, Krysta M. Svore, and Jianfeng Gao. Adapting boosting for information retrieval measures. *Inf. Retr.*, 13(3): 254–270, June 2010. ISSN 1386-4564.

Part VI
Information System

A Framework to Predict Software "Quality in Use" from Software Reviews

Issa Atoum and Chih How Bong

Faculty of Computer Science and Information Technology
Universiti Malaysia Sarawak
94300 Kota Samarahan, Sarawak, Malaysia

{ Issa.Atoum@gmail.com , chbong@fit.unimas.my }

Abstract. Software reviews are verified to be a good source of users' experience. The software "quality in use" concerns meeting users' needs. Current software quality models such as McCall and Boehm, are built to support software development process, rather than users perspectives. In this paper, opinion mining is used to extract and summarize software "quality in use" from software reviews. A framework to detect software "quality in use" as defined by the ISO/IEC 25010 standard is presented here. The framework employs opinion-feature double propagation to expand predefined lists of software "quality in use" features to domain specific features. Clustering is used to learn software feature "quality in use" characteristics group. A preliminary result of extracted software features shows promising results in this direction.

Keywords: "quality in use"; software reviews; opinion mining; ISO 25010; quality model; product reviews

1 Introduction

Online product reviews are a major information source of users' experience. Many online web sites give users the opportunity to share their experience and give ideas about software and possible enhancements. Reviews on popular software have increased dramatically; hence processing them is a laborious yet costly. Moreover, most of the time product reviews can be confusing and misleading. For example, comments like "I just don't like this product" and "The product took forever to be here" is lack of constructive expressions as the comments were not targeted on the product because some reviews consist of emotional expression and/or biases.

Software products are evaluated differently by different stakeholders' interests. For example, the publisher of the product may be interested in developing quality software while users care about the whole product while it is operational. Quality according to Gravin in the point view of a user is "meeting customer needs"[1]. If the software meets the needs then it is said to have good quality.

T. Herawan et al. (eds.), *Proceedings of the First International Conference on Advanced Data and Information Engineering (DaEng-2013)*, Lecture Notes in Electrical Engineering 285, DOI: 10.1007/978-981-4585-18-7_48, © Springer Science+Business Media Singapore 2014

Many software quality models such as McCall, Boehm, Dromey and FURUPS [2], [3] are built for quality from development perspective and does not fit to measure software quality from user point of view [4], [5]. For users, the purpose of using software is to help them achieve particular goals, as the *effectiveness*, *efficiency* and *satisfaction* with which users can achieve specified goals in specified environments. ISO/IEC 25010:2011(hereafter ISO 25010) covers the software quality by a model known as Systems and software Quality Requirements and Evaluation (SQuaRE). The ISO 25010 has the "quality in use" model, the focus of this work.

Opinion Mining can be used to identify important reviews and opinions to answer users' queries about quality. The more fine-grain works are on feature or aspect-based sentiment analysis where it determines the opinions on the features of the reviewed entity such as cell phone, tablet etc. More importantly, to our knowledge little research has been published in software reviews opinion mining. Mining software reviews can save users time and can help them in software selection process.

Table 1. definitions of "quality in use " characteristics from ISO 25010.

Characteristic	Definition
Effectiveness	Accuracy and completeness with which users achieve specified goals (ISO 9241-11).
Efficiency	Resources expended in relation to the accuracy and completeness with which users achieve goals (ISO 9241-11).
Freedom From Risk	Degree to which a product or system mitigates the potential risk to economic status, human life, health, or the environment.
Satisfaction	Degree to which user needs are satisfied when a product or system is used in a specified context of use.
Context Coverage	Degree to which a product or system can be used with effectiveness, efficiency, freedom from risk and satisfaction in both specified contexts of use and in contexts beyond those initially explicitly identified.

This paper proposes a framework to process software user reviews in order to extract one of the software quality indicators, "quality in use" as defined by the ISO 25010 (ISO, 2011). **Table 1** shows "quality in use" definitions. Note that these definitions has some intersection with product quality division of the the ISO standard, more precisely the *usability* characteristic.One major step of this research is to build a data set of software keywords or features. In the context of our study, software features are software properties that describe software "quality in use" such as the keyword *conform* and *resource* to describe satisfaction and efficiency characteristics respectively. Once data are in place, then a model is built utilizing topic modeling [6][7] and opinion mining [8] methods. Finally the model is evaluated against predefined criteria with users.

First, the problem is defined. Then related works are presented. After that, the proposed approach is explained. Finally, preliminary results are presented and the paper is concluded.

2   The problem

Although there are many software quality models such as McCall, Boehm, Dromey and FURUPS [2], [3], most of them target the software product or process characteristics and does not fit to measure software quality from user point of view [4], [5] (quality in use). Tweaking these models to allow measurement of "quality in use" can be cumbersome process and is outside the scope of this work.

Processing massive number of software product reviews is challenging due reviews subjectivity as it has many spam and unconstructive sentences.

3   Related Works

This section present the related works that are used to build the proposed framework grouped into topic modeling, and feature extraction and summarization.

3.1   Topic Modeling

Topic modeling methods can be instinctively viewed as clustering algorithms that cluster terms into *meaningful* clusters or subtopics. A famous topic modeling model is called LSI or LSA[9], [10]. LSA transforms text to low dimensional matrix and it finds the most common topics that can appear together in the processed text. Latent Dirichlet Allocation (LDA) is very famous topic modeling[7]. The model extends Probabilistic Latent Semantic Analysis (PLSA) model[6] to cover two problems: over fitting and the limitation of assigning probability to a document outside the training set.

3.2   Feature Extraction , Classification, and Summarization

Feature or topic extraction has been discussed in literature in many works such as [11], [12], [13], [14], [15]. Most of these works use the language semantics to extract features such as nouns and noun phrases along with their frequencies subject to predefined thresholds. Qiu et al. [15], [16] suggested to extract both features and opinion by propagating information between them using grammatical syntactic relations.

Leopairote, Surarerks, & Prompoon [17] proposed a model that can extract and summarize software reviews in order to predict software "quality in use". The model depends on a manually built ontology of ISO 9126 "quality in use" keywords and WordNet 3.0 synonyms expansion.

The Authors consider the work of [17] the most nearby to this paper. The difference from proposed work is that the proposed frameowrk employs word similarity and relatedness rather than rule based classification and ontologies.

4   Proposed Framework

This research proposes a framework that is composed of data preparation, feature extraction, sentiment orientation and "quality in use" overall scoring. First "quality in use" repository is built, and then software features are expanded. After that sentences polarity are calculated and sentence are grouped into its own "quality in use" characteristic. Finally the "quality in use" is scored to represent the overall "quality in use" value for a particular software.

It is worth to mention to emphasize that the framework is only targeting primary users while the role of developers will be limited to testing the framework during its development. The framework does not need to track quality in software development life cycle or replace any available current standard. **Fig. 1** illustrates the structure of the proposed framework. The methodology goes in these steps:

4.1   Data Preparation

To allow maximum coverage of "quality in use", different viewpoints are taken into consideration. These viewpoints are the ISO standard definition of "quality in use", and how often different keywords are used together. To achieve this goal a combination of: the ISO 25010 standard document, Google search results, the WordNet taxonomy and a sample of software product reviews are suggested.

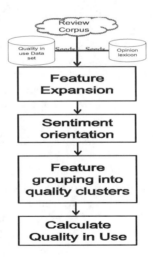

Fig. 1. Proposed methodology for predicting "quality in use" of softwaer user reviews

4.2   Feature Extraction

A feature is a target that the user is of interest (software entity name, or part of its attributes). In the context of this research, software features are software properties that describe software "quality in use" such as the keyword *conform* and *resource* to describe *satisfaction* and *efficiency* characteristics respectively. Features are extracted using double prorogation of opinion and features [15], [16], so it will allow to extract extra "quality in use" features that were not initially identified.

4.3   Sentiment Orientation

Knowing the polarity of each compared sentence is very important so that positive and negative opinions can be aggregated. The proposed approach suggests calculating the sentiment orientation of each sentence using a semi-supervised technique seeding it with list of combined opinion words list. The polarity of extracted sentences are calculated using Qiu method[16] and the SentiWordnet[1].

4.4   Feature Summarization

Given a set of reviews with many sentences, each sentence is grouped into a single "quality in use" characteristic such as *efficiency*, and *risk mitigation*. To resolve this issue the Zhai et al.'s Expectation Maximization algorithm [18] is borrowed. Assuming features and opinion words pairs are available from Qiu model (previous step), using a modified version of Expectation Maximization(EM) framework [19] , related software features are grouped and mapped to software's "quality in use" characteristics. In this work, the initialization of the EM algorithm will be features and opinion words extracted using Qiu algorithm. Finally quality is scored at the software level by aggregating polarity of related grouped software characteristics.

4.5   Overall "Quality in Use" Scoring

In this step the overall "quality in use" for the software is calculated. The final value of processed reviews can be shown to user in a percentage layout. For example user may get that certain software is 90% covering *efficiency*, 70% covering *effectiveness* and an overall of 80% "quality in use". One way to get the overall "quality in use" for a software is to average the "quality in use" for each of five characteristics using the formula (1). Other possible ways such as those that depends on user preferences or characteristic weighting could be considered in future research.

$$Quality_{in_{use}} = \sum_{c=1}^{5} Qc_i * Orientation(Qc_i)/5$$

Where Qc_i is the "quality in use" characteristics sentences classified as {*efficiency, effectiveness, risk mitigation, satisfaction, context coverage*}, $Orientation(Qc_i) \in \{0,1\}$ is a positive or negative orientation of each sentence.

5   Preliminary Results

Experiments have been carried out to prepare the data dictionary for "quality in use". **Fig. 2** illustrates the proposed approach.

---

[1] http://sentiwordnet.isti.cnr.it/

5.1   Data Preparation

Ten different domain software reviews have been crawled from CNET. The fields that have been extracted are: software name, pros, cons, summary and review rating. From these domains one domain were chosen at random for topic modeling (Section 5.2). Cross checking the validity of the framework can be applied using other software domains. The whole ISO document was also taken for topic modeling. Both documents were filtered from stop words, special chars and non-English words.

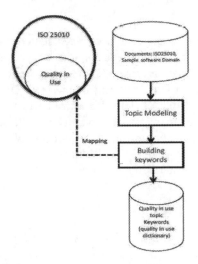

Fig. 2. Building "quality in use" dictionary

Other types of documents such as those from Google search are not considered at this stage. No synonym or word similarity using WordNet was applied.

5.2   Topic Modeling

The datasets prepared in section 5.1 were used to extract initial topics using a topic model algorithm known as Latent Dirichlet Allocation(LDA)[7]. Each data set were used alone to extract a list of possible topics. The perception of LDA is to discover topics of a given document. LDA was used with default parameters topics=10, α =5 (50/Topics), β=0.01 and 20 words per topic.

5.3   Finding Relevant Keywords

Given these initial topics from ISO document and the review sentences, the "quality in use" characteristic were constructed by taking one word at a time and verifying word's relevancy to the ISO 25010's "quality in use" characteristic definitions. Extracted topics were then categorized into three subtopics: *effectiveness*, *efficiency* and *risks*. Duplicate keywords were removed if found.

Table 2 shows a fragment of keywords categorized under the three subcategories that are manually extracted. First glance of the outcomes shows rather promising result. Col-locating the words with ISO document reveals that most words are indeed closely related to each subcategory. By detecting the semantics embedded in the user reviews through semantic analysis, different user experiences on a product are identified, which in turn allows better prediction of the software quality.

Table 2. Sample extracted topic keywords

"Quality in Use" Characteristic		
Effectiveness	*Efficiency*	*Risk Mitigation*
achieve	accessibility	alert
job	availability	bug
able	background	issue
automatic	capacity	confidentiality
control	compatible	lose
support	cpu	Crash
option	efficiency	Destroy
use	overload	Error

6   Conclusion

In this paper, a framework to detect software "quality in use" from online software reviews as defined by the ISO 25010 is proposed. The framework employs LDA topic modeling to build a data set of "quality in use", a semi-supervised learning to expand software "quality in use" features and predefined sets of opinion lexical to calculate polarity of several sentences. The framework is being developed and tested. Future work includes the application of the framework on other quality dimensions.

Acknowledgment

This study was supported in part by Universiti Malaysia Sarawak' Zamalah Graduate Scholarship and grant from ERGS/ICT07(01)/1018/2013(15).

7   References

1. Al-Qutaish, R. E. (2010). Quality models in software engineering literature: an analytical and comparative study. *Journal of American Science*, *6*(3), 166–175.
2. Blei, D. M. (2012). Probabilistic topic models. *Commun. ACM*, *55*(4), 77–84. doi:10.1145/2133806.2133826
3. Blei, D. M. D., Ng, A. Y. A., & Jordan, M. I. (2003). Latent dirichlet allocation. *J. Mach. Learn. Res.*, *3*, 993–1022. Retrieved from http://dl.acm.org/citation.cfm?id=944937

4. Deerwester, S., & Dumais, S. (1990). Indexing by latent semantic analysis. *Journal of the American society for information science*, *41*(6), 391–407.
5. Dempster, A., Laird, N., & Rubin, D. (1977). Maximum likelihood from incomplete data via the EM algorithm. *Journal of the Royal Statistical Society. Series B (Methodological)*, 1–38. Retrieved from http://www.jstor.org/stable/10.2307/2984875
6. Dromey, R. G. (1995). A model for software product quality. *Software Engineering, IEEE Transactions on*, *21*(2), 146–162. doi:10.1109/32.345830
7. Garvin, D. A. (1984). What does product quality really mean. *Sloan management review*, *26*(1), 25–43.
8. Ku, L., Liang, Y., & Chen, H. (2006). Opinion extraction, summarization and tracking in news and blog corpora. In *Proceedings of AAAI-2006 Spring Symposium on Computational Approaches to Analyzing Weblogs*.
9. Landauer, T. K., Foltz, P. W., & Laham, D. (1998). An introduction to latent semantic analysis. *Discourse Processes*, *25*(2-3), 259–284. doi:10.1080/01638539809545028
10. Leopairote, W., Surarerks, A., & Prompoon, N. (2012). Software quality in use characteristic mining from customer reviews. In *Digital Information and Communication Technology and it's Applications (DICTAP), 2012 Second International Conference on* (pp. 434–439). Ieee. doi:10.1109/DICTAP.2012.6215397
11. McCall, J. A., Richards, P. K., & Walters, G. F. (1977). *Factors in software quality*. General Electric, National Technical Information Service.
12. Mukherjee, A., & Liu, B. (2012). aspect extraction through Semi-Supervised modeling. *Proceedings of 50th anunal meeting of association for computational Linguistics (acL-2012)*, (July), 339–348.
13. Qiu, G., Liu, B., Bu, J., & Chen, C. (2009). Expanding domain sentiment lexicon through double propagation. In *Proceedings of the 21st international jont conference on Artifical intelligence* (pp. 1199–1204).
14. Qiu, G., Liu, B., Bu, J., & Chen, C. (2011). Opinion word expansion and target extraction through double propagation. *Computational linguistics*, *37*(1), 9–27.
15. Samadhiya, D., Wang, S.-H., & Chen, D. (2010). Quality models: Role and value in software engineering. In *Software Technology and Engineering (ICSTE), 2010 2nd International Conference on* (Vol. 1, pp. V1–320 –V1–324). doi:10.1109/ICSTE.2010.5608852
16. Wong, T.-L., Lam, W., & Wong, T.-S. (2008). An unsupervised framework for extracting and normalizing product attributes from multiple web sites. In *Proceedings of the 31st annual international ACM SIGIR conference on Research and development in information retrieval* (pp. 35–42). New York, NY, USA: ACM. doi:10.1145/1390334.1390343
17. Zhai, Z., Liu, B., Wang, J., Xu, H., & Jia, P. (2012). Product Feature Grouping for Opinion Mining. *Intelligent Systems, IEEE*, *27*(4), 37–44. doi:10.1109/MIS.2011.38
18. Zhang, L, & Liu, B. (2011). Identifying noun product features that imply opinions. In *Proceedings of the 49th Annual Meeting of the Association for Computational Linguistics: Human Language Technologies: short papers* (Vol. 2, pp. 575–580).
19. Zhang, Lei, Liu, B., Lim, S. S. H., & O'Brien-Strain, E. (2010). Extracting and ranking product features in opinion documents. In *Proceedings of the 23rd International Conference on Computational Linguistics: Posters* (pp. 1462–1470). Stroudsburg, PA, USA: Association for Computational Linguistics. Retrieved from http://dl.acm.org/citation.cfm?id=1944566.1944733

A Unified Framework for Business Process Intelligence

Abid Sohail , Dhanapal Durai Dominic

Department of Computer and Information Sciences,
Universiti Teknologi PETRONAS, Malaysia
abidbhutta@gmail.com

Abstract. Enterprises are striving to cut down their cost and the same time maintains an expectable level of quality in service delivery to keep the competitive edge. Integral to that is an analysis of business processes in order to identify inefficiencies in the design as well as execution of processes. Subsequently, based on the analysis improvement actions are taken. The techniques to identify inefficiencies in process are categorized into two types, a-priori (pre-execution) analysis and posterior (post-execution) analysis. This study focuses on posterior analysis, in which the data produced as a result of process execution are used to identify inefficiencies such as execution delay, and resource utilization. The aim of study is to analyze the existing work on business process improvement and build a unified framework for business process intelligence.

Keywords: Business Process Management, Process Improvement, Business Process Improvement, Workflow Management Systems

1 Introduction

Business process improvement is about analyzing the current behavior of process execution in order to identify the process inefficiencies such as agent assignment, resource utilization, and control flow etc. Dissatisfaction from the process execution is the beginning of improvement process. Improvement is a continuous process in determination of causes for inefficiency and finding way to eradicate them. In early 90's Business process reengineering (BPR) is initiated by (Michal Himmer, 1991 and Davenport, 1993) for business processes betterment. Business Process Reengineering (BPR) one way to identify weak point in process design and suggest a radical change method that is to make fundamental changes in process design for its optimal execution. Since then a lot of work has been done to provide radical changes for process improvement. Automation of workflow management systems (WMS) proposed by Georgakopoulos (1995). WMS are combination of technologies that provide methodologies and software support to business process, workflow specifications, optimization of specified processes and workflow automation [4]. Major expected goal in adoption of these technologies is to provide an optimized execution of organizational process and activities. An integrated tool that supports both business user and information technology can only provide flexible way to adopt changes for the improvement process. Automated WMS provide one way to suggest drastic change for process improvement. But a very weak diagnosis mechanism is provided to handle design complexity. Business process Management systems (BPM)

T. Herawan et al. (eds.), *Proceedings of the First International Conference on Advanced Data and Information Engineering (DaEng-2013)*, Lecture Notes in Electrical Engineering 285, DOI: 10.1007/978-981-4585-18-7_49,
© Springer Science+Business Media Singapore 2014

provide an overall monitoring and management of process. These systems are supposed to be the superset of WFMS [4]. Smith & Fingar (2003) suggest BPM provides a holistic process change management approach to reduce the time and cost of process structure and design change. BPM provides high flexibility in change management. Van der Aalst (2003) defines BPM as "Supporting business processes using techniques, methods and software to design, enact, control and analyze operational processes that are involving humans, organizations, applications, documents and other sources of information". BPM provides process centric approach that is the combination of information technology and business governance methodology [3].

Contribution of this research is to explore and categorized major and common components from improvement methods and organize them into a model. Developed framework is used to facilitate beginner researches to see major components and their order of execution in a single model. Unified model is proposed only for illustration purpose. Model components are gathered from the well known studies conducted in the domain of process improvement and combined with business process intelligence frameworks. Other contribution of this research is to show the research emphasis in using framework components. It is measured from the observation of model description and way of illustration. Research emphasis is noted on three scale that is, highly emphasis, major component and neglected component. Highly emphasized and major component is differentiated like if method is providing detail steps or referring to an establish method in absence of both it is considered as a major component. Which means research considering said sub component in method but its emphases is less. Observation is considered neglected in absence of component. In section 2 gives methodology section 3 presents related work, unified framework is presented in section 4, section 5 presents component usage by BPI researches and Conclusion and discussion is given in section 6.

2   Methodology

Major contribution of this research is to explore studies conducted on business process improvement, especially that are from the domain of business process intelligence. Other propose of this research is to identify most common components in designing a unified BPI framework. Common components are defined as most frequently used. Unified framework is designed to show the improvement process and relationship of components and their order of execution. Two research questions are establishes these are, RQ1: Create unified framework for BPI. Selection of component will do by mutual consciences of all authors. Answer to this question is provided in section 4 that shows the process improvement method with pre and post activates of process warehouse. RQ2: Identification of researches emphasis on identified procedural components of BPI unified framework. Answer to this research question will give you component categories and researchers consideration with highly emphasized, major component and neglected. Articles are selected randomly from online available digital databases that are ISI Thomson, Science direct, IEEE Explore and emerald. Levy and Timothy [7] provided guidelines are used for filtration process. This is followed by an iterative approach. In first iteration 90 articles were selected after abstract study. In second iteration 31 articles were filtered. Final selection is made by mutual consensus of all authors and finally selection 9 articles. Selected article are either discussing improvement framework components or providing complete framework.

3   Related Work

Business process improvement provides a way to monitor and improve the performance of business processes. Business processes redesign, business process reengineering (BPR), workflow management systems (WFMS), continuous process improvement (CPI), process restructuring and business process intelligence (BPI) are different term used for improvement process[7]. Identification and selection of improvement process is dependent of business user requirements [5]. Process evaluation and performance categorization for all behaviors including resources (i.e agent and resources reassignment) is crucial [4]. Griesberger [8] argue reorganization of process actors for optimistically recovery is equally important. Improvement is required at both stages i.e. before and after task execution [19]. Griesberger [8] in his work emphasized on identification and categorization of process quality dimensions through business key performance indicators. Whole process of improvement is divided into iterative executed steps. These steps provide analysis (finding weakness) through measurement of process log activities [18], change initiative (how to improve weakness) [4], agent and resource performance evaluations [4] and customer/user feedback [14]. A continuous improvement method is proposed by Liang (2012) that provides the incremental way to improve the business process. Lang proposed improvement method consists of four layered steps starts from the creation of workflow models lead to improved process. BPI implication guidelines are provided by Carsten et al. (2010). Based on these guidelines a morphological box for business process intelligence is provided. That shows the relationship of BPI components categorically. Tan et al. (2008) presents BPI model and also demonstrates its implementation with case study. Three steps are suggested for performance and management analysis that are measurement, analysis and response to user. Processes execution quality is measured through efficiency, speed, time, cost and scheduling strategies. Yan Li (2008) designed an intelligent business process system to adhere the process diagnosis, analysis, optimization, and prediction other than previous systems focusing on process definitions and running. Their proposed system is look like cockpit architecture provided by Castellanos (2004).

Dalmaris (2007) proposed a Knowledge intensive business process improvement (KBPI) framework. Which is consists of three parts functional theory of knowledge deduced from Karl Proper's epistemology, ontologies for business process representation and process audit evolution and improvements methods. Mutschler (2005) gives three layers BPI reference model. Process log data is maintained (produced during tasks execution) with special steps. After that clearing operation is performed on extracted log data. Syntactical and syntactically correction operation is performed on extracted data and is integrated with simulation data to process warehouse (PW). Schiefer (2004) gives an idea to merging the workflow produced data with produced activity log data through process information factory (PIF). PIF consists of these four components, i) Process warehouse (PWH), ii) Process data store (PDS), iii) Event Processing Container (EPC) and iv) PIF Builder. Analyzed and processed information is available to business user via dashboard. An integrated tool that supports both business user and information technology user is proposed by Daniela Grigoria et al ,(2004). Process engine provide multi dimensional view of information to business user through cockpit or dashboard.

4   Unified framework for business process intelligence

Unified framework for process improvement through business process intelligence is an ultimate objective of this research. Major portion of study is conducted through nine B selected PI methods/frameworks [10-18]. Selection is made randomly and by mutual consensus of authors. In development of unified framework first step is to explore the major and common components. Business intelligence based improvement methods and improvement methods without using business intelligence are two main categories. Both domains are considered in an initial study of gathering of all essential comments. Most of important evaluation criteria and established techniques for business process improvement are ignored in business process intelligence domain study [4]. For an instance optimal agent and resource assignment is ignored in BPI methods. Agent is the human resource and resource is the combination of tools or instruments used by the agent to perform process activities. Identified components are classified into four main layers. Functionality of each layer differentiates them from other layer. Unified frame work is generated after combining all selected component with respect to their functionality in each layer. Other part of this research is on the identification of researcher emphasized in using components in selected studies. BPI framework components are categories into four layers or step that are data sources, process warehouse creation process, measurement policies, decision making policies and user interfaces. Data sources defined as the collection sources that contain the process execution information. Process warehouse creation process leads to an integration of all data sources into another repository. Process warehouse differ from data warehouse in a case of having process activity log data and simulation data. Measurement policies are the combination of analytical tasks to perform on PW to extract inform for decision making. Guided way of improvement is required to ease the decision making for business user. Decision making policies are used for facilitation of business user in decision making. Dashboard or user interface is most important component for BPI provide user interaction with system.

Unified framework for business process intelligence provides a way to show clear view of structural flow of improvement process steps and activities. Flow of improvement activities derived from previous work. Provision of pre and post step of process warehouse in a single model is the most furious part of this research. Pre creation steps of process warehouse are already established in literature and have been sufficient investigation. Post step of process warehouse are very rarely investigated. For instance an example of how to use process warehouse an IT developer team developed the process warehouse then it needs to have strong policies to extract required information that provide effective navigation of information. Zellener [20] claims that no method is developed to provide improvement act. An effective and guided way of improvement is required. It can only be possible when a method have process to identify process weakness efficiently after measuring information from process warehouse. Intension of this research is to add measurement policies and decision making policies as a vital step in BPI framework.

General structure of BPI framework is presented in figure1 that shows complete relationship and step by step execution of all extracted components. General model is dived into three layers first is collection of data sources, second is creation of process warehouse after applying PW creating procedure and policies. Third layer is provision of interface after applying measurement and decision making policies on multi-dimensional data available in PW. Analyzed information is provided as an out put to business user in user friendly manners.

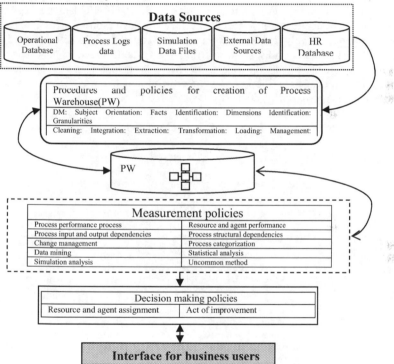

Fig.1 Unified framework for Business Process Intelligence

5 BPI framework components in relation with researcher emphases

Unified framework provides a descriptive framework to show the complete procedural and structural components for improvement process. This framework shows is very helpful for new research in finding all essential components in one model. Business intelligence based framework should have at least framework components. That are supposed major component and executed in order wise. Once the components are identified and gathered in a framework then our next aim it is to explore the BPI researcher emphasis for each procedural component. Study is made on nine selected method. Identification of emphasis and existence is perceived after studding and discussing with other authors. Finally usage is categorized as following highly emphases, major component and neglected component in studies model. First category is highly emphases, which mean existence of procedural component is highly considered by researchers. Second category is major component which means components are existed in model but focus is very little. Third category is neglected when existence is totally ignored. Table 1 shows the observational presence of procedural components from selected models. For an instance of study [10] we identify from the provided figure 3 "Architecture of the Business Process Intelligence tool suite" operational data sources, simulation data and HR data are neglected the data source components. In some cases complete illustration of method is studied to find response. Highly emphasized and major component is differentiated like if method is providing detail steps or referring to an establish method in absence of both cases it is noted as major component. Researcher considering said sub component in method but it emphases is less so, we categorized it as major component. Absence of component is noted as neglected

Table1. BPI framework components existence with reference to researcher emphasis

Framework components	Procedural components	Highly emphases	Major component	Neglected
Data sources	Operational data	[11,13]	[14,15,16,18]	[10,12,17]
	Process log data	[10-18]		
	Simulation data	[13,14]	[11,16,17]	[10,12,15,18]
	External source data	[10,11,12,13]	[14,15,16]	[17,18]
	HR data	[11,13]	[14,18]	[10,12,15,16,17]
Process warehouse procedures	Data warehouse perquisites	[13]	[11,14,15,16]	[10,12,17,18]
	Dimensional modeling	[10,11,12,16]	[13,18]	[14,15,17]
	PW repository management	[13]	[10,12,15,16,18]	[11,14,17]
Measurement policies	Process performance evaluation	[16]	[14,17,18]	[10,11,12,13,15]
	Process input and output dependencies		[14,18]	[10,11,12,13,15,16,17]
	Resource and agent performance			[10-18]
	Change management (redesign)		[14,15]	[10,11,12,13,16,17,18]
	Process categorization		[11,13,14,16]	[10,12,15,17,18]
	Process structural dependencies		[13,14,18]	[10,11,12,15,16,17]
	Data mining	[10,12,13,15,16]	[11]	[14,17,18]
	Simulation analysis	[10,12,13,15,16]	[11,17,18]	[14]
	Statistical analysis	[15]	[12,13]	[10,11,14,16-18]
	Uncommon method		[14,16]	[10,11,12,13,15,17,18]
Decision making policies	Resource and agent assignment			[10-18]
	Act of improvement			[10-18]
User interface	Cockpit	[10-18]		

It is observed that most of frameworks are ignoring major procedural components that are supposed to be essential by other improvement studies. Like agent and resource performance evaluation and reassignment, previous methods are also limited to provide user a guided way of making change i.e. how to make the decision and how to manage the change and reassignment of resources.

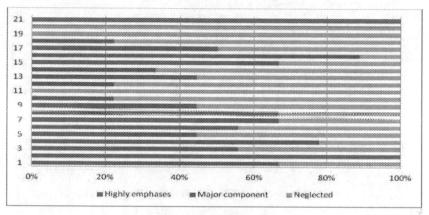

Fig.2. Comparisons of different parameters

6   Conclusions and Future work

This study concluded that measurement and analysis of information from raw data without the presence of PW is very time consuming and need more affords. Procedural components from 1 to 21 are analyzed with selected nine BPI frameworks and results with researcher's choice are shown in figure 2. These 21 components are altogether considering only sub components. Operational data sources, simulation data, process log data, external source data and HR data are the sub components of data sources. Process log data and user interface are considered highly emphasized components. Resource and agent performance assessment and their optimal assignment neglected in all studies. Many studied neglect and ignore existence of important components like how to guide user for decision making, change management and resources evaluation and job reassignment. Future directions from this work are to design a detailed agent and resource assignment method. Not only for agent and resource optimal assignments other neglected sub components also required a detail study.

7   References

1. Hammer, Michael. "Reengineering work: don't automate, obliterate." Harvard business review 68.4 (1990)
2. Georgakopoulos, Diimitrios, Mark Hornick, and Amit Sheth. "An overview of workflow management: from process modeling to workflow automation infrastructure." Distributed and parallel Databases 3.2 (1995): 119-153.
3. Hammer, Michael, and James Champy. "Business process reengineering." London: Nicholas Brealey (1993).

4. Sohail, A.; Dominic, P.D.D., "A gap between Business Process Intelligence and redesign process," Computer & Information Science (ICCIS), 2012 International Conference on , vol.1, no., pp.136,142, 12-14 June 2012 doi: 10.1109/ICCISci.2012.6297227

5. Smith, H., & Fingar, P. (2003). Business process management: the third wave (Vol. 1). Tampa: Meghan-Kiffer Press.

6. Van Der Aalst, Wil MP, Arthur HM Ter Hofstede, and Mathias Weske. "Business process management: A survey." Business process management. Springer Berlin Heidelberg, 2003.

7. Yair Levy and Timothy J. Ellis, A Systems Approach to Conduct an Effective Literature Review in Support of Information Systems Research, Informing Science Journal Volume 9, 2006

8. Griesberger, Philipp, Susanne Leist, and Gregor Zellner. "Analysis of techniques for business process improvement." (2011), ECIS 2011 Proceedings.

9. J. T. S. Ribeiro and A. J. M. M. Weijters. 2011. Event cube: another perspective on business processes. In Proceedings of the 2011th Confederated international conference on On the move to meaningful internet systems - Volume Part I (OTM'11), Vol. Part I. Springer-Verlag, Berlin, Heidelberg, 274-283.

10. Daniela Grigori, Fabio Casati, Malu Castellanos, Umeshwar Dayal, Mehmet Sayal, Ming-Chien Shan, Business Process Intelligence, Computers in Industry, Volume 53, Issue 3, April 2004, Pages 321-343, ISSN 0166-3615, 10.1016/j.compind.2003.10.007.

11. Schiefer, J.; Jun-Jang Jeng; Kapoor, S.; Chowdhary, P., "Process information factory: a data management approach for enhancing business process intelligence," e-Commerce Technology, 2004. CEC 2004. Proceedings. IEEE International Conference on , vol., no., pp.162,169, 6-9 July 2004

12. Castellanos, Malu, Fabio Casati, Umeshwar Dayal, and Ming-Chien Shan. "A comprehensive and automated approach to intelligent business processes execution analysis." Distributed and Parallel Databases 16, no. 3 (2004): 239-273.

13. B. Mutschler, M. Reichert, and J. Bumiller. An approach to quantify the costs of business process intelligence. In International Workshop on Enterprise Modeling and Information Systems Architectures (EMISA 05), pages 152–165, 2005.

14. Peter Dalmaris, Eric Tsui, Bill Hall, Bob Smith, (2007) "A framework for the improvement of knowledge-intensive business processes", Business Process Management Journal, Vol. 13 Iss: 2, pp.279 - 305

15. Yan Li; Shao-Ling Deng, "Design of Intelligent Business Process System and Process Remodeling," Intelligent Computation Technology and Automation (ICICTA), 2008 International Conference on , vol.1, no., pp.589,593, 20-22 Oct. 2008

16. Tan,W. Shen,W. Zhou,B. A business process Intelligence System for enterprise process performance management. IEEE Transactions on System, Man and Cybernetics, Part C.,36, (36), pp. 745-756 DOI: 10.1109/TSMCC.2008.2001571

17. Felden, Carsten, Peter Chamoni, and Markus Linden. "From Process Execution towards a Business Process Intelligence." In Business Information Systems, pp. 195-206. Springer Berlin Heidelberg, 2010.

18. Liang CHEN, X. L., Qing YANG (2012). "Continuous process improvement based on adaptive workflow mining technique." Journal of Computational Information Systems 8(7).

19. Markus Linden, Carsten Felden, and Peter Chamoni. Dimensions of Business Process Intelligence. M. zur Muehlen and J. Su (Eds.): BPM 2010 Workshops, LNBIP 66, pp. 208–213, 2011. © Springer-Verlag Berlin Heidelberg 2011

20. Zellner, G. (2011). A structured evaluation of business process improvement approaches. Business Process Management Journal, 17(2), 203-237.

Effective Risk Management of Software Projects (ERM): An Exploratory Literature Review of IEEE and Scopus Online Databases

[1]Uzair Iqbal Janjua, [2]Alan Oxley, [3]Jafreezal Bin Jaffer

[1]uzair_iqbal@comsats.edu.pk, [2]alanoxley@petronas.com.my
[3]jafreez@petronas.com.my
Department of Computer & Information Science
UNIVERSITI TEKNOLOGE PETRONAS, MALAYSIA

Abstract . The significance and the maturity level of software engineering have been increasing since 1968. However, software development, which falls under the umbrella of software engineering, is still evolving. Different software risk management process models, methods and techniques have been presented by researchers for the software industry to make the development of software projects more likely to succeed. However, a lot of software projects still fail to complete. The models used are general and based on the traditional technique of risk management. Hence, effective risk management (ERM) techniques are required for the development of mature software in order to increase the rate of success of software projects.

Keywords: Risks, Risk Management (RM), Effective Risk Management (ERM).

1   Introduction

Until the late 1980s, the theory of risk management (RM) was little discussed in the software development community. The need to identify risks in software projects and the avoidance of them has, though, always been appreciated. To handle risk factors researchers over the last three decades have proposed different software RM models and methods. However, still the success rate of software development is very low. Survey result [1] conducted in 2008 shows; out of 100, only 32 software projects succeeded, 24 projects were completely failed and 44 projects were seriously threatened by risks. Though some authors criticize the report of the Standish Group International, many other authors [2], [3] second the report. So, different researchers tried to find an ERM model to increase project success rate but still there is no model which can be called the de facto ERM model [4], [5], [6].

Researchers and practitioners, on the basis of their theoretical knowledge and practical observation have given many suggestions and advices for using different tools and methods to make the current practices of software RM more effective [4], [5], [6], [7].

T. Herawan et al. (eds.), *Proceedings of the First International Conference on Advanced Data and Information Engineering (DaEng-2013)*, Lecture Notes in Electrical Engineering 285, DOI: 10.1007/978-981-4585-18-7_50,
© Springer Science+Business Media Singapore 2014

Some authors and practitioners have suggested techniques to make the current practices of RM more effective even without the means of any effectiveness tools [8], [9], 10]. Whereas some others mentioned that ERM can be achieved by using intelligent software tools [2]. Even some authors believe that without information technology it is not possible to achieve ERM [11].

Kwak and Stoddard believe software ERM is a significant tool for a project manager to increase the probability of project success. This leads the authors to conduct a literature review study to find all the tools and suggestions to make RM of software development more effective to increase the success rate. So the objectives of the research work were:

a) To find the significance of effective risk management for the software development.

b) To find how current practices of risk management be make more effective.

The paper is structured as follows: section 2 explains the method used to explore literature of online databases. Section 3 discusses the important finding and result of current study; section 4 describes the conclusion and the future work.

2   Exploratory Literature Review of IEEE and Scopus Online Databases

In this literature review the authors have explored the Scopus and IEEE online databases from the year 1996 to 2012 with the key stream "software effective risk management". The aim of this research work is to explore the concepts and significance of ERM for software projects.

Two researchers selected papers for the primary study autonomously. Only those papers were selected, for the secondary study, which were directly related to software engineering. After the primary study, 34 papers were selected for the secondary study on the mutual understanding of both researchers. After the primary selection, a paper available at both data bases considered once for secondary study to avoid biasness. Authors have defined 5 criteria's for the selection of papers to achieve the objective of the current study. For the secondary study, only those papers were selected which at least fulfill 1 of the following 5 criteria (C).

C1.   Define the meaning of effective risk management.

C2.   Discuss or suggest any technique to make risk management more effective.

C3.   Discuss or suggest any tool to make risk management more effective.

C4.   Proposed any method or process model or framework or methodology to make risk management more effective.

C5.   Mention the need of effective risk management.

Details of selection process of studies are given in Table 1. Letter "S" is used to represent papers selected for secondary study (e.g. Study 1 is represented by S1). References of papers selected for secondary study, are given in Appendix A. Appendix A is presented in the end of paper, immediately after the reference section.

Table 1. Details of Search and Selection of Studies

Year Range	Search Result		No. of Papers Selected for Primary Study		No. of Papers Selected for Secondary Study
	Scopus	IEEE	Scopus	IEEE	
1995-1999	89	33	5	4	6
2000-2004	240	33	3	0	1
2005-2009	1301	112	14	19	11
2010-2012	1743	101	12	13	16
Total Paper	3397	298	34	38	34

3   Findings and Discussion of Exploratory Literature Review

Table 1 show that the proportion of papers selected for the secondary study, from the search results, is significantly larger for the two periods 2005 to 2009 and 2010 to 2012. Highest numbers of papers selected for secondary study are from year 2012. This indicates that the significance of software ERM has increased in recent years. Moreover, 47% of all the papers selected for the secondary study were published in the three years 2010 to 2012. The remaining papers of the secondary study were published over a period of 15 years. Last but not least, all of the papers from the last three years addressed the need for ERM either directly or indirectly. Overall, 79% of all the papers selected from the primary study addressed the need of change in current RM practices for ERM either directly or indirectly.

71% of the papers in the secondary study not only mentioned the need for ERM but also proposed a model/framework/methodology or discussed some tool or factors to make current practices of RM more effective.

18% of the papers in the secondary study not only give some suggestions but also recommend some tools for ERM. Only 41% of the papers in the secondary study proposed a framework or process model or method for the ERM. But all framework or process model or method is about one or two processes of risk management but not for whole processes of software RM.

50% of the papers in the secondary study defined ERM. The majority of the papers defined ERM directly whereas the rest defined ERM indirectly.

The authors explored papers in the IEEE and Scopus online databases dating from 1996. S1, written by Brekka L.T et al., is founded as a first study which enlightens the need of early identification of risks to make RM more effective. S4, written by Williams et al., was published in 1997 and will now be described. The paper discussed effective and ineffective RM methods on the basis of six years of practical experience by its authors. Some methods, within the eight different processes of RM, proved to be effective and some proved to be ineffective. At the end of the paper, the authors mentioned that when an organization has ERM in place then the organization can focus on avoiding future problems rather than solving current ones. In fact, the paper can act as motivation for any researcher to find the best practices of RM so as to make RM more effective.

Table 2. Quantitative Result of Exploratory Review of five Criterias

C1	C2	C3	C4	C5
Frequency =17 Percentage =50	Frequency =23 Percentage =68	Frequency =08 Percentage =23	Frequency =14 Percentage =41	Frequency =27 Percentage =79
S4,S5,S8-S12, S14-S16,S22, S24,S27-S29, S31,S32	S1-S4,S6,S8-S17, S19,S21,S24,S26, S27,S29,S31,S32	S2,S10, S16-S18,S22, S24,S27.	S1,S14,S16, S20,S23-S31, S34	S1,S2,S4,S5, S7-S10,S14, S16, S17, S19-S34

During the research, the authors came to realize that selection of the papers was the most difficult part of this study because, in the literature, the term "effective risk management," in the context of software engineering, is not clearly defined. Different authors have defined the term according to their knowledge or on the basis of their experience. Nevertheless, the majority believe that the success of a software project depends upon the effective management of risks, i.e. success depends on software ERM (S7, S13, S14, S30 and S31). Furthermore, software ERM is proactive in nature (S4).

68% of the papers in the secondary study recommend the manipulation or addition of some of the key steps or factors or principles in current RM activities in order to make the activities more effective, as compared with 23% of papers that recommend the use of tools for ERM. Some of key factors required to make RM more effective identified from papers selected for secondary study are: support of top management for RM, healthy culture among team members to promote RM, common goals, circulation of established guidelines of RM among all stakeholders, knowledge base of RM, sharing of experiences and best practices of RM, Entire project team involvement with clear roles and tasks, disciplined, open and continuous communication among all stakeholders, skilled manager for RM, Shared product vision, global perspective, forward-looking view and decision support system. Moreover, for ERM, RM processes should be well integrated, consistent, continuous, improvable and proactive in nature.

According to literature review an ERM process is proactive in nature, which enables organizations to achieve their goals by reducing current risks from their maximum levels through continuous monitoring. Furthermore ERM, process assists management with improved decision by making thorough up-to-date risk reporting, something that is only possible by clearly stating the responsibilities of each stakeholder, and by developing a healthy culture of communication among all stakeholders.

During search of related studies, authors also tried to find any other study which has conducted literature review about the ERM of software projects with similar parameters. But according to literature review of two databases there is no similar study.

4 Conclusion and Future Work

Evidence of the need for ERM and its positive impact on software projects is enough to advocate the worthiness of ERM. ERM in software development not only increase the success of software projects but also help in decision making and ensures the customers' satisfaction and overall improved financial performance of the organizations.

Following the literature survey that the authors undertook, it is believed by different researchers and practitioners that the current practices of software RM can be made more effective by adopting their suggestions.

Literature Review of current Study is limited to 2 online databases. In future, the authors will undertake a systematic literature review of ERM by scouring different electronic databases in order to identify further suggestions for ERM. Furthermore, during synthesis of selected papers for current study it is founded that many authors point out the involvement of "Risk Manager" in the RM activities. In future authors will try to find, is there any relationship between "Risk Manager" and ERM of software projects.

References

Standish Group International.: Chaos Summary 2009: 10 Laws of CHAOS. Technical Report (2009)

John Dhlamini, Isaac Nhamu, and Admire Kachepa.: Intelligent Risk management Tools for Software development. SACLA, Mpekweni Beach Resort, South Africa (2009)

Shikha , Dr. R. Selvarani.: An Efficient Method of Risk Assessment using Intelligent Agents. Second International Conference on Advanced Computing & Communication Technologies (2012)

Robert Stern, José Carlos Arias.: Review of Risk Management Methods. Business Intelligence Journal - Vol.4 No.1 (2011)

Mira Kajko-Mattsson, Jan Lundholm, Jonas Norrby.: Industrial Opinion on the Effectiveness of Risk management Methods. 33rd Annual IEEE International Computer Software and Applications Conference (2009)

George Holt.: Software Risk management From a System Perspective. Crosstalk, The Journal of Defense Software Engineering (2005)

Sven Roeleven, Michiel Jorna.: How to implement effective Enterprise Risk Management Building a sustainable Governance Risk & Compliance solution. Business White Paper, June (2011)

Daniel D. Galorath , Michael W. Evans.: Software Sizing, Estimation, and Risk Management. Auerbach Publications (2006)

Stephen. Ward.: Requirements for an Effective Project Risk Management Process. Project Management Journal, Vol. 30, No. 3, pp. 37-43 (1999)

Y.H. Kwak J. Stoddard.: Project risk management: lessons learned from software development environment.　　doi:10.1016/S0166-4972(03)00033-6,　　Elsevier　　Science　　(2003)

Appendix A: References of Papers Selected for Secondary Study

S1   Brekka L.T, Maksimovic V, Picardal C, Iftekharuddi, K.: Risk management and sys-
       tems engineering discipline. Aerospace and Electronics Conference, 1996. NAECON
       1996., Proceedings of the IEEE 1996 National, vol.2, no., pp.829,835 vol.2,
       doi: 10.1109/NAECON.1996.517748 (1996)

S2   Chittister, C.G, Haimes, Y.Y.: Systems integration via software risk manage-
       ment. Systems, Man and Cybernetics, Part A: Systems and Humans, IEEE Transactions
       on , vol.26, no.5, pp.521,532, Sep doi: 10.1109/3468.531900 (1996)

S3   Boehm, B.W., DeMarco, Tom.: Software risk management," Software, IEEE , vol.14,
       no.3,pp.17,19, doi: 10.1109/MS.1997.58922 (1997)

S4   Williams R.C, Walker J.A, Dorofee, A.J.: Putting risk management into prac-
       tice. Software, IEEE , vol.14, no.3, pp.75,82, doi: 10.1109/52.589240 (1997)

S5   Collofello J.S, Pinkerton, A.K.: Integrating risk management into an undergraduate
       software engineering course. Frontiers in Education Conference, 1997. 27th Annual
       Conference. Teaching and Learning in an Era of Change. Proceedings. , vol.2, no.,
       pp.856,860 vol.2, 5-8 , doi: 10.1109/FIE.1997.635987 (1997)

S6   ADLER, T.R., LEONARD, J.G. and NORDGREN, R.K.: Improving risk management:
       Moving from risk elimination to risk avoidance. Information and Software Technolo-
       gy, 41(1),pp.29-34, Elsevier Science (1999)

S7   KUMAR, R.L.: Managing risks in IT projects: An options perspective. Information and
       Management. 40(1), pp. 63-74., Elsevier Science (2002)

S8   Skelton, T.M.; Thamhain, H.J., "User-centered design as a risk management tool in
       new technology product development," Engineering Management Conference, 2005.
       Proceedings. 2005 IEEE International, vol.2, no., pp.690,694, doi:
       10.1109/IEMC.2005.1559237 (2005)

S9   Samad J, Ikram Naveed.: Managing the Risks: An Evaluation of Risk Management
       Processes. Multitopic Conference, 2006. INMIC '06. IEEE , vol., no., pp.281,287, 23-
       24, doi: 10.1109/INMIC.2006.358178 (2006)

S10   Skelton, T.M.; Thamhain, H.J., "Managing the Sources of Uncertainty in Technology
       Projects," Engineering Management Conference, 2006 IEEE International , vol., no.,
       pp.473,477, 17-20, doi: 10.1109/IEMC.2006.427991 (2006)

S11   Damian D, Chisan J.: An Empirical Study of the Complex Relationships between Re-
       quirements Engineering Processes and Other Processes that Lead to Payoffs in Produc-
       tivity, Quality, and Risk Management. Software Engineering, IEEE Transactions on ,
       vol.32, no.7,pp.433,453, doi: 10.1109/TSE.2006.61 (2006)

S12   Nyfjord J, Kajko-Mattsson M.: Communicating Risk Information in Agile and Tradi-
       tional Environments.Software Engineering and Advanced Applications,2007.33rd
       EUROMICRO   Conference   on ,vol.,no.,pp.401,408,28-31,   doi:
       10.1109/EUROMICRO.2007.22 (2007)

S13   Smite D.: Project Outcome Predictions: Risk Barometer Based on Historical Da-
       ta. Global Software Engineering, 2007. ICGSE 2007. Second IEEE International Con-
       ference on , vol., no., pp.103,112, 27-30, doi: 10.1109/ICGSE.2007.37 (2007)

S14   DEY, P.K., KINCH, J. and OGUNLANA, S.O.: Managing risk in software develop-
       ment projects: A case study. Industrial Management and Data Systems, 107(2), pp.
       284-303 (2007)

S15   Ye Tao.: A Study of Software Development Project Risk Management. Future Infor-
        mation Technology and Management Engineering, 2008. FITME '08. International
        Seminar on ,vol., no., pp.309,312, 20-20, doi: 10.1109/FITME.2008.125 (2008)
S16   Lai Yifei; Qianhua Zhang; Jia Junping.: Study on Project Risk Management Infor-
        mation System Based on Progress Schedule," Wireless Communications, Networking
        and Mobile Computing, 2008. WiCOM '08. 4th International Conference on , vol., no.,
        pp.1,5, 12-14 doi: 10.1109/WiCom.2008.2444 (2008)
S17   Al-Rousan, T, Sulaiman S, Salam R.A.: Project Management Using Risk Identification
        Architecture Pattern (RIAP) Model: A Case Study on a Web-Based Applica-
        tion. Software Engineering Conference, 2009. APSEC '09. Asia-Pacific , vol., no.,
        pp.449,456, 1-3, doi: 10.1109/APSEC.2009.42 (2009)
S18   Dapeng Liu, Qing Wang, Junchao Xiao.: The role of software process simulation mod-
        eling in software risk management: A systematic review. Empirical Software Engineer-
        ing and Measurement, 2009. ESEM 2009. 3rd International Symposium on , vol., no.,
        pp.302,311, 15-16 doi: 10.1109/ESEM.2009.5315982 (2009)
S19   Verma C, Amin S.A.: Significance of Healthy Organizational Culture for Superior Risk
        Management During Software Development. Developments in E-systems Engineering
        (DESE),2010 ,vol.,no.,pp.182,189,6-8, doi: 10.1109/DeSE.2010.37 (2010)
S20   Yong Hu, Xiangzhou Zhang, Xin Sun, Jing Zhang, Jianfeng Du, Junkai Zhao.: A Uni-
        fied Intelligent Model for Software Project Risk Analysis and Planning. Information
        Management, Innovation Management and Industrial Engineering (ICIII), 2010 Inter-
        national Conference on , vol.4, no., pp.110,113, 26-28, doi: 10.1109/ICIII.2010.504
        (2010)
S21   Wen-Hsien Tsai, Sin-Jin Lin, Jau-Yang Liu, Kuen-Chang Lee, Wan-Rung Lin, Jui-
        Ling Hsu.: Examining the implementation risks affecting different aspects of Enterprise
        Resource Planning project success. Computers and Industrial Engineering (CIE), 2010
        40th   International   Conference   on ,vol.,no.,pp.1,6,25-28,   doi:
        10.1109/ICCIE.2010.5668317 (2010)
S22   Azizi N, Hashim, K.: Enterprise level IT risks: An assessment framework and
        tool. Computer Science and Information Technology (ICCSIT), 2010 3rd IEEE Interna-
        tional Conference on ,vol.3,no.,pp.333,336,9-11, doi: 10.1109/ICCSIT.2010.5563565
        (2010)
S23   Ai-Guo Tang, Ru-long Wang.: Software project risk assessment model based on fuzzy
        theory. Computer and Communication Technologies in Agriculture Engineering
        (CCTAE), 2010 International Conference On , vol.2, no., pp.328,330, 12-13, doi:
        10.1109/CCTAE.2010.5544587 (2010)
S24   Avdoshin S.M, Pesotskaya E.Y.: Software risk management. Software Engineering
        Conference in Russia (CEE-SECR), 2011 7th Central and Eastern European , vol., no.,
        pp.1,6, doi: 10.1109/CEE-SECR.2011.6188471 (2011)
S25   Betz S, Hickl, S, Oberweis A.: Risk Management in Global Software Development
        Process Planning. Software Engineering and Advanced Applications (SEAA), 2011
        37th EUROMICRO Conference on , vol., no., pp.357,361, doi: 10.1109/SEAA.2011.64
        (2011)
S26   Tak Wah Kwan, Leung H. K N.: A Risk Management Methodology for Project Risk
        Dependencies. Software Engineering, IEEE Transactions on , vol.37, no.5, pp.635,648,
        doi: 10.1109/TSE.2010.108 (2011)
S27   Samer Alhawari, Louay Karadsheh, Amine Nehari Talet, Ebrahim Mansour.:
        Knowledge-Based Risk Management Framework for Information Technology Project.

International Journal of Information Management, Volume 32, Issue 1, Pages 50–65 (2012)

S28   BARATEIRO, J. and BORBINHA, J.: Integrated management of risk information, Federated Conference on Computer Science and Information Systems, FedCSIS , pp. 791-798 (2011)

S29   Khatavakhotan A.S, Siew Hock Ow.: Rethinking the Mitigation Phase in Software Risk Management Process: A Case Study. Computational Intelligence, Modelling and Simulation (CIMSiM), 2012 Fourth International Conference on , vol., no., pp.381,386, 25-27, doi: 10.1109/CIMSim.2012.62 (2012)

S30   Hashimi H, Hafez, A, Beraka M.: A Novel View of Risk Management in Software Development Life Cycle. Pervasive Systems, Algorithms and Networks (ISPAN), 2012 12th International Symposium on , vol., no., pp.128,134, 13-15, doi: 10.1109/I-SPAN.2012.25 (2012)

S31   Ying Qu, Meng-Jia Yuan, Feng Liu.: The risk factor analysis for software project based on the interpretative structural modelling method. Machine Learning and Cybernetics (ICMLC), 2012 International Conference on, vol.3, no., pp.1019,1024, 15-17,doi: 10.1109/ICMLC.2012.6359494 (2012)

S32   Lobato L.L, da Mota Silveira Neto P.A, do Carmo Machado I, de Alemida E.S, de Lemos Meira S.R.: Risk management in software product lines: An industrial case study. Software and System Process (ICSSP), 2012 International Conference on , vol., no., pp.180,189, 2-3, doi: 10.1109/ICSSP.2012.6225963 (2012)

S33   Bazaz Y, Gupta S, PrakashRishi O, Sharma L.: Comparative study of risk assessment models corresponding to risk elements. Advances in Engineering, Science and Management (ICAESM), 2012 International Conference on , vol., no., pp.61,66, 30-31 (2012)

S34   Zhang, Y., H. Yang, and X. Jiang.: Study of project risk continuous process pattern. Advances in intelligent and soft computing. Vol. 137 AISC (2012)

Adaptive Questionnaire Ontology in Gathering Patient Medical History in Diabetes Domain

Sherimon P.C.[1], Vinu P.V.[1] Reshmy Krishnan[2], Youssef Takroni[3],

Yousuf AlKaabi[4],Yousuf AlFarsi[5]

[1] M.S.University, India
[2] Muscat College, Muscat, Sultanate of Oman
[3] Arab Open University, Muscat, Sultanate of Oman
[4] SQU Hospital,Sultanate of Oman
[5] SQU Hospital,Sultanate of Oman

`@sherimon@aou.edu.om,`
`vinusherimon@yahoo.com,reshmy_krishnan@yahoo.co.in,`
`yst@aou.edu.om,aseer115@gmail.com,dryousufalfarsi@gmail.com`

Abstract. Clinical Decision Support System (CDSS) can be used to prepare diagnosis from different patient's details and hence physicians or nurses can review this diagnosis for improving the final decision. Due to the lack of CDSS in diabetes and related diseases in Sultanate of Oman, an Ontology based CDSS is proposed here. The deployed key components of the system are Adaptive Questionnaire Ontology, patient's semantic profile, guideline ontology and risk assessment reasoner. We here propose a model for gathering the patient medical history based on dynamic questionnaire ontology. Ontology is among the most powerful tools to encode medical knowledge semantically. It is an abstract model which represents a common and shared understanding of a domain. The model is explained and implemented for diabetes domain.
Keywords: Patient Semantic Profile, Questionnaire Ontology, OWL, Adaptive Questionnaire, Protégé

1   Introduction

When applying a Decision support system in a health domain, it is called Clinical decision support system(CDSS).This application can analyze data and can help healthcare providers to take fast and accurate clinical decisions. CDSS can be used to prepare diagnosis from different patient's details and doctors/nurses can review this diagnosis for improving the final decision [9]. Two main types of CDSS are there. CDSS based on knowledgebase and CDSS based on machine learning. In the first one, rules are applied to patient data using an inference engine and display the results to the end user.

T. Herawan et al. (eds.), *Proceedings of the First International Conference*   453
on Advanced Data and Information Engineering (DaEng-2013), Lecture Notes
in Electrical Engineering 285, DOI: 10.1007/978-981-4585-18-7_51,
© Springer Science+Business Media Singapore 2014

In most of the Health Care Information Management Systems, the patient information is collected on every visit to the hospital and thus the involvement of patients is minimal [13]. A traditional procedure is as follows: - When a patient visit a hospital, the nurse will first diagnose the patient and will record the preliminary observations such as readings of blood pressure, height, weight, body temperature etc. about the patient [1]. Other details are collected from the patient itself by the doctor. Here in most of the cases the patient medical history will be incomplete. It results from the following issues that can arise during the collection of patient details:

- Medical Personnel failed to ask the patient, relevant questions about allergies to any medicines, any family history of diabetes etc [12].
- Medical Staff didn't act according to the user context. For example the patient is suffering from severe diabetes, but they failed to refer to specialty unit.
- Patient doesn't want to disclose certain sensitive and personal health problems.
- Possibility of asking unwanted questions by the medical staff.
- Gathering of Patient information is sometimes a time-consuming task.

Many issues discussed here are the consequences of an inefficient patient history collection system. If we have an efficient system that automates the whole process, then it will save resources, time and also nurse can focus more on providing medical care to the particular patient. Also information collected in such a manner will be more structured/ organized and detailed than collected through traditional manual and personal interviews. An automated database system will be able to solve most of the issues presented before. The challenge is that, this automated system must be capable to record critical information from various patients also.

In this paper, Based on the analysis obtained from the survey conducted and from discussion with the experts, the proposed CDSS is designed .The architecture/model of CDSS is based on semantic web technology so that more trusted and reliable recommendations can be provided in response to patient data. In the proposed CDSS, the system uses an adaptive clinical questionnaire based on ontology. The users are required to answer the questions regarding the patient's medical history. The users can be clinicians or patients themselves [through online]. Depending on the adaptive nature of the questions, the user input will be checked against a list of potential answers [5]. This process is repeated till all the questions are done. So this ontology based adaptive information collection system iteratively captures precise and reliable patient information in each successive step. This intelligent questionnaire adapts itself as per patient's medical history, by asking relevant information which is significant to patient's profile. Using the semantic patient profile and the decision support ontology which is based on the diabetes and its associated diseases, the reasoner, predicts the risk assessment factor of hypertension with the help of a rule engine. The adaptive questionnaire ontology is developed using Protégé editor [14] and OWL API is used for the development of the patient history gathering system.

The paper is organized as follows: - Section 2 describes the comparison between early systems and their issues with the dynamic systems along with their benefits of using ontology. Model of Adaptive Questionnaire is presented in Section 3. The development of Questionnaire ontology and the adaptive nature of questions are explained here with different scenarios. Section 4 presents the user interfaces of patient, nurse and the doctor. The Implementation of Adaptive Questionnaire is presented in Section 5. Section 6 includes conclusion and future followed by acknowledgement and references.

2   Background

2.1 Early systems Vs Dynamic Systems in Medical Care

Information collections about patients such as past medical history, drug history, family medical history and social history have a vital role in clinical diagnosis system. A traditional system of interviewing by the doctor to the patient is not so efficient since getting required information from patients within the time limit is impossible. In almost all cases there is a chance of forgetting certain questions to be asked. Hence a traditional history collection by interview by doctors is incomplete and time consuming not only for collecting information but for documentation. With most patients, traditional interview is like piloting the plane without a checklist [2].
A computerized questionnaire can act like an efficient medium for collecting large and comprehensive patient's medical history without wasting the doctor's valuable time [1]. It can act as a preliminary survey of patients with total medical problems such as past medical history, drug history, family medical history and social history which are required for the further clinical diagnosis. It is easy to review than the conventional interview since data are systematically arranged [10]. When the questionnaire is made available in internet, patients from remote area also can fill the information without visiting the clinics for providing data and hence can save time and expenses. Moreover through automated questionnaire, we can make sure that no required data is left out. The questionnaire could make part of the medical record of that patient and doctor and can refer in between for further clarifications and add notations [6].

2.2 Ontology Based Approach

To overcome the lack of flexibility and adaptability to unexpected requirements and a general lack of intelligence, a layer of ontology is added on top of the functionalities in the questionnaire [7]. This causes for a pragmatic solution to implement a shift from a simple database management system into intelligent knowledgebase system in healthcare.[15].Since the ontology layer can be updated without the need for additional and costly software engineering work, modification of the questionnaire will be more convenient and cost effective[16]. Since a main advantage of ontology is knowledge sharing and reuse, adaptive questionnaire based on ontology makes the questionnaire more valuable for further usage. The top Ontology layer can enable the questionnaire system to perform decision support operations, and more efficient analysis can be obtained from the questionnaire.

3   Model of Adaptive Questionnaire Ontology

The model consists of three main components. The user interface, the Java engine, and the OWL file which contains the questions. The Java adaptive engine is implemented using OWL API [4]. The Java engine acts as intermediary between the user interface and the ontology. The main classes in ontology are questionnaire, sub questionnaire, start of questionnaire, question, the further question and answer. Different types of properties are used in the design of ontology [3]. Nature of the classes in the questionnaire is defined in Type properties. Structure of the questionnaire is described in structural properties. Composition properties will give idea how to combine classes [11]. Adaptive properties determine the adaptive (dynamic) behavior of the questionnaire. The main classes of the ontology are *Physical_History, Smoking_History, Alcolhol_History* etc.

Object Properties are used to define relationships between ontology classes [8]. Data properties are used to define relationships between an individual class and XML schema data type [8]. The object property *"has_diabetic_type"* associates *"Patient"* and *Type_Of_Diabetes* classes. The data property *"diabetic_type"* has values "Type1" and "Type2".

3.2 Adaptive Questions
The system asks domain specific questions. Most of them are designed to have Boolean type answers. For example, consider the question *"Do you Smoke?"* This question is an adaptive one which expects *Yes/No* answer. If the answer to the above question is given as *Yes*, then the system displays the sub question *"For How Many Years?"* otherwise, it displays questions from next category, say, for example, alcohol related questions. That means, if the answer is *No*, the system will not ask further questions related to smoking. It proceeds to the next category of questions, for example, alcohol. So the questionnaire adapts itself as per the patient's history and it asks only relevant questions according to the patient's context. Both the scenarios are given in Fig.1.

Adaptive Question – Scenario 1		Adaptive Question – Scenario 2	
>>System	Do you Smoke?	>>System	Do you Smoke?
<<User	Yes	<<User	No
>>System	For How Many Years?	>>System	Do you Drink Alcohol?
<<User	2	<<User	Yes
>> System	How many cigarettes do you smoke?	>> System	For How Many Years?
....		....	

Fig.1. Adaptive Question Scenario

4   User Interfaces

Protégé, OWL Java API, and pellet reasoner are the tools used in the implementation of adaptive patient history collection system.

4.1 Patient Interface

Prior to the hospital visit, a patient can enter his/her history through a web enabled patient interface. A web enabled interface provides the doctors with complete patient history before the start of diagnosis. Other advantages of a patient interface includes: patient can enter all the history according to their convenience from home, time can be consumed, the patient can answer some questions which are very sensitive/ personal, which are otherwise cannot be recorded during face-to-face consultation with the doctor. Patient will be given access to only patient's interface which is controlled by a login screen. They will not be allowed to view nurse or doctor interface. Personal History form and Diabetic History form are presented in Fig.2 and Fig.3

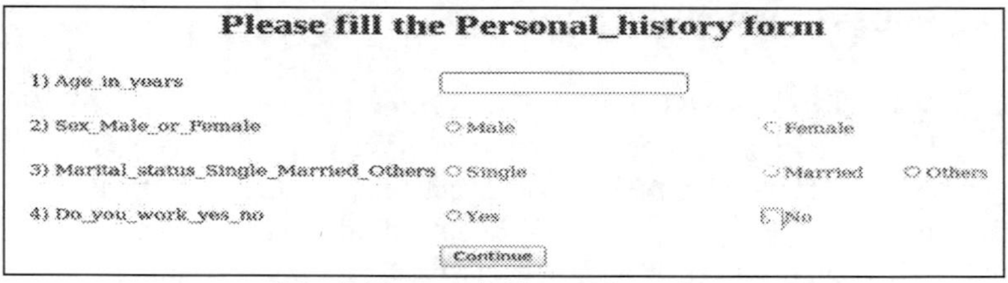

Fig.2 Personal History Interface

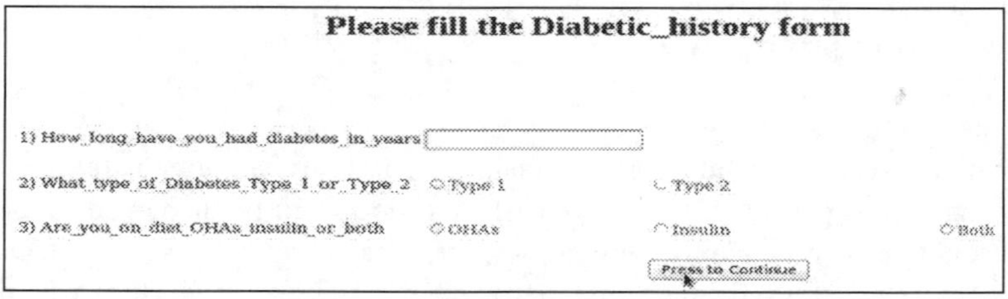

Fig.3. Diabetic History Interface

4.2 Nurse Interface

When the patient visits the hospital, nurse can view the information entered by the patient through the patient interface. This will help the nurse to arrive at a primary conclusion about the status of the patient. Now as per the recorded patient details, if the patient is at a high risk which requires immediate medical attention, s/he will be referred to intensive care unit. Otherwise, the nurse will examine the patient and records the preliminary observations about the patient such as readings of blood pressure, heart beat, height, weight, body temperature etc. These values are appended to the patient profile.

4.3 Doctor Interface

Similarly like nurse, the system provides complete information of the patient when s/he approaches the doctor for final diagnosis. Doctor can view the patient details entered earlier by the patient and the nurse. Further, the doctor will diagnose the patient and the observations are appended to the patient file. If the doctor suggests any lab tests, the test results are also incorporated in the patient file.

Please fill the General_Physical_examination form

Patient_general_condition

a_Pale	○ Yes	● No
b_jaundice	○ Yes	● No
c_clubbing	● Yes	○ No
d_Oedema	○ Yes	● No
e_Lymphadenopathy	○ Yes	● No
f_thyroid_swelling	● Yes	○ No

Fig.4. General Physical Examination – By Doctor

5   Implementation and Results

5.1 Using OWL API to parse Questionnaire Ontology

The Protégé-OWL API is an open-source Java library for the Web Ontology Language and RDF(S) [14]. The API provides classes and methods to load and save OWL files, to query and manipulate OWL data models, and to perform reasoning [14]. OWL API contains a set of interfaces used to manipulate ontologies. It consists of many functions to extract classes, sub-classes, object properties, data properties, and individuals of ontology. For example, *getSubClass( )* is a function used to extract all sub classes of a particular class.

5.2 Adaptive Questionnaire

Questions included in the ontology are of adaptive and non-adaptive nature. If a question is adaptive, as per the user input, further questions will be displayed. If the question does not have any adaptive properties, irrespective of the user input, the next question will be displayed. The first set of questions of adaptive nature is demonstrated in Smoking History Form.

Please fill the Smoking_history form

1) Do_you_smoke_Yes_No ● Yes ○ No

Continue...

Fig.5. Smoking History Form

The first adaptive question in this category is *Do_you_smoke_Yes_No* as shown in Fig.8. If the user input is **YES**, it triggers a call for additional questions, which is related to smoking as shown in Fig.6.

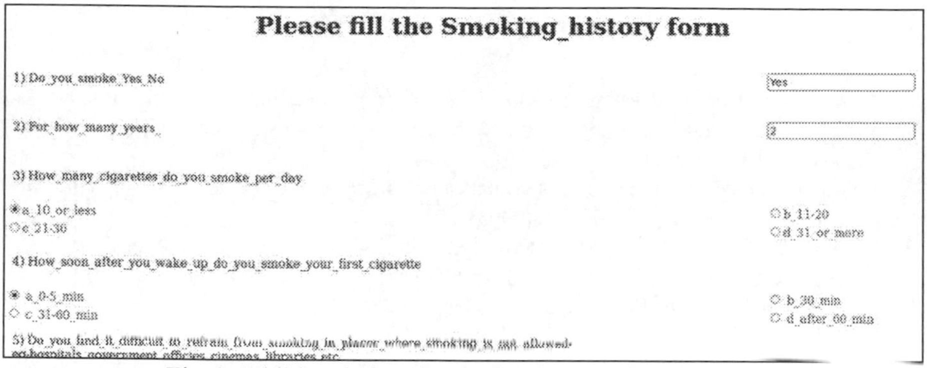

Fig.6. Additional Questions related to Smoking History

If the user input is *NO*, no additional questions related to Smoking will be displayed. Instead the system moves to the first question in the Alcohol history form.

Depending on the adaptive properties of the remaining questions, this process will continue until the end of questionnaire is reached.

6 Conclusion and Future

Patient history collection systems capture better information than face-to-face consultation. It also saves lot of time since the patients can enter their details according to their convenience. Our ultimate aim is to provide doctors, all relevant information related to a patient, so that they can perform a proper risk assessment. In a usual automated system, the number of questions asked to patients will be fixed. But here, since the questionnaire is adaptive, the number of questions asked to each specific patient varies. System has the intelligence to reduce the number of questions according to the user input. Ontology based systems can be extended and reused in a variety of problems in a similar domain. Any updates in the system can be applied directly by updating the questionnaire ontology without any software engineering work.

Generation of semantic patient profile and analysis and prediction of risk factors of diabetes are identified as the future work. Ontologies are semantic models for domains of reality. Ontology based reasoning makes a way to discover new knowledge, which can lead to new directions in research, particularly in medical domain.

Acknowledgement
This work is published as part of a project funded by The Research Council [TRC], Oman under Agreement No. ORG/ AOU/ ICT/ 11/ 015, Proposal No ORG/ICT/11/004 and Arab Open University, Oman Branch.

References

1. Ahmadian, L. Cornet, Ronald1; de Keizer, Nicolette F.1 Facilitating pre-operative assessment guidelines representation using SNOMED CT, Journal of Biomedical Informatics, Volume 43, Issue 6, pp. 883-890, (2010)
2. J. W. Bachman. The patient-computer interview: a neglected tool that can aid the clinician. Mayo Clinic Proceedings,78:67–78, (2003)
3. Sherimon P.C., Vinu P.V., Reshmy Krishnan, Development Phases of Ontology for an Intelligent Search System for Oman National Transport Company, In Proceeding(s) of the International Journal of Research and Reviews in Artificial Intelligence – IJRRAI, Vol 1, No.4, December, pp. 97-101, ISSN: 2046-5122, (2011)
4. Matt-Mouley Bouamrane, Alan Rector, Martin Hurrell, Ontology-Driven Adaptive Medical Information Collection System, Foundations of Intelligent Systems, Lecture Notes in Computer Science Volume 4994, pp 574-584, (2008)
5. Bouamrane M-mouley, Rector A, HurrellM (2008): Gathering Precise Patient Medical History with an Ontology-driven Adaptive Questionnaire. 2008:539-541.
6. Matt-Mouley Bouamrane, Alan Rector, Martin Hurrell, Development of an ontology for a preoperative risk assessment clinical decision support system. In Computer-Based Medical Systems, IEEE International Symposium, (2009)
7. Vinu P.V, Sherimon P.C., Reshmy Krishnan, Development Of Seafood Ontology For Semantically Enhanced Information Retrieval, International Journal of Computer Engineering and Technology, Volume 3, Issue 1, January- June, pp. 154-162 (2012)
8. Noy, N., McGuinness, D.: Ontology development 101: A guide to creating your first ontology, Technical Report SMI-2001-0880, Stanford Medical Informatics (SMI), Department of Medicine, Stanford University School of Medicine (2001)
9. Kamran Farooq, Amir Hussain, Stephen Leslie, Chris Eckl, Calum MacRae, Warner Slack, "An Ontology Driven and Bayesian Network Based Cardiovascular Decision Support Framework", Advances in Brain Inspired Cognitive Systems, Lecture Notes in Computer Science Volume 7366, 2012, pp 31-41.
10. Warner V. Slack, M.D., Phillip Hicks, Ph.D., Charles E. Reed et al. A computer Based Medical History System N Engl J Med; 274:194-198,(1966)
11. R.Subhashini, Dr. J. Akilandeswar, A Survey on Ontology Construction Methodologies, International Journal of Enterprise Computing and Business Systems, 2011
12. Sherimon P.C., Vinu P.V., Reshmy Krishnan, Youssef Takroni, Developing a Survey Questionnaire Ontology for the Decision Support System in the Domain of Hypertension, IEEE South East Conference, April 4-7, Florida, U.S. (2013)
13. Sherimon P.C., Vinu P.V., Reshmy Krishnan, Youssef Takroni, Ontology Based System Architecture to Predict the Risk of Hypertension in Related Diseases, Accepted in International Journal of Information Processing and Management (ISSN: 2093-4009), (2013)
14. http://protege.stanford.edu/plugins/owl/api/
15. Matt-Mouley Bouamrane,Alan Recter,Martin Hurrel(2008) :Using Ontologies for an intelligent Patient modeling ,Adaptation and management system:OTM 2008,part II,LNCS 5332,pp.1458-1470
16. S. Abidi, "Ontology-based knowledge modeling to provide decision support for comorbid diseases," Knowledge Representation for Health-Care, pp. 27-39, 2011.

Filtering of Unrelated Answers in a Cooperative Query Answering System

Maheen Bakhtyar[1,2], Lena Wiese[3], Katsumi Inoue[4], and Nam Dang[5]

[1] Asian Inst. of Technology Bangkok, Thailand Maheen.Bakhtyar@ait.asia
[2] CSIT, University of Balochistan, Pakistan MaheenBakhtyar@um.uob.edu.pk
[3] Inst. of CS, University of Göttingen, Germany wiese@cs.uni-goettingen.de
[4] National Inst. of Informatics, Tokyo, Japan ki@nii.ac.jp
[5] Tokyo Inst. of Technology, Tokyo, Japan namd@de.cs.titech.ac.jp

Abstract. A database system may not always return answers for a query. Such a query is called a *failing query*. Under normal circumstances, an empty answer would be returned in response to such queries. Cooperative query answering systems produce generalized and relevant answers when an exact answer does not exist, by enhancing the query scope and including a broader range of information. Such systems may apply various generalization techniques, also referred to as generalization operators, to relax certain conditions and obtain related answers. These answers are not exact but informative answers that potentially contain some of the information that the user needs. Therefore, we propose a method to filter out unrelated answers and return only related answers to the user. We also propose a mechanism to have a restricted and optimized generalized query space by limiting the number of queries produced. We determine the similarity between user query and the answer produced. Unrelated answers are pruned out, and only the related and informative answers are returned to the user.

Keywords: Cooperative Query Answering, Query Relaxation, Semantic Filtering, Unrelated Answers, WordNet, Inductive Conceptual Learning

1 Introduction

A query with no resulting answer is called a *failing query*. *Cooperative query answering systems* produce generalized and relevant results when an exact answer does not exist by enhancing the query scope and including a broader range of information. These systems apply various generalization techniques; also referred to as generalization operators; to relax certain conditions and to obtain related answers. These answers are not exact but informative that potentially contain partial information that users need.

Various generalization techniques have been developed to give a wider range of answers [1–5]. Inoue et al., discuss and analyze the properties of three of the generalization/relaxation techniques [6]. They also discuss the iterative combination of the three techniques (called operators).

T. Herawan et al. (eds.), *Proceedings of the First International Conference on Advanced Data and Information Engineering (DaEng-2013)*, Lecture Notes in Electrical Engineering 285, DOI: 10.1007/978-981-4585-18-7_52,
© Springer Science+Business Media Singapore 2014

We can observe that sometimes a number of answers produced after query generalization are not related to what the user originally asked. The reason is often that new structures are added to the query when generalizing the query. Therefore, we propose a framework to determine the similarity between the user's original query and the answers produced after query expansion. Unrelated answers can be pruned out so that only the related and informative answers are returned to the user.

2 Techniques for Query Generalization in Cooperative Query Answering Systems

Cooperative query answering systems generalize queries by enhancing the query scope and including a broader range of information. If the answer to a query is *null*, then cooperative query answering relaxes the query, employing various techniques. Relaxation of the query results in some informative answers instead of an empty set.

Deductive generalization of queries assists in providing informative answers to failing queries. Gaasterland et al., provide a formal definition of deductive generalization [5]. We consider conjunctive queries and the generalization operators for them. Conjunctive queries contain both positive and negative conjuncts, but for simplicity, we initially target only queries with positive conjuncts/literals. See [6] for a definition of a conjunctive query.

A generalization operator is a mechanism to generalize a query to enhance the query scope. When applied to a set of queries, it produces a set of relaxed and more general queries. See [6] for the formal definition. We consider three generalization operators for conjunctive queries and their iterative execution combined with each other, as discussed by Inoue et al., [6].

The **DC operator** relaxes a conjunctive query by dropping one of the conjuncts from the query at a time, making the query less restrictive.

The **AI operator** adds a new variable in the query and therefore introduces a more general query. This results in coverage of different values for newly added variable.

The **GR operator** allows replacing a sub-part of the failing query with the head of a rule in the knowledge base Σ (details in [6]). New constants and new conjuncts (but not variables) are potentially introduced in the generalized query, possibly removing some of the conjuncts, variables, and constants.

Inoue et al., state that it is sufficient to execute the three operators in a certain order when iteratively executing generalization steps. The authors apply the operators in breadth-first manner with the GR operator first, followed by DC and then AI. In some cases, GR is not applicable so we might apply DC then AI. When neither GR nor DC is applicable, we may apply only AI.

3   Similarity between User Query and the Returned Answer

We have discussed that the user may be disappointed if no answer is returned in response to a query. However, the user may be even more disappointed if unrelated answers are retrieved and returned in response. For example, if the query requires a *list of hotels in a particular town,* but the system provides a *list of hospitals in that town* after query generalization, the user would likely be unhappy, so such cases are best avoided. Similarity is a general notion of a metric based on the relatedness of two target inputs (e.g., concepts, terms, or documents). Similarity values are usually between 0 and 1, where 0 means not similar at all. We propose a method to find and remove unrelated answers from a set of generalized and expanded results.

In our approach, we make use of syntactic as well as semantic constructs to compare and match the original query with generalized queries or with answers obtained in response to those generalized queries. We make use of WordNet[6]; in particular, we use the similarity between various concepts in WordNet to understand semantics and prune out unrelated queries or answers. There are various methods for measuring similarity between a pair of words, several based on WordNet. We use the similarity metric by Wu and Palmer [9] to measure similarity between a pair of words. To better analyze, we express the operator tree by Inoue et al., in terms of the regular expression.

$$\underbrace{(AI)^+}_{I} \mid \underbrace{(DC)^+(AI)^*}_{II} \mid \underbrace{(GR)^+(DC)^*(AI)^*}_{III}.$$

We now examine each branch of the regular expression and develop similarity metrics for branches with expected dissimilar answers.

3.1   Anti-Instantiating Iteratively $(AI)^+$

The AI operator generates queries having new variables. Therefore, answers retrieved after introducing new variables according to a knowledge base may be unrelated to the original query. These answers might contain information entirely out of context for the query and the user's needs. To identify such situations, after AI execution, we match answers with the original query.

The AI operator and iterations of AI alone neither remove nor introduce any new predicate symbols. Therefore, the size (the sum of the arities of the predicates) of the query and the answer formulae should be equal, and it is possible to keep the answer predicates in the same order as in the query, with one-to-one matching of the query predicates and answer predicates. To measure

---

[6] WordNet[7, 8] is a lexical database and a useful tool for computational linguistics and natural language processing(NLP) that contains a formal hierarchical arrangement of English vocabulary. The concepts are interlinked on the basis of some relationship such as synonym sets(called synsets).

query-answer similarity for the AI^+ branch, we number the occurrences of variables and constants in a formula. Each such occurrence is called a position. For example, in the formula $\text{ill}(\text{mary}, X) \wedge \text{ill}(\text{peter}, X)$, mary occurs at position 1, X occurs at positions 2 and 4, and peter occurs at position 3. In AI^+, the number of predicates and the size of the parameter tuples in the query and the answer is equal. Similarity measurement functions are shown in Equations 1 to 6.

The overall similarity $SimQA_A(q, a)$ between the generalized query q and a candidate answer a is show in Equation 1.

$$SimQA_A(q,a) = \frac{\sum_{i=1}^{m} Sim_p(i,q,a)}{m}, \quad (1)$$

$$Sim_p(i,q,a) = \begin{cases} 1, & V(p_i(q)) \wedge \neg A(p_i(q)). \\ Sim_{wn}^h(i,q,a), & V(p_i(q)) \wedge A(p_i(q)) \\ & \text{producing } p_i(a) \wedge \neg PN(p_i(a)). \\ Sim_{pn}^h(i,q,a), & V(p_i(q)) \wedge A(p_i(q)) \\ & \text{producing } p_i(a) \wedge PN(p_i(a)). \\ 1, & C(p_i(q)) \wedge p_i(q) = p_i(a). \\ 0.5, & PN(p_i(q)) \wedge p_i(q) \neq p_i(a). \\ Sim_c(p_i(q), p_i(a)), & \text{otherwise.} \end{cases} \quad (2)$$

where, $Sim_p(i, q, a)$ is the similarity of the parameter (variables or constants) at position i and m is the total number of positions (sum of arities) in the query q (or the answer). Let $p_i(q)$ be the parameter, i.e., variable or constant, at the ith position in the query (and for a analogously). We must calculate the similarity $SimQA_A(q, a)$ for each answer $a \in A'$ with the original query q. The similarity value can be used to rank answers obtained in one AI step in the tree. Then irrelevant answers can be filtered based on a threshold. $Sim_p(i, q, a)$ is the similarity of the parameter at position i in the query and the answer formula. A description of each of the Boolean functions in Equation 2 is provided in Table 1.

Table 1. Boolean Functions used in Equation 2.

Function	Return Value
$V(arg)$	$True$ if arg is a variable, $False$ otherwise.
$A(arg)$	$True$ if arg is anti-instantiated, $False$ otherwise.
$PN(arg)$	$True$ if arg is a proper noun, $False$ otherwise.
$C(arg)$	$True$ if arg is a constant other than a proper noun, $False$ otherwise.

$Sim_{wn}^h(i, q, a)$ is the similarity of the symbol at an answer position that is not a proper noun having the same variable in the corresponding positions of the query. The similarity between two concepts is measured according to WordNet (as discussed earlier). $Sim_{wn}^h(i, q, a)$ works by iterating through all the positions in the answer and calculating the similarities (using WordNet) at the appropriate positions. We refer to this similarity as *horizontal* similarity, because the AI operation breaks bonds inside the query. For example, after AI, $\text{ill}(X, \text{asthma}) \wedge \text{allergic}(X, \text{inhaler})$ may become $\text{ill}(X, \text{asthma}) \wedge$

allergic(Y, inhaler); hence the bond created by common variable X is broken and the similarity needs to be checked after the answer is obtained. Similarly, $Sim_{pn}^h(i, q, a)$ is the similarity of the answer positions that are proper nouns having the same variable in the corresponding positions of the query. We assign a moderate similarity of 0.5 in the case of proper nouns, because we neither want to completely suppress the significance of proper nouns nor to completely ignore the notion of generalization. $Sim_c(r, s)$ is either the semantic similarity between the constants or becomes undefined. A query-answer pair is rejected or pruned out if the constants' similarity value is below a certain threshold T_c (0.5 in our case). This way we avoid having a huge cross product of two queries (leading to combinatory explosion) hence reducing and restricting the query space.

$$Sim_c(r, s) =$$

$$\begin{cases} Sim_{wn}(r, s), & \text{if } Sim_{wn}(r, s) >= T_c \\ \perp, & \text{otherwise.} \end{cases}$$

(3)

$$Sim_{wn}^h(i, q, a) = \frac{\sum\limits_{\substack{j=1,\ i \neq j \\ p_j(q) = p_i(q)}}^{m} Sim_{wn}(p_i(a), p_j(a))}{O\Big(p_i(q)\Big) - 1}$$

(4)

$$Sim_{pn}^h(i, q, a) -$$

$$\frac{\sum\limits_{\substack{j=1,\ i \neq j \\ p_j(q) = p_i(q)}}^{m} Sim_{pn}(p_i(a), p_j(a))}{O\Big(p_i(q)\Big) - 1}$$

(5)

$$Sim_{pn}(r, s) = \begin{cases} 1, & \text{if } r = s, \\ 0.5, & \text{if } r \neq s, \end{cases}$$

(6)

where r and s are two proper nouns.

For further optimization, we may get user feedback to decide if some variable is important and need not be anti-instantiated. This would reduce the query space by limiting the size of the cross product of two sub-queries.

Example 1 shows how the described functions can be used to measure similarity and how answers irrelevant to the original query can be filtered.

Example 1. $(AI)^+$.
Failing Query: $q = ill(X, \text{asthma}) \wedge ill(X, \text{fever}) \wedge \text{allergic}(X, \text{inhaler})$
Now, we analyze some answers obtained using SOLAR [10], in response to a few generalized queries.
Generalized Query: $q' = ill(X, \text{asthma}) \wedge ill(X, \text{fever}) \wedge \text{allergic}(\mathbf{V'}, \text{inhaler})$
Generalized Answer:

$a' = ill(\underbrace{\text{lisa}}_{1}, \underbrace{\text{asthma}}_{1}) \wedge ill(\underbrace{\text{lisa}}_{1}, \underbrace{\text{fever}}_{1}) \wedge \text{allergic}(\underbrace{\text{john}}_{\frac{0.5+0.5}{2}}, \underbrace{\text{inhaler}}_{1})[SimQA_A = .9]$

Explanation: The query is relaxed by replacing the third occurrence of X with a new variable V' as shown above. Since the overall similarity is still quite high, the answer may be related to user needs.
Generalized Query: $q'' = ill(X, \text{asthma}) \wedge ill(X, \text{fever}) \wedge \text{allergic}(V', \mathbf{V''})$
Generalized Answer:

$a'' = ill(\underbrace{\text{lisa}}_{1}, \underbrace{\text{asthma}}_{1}) \wedge ill(\underbrace{\text{lisa}}_{1}, \underbrace{\text{fever}}_{1}) \wedge \text{allergic}(\underbrace{\text{tonny}}_{\frac{0.5+0.5}{2}}, \underbrace{\text{fruit}}_{0.5})[SimQA_A = \mathbf{0.83}]$

Explanation: The relaxed query above is generated by replacing the constant `inhaler` with a new variable V''. This newly generated query results in a new constant `fruit` in the answer a''. The similarity between the concepts `inhaler` and `fruit` is calculated using the similarity algorithm [9] based on WordNet (the similarity is 0.5.)

Generalized Query: $q'''' = \text{ill}(X, \text{asthma}) \wedge \text{ill}(\mathbf{V'''}, \mathbf{V'''}) \wedge \text{allergic}(V', V'')$

Generalized Answer:

$$a'''' = \text{ill}(\underset{1}{\underbrace{\text{lisa}}}, \underset{1}{\underbrace{\text{asthma}}}) \wedge \text{ill}(\underset{\frac{0.5+0.5}{2}}{\underbrace{\text{peter}}}, \underset{0.3}{\underbrace{\text{bipolar\_disorder}}}) \wedge \text{allergic}(\underset{\frac{0.5+0.5}{2}}{\underbrace{\text{tonny}}}, \underset{0.26}{\underbrace{\text{sunlight}}})$$

$$[SimQA_A = \mathbf{0.59}]$$

Explanation: The similarity based on WordNet is calculated for positions 4 and 6 as described in case a''. We reject this answer based on Equation 3, as `sunlight` is not related to `asthma`. This query space reduction can be implemented beforehand during the generalization phase by only substituting constants that are related. Similarly, the value is calculated for position 3 as described previously in a'.

3.2 Zero or More AI Iterations after Iterative DCs $(DC)^+(AI)^*$

This branch can be further divided into the following sub-branches:

$(DC)^+$ (Dropping conditions iteratively) : Dropping a condition does not introduce any new variable or conjunct while generalizing a query. Therefore, no new constants/answers will be introduced. Only informative answers related to some cropped part of the original query are returned. However, the information lost with each generalization step depends on the constants and variables being dropped with the dropped literal. A function returning the similarity depending on each generalization step is given in Equation 7.

$$SimQA_D(q, a) = \frac{m'}{m} \tag{7}$$

where q is the original user query, a is one of the answers produced against some generalized query, m is the sum of arities of the predicates in the query, and m' is the sum of arities of the predicates in the answer (or in the generalized query). Example 2 shows how similarity is calculated when literals are dropped in generalization steps. We also suggest to take optional feedback from the user if one predicate is more important than the others. Then we can decide which literal to drop first. This optimization will help drop literals efficiently.

Example 2. $(DC)^+$.
Failing Query: $q = \text{patient}(X, \text{tokyo}, Y) \wedge \text{ill}(X, \text{asthma}) \wedge \text{allergic}(X, \text{inhaler})$
The generalized queries, their respective answers (obtained using SOLAR [10]), and the similarity values are:
Generalized Query: $q_1' = \text{ill}(X, \text{asthma}) \wedge \text{allergic}(X, \text{inhaler})$

Generalized Answer: $a_1' = \text{ill(peter, asthma)} \wedge \text{allergic(peter, inhaler)}$
$[\frac{4}{7} = \mathbf{0.57}]$
Explanation: Three positions dropped; hence, more information loss

We can see from the similarity values as well as the answers obtained that the similarity decreases with each step of generalization. We also notice that the similarity can be calculated before the actual answer is extracted in case of $(DC)^+$ because in this case, the similarity is not obtained on the basis of semantics but only considering the syntax of the query. This enables supervised and controlled answer generation. Supervised and controlled query generalization improves the efficiency of the system by omitting some queries for which an answer need not be calculated.

$(DC)^+(AI)^+$(Iterative dropping conditions followed by iterative AI) :
We already discussed that the DC operator alone does not introduce any new variables. However, if AI is applied after DC, then we do expect new variables in the generalized query. Therefore, we may obtain very dissimilar answers. DC will always reduce the size of the query, therefore reducing the answer size. In this case, one-to-one matching with the original query is not possible. Additionally, iterations over DC operator followed by AI increases the possibility of having queries with more dissimilar answers.

We propose a mechanism to first determine the amount of information retained after iteration over DC operations and then find out how similar the anti-instantiated part is when some information has already been cropped out during the DC iterations. A function returns the similarity between query and the answer is shown in Equation 8.

$$SimQA_{DA}(q,a) = SimQA_A(q^{DC^+}, a^{DC^+AI^+}) \times \frac{m}{m'}, \qquad (8)$$

where $SimQA_A$ is passed the query which is the result of the DC operator, along with the final answer obtained. The cropped query after DC is treated as the original query so that one-to-one matching is possible. We multiply this factor with the information retained after applying the DC operator. m is total number of positions in the original query, and m' is the number of positions remaining after DC operations have been carried out. This function first determines the information retained in the query after applying iterations of the DC operator and then finds how similar the retained information is to the query. Example 3 uses this function to find the similarity between the query and the answer with a single AI operation applied after multiple DC operations.

Example 3. Multiple DC followed by AI.
Failing Query: $q = \text{patient}(X, \text{tokyo}, Y) \wedge \text{ill}(X, \text{fever}) \wedge \text{allergic}(X, \text{inhaler})$
Explanation: $q^{DC.DC.AI} = \text{allergic}(X, Y)$ is the relaxed query produced after applying DC twice followed by single AI operation. Here, $\frac{m}{m'} = 0.28$ (Information Retained) and $a^{DC.DC.AI} = \text{allergic(peter, bronchodilator)}$ is the generalized answer.

Similarity: $SimQA_{DA}(q,a) = SimQA_A(q^{DC.DC}, a^{DC.DC.AI})$ of $0.28 = (1 + 0.3)/2$ of $0.28 = 0.18$

3.3 DC followed by AI after Iterative Goal Replacement $(GR)^+(DC)^*(AI)^*$

This branch can be further divided into sub-branches but we only focus on one of the branches to start with.

$(GR)^+$ **(Iterative goal replacement)**: Execution of the GR operator potentially adds new conjuncts, new variables, and new constants to create generalized queries. It replaces a sub-part of the failing query with the head of a matching *single-headed range-restricted (SHRR) rule* in the knowledge base (Σ). The answers returned against these queries might be extremely dissimilar.

Generalized queries in this case consist of two parts, a replaced part (we call it the body of the rule B) and an actual existing/preserved part E, which is not replaced. Logically speaking, the replaced part of the query should not semantically affect similarity directly because it is database dependent, reflecting how the knowledge base is defined. On the other hand, we also notice that some variables or constants may be dropped or new ones may be introduced; therefore, we have to analyze accordingly.

We realize that the newly introduced constants, variables, and conjuncts are database-dependent and are placed as a rule in the knowledge base; therefore, we do not consider them as irrelevant or dissimilar to the original query construct. For the same reason, we do not compare the body with the head of the rule so that the sense of generalization is retained. However, we do consider the constants and variables missing in the generalized queries because that is the information lost during generalization steps. We consider the inter-relationship R_{BE} between the body B and existing part E and then relate it with the inter-relationship R_{HE} between the head of rule H and the existing part E. Finding and analyzing the relevance R_{BE} and R_{HE} shows how much information has been lost during the relaxation process, and we also see how the bond between the body replaced and the existing part in the query is broken when the repeating variables or repeating constants are removed from the query.

We present a supervised and controlled answer generation mechanism for GR. We first assign a weight to each repeating variable and constant in E and B. By repeating variable (or constant), we mean a variable (or constant) that is present in both E and B. These variables and constants represent the bond or link between the body and the existing part, and we need to analyze the effects of breaking these links on query answer similarity. The weight for each repeating variable/constant is calculated using Equation 9.

$$w_e = \frac{O(e)}{m}, \quad (9) \qquad w^t = \sum (O(e^t) \times w_e), \quad (10) \qquad SimQA_G(q,a) = \frac{w^{q'}}{w^q} \quad (11)$$

where w_e is the weight for variable (or constant) e in E and B. $O(e)$ is the total number of occurrences of e, and m is the total number of positions in the original query. Once the weight for each repeating variable (or constant) is calculated, we find the total weight of the orignal query as well as that of the generalized query by Equation 10, where $t \in \{q, q'\}$ indexes the weight based on the number of repeating variables or constants in the original or relaxed query. The total similarity is calculated using Equation 11.

We know that w^q and $w^{q'}$ are the total weights of the repeating arguments in the original query and generalized query, respectively. w^q is the actual weight carried inside the query, whereas $w^{q'}$ is the weight retained after generalization. Therefore, the retained weight is calculated in Equation 11.

Example 4. $(GR)^+$.
Failing Query:

$$q = \overbrace{\texttt{ill}(X, \texttt{asthma}) \wedge \texttt{allergic}(X, \texttt{inhaler})}^{B} \wedge \overbrace{\texttt{gender}(X, \texttt{male}) \wedge \texttt{history}(X, \texttt{asthma})}^{E}$$

Weight of each Repeating Variable/Constt.: $w_X = \frac{4}{8} = .5, w_{\texttt{asthma}} = \frac{2}{8} = .25$
Generalized Query:
$$q' = \underbrace{\texttt{treat}(X, \texttt{injection})}_{H} \wedge \underbrace{\texttt{gender}(X, \texttt{male}) \wedge \texttt{history}(X, \texttt{asthma})}_{E}$$
SHRR rule: $\texttt{ill}(X, Y) \wedge \texttt{allergic}(X, Z) \rightarrow \texttt{treat}(X, \texttt{injection})$
Substitution: $\theta = \{\texttt{asthma}/Y, \texttt{inhaler}/Z, X/X\}$
Total Weight: $w^q = 4 \times 0.5 + 2 \times 0.25 = 2.5$
$w^{q'} = 3 \times 0.5 + 1 \times 0.25 = 1.75$
Total Similarity: $SimQA_G(q, a) = \frac{1.75}{2.5} = 0.7$

We have discussed and proposed a similarity metric for the similarity between a user query and generalized informative answers produced by a cooperative query answering system SOLAR [10]. We explained our similarity metrics for all three operators executed iteratively (i.e., DC, AI and GR) and the execution of AI followed by DC. We do not discuss the case of combination with GR yet in detail. DC applied after GR will keep the same variables, conjuncts, and constants as in the initial query, with new ones added by the GR operator. In iterative GR followed by iterative AI, the AI operator adds more variables by replacing constants and some variables. Therefore, similarity must be calculated for the answers generated against this case. We conclude that we need to check the similarity between the query and the answer only in case of AI and GR operators. Whenever these operators are applied in any iteration, they might introduce new variables or conjuncts and may result in unrelated answers. Once the similarity between the original query and the obtained answer is calculated, we filter out the answers with low similarity. A trial-and-error approach is required to define a threshold for deciding whether the answer is deemed related or not.

4 Conclusion, Limitation, and Future Work

We propose an approach to filter out unrelated answers in a cooperative query answering system and to return relevant answers to the user. We provide a similarity metric for all combinations of operators executed iteratively, except the GR operator combined with DC and AI. In the future, we intend to extend our approach and develop a similarity function for the remaining cases that are not covered yet. We also plan to evaluate our complete approach on a benchmark dataset.

5 Acknowledgements

We are thankful to Dr. Matthew N. Dailey for all his suggestions and help. We also thank Inoue lab, NII, Tokyo for the support.

References

1. Chu, W.W., Yang, H., Chiang, K., Minock, M., Chow, G., Larson, C.: Cobase: A scalable and extensible cooperative information system. Journal of Intelligent Information Systems **6** (1996) 223–259 10.1007/BF00122129.
2. Halder, R., Cortesi, A.: Cooperative query answering by abstract interpretation. In: Proceedings of the 37th international conference on Current trends in theory and practice of computer science. SOFSEM'11, Berlin, Heidelberg, Springer-Verlag (2011) 284–296
3. Shin, M.K., Huh, S.Y., Lee, W.: Providing ranked cooperative query answers using the metricized knowledge abstraction hierarchy. Expert Systems with Applications **32**(2) (2007) 469–484
4. Pivert, O., Jaudoin, H., Brando, C., HadjAli, A.: A method based on query caching and predicate substitution for the treatment of failing database queries. In: IC-CBR'10. LNCS, Springer (2010) 436–450
5. Gaasterland, T., Godfrey, P., Minker, J.: Relaxation as a platform for cooperative answering. Journal of Intelligent Information Systems **1**(3) (December 1992) 293–321
6. Inoue, K., Wiese, L.: Generalizing conjunctive queries for informative answers. In: Proceedings of the 9th International Conference on Flexible Query Answering Systems. Lecture Notes in Artificial Intelligence, Springer-Verlag (2011)
7. Miller, G.A.: Wordnet: A lexical database for english. Communications of the ACM **38** (1995) 39–41
8. Fellbaum, C., ed.: WordNet: An Electronic Lexical Database (Language, Speech, and Communication). The MIT Press (1998)
9. Wu, Z., Palmer, M.: Verbs semantics and lexical selection. In: Proceedings of the 32nd annual meeting on Association for Computational Linguistics. ACL '94, Stroudsburg, PA, USA, Association for Computational Linguistics (1994) 133–138
10. Nabeshima, H., Iwanuma, K., Inoue, K., Ray, O.: Solar: An automated deduction system for consequence finding. AI Commun. **23** (April 2010) 183–203

Measuring eGovernment Systems Success: An Empirical Study

Mohamed E. Edrees, Amjad Mahmood

Computer Science Department, Information Technology Collage
University of Bahrain, Bahrain
{mezzudien, amahmood}@uob.edu.bh

Abstract. eGovernment is connecting the government with citizens, businesses, and other stakeholders, via computers and internet. It has received fairly extensive attention in recent years. Many governments have realized the importance of using information and communication technologies to provide efficient and transparent government. It is therefore important to measure the success of government-to-citizen (G2C) eGovernment systems from the citizens' perspective. This study provides an empirical test of an adaptation of DeLone and McLean's IS success model in the context of G2C eGovernment. Data collected by questionnaire from 149 users of eGovernment gateway in Bahrain were analyzed. All hypothesized relationships between the six success variables are significantly or marginally supported. This paper concludes by discussing limitations, which should be addressed in future research.

Keywords: Electronic (e) Government, eGovernment system success, Systems success, Success/Impact Measurement, Delone and McLean

1   Introduction

Electronic government (eGovernment), in general, is connecting the government with citizens, businesses, and other stakeholders, via computers and internet. According to Organization for Economic Cooperation and Development, eGovernment is the use of information and communications technology, and particularly the internet, as a tool to achieve better government (Rose & Grant, 2010). Wang and Liao (2008) defined eGovernment as it is the government's use of ICT, particularly Web-based Internet applications, to enhance the access to and delivery of government information and services to citizens, business partners, employees, and other agencies and entities.

Electronic government is no longer just an option but a necessity for countries aiming for better governance (Gupta & Jana, 2003). New information and communication technologies offer the governments with new possibilities for providing citizens and businesses with better, more efficient services (Verdegem & Verleye, 2009). It allows interaction without the limitations of time and space that office hours and buildings impose, resulting in 24/7/365 access to, and potential e-democratic involvement with, government (Rose & Grant, 2010).

T. Herawan et al. (eds.), *Proceedings of the First International Conference on Advanced Data and Information Engineering (DaEng-2013)*, Lecture Notes in Electrical Engineering 285, DOI: 10.1007/978-981-4585-18-7_53,
© Springer Science+Business Media Singapore 2014

In general, eGovernment has received extensive attention in recent years. Many governments have realized the importance of using information and communication technologies to provide efficient and transparent government (Wang & Liao, 2008). Since late 1990s, governments at all levels have launched electronic government projects aimed at providing electronic information and services to citizens and businesses (Obi, 2009, Torres, Pina, & Acerete, 2005).

Wang and Liao (2008) noted that measuring and evaluating eGovernment progress and success have become a priority for decision makers in all countries and they emphasized on the need for evaluation efforts that assess the effectiveness of the eGovernment systems. Therefore, they empirically tested the adaption of DeLone and McLean's (2003) IS success model in the context of G2C eGovernment in Taiwan and they found that the data supported the hypnotized model. However, their empirical results need to be verified in different user populations and different eGovernment contexts and more empirical tests should be conducted in order to generalize the model validation in the context of eGovernment. Therefore, this research aims to revalidate a multidimensional G2C eGovernment systems success model based on the DeLone and McLean (2003) IS success model.

This paper is structured as follows. Section 2 reviews information systems success model. Section3 presents our research model and hypotheses. Section 4, section 5 and section 6 present and discuss methods, measures, and results of the study respectively. Finally, theoretical and managerial implications and directions for future research are discussed in section 7.

2   IS Success Models

DeLone and McLean (1992) reviewed the existing definitions of information systems success and their corresponding measures to provide a general and comprehensive definition of information systems success that covers different evaluation perspectives. They introduced their first IS success model that classifies information systems success and their corresponding measures into six major categories and described the interrelationships between them (see Fig. 1). Since then, empirical investigations of the multidimensional relationships among the measures of IS success were done by a number of studies. Empirical testing and validating of the D&M IS Success Model was the primary purpose of two research studies (DeLone & McLean, 2003).

Partial test of the DeLone and McLean (1992) model using a structural equation modelling was conducted by Seddon and kiew (1996). According to them, considerable support for the DeLone and McLean (1992) model was provided by the results. A goodness-of-fit test on the entire DeLone and McLean (1992) was performed by Rai, Lang, and Welker (2002) and they found that some goodness-of-fit indicators were significant but others were not. However, all of the path coefficients among success dimensions of the DeLone and McLean (1992) model were found to be significant (as cited, DeLone & McLean, 2003).

All researches that applied, validated, challenged and proposed enhancements to their original model and evaluated its usefulness were discussed by DeLone &

McLean in 2003. Based on the evaluation of those contributions, an updated information systems success model was proposed with minor refinements to the original model (see Fig. 2).

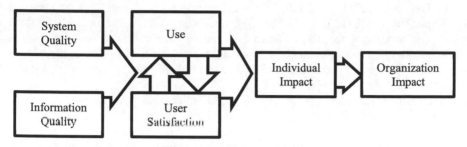

Fig. 1. DeLone and McLean's (1992) model

Within the G2C eGovernment context, citizens use an internet-based application to search information and conduct transactions. This Internet-based application is an IS phenomenon that lends itself to be studied using the updated IS success model. DeLone and McLean (2003) also suggest that further development, challenge, and validation of their model are needed. Thus, the updated IS success model can be adapted to the system success measurement in the G2C eGovernment context (Wang & Liao, 2008).

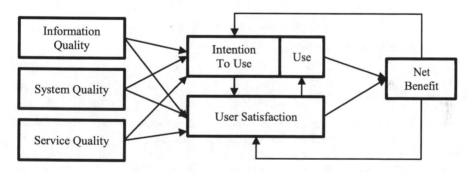

Fig. 2. Updated DeLone and McLean's (2003) model

3   Research model and hypotheses

This study adapted the research model proposed by Wang and Liao (2008) based on Delone and McLean (2003) information systems success model, which suggests that information quality, system quality, service quality, use, user satisfaction, and perceived net benefit are success variables in eGovernment systems. The model uses the term "use" instead of both "intention to use" and "use". Different stakeholders may have different opinions as to what constitutes a benefit to them. According to that, net benefits in this study refer to citizen-perceived net benefits as this study is focusing on measurement of eGovernment systems success from the perspective of the citizens' point of view. Moreover, as suggested by Wang and Liao (2008), to avoid model complexity and to reflect the cross-sectional nature of this study, the feedback links from net benefit to both use and user satisfaction were excluded.

As DeLone and McLean (2003) note, IS success is a multidimensional and interdependent construct and it is therefore necessary to study the interrelationships among, or to control for, those dimensions. Also, they conclude that the success model needs further development and validation before it could serve as a basis for the selection of appropriate IS measures. Thus, the following hypotheses were tested:

H1. Information quality will positively affect use in the eGovernment context.
H2. System quality will positively affect use in the eGovernment context.
H3. Service quality will positively affect use in the eGovernment context.
H4. Information quality will positively affect user satisfaction in the eGovernment context.
H5. System quality will positively affect user satisfaction in the eGovernment context.
H6. Service quality will positively affect user satisfaction in the eGovernment context.
H7. Use will positively affect user satisfaction in the eGovernment context.
H8. Use will positively affect perceived net benefit in the eGovernment context.
H9. User satisfaction will positively affect perceived net benefit in the eGovernment context.

4   Research methodology

4.1   Measures and Generation of Scale Items

The main purpose of this study is to revalidate a multidimensional G2C eGovernment systems success model based on the DeLone and McLean (2003) IS success model. This research is not concerning to develop new scale. Thus, the scale used by Wang and Liao (2008) was adapted in this research to ensure content validity of the scale. Likert scale (1-5) ranging from "Strongly disagree" to "Strongly agree" was used.

4.2   Sample and Procedure

The data used to test the research model were obtained from a sample of experienced users of Bahrain G2C eGovernment portal as Bahrain eGovernment authority provides citizens with single portal for all eGovernment services (www.bahrain.bh). Therefore, respondents were asked whether they had ever used Bahrain eGovernment portal, and if they replied in the affirmative, they were asked to participate in the survey. The questionnaire requested the respondents to relate their experience of using the eGovernment portal and to answer the questions accordingly. A total of 149 responses were obtained. Detailed descriptive statistics relating to the respondents' characteristics are shown in Table 1.

Table 1. Characteristics of the respondents

	Characteristic	Number	Percentage
Gender	Male	75	50.3
	Female	74	49.7
Age	< 20	12	8.1
	21 - 30	51	34.2
	31 - 40	38	25.5
	41 - 50	32	21.5
	> 51	16	10.7
Degree	High school	6	4.0
	Undergraduate	31	20.8
	Graduate	112	75.2
Industry	Student	11	7.4
	Manufacturing	8	5.4
	Service	23	15.4
	Government Agencies	84	56.4
	Education and Research	23	15.4

5 Results

In order to test the research model, two types of statistical analysis were applied using Statistical Package for Social Science (SPSS) version 20.0. Pearson correlation analysis was conducted to examine the relationship between the model factors. All relationship between model factors were found statistically significant at $p<0.01$. Moreover, all correlation values calculated exceeded 0.5, which were considered strong relationship. Table 2 summarizes the correlation values for the six factors.

Table 2. Pearson correlation results

Factor	USE	USERS	Perceived Net Benefit (NB)
Information Quality (IQ)	0.695**	0.744**	
System Quality (SQ)	0.694**	0.660**	
Service Quality (SERQ)	0.780**	0.755**	
USE		0.798**	0.724**
User Satisfaction (USERS)			0.701**

**Correlation is significant at the 0.01 level (2-tailed).

To examine hypothesized causal paths, multiple regressions analysis was conducted. Properties of the causal paths, including standardized path coefficients, p-values, and variance explained for each equation in the hypothesized model, are presented in Fig. 3. As hypothesized, both the information quality and service quality had a significant influence on both use and user satisfaction. As well as the influence of system quality were founded significant. Thus, H1, H4, H3, H6 and H2 were supported ($\beta=0.19$, $\beta=0.23$, $\beta=0.37$, $\beta=0.41$ and, $\beta=0.49$, respectively). The influences of system quality on use were not significant at $p<0.05$, but significant at $p<0.1$. Thus,

H5 were marginally supported (β=0.13). Consequently, system quality exhibited a stronger effect than information quality and service quality on use. However, the stronger effect on user satisfaction exhibited by service quality. In addition, use had a significant influence on both user satisfaction and perceived net benefit. H7 and H8 were supported (β=0.80 and β=0.45, respectively). Finally, user satisfaction appeared to be a significant determinant of perceived net benefit. H9 was supported (β=0.34).

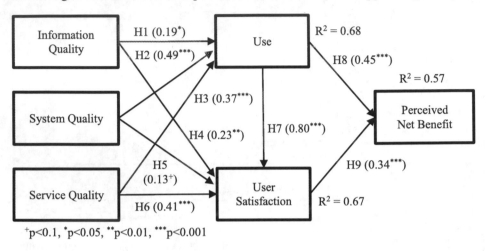

$^+$p<0.1, *p<0.05, **p<0.01, ***p<0.001

Fig. 3. Hypotheses testing results

Overall, this model accounted for 57% of the variance in perceived net benefit, with use exerting a stronger direct effect than user satisfaction on perceived net benefit. Sixty-seven percent of the variance in user satisfaction was explained by information quality, system quality, service quality, and use, while 68% of the variance in use was explained by information quality, system quality, and service quality. The direct and total effect of user satisfaction on perceived net benefit was 0.34. However, the direct and total effects of use on perceived net benefit were 0.45 and 0.72, respectively. Thus, use exhibited stronger direct and total effects on perceived net benefit than user satisfaction. Among the three quality-related constructs, both service quality and system quality had convergent total effect on perceived net benefit. Anyhow, services quality gained the strongest total effect on perceived net benefit. The direct, indirect, and total effects of information quality, system quality, service quality, use, and user satisfaction on perceived net benefit are summarized in Table 3.

6   Discussion

This study revalidated a multidimensional G2C eGovernment systems success model based on the DeLone and McLean (2003) IS success model, which captures the multidimensional and interdependent nature of G2C eGovernment systems success proposed by Wang and Liao (2008). The results supported their findings and indicated that information quality, system quality, service quality, use, user satisfaction, and perceived net benefit are valid measures of eGovernment system success. Moreover, all hypothesized relationships between the six success variables were significantly or marginally supported by the collected data. Except the link from system quality to use, our research results support the results obtained by Wang and Liao. However, this research found that hypothesized relationship between system quality and use significantly supported, while it was rejected by Wang and Liao results.

Table 3. The direct, indirect and total effect of dominants on perceived net benefit

	Direct effect			indirect effect			Total effect		
	USE	USERS	NB	USE	USERS	NB	USE	USERS	NB
IQ	0.19	0.23			0.15	0.22	0.19	0.38	0.22
SQ	0.49	0.13			0.39	0.40	0.49	0.52	0.40
SERQ	0.37	0.41			0.29	0.41	0.37	0.7	0.41
USE		0.80	0.45			0.27		0.8	0.72
USERS			0.34						0.34

Wang and Liao (2008) concluded with the importance of the effect of information quality on both use and user satisfaction which yield to the highest effects on perceived net benefit. The results of this research differ from that. The effect of system quality gained the highest influence on use and the effect of service quality gained the highest influence on user satisfaction. For the total effect on perceived net benefit, both system quality and services quality were gained convergent influence but the services quality gained the strongest effect. According to the model, in order to increase citizen-perceived net benefit, eGovernment authorities need to develop G2C eGovernment systems with good information quality, system quality, and service quality, which, in turn, will influence citizen system usage behavior and satisfaction evaluation, and the corresponding perceived net benefit. The results support Wang and Liao (2008) finding that the effect of system use found to have the strongest direct and total effect on perceived net benefit, indicating the importance of system use in promoting citizen-perceived net benefit. However, as they argued, simply saying that increased use will yield more benefits, without considering the nature of this use, is insufficient. Therefore, system use is a necessary condition of yielding benefits to citizens.

The results of this research emphasize on the importance of assuming a multidimensional, interdependent analytical approach in measuring the success of eGovernment systems and encourage eGovernment authorities to include measures for information quality, system quality, service quality, system use, user satisfaction, and perceived net benefit in their evaluation effort of eGovernment system success.

7 Conclusion

This research was conducted to revalidate a multidimensional G2C eGovernment systems success model based on the DeLone and McLean (2003) IS success model proposed by Wang and Liao (2008), which considers six success measures that are information quality, system quality, service quality, use, user satisfaction, and perceived net benefit. The findings of this study partially support Wang and Liao results. All hypothesized relationships between the six success variables were significantly or marginally supported by the data while rejected one of the relations. Wang and Liao emphasizes on the importance of the effect of information quality on both use and user satisfaction which yield to the highest effects on perceived net benefit while the results of this research show that the effect of both system quality and service quality are more important. Both of the results emphasize on the importance of assuming a multidimensional, interdependent analytical approach in measuring the success of eGovernment systems.

The results of this study were limited to a particular eGovernment system (Bahrain.bh) and targeted a specific citizen group in Bahrain. Thus, caution needs to be taken when generalizing these findings and in discussion of other eGovernment categories or user groups. Therefore, future research should be conducted to challenge and revalidate the model in different user populations and different eGovernment contexts, especially in government-to-business (G2B) and government-to-government (G2G) contexts and to enhance our understanding of the nature of eGovernment system success.

8   References

1. DeLone, W. H., & McLean, E. R. (1992). Information systems success: The quest for the dependent variable. Information Systems Research, 3(1), 60-95.
2. DeLone, W. H., & McLean, E. R. (2003). The DeLone and McLean Model of Information Systems Success: A Ten-Year Update. Journal of Management Information Systems, 9–30.
3. Gupta, M. P., & Jana, D. (2003). E-government evaluation: A framework and case study. Government Information Quarterly, 20, 365–387.
4. Irani, Z., Love, P. E., & Jones, S. (2008). Learning lessons from evaluating eGovernment: Reflective case experiences that support transformational government. Journal of Strategic Information Systems, 17, 155–164.
5. Isaac, W. C. (2007). Performance Measurement for the e-Government Initiatives: A Comparative Study. Dissertation Abstracts International.(UMI No: 3283471).
6. Obi, M. C. (2009). Development and Validation of a Scale for Measuring e-Government User Satisfaction. Dissertation Abstracts International.(UMI No. 3379230).
7. Rai, A., Lang, S. S., & Welker, R. B. (2002). Assessing the Validity of IS Success Models: An Empirical Test and Theoretical Analysis. Information Systems Research, 13(1), 50-69.
8. Rose, W. R., & Grant, G. G. (2010). Critical issues pertaining to the planning and implementation of E-Government initiatives. Government Information Quarterly, 26–33.
9. Seddon, P. B., & kiew, M. Y. (1996). A partial test and development of the DeLone and McLean model of IS success. Australasian Journal of Information Systems, 4, 90-109.
10. Torres, L., Pina, V., & Acerete, B. (2005). E-government developments on delivering public services among EU cities. Government Information Quarterly, 22(2), 217–238.
11. Verdegem, P., & Verleye, G. (2009). User-centered E-Government in practice: A comprehensive model for measuring user satisfaction. Government Information Quarterly, 26, 487–497.
12. Wang, Y.-S., & Liao, Y.-W. (2008). Assessing eGovernment systems success: A validation of the DeLone and McLean model of information systems success. Government Information Quarterly, 25, 717–733.

The Comparison of CRM Model : a Baseline to Create Enterprise Architecture for Social CRM

Meyliana[1] and Budiardjo, Eko K[2].

[1]School of Information System, Bina Nusantara University, Jakarta, Indonesia
meyliana@binus.edu
[2]Faculty of Computer Science , University of Indonesia, Depok, Indonesia
eko@cs.ui.ac.id

Abstract. Precise CRM implementation for companies certainly affects the business survival of a company because it will relate to the maintenance of good and right relationship with the customers which will make their loyalty improves. There are numerous models / frameworks of CRM which is widely known today. By making comparison between these models, we can make exact and useful deliverables to identify crucial points in building Social CRM model/ framework. To harmonize business with the Information Technology, we need to create enterprise architecture that suit the company's situation. The merger of Zachman framework and Architecture for Integrated Information System (ARIS) is viewed as the bridging of Social CRM model/framework with its technology development and information system which will bring success in the implementation process.

Keywords : CRM Model, Enterprise Architecture, Zachman Framework, Architecture for Integrated Information System (ARIS)

1   Introduction

Competition among business that is getting intense requires company's sharp observation especially in order to maintain their current customers and acquiring new ones, yet keep improving the value proposition to the customers. For that, any company has to focus on the application of Customer Relationship Management (CRM). CRM (CRM 1.0 or traditional CRM) is a philosophy and a business strategy supported by a system and a technology designed to improve human interactions in a business environment [7]. The application of CRM in a company requires integration from business process, internal functions as well as external networks that is supported by Information Technology [5]. Based on related literature, there are several models that were developed by some researchers. These models have been applied by several companies. Yet, in line with a highly dynamic business development, it certainly requires the latest and fast information technology to assist CRM implementation in the company. Without a proper information technology, a successful and good CRM implementation will be very difficult to reach. Information Technology is sig-

T. Herawan et al. (eds.), *Proceedings of the First International Conference on Advanced Data and Information Engineering (DaEng-2013)*, Lecture Notes in Electrical Engineering 285, DOI: 10.1007/978-981-4585-18-7_54,
© Springer Science+Business Media Singapore 2014

nificantly improving the CRM development, started from CRM 1.0 or well-known as traditional CRM, and continued by developing e-CRM (Electronic CRM) which is the management of CRM electronically, specifically, it handles all connections to the customers that utilizes information technology [10]. The vast information technology development has brought a significant CRM implementation development. After e-CRM, there comes m-CRM (mobile-CRM), it is an application type of CRM which is run, operated and accessed through mobile platform [13]. Not to mention that today, the development of social media has become overwhelming which the number of users exceed one billion people, a really fantastic number [20]. Therefore, the latest CRM development is the one above the social media, often mentioned as Social CRM or CRM 2.0, that is a philosophy and a business strategy, supported by a technology platform, business rules, processes, and social characteristics, designed to engage the customer in a collaborative conversation in order to provide mutually beneficial value in a trusted and transparent business environment. It's the company's response to the customer's ownership of the conversation [7]. As can be seen from the development of CRM (CRM → e-CRM → m-CRM → Social CRM), it proves that its implementation highly depends on information technology. Yet, it is often that the change in IT world does not always catch the change in business world. Therefore, an enterprise architecture that can bridge the business with IT and information system development is needed.

2   The Comparison of CRM Model

The CRM Model was first introduced by Tearcy & Wiersema, it is called the Value Discipline, model [4], [6]. This particular model explains three different focuses for companies on how they can generate good customer relation which has to be planned far in advance. It involves operational excellence, product leadership, and customer intimacy.

Another model from Peppers and Rogers describes that company should pursue four steps in building a close one-to-one relation with customers, it covers identify, differentiate, interact, and customize (IDIC), then this model is later called as the IDIC model [9], [6], [5].

Gartner Inc. develops a CRM model that is referred to as The Gartner Competency Model describes about CRM as a business strategy that maximizes profitability, revenue, and customer satisfaction by organizing around customer segments, fostering behaviour that satisfies customers and implementing customer-centric processes. CRM initiatives need a framework to ensure that programmers are approached on a strategic balanced and integrated basis. The framework, called the eight building blocks of CRM, helps organizations see the big picture, make their business case and plan their implementation [6], [5], [15].

QCi model is the product of a consultancy firm that describes this model as a customer management model. The heart of this model depicts a series of activities that companies need to perform in order to acquire and retain customers. The model features people performing processes and using technology to assist in these activities [6], [5], [16].

Fig. 1. (1) The Value Discipline Model [4]; (2) The IDIC Model [9]; (3) Eight Building Blocks of CRM and (4) The QCi Model (Source 3 & 4: Buttle, 2009, p.19-21)

This dynamic CRM framework was developed by Park dan Kim. Customers who are committed to a particular company is certainly a long-term benefit for the company itself. Their (the customers') commitments get built when their expectations are satisfied, and eventually they will give a fair value share on their relationship with the company they deal with. This framework is built by observing two perspectives, of the company and of the customers. From the company's perspective, this value will reflect the customer's equity, yet from the customer's perspective, it will represent the values that are felt by the customers from the relation. To manage a successful relation, company will need numerous information about customers, such as their data until what the company may offer to their customers or the feedback from the customers. Customer Information System (CIS) has a wide role in managing and distributing the customers' information. Yet, the gaps that exist between marketing strategy and IT strategy can be a significant barrier to achieve a successful CIS. What should be included in CIS may have database, communication channel and decision model of a relation management, which should be designed to facilitate two-way customers' relation exchange. This dynamic CRM framework involves the IT strategy into the framework, which finally enable to formulate the strategy results into the real business case [3].

Another model was developed by Adrian Payne. The model clearly identifies five core processes in CRM: the strategy development process, the value creation process, the multi-channel integration process, the performance assessment process, and the information management process. The first two represent strategy CRM, the multi-channel integration process represents operational CRM, the last two represent analytical CRM [6], [5], [8].

There was a research about dynamic capability approach done by Desai, Sahu and Sinha, which was basically broken down from strategic management theory. This research identifies sources of CRM competition performance in dynamic capability. What is "dynamic capability"? It is an organization *capability* to improve their skill continuously, making inovations and resetting their resources to keep pace with the changing environment needs. Information Technology competency is now considered as an important moderator in the relation of dynamic capability and competition performance. This model also explores the social network capacity influence toward dynamic capability [1].

Forrester develops the CRM model that is started from determining customer oriented strategy. Then this strategy will be separated into CRM processes that are integrated to each other with the help of technology and people who are involved inside [14].

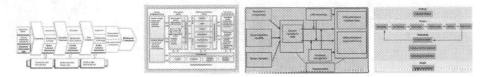

Fig. 2. (1) Dynamic CRM Framework [3]; (2) Payne's Five-Process Model (Source: Buttle, 2009, p.21), (3)Dynamic Capability CRM Model [1]; (4) Forrester CRM Model [14]

The CRM value chain is represented by Francis Buttle's model. The model consists of five primary stages and four supporting conditions leading forward the end goal of enhanced customer profitability. The primary stages of customer portfolio analysis, customer intimacy, network development, value proposition development, and managing the customer lifecycle are sequenced to ensure that a company with the support of its network of suppliers, partners, and employees, creates and delivers value propositions that acquire and retain profitable customers. The supporting conditions of leadership and culture, data and IT, people and processes enable the CRM strategy to function effectively and efficiently [6], [5].

FrontCRM framework is one that was developed by Budiardjo & Wira. This framework is designed to recognize or detect any business process that is running within every management aspect that is related to CRM, especially on marketing, sales and customer service (departmental plans). The framework is arranged based on the following core activities: strategic planning, marketing & sales, and the customer service of the company. The four core activities become the basic philosophy to apply CRM [12]. There are 3 steps to make the ontology of FrontCRM model, they are: (1) describing the top abstract, (2) the focus on business process distribution into 3 different business area (marketing, sales and customer service) and when it comes to this point, the business practice can be decomposed into sub-practice and CRM software features can be set as the automation practice, and (3) CRM software practice & features can be distributed into their respective business process [11].

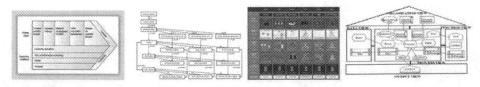

Fig. 3. (1) CRM Value Chain (Source: Buttle, 2009, p.20); (2) FrontCRM Framework [11]; (3) Zachman Framework [2]; (4) ARIS [2]

Based on the CRM models / framework above, we make comparison to get deliverables that can be seen at the following tables.

Table 1. The Comparison of CRM 1.0 Model

CRM 1.0 Model	References	Vision	Strategy	Focus	Process	Customer Experience	People & Organization	Information	Technology	Performance Appraisal (Measurement)	External Environment	Social Networking
The Value Discipline Model	Treacy & Wiersema, 1993			v	v	v						
The Identify, Differentiate, Interact, Customize (IDIC) Model	Peppers & Rogers, 1996				v			v				
Eight Building Blocks of CRM	Gartner Research, 2001	v (leadership, value proposition, social worth)	v (objective, segmentation, interaction)		v (customer life cycle, KM)	v	v (culture, organization chart, people skill, employee/partner/s supplier's attitudes)	v (data, analysis)	v (application, architecture, infrastructure)	v (Cost to serve, satisfaction, loyalty, social cost)		
QCi (Quality Competitiveness Index) Model	Hewson, et al, 2002		v (planning, value proposition)		v	v	v	v	v	v (cost to serve, satisfaction, loyalty)	v	
Framework of Dynamic CRM	Park & Kim, 2003		v		v		v	v	v			
Payne's Five-Process Model	Payne, 2005	v (business vision, industry & competitive characteristics)	v (customer choice, customer characteristics, segmentation, value proposition)		v			v (data, analysis tools)	v (IT systems, applications)	v (shareholder results, KPI)		
Dynamic Capability CRM Model	Desai, Suhu, & Sinha, 2007		v (market orientation)	v (customer, organization)	v		v (resource reconfigurability)	v (KM)	v	v (customer & organization's performance)	v (industry control)	v
Forrester CRM Model	Forrester Research, 2008		v (customer strategy)		v		v (people management)	v (data, analytic)	v (technology infrastructure)			
The CRM Value Chain	Buttle, 2009	v	v		v		v	v	v			
Framework FrontCRM	Budiardjo et al 2012		v		v		v (sales, marketing, customer service)	v	v			
Total		3	8	2	10	3	7	8	8	4	2	1

From the comparison above, we can see that all CRM models / framework result in processes that are followed by vision, information, and technology (each has a total of 8), and followed by people & organization (a total of 7). From this result we can take a temporary conclusion that all CRM models / framework are oriented into processes and must be supported by strategy, information, technology, as well as strong people & organization. The deliverables of this comparison can be made a foundation in making the models for Social CRM.

The strength of information technology has become one of the success points in implementing the CRM, therefore a detailed descriptions of Social CRM are needed for the construction of CRM information system. The description will meet its goal if company possess architecture enterprise for implementing this CRM.

3 The merge of Enterprise Architecture with Social CRM Model

Enterprise architecture is the whole enterprise, especially the business processes, technologies, and information systems of the enterprise [19]. Enterprise architecture methodologies are needed to bridge a fast business change with changes in information technology. There are numerous enterprise architecture methodologies that exist such as Zachman framework, The Open Group Architectural Framework (TOGAF), Federal Enterprise Architecture, The Gartner Methodology [19], Enterprise Architecture Planning (EAP), Architecture for Integrated Information System (ARIS), US Department of Defense Architecture Framework (DoDAF), etc [18]. This research employs a comparison of two enterprise architecture frameworks from Zachman Framework and Architecture for Integrated Information System (ARIS). The reasons for employing those 2 enterprise architecture frameworks are (1) both possess two-dimension structure and (2) Zachman framework speaks as ontology while ARIS explains how to build the information system [2].

Zachman Framework (ZF).

Zachman framework is a framework for modeling, evaluating, optimizing, managing and documenting of the entire business system. The basic characteristic of the Zachman architecture is related to the development and implementation from the business model into the information system model [2]. Zachman framework is the most comprehensive and mature framework to develop information system. This framework is introduced by John Zachman in 1987 which was inspired by engineering blue-print. Zachman defines architecture as a set of artifact design, which is recognized as the descriptive representative of numerous objects that are covered within a company. This framework consists of classification scheme with two-dimension to design artifact, where the horizontal axis provides taxonomy from several architecture artifacts and also consists of six abstract columns which each part has different focuses. They are WHAT focuses on data, HOW focuses on function, WHERE focuses on network, including location and connection, WHO focuses on people, WHEN focuses on time / event, and WHY focuses on motivation (see Fig 3 number 3).

Vertical axis provides numerous perspectives of the architecture as a whole. This axis describes buildtime lifecycle that consists of several steps of Business function modeling/Scope (level 1), Business process models/Business model (level 2), Logical models (IS model) System model (level 3), Physical models (IS model)/Technology model (level 4), "As-Built" IS model/Detailed representations (level 5), and Runtime phase that is related to the functioning enterprise (level 6).

Every perspective (line) is dedicated to one stakeholder within the information system, that possess Planner perspective (CEO and other top management executives), Owner (business manager and analyst), Designer (solution architect and implementation consultant), Builder (system designer), and Subcontractor (programmer) [17].

Architecture for Integrated Information Systems (ARIS).

ARIS emphasizes on inter artifact relation and how this artifact can be used properly as a foundation of business process analysis tools. ARIS frame-work is a process-centric and was introduced by Professor Scheer in 1992 [2], [17]. This framework is based on the business process model that is common and consists of five views, they are: Function view (goal, activities, software), Organization view (organizational units, computer hardware, machine resources), Data view (events, messages, environmental data), Output view (material input / output, services, financial resources), and Control / Process view (see Fig. 3 number 4).

View number 1 – 4 have static characteristics and used for modeling the internal relations, while view number 5 has a dynamic characteristic and used for modeling relations among elements that are possessed by different views [17]. Description levels are the second dimension of this framework. This description is broken down directly from a phase model that describes design aspects based on the software lifecycle concept [2]. These five description levels become the business strategy and goals, requirements definition, design specification, implementation description, and ICT.

The merge result of vertical and horizontal from Zachman framework and ARIS can be seen at the following figure.

To develop Social CRM model/framework (CRM 2.0), researchers utilize CRM 1.0 model/framework CRM 1.0 that generate several deliverables, which then the deliverables will be mapped into the merger of Zachman framework and ARIS. The mapping results are to be used for filling the framework cell business at the above Figure 12, where we can find the content at the Data column that has data & information about customer, transactional, and non transactional, Time column that has customer life cycle, Network column that has buildings, social media, networking infrastructure, collaboration with partners & suppliers, People column that has units which execute the CRM activities: marketing, sales, and customer service, Function column that has processes and focus on the CRM activities: the process of Acquisition, Retention, and Development, then Motivation column that has vision, strategy, and measurement CRM, as well as customer experience. The details of this explanation can be seen at the following figure, with red circle legend sign that shows business framework made for Social CRM (CRM 2.0) model/framework

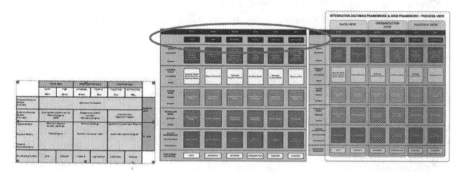

Fig. 4. (1) Vertical & horizontal comparison merge between ZF with ARIS [2]; (2 & 3) The result of Social CRM model/framework mapping into the Enterprise Architecture

The result of this mapping process is not the final one. There will be futher research that will put into details the business framework into a more comprehensive information for each stakeholder that is involved and will also generate another mapping into the industry value chain. This research is expected to generate a Social CRM (CRM 2.0) model/framework as well as able to describe the success of its implementation. This research aims to develop a Social CRM (CRM 2.0) model/framework for education, especially higher education.

4   Conclusion

Conclusion that we can take from this research activity is the comparison of CRM 1.0 model/framework can generate a very beneficial deliverables to identify important points of the Social CRM (CRM 2.0) model/framework. To harmonize business with information technology requires proper enterprise architecture. The merge of Zachman framework with Architecture for Integrated Information System is viewed able to bridge Social CRM (CRM 2.0) model/framework with the development of technology and information system. This developed model/framework will need to be made more detailed and specialized for particular industry to test the success of its implementation.

References

1. Desai, D., Sahu, S., Sinha, P. K.: Role of Dynamic Capability and Information Technology in Customer Relationship Management: A Study of Indian Companies. VIKALPA. 32(4), 45-62 (2007)
2. Kozina, M.: Evaluation of ARIS and Zachman Framework as Enterprise Architectures. Journal of Information and Organizational Sciences. 30(1), 115-136 (2006)
3. Park, C. H. and Kim, Y. G.: A Framework of Dynamic CRM: Linking Marketing with Information Strategy. Business Process Management Journal. 9(5), 652-671 (2003)
4. Treacy, M., Wiersema, F.: Customer Intimacy and Other Value Disciplines. Harvard Business Review. January – February, 83-93 (1993)
5. Buttle, F.: Customer Relationship Management, Concepts and Technologies, 2nd ed. Elsevier, Burlington, MA (2009)
6. Charantimath, P. M.: Total Quality Management, 2nd ed. Pearson, Dorling Kindersley, India (2011)
7. Greenberg, P.: Customer Relationship Management at the speed of light : social CRM strategies, tools, and techniques for engaging your customers, 4th ed. McGraw-Hill, USA (2010)
8. Payne, A.: Handbook of CRM, Achieving Excellence in Customer Management. Elsevier, Burlington, MA (2005)
9. Peppers, D., Rogers, M.: The One-to-one Future : Building Business Relationships One Customer at a Time. Doubleday, New York (1996)
10. Turban E., King D., Lang, J.: Introduction to Electronic Commerce, 3rd ed. Pearson, USA (2011)
11. Budiardjo, E. K., Perdana, W., Franshisca, F.: The Ontology Model of DrontCRM Framework. In: 2012 4th International Conference on Graphic and Image Processing Proceeding, pp.316-320. SPIE, Singapore (2012)
12. Budiardjo, E. K., Perdana, W.: FrontCRM : A Framework based on Theory of CRM, Penerapannya pada Toko Buku berskala UKM. In: Prosiding Seminar Nasional Aplikasi Teknologi Informasi 2008 (SNATI 2008), pp.A-149 – A-158. Yogyakarta, Indonesia (2008)
13. Camponovo, G., Pigneur, Y., Rangone, A., Renga, F.: Mobile Customer Relationship Management: An Explorative Investigation of the Italian Consumer Market. In: Proceedings of the 4th International Conference on Mobile Business, pp. 42-48. Sydney, Australia (2005)
14. Band, W.: Overview: Customer Relationship Management 2008. Executive Report, Forrester Research (2008) Available: http://experiencematters.wordpress.com/2008/12/15/forrester%E2%80%99s-2008 customer-experience-rankings

15. Close, W. S., Eisenfeld, B.: CRM at Work: Eight Characteristics of CRM Winners. Project Report. Gartner Research (2001) Available: http://www.gartner.com/resources/98800/98877/crm_at_work_eight_characteri_98877pdf
16. Hewson, W., Hicks, D., Meekings, A., Stone, Woodcock, N.: CRM in the Public Sector. Technical Report, Hewson Group (2002) Available: http//qci.uk/public_face
17. IDS Scheer: ARIS Design Platform. White Paper (2006) Available: http://www.arise-consulting.net/files/ARIS_Enterprise_Architecture_Solution_WP.pdf
18. ISO Architecture: Survey of Architecture Framework. Report (2013) Available: http://www.iso-architecture.org/ieee-1471/afs/frameworks-table.html
19. Sessions, R.: A Comparison of the Top Four Enterprise-architecture Methodologies. Technical Report (2007), Available: http://msdn.microsoft.com/en-us/library/bb466232.aspx
20. TechCrunch: ITU: There Are Now Over 1 Billion Users Of Social Media Worldwide, Most On Mobile. Blog (2012) Available: http://techcrunch.com/2012/05/14/itu-there-are-now-over-1-billion-users-of-social-media-worldwide-most-on-mobile/

About The Authors

Meyliana has been the faculty member of the School of Information Systems – Bina Nusantara University since 1997. Majoring in Business Process Management and Information System Development as professional track record. Graduated from Bina Nusantara University in 1996, holds Master in Management of Information Systems from Bina Nusantara University in 1999, and now, she is a Ph.D student in Computer Science, University of Indonesia. Currently she is the Deputy Vice Rector Operational of Alam Sutera Campus and Rector's Office Manager, Bina Nusantara University

Dr. Eko K. Budiardjo has been the faculty member of the Faculty of Computer Science - University of Indonesia since 1985. Teaching, research, and practical services are aligned; give result in a full spectrum of academic achievement. Majoring in Software Engineering as professional track record, he has made some scientific contribution such as Software Requirement Specification (SRS) patterns representation method, ZEF Framework, and FrontCRM Framework. Graduated from Bandung Institute of Technology (ITB) in 1985, holds Master of Science in Computer Science from the University of New Brunswick – Canada in 1991, and awarded Philosophical Doctor in Computer Science from the University of Indonesia in 2007. Currently he is the Vice Chairman of ICT Technical Committee of The National Research Council (DRN), and Chairman of The Indonesian ICT Profession Society (IPKIN).

Part VII
Mobile, Network, Grid and Cloud Computing

A Model for Client Recommendation to a Desktop Grid Server

Mohammad Yaser Shafazand, Rohaya Latip, Azizol Abdullah, and Masnida Hussin

Faculty of Computer Science and Information Technology,
Universiti Putra Malaysia, Malaysia
shafazand@aut.ac.ir
{rohayalt,azizol,masnida}@upm.edu.my
http://www.upm.edu.my

Abstract. A vast amount of idle computational power of desktop computers could be utilized throughout desktop grids. For an appropriate utilization, the scheduler, needs to determine clients which are best suited to deliver assigned jobs in time. Diversity of hosts (i.e. OS, hardware and network specifications) and intermittent availability of resources are known issues which complicate the schedulers work. As a solution to this problem, a client-server model consisting two modules for a desktop grid middleware is discussed: a module to forecast machine resource availability in the client side and a module in the server side that recommends clients to the scheduler that are the nearest to job expectations. Historic data, time-series analyses and machine learning are used for this purpose in the modules.

Keywords: Desktop Grid, Availability Prediction, Case-based Reasoning, Match-making, Recommender System.

1 Introduction

Desktop grids are a form of computational grids in which each resource is a desktop computer of an organization. The main goal is to utilize idle resources. Assuming the large amount of idle CPU time in each desktop computer, the desktop grid is potentially one of the biggest computing resource pools available for an organization. Therefore, a simple efficiency could do much contribution. To harness these resources, a middleware is needed to manage and match jobs and resources. One of the current focuses of desktop grid schedulers is to estimate if a resource or machine is available long enough for the task to finish and return its results [1]. The term availability could refer to both CPU or machine availability. CPU availability refers to the CPU resource of a machine being dedicated to perform computations for the benefit of the distributed systems [1]. We define machine availability as the machine system being in reach and available for the server to communicate with.

There are issues that make it difficult to guaranty the availability. For example, occasions when the resource owner initiates an activity, and needs a resource

T. Herawan et al. (eds.), *Proceedings of the First International Conference on Advanced Data and Information Engineering (DaEng-2013)*, Lecture Notes in Electrical Engineering 285, DOI: 10.1007/978-981-4585-18-7_55,
© Springer Science+Business Media Singapore 2014

that is being used by a job. The network used for interconnection between the machines could also have disruption and disconnection which also limits machine availability. Based on user usage needs, each client behaves differently in terms of resource availability each day of the week [2]. This behaviour could be predicted based on historic data which is not considered in some current desktop grid and volunteer computing systems.

Our model proposes a client-server solution which has two modules, the Availability Tracer module (AT) in the client side and the Recommender and Estimator module (RE) in the server side. The AT module uses historic availability data of a resource (i.e. CPU availability) to predict its availability for a short given time frame (e.g. one day). The RE module prioritizes and recommends the client machines based on availability and nearest resource matching for a received job to the scheduler. This process assists the scheduler and doesnt interfere with its inner functionality. Replicas for the job assigned could also be selected from the recommended list if needed. The rest of the paper is organized as follows: Section 2 explains the related works done in the area of forecasting for grid schedulers. Section 3 gives background information on CBR. Section 4 presents the overall architecture, 5 explains the forecasting module and its functionality and finally section 6 concludes and describes the future work.

2   Related Works

Most studies assume each system is dependable and traceable. This is in contrast with the desktop grid and volunteer computing environment which the machine hosts are not dependable and might not even be traceable. Also, the environments in the studies have less diversity than the desktop grid environment. Based on the approach for predicting, we have classified their studies.

2.1   Without Prediction

Solutions that ignore forecasting usually try to match the job needs with the available resources or performance characteristics on a client. As an example, matchmaking is a mechanism used in Condor (now named HTCondor) to match an idle job to an available machine [3]. This mechanism has the ability to define rules but it doesnt have a forecasting strategy. Most of the time these solutions suffer in resource utilization [4], [5]. The simplicity and low computational costs on the server is the main advantage of this category but still more efficient models could be proposed which could also have a low computational cost for the server.

2.2   Stochastic Based and Mathematical Modelling Prediction

Stochastic Based Prediction is known as using a combination of stochastic mathematical analysis and direct-numerical simulation to forecast the system performance [4]. This model tries to predict the performance of the system by creating a pattern or graph of performance or resource availability using probability model, statistical evaluations and mathematical analogies. One famous study which had a great impact on current studies is the network weather service (NWS) model discussed by [6] which does a short-term forecast of certain communication and computing resources. The forecasting systems are getting complex in this area and their monitoring systems might reduce the overall efficiency. The work done in [7] could be mentioned as a recent work to study availability, but their work doesnt consider a fine grained prediction of each machines availability to provide a more adequate feed to the server.

2.3 Machine Learning Based Prediction

This group is subjected to studies using learning algorithms to support the prediction mechanism mostly using historic data. In the category of regression computation we could name artificial neural networks (ANN) which some studies [8] have used in favor of the grid scheduler. The main disadvantage could be that ANNs learning process is quite complex [9] and it also lacks online learning (Learning throughout the process of functioning) by default. Some other studies [9] have used support vector machines (SVM) which is also another efficient subcategory of regression computation in the machine learning field. SVM based systems still lack the presence of online learning by default and need large learning sets of data with complicated learning. These studies do not assist the scheduler with a recommender system. Most of which replace the current systems with their own suggested functionality. Case-based Reasoning uses historic data and could be used where we have small learning datasets. It has also proven its benefits in studies benefiting the grid scheduler [10].

3 Case-Based Reasoning (CBR) for Machine Learning

CBR is based on the fact that new problems could be solved by remembering how similar problems have been solved in the past [11]. CBR is also known as a methodology for problem solving which could use a variety of technologies (e.g. Neural Networks, Genetic Algorithms, and Fuzzy Logic) for its purpose [11]. In CBR terminology, a case is usually referred to a previously experienced problem situation which could be saved and used for similar future problems [12]. This methodology is described as a cycle with four major activities:

1. *Retrieve* similar cases to the problem description from historic
2. *Reuse* (adapt) a solution suggested by a similar case. repository.
3. *Revise* (review) that solution to better fit the new problem if necessary.
4. *Retain* the new solution once it has been confirmed or validated.

In each of these activities any state of the art technology or algorithm that satisfies the problem could be used.

3.1   CBR Benefits

The following specifications of CBR benefit us in forecasting and recommending:

Domain model not available: CBR could be used where we dont have a deep domain model and we might only need to rely on a collection of historical problem-solving events [13].

Small Training sets: CBR could be used where the number of training examples is too small to obtain a generalization for future reference [13].

Simplified lazy learning: CBR learns throughout the process of problem solving and has a simple learning by example mechanism [12].

Incremental sustained learning: Each time a problem is solved, the experience will be saved for encountering future problems [12].

4   Architecture

One of the most important common practices considered is simplicity which mostly results in a lower computational load of a design. We have also considered accuracy as another important aspect. Simple systems mostly have lower accuracies, and thus lower performances to some extent. High accuracy might be over-qualifying, resulting to a unwanted computational load. Therefore, a trade-off between accuracy and simplicity must be considered. The overall model being central, client-server based or peer to peer also has an impact on the overall performance of the model. We must also note the role of the monitoring systems to record the historic traces. These systems must have the least resource usage load on the system they are running on.

An example of our models application on the BOINC architecture and on the Condor pool could be seen in Fig. 1 and Fig. 2 respectively. These sample models are fabricated based on the true BOINC architecture and Condor pool design [3] and the mentioned aspects. Other architectures could follow the same concept.

4.1   Job Scenario for BOINC architecture

Consider Fig. 1, when the client requests for a new job, it sends its request to the AT module. The AT module in turn predicts its resource availability in a needed time frame (e.g. one day) and sends this data and the request for a new job to the server (RE Module). The AT module also collects the availability history and updates its predictions and the RE time to time. The RE module in the server has a database of all the latest AT estimations. It creates a list of machines based on their relevant availability data and resource specifications and recommends best fitted machines to the scheduler. The scheduling server is feeded a list of machines by the RE module and continues its ordinary process. The internal process of the scheduler is left intact.

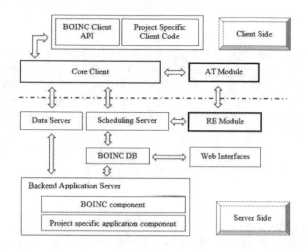

Fig. 1. Sample Client-Server Module Allocation in BOINC Architecture

4.2 Job Scenario for Condor pool

Another application example is on the Condor pool (Fig. 2). The same scenario described in the BOINC is considered here with the difference that in BOINC, the client requests jobs from the server. The RE module updates the collector with a list of recommended machines. When a job is submitted, the negotiator queries the collector to discover ready machines and jobs and uses this information to do the matchmaking and schedule the jobs. The rest of the procedure is the same described in the Condor running process [3].

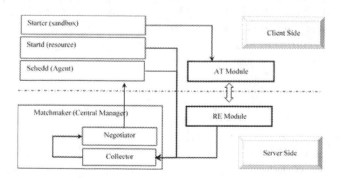

Fig. 2. Sample Client-Server Module Allocation in a Condor pool

5   Forecasting Model

For our model, we have considered analysing a dataset of actual desktop grid machines availability traces, gathered over six months [2].This dataset is part of the Failure Trace Archive [14]. We used an ARIMA (autoregressive integrated moving average) model to predict the CPU availability behaviour in our AT module. A Box-Jenkins [15] procedure has be followed for this goal. A case-based reasoning method is used for the server RE module.

5.1   Availability Tracer (AT) Module

We first packed our dataset traces of availability into frames each representing a day (24 hours). This is the granularity we assumed appropriate for the predictor. The AT has a database of these availability traces. Analysing the time series behaviour, we selected the simple to implement Auto Regressive (AR) model as in (1) for our predictor:

$$y_t = e_t + c_1 e_{t-1} \tag{1}$$

The formula in (1) is a *first-order auto regressive model* derived from analysing the availability patterns. y_t is our future prediction, the next 24 hours CPU availability. e_t is a constant. y_{t-1} is the availability of the t-1 day. c_1 is a factor for y_{t-1}. Both e_t and c_1 are calculated using methods discussed in the Box-Jenkins model. While each job is running on a machine, the resources availability is monitored and saved for further predictions.

5.2   Recommender and Estimator (RE) Module

Our selection of attributes for the CBR forecaster is based on the job and machine attributes used in studies [3], [5]. Each past submitted finished job has a relevant *Problem* , *Solution* and *Result* case(data structure) saved in the case base : some important attributes of the *Problem* case are: job name (JobName); required CPU (ReqCPU); required memory (ReqMem); required floating-point operations per hour to finish the job (Reqfloph); required maximum bandwidth quota (ReqBW); and DueTime. some attributes of the *Solution* data structure are: MachineId; number of mega floating-point operations a machine could finish in an hour (Mfloph); and the monitored CPU availability of the machine while finishing the job (CPUAvail). There are three main processes in this module :

Case retrieval process: This process receives a problem case as an input (job resource requirements). It is then responsible to fetch similar cases (machine profiles) to the problem case from the case-base repository using a global similarity. A local similarity is calculated between each case attribute considering each attributes weight in favour of the global similarity. These weights are predefined. The local similarity is defined by function f_i where i represents each attribute [5]. The case-base is initially filled with sample cases.

$$NumericAttributes : f_i(a,b) = 1 - \frac{|a-b|}{range} \tag{2}$$

It is shown in (2) a sample equation of local similarity for numeric attributes. a and b are attributes of different cases; the problem case and the past case saved in the case base. Also, *range* is the absolute value of the difference between the upper and lower set boundaries.

$$sim(A, B) = \sum_{i=1}^{p} w_i f_i(a_i, b_i) \tag{3}$$

In (3) the global similarity is calculated using local similarities. w_i is the weight of each attribute i and f_i is the local similarity for the attribute i which was described in (2). The most common algorithm used for the retrieval process is the nearest neighbour (NN) algorithm. The output of this process is a list of k most similar cases retrieved.

Solution adaptation process: The input of this process is a number of k similar cases to the problem case given from the case retrieval process. In this process, the solutions in the fetched cases are modified to fit the current situation. The *Mfloph* of the solution record is then modified (adapted) by the following formula based on the affection of the estimated CPU availability:

$$Mfloph_{adapted} = est(CPU_{Availability})\% * Mfloph_{current} \tag{4}$$

In (4), the percent of the estimated CPU availability is multiplied to the Mfloph of the machine to give the actual throughput the system is considered to provide. The results of this stage finally reach the scheduler/matchmaker.

Case retaining process: This process supports the learning mechanism by filling the case base with new cases and removing past cases which are obsolete based on their timestamp. After receiving the actual CPU (and/or machine) availability from the scheduler, these fields would be saved with the the Solution and Problem record in the case base. Also, obsolete cases are removed for more efficiency.

6   Conclusion

A model for a client availability predictor in cooperation with a server recommender system for a desktop grid or volunteer computing grid scheduler was described. Two modules were proposed, one in the server side (RE module) and one in the client side (AT module). Their combination makes it possible to recommend a client to the scheduler/matchmaker based on the predicted CPU or machine availability of the client machine. This is believed to decrease improper task distribution resulting to missing deadlines in desktop grid middleware. For more accurate job runtime estimation in jobs that have large amounts of data passed as input or received as the results, the network properties (e.g. bandwidth) could also be considered. In our future work, we would evaluate our model based on accuracy and efficiency of the treated system. Even a slight improvement in resource management in an environment with a large quantity of idle resources could make a major contribution.

References

1. D. Lázaro, D. Kondo, and J. M. Marquès, Long-term availability prediction for groups of volunteer resources, Journal of Parallel and Distributed Computing, vol. 72, no. 2, pp.281296, 2012.
2. B. Rood and M. J. Lewis, Multi-state grid resource availability characterization, in Proceedings of the 8th IEEE/ACM International Conference on Grid Computing, pp.4249, 2007.
3. T. Tannenbaum, D. Wright, K. Miller, and M. Livny, Condor – A Distributed Job Scheduler, in in Beowulf Cluster Computing with Linux, T.Sterling, Ed. MIT Press, 2002.
4. M. Wu and X.-H. Sun, Grid harvest service: A performance system of grid computing, Journal of Parallel and Distributed Computing, vol. 66, no. 10, pp. 13221337, 2006.
5. L. N. Nassif, J. M. Nogueira, A. Karmouch, M. Ahmed, and F. V. de Andrade, Job completion prediction using case-based reasoning for Grid computing environments, Concurrency and Computation: Practice and Experience, vol. 19, no. 9, pp. 12531269, 2007.
6. R. Wolski, Dynamically forecasting network performance using the Network Weather Service, Cluster Computing, vol. 1, no. 1, pp. 119132, 1998.
7. B. Javadi, D. Kondo, J.-M. Vincent, and D. P. Anderson, Discovering Statistical Models of Availability in Large Distributed Systems: An Empirical Study of SETI@home, Parallel and Distributed Systems, IEEE Transactions on, vol. 22, no. 11, pp. 18961903, 2011.
8. K. Singh, E. Ípek, S. A. McKee, B. R. de Supinski, M. Schulz, and R. Caruana, Predicting parallel application performance via machine learning approaches, Concurrency and Computation: Practice and Experience, vol. 19, no. 17, pp. 22192235, 2007.
9. L. Hu, X.-L. Che, and S.-Q. Zheng, Online System for Grid Resource Monitoring and Machine Learning-Based Prediction, Parallel and Distributed Systems, IEEE Transactions on, vol. 23, no. 1, pp. 134145, 2012.
10. E. Xia, I. Jurisica, J. Waterhouse, and V. Sloan, Runtime Estimation Using the Case-Based Reasoning Approach for Scheduling in a Grid Environment, in Case-Based Reasoning. Research and Development, vol. 6176, I. Bichindaritz and S. Montani, Eds. Springer Berlin / Heidelberg, pp. 525539, 2010.
11. I. Watson, Case-based reasoning is a methodology not a technology, Knowledge-Based Systems, vol. 12, no. 56, pp. 303308, 1999.
12. A. Aamodt, Case-based reasoning: Foundational issues, methodological variations, and system approaches, AI communications, vol. 7, pp. 3959, 1994.
13. J. A. Recio-García, P. A. Gonzlez-Calero, and B. Daz-Agudo, jcolibri2: A framework for building Case-based reasoning systems, Science of Computer Programming, no.0, p. -, 2012.
14. D. Kondo, B. Javadi, A. Iosup, and D. Epema, The Failure Trace Archive: Enabling Comparative Analysis of Failures in Diverse Distributed Systems, in Cluster, Cloud and Grid Computing (CCGrid), 2010 10th IEEE/ACM International Conference on, pp. 398407, 2010.
15. G. E. P. Box, G. M. Jenkins, and G. C. Reinsel, Time series analysis: forecasting and control. Prentice Hall, Englewood Cliffs, NJ, USA, 3rd edition edition, 1994, p. 598.

A Reliable Data Flooding in Underwater Wireless Sensor Network

Tariq Ali[1], Low Tang Jung[2], Ibrahima Faye[3]

[1]tariqhsp@gmaill.com [2]lowtanjung@petronas.com.my
[3]ibrahima_faye@petronas.com.my

[1,2]CIS DEPARTMENT, [3]FAS DEPARTMENT, UNIVERSITI TEKNOLOGI PETRONAS, TRONOH, MALAYSIA

Abstract. The problem of data gathering in the inhospitable underwater environment, besides long propagation delays and high error probability, continuous node movement also makes it difficult to manage the routing information during the process of data forwarding. Providing the better communication in UWSNs and maximize the communication performance in network is a significant issue due to volatile characteristics of water environment. This paper present our proposed a novel routing protocol called Layer by layer Angle Based Flooding (L2-ABF.), to handle the issue of horizontal communication between nodes on same depth level from the surface sinks, end-to-end delays and energy consumption. In L2-ABF, every node can calculate its flooding angle to forward data packet to the sinks, without using any explicit configuration and location information. The simulation results show that L2-ABF has some advantages over some existing flooding techniques and also can easily manage quick routing changes where node movements are frequent.

Keywords: Propagation delays, L2-ABF, Routing algorithms, UWSN, Underwater environment.

1 INTRODUCTION

A reliable Underwater Wireless Sensor Network (UWSN) can provide a better solution for the applications which operates under the constraints of volatile environments. These environments are not feasible for human presence due to unpredictable underwater activities. The manned explorations are not applicable in underwater environment due to some reasons, like high water pressure and vast area of ocean [1]. UWSN is a new branch of terrestrial base sensor networks. However UWSN has some similarities with terrestrial base wireless sensor networks, like large numbers of nodes and energy issues but still these are different from many aspects. Energy saving is a main concern in UWSNs because sensor nodes powered by batteries which are not easy to replace or recharge rather than in stationary networks. Due to water absorption, radio communications does not work well for underwater. Thus acoustic communication is employed as only viable solution for underwater sensor networks.

T. Herawan et al. (eds.), *Proceedings of the First International Conference on Advanced Data and Information Engineering (DaEng-2013)*, Lecture Notes in Electrical Engineering 285, DOI: 10.1007/978-981-4585-18-7_56,
© Springer Science+Business Media Singapore 2014

With this replacement, UWSN has to bear some cost in the form of speed which is five magnitudes less than the radio channel. However, the cost of a sensor node for underwater is higher than the ground base wireless networks. Another issue with underwater networks is that most of the sensor nodes can passively move with water currents [2]. The above-mentioned limitations show that terrestrial ad hoc network's protocols are not suitable for UWSN. The results of routing protocols for terrestrial networks are very poor in underwater networks. The limitations mention above demand some specifically new designed protocols for UWSN. Lot of researchers has been focusing on designing new routing protocols simply due to the key characteristics of underwater communications [3].

2   RELATED WORK

Underwater Sensor Networks are attracting the attention of industry and academia [4], [5]. At one side, it can enable a wide range of aquatic applications, and on the same time, adverse environmental conditions create a range of challenges for underwater communication and networking. The node mobility and sparse deployment can create problem for underwater sensor networks. Due to the continuous node movements with water currents, there may not be a persistent route from a source to a destination. This explains why an underwater sensor network can be viewed as a partially connected network [6], and the traditional routing protocols developed for terrestrial sensor networks, usually are not practical for such environment. Due to this intermittent connectivity, packets can be dropped when no routes are available to reach the destination.

2.1   Vector Based Forwarding (VBF)

A review of underwater network protocols till the year 2000 can be found in [7]. Several routing protocols have been proposed for underwater sensor networks. Vector Based Forwarding (VBF) protocol has been suggested in order to solve the problem of high error probability in dense networks. Here an idea of routing pipe like circuit switching, from the source to the destination is proposed, and all the flooding are carried out through this pipe. By using this approach, the retransmissions are decreased making VBF significantly improve energy efficiency. In [8], a two-phase flexible routing solution for long-term monitoring applications with an idea of centralized planning network routings and data paths has been proposed. Later on, the same authors proposed a protocol that can handle both delay sensitive and delay tolerant applications. In this protocol, a cross-layer approach was adopted to create an interaction between the routing functions and underwater characteristics.

2.2   Depth Base Routing (DBR)

For location based routing, most of the protocols require and manage full-dimensional location information of the sensor nodes in the network, which often is a challenge in

UWSNs [9]. Instead of requiring complete localized information the protocol in [10] needs only the local depth information. Obtaining the depth is not a problem, as the authors suggested to equip each node with an inexpensive depth sensor. Their simulation results showed that, DBR can achieve high packet delivery for dense networks with reduced communication cost. However the problem here is that, it doesn't show good results for sparse networks due to its greedy nature. In order to achieve same performance as for dense deployments, some recovery algorithms need to be explored when the greedy strategy fails.

2.3   Focused Bream Routing (FBR)

Without any prior location information of nodes, a large number of broadcast queries can burden the network which may result in reducing the overall expected throughput. In order to reduce such unnecessary flooding in [11], presented Focused Bream Routing (FBR) protocol for acoustic networks. Their routing technique assumes that, every node in the network has its own location information, and every source node knows about the location of the final destination. Other than this information, the location of intermediate nodes is not required. Routes are established dynamically during the traversing of data packet for its destination, and the decision about the next hop is made at each step on the path after the appropriate nodes have proposed themselves.

3   PROBLEM AND CHALLENGES

When considering underwater wireless sensor networks, a due consideration must be given to the possible challenges that may be encountered in the subsurface environment. Continuous node movement, deployment of networks and communication between the peer sensors nodes at same depth level are major issues posed by the host conditions. Most of the times, sensor nodes are considered to be static but sensor nodes underwater in fact can move up to 1-3/msec due to different underwater activities [12].

Following are the problem statements which investigate from literature review and related work.

1. The significant problem is horizontal communication between the sensor nodes on the same depth levels which cause for end-to-end delay, large data routing path from source node to destination node and also overhead for energy consumption.
2. Another problem is un-directional movement of nodes in underwater due to water current, which cause for rapid change in network topology and can also affect the efficiency of routing protocols.

3.1   Research Questions

Based on the known problems and issues in UWSNs, the following research questions are formulated for this research.

1. How can we minimize the effect of horizontal communication between sensor nodes on same depth levels with the help of routing decisions?
2. What are the deployment methods available for UWSN? How these methods can help to minimize the effect of node mobility?
3. How sensor nodes can manage and organize their locations and positions in an unstable underwater environment?

3.2   Research Objectives

Designed new routing protocol with the name of Layer by layer Angle Based Flooding (L2-ABF) for UWSN:

1. To reduce the horizontal communication between the sensor nodes on same depth levels
2. To overcome high end-to-end delay, large data routing path and energy constraint
3. To provide further directions in handling similar challenges in other areas of Underwater Wireless Sensor Network.

4   MODEL AND CONTRIBUTION

The horizontal communication between the sensor nodes on the same depth levels is known to cause an increase in data routing path from lower layer nodes to the surface sinks which are deployed on the water surface. The large data routing path is highly likely to increase end to end routing delay in the underwater data transmission and brings along with it the Power issues. The proposed Layer by Layer Angle Based Flooding (L2-ABF) protocol is aimed to reduce the horizontal communication between the peer sensors on the same depth levels in underwater sensor network. Angle based flooding architecture would be used in L2-ABF to achieve this goal. The anchored nodes flood the sensed data towards sink node through upper layer nodes. The packets forwarding mechanism is defined in section "A". The flooding zone is calculated by using the base angle $\Theta= 90 \pm 10K$ and the size of the flooding zone is depending on the value of variable K. The maximum angle range for flooding zone is always greater than 0 and less than π as shown in Fig.1.

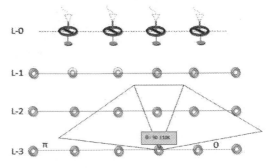

Fig. 1. L2-ABF angle base architecture

4.1 Packet Forwarding Mechanism

The angle base flooding approach is used in our proposed routing protocol. This routing mechanism does not based on sensor node location information and is to be designed for delay and Power-efficient multi-layer communication in underwater acoustic networks. In this routing mechanism, there is no need for sender node to know its own location and the location of final destination (Sink) before transmitting the data packets. Anchor nodes flood the sensed data towards surface sinks via the upper layer nodes. The forwarder node will define the flooding zone by using the initial angle Θ =90 $\pm10K$. Here K is a variable and has a finite set of value K $\{1, 2.....8\}$. After defining the flooding zone, node will send Hello Packets (HP) within the defined zone and wait for Hello Reply (HR). If there is no HR received, node increase the value of K in initial angle, to increase their flooding zone until the basic condition is meet (k $\leq$ 8). To increase the reliability of network, here we assume that the nodes can pick random value of variable K, to increase the size of flooding zone and it depends on the movement of nodes. We also assume that if nodes did not receive any Hello Reply at the last value of k, so node can move up by using hydraulic pressure and directly forward data packets to the sink.

1. If node receive very small number of Hello Reply. Therefore, nodes will consider the movement is slow. So nodes will use large value of variable k to increase the flooding zone.
2. If node receive more number of Hello Reply. Therefore, nodes will consider the movement is fast. So nodes will use small value of variable k to increase the flooding zone.

4.2 Algorithm 1: Data packet forwarding

Algorithm {Forward Data Packet (DP), Hello Packet (HP), Hello Reply (HR)}

1. Node has DP and ready to send
2. Check the status of DP in Q2   //Q2 is a buffer history
3. If (DP in Q2)
4.   Discard the packet // Packet already sent
5. END if
6. If (DP not in Q2)
7. Calculate the Flooding Zone using $\theta = 90\pm10K$ // K is variable has set of values [1,2.....8]
8. Send HP   // Inside the define flooding Zone
9.   Wait for HR
10. If (HR received) = yes
11. Forward "DP"
12. Go to rest mode   //Acknowledge Node is Qualified for the further flooding
13. Else
14. If (K $\leq$8)

15. Go on step 7   //to increase the size of flooding zone with increasing the value of
    "K"
16. Else
17. Send DP directly // By using hydraulic pressure
18. End if
19. End

4.3   Priority and Layers Assigning Method

Figure 2 shows the architecture of UWSN that L2-ABF assume. In which nodes are
deployed from surface to the bottom of sea. Sinks are deployed on water surface and
consider being static after deployed and as well as luxury with acoustic and radio
communication [13]. The floating nodes are deployed with different levels of depth
using the bouncy control mechanism. We only consider horizontal movements on
floating nodes within the define region. The vertical variation is very little and normally ignored [14]. We assume that all the sensor nodes can compute the priority to
become next forwarder. The process is as follow: The sink nodes will assign layer
numbers to the floating nodes through Hello Packets. The nodes will consider in first
layer those received Hello packets first and rebroadcast Hello Packet with decrement
one in layer number and so on. This process will continue until the value of Layer-ID
becomes zero. Layer number of nodes is changing when nodes will change its layer
and each node on same layer has same layer number but different IDs. On the basics
of layer number every nodes can measure, how mush far from the upper layer nodes
and surface sinks. When multiple nodes are inside the flooding zone and they want to
respond of Hello Packets. Therefore, every node in zone will calculate the priority on
the basics of layer number and residual energy. After calculation the priority, nodes
broadcast this information to the sender. The sender node will save information of the
five highest priority nodes into its priority queue. The sender node will select the next
forwarder for these data packets that have highest priority values in his queues.

4.4   Waiting Time Calculation

When a forwarding nodes cannot find the next nodes from upper layers after the first
attempt. The source node will make two attempts as increasing its flooding zone angle
to send the current data packets towards upper layer nodes before move up to forward
data packet directly. We defined the range of values between (0 , 9) [15], to calculate
the maximum waiting time to forward the data packets toward upper layers, where 0
mean no wait and 9 can be the maximum waiting time in worst case. We assume that
node will wait t1 time, before going for other attempt. It depends on the number of
nodes replied in first attempt, t1 can be calculated as:

$$t_1 = \frac{k}{n_1+1} \tag{1}$$

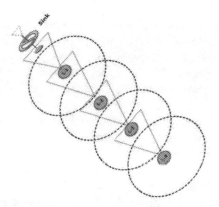

Fig. 2. Diagram illustrates the concept of layer number from Sink-to-Sensor nodes

Where K is a constant that has maximum value of the waiting time and n_1 is the number of nodes replied in first attempt. If there is no reply from the upper layer nodes, it will wait t_2 time, to go for second attempt as increasing the angle of flooding zone. The t_2 time is:

$$t_2 = \frac{k}{n_2+2} \tag{2}$$

Where K is a constant that has maximum value of the waiting time and n_2 is the number of nodes replied in 2nd round. Every times depend on two factors. First, the number of nodes replied after the other attempt and second, the difference between number of nodes in first and second attempt. So the average of these times is:

$$T = \frac{\left[\frac{K}{|n_2-n_1+1|} + \frac{K}{|n_2+1|} + \right]}{2} \tag{3}$$

From the equation '3', it is clear that the waiting time can be calculated by the availability of neighbor's nodes and the frequency of change in them.

5   PERFORMANCE EVALUATIONS

NS-2 was used to evaluate the performance of L2-ABF. In our simulation we take 300 sensor nodes (both sink and floating nodes) deployed in 3D area of (800×800×500m). We use single sink at the center of water surface and the sink is static but luxury with both (Radio, Acoustic) types of communication. The distance between the layers of floating nodes can be up to 100m. In this experiment, sink nodes consider being static after deployed but remaining nodes are floating in nature and we ignore the vertical movement on floating nodes. The only horizontal movements between nodes was considering with different water current up to 1-4m/s at fixed motions.

5.1   Mobility of Nodes

Fig.3: (a) shows the data delivery ratio at different speed of nodes movements, while we considered two type of nodes movements 2m/s, 4m/s and also static nodes. The data delivery ratios are 100% with the suggested number of nodes in the network and remain static, even not serious effect on these delivery ratios with decreasing the density of nodes. We can achieve around 90% data delivery ratios if 30% nodes are not available in the network, even if we see the sparseness of nodes where 50% of nodes are not available, still we can receive 85% data packets on average nodes movements. Now we analyzed the end-to-end delays and energy consumptions with different nodes movements as shown in Fig.3: (b) and (c).Here we can see, little bit variations in results on nodes movements as compare to static nodes. This difference is minor at beginning, after that it start to increase as number of nodes start to decrease, but still we can see this difference is not so high until 50% nodes are part of networks.

In the whole scenario, it is clear that, node mobility has no serious effect on end-to-end delays and energy consumptions as only minor difference on different nodes speeds. It is only due to that there is no need to maintain complex routing table for location information of sensor nodes, even when node change its position.

The performance of L2-ABF was checked with different packets load, we analyzed the delivery ratios and end-to-end delays by producing more data packets in the network.

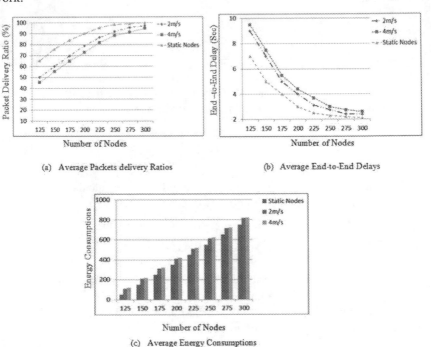

(a) Average Packets delivery Ratios   (b) Average End-to-End Delays

(c) Average Energy Consumptions

Fig. 3. Simulations Results

In normal case, network generates 1 packet per second, but here we consider the cases where first the network generates 3 packets and then 2 packets per second. Fig. 4(a) presents the delivery ratio with different offered loads. It shows that with dense nodes, the delivery ratios are almost the same and the difference starts when the number of nodes starts to become sparse. At high offered load and with fewer nodes in the network, sometime a node cannot find next hop and packets start to increase in the buffer which results in discarding them. Fig. 4(b) shows the variations in end-to-end delays when we increase the number of packets in the network. It shows that, the network can handle easily when 50% more packets becomes the part of the network; even these delays are affordable when double packets have been generated in the network.

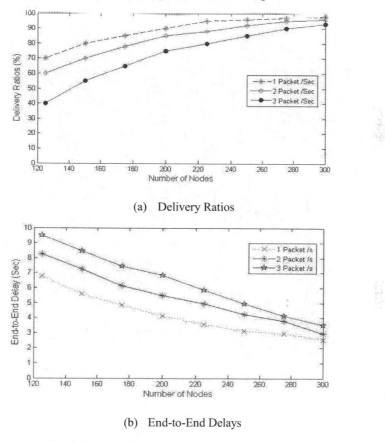

(a) Delivery Ratios

(b) End-to-End Delays

Fig. 4. Performance of L2-ABF with different loads

5.2 Analysis with different K's Values

Now we analyzed the end-to-end delays and data delivery ratios with different values of variable K which is used by nodes to forward data packets toward sink nodes as shown in Fig.5 (a & b). Here we can see, little bit variations in results with different values of variable K. The result shows that when node use small value of variable K to forward data packets, so there is very little effect on delays as well as delivery ratios. This difference is minor at beginning, after that it start to increase as number of nodes start to decrease and node use large value of variable K, but still we can see this difference is not so high until 50% nodes are part of networks.

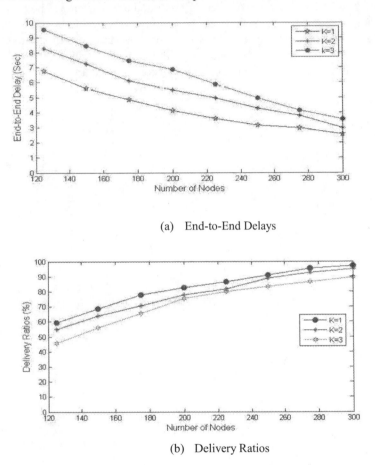

(a)   End-to-End Delays

(b)   Delivery Ratios

Fig. 5. Performance with different values of K

6   COMPARISON WITH VBF

L2-ABF was compared with VBF so that their performance could be evaluated. The data delivery ratios with a single sink was compared. A surface sinks was placed at the center of the network for the data delivery ratios with a single sink to provide the comparison of both L2-ABF and VBF. The delivery ratios of L2-ABF were less affected than the VBF delivery ratios as shown in Fig 6 (a). This was caused by routing pipe is used in VBF [16]. In this case, data packets could only be forwarded by the nodes to the surface of the water without any consideration being taken as to the position of the sink node.

The end-to-end delay with various layer counts, 2 to 9, from the source to the sink was measured. A larger end-to-end delay is the result of VBF due to 3-way handshake process before the data packets are transmitted. Additionally, this difference is made more significant by a larger number of layers due to the control packets being re-

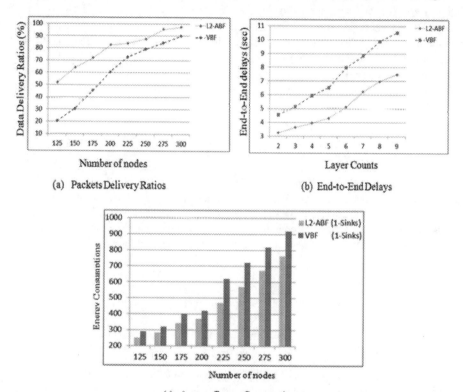

(a) Packets Delivery Ratios   (b) End-to-End Delays

(c) Average Energy Consumptions

quired to be exchanged at every layer. On the other hand, as shown in Fig. 6 (b), the end-to-end delay is increased linearly with L2-ABF according to the number of lay-ers.

Fig. 6. Comparisons with VBF

Now we evaluate energy consumptions with different number of nodes as shown in Fig.6: (c). Here we can see, the variations in results. The difference is minor at begin-ning, after that it start to increase as number of nodes start to increase, but still we can see this difference is not so high in L2-ABF even all nodes are part of networks.

7   CONCLUSION

This paper presents our proposed routing algorithm with the name of L2-ABF for UWSN. To prevent the flooding on whole network, L2-ABF used the calculated flooding zone to forward the data packets toward upper layer nodes. The novelty of this algorithm is that the routing does not depend on any type of geographical information or special type of hardware and configuration. We tried to keep the routing overhead minimal with low bandwidth and extremely low data bit rates for UWSN. In this algorithm every nodes have the capability to calculate its flooding zone to forward data packets with control flooding. L2-ABF uses the idea of angle based routing, instead of distance routing or dynamic addressing base routing according to the requirements of underwater environments. The main advantage of L2-ABF is that the delivery ratios do not depend on the sparseness or the density of the sensor nodes. The simulation results shows, that it can achieve high delivery ratios around 90% for both sparse and dense network. The enhancement of simulation results and angle zone optimization can be the future plan for this algorithm.

REFRENCES

1. Ayaz, M., A. Abdullah.: An efficient Dynamic Addressing based routing protocol for Underwater Wireless Sensor Networks. Comput. Commun., 2012. 35(4): p. 475-486 (2012)
2. Heidemann, J., Li, Yuan.: Research challenges and applications for underwater sensor networking. in Wireless Communications and Networking Conference, 2006. WCNC 2006. IEEE (2006)
3. Ayaz, M., A. Abdullah.: Hop-by-Hop Dynamic Addressing Based (H2-DAB) Routing Protocol for Underwater Wireless Sensor Networks, in Proceedings of the 2009 International Conference on Information and Multimedia Technology. IEEE Computer Society. p. 436-441 (2009)
4. Wahid, A., L. Sungwon, K. Dongkyun.: An energy-efficient routing protocol for UWSNs using physical distance and residual energy. in OCEANS, 2011 IEEE - Spain (2011)
5. Kai Chen, Y.Z., Jianhua He.: A Localization Scheme for Underwater Wireless Sensor Networks. International Journal of Advanced Science and Technology, Vol. 4 (2009)
6. Zheng, G., Colombi, G.: Adaptive Routing in Underwater Delay/Disruption Tolerant Sensor Networks. in Wireless on Demand Network Systems and Services, 2008. WONS 2008. Fifth Annual Conference on (2008)
7. Xie, P., J.-H. Cui, L. Lao.: VBF: Vector-Based Forwarding Protocol for Underwater Sensor Networks, in Networking 2006. Networking Technologies, Services, and Protocols; Performance of Computer and Communication Networks; Mobile and Wireless Communications Systems. Springer Berlin / Heidelberg. p. 1216-1221 (2006)
8. Daeyoup, H., K. Dongkyun. DFR.: Directional flooding-based routing protocol for underwater sensor networks. in OCEANS (2008)
9. Shin, D., D. Hwang, D. Kim, DFR.: An efficient directional flooding-based routing protocol in underwater sensor networks. Wireless Communications and Mobile Computing (2011)
10. Yan, H., Z.J. Shi, and J.-H. Cui, DBR.: depth-based routing for underwater sensor networks, in Proceedings of the 7th international IFIP-TC6 networking conference on AdHoc and sensor networks, wireless networks, next generation internet. 2008, Springer-Verlag: Singapore. p. 72-86.
11. Jornet, J.M., M. Stojanovic, M. Zorzi.: Focused beam routing protocol for underwater acoustic networks, in Proceedings of the third ACM international workshop on Underwater Networks. 2008, ACM: San Francisco, California, USA. p. 75-82.

12. Zhou, Z., Peng, Z.: Efficient multipath communication for time-critical applications in underwater acoustic sensor networks. IEEE/ACM Trans. Netw., 2011. 19(1): p. 28-41 (2008)
13. Ali, T. and L.T. Jung.: Flooding control by using Angle Based Cone for UWSNs. in Telecommunication Technologies (ISTT), 2012 International Symposium on (2012)
14. Basagni, S., Petrioli, C.: Optimizing network performance through packet fragmentation in multi-hop underwater communications. in OCEANS 2010 IEEE - Sydney (2010)
15. Wang, D., F.J.Pierre.: Acoustically focused adaptive sampling and on-board routing for marine rapid environmental assessment. Journal of Marine Systems, 2009. 78(Supplement 1): p. S393-S407 (2009)
16. Ayaz, M., A. Abdullah.: A survey on routing techniques in underwater wireless sensor networks. Journal of Network and Computer Applications, 2011. 34(6): p. 1908-1927 (2011)

Analyzing the Performance of Low Stage Interconnection Network

Mehrnaz Moudi and Mohamed Othman

Department of Communication Technology and Network,
Universiti Putra Malaysia, 43400 UPM, Serdang, Selangor D.E., Malaysia.
mehrnazmoudi@gmail.com, mothman@upm.edu.my

Abstract. In order to avoid crosstalk, a new architecture is proposed for Optical Multistage Interconnection Networks (OMINs). In the new architecture, two switches are replaced by one switch in each row. Reduction in the number of switches makes the considerable reduction in the execution time. To study the performance of the new architecture, analytical techniques also can be used effectively. The theory of probability is used to derive mathematical equation for network bandwidth allocation of a unit load. The obtained results show the improvement in the network performance. By increasing load, the bandwidth is reduced. In addition the simulation is applied to validate the new architecture and show improvement in the performance by approximately 30 % reduction in the execution time.

Keywords: Bandwidth; Crosstalk; Low Stage Interconnection Network; Switch Element.

1 Introduction

Multistage Interconnection Networks (MINs) are key elements in switching, communication applications, telecommunication and parallel computing systems [8]. MIN consists of N inputs, N outputs, and n stages ($n = log_2 N$). The connection among the Switching Elements (SEs) in these stages has a certain pattern which any network input port can be connected to any output port. Although the electronic MINs were the most widely used networks, with growing demand for bandwidth, optical technology is an interesting procedure to implement MINs. Also it used to transmit the messages. The two fundamental differences between the electronic MINs and the optical MINs (OMINs) are the optical-loss during switching and the crosstalk problem in the optical switches [19], [15]. Fig. 1 shows a general model of OMIN.

Several proposed architectures for OMINs are Delta networks, Omega networks, Generalized Cube networks, Clos networks, Crossbar networks, Benes networks and so on [18]. In this research, we focused on the Omega network architecture. In Omega networks, transferring messages between every output to every input is fully accessible [6], [2], [14]. The crosstalk in Omega network, the result of interacting two signal channels with each other in each SE, is the

T. Herawan et al. (eds.), *Proceedings of the First International Conference on Advanced Data and Information Engineering (DaEng-2013)*, Lecture Notes in Electrical Engineering 285, DOI: 10.1007/978-981-4585-18-7_57,
© Springer Science+Business Media Singapore 2014

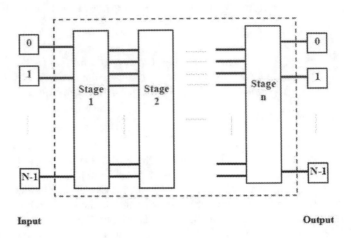

Fig. 1: General Model of OMIN

main drawback of optical switches, although there are many significant interests by using optical technology to implement interconnection networks and switches [1], [4], [5]. There are two types of logic states for SE connection. The straight or cross connection in the SE are shown in Fig. 2.

a) Straight Connection   b) Cross Connection

Fig. 2: Straight and Cross Logic State of a 2 × 2 SE

According to the previous studies, there are the crosstalk-free scheduling algorithms to route in OMINs to solve the crosstalk. These algorithms are such as four heuristic algorithms, Genetic Algorithm (GA), Simulated Annealing algorithm (SA), Zero algorithm and Ant Colony Optimization algorithm (ACO) [12]. As the performance of communication networks are affected by efficient message routing algorithms directly, the first goal for all of these routing algorithms is to avoid crosstalk [3],[10], [14]. In this paper, we will evaluate the performance of the new architecture for Omega network by doing probabilistic analysis to find out the bandwidth of the network in different loads. The main advantage of the proposed architecture is making fast crosstalk-free network with lower cost in terms of hardware. The new architecture is designed to improve the network performance in terms of reducing the execution time and improving bandwidth.

The remainder of the paper is organized as follows: Section 2 introduces the new architecture. Performance setup, simulation and analysis are presented in Section 3. In Section 4, the analytical results are discussed. Finally, Section 5 concludes this paper.

2 Low Stage Interconnection Network

The Low Stage Interconnection Network is new architecture for OMIN. In each row of the Low Stage architecture, two SEs converts to one SE. The number of SEs is reduced and the number of stages will be less than the original. Three models are introduced for this architecture. Although, the transformation in Low Stage can be applied to Omega networks with even number of stages. The new number of stages for Low Stage Interconnection Network is presented in Table 1. After calculating the new number of stages and grouping switches according to the configuration of these models, we investigate routing in the new architecture to improve the bandwidth in the less execution time.

Table 1: Number of Stages in Low Stage Models

Models	Number of stages
First Low Stage	$(\log_2 N) - 1$
Second Low Stage	$(\log_2 N)/2 + 1$
Third Low Stage	$(\log_2 N) - 1$

An $N \times N$ new architecture can be divided into three models. The construction of Low Stage Interconnection Networks is described in Figs. 3, 4 and 5.

In the First Low Stage (FLS), the switches in each row for two middle stages are merged as one group, Group 1, which applies like one switch. Then new number of stages for this model has been formulated in Table 1. Fig.3 shows the FLS, which is the first model of the new architecture.

For the Second Low Stage (SLS) model, all of the switches except the switches in the first and last stage in each row are merged together. In this way, merging starts from the second stage and merges this switch with the next switch in the third stage; the switch in fourth stage will be merged with the switch in fifth stage and it will continue until finishing the switches except the switches in the first and the last stage for each row. The number of groups of switches for each row is $(n/2) - 1$ groups except the switches in the first and last stage. This model is shown in Fig.4.

The Third Low Stage (TLS) model is applied on the switches in each row from left to right except the first and last stage, and then makes $(n - 3)$ groups of

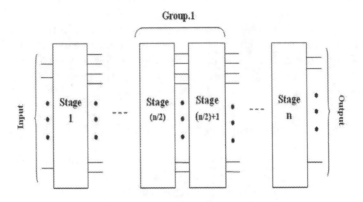

Fig. 3: First Low Stage Model of the OMIN

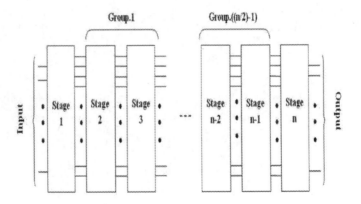

Fig. 4: Second Low Stage Model of the OMIN

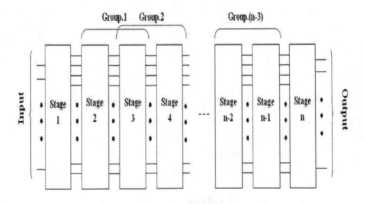

Fig. 5: Third Low Stage Model of the OMIN

switch for each row. This model is useful for big size networks. Fig.5 illustrates the construction of this model. The new number of stages is calculated like the first model.

3 Performance Analysis

In this paper, we study the performance of OMIN with respect to crosstalk problem by applying new architecture for the Omega network [7]. The scheduling messages without crosstalk in the Low Stage Interconnection Networks and OMIN are similar together. At each time slot, one group of messages is sent through the network to avoid the effect of crosstalk. The goal is proposing a new architecture for an interconnection network to quickly route all messages without crosstalk and lower cost. Two used parameters for performance evaluation are bandwidth and execution time. Throughout this research, for the new architecture, the focus was on the bandwidth to improve the performance of Low Stage Interconnection Network.

First parameter is bandwidth for evaluation in this paper. The bandwidth for the low Stage Interconnection Network will be evaluated by doing probabilistic analysis in different loads. It can be a good way to analyze the performance by calculating the bandwidth under a load l which is the probability for an input that is active or otherwise [9].

Fig.6 shows three different models of switches. They can be 2-active, 1-active or 0-active. It is clear that 0-active and 1-active switches do not allow any crosstalk whereas the 2-active switch produces crosstalk [9]. P is defined as the probability for existed request at an input link of a switch then:

P_2 = probability for 2-active switch,
P_1 = probability for 1-active switch,
P_0 = probability for 0-active switch.

The following equations were derived to find the bandwidth.

$$P_2 = l \times l = l^2 \tag{1}$$

$$P_1 = l \times (1 - l) + (1 - l) \times l = 2l \times (1 - l) \tag{2}$$

$$P_0 = (1 - l) \times (1 - l) = (1 - l)^2 \tag{3}$$

Eqs (1) to (3) are utilized to obtain the value of P_n. The bandwidth of n-stage for the network ($N = 2^n$) is expressed as:

$$BW = P_n \times 2^n \tag{4}$$

Another parameter is execution time which is elapsed time between the beginning and the end of scheduling algorithm. It is calculated for each algorithm including from the source and destination generation time until scheduling the messages without crosstalk for each permutation set [1],[3]. To route the messages, we find list of messages by using Bitwise Window Method (BWM) [1] to identify any crosstalk. This method is the last improvement for Window Method (WM)[1] that is a used method between messages, which should not be in the same group because of crosstalk. The advantage of BWM compared with WM is saving in time to find crosstalk. Next step is drawing a conflict matrix with new number of stages. It is a square matrix with $N \times N$ entry where N is the network size. Based on the conflict matrix, the routing algorithm is used to sort and select the messages for scheduling in different independent subsets to avoid the crosstalk in the new architecture.

The used scheduling algorithm in this paper is the Sequential Decreasing. The main reason to select this algorithm is scheduling the messages in different independent subsets in decreasing order to avoid the crosstalk. The scheduling algorithm is executed 1000 times to attain the accurate results for various network sizes. The speed increment for the large network size is obvious. Based on the evaluation, the results show that the Low Stage Interconnection Network can achieve high performance with a reduction in execution time and improvement for bandwidth.

4   Analytical Results

In Fig.6, obliviously the execution time is reduced in the Low Stage Interconnection Network. While this reduction for the SLS and TLS models are more than the FLS model. Approximately this reduction in SLS is more than all of them. The reason is related to the calculation way of the number of stages. In this model messages can be scheduled faster than the other two models. Although, the execution time for SLS is roughly similar to TLS model. Overall, the execution time of Low Stage Interconnection Network is approximately 30% less than the OMIN.

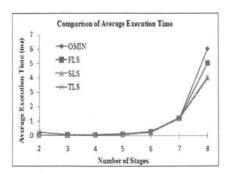
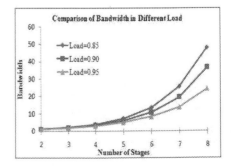

Fig. 6: Execution Time for Low Stage Interconnection Network

Fig. 7: Bandwidth versus Number of Stages for SLS Model

As SLS model obtained better result in terms of execution time, we evaluate bandwidth in the selected model. In Fig.7, bandwidth for SLS model of the Low Stage Interconnection Network has been demonstrated in l=0.85, l=0.90 and l=0.95. For small size of Low Stage Interconnection Networks, provided bandwidth is nearly equal for different loads, where in the large network size by increasing load, the bandwidth will decrease. In different load, the difference among bandwidth proves the significant role for the load on the bandwidth. This role shows that by increasing the load, there is a reduction in bandwidth.

5 Conclusion

The contribution of this research is studying the performance of the proposed architecture by analytical techniques. A new Interconnection Network architecture used to schedule permutation connections without crosstalk in terms of the two performance metrics; bandwidth and execution time. Low Stage Interconnection Network is name of the new architecture based on the Omega network. It has three models that used to compare with original OMIN. There is approximately 30% reduction in the execution time for these three models. Another performance metric, the bandwidth, was considered in the second model of Low Stage Interconnection Network. The similar bandwidth for small network size was obtained in different loads, while the bandwidth was reduced in bigger loads obviously.

Acknowledgments. The study was supported by the Research University Grant Scheme (RUGS), Universiti Putra Malaysia (RUGS Number: 05/01/11/12 50RU).

References

1. Abed, F., Othman, M.: Efficient window method in optical multistage interconnection networks. In: IEEE International Conference on Telecommunications and Malaysia International Conference on Communications (ICT-MICC), pp. 181-185. IEEE Press, Malaysia (2007).
2. Almazyad, A.S.: Optical omega networks with centralized buffering and wavelength conversion. Journal of King Saud University - Computer and Information Sciences. 23(1), 15-28 (2010).
3. Al-Shabi, M., Othman, M.: A New Algorithm For Routing And Scheduling In Optical Omega Network. International Journal of the Computer, the Internet and Management. 16(1), 26-31 (2008).
4. Atiquzzaman, M., Akhtar, M.S.: Performance of buffered multistage interconnection networks in a nonuniform trafc environment. Journal of Parallel and Distributed Computing. 30(1), 52-63 (1995).
5. Bashirov, R., Karanller, T.: On path dependent loss and switch crosstalk reduction in optical networks. Information Sciences, 180(6), 1040-1050 (2010).

6. Borella, A., Cancellieri, G., Mantini, D.: Space Division Architectures for Crosstalk Reduction in Optical Interconnection Networks. In: QoS-IP, LNCS, vol. 2601, pp. 460-470. Springer, Heidelberg (2003).
7. Brenner , M., Tutsch, D., Hommel, G.: Measuring transient performance of a multistage interconnection network using ethernet networking equipment, In: International Conference on Communications in Computing (CIC02). pp. 211216. USA, (2002).
8. Garofalakis, J., Stergiou, E.: Analytical model for performance evaluation of Multilayer Multistage Interconnection Networks servicing unicast and multicast traffic by partial multicast operation. Performance Evaluation. 67, 959-976 (2010).
9. Katangur, A.K., Akkaladevi, S., Pan, Y.: Analyzing the performance of optical multistage Interconnection networks with limited crosstalk. Cluster Computing. 10, 241-250 (2007).
10. Katangur, A.K., Akkaladevi, S., Pan, Y., Fraser, M.D.: Applying Ant Colony Optimization to Routing in Optical Multistage Interconnection Networks with Limited Crosstalk, In: 18th International Parallel and Distributed Processing Symposium, IPDPS04 (2004).
11. Lu, E., Zheng, S. Q.: Parallel Routing and Wavelength Assignment for Optical Multistage Interconnection Networks, In: Proceedings of the International Conference on Parallel Processing, ICPP04 (2004).
12. Moudi, M., Othman, M.: A Challenge for Routing Algorithms in Optical Multistage Interconnection Networks. Journal of Computer Science. 7(11), 1685-1690 (2011).
13. Nitin, Kumar Sehgal, V., Sharma, R., Singh Chauhan, D., Srivastava, N., Garhwal, S.: Modified fault tolerant Combining Switches Multistage Interconnection Networks with chaining: Algorithm, design and cost issues. TENCON 2008 - IEEE Region 10 Conference. pp. 1-5 (2008).
14. Othman, M., Shahida, T.D.: The Development of Crosstalk-Free Scheduling Algorithms for Routing in Optical Multistage Interconnection Networks, Trends in Telecommunications Technologies, Christos J Bouras (Ed.), ISBN: 978-953-307-072-8, InTech (2010).
15. Pan Y., Qiao, C., Yang, Y. Wu, J.: Recent developments in optical multistage networks. In: Ruan, L., Du, D.-Z. (Eds.) Optical NetworksRecent Advances, pp. 151185. Kluwer Academic, Norwell (2001).
16. Pan, Y., Qiao, C., Yang, Y.: Optical Multistage Interconnection Networks: New Challenges and Approaches. IEEE Communications Magazine, Feature Topic on Optical Networks, Communication Systems and Devices. 37(2), 50-56 (1999).
17. Shanmugam, G., Ganesan, P., Vanathi, P.T.: Metaheuristic algorithms for vehicle routing problem with stochastic demands. J. Comput. Sci. 7, 533-542 (2011).
18. Tian, H., Katangur, A. K., Zhong, J., Pan, Y.: A Novel Multistage Network Architecture with Multicast and Broadcast Capability. Journal of Supercomputing. 35(3), 277-300 (2006).
19. Vaez, M.M., Lea, C.-T.: Strictly nonblocking directional-coupler-based switching networks under crosstalk constraint. IEEE Transaction Communication. 48, 316323 (2000).

Concurrent Context-Free Grammars

Gairatzhan Mavlankulov[1,*], Mohamed Othman[1,**], Mohd Hasan Selamat[1],
and Sherzod Turaev[2]

[1]Department of Communication Technology and Network, Universiti Putra Malaysia
43400 UPM Serdang, Selangor D.E., Malaysia
[2]Department of Computer Science, International Islamic University Malaysia 53100
Gombak, Selangor D.E., Malaysia
{gairatjon@gmail.com,mothman@upm.edu.my,
hasan@upm.edu.my,sherzod@iium.edu.my}

Abstract. In this paper we study some properties of context-free concurrent grammars which are controlled by Petri nets under parallel firing strategies, i.e., the transitions of a Petri net fire simultaneously in different modes. These variants of control increase the computational power of generative devices. Moreover, they can be used as theoretical models for parallel computing devices.

Key words: Petri nets, parallel firing, controlled grammars, parallel computing

1   Introduction

Recently in [2,3,4,5,11] were introduced different variants of a Petri net controlled grammar, which is a context-free grammar equipped with a Petri net, whose transitions are labeled with rules of the grammar or the empty string, and the associated language consists of all terminal strings which can be derived in the grammar. The sequence of rules in every terminal derivation corresponds to some occurrence sequence of transitions of the Petri net which is enabled at the initial marking and finished at a final marking of the net. It can be considered as mathematical models for the study of concurrent systems appearing in systems biology and automated manufacturing systems. The distinguished feature of all of these variants is that the transitions of a Petri net fire sequentially. In this paper we introduce a new variant of theoretical models for parallel computation using Petri nets under parallel firing strategies, called grammars controlled by Petri nets under parallel firing strategies (concurrent grammars), i.e. the transitions of a Petri net fire simultaneously in different modes. We show two examples of context free concurrent grammars which can generate non-context free languages. Noted, these languages can not be generated by Petri Net controlled grammars in the sequential case.

---

* Corresponding author
** The author is also an associate researcher at Computational Science and Mathematical Physics Lab., Mathematical Science Ins., UPM.

T. Herawan et al. (eds.), *Proceedings of the First International Conference
on Advanced Data and Information Engineering (DaEng-2013)*, Lecture Notes
in Electrical Engineering 285, DOI: 10.1007/978-981-4585-18-7_58,
© Springer Science+Business Media Singapore 2014

2   Preliminaries

2.1   Grammars and languages

A *context-free grammar* is a quadruple $G = (V, \Sigma, S, R)$ where V and Σ are the disjoint finite sets of *nonterminal* and *terminal* symbols, respectively, $S \in V$ is the *start* symbol and $R \subseteq V \times (V \cup \Sigma)^*$ is a finite set of *(production) rules*. Usually, a rule (A, x) is written as $A \to x$. A rule of the form $A \to \lambda$ is called an *erasing rule*. $x \in (V \cup \Sigma)^+$ *directly derives* $y \in (V \cup \Sigma)^*$, written as $x \Rightarrow y$, iff there is a rule $r = A \to \alpha \subset R$ such that $x = x_1 A x_2$ and $y = x_1 \alpha x_2$. The rule $r : A \to \alpha \in R$ is said to be *applicable* in sentential form x, if $x = x_1 A x_2$, where $x_1, x_2 \in (V \cup \Sigma)^*$ The reflexive and transitive closure of $\Rightarrow$ is denoted by $\Rightarrow^*$. A derivation using the sequence of rules $\pi = r_1 r_2 \cdots r_n$ is denoted by $\xrightarrow{\pi}$ or $\xrightarrow{r_1 r_2 \cdots r_n}$. The *language* generated by G is defined by $L(G) = \{ w \in \Sigma^* \mid S \Rightarrow^* w \}$. The family of context-free languages is denoted by $\mathcal{CF}$.

2.2   Multisets

A *multiset* over an alphabet Σ is a mapping $\mu : \Sigma \to \mathbb{N}$. The set Σ is called the *basic set* of a multiset ν and the elements of Σ is called the *basic elements* of a multiset μ. A multiset μ over an alphabet $\Sigma = \{a_1, a_2, \ldots a_n\}$ can be denoted by

$$\mu = (\mu(a_1)a_1, \mu(a_2)a_2, \ldots, \mu(a_n)a_n)$$

where $\mu(a_i)$, $1 \le i \le n$, is the multiplicity of a_i, or as a vector

$$\mu = (\mu(a_1), \mu(a_2), \ldots, \mu(a_n)),$$

or as the set in which each basic element $a \in \Sigma$ occurs $\mu(a)$ times

$$\mu = \{\underbrace{a_1, \ldots, a_1}_{\mu(a_1)}, \underbrace{a_2, \ldots, a_2}_{\mu(a_2)}, \ldots, \underbrace{a_n, \ldots, a_n}_{\mu(a_n)}\}.$$

The empty multiset is denoted by ϵ, that is $\epsilon(a) = 0$ for all $a \in \Sigma$. The set of all multisets over Σ is denoted by $\Sigma^\oplus$. Since Σ is finite, $\Sigma^\oplus = \mathbb{N}^{|\Sigma|}$. The power (or cardinality) of a multiset $\mu = (\mu(a_1), \mu(a_2), \ldots, \mu(a_n))$ denoted by $|\mu|$, is $\sum_{i=1}^{n} \mu_i$. A multiset μ is a *set* if and only if $\mu(a) \le 1$ for all $a \in \Sigma$. For two multisets μ and ν over the same alphabet Σ, we define

– the *inclusion* $\mu \sqsubseteq \nu$ by

$$\mu \sqsubseteq \nu \text{ if and only if } \mu(a) \le \nu(a) \text{ for all } a \in \Sigma;$$

– the *sum* $\mu \oplus \nu$ by

$$(\mu \oplus \nu)(a) = \mu(a) + \nu(a) \text{ for each } a \in \Sigma,$$

and we denote the sum of multisets $\mu_1, \mu_2, \ldots, \mu_k$ by $\sum_{i=1}^k \mu_i$, i.e.,

$$\sum_{i=1}^k \mu_i = \mu_1 \oplus \mu_2 \oplus \cdots \oplus \mu_k;$$

– the *difference* $\mu \ominus \nu$ by

$$(\mu \ominus \nu)(a) = \max\{0, \mu(a) - \nu(a)\} \text{ for each } a \in \Sigma.$$

2.3 Petri nets

A Petri net is a triple (P, T, δ) where P and T are finite disjoint sets of *places* and *transitions*, respectively, a mapping $\delta : T \to P^{\oplus} \times P^{\oplus}$ is a mapping which assigns to each transition $t \in T$ a pair $\delta(t) = (\alpha, \beta)$ where α and β. Graphically, a Petri net is represented by a bipartite directed graph with the node set $P \cup T$ where places are drawn as *circles*, transitions as *boxes*. For each transition $t \in T$ with $\delta = (\alpha, \beta)$, the multiplicities $\alpha(p)$, $\beta(p)$ of a place $p \in P$, give the number of arcs from p to t and from t to p, respectively. A multiset $\mu \in P^{\oplus}$ is called a *marking*. For each $p \in P$, $\mu(p)$ gives the number of *tokens* in p. Graphically, tokens are drawn as small solid *dots* inside circles.

A *place/transition net* (p/t net for short) is a quadruple $N = (P, T, \delta, \mu_0)$ where (P, T, δ) is a Petri net, $\iota \in P^{\oplus}$ is the *initial marking*.

A transition $t \in T$ with $\delta(t) = (\alpha, \beta)$ is *enabled* at a marking $\mu \in P^{\oplus}$ if and only if $\alpha \sqsubseteq \mu$. In this case t can *occur* (*fire*). Its occurrence transforms the marking μ into the marking $\mu' \in P^{\oplus}$ defined by $\mu' = \mu \ominus \alpha \oplus \beta$. We write $\mu \xrightarrow{t}$ to denote that t may fire in μ, and $\mu \xrightarrow{t} \mu'$ to indicate that the firing of t in μ leads to μ'. A finite sequence $t_1 t_2 \cdots t_k$, $t_i \in T, 1 \leq i \leq k$, is called *an occurrence sequence* enabled at a marking μ and finished at a marking μ_k if there are markings $\mu_1, \mu_2, \ldots, \mu_{k-1}$ such that

$$\mu \xrightarrow{t_1} \mu_1 \xrightarrow{t_2} \ldots \xrightarrow{t_{k-1}} \mu_{k-1} \xrightarrow{t_k} \mu_k.$$

In short this sequence can be written as $\mu \xrightarrow{t_1 t_2 \cdots t_k} \mu_k$ or $\mu \xrightarrow{\nu} \mu_k$ where $\nu = t_1 t_2 \cdots t_k$. For each $1 \leq i \leq k$, marking μ_i is called *reachable* from marking μ. $\mathcal{R}(N, \mu) \subseteq P^{\oplus}$ denotes the set of all reachable markings from a marking μ.

Let $N = (P, T, \delta, \iota)$ be a p/t net and $F \subseteq \mathcal{R}(N, \iota)$ be a set of markings which are called *final markings*. An occurrence sequence ν of transitions is called *successful* for F if it is enabled at the initial marking ι and finished at a final marking τ of F. If F is understood from the context, we say that ν is a *successful occurrence sequence*.

A *labeled Petri net* is a tuple $K = (\Delta, N, \gamma, F)$ where Δ is an alphabet, $N = (P, T, \delta, \iota)$ is a p/t net, $\gamma : T \to \Delta \cup \{\lambda\}$ is a transition labeling function and $F \subseteq \mathcal{R}(N, \iota)$.

2.4 Context-Free Petri nets

A context-free Petri net is a Petri net $N = (P, T, F, \phi, \beta, \gamma, \iota)$ where
- labeling function $\beta : P \to V$ and $\gamma : T \to R$ are bijections;
- there is an arc from place p to transition t if and only if $\gamma(t) = A \to \alpha$ and $\beta(p) = A$. The weight of the arc (p, t) is 1;
- there is an arc from transition t to place p if and only if $\gamma(t) = A \to \alpha$ and $\beta(p) = \chi$ where $|\alpha|_\chi > 0$. The weight of the arc (t, p) is $|\alpha|_\chi$;
- the initial marking ι is defined by $\iota(b^{-1}(S)) = 1$ and $\iota(p) = 0$ for all $p \in P - \{\beta^{-1}(S)\}$

Example 1. Let G_1 be a context-free grammar with the rules:

$$r_0 : S \to AB,\ r_1 : A \to aAb,\ r_2 : A \to ab,\ r_3 : B \to cB,\ r_4 : B \to c$$

(the other components of the grammar can be seen from these rules). Figure 1 illustrates a cf Petri net N_1 with respect to the grammar G_1. Obviously, $L(G_1) = \{a^n b^n c^m \mid n, m \geq 1\}$.

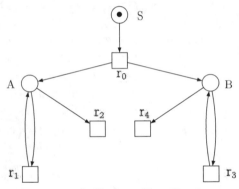

Figure 1. A Context-Free Petri net

3 Definitions, Examples, Results

3.1 Multisteps

Definition 1. *Let* $K = (\Sigma, N, \gamma, F)$, $N = (P, T, \delta, \iota)$, *be a labeled Petri net and let* $A = \{t_1, t_2, \ldots, t_k\} \subseteq T$ *with* $\delta(t_i) = (\alpha_i, \beta_i)$ *for* $1 \leq i \leq k$. *The transitions of a multiset* $\nu \in A^{\oplus}$ *are simultaneously/parallelly enabled/firable at a marking* $\mu \in \mathcal{R}(N, \iota)$ *if and only if*

$$\sum_{i=1}^{k} \nu(t_i)\alpha_i \sqsubseteq \mu.$$

Then the transitions of ν *parallelly fire* resulting in the new marking μ' defined by

$$\mu' = \mu \ominus \sum_{i=1}^{k} \nu(t_i)\alpha_i \oplus \sum_{i=1}^{k} \nu(t_i)\beta_i.$$

A multiset ν whose transitions fire parallelly is called a *multistep*. We write $\mu \xrightarrow[m]{\nu} \mu'$ to denote that the a multistep ν at μ leads to μ'.

Further, we define some special types of multisteps with respect to the basic sets and multisets.

Let $A = \{t_1, t_2, \ldots, t_k\} \subseteq T$ with $t_i = (\alpha_i, \beta_i)$, $1 \leq i \leq k$, and let a multistep $\nu \in A^\oplus$ be enabled at a marking $\mu \in P^\oplus$.

- The multistep ν is called *global* if $\nu(t) > 0$ for all $t \in A$ and
  - $A = T$ or
  - for all B where $A \subset B \subseteq T$, and for all extensions $\bar{\nu} \in B^\oplus$ of ν,

$$\sum_{t \in B} \bar{\nu}(t)\alpha \not\sqsubseteq \mu.$$

- The multistep ν is called *labeled* if $A = T_x = \{t \in T : \gamma(t) = x\}$ for some $x \in \Sigma$.
- The multistep ν is called *maximal* if and only if for all $\eta \in A^\oplus$,

$$\nu \sqsubseteq \eta \text{ and } \sum_{i=1}^{k} \eta(t_i)\alpha_i \sqsubseteq \mu \text{ imply } \eta = \nu.$$

- The multistep ν is called a *step* if ν is a set, i.e., $\nu \subseteq A$.

We denote a multistep, a global multistep, a labeled multistep, a maximal multistep, a step and a maximal step by m, g, γ, $\widehat{m}$, s and $\widehat{s}$, respectively. We also use general notation a *multistep in f-mode*, $f \in \{m, g, s, \gamma, \widehat{m}, \widehat{s}\}$ taking into consideration above defined types of multisteps. It is also of interest to consider combined variants of multisteps such as a *maximal labeled multistep*, a *maximal labeled step*, a *global and maximal multistep* and a *global and maximal step* denoted by $\widehat{m}_\gamma$, $\widehat{s}_\gamma$, $\widehat{m}_g$, and $\widehat{s}_g$, respectively.

We write $\mu \xrightarrow[f]{\nu} \mu'$ to indicate that the multistep ν in f-mode, at μ leads to μ', where $f \in \{m, s, \gamma, \widehat{s}, \widehat{m}, \widehat{m}_\gamma, \widehat{s}_\gamma, \widehat{m}_g, \widehat{s}_g\}$.

Let $A_i \subseteq T$, $1 \leq i \leq n$. A finite sequence $(\nu_1, \nu_2, \ldots, \nu_n)$, $\nu_i \in A_i^\oplus$, $1 \leq i \leq n$, is called a *multistep sequence* leading from a marking μ to a marking μ' if there are markings $\mu_1, \mu_2, \ldots, \mu_{n-1}$ such that

$$\mu \xrightarrow[f]{\nu_1} \mu_1 \xrightarrow[f]{\nu_2} \mu_2 \cdots \mu_{n-1} \xrightarrow[f]{\nu_n} \mu'$$

where $f \in \{m, s, \gamma, \widehat{s}, \widehat{m}, \widehat{m}_\gamma, \widehat{s}_\gamma, \widehat{m}_g, \widehat{s}_g\}$. In short the multistep sequence can be written as $\mu \xrightarrow[f]{\nu_1 \nu_2 \cdots \nu_n} \mu'$.

3.2 Concurrent context-free grammar and language

Let $K = (\Sigma, N, \gamma, F)$, $N = (P, T, \delta, \iota)$, be a labeled Petri net and let $\nu \in A^\oplus$, for some subset $A = \{t_1, t_2, \ldots, t_k\} \subseteq T$. We define a multiset $\gamma(\nu)$ by

$$\gamma(\nu) = \{\underbrace{\gamma(t_1), \ldots, \gamma(t_1)}_{\nu(t_1)}, \ldots, \underbrace{\gamma(t_k), \ldots, \gamma(t_k)}_{\nu(t_k)}\}.$$

We use the notion $\gamma^{-\lambda}(\nu)$ to denote the multiset of λ-free elements of $\gamma(\nu)$, i.e.,

$$\gamma^{-\lambda}(\nu) = \{\gamma(t) \in \gamma(\nu) : \gamma(t) \neq \lambda\}.$$

Definition 2. *A concurrent grammar is a tuple* $\mathcal{G} = (V, \Sigma, S, R, N, \gamma, F)$ *where* $G = (V, \Sigma, S, R)$ *is a context-free grammar and* $N = (P, T, \delta, \iota)$ *is a p/t net,* $\gamma : T \to R \cup \{\lambda\}$ *is a transition labeling function and* F *is a set of final markings.*

Definition 3. *We say that* $x \in (V \cup \Sigma)^*$ *directly derives* $y \in (V \cup \Sigma)^*$ *in f-mode,* $f \in \{m, s, \gamma, \widehat{s}, \widehat{m}, \widehat{m}_\gamma, \widehat{s}_\gamma, \widehat{m}_g, \widehat{s}_g\}$*, written as* $x \underset{f}{\Longrightarrow} y$ *if and only if the following conditions are satisfied:*

(1) $x = x_1 A_{i_1} x_2 A_{i_2} \cdots x_k A_{i_k} x_{k+1}$ *and* $y = x_1 u_{i_1} x_2 u_{i_2} \cdots x_k u_{i_k} x_{k+1}$*, where*
  (a) $\rho = \{r_{i_j} : A_{i_j} \to u_{i_j}\} \in R'^{\oplus}$ *for some* $R' \subseteq R$*,*
  (b) $x_i, y_i \in (V \cup \Sigma)^*$ *for* $1 \leq j \leq k+1$*;*
(2) there is a multiset $\nu \in T'^{\oplus}$ *for some* $T' \subseteq T$*, the transitions of which are firable at a marking* $\mu \in P^{\oplus}$ *in f-mode, such that* $\rho = \gamma^{-\lambda}(\nu)$*.*

Definition 4. *The language* $L(\mathcal{G})$ *generated by the concurrent grammar* $\mathcal{G}$ *in f-mode consists of all words* $w \in \Sigma^*$ *such that there is a derivation*

$$S \underset{f}{\overset{\rho_1}{\Longrightarrow}} w_1 \underset{f}{\overset{\rho_2}{\Longrightarrow}} \cdots \underset{f}{\overset{\rho_n}{\Longrightarrow}} w_n = w$$

where $\rho_i \in R_i^{\oplus}$*,* $R_i \subseteq R$*,* $1 \leq i \leq n$*, and there is an multistep occurrence in f-mode in* (Σ, N, γ, F)

$$\iota \underset{f}{\overset{\nu_1}{\longrightarrow}} \mu_1 \underset{f}{\overset{\nu_2}{\longrightarrow}} \cdots \underset{f}{\overset{\nu_n}{\longrightarrow}} \mu_n, \mu_n \in F,$$

where $\nu_i \in T_i^{\oplus}$*,* $T_i \subseteq T$*,* $1 \leq i \leq n$*, such that* $\rho_i = \gamma^{-\lambda}(\nu_i)$*.*

3.3   Examples and Results

In this section we show some examples where parallel firing strategy in Petri Net controlled grammars can generate some *non context free languages* which can not be generated in a sequential case. These facts show that controlling by parallel firing strategy has more computational power than controlling by sequential case.

Example 2. Let $G_2 = (V, \Sigma, S, R)$ is context-free grammar , where $V = \{S, A\}$, $\Sigma = \{a, b\}$, $R = \{S \to AA, A \to aA, A \to a\}$,for all $a \in \Sigma$.

If we consider the maximal labeled mode, it can be easily seen that the grammar generate the language $L(G_2) = \{ww : w \in \Sigma\}$ which is not context-free language. For example, if $\Sigma = \{a, b\}$, the set of labeled rules will be as

r_1: $S \to AA$
r_2: $A \to aA$
r_3: $A \to bA$
r_4: $A \to a$
r_5: $A \to b$.

The derivation steps for generating word **aaabaaab** would be like

$S \overset{r_1}{\Rightarrow} AA \overset{r_2}{\Rightarrow} aAaA \overset{r_2}{\Rightarrow} aaAaaA \overset{r_2}{\Rightarrow} aaaAaaaA \overset{r_5}{\Rightarrow} aaabaaab$

Example 3. Let $G_3 = (V, \Sigma, S, R)$ is context-free grammar , where $R=\{r_1 : S \to SS$, $r_2 : S \to a\}$ and $\Sigma = \{a\}$.

Here also if we consider maximal labeled mode, it is obvious, using r_1, we get doubled number of **S**'s in each derivation step.
$S \overset{r_1}{\Rightarrow} S^2 \overset{r_1}{\Rightarrow} S^4 \overset{r_1}{\Rightarrow} S^8 .... \overset{r_1}{\Rightarrow} S^{2^k}$.
Application of the r_2 rule in any step replaces all **S**'s with a's, consequently $S \overset{*}{\Rightarrow} a^{2^k}$. Therefore $L(G_3) = \{a^{2^n} : n \geq 0\}$ which is not context-free.

Example 4. Let $G_4 = (V, \Sigma, S, R)$ is context-free grammar , where $R=\{r_1 : S \to SS$, $r_2 : S \to AAA, r_3 : S \to \lambda, r_4 : A \to a, r_5 : A \to b, r_6 : A \to c\}$ and $\Sigma = \{a, b, c\}$.

Now we will consider concurrent grammar in global mode. Let us consider derivation steps of this grammar. In the starting point we choose one of the rules r_1, r_2 or r_3. If we choose r_3, obviously we get an empty string. $S \overset{r_3}{\Rightarrow} \lambda$. If we choose r_2, first derivation step will be as $S \overset{r_2}{\Rightarrow} AAA$. In the next step we must apply r_4, r_5 or r_6 rules in one step(choosing nondeterministically), so we get strings $S \overset{r_2}{\Rightarrow} AAA \xrightarrow{\{r_4,r_5,r_6\}} \{abc, acb, bac, bca, cab, cba\}$. Other strings will be generated by using r_1 in the starting point $S \overset{r_1}{\Rightarrow} SS$. In the next step we must choose two different rules from the set of rules $\{ r_1, r_2, r_3 \}$. Let us consider all cases. If we choose rules r_1, r_3 and apply them any times it can be easily seen that derivation will not be changed $S \overset{r_1}{\Rightarrow} SS \xrightarrow{\{r_1,r_3\}^+} SS$. If we choose and apply rules r_2, r_3 we get $S \overset{r_1}{\Rightarrow} SS \xrightarrow{\{r_2,r_3\}} AAA$ and consequently $S \overset{r_1}{\Rightarrow} SS \xrightarrow{\{r_2,r_3\}} AAA \xrightarrow{\{r_4,r_5,r_6\}} \{abc, acb, bac, bca, cab, cba\}$. By applying the rules r_1, r_2 n times we get derivation
$S \overset{r_1}{\Rightarrow} SS \xrightarrow{r_1,r_2} \{SSAAA, AAASS\} \xrightarrow{\{r_1,r_2\}^n} A^k SSA^l$, where $k, l \geq 0$ and $k + l = 3n$. If inside of these steps we apply the rules r_1, r_3 derivation will not be changed as we showed before. Therefore another combination of applicable rules will be r_2, r_3. $A^k SSA^l \xrightarrow{\{r_2,r_3\}} A^{3(n+1)}$, for generate strings $w : w \in \Sigma^*$ it is necessary go to place A. Otherwise if transition will go to place S the derivation will be stopped, because there is no S in derivation. So , using rules r_4, r_5, r_6 $n+1$ times we get $A^{3(n+1)} \xrightarrow{\{r_4,r_5,r_6\}^{n+1}} \{w : w \in \Sigma^*, \Sigma = \{a, b, c\}, where \#a = \#b = \#c\}$

4 Conclusion

We have introduced a new variant of theoretical models for parallel computation using Petri nets under parallel firing strategies, called grammars controlled by Petri nets under parallel firing strategies (i.e., concurrent grammars), which are natural formal models of concurrent, asynchronous, distributed, parallel, nondeterministic and stochastic systems. We have defined various concurrent grammars with respect to classes of Petri nets, firing modes, labeling strategies and final marking sets. Also we have shown some examples of concurrent grammars which can generate non-context free languages, for instance, the languages $L_1 = \{ww : w \in \Sigma\}$, $L_2 = \{a^{2^n} : n \geq 0\}$, $L_3 = \{w : w \in \Sigma^*, \Sigma = \{a, b, c\}, \#a = \#b = \#c\}$. Comparing the generative power of concurrent grammars with that of grammars in Chomsky hierarchy and other parallel grammars is our future works.

Acknowledgments. This work has been supported by Ministry of Higher Education of Malaysia via Fundamental Research Grant Scheme FRGS /1/11/SG/UPM/01/1

References

1. Hans-Dieter Burkhard. Ordered Firing in petri nets. Information Processing and Cybernetcs, 2/3:71-86, 1989.
2. J. Dassow and S. Turaev. k-Petri net controlled grammars. Language and Automata Theory and Applications. Second International Conference, LATA 2008. Revised Papers, volume 5196 of LNCS, pages 209-220. Springer, 2008.
3. J. Dassow and S. Turaev. Grammars controlled by special Petri nets. Language and Automata Theory and Applications, Third International Conference, LATA 2009, volume 5457 of LNCS, pages 326-337. Springer, 2009.
4. J. Dassow and S. Turaev. Petri net controlled grammars: the power of labeling and final markings. Romanian Jour. of Information Science and Technology, 12(2):191-207, 2009.
5. J. Dassow, G. Mavlankulov, M. Othman, S. Turaev, M.H. Selamat and R. Stiebe, Grammars Controlled by Petri Nets, In: P. Pawlewski (ed.) Petri nets, INTECH, 2012, ISBN 978-953-51-0700-2.
6. B. Farwer, M. Jantzen, M. Kudlek, H. Rolke, and G. Zetzsche. Petri net controlled finite automata. Fundamenta Informaticae, 85(1-4):111-121, 2008.
7. B. Farwer, M. Kudlek, and H. Rolke. Concurrent Turing machines. Fundamenta Informaticae, 79(3-4):303-317, 2007.
8. M. Hack. Petri net languages. Computation Structures Group Memo, Project MAC 124, MIT, Cambridge Mass., 1975.
9. M. Jantzen and G. Zetzsche. Labeled step sequences in Petri Nets. In Petri Nets 2008, volume 5062 of LNCS, pages 270-283. Springer, 2008.
10. J.L. Peterson. Petri net theory and modeling of systems. Prentice-Hall, Englewood Clis, NJ, 1981.
11. S. Turaev. Petri net controlled grammars. In Third Doctoral Workshop on Mathematical and Engineering Methods in Computer Science, MEMICS 2007, pages 233-240, Znojmo, Czechia, 2007. ISBN 978-80-7355-077-6.

Effect of the Spin-orbit Interaction on Partial Entangled Quantum Network

Abdel-Haleem Abdel-Aty[1], Nordin Zakaria[1], Lee Yen Cheong[2], and Nasser Metwally[3]*

[1] Computer and Information Science Department, Universiti Teknologi Petronas,
31750 Tronoh, Perak, Malaysia
email: amahmedit@yahoo.com
[2] Fundamental and Applied Science Department, Universiti Teknologi Petronas,
31750 Tronoh, Perak, Malaysia
[3] Mathematics Department, University of Bahrain, 32034 Kingdom of Bahrain

Abstract. Dzyaloshiniskii–Moriya (DM) interaction is used to generate entangled network from partially entangled states in the presence of the spin–orbit coupling. The effect of the spin coupling on the entanglement between any two nodes of the network is investigated. It is shown that the entanglement decays as the coupling increases. For larger values of the spin coupling, the entanglement oscillates between upper and lower bounds. For initially entangled channels, the upper bound does not exceed its initial value, whereas for the channels generated via indirect interaction, the entanglement reaches its maximum value.

Keywords: Entengled quantum network, entanglement, spin-orbit interaction

1 Introduction

Quantum Information Technology (QIT) promises faster, more secure means of data manipulation by making use of the quantum properties of matter [1, 2]. Nowadays, much applications of information science are converted from classical to quantum information and proved that the quantum information give a better results compared with classical one. An examples of QIT , Quantum Neural Network QNN [3], Quantum Computing [4], Quantum Artificial Intelligence QAI [5, 6] and Quantum Communications [7]. One of the most important topics in QIT is the generation of the entangled quantum networks. Quantum networks have been implemented experimentally [7–10] and theoretically [11–15].

The Dzyaloshiniskii–Moriya (DM) interaction is a natural phenomena discovered in 1960 by Moriya (Dzyaloshinskii–Moriya (DM) as an antisymmetric and anisotropic exchange coupling between two spins [16, 17]. It is found that

---

* Please note that the LNCS Editorial assumes that all authors have used the western naming convention, with given names preceding surnames. This determines the structure of the names in the running heads and the author index.

T. Herawan et al. (eds.), *Proceedings of the First International Conference on Advanced Data and Information Engineering (DaEng-2013)*, Lecture Notes in Electrical Engineering 285, DOI: 10.1007/978-981-4585-18-7_59,
© Springer Science+Business Media Singapore 2014

the DM interaction strengthen the entanglement among particles which implies
that the DM interaction plays an important role in the field of quantum networks [18]. The quantum correlation as a result of the DM interaction between
two particles is investigated by many authors(see for examples [19–23]). The
thermal entanglement between two qubits in the Heisenberg XYZ model and
the effect of the DM interaction and its strength are discussed by Da Chuang
and Z. Liang Cao [24]. The effect of the intrinsic decoherence in the teleportation
of two qubits XYZ model is studied in the presence of DM interaction [25].

Metwally [11] introduced a theoretical protocol to generate multi–nodes quantum network by using maximum entangled states, where the terminals of each
disconnected node are connected via DM interaction. The possibility of generating entangled network by using a class of partially entangled network is discussed
by Abdel–Aty et. al, [12]. Therefore, we are motivated to investigate the effect
of the spin–orbit and the efficiency of the generated entangled network in the
presence of DM interaction.

This paper is organized as follows: in section (2) the model and its evolution
are introduced. The entanglement between the different nodes is quantified for
different values of the spin–orbits coupling and the DM's strength are discussed
in section (3). We discuss our results in section (4).

2   The Model

It is assumed that entangled network which consists of four nodes is generated by
using partial entangled states of Werner type [26, 12]. In the suggested quantum
network the node 1 and node 2 is initially connected as well as node 3 and 4 (solid
line). The dash line between the node 2 and node 3 represents the connections
which generated by the DM interaction as in Fig. 1.

Consider a source generates a partial entangled state of the form

$$\rho_{ij} = \frac{1 - F_w}{3} I_4 + \frac{4F_w - 1}{3} |\psi^-\rangle\langle\psi^-| \tag{1}$$

where $ij = 12, 34$, $|\psi^-\rangle = \frac{1}{\sqrt{2}}(|01\rangle - |10\rangle)$ is the singlet Bell state, and F_w is
the maximal fraction corresponding to the Werner-state. The initial state of the
total system is given by

$$\rho_{1234}(0) = \rho_{12} \otimes \rho_{34}. \tag{2}$$

The Hamiltonian describing the evolution of the system for a two-qubit spin-
orbit chain with the DM interaction is switched on the x-axis can be written
as

$$\mathcal{H} = J_x \sigma_x^{(k)} \sigma_x^{(l)} + J_y \sigma_y^{(k)} \sigma_y^{(l)} + J_z \sigma_z^{(k)} \sigma_z^{(l)} + D_x(\sigma_y^{(k)} \sigma_z^{(l)} - \sigma_z^{(k)} \sigma_y^{(l)}) \tag{3}$$

where k, l represent the nodes which are connected together via DM interaction
with x−component of strength D_x; J_x, J_y and J_z are the x, y and z-component
of the real coupling coefficients, respectively, and the $\sigma^{(k)}, \sigma^{(l)}$ are the Pauli

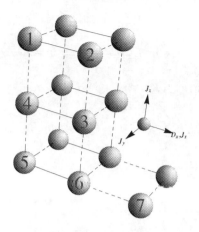

Fig. 1. The suggested quantum network.

matrices ($\sigma_x = |1\rangle\langle 0| + |0\rangle\langle 1|$, $\sigma_y = i(|1\rangle\langle 0| - |0\rangle\langle 1|)$ and $\sigma_z = |0\rangle\langle 0| - |1\rangle\langle 1|$).
In our case k and l represent the second and third qubit, respectively.

The final density operator of the network is given by

$$\rho_{1234}(t) = \mathcal{U}(t)\rho_{1234}(0)\mathcal{U}^\dagger(t), \tag{4}$$

where

$$\mathcal{U}(t) = e^{-i\mathcal{H}t} \tag{5}$$

and $\mathcal{U}^\dagger(t)$ is the conjugate of $\mathcal{U}(t)$

As we mentioned above that the node 1 and node 2 and node 3 and 4 are initially connected, so the connection will be generated between the second and third node. The unitary operator is defined by

$$\mathcal{U}^{(23)}(t) = \left[\cos(J_x t) - i\sin(J_x t)\sigma_x^{(2)}\sigma_x^{(3)}\right] \times \left[\cos(J_y t) - i\sin(J_y t)\sigma_y^{(2)}\sigma_y^{(3)}\right]$$
$$\times \left[\cos(J_z t) - i\sin(J_z t)\sigma_z^{(2)}\sigma_z^{(3)}\right] \times \left[\cos^2(D_x t) + \sin^2(D_x t)\sigma_x^{(2)}\sigma_x^{(3)}\right.$$
$$\left. - \frac{i}{2}\sin(2D_x t)(\sigma_z^{(2)}\sigma_y^{(3)} - \sigma_y^{(2)}\sigma_z^{(3)})\right]. \tag{6}$$

In matrix form, the unitary operator Eq. (6) can be written as

$$\mathcal{U}^{(23)}(t) = \begin{pmatrix} u_{ee,ee} & u_{ee,eg} & u_{ee,ge} & u_{ee,gg} \\ u_{eg,ee} & u_{eg,eg} & u_{eg,ge} & u_{eg,gg} \\ u_{ge,ee} & u_{ge,eg} & u_{ge,ge} & u_{ge,gg} \\ u_{gg,ee} & u_{gg,eg} & u_{gg,ge} & u_{gg,gg} \end{pmatrix}, \tag{7}$$

where

$$u_{ge,eg} = [i\sin J_z t + \cos J_z t]\Big[\sin^2 D_x t(-\sin J_x t\sin J_y t + \cos J_x t\cos J_y t)$$

$$-i\cos^2 D_x t\sin(t(J_x + J_y))\Big],$$

$$u_{ge,ge} = [i\sin J_z t + \cos J_z t]\Big[\cos^2 D_x t(-\sin J_x t\sin J_y t + \cos J_x t\cos J_y t)$$

$$-i\sin^2 D_x t\sin(t(J_x + J_y))\Big],$$

$$u_{ge,gg} = \frac{1}{2}\sin D_x t(i\sin J_x t + \cos J_x t)(i\sin J_y t + \cos J_y t)(i\sin J_z t + \cos J_z t),$$

$$u_{gg,ee} = [\cos J_z t - i\sin J_z t]\Big[\sin^2 D_x t(\cos J_x t\cos J_y t + \sin J_x t\sin J_y t)$$

$$-i\cos^2 D_x t\sin t(J_x - J_y)\Big],$$

$$u_{gg,ge} = -\frac{1}{2}\sin D_x t(i\sin J_x t + \cos J_x t)(i\sin J_y t - \cos J_y t)(i\sin J_z t - \cos J_z t),$$

$$u_{gg,gg} = [\cos J_z t - i\sin J_z t]\Big[\cos^2 D_x t(\cos J_x t\cos J_y t + \sin J_x t\sin J_y t)$$

$$-i\sin^2 D_x t\sin t(J_x - J_y)\Big],$$

and the rest of components $u_{ge,ee} = -u_{ge,gg} = u_{eg,ee} = -u_{ge,gg}$, $u_{gg,eg} = -u_{gg,ge} = u_{ee,eg} = -u_{ee,ge}$, $u_{eg,eg} = u_{ge,ge}$, $u_{eg,ge} = u_{ge,eg}$, $u_{ee,gg} = u_{gg,ee}$ and $u_{ee,ee} = u_{gg,gg}$. Using Eqs. (4) and (7), one gets the final entangled network between four nodes. Since we are interested to quantify the degree of entanglement between different nodes, one can obtain the required density operator between each pair of nodes by tracing out the other two. For example, the density operator between the first and the second node is given by $\rho_{12} = \mathrm{tr}_{34}\{\rho_{1234}(t)\}$.

3   Results and Discussion

In this section, we quantify the entanglement between each pair of nodes. Practically, we consider the channels ρ_{ij}, where $ij = 12, 13$ and 14. For this aim, we use Wootters's concurrent as a measure for entanglement [27] which is defined as

$$\mathcal{C} = \max\{\sqrt{\lambda_1} - \sqrt{\lambda_2} - \sqrt{\lambda_3} - \sqrt{\lambda_4}, 0\}, \tag{8}$$

where λ_k, $(k = 1..4)$ is the eigenvalue of the matrix $\rho_{ij}(\sigma_y^{(i)} \otimes \sigma_y^{(j)})\rho_{ij}^*(\sigma_y^{(i)} \otimes \sigma_y^{(j)})$.

The entanglement behavior (concurrence) of the entangled state between nodes "1" and "2", ρ_{12}, is described in Fig. 2 for different coupling values J_i, $(i = x, y, z)$, and the strength of DM is assumed to be fixed, $D_x = 0.2$. Fig. 2a describes the evolution of the concurrence C in the presence of zero coupling or one and only one non–zero coupling. It is clear that for $j_x = J_y = J_z$, the concurrence decays gradually to its minimum bound ($C = 0.4$) and then increases to maximum bound without exceeding the initial bounds. This shows that the decay is coming from the interaction between the second and third node, and consequently some correlations are lost. However, the upper and lower bounds will depend on the non-zero coupling when it is switched on. This behavior shows that the minimum bound of C for $J_x \neq 0$ is always larger than that depicted for $J_y \neq 0$ or $J_z \neq 0$. On the other hand, the concurrence vanishes completely for $J_y \neq 0$ or $J_z \neq 0$, i.e., $C = 0$ as the scaled time increases without exceeding the initial upper limit [11, 12]. In Fig. 2c, we investigate the behavior of the concurrence in which $J_x - J_y - J_z \neq 0$ is considered. It can be seen that when the coupling parameters are small, $J_x = J_y = J_z = 0.1$, the concurrence C decays gradually and vanishes when $t \in [7, 8]$. For larger values of J_i, $i = x, y, z$, the concurrence decays comparably faster.

Figure 2 b describes the behavior of concurrence for the entangled state ρ_{12}, where two non-zero couplings are considered. The general behavior is similar to that of depicted in Fig. 2, but the number of oscillations increases between the upper and lower bounds. If we compare the solid curves in Figs. 2 a & 2 b, we see that the presence of the coupling causes the concurrence to decay quicker.

Figure 3 describe the behavior of the concurrence for the entangled state which is generated between the nodes "1" and "3" via direct interaction. Fig. 3 a describes the behavior of C for only one non-zero coupling where the same values of the coupling and DM's strength as considered in Fig. 1 are being used. Since the two nodes are initially disentangled, the concurrence $C = 0$ at $t = 0$. As soon as the interaction is switched on an entangled state is generated between the first and the third node, and consequently the concurrence increases to reach its upper bound ($C = 0.4$). However, as t continues to elapse, the concurrence decays and vanishes completely. This behavior is repeated periodically.

The dynamics of concurrence, C, for the channel ρ_{13} when two non-zero couplings are considered is depicted in Fig. 3 b. The values of J_i, where $i = x, y, z$ are the same as in Fig. 2b. This figure shows that the concurrence oscillates between its lower and upper bounds quickly. The phenomena of the sudden–death and sudden–birth appear of the entanglement are appeared clearly.

Figure 3c describes the behavior of C for the state ρ_{13} when $J_x = J_y = J_z \neq 0$ have the same values as in Fig. 2c. It is clear that the general behavior is similar to that as depicted in Fig. 3b, but the number of oscillations increases as the value of the coupling increases. We note that the upper bound is slightly larger for large J_i.

We investigated the behavior of entanglement for the state ρ_{14}, which is generated via indirect interaction, is given in Figs. 4a–4c. It is clear that the behavior of C is similar to that displayed for ρ_{13}. However, the upper bound for the state ρ_{14} is much larger and reaches maximum value at $C = 1$. The number of oscillations of the concurrence increases as the spin–orbit coupling increases.

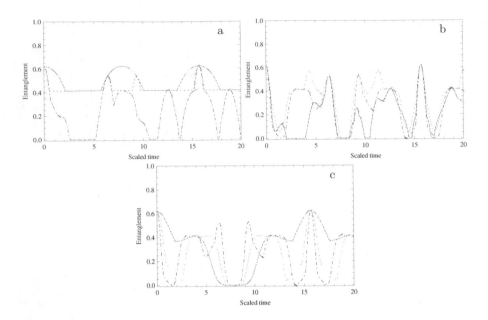

Fig. 2. The dynamics of the entanglement between node 1 and node 2, $\mathcal{C}_{12}$, (a): are given by the red line for $J_x = J_y = J_z = 0$ (without the effect of spin), the black dotted line represents the entanglement for $J_x = 0.5$ and $J_y = J_z = 0$, the green line for $J_y = 0.5$ and $J_x = J_z = 0$, and the blue dashed line represents the entanglement for $J_x = J_y = J_z = 0$ and $D_x = 0.2$. (b)Similar to a but the red line is for $J_x = J_y = 0.5$ and $J_z = 0$, black dotted line represents the entanglement for $J_x = J_y = 0.5$ and $J_z = 0$, the green line for $J_x = 0$ and $J_y = J_z = 0.5$, the blue dashed line represents the entanglement for $J_x = J_y = J_z = 0.5$ and $D_x = 0.2$. (c) The red line for $J_x = J_y = J_z = 0.1$, the green dashed line for $J_x = J_y = J_z = 0.3$ and the blue dashed line for $J_x = J_y = J_z = 0.5$ and $D_x = 0.2$.

On the other hand, the oscillation increases if all the couplings have non–zero values.

4   Conclusion

We discussed the effect of the spin–orbit coupling on the entanglement between different nodes in the quantum network. In general, the entanglement decays for non–zero coupling. The phenomena of the sudden–death and sudden–birth appeared for larger coupling values.

It shown that initially in the entangled channel the coupling constant has no effect on the upper bound of entanglement. However, the lower bound of entanglement do not vanish for non–zero couplings. The number of oscillation increases as the the coupling increases. For entangled channels which are gener-

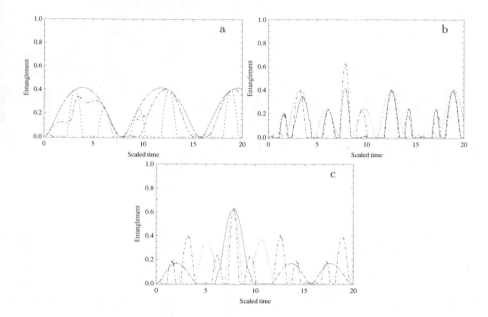

Fig. 3. The dynamics of the entanglement between node 1 and node 3, $\mathcal{C}_{13}$, (a), (b) and (c) is the same with figure 2 but for the channel 13.

ated either via direct or indirect interaction, the concurrence and the number of oscillation increase as the coupling increases.

Finally, it also shown that the generated entangled channel between any two nodes via indirect interaction has a large degree of entanglement. The upper bound exceeds the initial entangled state. Therefore, one can generate maximum entangled state from the less entangled state by controlling the spin–orbit coupling. This means that one can use the terminals of the generated entangled network to perform quantum information task with high efficiency.

References

1. Chuang, Isaac L., Nielsen, Michael A.: Quantum computation and quantum information. In Cambridge University Press, (2000)
2. Zhang, Jian-Song., Chen, Ai-Xi.: Review on quantum discord of bipartite and multipartite systems. Quant. Phys. Lett. 1, 69-77 (2012)
3. Ezhov, A.A.: Pattern Recognition with Quantum Neural Networks. Lecture Notes in Computer Science. 2013, 60–71 (2001)
4. Caraiman, S. and Manta, V.: Image processing using quantum computing. System Theory, Control and Computing (ICSTCC), IEEE Press. 1–6 (2012)
5. Seth Lloyd, Masoud Mohseni, Patrick Rebentrost.: Quantum algorithms for supervised and unsupervised machine learning. http://arxiv.org/abs/1307.0411 (2013)
6. Ahmed Younes, Database Manipulation Operations on Quantum Systems, Quantum. Information. Review. 1, 9-17 (2013) (Jan. 2013), PP:9-17

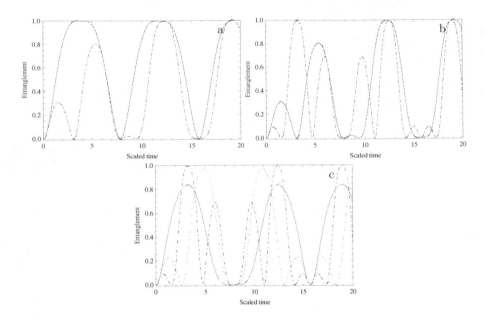

Fig. 4. The dynamics of the entanglement between node 1 and node 4, C_{14}, (a), (b) and (c) is the same with figure 2 but for the channel 14.

7. Elliott, Y.C.: Building the quantum network. New J. Phys. 4(3), 46-53 (2002)
8. Mink, A., Hershman, B.J., Nakassis, A., Tang, X., Lu, R., Su, D.H., Clark, C.W., Williams, C.J., Hagley, E.W., Wen, J., Bienfang, J., Gross, A.J.: Quantum key distribution with 1.25 gbps clock synchronization. Optics Express. 12(9), 20112016 (2004)
9. Buntschu F. :Long-term performance of the swissquantum quantum key distribution network in a field environment. :New J. Phys, 13:123001, 2011.
10. Ishizuka H., et al.: Field test of quantum key distribution in the tokyo qkd network. Optics Express. 19(11), 1038710409, (2011)
11. Mtwallay, M.: Entangled network and quantum communications. Phys. Lett. A. 375(3), 426853 (2011)
12. Abdel-Aty, Abdel-Haleem., Cheong, LeeYen., Zakaria, Nordin., Metwally, Nasser,. Quantum network via partial entangled state. In: International Conference in Quantum Optics and Quantum Information (icQoQi), IOP Press (2013).
13. Abdel-Haleem Abdel-Aty, LeeYen Cheong, Nordin Zakaria, and Nasser Metwally. Entanglement and teleportation via partial entangled-state quantum network. (Submitted) Quant. Inf. Process. , 2013.
14. Darwish, M., Obada, A.-S.F,. El-Barakaty, A.: Purity loss for a cooper pair box interacting dispersively with a nonclassical field under phase damping. Appl. Math. Inf. Sci, 5, 122 (2011)
15. Hessian, H.A.: Entropy Growth and Formation of Stationary Entanglement due to Intrinsic Noise in the Two-Mode JC Model. Quant. Phys. Lett. 2, 1-9 (2013)
16. Moriya, T.: New mechanism of anisotropic superexchange interaction. Phys. Rev. Lett, 4, 228 (1960)

17. Friesen, M., Chutia, S., Joynt, R.: Detection and measurement of the dzyaloshinskii-moriya interaction in double quantum dot systems. Phys. Rev. 73, 241304 (2006)

18. Zhang, Guo-Feng.: Thermal entanglement and teleportation in a twoqubit heisenberg chain with dzyaloshinski-moriya anisotropic antisymmetric interaction. Phys. Rev. A. 75, 034304 (2007)

19. Sha-Sha Li, Ting-Qi Ren, Xiang-Mu Kong, and Kai Liu.: Thermal entanglement in the heisenberg "XXZ" model with dzyaloshinskiimoriya interaction. Physica A: Statistical Mechanics and its Applications, 391(12), 35–41 (2012)

20. Yeo, Y.: Teleportation via thermally entangled states of a two-qubit heisenberg XX chain. Phys. Rev. A, 66, 062312 (2002)

21. Ming-Yong Ye, Wei Jiang, Ping-Xing Chen, Yong-Sheng Zhang, Zheng- Wei Zhou, and Guang-Can Guo. Local distinguishability of orthogonal quantum states and generators of su(N). Phys. Rev. A, 76:032329, Sep 2007.

22. Ahmed, A-H.M., Zakaria, M.N., Metwally, N.: Teleportation in the presence of technical defects in transmission stations. Appl. Math. Inf. Sci. 6(3), 781 (2012)

23. Mohamed, A.-B.M.: Quantum discord and its geometric measure with death entanglement in correlated dephasing two qubits system. Quantum. Information. Review. 1, 1-7 (2013)

24. Da-Chuang Li and Zhuo-Liang Cao. Thermal entanglement in the anisotropic heisenberg "XYZ" model with different inhomogeneous magnetic fields. Optics Communications, 282(6), 1226–1230, (2009)

25. Zhenghong He, Zuhong Xiong, and Yanli Zhang. Influence of intrinsic decoherence on quantum teleportation via two-qubit heisenberg fXYZg chain. Physics Letters A, 354(12):79 83, 2006.

26. Majumdar, A.S., Adhikari, S., Ghosh, B., Nayak, N., Roy, S.: Teleportation via maximally and non-maximally entangled mixed states. Quant. Info. Comp. 10, 0398 (2010)

27. Hill, S., Wootters, W.K.: Entanglement of a pair of quantum bits. Phys. Rev. Lett. 78, 50225025 (1997).

Enhancement of the RFID Data Management using EPCIS and e-Transaction

Worapot Jakkhupan[1] and Somjit Arch-int[2]

[1] Information and Communication Technology Programme, Faculty of Science,
Prince of Songkla University, Songkla 90112, Thailand
worapot.j@psu.ac.th
[2] Department of Computer Science, Faculty of Science, Khon Kaen University,
Khon Kaen 40000, Thailand
somjit@kku.ac.th

Abstract. One of the challenges in RFID adoption is to deal with massive RFID data. The large amount of data continuously generated from RFID devices come with noise and redundant, and the data may be missing. This study proposes the enhancement of a high level RFID data management using the data from EPCIS and e-Transaction as pre-processing data sources. The results reveal that the enhanced features improve the performance of the system by eliminating noise and redundant, therefore, the operation time is significantly reduced. The missing data can be detected at the end of the operation.

Keywords: RFID, Data Management, Supply Chain, EPCglobal Network Standards, Electronic Transaction

1 Introduction

Radio frequency identification (RFID), an automatic object identification technology, has become growing interested from the supply chain and the academic researchers [1, 2]. RFID uses the radio-frequency waves to transfer data from RFID tags, attached on the physical objects, to the system. RFID automatically captures the data from movable and invisible objects, which is useful to monitor the movement of the products [3]. Although RFID has been introduced over a decade, but there are still many barriers in a large-scale RFID adoption such as high investment, long term implementation, and the unreliability of RFID data [4].

Since RFID generates high volume of data, enterprise application (EA) has to deal with massive data such as redundant and noise, as well as missing data [4, 5]. In the RFID operation, RFID reader interrogates all tags in range and generates the report, approximately one second per report. Each report may contain noise and redundant data. The huge amount of redundant and noise data causes high traffic in the network, which can reduce the performance and burden the system [6]. Likewise, missing data cause system unreliable. Therefore, the RFID system requires the features to reduce the redundant and noise data, as well as the feature to define the missing data.

T. Herawan et al. (eds.), *Proceedings of the First International Conference
on Advanced Data and Information Engineering (DaEng-2013)*, Lecture Notes
in Electrical Engineering 285, DOI: 10.1007/978-981-4585-18-7_60,
© Springer Science+Business Media Singapore 2014

This study aims to enhance the high level RFID data management by providing RFID system the pre-processing data taken from e-Transaction and EPCIS. The proposed system was experimented by simulating a virtual RFID-enabled supply chain based on realistic supply chain operations. When comparing with the traditional system, the proposed system generates a smaller quantity of noise and redundant data, thus, the operations perform faster, and the missing data can be defined as well.

This paper is organized as follows. Sect. 2 introduces the background of this study. Sect. 3 describes the system design and experimentation. Sect. 4 reveals the results and discussions. Finally, Sect. 5 summarizes the conclusions.

2 Literature Review

2.1 RFID Data Types and Problems

Xingyi et al. [7] has categorized the RFID data into three types, Primitive Event Data (PED), Basic Event Data (BED), and Complex Event Data (CED). Likewise, Bai et al. [8] has classified the RFID data into three types, duplicate, false positive (or noise), and false negative (or missing). The types of the RFID data and the concepts of this study are drawn in Fig. 1.

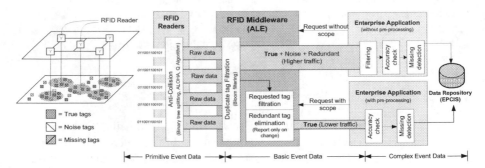

Fig. 1. RFID data types

Primitive Event Data (PED). The binary data are transferred from all of the RFID tags to the RFID readers cause the data collision. The anti-collision algorithms embedded in the RFID reader are used to deal with the data collision, such as binary tree splitting and framed-slotted ALOHA [6, 9]. In each round of generating report, the duplicate data should be eliminated. The algorithms embedded in the RFID middleware, such as Bloom Filters, are used to filter the duplication [10, 11].

Basic Event Data (BED). PED, generated from RFID middleware, may include noise, redundant, and missing data. The noise data are the data that generated from middleware but unwanted from EA. The redundant data were already reported in the previous cycle but still continuously appear in the report. The missing data are the tags that should be found in the operation but missing from the report [12, 13].

The best solution to correct the RFID data is to provide the pre-processing data to the RFID middleware [14]. The list of tags stored in the data repository or the related transaction can be used to define the pre-processing data [12, 13, 15]. Therefore, this study adopts EPCIS and e-Transaction as the pre-processing data sources aims to solve the BED problems.

Complex Event Data (CED). CED contains full context of RFID event stored in the data repository, in this case, EPCIS. CED says "At the location A, at time T, the following contained EPCs were aggregated on the following SSCC, as referred to the transaction message M" in the pallet preparation process.

2.2 EPCglobal Network Standards

An organization named EPCglobal[1] has been developing the critical elements of the new network concept called the EPCglobal Network. The goals of EPCglobal Network are to standardize the RFID infrastructure and to enable the product visibility throughout the supply chain. The architecture of the EPCglobal Network consists of a set of standards, which serve the particular operations.

Tag Data Standard (TDS). TDS standardizes the object identification, called Electronic Product Code (EPC). TDS also provides patterns to instruct RFID middleware how to filter the RFID tags. The examples of using TDS to identify objects and to define the filtering patterns are shown in Table 1.

Table 1. Patterns and examples of TDS identification and filtering

Objects		Identification	Filtering
SSCC	Pattern	urn:epc:id:sscc:CompPrefix.Serial	urn:epc:idpat:sscc:CompPrefix.Serial
	Example	urn:epc:id:sscc:88511111.000001111	urn:epc:idpat:sscc:88511111.000001111
SGTIN	Pattern	urn:epc:id:sgtin:CompPrefix.ItemRef.Pattern	urn:epc:idpat:sgtin:CompPrefix.ItemRef.Pattern
	Example	urn:epc:id:sgtin:88511111.1111.1 urn:epc:id:sgtin:88511111.1111.2	Individual urn:epc:idpat:sgtin:88511111.01111.1 urn:epc:idpat:sgtin:88511111.01111.50
		... urn:epc:id:sgtin:88511111.1111.50	Wildcard urn:epc:idpat:sgtin:88511111.01111.* urn:epc:idpat:sgtin:88511111.01111.[1-50]
SGLN	Pattern	urn:epc:id:sgln:CompPrefix.LocID.extendID	urn:epc:idpat:sgln:CompPrefix.LocID.extendID
	Example	urn:epc:id:sgln:88511111.9999.1	urn:epc:idpat:sgln:88511111.9999.1

Application Level Events (ALE). ALE separates RFID readers from enterprise applications. ALE filters the RFID tags according to the filtering patterns defined by TDS. Each round of the RFID interrogation is called event cycle (EC), and the report generated from EC is called ECReport. The tags that are already reported in the previous ECReport should not be reported again. In order to eliminate noise tags, EA defines the filtering patterns of the included or excluded tags in each operation. ALE provides *reportOnlyOnChange* function to eliminate the redundant data.

---

[1] http://www.gs1.org/epcglobal

Electronic Product Code Information Service (EPCIS). EPCIS is a standard RFID data repository that stores the CED. EPCIS is composed of three main features; EPCIS data repository, EPCIS capture interface, and EPCIS query interface. The data stored in EPCIS are called EPCIS events, which represent the business context.

2.3   Electronic Transaction Management System

There are four transactions related in the product movement; Purchase Order (PO), Purchase Order Confirmation (POC), Despatch Advice (DA), and Receiving Advice (RA). In this study, GS1 XML Standard was adopted to run e-Transaction system. The collaboration between e-Transaction and the EPCglobal Network Standards simulated in this study is drawn in Fig. 2.

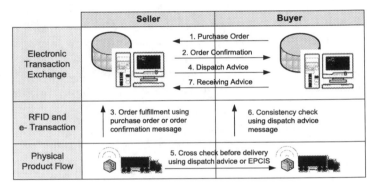

Fig. 2. The collaboration between e-Transaction and EPCglobal Network Standards

3   System Design and Experimentation

3.1   A Case Study

This study simulated the pallet delivery scenario, which is including pallet preparation, pallet sending, and pallet receiving. Two types of data filtering patterns will be compared; wildcard and individual.

Wildcard Data Filtration. Wildcard data filtration used in pallet preparation, allows the operating * or [range of number] to identify the filtering pattern. For example, buyer purchases product number 8851111.11111. RFID system should allow only items that the identifier starts with 88511111.1111 to be accumulated on the pallet. In the pallet preparation, the filtering pattern can be retrieved from PO or POC.

Individual Data Filtration. is used to check the consistent of pallet and products one-by-one. The list of tags can be retrieved from the aggregation event in EPCIS or from DA document. Since RFID system has exactly the list of EPCs, the system can define the missing tags at the end of the process.

3.2 System Architecture

The system architecture was designed based on real operations. The EPCglobal Network components were set up, and the additional components were developed. The architecture of the proposed system is shown in Fig. 3. The Enterprise Application (EA) operates the common business process. In our simulation, EA includes pallet preparation system, pallet shipping system, and pallet receiving system. The related EPCglobal Network Components, include EPCIS and ALE, operate the standard RFID data management. The e-Transaction Management system communicates between business partners. The Connectors provide interfaces to connect the EPCglobal Network Components with other applications.

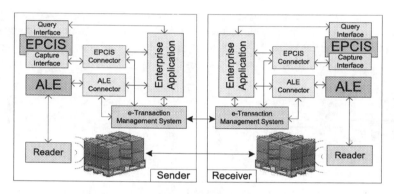

Fig. 3. Architecture of the proposed system

3.3 Experimentation

This study simulated the RFID-enabled pallet movement. As described in Fig. 4, assume that buyer orders 100 items of product number 88511111.1111 and 100 items of product number 88511111.2222 from seller. In the pallet preparation, the pallet number urn:epc:id:sscc:88511111.000000001 contains 200 EPCs of the purchased products. From our site survey, in a small and untidy business location with full of RFID tags attached on the objects, such as in the warehouse, the staffs cannot remove the noise tags from the RFID reading zone. Thus, we assume that there are 200 noise tags in the RFID reading zone. All of the tags are functioning and can be interrogated by the RFID readers.

In the pallet preparation process, the system allows only the tags that the identifier starts with 88511111.1111 and 88511111.2222 to be accumulated in the pallet. The system finished after the quantity of each product reaches 100. In the pallet shipping and receiving, the systems match the interrogated tags one-by-one with the list of tags extracted from EPCIS or DA document. At the start of the operation, the status of tag is set as missing, after the tag has been found, the system changes the status of tag as found. Thus, at the end of the operation, if the tag still not be found, the system will identify that tag as missing and report to the staffs. Finally, the staffs can manually find out and fix the missing problem.

3.4   Research Materials.

This research used the tools, including hardware and software, to simulate the RFID system. The existing software compliant with the EPCglobal Network Standards were selected to implement the system. The selected tools are as Rifidi[2], ALIEN 9900, LogicAlloy[3], and Fosstrak[4].

4   Results and Discussions

4.1   Results

The average size of ECReport and the average time of each process were collected. The performances of the system with and without the enhancements are compared. Concurrently, the performance of the wildcard and individual tag filtration are revealed. The results are shown in Fig. 4 and Fig. 5.

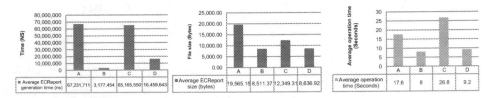

(a) Average Report Generation time   (b) Average ECReport size   (c) Average operation time

Fig. 4.   The experimental results (A=Pallet preparation without filtration, B=Pallet preparation with wildcard filtration, C=Pallet sending and receiving without filtration, D=Pallet sending and receiving with individual filtration)

4.2   Discussions

The results in Fig. 5a and 5c reveal the performance of the system without tag filtration. Since EA receives all tags, the number found tags, which are including noise and redundant tags, is very high. Therefore, EA has to filter and eliminate the redundant and noise tags by itself, thus, the operation takes longer time. On the other hand, the results in Fig. 5b and 5d reveal the performance of the enhanced system with the tag filtration. Since the redundant and noise tags are prior eliminated, EA receives only the tags that meet the requirement; therefore, the noise tag is equal to zero. Since EA does not have to filter the noise tags, the operation takes shorter time. Finally, the comparison between the wildcard filtration (Fig. 5b) and individual filtration (Fig. 5d) shows that the wildcard filtration takes shorter time than the individual filtration.

---

[2]   http://www.rifidi.org
[3]   http://www.logicalloy.com
[4]   https://code.google.com/p/fosstrak

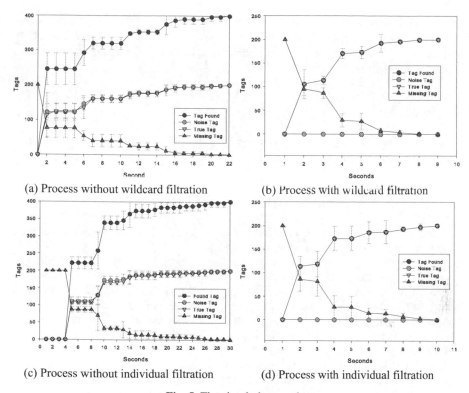

(a) Process without wildcard filtration   (b) Process with wildcard filtration

(c) Process without individual filtration   (d) Process with individual filtration

Fig. 5. The simulation results

5   Conclusions

In order to manage the huge amount of massive RFID data automatically generated from RFID devices, RFID system requires the pre-processing data to filter the RFID tags in the particular operations. This research proposes an enhancement of the high level RFID data management by including the pre-processing data acquired from EPCIS and e-Transaction, which were experimented in the pallet preparation, pallet sending and pallet receiving processes. The comparison between with and without data filtration in the high level RFID data management reveals that, the data filtration reduces the size of data in the whole network because the redundant and noise data were eliminated before being propagated into the network. The enterprise application receives only the filtered data, thus, the operations perform faster. The individual tag filtering pattern can be applied to enhance the missing tag detection feature, which improves the reliability of the RFID system.

Acknowledgments. The authors gratefully thank the Global Standards 1 (GS1) Thailand for supporting the data for this study. Financial support for this research was provided by the Thailand Research Fund (TRF) through the Royal Golden Jubilee

(RGJ) Ph.D. Program (Grant No.PHD/0154/2550), which is sincerely acknowledged. The authors wish to acknowledge the support of the Khon Kaen University Publication Clinic, Research and Technology Transfer Affairs, Khon Kaen University, for their assistance.

Reference

1. Wu, D.L., Ng, W.W.Y., Yeung, D.S., Ding, H.L.: A brief survey on current RFID applications. In: Internation Conference on Machine Learning and Cybernetics, pp. 2330-2335. (2009)
2. Ngai, E, Moon, K.K., Riggins, F.J., Yi, C.Y.: RFID research: An academic literature review (1995-2005) and future research directions. Int. J. Prod. Econ. 112(2), 510-520 (2008)
3. Wang, F., Liu, P.: Temporal management of RFID data. In: 31st International Conference on Very Large Data Bases, pp. 1128-1139. (2005)
4. Wu, N., Nystrom, M., Lin, T., Yu, H.: Challenges to global RFID adoption. Technovation. 26(12), 1317-1323. (2006)
5. Cheong, T., Kim, Y.: RFID data management and RFID information value chain support with RFID middleware platform implementation. In: International Conference on On the Move to Meaningful Internet Systems, pp. 557-575. Springer-Verlag, Berlin (2005)
6. Lee, C.H., Chung, C.W.: An approximate duplicate elimination in RFID data streams. Data. Knowl. Eng. 70(12), 1070-1087 (2011)
7. Xingyi, J., Xiaodong, L., Ning, K., Baoping, Y.: Efficient Complex Event Processing over RFID Data Stream. In: 7th IEEE/ACIS International Conference on Computer and Information Science, pp. 75-81. IEEE Press, New York (2008)
8. Bai, Y., Wang, F., Liu, P.: Efficiently filtering RFID data streams. In: 1st International VLDB Workshop on Clean Databases, pp. 50-57. (2006)
9. Zhu, L., Yum, T.-S.P.: A Critical Survey and Analysis of RFID Anti-Collision Mechanisms. IEEE. Commun. Mag. 49(5), 214-221 (2011)
10. Lee, J., Kwon, T., Choi, Y., Das, S.K., Kim, K.-A.: Analysis of RFID anti-collision algorithms using smart antennas. In: 2nd International Conference on Embedded Networked Sensor Systems, pp. 265-266. ACM, New York (2004)
11. Mahdin, H., Abawajy, J.: An Approach to Filtering RFID Data Streams. In: 10th International Symposium on Pervasive Systems, Algorithms, and Networks, pp. 742-746. (2009)
12. Tan, C.C., Sheng, B., Li, Q.: Efficient Techniques for Monitoring Missing RFID Tags. IEEE T. Wirel. Commun. 9(6), 1882-1889 (2010)
13. Li, T., Chen, S., Ling, Y.: Identifying the missing tags in a large RFID system. In: 11th ACM International Symposium on Mobile ad hoc networking and computing, pp. 1-10. ACM, New York (2010)
14. Anagnostopoulos, A.P., Soldatos, J.K., Michalakos, S.G.: REFiLL: A lightweight programmable middleware platform for cost effective RFID application development, Pervasive and Mobile Computing. 5(1), 49-63 (2009)
15. Zhang, R., Liu, Y., Zhang, Y., Sun, J.: Fast Identification of the Missing Tags in a Large RFID System. In: 8th Annual IEEE Communications Society Conference on Sensor, Mesh and Ad Hoc Communications and Networks, pp. 278-286. IEEE Press, New York (2011)

Multi-Criteria Based Algorithm for Scheduling Divisible Load

Shamsollah Ghanbari[1], Mohamed Othman [*2] ,Wah June Leong[1], and Mohd Rizam Abu Bakar[1]

[1] Laboratory of Computational Science and Mathematical Physics
Institute For Mathematical Research
[2] Dept of Communication Tech and Network
Faculty of Computer Science and Information Technology
Universiti Putra Malaysia 43400 UPM Serdang, Selangor D.E., Malaysia
myrshg@gmail.com, mothman@upm.edu.my

Abstract. Divisible load theory has become a popular area of research during the past two decades. Based on divisible load theory the computations and communications can be divided into some arbitrarily independent parts and each part can be processed independently by a processor. Existing divisible load scheduling algorithms do not consider any priority for allocating fraction of load. In some situation the fractions of load must be allocated based on some priorities. In this paper we propose a multi criteria divisible load scheduling algorithm. The proposed model considers several criteria with different priorities for allocating fractions of load to processors. Experimental result indicates the proposed algorithm can handle the priority of processors.

Keywords: Divisible load scheduling, priority, multi criteria, AHP.

1 Introduction

The first article about divisible load theory(DLT) was published in 1988[9]. Based on DLT, it is assumed that the computation can be partitioned into some arbitrary sizes, and each partition can be processed independently. In the past two decades, DLT has found a wide variety of applications in parallel processing area such as data intensive applications [7], data grid application [8] , image and vision processing[6] and so on. Ten important advantages of DLT has been listed in[10]. DLT was applied for various network topologies including chain, star, bus, tree [4], three-dimensional mesh[13]. Existing divisible load scheduling algorithms do not consider any priorities for allocating fractions of load to processors. In some situation the fractions of load must be allocated based on some priorities. For example when we have limitation on memory or buffers, existing divisible load scheduling algorithms may have NP

---

[*] The author is also an associate researcher at Computational Science and Mathematical Physics Lab., Institute for Mathematical Research, UPM.

T. Herawan et al. (eds.), *Proceedings of the First International Conference on Advanced Data and Information Engineering (DaEng-2013)*, Lecture Notes in Electrical Engineering 285, DOI: 10.1007/978-981-4585-18-7_61,
© Springer Science+Business Media Singapore 2014

547

complexity[14]. In this case a multi criteria based algorithm would be very useful. This paper proposes a multi criteria divisible load scheduling algorithm. The proposed model considers several criteria with different priorities for allocating fractions of load. The proposed algorithm is based on analytical hierarchy process theory(AHP). AHP is a multi-criteria decision-making(MCDM)/ multi-attribute decision-making (MADM) model which was developed by T.Saaty[2]. Over the past three decades, AHP has been found a number of applications in various fields[1,2,3]. AHP is a suitable method for solving priority-based problems such as scheduling with various attributes and alternatives as well[5]. The rest of this paper is organized as following sections: section two explains the basic concepts of divisible load scheduling, section three explains briefly hierarchy process theory, in section four we propose a new multi criteria based divisible load scheduling algorithm. In section five we analyze the proposed model. Finally in section six, we provide some experimental results for supporting the proposed model.

2   Divisible Load Scheduling

In general, DLT assumes that the computation and communication can be divided into some parts of arbitrary sizes and these parts can be independently processed in parallel(see Fig.1). DLT assumes that initially amount of load is

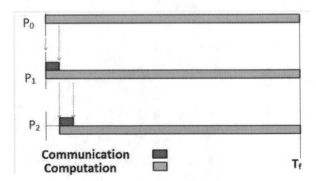

Fig. 1: Gantt chart-like timing diagrams for divisible load

held by the originator p_0. The originator does not do any computation. It only distributes $\alpha_1, \alpha_2, ..., \alpha_m$ fractions of load on worker processors $p_1, p_2, ..., p_m$. Condition for optimal solution is that all processors stop processing at the same time[9]. The goal is to calculate $\alpha_0, \alpha_1, ..., \alpha_m$; where $\alpha_0 + \alpha_1 + ... + \alpha_m = 1$. Eq.(1) shows a general timing equation(closed form) for DLT[4].

$$T_i(\alpha) = \begin{cases} \alpha_0 w_0 & i = 0 \\ \Sigma_{j=1}^{i} z_j \alpha_j + \alpha_i w_i & i = 1, 2, ..., m \end{cases} \tag{1}$$

3 Analytical Hierarchy Process

Analytical Hierarchy Process(AHP) is a multi-criteria(attribute) decision-making. A simple form of AHP is consisted of three levels including objective level, attributes level and alternatives level. Each level uses comparison matrixes for comparing the priorities. Assume that $A = [a_{ij}]$ is a comparison matrix. Each entry in matrix A is positive. In this case A is a square matrix $(A_{n \times n})$. There is only a vector of weights such as $v=(v_1,..., v_n)$ associated with any arbitrary comparison matrix such as A. Eq.(2) indicates relationship between the elements of comparison matrix A and its vector of weights $v[1,2]$.

$$a_{ij} = \begin{cases} \frac{v_i}{v_j} & i \neq j \\ 1 & i = j \end{cases} \tag{2}$$

An essential step in AHP is to calculate vector of weights(v). Based on [1,2,3,11] Vector of weights can be computed by solving Eq.(3).

$$Av = \lambda_{max}.v \tag{3}$$

Where λ_{max} denotes the principal eigenvalue of A and v denotes the corresponding eigenvector. If A is absolutely consistent then $\lambda_{max} = n[2]$. A metric for evaluating consistency of comparison matrix is named consistency rate(CR), it can be calculated by Eq.(5). According to[11,1,2] if $CR < 0.1$ then comparison matrix will be consistent. In Eq.(5), RI and CI denote the random index and consistency index respectively. CI can be calculated by Eq.(4) and RI can be obtained by using table 1[2]. Other methods for getting RI are available in [1,3,11].

$$CI = \frac{\lambda_{max} - n}{RI} \tag{4}$$

$$CR = \frac{CI}{RI} \tag{5}$$

Number of rows	2	3	4	5	6	7	8	9	10
Random Index	0.00	0.58	0.90	1.12	1.24	1.32	1.41	1.45	1.49

Table 1: Random Index(RI) vs number of rows(columns) of matrix

4 Proposed Algorithm

Our proposed model consists of a scheduler, some processors, a multi criteria(attributes) decision maker and a matrix solver see Fig.2(a). A decision maker consists of three levels of priorities including divisible load level (objective level), criteria level (attribute level) and processors level (alternative level), see fig.2(b). In general the proposed model can be sketched as algorithm 1.

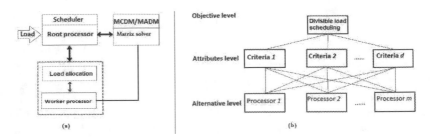

Fig. 2: (a)Framework of Proposed Model (b) Three Levels of a Decision Maker

---

Algorithm 1 A general multi-criteria based divisible load scheduling algorithm

Input: $\psi = \{P_1\ ,P_1,..,P_m\}$ a set of processors(T_{cp},T_{cm} and for each processor w_i,z_i);C=$\{C_1\ ,C_2,..,C_d\}$ a set of criteria;

Output:fraction of load $\alpha_1,\alpha_2,...,\alpha_m$

Description:

1: **let** k←1
2: **while** $k \le d$ **do**
3:     $Q^{c_k} \leftarrow$ Comparison_Mat(ψ,C_k);
4:     **For** Q^{c_k} **compute** CR; **Note:**This parameter can be calculated by Eq.(5)
5:     **if** $CR < 0.1$ **then**
6:        **let** k← k+1
         ELSE exit (the matrix is inconsistent.)
7:     **end if**
8: **end while**
9: **let** R←Comparison_Mat(C,Priority of criteria);
10: **For** R **compute** CR; **Note:**This parameter can be calculated by Eq.(5)
11: **if NOT**($CR < 0.1$) **then**
12:     **exit** (the matrix is inconsistent.)
13: **end if**
14: **let** $\Delta \leftarrow$ nil ;**let** k←1
15: **while** $k \le d$ **do**
16:     **solve** $Q^{c_k}v^{(k)}=\lambda_{max}v^{(k)}$;
17:     **attach**($\Delta,v^{(k)}$);
18: **end while**; **Note:** at the end of this loop we have $\Delta=[v^{(1)}\ v^{(2)}...v^{(d)}\ ]$
19:     **solve** $R\Lambda=\lambda_{max}\Lambda$;
20: **let** PVD←$\Delta.\Lambda$; Note that PVD is an array with m elements
21: **let** i←1;
22: **while** $i \le m$ **do**
23:     **assign** processor P_i with i^{th} element of PVD;
24: **end while**;
25: **sort** processors based on their PVD value;
26: **allocate** fraction of load to the sorted processors based on Eq.(1) and **compute** $\alpha_1,\alpha_2,...,\alpha_m$

Suppose that $\psi = \{p_1, p_2, ..., p_m\}$ is a set of worker processors that requests load in a divisible load scheduling model. We assume that $\Re = \{c_1, c_2, ..., c_d\}$ is a set of criteria for choosing a processor based on a decision maker such as root processor as well. In this algorithm $Comparison\_Mat()$ is a procedure that makes a comparison matrix. This procedure compares elements of a vector based on a particular criterion. Assume that the ratio of priority of p_i to p_j for getting a fraction of load is $\frac{v_i}{v_j}$. This procedure makes a square matrix with m rows and columns. In this case m is the number of worker processors. Steps 1-8 of algorithm 1 make d comparison matrixes which are Q^{c_1}, Q^{c_2},..., Q^{c_d} based on a criterion such as c_g. Step 9 of algorithm 1 compares priorities of criteria. In this step each criterion will be compared with other criteria. Steps 4-7 and 10-13 of algorithm investigate the consistency of comparison matrixes. Consistency can be tested by using Eq.(5).The next step of algorithm is to compute vector of weights for produced matrixes. These processes have been done in steps 14-19 algorithm 1. Assume that $v^1, v^2, ..., v^d$ are corresponding vector of weights of $Q^{c_1}, Q^{c_2}, ..., Q^{c_d}$. In this case we can define Δ as a priority matrix of processors. Matrix (Δ) is shown in Eq.(6). Vector of weights for R is denoted by Λ in algorithm 1. It is calculated in step 19.

$$\Delta = [v^1 v^2 ... v^d] \tag{6}$$

Step 20 calculates $priority\,vector\,of\,distribution(PVD = \Delta.\Lambda)$. Each element of PVD denotes the priority of the corresponding processor to get a fraction of load. Steps 22-24 assign the PVD values to the corresponding processors. Step 25 sorts the worker processors based on their PVD-values. Finally the load is allocated to the sorted processors in step 26.

5   Analysis of Proposed Algorithm

This section mainly discusses about two important issues related to the proposed algorithm including complexity and consistency. The complexity of proposed algorithm(denoted by Ω_{total}) can be calculated by Eq.(7). In Eq.(7), $\Omega_{priority}$ can be calculated as Eq.(8) which denotes the complexity of computing the priority vectors of comparison matrixes. Meanwhile, Ω_{DLT} denotes the complexity of solving DLT. Moreover, $\Omega_{consistency}$ can be calculated as E.(9) which denotes the complexity of investigating consistency of matrixes. In this case according to[12] we assume that a matrix multiplication takes approximately $m^{2.81}$ arithmetic operations (additions and multiplications). Where, m and d are the number processors and criteria respectively. Consistency indicates that each of the comparison matrixes has a logically reasonable value.

$$\Omega_{total} = \Omega_{DLT} + \Omega_{priority} + \Omega_{consistency} \tag{7}$$

$$\Omega_{priority} = d^{2.81} + d.m^{2.81} \tag{8}$$

Consistency of proposed algorithm mainly depends on the decision makers. In other word, if the decision makers adjust elements of comparison matrix based on the real priority of scheduling, they can make consistent comparison matrixes. Consistency can be calculated by Eq.(5) of comparison matrixes. We should solve Eq.(3) for calculating the consistency of comparison matrix. If any comparison matrixes are not consistent we should make a comparison matrix again. This inconsistency increases the complexity of algorithm. Therefore, $\Omega_{consistency}$ can be calculated by Eq.(9). In Eq.(9),k denotes the number of comparison matrixes that have been rejected(recomputed) because of inconsistency.

$$\Omega_{consistency} = k[max(d, m)]^{2.81} \tag{9}$$

6 Experimental Results

In this section we have provided some experimental results. For this purpose we consider two separate cases, in the $first\ case$ we assume four worker processors $(p_1,p_2,p_3$ and $p_4)$. For each processor three criteria(c_1,c_2,c_3) are considered. Table 2 indicates comparison matrixes based on criteria c_1 ,c_2 , c_3 and their corresponding priority vectors$(Q^{c_1}$,Q^{c_2} , $Q^{c_3})$. Comparison matrix for criteria based on decision maker is denoted by matrix C as well. In this case we investigate consistency of comparison of criteria matrix C.

$$C = \begin{pmatrix} 1 & 3 & 5 \\ \frac{1}{3} & 1 & 3 \\ \frac{1}{3} & \frac{1}{3} & 1 \end{pmatrix}$$

criteria	c_1					c_2					c_3				
Processors	p_1	p_2	p_3	p_4	Q^{c_1}	p_1	p_2	p_3	p_4	Q^{c_2}	p_1	p_2	p_3	p_4	Q^{c_3}
p_1	1	$\frac{1}{4}$	4	$\frac{1}{6}$	0.17	1	$\frac{1}{2}$	$\frac{1}{3}$	5	0.20	1	$\frac{1}{5}$	1	$\frac{1}{2}$	0.12
p_2	4	1	4	$\frac{1}{4}$	0.29	2	1	$\frac{1}{2}$	7	0.31	5	1	2	2	0.43
p_3	$\frac{1}{4}$	$\frac{1}{4}$	1	$\frac{1}{5}$	0.05	3	2	1	9	0.44	3	$\frac{1}{2}$	1	3	0.33
p_4	6	4	5	1	0.49	$\frac{1}{5}$	$\frac{1}{7}$	$\frac{1}{9}$	1	0.05	1	$\frac{1}{2}$	$\frac{1}{3}$	1	0.12

Table 2: comparison matrix and their corresponding priority vectors

By using Eq.(3) we have:λ_{max} =3.039 ,$CI = \frac{3.039-3}{3}$=0.0195,$CR = \frac{0.0195}{0.58}$=0.033. Matrix C is consistent because CR <0.1. Consistency of comparison matrixes in table 2 can be investigated similarly. Based on steps 15-20 of proposed algorithm PVD for priority of processors can be formed as follow:

$$PVD = \Delta.\Lambda = \begin{pmatrix} 0.17 & 0.20 & 0.12 \\ 0.29 & 0.31 & 0.43 \\ 0.05 & 0.44 & 0.33 \\ 0.49 & 0.05 & 0.12 \end{pmatrix} \begin{pmatrix} 0.57 \\ 0.31 \\ 0.12 \end{pmatrix} = \begin{pmatrix} 0.17 \\ 0.32 \\ 0.20 \\ 0.31 \end{pmatrix}$$

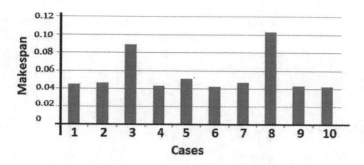

Fig. 3: possible value for makespan with different order of processors

We assume that $T_{cp}=0.1$, $T_{cm}=.01$, $w_0=0.1$, $z_0=0$, $w_1=0.1$, $z_1=.01$, $w_2=0.2$, $z_2=0.02$, $w_3=0.3$, $z_3=0.03$. At first we consider the root processor allocates fractions of load to worker processors which have been sorted based on priorities. In this case priority value of p_1, p_2, p_3 and p_4 are 0.17, 0.32, 0.20 and 0.31 respectively. Therefore, processor p_1 gets its fraction of load first, and then p_3, p_4 and p_2 get their fractions of load respectively. Thus fractions of load would be: $\alpha_0 =0.326517$, $\alpha_1 = 0.326517$, $\alpha_2 = 0.10715$, $\alpha_3 =0.16001$, $\alpha_4 = 0.07921$ and makespan would be 0.003265. On the other hand we consider the system without any priority, in this case $\alpha_0 = 0.32634$, $\alpha_1 = 0.32634$, $\alpha_2 = 0.16154$, $\alpha_3 =0.10662$, $\alpha_4 = 0.07917$ and makespan=0.003263.

In the *second case* we examined the proposed model on a multi level tree network. For this purpose we assume that the processors are connected to each other as a binary tree. We assume p_0, p_1 and p_2 are root processor, left-child(in first level) and right-child(in first level) respectively. Processor p_3, p_4,p_5 and p_6 are nodes of second level. However, p_7, p_8,...,p_{14} are nodes of third level. We assume that $w_i = 0.1 * i$ and $z_i = 0.001 * i$ for i_{th} processor and $T_{cp}=1$ and $T_{cm}=1$ for all processors as well. In this case makespan would be 0.0427524. We examined the algorithm for several orders of priorities in the binary tree. It can be seen that different priority of processors results different makespan. Fig.3 shows the makespan in 10 separate order of priorities. In case 1 processor p_8 obtains its fraction of load before p_7. In case 2,3 and 7; p_7 obtains its fraction of load before p_3, p_1 and p_2 respectively. In case 4,5 and 6; p_{13} obtains its fraction of load before p_6, p_2 and p_{14} respectively. Finally in case 8 and 9; p_{14} obtains its fraction of load before p_2 and p_7 respectively. Meanwhile case 10 is optimal. The results show that the proposed algorithm slightly increases makespan. Therefore we need to define a priority based closed form to deal with the problem.

7 Conclusion

We proposed a multi-criteria based algorithm for divisible load scheduling. The advantage of the model is being able to handle priorities of processors to get fractions of load. Comparing with the previous DLT, the algorithm has a small ignorable increase in complexity. Another problem is related to optimality. We indicated that the proposed model may obtain different makespan. It means that the makespan should be defined under priority condition. Improving the proposed model in order to calculate a closed form formula for multi criteria divisible load scheduling has been considered as future work.

ACKNOWLEDGMENTS

This work has been supported by UPM Research University Grant Scheme RUGS 05-01-10-0896RU/F1 and MOHE Fundamental Research Grant Scheme FRGS/1/11/SG/UPM/01/1.

References

1. Saaty, Thomas L. What is the analytic hierarchy process?. Springer Berlin Heidelberg, 1988.
2. Saaty, Thomas L. "How to make a decision: the analytic hierarchy process." European journal of operational research 48(1), 9-26 (1990).
3. Saaty, Thomas L. "The Modern Science of Multi criteria Decision Making and Its Practical Applications: The AHP/ANP Approach." Operations Research (2013).
4. Robertazzi, Thomas G., ed. Networks and Grids: Technology and Theory. Springer, 2007.
5. Ghanbari, Shamsollah, and Mohamed Othman. "A Priority based Job Scheduling Algorithm in Cloud Computing." Procedia Engineering 50, 778-785 (2012).
6. Li, Ping, Bharadwaj Veeravalli, and Ashraf A. Kassim. "Design and implementation of parallel video encoding strategies using divisible load analysis." Circuits and Systems for Video Technology, IEEE Transactions on 15(9), 1098-1112 (2005).
7. Ko, Kwangil, and Thomas G. Robertazzi. "Equal allocation scheduling for data intensive applications." Aerospace and Electronic Systems, IEEE Transactions on 40(2), 695-705 (2004).
8. Abdullah, Monir, Mohamed Othman, Hamidah Ibrahim, and Shamala Subramaniam. "Optimal workload allocation model for scheduling divisible data grid applications." Future Generation Computer Systems 26(7), 971-978 (2010).
9. Cheng, Yuan-Chieh, and Thomas G. Robertazzi. "Distributed computation with communication delay [distributed intelligent sensor networks]." Aerospace and Electronic Systems, IEEE Transactions on 24(6), 700-712 (1988).
10. Robertazzi, Thomas G. "Ten reasons to use divisible load theory." Computer 36(5), 63-68 (2003).
11. Patrick S. Chen, "On Vargas's Proof of Consistency test for 3 x 3 Comparison Matrices in AHP" Journal of the Operations Research Society of Japan 45, 233-241 (2002).
12. Strassen, Volker. "Gaussian elimination is not optimal." Numerische Mathematik 13(4), 354-356 (1969).
13. Drozdowski, Maciej, and Wodzimierz Gazek. "Scheduling divisible loads in a three-dimensional mesh of processors." Parallel Computing 25(4), 381-404(1999).
14. Berliska, J., and Maciej Drozdowski. "Heuristics for multi-round divisible loads scheduling with limited memory." Parallel Computing 36(4), 199-211 (2010).

Performance Analysis of Mobile Ad Hoc Networks using Queuing Theory

Aznida Hayati Zakaria[1], Md.Yazid Mohd Saman[1], Ahmad Shukri M Noor[1], Ragb O. M. Saleh[1]

[1]Department of Computer Science, Faculty of Science & Technology
University Malaysia Terengganu
Kuala Terengganu, Terengganu, Malaysia
aznida76@gmail.com, {yazid, ashukri@umt.edu.my}

Abstract. Delays and queuing problems are common features in our daily lives. We may also find it in technical environments, such as in manufacturing, computer networking and telecommunications. Queuing theory is the mathematics of waiting lines. It has been extensively applied to represent and analyse resource sharing systems, such as communication and computer systems. Unlike traditional wireless networks that need an expensive infrastructure to support mobility, Mobile Ad hoc Networks (MANETs) are completely infrastructure-less. In MANET, each node is connected through wireless links with other nodes. Because of the limited transmission range of mobile nodes, more than one hop may be needed to send and receive data across a wireless network. Therefore in order to facilitate communication within the network, a routing protocol is used to discover paths between the nodes. Nodal mobility can cause unpredictable network topology changes in MANETs. This paper discusses the formulation of queuing theory for analysing the performance MANET protocols by determining its arrival times, average waiting times, and response times.

Keywords: Queuing Theory, MANET, Routing Protocols

1 Introduction

In recent years; wireless or mobile networking is becoming easily accessible and affordable to the public. The proliferation of mobile computing and communications devices (e.g. cell phones, laptops, handheld digital devices, personal digital assistants) is driving a revolutionary change in our information society [1],[2],[8],[9]. Mobile users can use cellular phones for multiple purposes such as checking e-mails and browsing the Internet. Travellers with portable computers can surf the Internet in places like coffee shops and airports or other public locations; while researchers can exchange files and other information by connecting portable computers via wireless LANs at any places. Future information technology will be mainly based on this technology [1],[8].

T. Herawan et al. (eds.), *Proceedings of the First International Conference on Advanced Data and Information Engineering (DaEng-2013)*, Lecture Notes in Electrical Engineering 285, DOI: 10.1007/978-981-4585-18-7_62,
© Springer Science+Business Media Singapore 2014

A MANET is a type of ad hoc network that can change locations and configure itself rapidly. Because MANETs are mobile, they use wireless connections to connect to various networks [8]. There are no routers, servers, access points or cables in a MANET. Nodes can move freely and in arbitrary ways, so it may change its location from time to time. Each node may be a sender or a receiver, and any node may work as a router and do all router functions. This means that it can forward packets to other nodes [2]. Examples for applications of MANETs include meeting or conferences; military operations; and search and rescue operations.

Performance studies of MANET usually employ either analytical modelling or simulations. Simulation can evaluate complex system models at several levels of details [19]. However, simulations are costly in terms of programming and computing time. Meanwhile, analytical modelling involves constructing a mathematical model of the system behaviour [3]. Mathematical model can be used to abstract the essential characteristics of a computer system and analyse the system behaviour [6]. Simulation and analytical modelling often use queuing network system as the model for analysing the performance of a MANET. Currently, the analytical model for MANET using queuing theory has never been put forward and published. Therefore, this is the key motivating factor to develop an analytical model based on queuing system in this research. Our primary purpose in this study is to determine the arrival times, average waiting times, and response times when delivering data between mobile nodes.

The rest of the paper is organized as follows. A general description of MANET is depicted in Section 2. The queuing model for MANET is discussed in Section 3 while the performance analysis of MANET using queuing theory is described in Section 4. Finally, we conclude the paper in Section 5.

2 Mobile Ad Hoc Network (MANET)

A Mobile Ad hoc Network (MANET) is an ad hoc wireless network, and is a self-configuring network of mobile nodes connected by wireless links [4]. The nodes in MANET are free to move about arbitrarily thus the topology may change rapidly and unpredictably [1]. Each node functions as both a host and a router [14]. The nodes in MANET are equipped with wireless transmitters [8]. If there are only two nodes in MANET that want to communicate with each other and are located very closely to each other; then no specific routing protocols or routing decisions are necessary. On the other hand, if there are a number of mobile hosts wishing to communicate; then the routing protocols are important. Some critical decisions have to be made such as which is the optimal route from source to the destination.

Based on Figure 1, suppose that node 1 wants to send data to node 3 but node 3 is not in the range of node 1. Then in this case, node 1 may use the services of node 2 to transfer data since node 2's range overlaps with both the node 1 and node 2. Node 2 act as a router in this case, which forward the data from node 1 to node 3. Indeed, the routing problem in a real ad hoc network may be more complicated than this example suggests due to the inherent non uniform propagation characteristics of wireless transmissions and due to the possibility that any or all of the nodes involved may move at any time [4].

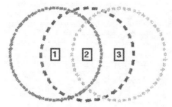

Fig. 1 A Simple MANET [2]

One of the main difficulties in MANET is the routing problem. Many routing protocols have been proposed in the past. These protocols have been designed for accurate, fast, reliable routing for a high volume of changeable network topology [2]. The classification of routing protocols differentiate according to their technique, their hop count, link state, and source routing in a route-discovery mechanism [5],[14],[15].

The Dynamic Source Routing (DSR) protocol is used in this study. It is one of the examples of an on-demand routing protocol that is based on the concept of source routing. It is designed for the use in multi hop ad hoc networks of mobile nodes [4].   DSR uses no periodic routing messages thus reduces network bandwidth overhead, conserves battery power and avoids large routing updates [12],[16]. This protocol is composed of the two mechanisms of Route Discovery and Route Maintenance [13] and each operate entirely *on demand*.

3 Queuing Model for MANET

A MANET topology can be defined as a dynamic (arbitrary) multi-hop graph $G^2(N, r_0(N))$, where N is a finite set of mobile nodes $(1,2,....,N)$ and the nodes are independently placed on a two-dimensional area A. Each node is assumed to have the transmission range, defined by $r_0(N)$. Let r_{ij} denote the distance between nodes i and j. Nodes i and j are said to be neighbors if they can directly communicate with each other, that is if $r_{ij} \leq r_0(N)$. A circle area of $\pi r_0^2(N)$ is termed the communication area of a node. We assume the number of nodes that are neighbors of node i, if the node is within the communication area of $\pi r_0^2(N)$ in the node deploy area of a two-dimensional area A. A is a rectangular area of size a x b for $a \geq b$. This network model is depicted in Figure 2 [7],[9].

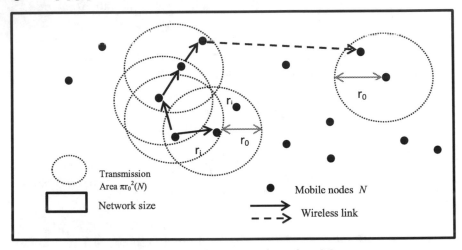

Fig. 2 The model of network topology [7]

Assume that there are N nodes in the ad hoc network where node S is the source and node D is the destination for the packets of data. Figure 3 shows node S and node D as queuing nodes. An arriving transaction consists of jobs. Communication among nodes is via the process of sending RREQ packets by node S and returning RREP packet by node D. First, we assume that the packets passing to the destination node is abstracted to a queuing system with single server, single queue.

Fig. 3 MANET Queuing Model

In this case, we study the performance analysis when packets of data from node S are being sent to its destination node D as Figure 3 shows. We consider that, each node has a service at any point in time. In other words, each node corresponds to one job/task at any one time.

4 Performance Analysis of MANET

This section presents the mathematical implementation of queuing network model based on MANET.

4.1 Model Input : Workload, Transaction Processing, and Arrivals Rates

The network consists of N interconnected nodes. In this model, transaction arrives according to a Poisson process and consists of jobs. Each node corresponds to one job at any one time. Each transaction has arbitrary capacity of nodes. Consequently, the service time given to transaction is arbitrary. Thus, we model the nodes by queuing network in which each of the n identical site is represented by an M/G/1 system. The letters M and G indicate the probability distribution n of user inter arrival times and of service times respectively, and 1 is the number of identical parallel server in the queuing system.

Transaction arrivals to the nodes are modelled with parameter λ. The arrival of transactions from every node is assumed to be a Poisson process, and then their sum is also Poisson. Transactions are also assumed to be executed at a single node. Thus, the arrival rate of transaction at a single node amount to

$$\lambda = \ell.\lambda + (n-1).(1-\ell).\lambda/((n-1)) \tag{1}$$

The first term describes a share of ℓ of the incoming λ transactions can be executed locally, and whereas the remaining $(1-\ell) \cdot \lambda$ transactions are forwarded to nodes where appropriate data is available. The other n-1 nodes also forward $(1-\ell)$ of their λ transactions, which are received by each of the remaining nodes with equal probability 1/(n-1).

4.2 Model Output : Performance Criteria

This paper considers the average response times as the performance criteria. The average waiting time $\overline{W_c}$ at a local node can be obtained using the Pollaczek-Khinchin formula for general M/G/1 systems [6].

$$\overline{W}_c = \frac{\lambda.E(B_c^2)}{2(1-\rho)} \tag{2}$$

Similar to the calculation of $\overline{W}_c$, the mean waiting time $\overline{W}$ at a local node in MANET is found to be :

$$\overline{w} = \frac{\lambda.t^2}{1-\lambda.t} \tag{3}$$

where λ is the arrival rate per node and t is the mean service time. By symmetry, the average queuing delays at all stations are the same. The expected queuing delay that a packet waiting for transmission in the entire network from source to destination is given by

$$E(D_Q) = \overline{W}.E(H)$$

(4)

where $\overline{W}$ is mean waiting time at local node and $E(H)$ is the expected hop count between a source and destination pair.

The mean waiting time at local node is the time that user or transaction spends in a queue waiting to be serviced. Meanwhile, the response time is the total time that a job spends in the queuing system [6]. In other word, the response time is equal to the summation of the waiting time and the service time in the queuing system [11]. On average, a transaction needs to wait for $\overline{W}$ seconds at a node to receive a service of t seconds. Thus, the response time is given by

$$\overline{R} = \overline{W} + t$$

(5)

where $\overline{W}$ is mean waiting time at local node and t is the mean service time.

The average response time over all transactions types in the entire network from source to destination results in

$$\overline{R}_{total} = \overline{R}.E(H)$$

(6)

where $\overline{R}$ is the response time at local node and $E(H)$ is the expected hop count between a source node and destination node.

4.3 Results

Figure 4 presents the results of waiting times with three transaction arrival rates (0.5s, 1.0s, 2.0s) for hop count three, five and eight. The curve in Figure 4 shows a positive relationship where an increase in the waiting time is associated with an increase in the hop count from source node to destination node. The increasing hop count caused the services take much longer time to response to all requests in the entire network. Figure 4 indicates that the minimum value of transaction arrival rate results in minimum waiting times for transmission in the entire network, from source node to destination.

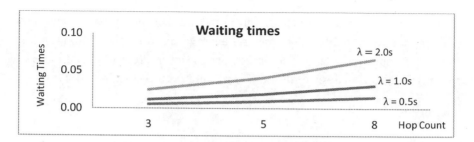

Fig. 4 Performance of MANET (Waiting Times vs Hop Count)

Variation of the values of the average response times based on the hop count (three, five and eight) is shown in Figure 5. Note that from the figure, the curve shows a positive relationship where an increase in average response time is associated with an increase of hop counts. In other words, every increment of hop count causes the average response time to increase. This directly reflects the definition of the response time, ie. elapsed time from the initiation to the completion of the transaction, and with the inclusion of the waiting time and the service time.

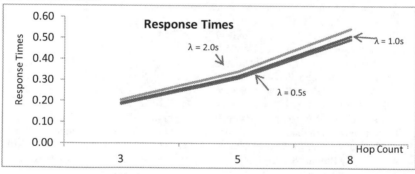

Fig. 5 Performance of MANET (Response Times vs Hop Count)

5 Conclusion

This paper proposed a queuing network model based on dynamic source routing protocol to analyse minimum waiting times and response times in MANETs. This study evaluates the performance of ad hoc routing based on dynamic source routing protocol (DSR). The novelty of this research is in providing an analytical queuing model for mobile ad hoc network based on M/G/1 queuing system. We model MANET by closed queuing network in which each of the n number of hop counts are represented by a M/G/1 queuing system. It is noted that every increment in the number of hop count caused the increasing values in the waiting time and average response time. In the future, extensive complex simulations could be carried out in order to gain a more in-depth performance analysis of the ad hoc routing protocols in MANET.

6 References

1. Chlamtac, I., Conti, M., & Liu, J. J. N. (2003). Mobile ad hoc networking: imperatives and challenges. *Ad Hoc Networks*, *1*(1), 13-64.
2. Loo, J., Mauri, J. L., & Ortiz, J. H. (Eds.). (2012). *Mobile Ad Hoc Networks: Current Status and Future Trends*. CRC Press.
3. Gokhale, S. S., & Trivedi, K. S. (1998). Analytical Modeling. *Computer Engineering*, *0291*.
4. Ramesh, V., Subbaiah, D. P., Rao, N. K., & Raju, M. J. (2010). Performance Comparison and Analysis of DSDV and AODV for MANET. *IJCSE) International Journal on Computer Science and Engineering*, *2*(02), 183-188.
5. Gorantala, K. (2006). Routing protocols in mobile ad-hoc Networks. *Master's Thesis in Computing Science, June, 15*.
6. Klienrock, L. (1975). Queuing Systems, Volume 1: Theory. *John Wiley & Sons*.
7. Lee, A., & Ra, I. (2012). A Queuing Network Model Based on Ad Hoc Routing Networks for Multimedia Communications. *Appl. Math*, *6*(1S), 271S-283S.
8. Basagni, S., Conti, M., Giordano, S., & Stojmenovic, I. (Eds.). (2004). *Mobile ad hoc networking*. Wiley-IEEE press.
9. Erciyes, K., Dagdeviren, O., Cokuslu, D., Yılmaz, O., & Gumus, H. Modeling and Simulation of Mobile Ad hoc Networks.
10. Hofmann-Wellenhof, B., Lichtenegger, H., & Collins, J. (1993). Global positioning System. Theory and Practice. *Global Positioning System. Theory and practice., by Hofmann-Wellenhof, B.; Lichtenegger, H.; Collins, J.. Springer, Wien (Austria), 1993, 347 p., ISBN 3-211-82477-4*
11. Deris, M. M., Aznida, Z., Saman, M. Y., Suryani, W. A., & Zarina, M. (2007). Performance modelling of asynchronous replica distribution technique on distributed database systems. *International Journal of Computer Mathematics*, *84*(2), 183-192.
12. Bansal, M., Rajput, R., & Gupta, G. (1999). Mobile ad hoc networking (MANET): Routing protocol performance issues and evaluation considerations.
13. Raut, S. H., Tech-II, M., & Ambulgekar, H. P. (2012). Routing Protocols in Mobile Adhoc Network (MANET). *International Journal of Engineering*, *1*(9).
14. Roy, R. R. (2011). Mobile Ad Hoc Networks. In *Handbook of Mobile Ad Hoc Networks for Mobility Models* (pp. 3-22). Springer US.
15. Taneja, S., & Kush, A. (2010). A Survey of routing protocols in mobile ad hoc networks. *International Journal of Innovation, Management and Technology*, *1*(3), 2010-0248.
16. Tyagi, S. S., & Chauhan, R. K. (2010). Performance analysis of proactive and reactive routing protocols for ad hoc networks. *International Journal of Computer Applications*, *1*(14), 27-30.

Service Dynamic Substitution Approach Based on Cloud Model

Yan Gong[1], Lin Huang[1,2], and Ke Han[1]

[1] Chinese Electronic Equipment System Engineering Corporation, Beijing, China
[2] State Key Laboratory of Networking and Switching Technology,
Beijing University of Posts and Telecommunications, Beijing, China

gongyan@bupt.edu.cn, huanglin1204@gmail.com,
kehan@yahoo.com.cn

Abstract. Performance of web service may fluctuate due to the dynamic Internet environment, which makes the Quality of Service (QoS) inherently uncertain. It is necessary to reconfigure the web service to enable its QoS values to meet users' demands. In this paper, we propose an efficient and effective dynamic substitute approach based on cloud model. Our approach employs cloud model to compute the QoS uncertainty to determine dynamic substitute targets. By targeting substitutions, the reconfigured web service will better satisfy users' requirements. Both theoretical analysis and experiment results have proven the feasibility and effectiveness of the approach.

Keywords: Web Service, Cloud Model, QoS.

1 Introduction

Web service is the core technology to implement Service Oriented Architecture (SOA) and the basic functional entity to build service-oriented software. Running in a dynamic environment, the Web service is faced with various changes. The QoS of the existing Web services may be unstable, but new Web services will continuously appear with probably better QoS. Therefore, after composing the Web services, continuous maintenance should also be conducted to ensure that the QoS of the service composition will meet the requirements continually.

The pervious concerns about the service QoS tend to focus on the users' changes in their requirements towards it. By determining what service needs to be replaced and by conducting the dynamic replacement, the SOA-based software will re-achieve the users' QoS requirements. However, it should be noted that the QoS value of the Web service that the software is made up of is with significant fluctuations, which will impact on the stability of the entire software's QoS value and lead to its failure in meeting the users' needs. Thus, a way to monitor and provide dynamic replacement to the Web services of the software is needed to improve its QoS stability.

This paper presents a service dynamic replacement method based on cloud model. For QoS value fluctuations of the software's member services, a monitoring module is

T. Herawan et al. (eds.), *Proceedings of the First International Conference on Advanced Data and Information Engineering (DaEng-2013)*, Lecture Notes in Electrical Engineering 285, DOI: 10.1007/978-981-4585-18-7_63,
© Springer Science+Business Media Singapore 2014

introduced in to do real-time monitoring. It uses backward cloud algorithm of the cloud model to turn quantitative QoS into qualitative QoS and assesses the level of its QoS instability according to the threshold parameter that has been set up in advance. Once a member service is found with great QoS value fluctuation, another service from its candidate list will be selected to replace it, stabilizing the QoS of the software.

The remainder of this paper is organized as follows. Section 2 introduces the related works of service substitution. Section 3 describes the service dynamic substitution based on cloud model. Section 4 shows the experiments and Section 5 concludes the paper.

2   Related Work

Paper [1] introduces an approach to service substitution based on user preferences over non-functional properties of services. The approach utilizes preference networks for representing and reasoning about preferences over non-functional properties. Service substitution based on the functional properties of components has been addressed by many authors [2-4]. These papers focus on the fixing of the software. Differently, this paper focuses on the instability of the software's non-functional properties during its running and aims to conduct unstable dynamic replacement to the member services to meet the users' requirements.

Paper [5] puts forwards a dynamic substitution approach of QoS-driven services. It records QoS reference value of each member service in a list, extracts their real time QoS values once over a period of time and compares them with the reference values. If the difference is greater than the given threshold value, the counter of the member service will add one. This method has the following two disadvantages: first, it can not effectively detect the uncertainty of QoS, leading to the inaccuracy in selecting the services that need replacement; second, it can not select the service of stable QoS values from the candidate list or ensure the success of the service replacement. This paper will compare it with the research method we propose through experiments.

3   Dynamic Substitution Approach Based on Cloud Model (DSCM)

In the service-oriented software, the main operating of the non-functional dynamic replacement is service replacement. The dynamic replacement of non-functional properties does not change the logical structure of the software. It only hopes to stabilize the software by replacing one or more of its member services. The service dynamic replacement process shows in Figure 1. The proposed DSCM approach contains two phases. The first phase is QoS instability detection (Section 3.1), in which we adopt cloud model to transform the quantitative QoS to qualitative QoS for the QoS uncertainty computation. The second phase is service substitution (Section 3.2), which is to find the service that meets the requirements from the candidate services and use it as replacement.

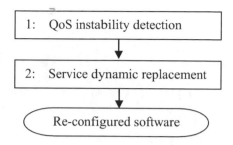

Fig. 1. Procedures of DSCM approach

3.1   QoS instability detection

Most SOA-based software are composed of a large number of Web services, which have the open, dynamic and changeable features. Web service might show instability of non-functional attributes, affecting the user experience. To solve the instability, this paper has designed a module to test the non-functional attributes of the SOA-based software. The module will examine the QoS value of each service in the service-oriented software. When it detects great changes in service QoS value, it will decide to replace the member service with another one from its candidate list with stable QoS values in order to stabilize the QoS values of the software.

To evaluate the degree of the service QoS uncertainty and ensure the reliable service replacement, DSCM adopts cloud model to compute the uncertainty by transforming quantitative QoS values (transaction logs) to qualitative QoS concept (uncertainty level). According to the uncertainty level, a web service with consistently bad QoS can be distinguished from those services with a large QoS variance.

Cloud model.

Cloud model [6] is a model of uncertainty transition between a linguistic term of a qualitative concept and its numerical representation. It can be employed for the uncertainty transition between qualitative concept and quantitative description.

The overall characteristics of cloud model may be reflected by its three numerical characteristics: Expected value (Ex), Entropy (En) and Hyper-Entropy (He). In the discourse universe, Ex is the position corresponding to the center of the cloud gravity, whose elements are fully compatible with the linguistic concept; En is a measure of the concept coverage, i.e., a measure of the fuzziness, which indicates how many elements could be accepted to the qualitative linguistic concept; and He is a measure of the dispersion on these cloud drops, which can also be considered as the entropy of En. Then, the vector $NC = \{ Ex , En , He \}$ is called the eigenvector of cloud model.

In this study, we apply these three numerical characteristics of backward cloud generator (Algorithm 1) to denote the uncertainty of QoS by transforming QoS quantitative values to qualitative concept.

Algorithm 1. Backward cloud generator.
Input: n transactions of a web service, i.e., n cloud
drops $\{x_1, x_2, \ldots, x_n\}$.
Output: the three numerical characteristics Ex, En, and
He of the n cloud drops.

1: $Ex = \overline{X} = \dfrac{1}{n}\sum\limits_{i=1}^{n} x_i$;

2: $T^2 = \dfrac{1}{n-1}\sum\limits_{i=1}^{n}\left(x_i - \overline{X}\right)^2$;

3: $En = \dfrac{\sqrt{\pi/2}}{N}\sum\limits_{i=1}^{N}\left|x_i - Ex\right|$;

4: $He = \sqrt{T^2 - En^2}$.

The working process of algorithm 1 is as follows:

- Line 1: According to x_i, compute the sample mean $\overline{X}$. The expected value of the web service on its QoS can be calculated by $Ex = \overline{X}$;
- Line 2: Compute the sample variance T^2 ;
- Line 3: Compute estimated value of *En*;
- Line 4: Compute estimated value of *He*.

Application of the cloud model.

Detection module works as follows: Every once in a certain period of time, it extracts the QoS value of each member service. In order to use cloud model to compute the uncertainty of the Web service QoS, set the parameters λ and h, respectively as the threshold values of *He* and *En*, for evaluating the stability of the Web service. If the monitored Web service satisfies the following conditions:

$$En \leq \lambda \quad \text{and} \quad He \leq h \tag{1}$$

It indicates that its service QoS is stable. Otherwise, the QoS of the service is unstable, and of large variational amplitude and poor reliability. And the member service will be added to list of services that need update and the service replacement operation will be triggered. Through constant monitoring of the module, member services of unstable QoS values will be found, ensuring the QoS values of the software stay in a stable state.

```
Algorithm 2. Monitor.
Input: an application S = {S₁,S₂,...,Sₙ}.
Output: needReplaceList with unstable QoS or Null.
```

```
1:   List substitutionList;
2:   SET table A with {a₁,a₂,...,aₘ} of Sₙ;
3:   SET table T with {t₁,t₂,...,tᵢ} of Tₙ;
4:   Every time unit tᵢ{
5:   FOR j = 0 to Size(A) {
6:   Get every QoS of Sₙ{a₁, a₂,...,aₘ};
7:   };
8:   };
9:   By using En and He to monitor QoS transactions;
10:  SET temp1 = En and temp2 = He;
11:  Get every temp1 and temp2 of Sₙ{a₁,a₂,...,aₘ};
12:  IF( temp1 > λ ) and ( temp2 > h );
13:  SubstitutionList. add(Sₙ);
14:  Return substitutionList.
```

The working process of algorithm 2 is described as follows:

- Line 1 - Line 2: demonstrate the QoS value list of member services.
- Line 3 - Line 8: gain and store the QoS attribute values of all the member services in each unit of time.
- Line 9 - Line 11: calculate En and He through the backward cloud algorithm's quantitative - qualitative conversion function of the cloud model.
  Line 12 - Line 14: compare the En and He with the given thresholds λ and h. If the comparison result is greater than a defined given threshold, the service will be added to the list of services need to be replaced.

3.2 Substitution Algorithm

The input of replacement algorithm is composed of a set of services with unstable QoS values. The algorithm is intended to replace the member services of the collection. It will select the suitable services from the candidate list serve the same function to take the place.

```
Algorithm 3. Service Substitution.
Input: a request of a new QoS S'ₙ = {a'₁, a'₂, ..., a'ₘ}, a set
Mᵢ of all Candidate services of Sᵢ, substitutionList with
unstable QoS.
Output: The new application S' or Null.
```

```
1:  IF(substitutionList! = Null) {
2:  For i = 0 to Size(substitutionList) {
3:  IF(exist Sji ∈ Mi and the QoS of Sji
        satisfy(a'1, a'2, ..., a'm) and En ≤ λ, He ≤ h){
4:  replace Sj with Sji; Return S';
5:  } ELSE{
6:  Return Null;
7:  };
8:  }.
```

- Line 1 - Line 2: The input of replacement algorithm is composed of a set of services with unstable QoS values.
- Line 3 - Line 8: Find the candidate services satisfying requirements and make replacement.

4   Experiment

All the experiments are conducted on the same computer with Pentium 3.0 GHz processor, 4.0GB of RAM, Windows. The experiments are based on the following assumptions: a software contains six member services; the entropy and hyper entropy threshold values of the QoS uncertainty calculation are set as $\lambda = 3.2$, $h = 5.4$. Monitoring information of each service's execution time is shown in Figure 2.

As can be seen from Figure 2, when some member services of the software show fluctuation in QoS, such as the service S3 shows great fluctuations in service execution time in the middle stage, our approach will replace them dynamically to stabilize the QoS values of the software and to meet users' expectations.

As can be seen from Figure 3, regardless of how many candidate services there are, the success rate of the DSCM is significantly better than the QDSDS. The average success rate of DSCM is up to 97.5% while with QDSDS, it is only 83.1%. The reason for the DSCM's significant success rate lies in the cloud-based QoS model used in the service replacement, which excludes the candidate services of great QoS fluctuations and adopts those with stable QoS values. Thus the success rate of the service replacement is improved.

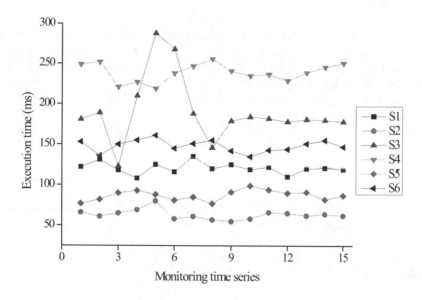

Fig. 2. Monitoring information

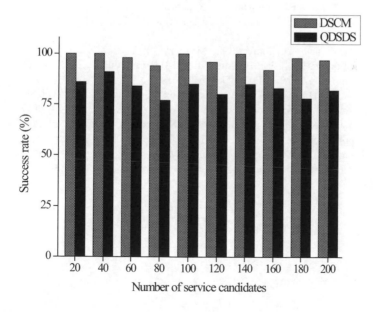

Fig. 3. Success rate

5   Conclusion

Service-oriented software is composed by a combination of services. Because of the dynamic nature of Web service environment, the QoS often fluctuates with inherent uncertainty and therefore needs to be dynamically replaced. Based on the changes in the QoS value of the software itself and users' requirements for the software, this paper proposes a dynamic replacement method based on cloud model. It adopts cloud model to transform the quantitative QoS to qualitative QoS for the uncertainty computation and judges the uncertainty level by setting the corresponding service QoS parameters. When a member service with unstable QoS values is found, it will be replaced with the candidate service with the same or better QoS values to stabilize the QoS values of the entire software.

Acknowledgments. This work is supported by the China Postdoctoral Science Foundation funded project (No.2012M521839).

References

1. Santhanam, G., Basu, S., Honavar, V.: Web Service Substitution Based on Preferences Over Non-functional Attributes. In: IEEE International Conference on Services Computing, pp.210-217. IEEE Press, Bangalore (2009)
2. Benatallah, B., Casati, F., Toumani, F.: Representing, Analysing and Managing Web Service Protocols. Data and Knowledge Engineering. 58(3), pp.327-357. Elsevier Press, (2006)
3. Taher, Y., Benslimane, D., Fauvet, M., Maamar, Z.: Towards an Approach for Web Services Substitution. In: 10th IEEE International Database Engineering and Applications Symposium, pp.166-173. IEEE Press, Delhi (2006)
4. Pathak. J., Basu. S., Honavar. V.: On Context-specific Substitutability of Web Services. In: IEEE International Conference on Web Services, pp.192-199. IEEE Press, (2007)
5. Song, Z., Zhang, X., Yin, Y.: QoS Driven Service Dynamic Substitution Method. Computer Applications and Software. 1(9), pp.27-30. (2012)
6. Li, D., Cheung, D., Shi, X., Ng, V.: Uncertainty Reasoning Based on Cloud Models in Controllers. Computers and Mathematics with Applications. 35(3), pp.99-123. Elsevier Press, (1998)

Text Summarization in Android Mobile Devices

Oi-Mean Foong[1], Suet-Peng Yong[2] and Ai-Lin Lee[3]

Computer and Information Sciences Department, Universiti Teknologi PETRONAS
Bandar Seri Iskandar, 31750 Tronoh, Perak, Malaysia
[1,2]{foongoimean, yongsuetpeng}@petronas.com.my, [3]mellissalee90@gmail.com

Abstract. This paper presents a text summarization in Android mobile devices. With the proliferation of small screen devices and advancement of mobile technology, the text summarization research has been inspired by the new paradigm shift in accessing information ubiquitously at anytime, anywhere and anyway on mobile devices. However, it is a challenge to browse large documents in a mobile device because of its small screen size and information overload problems. In this paper, a semantic and syntactic based summarization was attempted and implemented in a text summarizer. The objectives of the paper are two-fold. (1) To integrate WordNet 3.1 into the proposed system called TextSumIt which condenses single lengthy document into shorter summarized text. (2) To provide better readability to Android mobile users by displaying the salient ideas in bullets points. Documents were collected from DUC 2002 and Reuter news datasets. Experimental results show that the text summarization model improves the accuracy, readability and time saving in the text summarizer as compared with MS Word AutoSummarize.

Keywords: Android mobile devices, text summarization, information overload.

1 Introduction

With the rapid advancement of mobile technology and affordable handheld devices, users could access information from internet ubiquitously at anytime, anywhere and anyway on mobile devices. However, it poses many challenges such as low network bandwidth, limited small screen size, low memory capacity and high power consumption to load and visualize large documents on handheld mobile devices [1]. Yet, Android has gained popularity for information retrieval on-the-fly among users because many developers use the open source Eclipse IDE with Android SDK and ADT development kit that write mobile applications to extend the functionality of the mobile devices.

On the contrary, large internet articles or journals are often cumbersome to read as well as comprehend. It is time-consuming and we have limited time to assimilate all of the articles which are at times, exceeding our capability to perceive. In addition, time is precious as today's fast paced era has caused people to demand for quick results. As the information continues to grow exponentially, there exists a need to retrieve and filter the overloaded information. In fact, information overload is an increasing problem both in the workplace and in our general daily life. Readers either

T. Herawan et al. (eds.), *Proceedings of the First International Conference on Advanced Data and Information Engineering (DaEng-2013)*, Lecture Notes in Electrical Engineering 285, DOI: 10.1007/978-981-4585-18-7_64,
© Springer Science+Business Media Singapore 2014

skip reading the content or unable to process the information well, leading to wrong decisions made.

In this context, there is an apparent need for an Automatic Text Summarization (ATS) as a solution to accommodate the growing information while saving time in producing it manually. ATS extracts content of the document using an algorithm, produces coherent and correctly-deliberated summaries, and displays the most important points of the original text to the user in a more condensed way and in accordance to each user's needs [2]. In this paper, the objectives are two-fold. (1) To integrate syntactic and semantic techniques in Text Summarizer. (2) To provide a better readability of shorter text for mobile users in bullet points. The proposed technique would identify and rank the top 20% of importance sentences from original document to produce shorter summarized text.

Most researchers applied extractive summary as it is more difficult to develop abstractive summary due to its implementation of deep natural language processing. The main challenge would be to develop on an Android mobile platform to generate a shorter but precise summary text.

This paper is structured as follows. Section 2 presents the related work. Section 3 proposes the system architecture of the text summarizer. Section 4 presents the results of quantitative and qualitative evaluation of the text summarizer prototype. The last section states the conclusion of the research.

2 Related Work

The earliest works of text summarization started in 1950s [3]. The initial work started off with the implementation of statistical techniques in text summarization and gradually improved towards using natural language process (NLP), semantic analysis, fuzzy logic, swarm intelligence, genetic algorithm [4] and lastly hybrid fuzzy swarm [5]. The challenge to text summarization is the summary quality which remains a key issue in many researches.

There are two categories of text summarizers namely statistical and linguistic. Statistical summarizers operate by finding the important sentences using statistical techniques such as word frequency, text position, cue words, heading and sentence position [6], [7]. On the other hand, linguistic summarizers use knowledge about the language such as syntax, semantics usage and key concepts to summarize a document [8],[9]. According to Saggion and Poibeau, summaries can be classified into three categories: (1) Indicative (2) User-oriented or Generic and (3) Extractive or Abstractive [10]. Firstly, the Indicative summaries usually provide the general concepts of the text document without showing specific content. Secondly, the user-oriented summaries focus on the interest of the readers on certain topics. It favors specific themes or aspects of the text. Generic summaries convey the point of view of the authors on the input text. Thirdly, the *Extractive Summary* uses a fragment of the source text (key clauses, key phrases, sentences, etc.) to structure the summary, i.e., summary copied from input [11]. These summaries lack coherence as compared to abstractive summaries as it only conveys an approximate content of the source text.

The *Abstractive Summary* reconstructs the extracted sentences, i.e., paraphrasing sentences to form a more cohesive and coherent summary. This method can condense text more strongly as compared to extraction by developing an understanding and expressing main concept of documents in clear natural language [12].

Semantic extraction involves the understanding of the structure and meaning of the natural language to produce semantic information from text documents [13]. However, the critical issue in extracting text semantically is the ambiguity and uncertainty of the meaning of texts. The improvement in summaries is achieved by applying lexical knowledge (e.g. WordNet) towards the text summaries to build a more comprehensive text. In addition, the cohesiveness of sentences can be streamlined by mapping the terms within the sentence to similar concepts using a lexical database called WordNet [14]. It connects four types of Part-of-Speech (POS); nouns, verbs, adjectives, and adverbs in which it groups the words into sets of synonyms called synset. Each synset consists of the word, its explanation and its synonyms. The main idea of WordNet is to combine the usage of a dictionary and thesaurus which can support text analysis and the implementation of artificial intelligence applications such as word sense disambiguation, automatic text summarization, text categorization and information retrieval.

3   TextSumIt System Architecture

The proposed TextSumIt model which was proposed in authors' previous research focused on preprocessing and features selection as shown in Fig. 1 [15]. Users will input source text documents to be summarized. During the preprocessing text stage, the text document undergoes a process of tokenization, stop words removal, WordNet stemming and finally Part-of-Speech (POS) tagging. The preprocessing is done to make it easier for the process of learning algorithm.

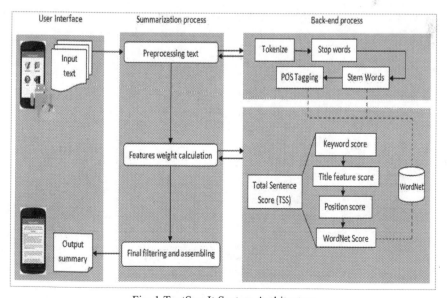

Fig. 1 TextSumIt System Architecture

Features weight computation of the Total Sentence Score (TSS) which consists of keyword score, title score, position score, and WordNet score. These features weight will determine the rank of the whole sentences from the source text. The sentence with the highest weight will be selected for the summary output. It is enhanced with the semantic extraction algorithm using WordNet.

The last stage of the text summarization process is final filtering and assembling. At this stage, undefined references of sentences are filtered and the rest are assembled to form a complete summary as output to users.

3.1 Features weight calculation

- **Total Sentence Score (TSS)**

A final total sentence score (*TSS*) is calculated for each sentence using Eq. 1.

$$TSS(Si) = w_1 KW(Si) + w_2 Title(Si) + w_3 Pos(Si) + w_4 WN(Si) \tag{1}$$

where *TSS* is the Total Sentence Score, *KW* is the keyword score, *Pos* as the position score and *WN* as the WordNet score. w_1, w_2, w_3, w_4 are the weights for each feature. The top scoring sentences will be selected as the output summary according to the compression rate set by users.

- **Keyword Score**

Keyword score is based on the Term Frequency (*TF*) and Inverse Sentence Frequency (*ISF*) algorithm [16]. It is computed by Eq. 2.

$$TF - ISF(w,s) = TF(w,s) * ISF(w) \tag{2}$$

where the term frequency *TF (w, s)* is the number of times the word *w* occurs in sentence s, and inverse sentence frequency *ISF(w)* is computed in Eq. 3.

$$ISF(w) = \log\left(\frac{S}{SF(w)}\right) \tag{3}$$

where sentence frequency, *SF(w)* is computed as the number of sentences in which the word *w* occurs. *S* is the total number of sentences in the document.

- **Title Feature Score**

This algorithm emphasizes on the sentences with keywords that are present in the title which resemble the theme of the document. It can be calculated using Eq. 4 [17].

$$Title(Si) = \frac{Keywords\ in\ Si \cap Keywords\ in\ Title}{Keywords\ in\ Si \cup Keywords\ in\ Title} \tag{4}$$

- **Position Score**

The position score in this study is based on its similarity to the first or last sentence of the document. The average score is calculated using Eq. 5, 6, 7.

$$P_{First}(Si) = \frac{Keywords\ in\ Si \cap Keywords\ in\ S_{First}}{Keywords\ in\ Si \cup Keywords\ in\ S_{First}} \tag{5}$$

$$P_{Last}(si) = \frac{Keywords\,in\,S_i \;\cap\; Keywords\,in\,S_{Last}}{Keywords\,in\,S_i \;\cup\; Keywords\,in\,S_{Last}} \tag{6}$$

$$Pos(si) = \frac{P_{First}(si) + P_{Last}(si)}{2} \tag{7}$$

- **WordNet Score**

The algorithm focuses on keyword semantics extraction was illustrated by Barrera and Verma [18]. The model's objective is to select the sentences which contain words that have the closest meaning to the keywords extracted. The score allocated for each word w is computed using Eq. 9:

$$score(w) = \frac{1}{2^l} \tag{9}$$

where l refers to the minimum level determined when the word w is compared with the synsets. The overall equation for WordNet score is shown in Eq. 10:

$$WN_{syn}(si) = \sum_{w \in S_i} scoresyn(w) \tag{10}$$

The closer the word w in a sentence, S_i to the synsets, the higher the WordNet score.

4 Result and Discussion

Experiments were set up with the aim of testing the system performance on accuracy (F-Score), readability and time saving in the text summarizer using DUC 2002 and Reuter news datasets [19]. In the following GUI section, screen shots were captured to showcase the block design in the TextSumIt mobile application which addressed the readability criteria.

4.1 Graphical User Interface (GUI)

The interfaces used for users' navigation within the TextSumIt application on Android are shown in Fig. 2(a) and 2(b). The output summary in Fig. 2(c) is arranged into three blocks: (1) The first block consists of the title of an article (2) The second block includes keywords in the article (3) The third block extracts important sentences into bullet points to make it easier for reading. The compression rate is scalable such that user can select various compression rates of 10%, 20%, 30%, 40% or 50% from the original text. Only top 20% of the important sentences which contain main ideas will be shortlisted for display.

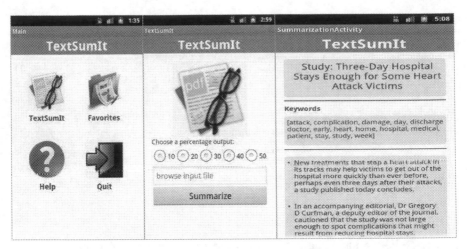

Fig2(a) Home Screen   Fig.2(b) Text Input   Fig.2(c) Summary Output

4.2 System Performance Evaluation

The evaluation was performed on 30 random articles from DUC 2002 based on 20% compression rates which was used by previous researchers [5]. Each 100-word article was selected as an input for the system summary and MS Word AutoSummarize.

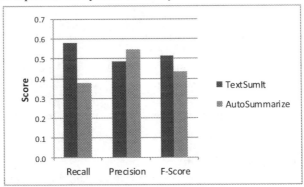

Fig.3 Overall average performance comparison between TextSumIt and AutoSummarize

Fig. 3 shows the average recall, average precision and average F-Score of the DUC 2002 articles in the system prototype and the MS Word AutoSummarize. Both summaries were compared to the reference summaries which are human-generated by experts. In general, TextSumIt has a higher F-score of 0.513 as compared to MS Word AutoSummarize which has a lower F-Score of 0.432. The significant difference in the results is due to the improved algorithm used in the sentence extraction.

TextSumIt which used a semantic based algorithm was able to extract sentences more precisely as compared to MS Word AutoSummarize which only used frequency-based algorithm. In spite of the higher average precision score of 0.547

obtained by the MS Word AutoSummarize, the proposed TextSumIt model yields an average F-score of 0.513 satisfactorily as compared to the alternative model.

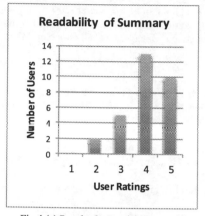

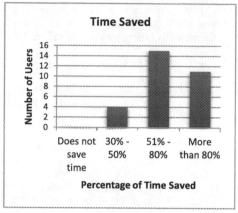

Fig.4 (a) Results for Readability   Fig.4 (b) Results for Percentage of Time Saved

In this section, system evaluation was conducted on readability and time saved criteria. There were 30 users who were randomly selected as the target group. These users were mainly university students who used to read lengthy documents such as research papers, journals or Reuter news on Android mobile devices. The results of the assessment on readability and time criteria are as shown in Fig. 4(a) and Fig. 4(b) respectively.

It was observed that the mobile TextSumIt application provided high readability (rating 4) and more than 80% of users were able to save 51%-80% of their time assimilating lengthy documents. In general, the response was positive and majority of them will use mobile application either for personal use or at workplace.

5   Conclusion

In this paper, an attempt was made to extract important sentences of a document by integrating syntactic and semantic of words which have similar meaning to title or keywords in the document. From the system evaluation, the proposed TextSumIt model yields an average F-score of 0.513 satisfactory as compared to the alternative model. The mobile application also provides better readability by displaying main ideas of original article in bullets points after final sentence filtering. Most mobile users found that it would save time in browsing shorter summarized text. For future work, it is recommended to incorporate text coherence in text summarization in Android mobile devices.

Acknowledgment

We would like to thank Research and Innovation Office (RIO) for the support in the research.

References

1. Yang, C. C., Wang, F. L.: An Information Delivery System with Automatic Summarization for Mobile Commerce. *Decision Support Systems*, 43(1), pp. 46–61, Elsevier (2007)
2. Mani, I.: Recent Developments in Text Summarization. In: Proceedings of the tenth International Conference on Information and Knowledge Management, CIKM, New York (2001)
3. Luhn, H.P.: The Automatic Creation of Literature Abstracts. *IBM Journal,* pp. 159-165 (1958)
4. Abuobieda, A., Salim, A., Albaham, N., Osman, A.T., Kumar, Y.J.: Text Summarization Features Selection Method using Pseudo Genetic-based Model. In: Proceedings International Conference on Information Retrieval Knowledge Management, pp.193-197. (2012)
5. Binwahlan, M.S., Naomie, S., Suanmali, L.: Fuzzy Swarm Diversity Hybrid Model for Text Summarization. Information Processing and Management , vol. 46, pp. 571-588. (2010)
6. Baxendale, P.: Machine-made Index for Technical Literature - An Experiment. IBM Journal of Research Developmen 2(4), (1958)
7. Edmundson, H.P.: New Methods in Automatic Extracting. Journal of the ACM, 16(2), pp. 264-285, New Work (1969)
8. Miranda-Jimunez, S., Gelbukh, A., Sidorov, G.: Summarizing Conceptual Graphs for Automatic Summarization Task. LNAI 7735, pp. 245-253, Springer-Verlag Berlin Heidelberg (2013)
9. Tonelli, S., Planta, M.: Matching Documents and Summaries uses Key-Concepts. In: Proceedings of the French Text Mining and Evaluation Workshop, pp. 1-6 (2011)
10. Saggion, H., Poibeau, T.: Automatic Text Summarization: Past, Present and Future. Multi-source, Multilingual Information Extraction and summarization II, Natural Language Processing, pp. 3-21, Springer-Verlag Berlin Heidelberg (2013)
11. Foong, O. M., Oxley, A., Sulaiman, S.: Challenges and Trends of Automatic Text Summarization. International Journal of Information and Telecommunicationi Technology 1(1), pp. 34-39 (2010)
12. Gupta, V. : A Survey of Text Summarization Extractive Techniques. Journal of Emerging Technologies in Web Intelligence 2(3), pp. 258-268. (2010)
13. Jusoh, S., Fawareh, H.M.: Semantic Extraction From Texts. In Proceedings of International Conference on Computer Engineering and Applications IPCSIT (2011)
14. WordNet, http://wordnet.princeton.edu
15. Foong, O.M., Lee, M. : TextSumIt: A Semantic Single Document Summarization Model on Android Mobile Devices. *Applied* Mechanics & Materials(263-266), IT Applications in Industry, pp. 1902-1909. (2012)
16. Yu, L., Duan, X., Tian, S., Guo, H.: Topic Extraction based on Product Reviews. Journal of Computational Information System, 9(2), pp. 773-780 (2013)
17. Foong, O.M., Oxley,A.: A Hybrid PSO Model in Extractive Text Summarizer.In Proceeding *IEEE Symposium on Computers & Informatics*, pp.130-134 (2011)
18. Barrera, A., Verma, R.: Combining Syntax and Semantics for Automatic Extractive Single-Document Summarization. In: Proceeding CICLing'12 of the 13th international conference on Computational Linguistics and Intelligent Text Processing, vol. Part II, pp. 366-377 (2012)
19. Lin, C.-Y.: ROUGE: A Package for Automatic Evaluation of Summaries. In: *Proceedings of Workshop on Text Summarization Post-Conference Workshop (ACL 2004)*, Barcelona, Spain (2004)

The Design of Android Metadata based on Reverse Engineering using UML

Zahidah Iskandar Shah[1] and Rosziati Ibrahim[2]

[1] Universiti Tun Hussein Onn Malaysia, xaza89@gmail.com,
[2] Universiti Tun Hussein Onn Malaysia, rosziati@uthm.edu.my

Abstract. The UML is a modelling language common use in software development. UML is the de-facto standard language used to analyse and design object-oriented software systems. The UML is a modeling language common use in software development. However, it cannot describe the Android platform, because UML is a general purpose, tool supported, modeling language that can be applied to all domains and platforms. It does not provide delicate concepts to express peculiar features of a specific system The Android application can be develop either using C or Java language, both of those language programming are object-oriented. In the context of object oriented systems, one of the techniques that can used to improve quality of software application is by using reusability technique. This paper presents how to design android application using extension UML model and software reuse that can produce a high quality Android application.

Keywords: Unified Modeling Language (UML), Android, Reusability

1 Introduction

Smart phone is built with advanced computing ability. There are many features evolved from simple devices used to powerful devices such as Wi-Fi connection, latest generation web browser and hardware devices etc. as compared to traditional phones in market. This mobile phone provides development platforms and software stores such as Android Market, Apple App Store where developers can distribute their applications. Most of these devices run Android Operating System. Since the Android operating system is open source and free it dominates among the smart phone market. According to the Android statistic by AppBrain shows the number of available applications usage [1] on the Android market. For example, the Android Market started with 2,300 applications in March 2009, and the last update, in January of 2012, indicates that there are currently more than 380,000 applications available [2]. There are various potential reasons for this current explosion in the number of available mobile applications. One of them could be the large increase in the number of developers for these platforms [3], as well as the availability of decentralized mobile applications stores [4], or the ease of building new applications [5], which attracts many new developers to

T. Herawan et al. (eds.), *Proceedings of the First International Conference on Advanced Data and Information Engineering (DaEng-2013)*, Lecture Notes in Electrical Engineering 285, DOI: 10.1007/978-981-4585-18-7_65,
© Springer Science+Business Media Singapore 2014

develop an application. A more fundamental reason why so many mobile applications are developed, might be the use of proven software engineering practices, such as code reuse [6]. The software reuse is prevalent (compared to regular open source software) in mobile applications of the Android Market [7]. Research has shown that the judicious usage of code reuse tends to build more reliable systems and reduce the budget of the total development [6][8]. Various types of software reuse exist, like inheritance, code, and framework reuse [8], each having its own advantages and disadvantages.

So far, a lot of Android application has low quality in terms of stability and performance. The Android application can be develop either using C or Java language, both of those language programming are object-oriented. In the context of object oriented systems, one of the techniques that can used to improve quality of software application is by using reusability technique. Victor et. al research have showed the strong impact of reuse on product productivity and especially, on product quality, or defect density and rework density, in the context of object oriented systems[6]. Software reuse is the process of using existing software artifacts instead of building them from scratch which involves three steps (a) selecting a reusable artifacts, (b) adapting it to the purpose of the application (b) integrating it into the software product under development.

One important step in developing a software application based on Model Driven Engineering (MDE) is analysis and design. Unified Modeling Language (UML) is a powerful communication between analysts and designers. According to Object Management Group [9] UML is the de-facto standard language used to analyse and design object-oriented software systems. The UML is a modeling language common use in software development. It provides well-defined modeling concepts and notations in order to state the behaviour of the system [10]. However, it cannot describe the Android platform, because UML is a general purpose, tool supported, modeling language that can be applied to all domains and platforms. It does not provide delicate concepts to express peculiar features of a specific system [11]. This paper presents how to design android application using extension UML model and software reuse that can produce a high quality Android application.

Organization of paper, in section 1 we present the introduction of building Android application. In Section 2, we review the related work in the area of reusability technique and UML Model. Section 3 discuss about the features of Android application. In Section 4 we present the modeling of UML Diagram, and in Section 5 we provide the proposed system architecture for Android application development, and we close with some concluding remarks.

2   Related Work

A UML supports the extension mechanisms [12]. According to Aldawud et. al the extension mechanism of UML support by UML Profile whereas the profile enable specifying, visualizing, and documenting the artifacts of software systems based on Aspect Orientation [13]. An extending UML Meta-model has been discover by Minhyuk et. al. but it just for the main program and does not includes the reusable technique in the development [11]. There are a huge number of tools that provide software modeling support using graphical notations as UML, the standard object-oriented modeling language. Some of these are able to generate code from UML models, as UModel [14], and Artisan Studio [15]. However, these tools focus on traditional software development, and do not completely support Android applications development. IBM Rational Rhapsody [16] is a commercial tool for software modeling, and recently, it was extended to support modeling and code generation for Android applications. The use of UML activities, like building blocks to construct applications, is proposed by Arctis [17]. These activities are after translated to a state machine in order to generate code and other files needed to wrap the state machines into an executable Android application. Abilio [18] use MDE approach for Android applications development, which addresses how to model specific aspects of Android applications, as intent and a data/service request, by using standard UML notations.

So far there are no researches discovering about the reusability module in Android mobile operating system as many of Android application using reusable module [7]. It has even been stated that there are few alternatives to software reuse that are capable of providing the gain of productivity and quality in software projects demanded by the industry [19]. In this paper we are focusing the reusable module in Android repository library and resources.

3 Android Features

There are four application components, activities, service, the content providers, and Broadcast Receiver. Each component is playing different role depending on our applications needs, but each one exist in unique building block that listed in a file called `AndroidManifext.xml`. It shows in the Figure 1 below: Activity is the most common one in building blocks. An Activity is usually represents single screen with user interface. Each activity is implemented as a single class that extends the Activity base class. A Service is a component that runs without user interface. For example is media player application. The music is played as a background without blocking user interaction with an activity. They can notify the user via the notification framework in Android. Each service class has a corresponding <service> declaration in its package's `AndroidManifest.xml`. Content Provider is a class that implements standard methods to let other applications store or retrieve data handheld by the content provider. Content Provider gives permission to read and write information about particular person. It is useful if we want our application's data to be shared with other applications. Content Provider is implemented as subclass of ContentProvider. It is write in standard API to enable other application to perform transactions. Broadcast Receiver is a class that handles execution of code in the application in reaction of external events. It is responds to system wide broadcast announcements such as a broadcast announcing the screen was turned off, phone rings or picture is being captured. They may display notification to alert user but does not display the

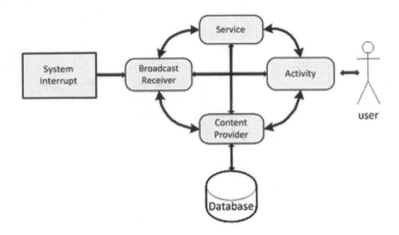

Fig. 1. Android components

user interface. Broadcast Receiver also registered in `AndroidManifest.xml`. It is
implemented as a subclass of BroadcastReceiver and each broadcast is delivered
as an intent object.

In activity has a unique life cycle which provides some callback methods
that handle some aspects of the service life cycle. There are seven operations
in activity feature which are `onCreate()`, `onStart()`, `onResume()`, `onPause()`,
`onStop()`, `onDestroy()` and `onRestart()`. Each activity can then start another
activity in order to perform different actions. Each time a new activity starts, the
previous activity is stopped, but the system preserves the activity in a stack [19].
For service also have the life cycle operations different from activity. The most
important callback methods in service includes `onStartCommand()`, `onBind()`,
`onCreate()` and `onDestroy()`. The services can be either in two forms of started
or bound. A service is started when an application component (such as an activity) starts it by calling `startService()`. Another application component can
start a service and it will continue to run in the background even if the user
switches to another application. Additionally, a component can bind to a service
to interact with it and even perform interprocess communication (IPC) [16].

4   Modeling UML for Android

In this section, we discuss about the UML meta-model formed through the
extended mechanism mentioned in previous section. The concepts of Android
meta-model can be understood but need to realize the relationship between User
Interface and System Resources. The user interface is using the MVC pattern
proposed by Bup Ki Min et. al It consists of Model (contents), View (showing
contents on user interface), and Controller (logics controlling contents and user
interface). Next, the System Resource consists of Resource Controller, Content

Resource and Hardware Resource [17]. It can be shown in Figure 2 below. In

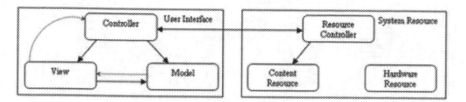

Fig. 2. Relationship between Interface and System Resource

this paper we used stereotypes and tagged values as modeling elements to model Class Diagram of Android platform. They enable the modeling element to attach a new semantic meaning. The UML meta-model is naturally extended as stereotype is defined to describe the modeling elements of Android platform. The modeling elements expressed by the stereotypes are separated according to the Android features in previous section. Next, the tagged values are used to for separated modeling elements which should be more accurately divided.

4.1 Structure Fundamental

Android fundamentals are shown in Fig. 3. From standard UML elements Android fundamentals are extend through the stereotypes.

The particular components in the Android fundamentals are expanded from the Class element that is one of the standard UML elements through the stereotypes. The Intent expands the Attribute which is a modeling element in the UML. The Intent is used in Activity components Different from other components, the Intent is expressed with the additional element Navigation. The Navigation expands the Association which is a modeling element in the UML. The components within the Android fundamentals are divided finely by the tagged values shown in Table 1.

Table 1. Tagged values relevant to the meta-model in Fig. 3

Stereotype	Tag	Tagged Value	Description
Activity	Implementation type	[Code—XmlLayout]	Way to implement user interface
Service	Service type	[Music—FileIO —ContentProvider]	Kinds of services
Content Provider	Content type	[SharedPreference —InternalStorage—]	Way to store data

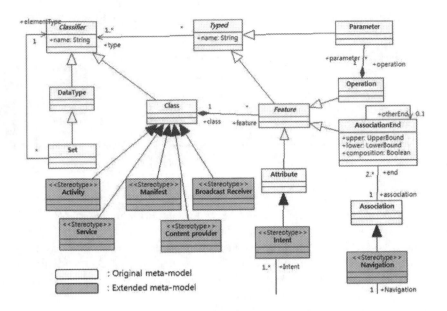

Fig. 3. The Meta-model for Android Fundamentals

4.2 Dynamic Fundamental

The stereotype <<LifeCycle>> is used to represent the operations related with the Android life cycles. The dynamic features in an Android application are modelled by the lifecycles. The stereotype <<LifeCycle>> is represented in Activity and Service components.

4.3 User Interface

In designing user interface, the MVC (Model-View-Control) pattern is applied. Therefore, the modeling elements for user interface are classified into the Model, the View, or the Control groups. For changeability, the View group is classified into UIControlButton, UIControlText, UIControlSelection, and UIControlViewer and UIControlLoading. In a design level, a developer might frequently change a view as Android provides various similar views used in a same role. Moreover, it will be helpful to explain what kinds of views are used, when a developer communicates with other developers. The Control and the Model elements are indicated by the stereotypes, <<Controller>> and <<Model>> respectively. They are expanded from the Class element.

5 Proposed System Architecture

In this section, the framework will briefly describe. The formation of UML Android based on component library reused. Figure 4 shows the proposed framework. First,we select an Android application source code that support to create

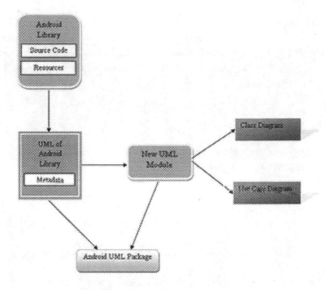

Fig. 4. New android UML package

a new Android application. To this Android application we do reverse engineering to form models that called as RE Model. For example, we want to create calculator application by adding new features that can be used to calculate min and standard deviation. This function has not yet been implement in basic calculator application. First we do reverse engineer to convert from source code of calculator to form a basic calculator model. Then, we create new component models that represents new features of calculator. After that combine both of models to form a model of new calculator application. The benefit from the proposed framework we can improve the Android application based on basic Android application.

By using this framework hopefully a high quality Android application can be produced.

6   Conclusion

This paper describes the modeling UML diagram for Android platform based on software reuse. The extended UML meta-model will provide the way to describe more specifically an Android application in modeling. It will help the developer to create Android source code based on modeling UML. The design and implementation phase will be more precise as it discovers the reusable module in Android source code.

For future work, Since Android has a many of powerful features and it is an open source platform based on Java, Android will add new concepts and elements to describe the behavior of the Android application.

7   Acknowledgements

This research is supported by graduate research incentive grants (GIPS) under project Vot No. 1254. The authors would like to thanks to Universiti Tun Hussein Onn (UTHM) for this grant.

References

1. Appbrain, http://www.appbrain.com/stats/number-of-android-apps
2. Tiuri van Agten, http://www.distimo.com/blog /2012_01_google-android-market-tops-400000-applications
3. Beth Stackpole, http://www.computerworld.com/s/article/9217885/ Your_next_job_Mobile_app_developer_
4. Dan Rowinski, http://readwrite.com/2012/02/06/infographic_history_of _mobile_app_stores
5. Steve Lohr, from http://www.nytimes.com/2010/07/12/technology/ 12google.html?_r=0 (2010)
6. Basili, V. R., Briand, L. C., Melo, W. L.: How reuse influences productivity in object-oriented systems. Communications of theACM, 39(10), 104-116 (1996).
7. Ruiz, I. J. M., Nagappan, M., Adams, B., Hassan, A. E.: Understanding reuse in the android market. In Program Comprehension (ICPC), 2012 IEEE 20th International Conference on (pp. 113-122). IEEE (2012).
8. Frakes, W. B., Kang, K.: Software reuse research: Status and future. Software Engineering, IEEE Transactions on, 31(7), 529-536 (2005).
9. Object Management Group, http://www.omg.org/news/releases/pr2003/04-17-03.htm.
10. Lavagno, L., Martin, G., Selic, B. V. (Eds.).: UML for real: design of embedded real-time systems. Springer (2003).
11. Ko, M., Seo, Y.-J., Min, B.-K., Kuk, S., Kim, H. S.: Extending UML Metamodel for Android Application. Computer and Information Science (ICIS), 2012 IEEE/ACIS 11th International Conference on (pp. 669 674) (2012)
12. Farhad, J.: The UML Extension Mechanisms Department of Computer Science, University College London
13. Aldawud, O., Elrad, T., Bader, A.: UML profile for aspect- oriented software development. In Proceedings of Third International Workshop on Aspect-Oriented Modeling (2003).
14. Altova, http://www.altova.com /umodel/uml-code-generation.html
15. Atego, http://www.atego.com/products/artisan-studio/
16. IBM Software, http://www-03.ibm.com/software/products/us/en/ratirhapfami
17. Kraemer, F. A.: Engineering android applications based on UML activities. In Model Driven Engineering Languages and Systems (pp. 183-197). Springer Berlin Heidelberg (2011).
18. Parada, A. G., Brisolara, L. B. D.: A Model Driven Approach for Android Applications Development. In Computing System Engineering (SBESC), 2012 Brazilian Symposium on (pp. 192-197). IEEE.
19. Mili, H., Mili, F., Mili, A.: Reusing software: Issues and research directions. Software Engineering, IEEE Transactions on, 21(6), 528-562 (1995).
20. Creative Commons Attribution, http://developer.android.com/guide/components /services.html.

The Idle Resource Items Workload and Implication on Different Weight Balance Rate in Grid Scheduling

Bakri Yahaya, Rohaya Latip, Azizol Abdullah and Mohamed Othman

Department of Communication Technology and Network,
Faculty of Computer Science & Information Technology,
Universiti Putra Malaysia, 43400 UPM, Serdang, Selangor, Malaysia
bakriy@gmail.com, {rohaya, azizol, mothman}@fsktm.upm.edu.my

Abstract. This paper discusses the impact of background workload for idle resource items of computing elements in grid computing. Most of the previous research did not consider this factor. A resource item may be processing local system operations when the grid perceives them to be idle, thus upsetting grid processing activities. The introduction of the resource items and background load factor in this study will reveal the true computing capability of computing elements. This background load factor, represented in the form of weightage on resource item, is tested to seek overall grid performance. By allocating the right balance of workload weightage of resource item in a computing element, a significant improvement in processing performance is achieved.

Keywords: Scheduling, resource item, background load, grid computing, idle, workload.

1   Introduction

Resource management is responsible in making the arrangement and control of activities that related to "sharing" capabilities. It serves to determine the right match between the source and workload based on environment capacity, workload capacity and also based on the policies used in the system. In general, studies on "resource management" centers on the authentication, job submission, scheduling and load balancing discipline. In line with that, this paper will explore the ideas that related to scheduling and background load issues in the hierarchical grid topology environment.

The focus was deepened into the workload issue on idle resource state items. In the meantime, the most fitting resource item weight balance rate is studied too. For the purposes of this study, there are two domains that need to be clarified before further discussions are made. Due to the focused study on resource items and background load, the initial decomposition relates to resource items that usually become preferred choices in grid environment and followed by an elaboration of background load based on previous studies. Unfortunately, the background load study did not impress the researches or were marginalized.

T. Herawan et al. (eds.), *Proceedings of the First International Conference on Advanced Data and Information Engineering (DaEng-2013)*, Lecture Notes in Electrical Engineering 285, DOI: 10.1007/978-981-4585-18-7_66,
© Springer Science+Business Media Singapore 2014

Findings from previous studies, clearly illustrate that there is diversity in the choices of the resources items. Those resource items are CPU speed, memory, bandwidth and other capacity such as I/O overhead and the number of CPUs in each unit and other related hardware connected to the grid environment. The most used resource items in a grid study are the CPU, followed by the memory, while network computing resources such as network protocol [1,2] and network bandwidth [5] were either selected as complementary or as the key factor, based on the study requirement. Overall, the result of these observations is consistent with the basic understanding of grid computing written in 2001 [8].

The question is what does the pattern of the workload dissemination of resource items in the computing elements is like. Let's include the CPU, memory and network bandwidth as a research resource items. Should the system, consider the current or idle load from the three resource items independently. So, how does the system allocate the new value or increment value for the CPU, memory and network bandwidth after receiving a job? What is the appropriate strategy to be implemented in a way to overcome this issue? This is very important to understand as we are going to create a decision path for a job migration and load distribution.

Furthermore, the discussion will touch about the resource items workload for the idle computing element, associating to the implication on different resource structure in grid computing. Idle resource item workload pattern on different structures and computing element capacity will be captured to seek a variety of information. This information will enable the research to go further. Then, all of the information found will undergo the testing based on the selected algorithm. Different sets of tests have been developed to compare the obtained results and the impact on simulation performance.

This paper will contribute to the resource items workload or background load for idle computing element pattern. Besides that, the set of testing will lead to the useful indication of computing element structure, capacity and utilization information on idle resource. Finally, the suitable background workload set or rate will be chosen for grid simulation. The rest of this paper is organized as follows. Section 2 illustrates the related works. The discussion in background load and computing power is discussed in Section 3 and followed by Section 4 which presents the simulation preparation and Section 5 for the result and conclusion.

2   Related Work

2.1   CPU and Other Resource Items

The substances of discussion are centered in the use of resource item, then the manipulation techniques used on resource item and finally the involvement of the background load in the research. In a nutshell, the exploration will touch on the hierarchical grid structure and resource items used. There are many studied based on a clustered environment previously that adopted multiple resource item, they are [6, 7, 9, 10, 11, 12] which undertaken in line with their study objectives. In [7] the implementation is through neighbors concept and delay parameter used as a condition for the partner's survival and filtered by the threshold value. The CPU utilization, communication delay, availability of nodes and the queue length are the sources of load and they contribute to the total cluster load. This implementation causes the information exchange rate to be relatively high, because the scheduler and system monitoring reside on both of the grid manager and cluster manager.

Studies by [6] also introduced a decentralized grid model and driven by the dynamic load balancing algorithm. The resource items used in the research are the CPU utilization, memory usage and queue length. The queue length from the node was the only source of load index value. The load factor used is not comprehensive and need to be expanded because the study is inclusive of the CPU and memory utilization. Research and development of algorithm through the concept of dual layer scheduling has been implemented by [9]. Top-layer scheduling used to choose an engine to parse the workflows meanwhile the bottom-layer scheduling works for task-based workflow. The mathematical calculation techniques used to determine the current status of the existing engine and the threshold applied to manage the workload balancing. Resource item's information that were recorded are the total CPU numbers, CPU speed, operating system, load status and network bandwidth. However, as a whole, this strategy will not work properly if there is no high imbalance. In fact, this study is only suitable and limited to this research.

Furthermore, research by [10] has emphasized the use of statistics as a solution to the problems of scheduling and load balancing. The resource items information used are the CPU speed, CPU utilization, number of active processes, memory and I/O usage statuses. In short, this study developed an algorithm with fully automatic capability and the information obtained allows the user to plan the size of the threshold before the workload is being processed. Research in [11] based on three section grid model structures, which are computing element, cluster manager and grid manager. Essentially, this study adopts the parameters like task ID, task size, CPU speed, workload index, the size of bandwidth and some other parameters derived from a mathematical formula based on possibility and estimation. In brief, this study did not significantly differ with other studies discussed above.

Finally, research in [12] emphasis on the grid the computing challenges inclusive of heterogeneity, scalability and adaptability. This study relies on two sources of data, first the premier source that came from the computing element information and the secondary data in accordance to the policies and mathematical formulations. In short, this study needs revision on a more stable platform to determine the validity of the results. The background load is neither considered as a basic load of computing element nor used in the research evaluation factor.

2.2   Background Load

As discussed at the beginning of this paper, background load is not an impressive research area to the extent of being marginalized by the researchers. However, there are researchers who are still aware about the benefits or implications of background load for the resource performance or computing resources generally. To evaluate the impact of multitasking on multiprogramming environments and to ensure reproducible results, there is need to have a well-defined and constant background load [21]. Research deployment in the grid or distributed environment cannot prevent the unpredictable load changes. Therefore a dynamic balancing system is needed to react to such load oscillations and improve performance [20].

Research on host load predictions made by [19] is based on free load profiles that aim to improve the overall performance. They did the experiment successfully and were able to obtain the expected results. There is a research working from Large-Scale HLA-based simulation and according to [20] these simulations are highly dependent on a distributed environment, so much so that they can undergo substantial performance loss due to heterogeneity of resources, presence of external background load, and improper placement of simulation elements or dynamic simulation. Their findings on the implication of the background load onto HLA performance is one of the evidence of the importance of background load to be considered in grid research.

Additionally, researches done by [5] were based on the similar concept that has been discussed above, but the mathematical techniques and methods of measurement of workload on the grid system are different. Each accepted jobs for processing will result in the computing element utilization increase. This study measured and distributed the workload on the resource items such as the CPU, memory and network. Meanwhile, each of the involved resource items had been allocated with a specific weight value to offset the processing space. This allows the resource to completely process the received workload before switching to ready status to receive a new job. This study achieved good simulation result, able to balance the workload at grid level and taking the background load of the computing element as the basic load before absorbing a new job.

3   Discussion

3.1   Background Load in Grid Simulation

As previously discussed, the background load of the computing element was never taken into consideration in the previous papers except [5, 20]. The assumption that idle resources do not contain the workload value of every resource item is improper. This is because the idle resources may be in a ready position to perform a computational task, partly used by the other task or used by the operating system file only. Evidently, the resource items on idle computing element contain its own weight that should be refined and can be categorized as a background load.

Therefore, to update the resource item values, the current values held by the background load on every resource item should be taken into account. Research conducted by [14, 15] has divided the simulation testing into three conditions based on different workload criteria. They discovered that, different application running on computing element required different services and will be given different priority. They had proven that the implementation with background load attributes contributed to the different results, or in other words, different level of performance. On the whole, the generated information is very useful for scheduling purposes and also for load balancing practices

The use of background load is not contradicted by later researchers and the background load is an important factor to be addressed in a heterogeneous computing environment or simulation [13, 20]. According to [16] the background load value should constantly be at thirty percent of the whole utilization rate, if otherwise not specified. Meanwhile, research done by [17] agreed that if the background load reaches thirty percent of the utilization rate and exceeded the stipulated period then the experiment is not valid. On the other hand, [17] also describes that the normal range of background load is from ten percent to thirty percent of overall utilization.

3.2 Background Load and Available Computing Power

When the background load of computing elements are taken into account then the processing space for each unit of computing elements will decline at various rates. This inconsistent reduction happens due to the various possibilities including the infrastructure capability, computing power capability, the memory size, the network capability and the activities running on the particular computing element. In spite of the possibilities explained above which makes the computing element seem to work, the computing element is in an idle status. The idle status means that the computing element is not on demand from the grid systems request. In other cases, no doubt there is a possibility that the computing element is in fully idle status.

Conceptually and logically, when a piece of the idle processing space is used by other components or applications, then the available processing space should be reduced. This also means that the remaining power of resource items has to be well managed. As explained above, this study took into account the sources of three resource items which are CPU, memory and network bandwidth. Thus, each of these resource items should be set with a permanent weight value. By adopting this technique the resource item value adjustments are easy to be fulfilled and implemented. Studies by [5] had run a resource item weights selection process for the idle computing element or active nodes. However, they have suggested a more suitable weight to be implemented which holds 60 percent to the CPU, network bandwidth has 30 percent of the weight and the balance allocated to the memory at 10 percent.

The bigger number of workload information involved in the research will cause the processing time and a message exchange rate increase. Then, the high rate of message exchange will result in the increment of overhead costs. Various types of workload information available, used to complement the implementation of simulation research. Generally, the type of information is for example JobID, delivery time, waiting time, run time, the number of processors allocated the percentage of memory usage and other information too. [18]. There are two simple strategies to choose in the workload dissemination direction, which are top to bottom or bottom to top dissemination. On the whole, this study will adopt the distribution from top to bottom fashion. This strategy begins with the consideration of grid level and moves toward the cluster level and finally reach the computing element level.

4   Simulation Preparation

4.1   Simulation Setup and Testing Algorithm

This research work has been implemented through GridSim and simulation parameters properties are as follows:

Table 1. Parameter Properties

Number	Parameter	Value
1	Number of tasks	2000
2	Size of task (MI)	300000-500000
3	Number of nodes per cluster	10
4	Number of clusters	10
5	Processor Speed (MIPS)	500-5000
6	Memory size (MB)	500-1000

The simulation algorithm steps are as follows:

Table 2. Algorithm Steps

Steps	Steps Detail	Sub Steps	Micro steps
1	The user submits request to the grid system,		
2	Grid initiates the infrastructure and parameters,		
3	The grid manager,	Check the job, Send job to queue,	
4	The job scheduler will determine the destination	Check the job size. Compare with the threshold value.	
		Select the cluster	Removes the overload clusters. Select the suitable cluster
		Select the computing element or node.	Removes the occupied nodes. Select the suitable node
5	The grid broker will send the job.		
6	Grid system gets the background load from the node		
7	Calculate the processing time.		
8	Change the node status to busy.		
9	Update the status and repeat until end of job.		

The simulation experiment scenarios are as follows:

Table 3. Experiment Scenarios

Scenario	CPU Weight %	Network Weight %	Memory Weight %
Set A	60	30	10
Set B	60	20	20
Set C	60	10	30
Set D	70	10	20

4.2   Rules of Simulation Equations

This research used a number of parameters for filtering purposes and decision making. These are as follows:

Table 4. Simulation Parameters and Equations Summary

Num	Parameter	Value	Detail
1	BGL CPU	Utilization	Background load for CPU
2	BGL Mem	Utilization	Background load for Memory
3	BGL Net	Utilization	Background load for Network.
4	Load	CF Load	Current load hold by the each computing unit.
5	ACL	Cluster load	Average cluster load in percentage
6	AL	System load	Average system load.
7	Sigma	Standard Deviation	The workload distribution value
8	Threshold	Simulation upper limit	Setting up the limitation of simulation.

All of the detail calculations and mathematical formulas are referring to the research work done by [5].

5   Results and Conclusion

5.1   Resource Items Weight Distribution Results

This test ties with 4 groups of background loads or utilization ratings. They are BGL A with 10 percent idle load, followed by BGL B with 20 percent idle load and lastly BGL C with 30 percent idle load. These idle background load consideration is in line with the suggestion by [16, 17]. However, bigger idle background workload BGL D with a 50 percent rate is chosen for the purpose of comparison.

Table 5. Background Load vs Scenarios   **Table 6.** Scenarios vs Background Load

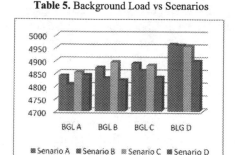
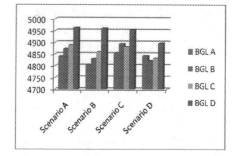

Figure 1 and 2 describe the results from the idle resource background load experiment. As explained before, the created scenarios deal with four groups of background load settings, specifically are 10, 20, 30 and 50 percent idle background load from the overall computing element power. After 80 runs of the experiment, results show that scenario B has proven suitable of being chosen as the best weight balance pattern for low idle resource background load rating. Although the recorded gap results are small, about 0.25%, but it is sufficient to describe the advantages of scenario B over the other scenarios.

Scenario D's capabilities are beyond expectations and it has outperformed the other scenarios on the rest of the idle resource background load group. For this scenario, weight balance pattern is 70% representing the CPU, 20% for the memory and 10% for the network bandwidth. The experiment results recorded were about 0.36%, 0.29% and 0.33% difference for scenario D compared to the last position on each idle resource background load group. The resulting differences are quite small due to the small ranges of test parameters.

In brief, the allocation of CPU weight should be bigger to support the position as the main processing unit. From the experiment, the results for scenario D which represents 70% CPU weight allocation were shown to be the best balance in the combination. As for the rest of the resource items, the memory should be allocated with bigger weight balance than for network bandwidth.

Based on the observation from the experiment, we can conclude that:

1.   The CPU should have a bigger portion of the weight balance to process the jobs. Technically, computing elements also need some extra space to smoothly run the job or task.

2.   The memory should be allocated with more weight balance ratio compared to network bandwidth. As observed from the result pattern, the scheduling performance is in the decline if the memory was only allocated with 10% weight.

3.   By allocating more weight balance to memory and reducing the weight balance for the CPU will cause disturbance in major processing areas, thereby reducing the overall scheduling performance.

4.   The combination of weight balance under the scenario D for 70%, 20% and 10% are the most suitable allocation to overall weight balance pattern.

This background load matter would be almost endless when considering its vast sources. For example, discussing the operating system alone would involve activities such as updates, processes, application management, security patches and etc. However, distinguishing background load behavior as opposed to identifying the sources makes it more manageable. This behavioral approach would be expressed through its rapid changes, periodical distribution of the load and finally the probability of its changes. In the end, whatever the approach the discussion centers upon, it will depend on the right weightage or the right ratio of background load assigned to the resource items that should influence the course of the research. In the future, this work will explore on the scalability, load balancing and scheduling improvement.

References

[1]   Rich Wolski, Neil Spring, Jim Hayes, "Predicting the CPU availability of time-shared Unix systems on the computational grid," Cluster Computing 3, Baltzer Science Publisher BV, pp. 293-301, 2000.

[2]   Lanier Watkins, William H. Robinson, Raheem A. Beyah, "A passive solution to the CPU resource discovery problem in cluster grid networks," IEEE Transactions on Parallel and Distributed System, vol. 22, no. 12, 2011.

[3]   Hyojeong Kim, Kihong Park, "Tackling the memory balancing problem for large-scale network simulation," IEEE International Symposium on Modelling, Analysis and Simulation of Computers and Telecommunication Systems, MASCOTS, pp. 1-10, 2008.

[4]   Amrik Singh, Lalit K. Awasthi, "Performance comparisons and scheduling of load balancing strategy in grid computing," IEEE International Conference on Emerging Trends in Networks and Computer Communications, ETNCC, 2011.

[5]   Yun-Han Lee, Seiven Leu, Ruay-Shiung Chang, "Improving jobs scheduling algorithms in a grid environment," Future Generation Computer Systems 27,Elsevier, pp. 991-998, 2011.

[6]   P. K. Suri, Manpreet Singh, "An efficient decentralized load balancing algorithm for grid," IEEE 2nd International Advance Computing Conference, pp. 10-13, 2010.

[7]   Jasma Balasangameshwara, Nedunchezhian Raju, "A decentralized recent neighbour load balancing algorithm for computational grid," The International Journal of ACM Jordan, ISSN 2078-7952, vol. 1, no. 3, pp. 128-133, September 2010.

[8]   Ian Foster, Carl Kesselman, Steven Tuecke, "The anatomy of the grid:Enabling scalable virtual organizations," International Journal of High Performance Computing Applications, vol. 15, issue 3, ACM, pp. 200-222, 2001.

[9]   Yan Ma, Bin Gong, "Double-layer scheduling strategy of load balancing in scientific workflow," IEEE 15th International Conference on Parallel and Distributed Systems, pp. 671-678, 2009.

[10] Bin Lu, Hongbin Zhang, "Grid load balancing scheduling algorithm based on statistics thinking," IEEE The 9th International Conference for Young Computer Scientists, pp. 288-292, 2008.

[11] Malarvizhi Nandagopal, Rhymend V. Uthariaraj, "Hierarchical load balancing approach in computational grid environment," International Journal of Recent Trends in Engineering and Technology, vol. 3, no. 1, ACEEE, pp. 19-24, 2010.

[12] B. Yagoubi, Y. Slimani, " Task load balancing strategy for grid computing," Jounal of Computer Science 3 (3), pp. 186-194, 2007.

[13] Hui Li, Rajkumar Buyya, "Model-based simulation and performance evaluation of grid scheduling strategies," Future Generation Computer Systems 25, Elsevier, pp. 460-465, 2009.

[14] Piyush Maheshwari, "A dynamic load balancing algorithm for a heterogeneous computing environment," IEEE, Proceedings of the 29th Annual Hawaii International Conference on System Sciences," pp. 338-346, 1996.

[15] Sridhar Gopal, Sriram Vajapeyam, "Load balancing in a heterogeneous computing environment," IEEE, 1998.

[16] J.H. Abawajy, "Adaptive hierarchical scheduling policy for enterprise grid computing systems," Journal of Network and Computer Applications 32, Elsevier, pp. 770-779, 2009.

[17] Omer Ozan Sonmez, Hashim Mohamed, Dick H.J. Epema, "On the benefit of processor coallocation in multicluster grid systems," IEEE Transactions on Parallel and Distributed Systems, vol. 21, no.6, June 2010.

[18] A.Iosup, H.Li,C. Dumitrecu, L. Wolters, D.H.J. Epema, "The Grid Workload Format," 27 November 2006. "unpublished".

[19] Sena Seneviratne and David Levy, "Host load prediction for grid computing using free load profiles," Lectures Notes in Computer Science 3719, Springer-Verlag, pp. 336-344, 2005.

[20] Robson Eduardo De Grande and Azzedine Boukerche, "Predictive dynamic load balancing for Large-Scale HLA-based simulations," 15th IEEE/ACM International Symposium on Distributed Simulation and real Time Applications," IEEE, pp. 4-11,2011.

[21] Wolfgang E. Nigel and Markus A. Linn, "Parallel programs and background load: Efficiency studies with the PAR-Bench System," ACM, pp. 365-375, 1991.

[22] Maciej Smolka and Robert Schaefer,"Computing MAS dynamics considering the background load," Lectures Notes Computer Science 3993, pp. 799-806, 2006.

Two-Level QoS-Oriented Downlink Packet Schedulers in LTE Networks: A Review

Nasim Ferdosian and Mohamed Othman

Department of Communication Technology and Network,
Universiti Putra Malaysia, 43400 UPM, Serdang, Selangor D.E., Malaysia
Email: n.ferdosian@gmail.com, mothman@ upm.edu.my

Abstract. The Long Term Evolution (LTE) as a mobile broadband technology supports a wide domain of communication services with different requirements. Therefore scheduling of all flows from various applications by the same strategy of resource allocation would not be efficient. Accordingly it is necessary to design new scheduling algorithms by considering the Quality of Service (QoS) requirements defined for each application. In this regard, this paper will provide a brief overview of previously reported schedulers for QoS support in LTE cellular networks by taking into account the various important QoS parameters such as delay, packet loss ratio and data rate. The resource distribution problem in LTE networks and solutions from the numerous numbers of previous resource allocation approaches are outlined and compared through a comparative table. The future direction for solving these problems will be stated. Overall this study summarises the current state of knowledge on the QoS-oriented scheduling algorithms for LTE networks.

Keywords: Long-term Evolution (LTE), scheduling, QoS, GBR and non-GBR.

1 Introduction

Mobile communication technologies are becoming constantly prevalent in our daily life with an expected increasing growth rate of mobile data higher than the fixed data. Nowadays mobile devices are able to support a wide range of different real-time and non-real time applications from call services to mobile Internet browsing, online TV, games, video-conference calls and so forth. According to the studies by Cisco [1] the monthly world's mobile data traffic growth will reach 11.2 Exabyte by 2017 and video traffic will settle two-thirds of overall foreseen mobile data traffic. The Third Generation Partnership Project (3GPP) introduced the LTE [2] intending to design a system that can provide an efficient improvement in throughput over the older mobile standards and support multiple classes of QoS. In fact the most important novelty introduced by LTE technology is the enhanced support of QoS constraints among all its performance targets.

With the aim of efficient support of current high variety of services, the efficient use of limited share bandwidth is essential. So the purpose of effective resource allocation strategies is crucial to meet the LTE targets (maximum spectral

T. Herawan et al. (eds.), *Proceedings of the First International Conference on Advanced Data and Information Engineering (DaEng-2013)*, Lecture Notes in Electrical Engineering 285, DOI: 10.1007/978-981-4585-18-7_67,
© Springer Science+Business Media Singapore 2014

efficiency, fairness, and QoS). However, it is typically impossible to accomplish all three intended goals at the same time [3] . Each factor can be supplied always at the cost of reducing another one. In this sense the main challenge is designing an allocation strategy to create a trade-off among the system performance and other desired targets of the network.

Several studies have conceived the concept of QoS provisioning in LTE networks [4] . However, there have been no controlled studies to compare differences in QoS aware schedulers based on the QoS characteristics. In [5] the authors provided an inclusive overview of the common presented techniques in literature by representing a performance comparison of them with focus on all design aspects of the downlink packet scheduling in LTE. Another performance evaluation of representative scheduling strategies can be found in [6] limited to the uplink. The presented comparison of schedulers is in terms of their fairness, throughput and spectral efficiency. The authors in [7] considered interference mitigation and scheduling policy as two key design factors of LTE systems for achieving the performance goals and reviewed the scheduling methods proposed based on interference mitigation with the aim to increase the QoS of cell-edge nodes.

In this paper we conducted a literature survey on the previously reported schedulers which have focused on support for QoS differentiation in LTE cellular networks, highlighting the important features of them, and classify them according to the QoS parameters considered as scheduling metrics for developing schedulers. In view of the fact that the most of the policy considerations are valid for both uplink and downlink resource allocation algorithms, therefore we focus on the downlink scheduling as the subject of our study. The rest of this paper is organized as follows. In section 2 we can overview the architecture of the LTE focusing on the issues of the resource sharing. Section 3 represents different characteristics of QoS in LTE networks pertain to design of resource sharing schedulers. In the fourth section we discuss approaches that are usually to use in this case and compare them in the fifth section. Finally in section 6, conclusions are drawn with particular attention to the new research directions.

2   Overview of Resource Allocation in LTE Downlink

Selecting an appropriate scheduling scheme is not standardized by the 3GPP specification for LTE [8] . Alternatively it is left to the vendors as an implementation decision, to adaptively configure and implement a well suitable algorithm according to the desired concerns of the system. It is worthy to mention that the responsibility of providing all these targets is up to the implementation of eNodeB residing in the MAC layer. The eNodeB assigns each active user a fraction of the total system bandwidth to share available resources among them by using a multiple access technique [9] . LTE downlink 3GPP adopted Orthogonal Frequency Division Multiple Access (OFDMA) as an access technique to accommodate a wide number of user equipment with different QoS application requirements and in different channel conditions. OFDMA allows multiple accesses by allocating disjoint selective collection of sub-carriers to each individual user to leverage multi user diversity and provide high scalability and robustness.

The LTE radio resources are distributed in time and frequency domains. The LTE resource grid structure is displayed in 1. Each OFDMA frame is constructed of ten 1ms sub-frames in time domain and a sub-channel of 12 consecutive same size sub-carriers that cover 180 kHz of the frequency domain. The basic resource unit for mapping sub-carriers to active users is called Resource Block (RB). Each RB spans over a 0.5 ms time extent and one sub-channel.

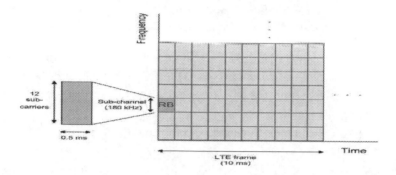

Fig. 1. LTE radio resource grid structure

3   Quality of Service in LTE Networks

In the field of telecommunication network, the term Quality of Service indicates a measure of how efficient and reliable a network can fulfill a guaranteed level of satisfaction for its diverse services from real-time to non real-time services. Each application data flow is associated with a bearer. A bearer can be considered as a virtual connection established between Packet Data Network Gateway and User Equipment. Typically, the service bearers are categorized into two main classes, Guaranteed Bit Rate (GBR) and non-Guaranteed Bit Rate (non-GBR). On the other hand each bearer is characterized by a set of QoS attributes which is called QoS Class Identifier (QCI) [10] . QCI defines the QOS class to which the bearer belongs with the parameters associated with each data flow such as:

Priority. The number given to each class of QoS which defines how emergency the bearer need to be resource allocated.

Bearer Type. This parameter specifies the kind of the connected bearer: GBR bearer or non-GBR one.

Packet Delay Budget (PDB). The maximum delay allowed for the transmission of queued packets.

Packet Loss Rate. The maximum tolerable number of erroneous bits associated to a traffic flow of a given QCI.

The dedicated QCI label of bearer is one of the main factors that determine the behavior adopted by eNodeB for scheduling that particular type of bearer.

The scheduler algorithms drive the specific allocation decision for each user based on a comparative metric per a specific RB. This metric is a scalar value which indicates a transmission priority for each user equipment founded on the information related to each stream of users. The related information is interpreted as the key parameters that construct the evaluating metric.

Particularly in every kind of broadband wireless network there are a number of parameters that can be used to quantify the QoS application requirements and consequently enabling differentiation among flows carrying application data. These parameters consist of throughput, queue size, packet delay, packet loss rate and so on. The choice of these parameters for making priority metric is rather controversial, and there is no agreement about the number and kind of critical parameters which should be taken into account, however there are some issues we have to consider in choosing them:

- Depends on the kind of traffic (real or non-real time)
- The level of the complexity resulted from the created priority function

4 Downlink QoS-Oriented Scheduling Algorithms under Study

To date various scheduling techniques have been developed and introduced to serve downlink packet scheduling in LTE networks. In the following we go over the main points of the studied algorithms. The purpose of these scheduling algorithms is to distribute system resources among service flows by focusing on influential QoS factors in two main stages of time and frequency. We selected these specific methods due to their successful supporting of divers QoS requirements and their presented results.

4.1 Target Bit Rate

Authors in [11] introduced a QoS-oriented scheduler for Best Effort and Constant Bit Rate traffic focusing on the guaranteed bit rate requirements. To the end of controlling the signaling overhead and consequently decreasing the complexity, this solution is decoupled between time and frequency schedulers in the way that through each domain just a limited number of users are passed. The total users are grouped into two sets. First set contains users below their target bit rate and second set compromises remaining users. Then the prioritization of users within set 1 and 2 is done using common well-known (BET) Blind Equal Throughput and Proportional Fair (PF) approaches in time domain respectively. After selecting a number of candidate users by the Time Domain (TD) scheduler, they will be allocated resources in the frequency domain (FD) by using the PF scheme. The BET and PF metrics are expressed as:

$$M_{BET} = 1/(R[n]) \tag{1}$$

$$M_{PF} = (\bar{D}[n])/(R[n]) \tag{2}$$

where n is the user index, $\bar{D}[n]$ is the wideband throughput expected for the user n over all the bandwidth and $R[n]$ is the past average throughput of user n which is updated every Transmission Time Interval (TTI) [12] .

The main focus of this method is on improving total throughput along with considering guaranteed bit rate measurements as the only one QoS parameter.

4.2   Delay-throughput

A flexible QoS oriented scheduler based on two stages (time and frequency domains) proportional fair scheduling principle described in [13] for real time video traffic. The proposed algorithm considers arrival rate and head of line packet delay as QoS constraints. The proposed algorithm uses a metric which is a combined function of delay, throughput and Channel Quality Identifier (CQI) factors as follows:

$$M = F_D \times F_{CQI} \times F_T \tag{3}$$

where F_D is a function QoS delay factors, F_{CQI} is a function of CQI indicating channel state information of each user, and F_T is factor of corresponding throughput calculation.

In the time domain step which is the first phase of scheduling, all user equipments are sorted according to a new proposed ranking metric and in the next step, frequency domain, the actual RBs are allocated to the users with pending retransmissions, and then to the delay sensitive users. Finally the remaining RBs are given to the rest of priority users. This algorithm also cannot be considered as a strong QoS provisioning scheduler because of the ignoring other QoS factors such as the minimum data-rate requirements.

4.3   QCI-throughput

The authors in [14] applied a self-optimization method to the LTE network scheduler in response to the active changes of network conditions and traffic over time and proposed an Optimized-Service Aware (OSA) scheduler. To simplify the complexity of the resource allocation procedure, it has been partitioned into three separate stages, QoS classes identified classification, time domain and frequency domain scheduling. At the first step each bearer is classified into different QoS classes based on its CQI factor. Then the TD scheduler prioritizes the classified bearers according to their QoS requirements and categorizes them into separate prioritized candidate bearers: GBR and non-GBR. GBR bearers typically carry real time applications sensitive to delay and need to be served with a guaranteed bit rate. OSA algorithm sorts each GBR bearer according to the Head of Line (HOL) packet delay in the buffer of related bearer. On the other hand the non-GBR bearer list is ordered according to the following priority metric:

$$M_{OSA} = (\bar{D}[n])/(\theta[n])W_{QoS} \tag{4}$$

where $\theta[n]$ is the normalized average channel condition estimate of bearer n and W_{QoS} is the QoS weight. Two created sorted candidate groups are passed through the FD scheduler to be assigned optimal spectrum. The FD scheduler allocates the best RB to the highest GBR priority bearer. After giving enough resources to all GBR bearers, if any RB is still remained, FD scheduler assigns them to non-GBR bearers. The OSA algorithm can be demonstrated to be unsuitable for dealing with bounded losses as another factor of QoS support.

4.4   Combination of Multiple QoS Parameters

The ranking function of traditional scheduling algorithms which are only based on the queue's priority, ignoring other metrics, would impose a lack of sufficient intellect over the resource allocation process. In response to this challenging problem, the authors in [15] introduced a new TD scheduling algorithm with emphasis in overload states. This overload state allocation algorithm supports QoS constraints by ordering the bearers using a ranking function of multiple metrics including priority, loss, delay, and queue depth. A priority value is considered for each metric, in order to make emphasis over any critical measurement. This ranking function is a combination of normalized metrics, individually multiplied in their corresponding priority as follows:

$$M_{Knapsack} = \sum_i R_i \tag{5}$$

where R_i indicates normalized prioritized metric i and i can be any subset of afore mentioned QoS measurement parameters such as packet loss and delay.

Eventually an FD scheduler as an open complementary option can be selected among the existed schedulers to assign appropriate radio RBs to the optimal set of data packets for the sake of throughput.

5   Discussion

Providing the required QoS is vital to deliver a good user experience over the mobile Internet. The notion of QoS is becoming even more important as device capabilities have revealed the desire for consumers to use more rich media content such as video. The new scheduling services enhance the capabilities in providing the required QoS for next-generation mobile Internet applications. Through this study, we concisely explained a selection of downlink resource allocation approaches and explored their ability to efficiently support diverse QoS requirements. Different QoS mechanisms studied in this paper followed a network initiated QoS control based on the GBR and non-GBR bearers, which is a class-based packet forwarding treatment for delivering real-time and non-real-time traffics. Likewise they followed a two level framework where the resource allocation procedure is divided between TD and FD schedulers. Accordingly this separation can make it easier to optimize each step of scheduling independently and result in significant reduction of computational complexity.

In order to comply with the QoS requirements, the aforementioned algorithms present a metric based on a single or multiple QoS factors. The knapsack algorithm introduced an adaptive ranking function that can be generalized and make possible adding any desired traffic metric. It also allows the operator to define the critical level of each metric by assigning a specific priority level.

Table 1. Comparison of studied QOS-oriented scheduling algorithms

Scheduler Name	QoS Parameters	Type of Traffic
[11]	Target Bit Rate	Best Effort and Constant Bit Rate traffic
[13]	Delay	Video traffic
OSA [14]	QCI	Real time and non- real time
Knapsack [15]	Delay, Loss, Queue Depth, Priority, MBR/AMBR	VoIP and data connections

On the other hand, from the viewpoint of the performance enhancement target of LTE systems, the studied algorithms take into account either the expected or achievable throughput of each user, except the last algorithm which responses to the need of high performance by mapping the LTE resource allocation scenario to the fractional knapsack problem and solving it by utilizing a greedy algorithm. Intuitively there is no need to use channel feedback for calculating user throughput. Therefore as a result, it would cause a reduced signaling overhead. Table 1 demonstrates the comparison of these different downlink scheduling.

6 Conclusion

Despite the huge amount of resource allocation approaches proposed so far, it is difficult to have a simple and fair weighting policy that can meet all QoS requirements of all connections in a time-varying wireless network. Therefore it is needed to introduce a principle of metric decoupling between time domain and frequency domain schedulers that is fundamental for maximizing throughput control. In this regards, in order to obtain a deep understanding and conceptual view of the QoS based scheduling concept, we discussed and classified the most interested and relevant techniques, from our point of view, presented in the literature according to their leveraging parameters of QoS. By making the allocation decision depending on the QoS attributes of each flow, the service provider significantly can increase the system revenue by providing their customers with an integrated variety of service plans.

Consequently, this comparative study would reveal the need for resource allocation schemes in the way that obtain all LTE concerned targets simultaneously. Therefore channel aware approaches must be used with QoS aware strategies to provide a good balance between multi-QoS provisioning to support mixes of real-time/non-real-time traffic and system performance maximization.

As another future direction of research we can investigate the reasons of the issue that in practice the proposed schedulers so far are almost never implemented by switching and router manufacturers and QoS services are virtually never implemented and billed by Internet Service Providers.

Acknowledgments. This work has been supported by the Malaysian Ministry of High Education under the Fundamental Research Grant Scheme FRGS/1/11/SG/UPM/01/1.

References

1. Cisco. 2012. Cisco Visual Networking Index: Global Mobile Data Traf?cForecast Update,2011-2016, http://www.cisco.com/en/US/solutions/collateral/ns341/ns525/ns537/ns705/ns827/white-paper-c11-520862.pdf.
2. 3GPP, Tech. Specif. Group Radio Access Network-Requirements for Evolved UTRA (E-UTRA) and Evolved UTRAN (E-UTRAN), 3GPP TS 25.913 V9.0, (2009)
3. Gui X., Ng T. S.: Performance of Asynchronous Orthogonal Multicarrier CDMA System in a Frequency Selective Fading Channel, IEEE Transactions on Communications, 47(7), 1084-1091, (1999)
4. Ekstrom, H.: QoS control in the 3GPP evolved packet system. IEEE Commun. 47, 76-83, (2009)
5. Capozzi F., Piro G., Grieco L. A., Boggia G., Camarda P.: Downlink Packet Scheduling in LTE Cellular Networks: Key Design Issues and a Survey. IEEE Commun. Surveys and Tutorials. 15(2), 678 - 700, (2012)
6. Elgazzar K., Salah M., Taha A. M., Hassanein H.: Comparing Uplink Schedulers for LTE. In: 6th International Wireless Communications and Mobile Computing Conference. pp. 189-193, ACM, (2010)
7. Kwan, R., Leung, C. : A survey of scheduling and interference mitigation in LTE. J. Elect. Comput. Eng. 2010, 1-10, (2010).
8. 3GPP. Tech. Specif. General Packet Radio Service (GPRS) enhancements for Evolved Universal Terrestrial Radio Access Network (E-UTRAN) access, TS 23.401 v9.4.0, (2010)
9. Ghosh A., Zhang J., Andrews J., Muhamed R.: Fundamentals of LTE. Prentice Hall, (2010)
10. 3GPP, Tech. Specif. Group Services and System Aspects - Policy and charging control architecture (Release 9). 3GPP TS 23.203, V9.4.0, (2010)
11. Monghal G., Pedersen K. I., Kovacs I. Z., Mogensen P. E.: QoS oriented time and frequency domain packet schedulers for the UTRAN long term evolution. In: Proc. of IEEE Veh. Tech. Conf., VTC-Spring, Marina Bay, Singapore, (2008)
12. Xalali R., Padovani R., Pankaj R.: Data throughput of CDMA-HDR a high efficiency-high data rate personal communication wireless system.In: Vehicular Technology Conference Proceedings,(VTC Tokyo). Vol.3,pp.1854-1858,IEEE,(2000)
13. Nonchev S., Valkama M.: QoS-oriented packet scheduling for efficient video support in OFDMA-based packet radio systems. In: Multiple Access Communications. pp. 168-180. Springer Berlin Heidelberg, (2011)
14. Zaki Y., Weerawardane T., Gorg C., Timm-Giel A.: Multi-qos-aware fair scheduling for lte. In: Vehicular Technology Conference (VTC Spring). pp.1-5. IEEE,(2011)
15. Brehm M., Prakash R.: Overload-state downlink resource allocation in LTE MAC layer. Wireless Networks Springer. 1-19, (2012)

Using search results metadata to discover effective learning objects for mobile devices

Rogers Phillip Bhalalusesa[1] and Muhammad Rafie Mohd Arshad[1]

[1] School of Computer Science , Universiti Sains Malaysia, 11800 Pulau Pinang, Malaysia
rpb11_com071@student.usm.my, *rafie@cs.usm.my*

Abstract. In mobile learning, limitations of most mobile devices to access rich multimedia contents demands the use of small and interactive learning objects. The selection of learning objects for mobile devices from repositories is based on learning objects metadata. However, most of the metadata are not readily available without downloading the learning objects first. Web crawlers can be used to retrieve the metadata but unfortunately some repositories have policies of blocking the crawling software agents. On the other hand the search engines which can be crawled publicly such as Google contains summary of search results which can be used to generate metadata for rating and identifying the learning objects that may be effective for mobile devices. This paper therefore presents a mechanism of crawling the search engines for metadata as used in Automatic Mobile Learning Objects Compilation (AMLOC) model to facilitate the discovery of learning objects for mobile devices.

Keyword: Learning Objects Metadata, Web Crawler, Search Engine, Mobile Learning

1 Introduction

Mobile devices are increasingly being adopted in many applications in our lives. The learning community is also slowly adopting mobile devices to ease learning activities [1]. When mobile devices are used in learning activities it is called Mobile learning (m-learning). The spread of m-learning is currently limited by the unavailability of effective learning materials that can be used in mobile devices. The mobile devices have many limitations that make it difficult to access the learning materials that have been developed for computers [2]. Computers use learning objects (LO) as sources of learning materials. LO can be considered as any digital resource that can be used to build learning materials (courses)[3]. LO can be combined with other LO to form larger LO and they can also be reused in other domains. Because of their reusability LO are commonly known as Reusable Learning Objects (RLO). The RLO are stored in the internet repositories and can be easily found by using search engines such as Google[4]. However since not all RLO can be used in mobile devices then information systems have to filter out the RLO when deploying the RLO for mobile

T. Herawan et al. (eds.), *Proceedings of the First International Conference on Advanced Data and Information Engineering (DaEng-2013)*, Lecture Notes in Electrical Engineering 285, DOI: 10.1007/978-981-4585-18-7_68,
© Springer Science+Business Media Singapore 2014

devices. The RLO are stored in the repositories with information describing their nature. This information is known as learning Objects Metadata (LOM) [5]. LOM facilitate the discovery and filtering of the RLO. Therefore in order to filter out the RLO, their metadata are compared to the features of mobile devices and only those RLO whose metadata match mobile device features are selected.

In order to retrieve LOM, web crawling techniques can be used. Web crawling is an important process to get a subset of information that can be ported in mobile devices. However not all repositories allow web crawler tools such as robots to access their websites and therefore using web crawlers directly on the RLO repositories may not bring all the results[6]. On the other hand since search engines usually contain results from most of renowned RLO repositories, they give a better choice for sending the crawler agent to get the LOM. Using the search engines as LOM generators imposes another challenge since search engines house the summary of search results and not all of them are true LOM. Therefore there is a need of a mechanism to map the search results summary into acceptable LOM for mobile devices.

The paper is organised as follows. Section two introduces Automatic Mobile Learning Objects Compilation (AMLOC). Section three explains the mechanism of using Google search results summary to create LOM and Section four gives the related work while Section five concludes the paper and gives recommendations for ways forward.

2 Automatic Mobile Learning Objects Compilation (AMLOC)

AMLOC is a model that is currently being developed at Universiti Sains Malaysia that automatically searches, identifies and compiles Reusable Learning Objects that can be used in mobile devices from RLO Repositories. The model identifies the effective m-learning RLO based on the summary of search results in accordance to the learning templates of a course from a Learning Management System (LMS) such as Moodle. Once all the RLO have been identified the model compiles all the RLO into complete sets of learning units as a Sharable Content Object Reference Model (SCORM) package that is uploaded back to the LMS. The compiled learning unit can be ported in mobile devices since all the RLO that have been compiled can be used in mobile devices or it can also be accessed in the LMS.

2.1 Limitations of Mobile Devices in Access Learning Materials

Mobile devices comprises of small, handheld computing devices such as smartphones, mobile phones and tablets that can be used access electronic components[1]. Being originally designed to be portable and for personal communication purposes most of the mobile devices face a lot of challenges when it comes to accessing the learning materials. These challenges come as a result of the limitation that these mobile devices have in terms of storage size, processing power, screen size and battery life, connectivity and cost compared to other high end devices such as Computers and Laptops. The table below describes in details the limitations of mobile devices [7].

Table 1. Limitations of Mobile Devices.

No	Item	Description
1	Power supply	Mobile devices use rechargeable battery which for most of the devices cannot last for a long time.
2	Processing Power	Processing power of most mobile device is lower compared to computers. It may not handle efficiently high computation tasks.
3	Screen Display	Most of the features of most mobile device including the screen are small in order to make them portable.
4	Data Storage	Most of the mobile devices are designed with small storage size and allowed to export most of its data to external storages
5	Connectivity	For a mobile devices the connectivity can vary from one location to the next
6	Cost	The mobile devices are expensive to own and the internet subscription can also be very high

2.2 Challenges of Using Mobile Devices in Accessing Learning Materials

The challenges of using the mobile devices to access the learning materials can be categorized in social economical, pedagogical and technological challenges.

Social economic challenges are based on costs and the lifestyle. Mobile devices that have the capability of accessing and deploying learning materials are expensive to own[8]. The data plans that can allow the mobile devices to use internet to get the learning materials are also expensive to subscribe to. Also very few companies create mobile learning materials and they sell them at high prices. Lastly in our societies, the mobile devices are not trusted to support sensitive activities such as learning.

Pedagogical challenges are concerned with getting quality learning from mobile devices. Because of small screen, interactive instructional materials are difficult to port and sequence into the mobile devices. In addition to that mobile limitations makes it difficult to embed learning quality features such as learning style, Instruction Systems Design principles, learning theories and leaner's' preferences in the learning materials[2].

Technological challenges are more to do with using the mobile devices to access the learning materials[2]. The mobile devices have low storage which means that the learning materials can not be stored in high quantity. Apart from that the processing power of mobile devices is low which makes it hard to process multimedia materials. Not only that but also the low battery life makes it hard to stay connected to the internet or access multimedia RLO.

2.3 Effective RLO for Mobile Devices

In order to deploy effective learning materials for mobile devices special kind of RLO have to be used. The atomic RLO can be easily downloaded or streamed in less time using less resources and therefore they can be used as mobile device materials. In addition to that, The multimedia RLO (such as lecture notes, video, presentations and simulations) are better suited in mobile devices because they are rich in interactivity and express the knowledge domain better. Hence atomic multimedia RLO can be used to achieve better learning materials for mobile devices as they give room to incorporate learning theories, styles and preferences[9]. Furthermore, in order to decrease the cost of RLO for mobile devices, the Open Education Resources (OER) can be used. With OER there will not be need to purchase RLO. Therefore atomic multimedia OER materials can inferred as the effective RLO for mobile devices.

3 Generating LOM from Search Results Summary

In order to show the mechanism of generating the LOM from the search results the description of how the search results summary can be mapped will be given. This will be followed by a description of the type of crawler that is used and lastly the architecture of the AMLOC web crawler will be described.

3.1 Mapping of Search Results Summary to Mobile LOM

AMLOC uses a subset of standard LOM developed by the Institute of Electrical and Electronics Engineers (IEEE) in 2002 [10]. The concept deducing a set of LOM based on IEEE LOM is known as Application Profile. Therefore AMLOC uses an Application profile containing Technological, Pedagogical and Social Economic Metadata based on the limitations of the mobile devices given in section 2. The LOM of the mobile RLO can be summarised in the following table.

Table 2. Mapping Mobile LOM to Search Results.

No	Type of LOM	LOM	Search Results Summary
1	Technological Metadata	Size	Size
		Format	Format
		Storage	Size
2	Pedagogical Metadata	Learning Space	Page Rank
		Coverage	Search Keyword
		Start and Stop	Format
		Total Time	Length
3	Social Economic	Source	Source (OER Repositories)
		Modifiability	Source (OER Repositories)
		Sharing and Reuse	Source (OER Repositories)
		Context	Source

3.2 Data Mining Tool to extract Search Results Summary

The process of data mining internet sites is called web scraping. Web Scraping, also known as web harvesting or web data extraction, is the process of extracting information or data from websites using software programs such as crawlers, spiders and bots[6]. Of all the data mining tools the web crawlers are the most effective because they can get all the contents. In addition to that the web crawlers can be used in any web server because they do not contain robotic programs that will execute in the remote server. The web crawler accesses the search results and for each link of the results found, it crawls that link. When a web crawler only follows a few hyperlinks according to predetermined conditions it is called a focused crawler[11]. That means, if a search result does not contain LOM, the web crawler can visit the search result link to search the for LOM. Since AMLOC relies on getting all the LOM from search results and follow only a few links for those results without LOM, then AMLOC uses focused web crawler. The AMLOC focused crawler uses breadth first algorithm since once it goes into the search results and does not find the required LOM it goes further into that search results. Once it finds the required LOM it moves on to the next link in the search results page

3.4 The AMLOC Web Crawler Implementation

The AMLOC Web Crawler is built to run entirely in Object Oriented Environment. Being a model that searches for learning objects for mobile devices it has to make sure that it can be used in mobile devices. The focused Web crawler of AMLOC is therefore implemented using Jsoup because Jsoup API can be used well in mobile operating systems such as Android[12]. The user starts by sending the search term of the learning object in the web browser. The search is conducted by the search engine and the results are listed in the search results page. The web crawler crawls the search results page and gets the search summary for each search result. The architecture of the Web crawler used in AMLOC is depicted in the figure 1 below.

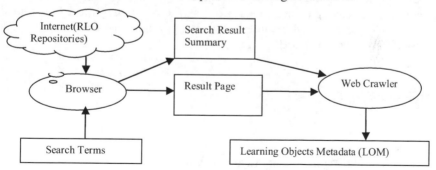

Fig. 1. Web Crawler Architecture

The UML Class Diagram is a standard Modelling language for Information Systems. The top level UML class diagram of the focused Web Crawler is shown below. As noted from the UML Diagram the package of AMLOC is separated from the Internet which shows that all the programming of the crawler is done locally in the AMLOC. This is important so as to limit the crawler from executing in the remote server.

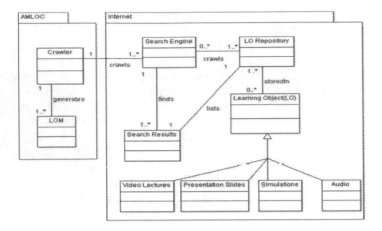

Fig. 2. AMLOC Web Crawler UML Diagram

3.2 Challenges of Crawling Search Results for Mobile LOM

The search results give only a summary of the LOM and therefore not all the LOM can be deduced. In addition, some repositories do not allow the LOM to be found by search engine. This may result in some RLO being missed by the crawler.

The use of web crawlers to filter the RLO introduces a mechanism of checking the metadata of the RLO. This means that some computation power will be used to send the crawler agent to the search engine. This step is not present when users use legacy process of downloading the RLO from the repositories straight away.

Furthermore, the latency of the whole process of downloading the RLO from the repositories will be increased because of the first process when collecting the LOM whenever this mechanism is used.

4 Related Work.

4.1 Using Search Engine to Rate Learning Objects

Most of studies that rate the learning objects use the public available metadata from the search results summary. The studies [13-16] used Google Page Rank technique to evaluate the effectiveness of the RLO. Another study by Yen and company introduced an algorithm to measure the quality of RLO returned from search results based on Google Search algorithm and e-commerce guided search techniques [13]. The studies [14, 15] use Fuzzy Algorithm to rank RLO based on their ontological and contextual information provided in the search results. The studies above imply that one can use the information surrounding the search results to rank and recommend the RLO but however they do not focus on mobile devices.

4.2 Using Web Crawling to Search Learning Objects

The study [11] describes how a focused crawler can be used to extract RLO from the internet using breadth first. The study uses the solution of following up the links starting from the search pages and collecting resources showing up. Another study [17] uses focused web crawler to iterate through a search engine instance of its repository and provide users with only the RLO they are interested in. These studies show that focused web crawlers can be used to capture LOM from the repositories.

4.3 Using Application Profiles of IEEE LOM

LOM have been standardized by IEEE in 2002 and contain nine areas which are General, Life Cycle, Meta-Metadata, Technical, Educational, Rights, Relation, Annotation and Classification Characteristics [10]. Some researchers use all IEEE LOM while others tend to add-on and/or remove some of IEEE LOM. The Singapore Application Profiles(SingPore) and SCORM utilize all the standard IEEE LOM[10]. Canadian application profile (CanCore) uses a subset of IEEE LOM[10]. The studies [18-20] used subsets of IEE LOM and added their own LOM in order to address other requirements. These studies confirm that the IEEE LOM can be modified to address specific needs and such an approach has been used in AMLOC.

5 Conclusion and Recommendation

Mobile devices are increasingly being used in learning activities but still lack features to access all the learning materials[21]. It is important therefore to filter the RLO that will be effective for mobile devices. Currently there is Open Archives Initiative - Protocol for Metadata Harvesting (OAI-PMH) to share the Dublin Core metadata of publications and as a result we get tools like BibTex and Endnote to harvest the publications metadata[22]. However such protocols do not apply to teaching materials. As there are no standard protocols for harvesting LOM for teaching materials, repositories owners should increase the accessibilies of LOM through web services so that anyone can use LOM to identify teaching RLO before downloading.

References

1.   Muyinda, P.B.: Deploying and Utilizing Learning Objects on Mobile Phones vol. (2010).
2.   Ryu, H.,Parsons, D.: Innovative mobile learning: Techniques and technologies: Information Science Reference. (2008).
3.   Wiley, D., Waters, S., Dawson, D., Lambert, B., Barclay, M., Wade, D., et al.: Overcoming the limitations of learning objects. Journal of Educational Multimedia and Hypermedia. vol. 13(4): pp. 507-521. (2004).
4.   Tony, K., Lisa, B.,Neil, L.: Institutional Use of Learning Objects: Lessons Learned and Future Directions. Journal of Educational Multimedia and Hypermedia. vol. 13(4). (2004).

5. Churchill, D.: Towards a useful classification of learning objects. Educational Technology Research and Development. vol. **55**(5): pp. 479-497. (2007).
6. Jacob, G., Kirda, E., Kruegel, C.,Vigna, G.: PUBCRAWL: protecting users and businesses from CRAWLers. In: Proceedings of the 21st USENIX conference on Security symposium, USENIX Association: Bellevue, WA. pp. 25-25. (2012).
7. Uden, L.: Activity theory for designing mobile learning. International Journal of Mobile Learning and Organisation. vol. **1**(1): pp. 81-102. (2007).
8. Lim, T.S.K., Mansor, F.,Norziati, M.: Mobile learning via SMS at Open University Malaysia: Equitable, effective, and sustainable. International Review of Research in Open and Distance Learning. vol. **12**(2). (2011).
9. Sharples, M., Taylor, J.,Vavoula, G.: Towards a theory of mobile learning. Proceedings of mLearn 2005. vol. **1**(1): pp. 1-9. (2005).
10. Mason, R.T.: Interoperability Gap Challenges for Learning Object Repositories & Learning Management Systems. Nova Southeastern University: United States -- Florida. pp. 257. (2011).
11. Tane, J., Schmitz, C.,Stumme, G.: Semantic resource management for the web: an e-learning application. In: Proceedings of the 13th international World Wide Web conference on Alternate track papers & posters. ACM. (2004).
12. Dinh, P.C.,Boonkrong, S.: The Comparison of Impacts to Android Phone Battery Between Polling Data and Pushing data. In: IISRO Multi-Conferences Proceeding. Thailand. (2013).
13. Yen, N.Y., Shih, T.K., Chao, L.R.,Jin, Q.: Ranking metrics and search guidance for learning object repository. Learning Technologies, IEEE Transactions on. vol. **3**(3): pp. 250-264. (2010).
14. Ochoa, X.,Duval, E.: Use of contextualized attention metadata for ranking and recommending learning objects. In: Proceedings of the 1st international workshop on Contextualized attention metadata: collecting, managing and exploiting of rich usage information. ACM. (2006).
15. Tsai, K.H., Chiu, T.K., Lee, M.C.,Wang, T.I.: A learning objects recommendation model based on the preference and ontological approaches. In: Advanced Learning Technologies, 2006. Sixth International Conference on. IEEE. (2006).
16. Han, K.: Quality rating of learning objects using Bayesian Belief Networks. Simon Fraser University. (2004).
17. Biletskiy, Y., Wojcenovic, M.,Baghi, H.: Focused crawling for downloading learning objects. Interdisciplinary Journal of E-Learning and Learning Objects. vol. **5**(1): pp. 169-180. (2009).
18. Quemada, J., Huecas, G., de-Miguel, T., Salvachúa, J., Fernandez, B., Simon, B., et al.: Educanext: a framework for sharing live educational resources with isabel. In: Proceedings of the 13th international WWW conference on Alternate track papers & posters. ACM. (2004).
19. Puustjärvi, J.: Syntax and Semantics of Learning Object Metadata. In: Learning Objects 2: Standards, Metadata, Repositories, and Lcms. pp. 41. (2007).
20. Verbert, K., Gašević, D., Jovanović, J.,Duval, E.: Ontology-based learning content repurposing. In: Special interest tracks and posters of the 14th international conference on World Wide Web. ACM. (2005).
21. Pathmeswaran, R.,Ahmed, V.: SWmLOR: Technologies for Developing Semantic Web based Mobile Learning Object Repository. The Built & Human Environment Review. vol. **2**(1). (2011).
22. Lagoze, C.,Van de Sompel, H.: The making of the open archives initiative protocol for metadata harvesting. Library hi tech. vol. **21**(2): pp. 118-128. (2003).

Part VIII
Web Data, Services and Intelligence

Analysis of Trust Models to Improve E-Commerce Multi-Agent Systems

Elham Majd[1], Vimala Balakrishnan[1]

[1] Faculty of Computer Science and Information Technology,
University of Malaya, Kuala Lumpur, Malaysia
elham.majd@siswa.um.edu.my

Abstract. Agents in highly dynamic multi-agent environments can break contracts. Hence, trust models play a critical role in making better evaluations of unknown candidates, and better decisions about when, and who, to interact, especially in e-commerce. However, current trust models do not contain any standard components. It is noted that finding standard components are effective in designing a trust model which can support heterogeneous information related to multi-agent systems. This paper compares and analyzes some recent trust models based on three important components: reliability, similarity, and satisfaction. In addition, a comparison is made between the most representative models to point out significant strengths and weaknesses that exist in current trust models. The ultimate goal of this study is to define the relationship between these components. This analysis helps to identify the standard components of trust models for e-commerce multi-agent systems.

Keywords: E-commerce, Multi-agent systems, Reliability, Satisfaction, Similarity, Trust model

1 Introduction

Trust, is one the factors that helps agents to reduce some of the uncertainty in their interactions, and achieve expectations about the behaviors of others which can increase or decrease with more experiences, (i.e., interactions or observations). Recent experiences are more important than old ones since old experiences may become outdated or irrelevant with the passing of time [1, 2]. Trust can be defined as "a particular level of the subjective probability with which an agent assesses that another agent or group of agents will perform a particular action"[3].

Generation of economic activities via electronic transactions, in particular, needs the presence of a trust system to ensure the fulfillment of a contract [4, 5]. Indeed, establishment of trust between unknown agents enables extension of a successful transaction to a much broader range of participants in an e-commerce multi-agent environment [6].

Different trust models have been suggested for multi agent environments, such as "Bayesian Network Trust Model"[7], "Probability Certainty Distribution Model" [8], "Trust based Recommender System for Peer Production Services Model (TREPPS)"

T. Herawan et al. (eds.), *Proceedings of the First International Conference on Advanced Data and Information Engineering (DaEng-2013)*, Lecture Notes in Electrical Engineering 285, DOI: 10.1007/978-981-4585-18-7_69,
© Springer Science+Business Media Singapore 2014

[9], and "Dynamic Trust Model" [10]. These models differ from one another based on several components which they were built on. Therefore, a comparative analysis of these models based on the identified components helps to better understand the existing and improve the future trust models. This paper intends to compare and analyze some of the existing trust models based on three standard components: similarity, satisfaction and reliability.

The rest of the paper is organized as follows: Section Two describes the concept of multi-agent systems. Section Three introduces the three main components of current multi-agent trust models. Section Four analyses the most representative of trust models in multi-agent environment. In Section Five the analyzed models are compared in order to define the strengths and weaknesses of components of current trust models. In Section Six the interfaces of the identified standard components are presented and finally, Section Seven contains the conclusions and future work in relation to multi-agent trust models in e-commerce.

2 Multi Agent Systems

Multi-Agent Systems are complex systems which are combined of several autonomous agents with limited abilities, but they can interact in order to achieve a global objective [11, 12]. The multi-agent systems are autonomous in that they work without human interventions. The multi-agents also react to environment [13].

However in e-commerce multi-agent environment some critical transactions such as those related to personal online shopping must be securely performed [14, 15], trust plays a critical role to improve the transaction in e-commerce. Thus, in order to extend a successful transaction to a much broader range of participants, different trust models have been suggested.

In the following section, we will define three main components which should be considered for designing of trust models in e-commerce multi-agent systems.

3 Standard Components

From the analysis criteria of the existing trust models, three main components are determined and any standard model to be designed should have these components in mind. These proposed standard components are presented in Table 1.

Table 1. The standard components

Components	Definition
Similarity	The likeliness between the preferences experienced by the providing agent and recommending agents
Satisfaction	Degrees of consent from the consumption of services provided by provider agents
Reliability	The ability of an agent to consistently provide services properly

3.1 Similarity

In a heterogeneous environment, if a requesting agent has insufficient experience about the providing agents or try to buy items from unknown providers, they should make a decision according to recommendations of other agents who have sufficient information about the providers. In this case, they should consult their acquaintances to find out which one of these providers can provide the items according to their highest value of preferences. So, if agent A wants to trust agent B, it should evaluate similarity between the preferences [16].

3.2 Satisfaction

Agents are autonomous, and any two agents may have different preferences on the same item. Thus, they have various degrees of satisfaction from the using the same item [16]. Hence, when completing an interaction, service requestor needs to rate provider's performance through the feedback interface in order to report its satisfaction of current interaction [9]. This information is vital as satisfaction rating represents the confidence level of users in the services and resources provided by the agents [17].

3.3 Reliability

Finally, the reliability of a system has generally been defined as the probability that a system will perform as per its specification for a specified duration of time [18]. However, when agents cooperate, possibility arises that an agent may deceive its partner for its own benefit. In order to ensure selection of a reliable partner, it is necessary to investigate reliability among agents [19], especially an e-commerce agent might be reliable if the price of the transaction is low enough, while its reliability might significantly decrease in the case of very high price [20].

4 Trust Models

This section describes some of the existing trust models, focusing only on their main characteristics.

One of the proposed models is "Bayesian Network trust Model" [7] which uses probability method to identify the relationship between different agents. In this model, the requesting agent evaluates providers' trustworthiness based on its own experiences and recommendations that it obtains from its peers [21].

The principle of this model is based on Bayesian network. In fact, each Bayesian network has a root node, T, with two branches called satisfying and unsatisfying, denoted by 1 and 0, respectively. The agents' overall trust in the provider's competencies is represented by $P(T=1)$, which is the proportion of interactions with satisfactory results out of all the interactions with the same provider [22]. The model then measures the overall degree of agents' satisfaction of interactions by indicating the importance of each aspect such as product quality. This model is one of the few models which compute the reliability of each agent according to different service aspects.

"Probability Certainty Distribution Model" [8] presents a formula to calculate the trust degree between recommending agents and then calculates the reliability of each one. This model evaluates reliability of each agent according to a particular task which requesting agent is asked to solve. The model composes a trust value of the agent considering the probability of a positive/negative outcome, where the pair (r,s) of past interactions denotes positive (satisfying) and negative (unsatisfying) interactions, respectively, and also the certainty placed in the probability. This model uses two operators for combining the trust values from multiple sources [23].

"TREPPS model" measures reliability and similarity among agents. This model provides a recommender system based on the trust of the social networks. Trust in this model is computed by means of fuzzy logic applications to appropriately assess the quality and the reliability of peer production services [24].

"Dynamic trust model" [10, 25] evaluates reliable transactions in multi-agent environment. This model proposes an exponential averaging update function to store the value of satisfaction instead of storing all the satisfaction levels of all interactions. Thus, this model successfully reduces the storage overhead. On the other hand, this model provides a simple formula to evaluate the similarity between preferences of two agents, with each agent's predetermined preferences, in the range of -1 to 1.

5 Comparisons and Discussion

The trust models in the previous section are compared and analyzed based on the three identified components. The results are summarized in Table 2. Generally it can be concluded that reliability makes up the strong component, whereas similarity and satisfaction are the weak components of these models.

Table 2. The Comparative analysis of Trust models

Models	Similarity	Satisfaction	Reliability
Bayesian Network Trust Model	N/A	Importance of each aspect	Proportion of satisfying interactions
Probability Certainty Distribution Model	N/A	N/A	Number of satisfying and unsatisfying past interactions along with uncertainty
TREPPS model	Calculated by an important weight	Rating of provider via buyer t after each interaction	Requester's satisfaction of past interactions and weight of time
Dynamic Trust Model	Average sum of preferences items until the time of interaction	N/A	Satisfaction of past interactions to provide a particular service

N/A: not available

As illustrates in Table 2, all models used the amount of satisfying rates of past interactions for calculating reliability of each agent. However, the above mentioned models did not give any clear explanation on how to calculate satisfaction. Ambiguous methods were presented by "TREPPS model", in which requester evaluates provider after each interaction to determine whether the interaction have provided satisfaction or not. Besides, "Bayesian Network Trust Model" considers the satisfaction in relation to the provider who can provide services according to service aspect of requester preferences. This reveals the fact that the reliability of each provider should be computed according to the evaluation results of satisfying previous interactions between requester and provider agents. Meanwhile, one of the models that involved uncertainty in computing reliability is "Probability Certainty Distribution Model".

On the same note, similarity is a concept which requires more investigation. "TREPPS model" defines a complex formula but without a comprehensive explanation for applying this formula to find the similarity between two agents. On the other hand, "Dynamic Trust Model" can calculate similarity by taking into consideration the average sum of preferences items of each agent.

6 Three Standard Components Interfaces

In this section, the interfaces of identified components are described.

When a requester has insufficient or none of information about providers, it should make a decision according to recommendations of recommender agents.

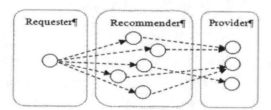

Fig. 1. A multi agent environment contain Requester, Recommenders, and Providers

In this case, the requester agent should select recommender agents which share similar preferences' aspects, so in this step requester agent calculates the similarity of each recommender agents according to average sum of preferences items.

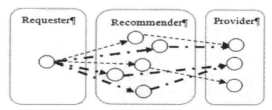

Fig. 2. Select agent B, C, and D as similar agents to requester

According to Fig. 2, requester agent A calculates the similarity of each agent considering the existing trust models noted in previous section, it seems that

"Dynamic Trust Model" presented an appropriate formula to calculate the similarity between preferences of two agents, with each agent's predetermined preference, fa_i on each item a, in the range of -1 to 1. The similarity, $w_{i,j}$, between preferences of two agents, i and j, is computed as:

$$w_{i,j} = \sum_a (1 - |fa_i - fa_j|) \tag{1}$$

where a, shows different items.

Then, the requester should proceed to check the reliability of each recommender considering the evaluation of satisfying interactions acquired from previous interactions with that particular recommender.

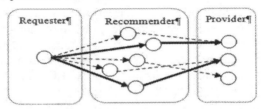

Fig. 3. Select agent B and D as similar and reliable agents

According to Fig. 3, requester agent A, finally selected recommender agent B and D as similar and reliable agents. It seems that "Bayesian Network Trust Model" presented a stronger formula to evaluate satisfaction value. It measures the overall degree of agents' satisfaction of interactions as follows:

$$s = w_{ds} * s_{ds} + w_{fq} * s_{df} \tag{2}$$

where ds and fq denotes the each aspect of services, such as product quality. Moreover, w_{ds} and w_{fq} reflect the weights, which indicate the importance of each aspect such as delivery time depending on the user's preferences. Each agent has a satisfaction threshold, s_t. If $s > s_t$, the interaction is considered satisfying, otherwise, it means it is unsatisfying. "Probability Certainty Distribution Model" then calculated the reliability by considering satisfaction value of previous interactions and the uncertainty placed in its probability formula. The reliability of each agent is then computed based on the binary event (r,s), where r denotes the satisfying interaction and s is un satisfying interaction, by a belief as $B = \{(a,b,u) | a > 0, b > 0, u > 0, a + b + u = 1\}$.

where $a = c\dfrac{r+1}{r+s+1}$ and $b = c\dfrac{s+1}{r+s+1}$ reflect the probability of positive and negative outcomes and $u = 1 - c$ represents the uncertainty placed in the probability, where Eq.3 denotes the certainty of future interaction outcomes.

$$c(r,s) = \frac{1}{2} \int_0^1 \left| \frac{x^r(1-x^r)^s}{\int_0^1 x^r(1-x^r)^s} - 1 \right| dx \tag{3}$$

7 Conclusion

In this paper, we defined the three main components of trust models for multi agent systems in e-commerce according to analyses of the most representative trust models. The comparison between existing models showed the computing of reliability is the strong point, while the calculating similarity and satisfaction are the weak points of these models. Moreover, we determined the relationship between these trust components. Before sending a query to recommenders in multi-agent environment, the similarity between recommenders and requester should be investigated; then the reliability of each recommender would be evaluated according to the satisfaction of their past interactions. This relationship represents how these three trust components should be implemented and measured in future e-commerce multi-agent trust models. For future work, we intend to provide other standard components to improve e-commerce multi-agent systems.

References

1 Aref, M. M.: A multi-agent system for natural language understanding, in Proceedings of the International Conference on Integration of Knowledge Intensive Multi-Agent Systems, Cambrige MA, pp. 36--40 (2003).
2 Wang, Y., Vassileva, J.: A review on trust and reputation for web service selection, presented at the 27th International Conference on Distributed Computing Systems Workshops (ICDCSW'07), Toronto, Ontario, Canada (2007).
3 Xin, L.: Trust beyond reputation: Novel trust mechanisms for distributed environments, Nanyang Technological University (2011).
4 Zhou, L.: Organizational Knowledge Management Based on Social Network, in Second International Conference on Information and Computing Science (ICIC'09), pp. 289--292, Beijing, China (2009).
5 Walter, F., Battiston, S., Schweitzer, F.: Coping with information overload through trust-based networks, in Managing Complexity: Insights, Concepts, Applications, D. Helbing, Ed., ed Berlin Heidelberg: Springer Berlin Heidelberg, pp. 273--300 (2008).
6 Yu, T., Winslett, M., Seamons, K..: Automated trust negotiation over the internet, Proc. of Multiconference on Systemics, Cybernetics and Informatics (2002).
7 Wang, Y., Vassileva, J.: Bayesian network-based trust model, in Proceedings of IEEE/WIC International Conference on Web Intelligence (WI), pp. 372--378, Halifax, Canada (2003).
8 Lorenzi, F., Baldo, G. , Costa, R., Abel, M., Bazzan, A., Ricci, F.: A Trust Model for Multiagent Recommendations, J. Emerging Technologies in Web Intelligence, 2, pp. 310--318 (2010).
9 Li Y. M., Kao, C. P.: TREPPS: A trust-based recommender system for peer production services, Expert systems with applications, 36, pp. 3263--3277 (2009).
10 Das, A., Islam, M. M., Sorwar, G.: Dynamic trust model for reliable transactions in multi-agent systems, in 13th International Conference on Advanced Communication Technology (ICACT), pp. 1101-1106, Korea (2011).
11 Kulasekera, A., Gopura, R., Hemapala, K., Perera, N.: A review on multi-agent systems in microgrid applications, in Innovative Smart Grid Technologies (ISGT India), pp. 173--177, IEEE Press, India (2011).
12 Shoham Y, Leyton-Brown, K..: Multiagent systems: Algorithmic, game-theoretic, and logical foundations: Cambridge University Press (2009).

13 Pipattanasomporn, M., Feroze, H., Rahman, S.: Multi-agent systems in a distributed smart grid: Design and implementation, in Power Systems Conference and Exposition, PSCE'09, pp. 1--8, IEEE Press (2009).

14 Jung, Y., Kim, M., Masoumzadeh, A., Joshi, J. B. D.: A survey of security issue in multi-agent systems, Artificial Intelligence Review, pp. 1--22 (2012).

15 Zacharia, G., Moukas, A., Maes, P.: Collaborative reputation mechanisms for electronic marketplaces, Decision support systems, 29, pp. 371--388 (2000).

16 Battiston, S., Walter, F. E., Schweitzer, F.: Impact of trust on the performance of a recommendation system in a social network, in Proceedings of the Workshop on Trust in Agent Societies at the Fifth International Joint Conference on Autonomous Agents and Multi-Agent Systems (AAMAS'06), pp. 1--15, United Kingdom (2006).

17 Woo, J. W., Hwang, M. J., Lee, C. G., Youn, H. Y.: Dynamic role-based access control with trust-satisfaction and reputation for multi-agent system, in Advanced Information Networking and Applications Workshops (WAINA) at IEEE 24th International Conference on Advanced Information Networking and Applications, pp. 1121--1126, Perth, Australia (2010).

18 Sundresh, T. S.: Semantic reliability of multi-agent intelligent systems, J. Bell Labs Technical, 11, pp. 225--236 (2006).

19 Li, L., Li, H., Lu, G., Yao, S. W.: A Quantifiable Trust Model for Multi-agent System Based on Equal Relations, in International Conference on Computational Intelligence and Security, pp. 291--295, China (2007).

20 Garruzzo, S., Rosaci, D.: The roles of reliability and reputation in competitive multi agent systems, in On the Move to Meaningful Internet Systems: OTM, ed Berlin: Springer Berlin pp. 326--339, Heidelberg (2010).

21 Lu, J., Li, R., Lu, Z., Ma, X.: Primary-Backup Access Control Scheme for Securing P2P File-Sharing Systems, International J. Information Technology, 1, pp. 8--15 (2009).

22 Firdhous, M., Ghazali, O., Hassan, S.: Trust Management in Cloud Computing: A Critical Review, International J. Advances in ICT for Emerging Regions (ICTer), 4, pp. 24--36 (2012).

23 Hang, C. W., Wang, Y., Singh, M. P.: An adaptive probabilistic trust model and its evaluation, in Proceedings of the 7th international joint conference on Autonomous agents and multiagent systems, pp. 1485--1488, Estoril, Portugal (2008).

24 Bustos, F., López, J., Julián, V., Rebollo, M.: STRS: Social Network Based Recommender System for Tourism Enhanced with Trust, in International Symposium on Distributed Computing and Artificial Intelligence 2008 (DCAI 2008), pp. 71--79 (2009).

25 Das, A., Islam, M. M.: SecuredTrust: A Dynamic Trust Computation Model for Secured Communication in Multiagent Systems, Dependable and Secure Computing, IEEE Transactions on, 9, pp. 261--274 (2012).

Augmenting Concept Definition in Gloss Vector Semantic Relatedness Measure using Wikipedia Articles

Ahmad Pesaranghader[1], Ali Pesaranghader[2], and Azadeh Rezaei[1]

[1] Multimedia University (MMU), Jalan Multimedia, 63100 Cyberjaya, Malaysia
[2] Universiti Putra Malaysia (UPM), 43400 UPM, Serdang, Selangor, Malaysia
{ahmad.pgh, ali.pgh, azadeh.rezaei}@sfmd.ir

Abstract. Semantic relatedness measures are widely used in text mining and information retrieval applications. Considering these automated measures, in this research paper we attempt to improve Gloss Vector relatedness measure for more accurate estimation of relatedness between two given concepts. Generally, this measure, by constructing concepts definitions (Glosses) from a thesaurus, tries to find the angle between the concepts' gloss vectors for the calculation of relatedness. Nonetheless, this definition construction task is challenging as thesauruses do not provide full coverage of expressive definitions for the particularly specialized concepts. By employing Wikipedia articles and other external resources, we aim at augmenting these concepts' definitions. Applying both definition types to the biomedical domain, using MEDLINE as corpus, UMLS as the default thesaurus, and a reference standard of 68 concept pairs manually rated for relatedness, we show exploiting available resources on the Web would have positive impact on final measurement of semantic relatedness.

Keywords: Semantic Relatedness; Biomedical Text Mining; Web Mining; Bioinformatics; UMLS; MEDLINE; Wikipedia; Natural Language Processing.

1 Introduction

How do you relate "*mouse*" to "*computer*"? What about "*mouse*" and "*medical testing*"? Humans, because of cognitive knowledge possession regarding to the word meanings, have an ability to answer above questions through semantic judgment of relatedness for the involved concepts. An intelligent algorithm that computationally imitates this ability is goal of many natural language processing studies. This quantification of lexical semantic relatedness has a broad application. Muthaiyah and Kerschberg [1] used semantic relatedness measures for ontology matching. Pekar et al. [2] employed them to enhance a question answering system in the tourism domain. Chen et al. [3] applied these measures in machine translation. For the biomedical domain, Bousquet et al. [4] investigated codification of medical diagnoses and adverse drug reactions using semantic relatedness and similarity measures.

The output of a relatedness measure is a value, ideally normalized between 0 and 1, indicating how semantically related two given terms (words) are. Mainly, with respect

T. Herawan et al. (eds.), *Proceedings of the First International Conference on Advanced Data and Information Engineering (DaEng-2013)*, Lecture Notes in Electrical Engineering 285, DOI: 10.1007/978-981-4585-18-7_70,
© Springer Science+Business Media Singapore 2014

to the semantic relatedness measurement, there are two models of computational technique available: taxonomy-based model and distributional model.

In taxonomy-based model, proposed measures take advantage of lexical structures such as taxonomies. Measures in this model largely deal with semantic similarity measurement as a specific case of semantic relatedness. For general English text, studies on measuring similarity rely on WordNet, a combination of dictionary and thesaurus, designed for supporting automatic text analysis and artificial intelligence applications. In clinical and biomedical studies, researchers employ the Unified Medical Language System (UMLS), a large lexical and semantic ontology of medical vocabularies maintained by the National Library of Medicine.

The notion behind distributional model comes from Firth idea (1957) [5]: "a word is characterized by the company it keeps". In these measures, words specifications are derived from their co-occurrence distributions in a corpus. These co-occurred features will be represented in vector space for the subsequent computation of relatedness.

Gloss Vector semantic relatedness measure is a distributional-based approach that by constructing definitions (Glosses) for the concepts from a predefined thesaurus, estimates semantic relatedness of two concepts through calculation of the angle between those concepts' Gloss vectors. However, as thesauruses do not provide expressive definitions for all the concepts, particularly specialized ones, this definition construction task can be tricky. In this paper, by considering Wikipedia articles and other resources, we try to enrich concepts' definitions further. We see exploitation of the resources on the Web enhances reliability of semantic relatedness measurement.

The remainder of the paper is organized as follows. Section 2 presents proposed semantic similarity and relatedness measures. In Section 3, we list data and resources employed for our experiments in the biomedical domain. In Section 4, our method for augmenting definition of a concept is described. In Section 5, experiments and analysis are given. Finally, the conclusions and future studies are stated in Section 6.

2   Related Works

The proposed semantic measures in the literature either address relatedness or similarity. Similarity in words is specific case of relatedness; as an example "*dissection*" and "*anatomization*" are similar words but "*dissection*" and "*scalpel*" are just related. This study we distinct semantic similarity from semantic relatedness.

2.1   Semantic Similarity Measures

Measures of semantic similarity are dependent on a hierarchical structure of concepts derived from taxonomies or thesauruses. In this regard, a concept is specific sense of a word (term), which would denote polysemy. One concept can also have different representative words (terms), which would cause synonymy. The hierarchical structures of the concepts may get equipped with relations including *is-a*, *has-part*, and *is-a-part-of* or any other types of relationship. Semantic similarity measures employ positional information of the concepts in the taxonomy in order to estimate how similar two input concepts are. Figure 1 illustrates part of a life-form taxonomy.

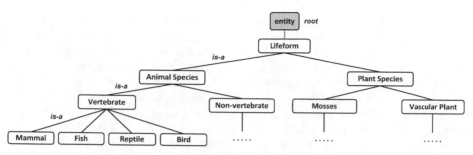

Fig. 1. A small portion of a lifeform taxonomy.

Path measure [6] is a simple similarity technique based on the calculation of the reciprocal of the shortest path from one input concept to another in the taxonomy. In path measure for finding shortest path between a pair of concepts we count nodes (number of jumps/paths incremented by 1). Path measure does not consider growth in specificity for concepts when we go down from the root towards leaves.

In order to deal with the problem of path measure, Wu and Palmer [7] propose a method which is based on both path and depth of the concepts. By using least common subsumer (LCS) of the two concepts they determined how specific in meaning the two input concepts can be. From information retrieval points of view, when one concept is less frequent than the other one, it is more informative. This informativeness quality denotes abstractness or concreteness in meaning for the concepts, so it is important for semantic similarity measurement. Path and depth based similarity measures lack consideration of this distinctive attribute.

In order to measure how informative one concept is, we need to know frequency of that concept in an external corpus. Resnik introduced this characteristic of a concept as its information content (IC) [8] and quantified it by the negative log of the probability of the concept in a corpus. He also measured two concepts similarity through calculating IC value of their LCS in a taxonomy. Jiang and Conrath [9] and Lin [10] in their studies improved accuracy of estimated similarity based on concepts' ICs. However, the process of IC calculation for a concept is a challenging task. Absence of the concept's representative word(s) in the corpus and ambiguity of the words in the corpus are just some problematic issues for IC computation.

Pesaranghader and Muthaiyah [11] introduced IC vectors which rely on the definitions of the concepts and their constituent words. They propose that the cosine of the angle between IC vectors of the two input concepts put together and IC vector of their LCS is a more reliable and effective way in semantic similarity estimation.

In measures of semantic similarity the dependency on taxonomy relations can be a disadvantage as taxonomies tend to be static and cannot keep up with the rapidly changing structure of knowledge in a given discipline.

2.2 Semantic Relatedness Measures

Measures of semantic relatedness mostly rely on distributional properties of concepts in large text corpora as an easier way to keep track of changes in a given knowledge domain. Nonetheless, Lesk [12] calculate the strength of relatedness between a pair of concepts as a function of the overlap between their definitions by considering their constituent words. Banerjee and Pedersen [13] proposed the Extended Gloss Overlap measure (also known as Adapted Lesk) which augment concepts definitions with the definitions of senses that are directly related to it in WordNet for an improved result. The main drawback in these measures is their strict reliance on concepts definitions and negligence of other knowledge source.

To address forging limitation in the Lesk-type measures, Patwardhan and Pedersen [14] introduced the Gloss Vector measure by joining together both ideas of concepts definitions from a thesaurus and co-occurrence data from a corpus. In their approach, every word in the definition of the concept from WordNet is replaced by its context vector from the co-occurrence data from a corpus and relatedness is calculated as the cosine of the angle between the two input concepts associated vectors (gloss vectors). This Gloss Vector measure is highly valuable as it: 1) employs empirical knowledge implicit in a corpus of data, 2) avoids the direct matching problem, and 3) has no need for an underlying structure. Therefore, in another study, Liu et al. [15] extended the Gloss Vector measure and applied it as Second Order Context Vector measure to the biomedical domain. The UMLS was used for driving concepts definitions and biomedical corpuses were employed for co-occurrence data extraction. In brief, this method gets completed by five steps which sequentially are, 1) constructing co-occurrence matrix from a biomedical corpus, 2) removing insignificant words by low and high frequency cut-off, 3) developing concepts extended definitions using UMLS, 4) constructing definition matrix by results of step 3 and 4, and 5) estimating semantic relatedness for a concept pair. For a list of concept pairs already scored by biomedical experts, they evaluated Second Order Context Vector measure through Spearman's rank correlation coefficient. Figure 2 illustrates the entire procedure.

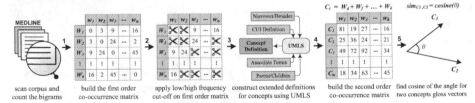

Fig. 2. 5 steps of Gloss Vector relatedness measure

Even though Gloss Vector semantic relatedness measure has proven its application in wide variety of intelligent tasks, it has always suffered from a constraining feature indicating this measure's extreme reliance on a thesaurus by which concepts get defined. This issue is even more intense when we apply this measure into a specialized area of knowledge for which accordingly we need a specialized thesaurus. In this type of thesauruses all involved concepts need to be expressively and precisely described. Moreover, these sources need to get updated constantly to keep up with new evolutions in the given discipline. Nonetheless, all available thesauruses do not deliver all the concepts, and even if they do, for many of concepts the provided definitions are short and undependable. Here is the checking point where the (extended) definition construction step curbs the procedure of relatedness calculation.

In Section 4, we explain our solution for resolving the problem of definition construction in Gloss Vector relatedness measure by relying on external resources on the Web which are built mainly based on shared public knowledge. We examine impact of exploiting these resources alone as well as in tandem with the thesaurus in the Gloss Vector measure which gets defined in the first place. The results of experiments along with the discussion get presented in Section 5. In the following section we list experimental resources of our experiments in the biomedical domain.

3 Experimental Data

3.1 Unified Medical Language System (UMLS)

The Unified Medical Language System[1] (UMLS) is a knowledge representation framework designed to support biomedical and clinical research. Its fundamental usage is provision of a database of biomedical terminologies for encoding information contained in electronic medical records and medical decision support. It comprises over 160 terminologies and classification systems, and Coding Symbols for Thesaurus of Adverse Reaction Terms (COSTART) is one of them. COSTART was developed by the United States Food and Drug Administration (FDA) for the coding, filing and retrieving of post-marketing adverse reaction reports. The scope of the research is limited to the COSTART vocabulary and all tested concepts belong to this resource. We employed 2012AB release of the UMLS containing the last version of COSTART.

3.2 MEDLINE Abstract

MEDLINE[2] contains over 20 million biomedical articles from 1966 to the present. The database covers journal articles from almost every field of biomedicine including medicine, nursing, pharmacy, dentistry, veterinary medicine, and healthcare. For the current study we used MEDLINE article abstracts as the corpus to build a first order term-term co-occurrence matrix for later computation of second order co-occurrence matrix. We used the 2013 MEDLINE abstract.

3.3 Reference Standard

The reference standard[3] in our experiments is a set of scored medical concept pairs from University of Minnesota Medical School as a result of their experimental study [16]. Eight medical residents were invited for participation in scoring the relatedness of these concept pairs in order to have them voted based on human judgment. Most of the concepts from original reference standard are not included in the COSTART ontology applied in our experiments. Therefore, after removing them from the original dataset, a subset of 68 concept pairs for testing Gloss Vector semantic relatedness measure was available. It should be mentioned here that the main reason for selecting COSTART with this rather small coverage of concept pairs from the original data set was its

---

[1] http://www.nlm.nih.gov/research/umls
[2] http://mbr.nlm.nih.gov/index.shtml
[3] http://rxinformatics.umn.edu/data/UMNSRS_relatedness.csv

potential to represent the vulnerability of Gloss Vector measure in demanding for a dependable thesaurus which emphasizes a need for alternative resources.

4   Methods

Totally, COSTART comprises 1501 biomedical concepts in its ontology (excluding root). For this quantity of concepts more than thirds of them (to be precise 619 concepts) does not have definition assigned in any vocabulary included in the UMLS. This means no chance for measuring semantic relatedness for this type of concepts as the gloss vectors of them will be unavailable. Fortunately, all the 74 involved concepts in our 68 concept pairs of reference standard have the definition of their own in the UMLS. However, for many of these tested concepts the provided definitions are short and unexpressive. Moreover, as it is shown in the [15] the best result from Gloss Vector measure would be achieved when we consider extended form of definitions for concepts (by adding definitions of the parents and children of a concept to the definition of its own). The absence of these extra definitions also hinders calculation of more reliable relatedness estimations.

In order to avoid this drawback we will augment extended definitions of the concepts even further. For this purpose we manually extract definitions of the concepts from free online resources especially Wikipedia articles and in a very rare cases "www.thefreedictionary.com/medicine" website. Even though, by using some automated techniques we could exploit definitions of the concepts from the Web, these definitions are meticulously hand-selected in order to estimate the most accurate impact of the online resources helping for relatedness measurement used in specific text mining task or other NLP-related applications in general.

These extracted definitions from Wikipedia articles can be employed alone for those concepts which do not have any definitions in the UMLS vocabularies or can be concatenated to the definition of those concepts for which definitions are assigned. As an example, the computed definition for the concept "*earache*" with the concept unique identifier (CUI) of *C0013456* from the UMLS is "*A disorder characterized by a sensation of marked discomfort in the ear; Painful sensation in the ear region; Pain in the ear*". By extracting corresponding Wikipedia article for this concept we will have an additional definition which is "*Otalgia or an earache is ear pain. Primary otalgia is ear pain that originates inside the ear. Referred otalgia is ear pain that originates from outside the ear. Otalgia is not always associated with ear disease. It may be caused by several other conditions, such as impacted teeth, sinus disease, inflamed tonsils, infections in the nose and pharynx, throat cancer, and occasionally as a sensory aura that precedes a migraine*". It is noticeable that the provided definition for this concept in Wikipedia is much more expressive than the available definition in the UMLS. As mentioned, these additional definitions can be used alone or combined with the presented concepts' definitions in the UMLS.

For doing experiments of the study all the 74 concepts including in the reference standard as well as their direct parents and children (PAR/CHD) and broader and narrower (RB/RN) concepts in the COSTART taxonomy are provided with expressive definitions from free online resources particularly from the Wikipedia articles.

5   Experiments and Discussions

The experiments of the study mainly deals with the different ways of definition construction whether by applying definitions developed from the UMLS as the well-known meta-thesaurus in the field of biomedical or by using free online resources especially Wikipedia articles elaborating involved medical and biomedical concepts.

For the experiment results evaluation, Spearman's rank correlation coefficient for assessing the relationship between the reference standards and the auto-generated semantic relatedness results is employed. Spearman's rank correlation, ρ, is a non-parametric measure of statistical dependence between two variables. Here we assume there is no relationship between the two sets of data. This algorithm sorts data in both sets from highest to lowest, and then subtracts the two sets of ranks and gets the difference d. The Spearman's correlation between the ranks is attainable from:

$$\rho = 1 - \frac{6 \times \sum_{i=1}^{n} d_i^2}{n(n^2 - 1)} \tag{1}$$

Table 1 shows the calculated amount of Spearman's rank correlation for the different types of definition construction. For this purpose instead of considering the definitions of the concepts alone, for calculation of their equivalent gloss vectors, the extended definitions of the concepts by appendage of their parents and children's definitions to their own definition is considered.

Table 1.   Spearman's rank correlation of semantic relatedness for different concept definition construction types

Definition Type	Spearman's Rank Correlation
UMLS	0.5024
Wikipedia	0.4998
UMLS and Wikipedia	**0.5402**

We can see, based on Spearman's rank correlation results, when we use Wikipedia articles along with the concepts definitions from the UMLS the most accurate semantic relatedness gets calculated. Moreover, the result of Spearman correlation indicates when we do not consider concepts definitions from the UMLS and just take into account collected definitions from Wikipedia, the estimated semantic relatedness are very close. This shows for those concepts without any definitions from the UMLS the Wikipedia articles and other resources on the Web are dependable alternatives.

6   Conclusion

Considering wide applications of Gloss Vector semantic relatedness measure, by considering Wikipedia articles as an extra choice for developing definitions of the concepts, we tried to avoid common drawback of concept's definition absence in the applied thesaurus. In the future works, for augmenting concepts definitions other social and shared resources solely based on public knowledge can be investigated. Moreover, for collecting additional definitions from these sources, different automatic techniques can get examined. Besides, our proposed approach can be tested on extrinsic tasks such as word sense disambiguation, or get evaluated in other domains.

References

1. Muthaiyah, S., Kerschberg, L.: A Hybrid Ontology Mediation Approach for the Semantic Web. International Journal of E-Business Research. 4, 79--91 (2008)
2. Pekar, V., Ou, S., Constantin Orasan, C., Spurk, C.,Negri, M.: Development and alignment of a domain-specific ontology for question answering," In: Proceedings of the 6th Edition of the Language Resources and Evaluation Conference (LREC-08), May. (2008)
3. Chen, B., Foster, G., Kuhn, R.: Bilingual Sense Similarity for Statistical Machine Translation. In: Proceedings of the ACL, pp. 834--843 (2010)
4. Bousquet, C., Lagier, G., LilloLe, L.A., Le Beller, C., Venot, A., Jaulent, M.C.: Appraisal of the MedDRA Conceptual Structure for describing and grouping adverse drug reactions. Drug Safety, vol. 28, no. 1, pp. 19--34 (2005)
5. Firth, J.R.: A Synopsis of Linguistic Theory 1930-1955. In Studies in Linguistic Analysis, pp. 1--32 (1957)
6. Rada, R., Mili, H., Bicknell, E., Blettner, M.: Development and Application of a Metric on Semantic Nets. IEEE Transactions on Systems, Man and Cybernetics. 19, 17--30 (1989)
7. Wu, Z., Palmer, M.: Verb Semantics and Lexical Selections. In: Proceedings of the 32nd Annual Meeting of the Association for Computational Linguistics, (1994)
8. Resnik, P.: Using Information Content to Evaluate Semantic Similarity in a Taxonomy. In: Proceedings of the 14th International Joint Conference on Artificial Intelligence, pp. 448--453 (1995)
9. Jiang, J.J., Conrath, D.W.: Semantic Similarity based on Corpus Statistics and Lexical Taxonomy. In: International Conference on Research in Computational Linguistics. (1997)
10. Lin, D.: An Information-theoretic Definition of Similarity. In: 15th International Conference on Machine Learning. Madison, USA, (1998)
11. Pesaranghader, A., Muthaiyah, S.: Definition-based information content vectors for semantic similarity measurement. In: Proceedings of the 2nd International Multi-Conference on Artificial Intelligence Technology (M-CAIT), pp 268--282 (2013)
12. Lesk, M.: Automatic Sense Disambiguation Using Machine Readable Dictionaries: How to Tell a Pine Cone from an Ice-cream Cone. In: Proceedings of the 5th Annual International Conference on Systems Documentation, pp. 24--26. New York, USA (1986)
13. Banerjee, S., Pedersen, T.: An Adapted Lesk Algorithm for Word Sense Disambiguation using WordNet. In: Proceedings of the 3rd International Conference on Intelligent Text Processing and Computational Linguistics. Mexico City (2002)
14. Patwardhan, S., Pedersen, T: Using WordNet-based Context Vectors to Estimate the Semantic Relatedness of Concepts. In: Proceedings of the EACL 2006 Workshop, Making Sense of Sense: Bringing Computational Linguistics and Psycholinguistics together. pp. 1--8. Trento, Italy (2006)
15. Liu, Y., T. McInnes, B.T., Pedersen, T., Melton-Meaux, G.,Pakhomov. S.: Semantic relatedness study using second order co-occurrence vectors computed from biomedical corpora, UMLS and WordNet. In: Proceedings of the 2nd ACM SIGHIT IHI, pp. 363–371 (2012)
16. Pakhomov, S., McInnes, B., Adam, T., Liu, Y., Pedersen, T., Melton, G.: Semantic Similarity and Relatedness between Clinical Terms: An Experimental Study. In: Proceedings of AMIA, pp. 572--576 (2010)

Development of Web Services Fuzzy Quality Models using Data Clustering Approach

Mohd Hilmi Hasan, Jafreezal Jaafar, and Mohd Fadzil Hassan

Computer & Information Sciences Department, Universiti Teknologi PETRONAS, 37150 Tronoh, Perak, MALAYSIA

{mhilmi_hasan, jafreez, mfadzil_hassan}@petronas.com.my

Abstract. This paper presents the fuzzy clustering of web services' quality of service (QoS) data using Fuzzy C-Means (FCM) algorithm. It was conducted based on actual QoS data gathered from the network. The work involved three data sets that represented three different QoS parameters. Each data set contained 1500 data points. The clustering was validated using Xie-Beni index to ensure that it performed optimally. As a result, three fuzzy quality models were produced that represented the three QoS parameters. The work implies potential new findings on fuzzy-based web services' applications, mainly in reducing computational complexity. The work also benefits the less technical-knowledgeable requestors as the fuzzy quality models can guide them to find services with realistic QoS performance. For future work, the fuzzy quality models will be employed in web services' QoS monitoring application. They will also be equipped with an adaptive mechanism that supports the dynamic nature of web services.

Keywords: Clustering, fuzzy clustering, Fuzzy C-Means, QoS clustering, web services clustering.

1 Introduction

The discovery and selection processes of web services are very crucial and hence require cautious consideration from requestors [1]. That can ensure the right services are subscribed, thus granting satisfaction to the requestors as their expected functional and non-functional requirements are delivered. However, the growing number of available web services candidates has made these processes more complicated. As a result, numerous researches have been conducted on web services categorization as one of the ways to assist requestors to identify suitable services for them [2]. Categorization is a way to group web services based on their features. For example, web services can be differentiated into different groups according to their functionalities. By having this grouping, it is evident that the discovery and selection processes can be improved since the computations for discovery and selection requests are executed based on the matched groups only rather than the whole web services candidates. Hence, it reduces the computational time and complexity [3]. In web services categorization area, there are two terms that are highly being investigated namely clustering

T. Herawan et al. (eds.), *Proceedings of the First International Conference on Advanced Data and Information Engineering (DaEng-2013)*, Lecture Notes in Electrical Engineering 285, DOI: 10.1007/978-981-4585-18-7_71,
© Springer Science+Business Media Singapore 2014

and classification. Clustering is an unsupervised method in which the categorization is performed using unlabelled data and without the predefined desired output. In contrast, classification involves categorization using labeled data with proper descriptions on the expected output. Hence, classification is considered as a supervised method [4, 5]. Apparently, there are various kinds of implementation that solve their web services categorization problems using clustering [6-8] and classification [9-11] methods. Most of the works, however, focus on categorizing web services based on functional criteria. This paper argues that it is also equally important to categorize web services based on their non-functional aspects. That will result in the grouping of web services can be made according to their quality of service (QoS). Apart from having less computational time and complexity, this kind of web services categorization also helps requestors especially those with less technical experience to understand the reasonable QoS values that they should expect the providers to offer. Mobedpour and Chen [12] state in their work that the general assumption that requestors can create their own QoS queries with exact parameter values is dubious since they may not even able to distinguish between the realistic and unrealistic values. If the requestors specify QoS requirements that are lower than the realistic values, they will get the underperforming services. On the other hand, specifying QoS requirements which are higher than the realistic values may turn out to be no matching service is available for selection. The QoS-based categorization is able to solve the issue by providing information on what performance criteria are considered as good, moderate and poor, for instance. Furthermore, it also provides a significant benefit for the requestors and providers as it can become a reference during contract negotiation and specification. This issue has become the main motivation of this work.

In another perspective, this paper also proposes that the categorization is carried out based on fuzzy procedure. This kind of categorization can provide facilitation to the development of fuzzy-based web services solutions. As apparent in literature, a growing interest have been put on the development of this kind of solutions which covers areas like service selection [12, 13], service reputation [14] and overhead forecasting [15]. Fuzzy-based applications contain inference mechanism which is generated from either expert knowledge or data. The former, however, may result in loss of accuracy [16] and consume a lot of time. Moreover, in certain circumstances, the expert knowledge may not always be available [17]. Thus, inference from data has become the preferred alternative for the development of fuzzy-based applications. Apparently, there are numerous methods that have the capability to carry out the inference task. However, it is evident that all of the available methods have their advantages and disadvantages, hence none of them is able to offer the optimal solution for all kinds of data sets [18]. In this work, the use of the well-known fuzzy clustering method, namely Fuzzy C-Means (FCM) was proposed. On the whole, this paper focuses on fuzzy clustering of web services' QoS data to generate fuzzy quality models. These models can be used as the reference in requirements negotiation and subscription. They can also be used as the inference components of the fuzzy-based applications.

In a nutshell, the objective of this paper is to present the development of web services' fuzzy quality models using data clustering approach. The process was performed upon the actual web services QoS data using FCM algorithm. Section 2 describes the FCM algorithm as well as the QoS data sets used in the experiments. This section also presents the validity index that was used to measure the performance of the proposed clustering. Next, the experimental processes and their results are presented in section 3. The experiments comprise the validation tests and the clustering processes. Finally, section 4 contains conclusion that summarizes the outcomes of the work and outlines some potential future directions.

2   Material and Methods

2.1   Fuzzy C-Means

FCM is a data clustering method that categorizes a collection of data into a number of different clusters. It assigns each data point to a specific cluster with a membership degree [19]. That means, each data point can belong to two or more clusters with different degrees of membership. FCM performs the clustering of data points through an iterative process that minimizes the objective function. Assume that there are n data points represented by $\{X_1, X_2, ..., X_n\}$ and c number of clusters. During its execution, initially, FCM generates the center of each cluster, c_i, $i=1, 2, ..., c$, through guessing. Hence, in general this initial set of centers' values does not represent the optimal clustering condition. Then, FCM performs the computation to assign each data point a membership degree for each cluster [20]. The computation is performed based on the following equation, in which all of the produced membership degrees are stored in matrix U.

$$u_{ij} = \frac{1}{\sum_{k=1}^{c} \left(\frac{d_{ij}}{d_{kj}}\right)^{2/(m-1)}} \tag{1}$$

where $d_{ij} = ||c_i - x_j||$ is the value of Euclidean distance from jth data point to the ith cluster center. The index of fuzziness is represented by m [20].

Next, FCM calculates the objective function which is defined as the following equation:

$$J(U, c_1, ..., c_c) = \sum_{i=1}^{c} J_i = \sum_{i=1}^{c}\sum_{j=1}^{n} u_{ij}^m d_{ij}^2 \tag{2}$$

The computation of FCM will be terminated when the value of this objective function reaches certain threshold value, which is the minimum value [20]. If it still does not reach this threshold value, the process is continued by computing the new value for the center of each cluster. These new values will replace the previous ones which were initiated by guessing. The clusters' centers are computed based on the following:

$$c_i = \frac{\sum_{j=1}^{n} u_{ij}^{m} X_j}{\sum_{j=1}^{n} u_{ij}^{m}} \quad (3)$$

Next, the same process of computing membership degrees using Eq. (1) will take effect. The whole processes are then iteratively computed until the objective function reaches its threshold value. In each repetition, FCM will minimize the objective function as well as improve the values of the cluster centers. In this work, the FCM computations were executed using Matlab's Fuzzy Clustering and Data Analysis Toolbox.

2.2 QoS Data Sets

The experiments were conducted using the QoS data set provided by Al-Masri and Mahmoud [21, 22]. The data set comprised the data that were collected from 365 real web services using Web Service Crawler Engine (WSCE). In this work, three QoS parameters were chosen namely response time, latency and availability. For each parameter, 1500 data points were used to generate the clustering results.

2.3 Clustering Validity Indices

Numerous researches have been conducted to produce assessment indices that can evaluate the optimization of clustering [23]. These indices are known as clustering validity indices. In finding the optimal clustering solution, two criteria are considered for validation process namely compactness and separation [24, 25]. The compactness emphasizes on the closeness among cluster members. Therefore, a more compact clustering means the members of each cluster are closer to each other. It is normally measured as variance, in which the minimum value shows the better compactness. On the other hand, separation emphasizes on the distance among clusters. That means clusters that are widely spaced produce a good degree of separation. Separation can be calculated based on one of these three measurements; the distance between the closest member of clusters, the distance between the most distant member of clusters, or the distance between the clusters' centers [25]. On the whole, the validity indices are used to measure how good the clustering is by determining the optimal number of clusters that has high degrees of compactness and separation.

There are a number of validity indices that are available for use. To select the best index is a daunting task since every index has not been proven to be the best choice for all conditions [23, 26]. Furthermore, there are some of the indices that only consider the membership values in their calculation. On the other hand, some other indices consider both the membership values as well as the data set itself, which can produce more representing results [25, 27]. This paper is interested in the latter and hence, proposes the use of Xie-Beni (XB) validity index. The XB validity index is proposed due to its well performance in the conditions when the number of clusters is in the range of two to ten and fuzzy weighting component is between 1.01 to 7, as reported by Pal and Bezdek [28].

To define XB validity index, assume $X = \{x_1, x_2, ..., x_n\}$ be a set of n data points and $V = [v_1, v_2, ..., v_c]$ be a matrix of cluster centers. Besides, c is the number of clusters and μ_{ij} is the degree of membership of x_j that belongs to v_i [29]. Let m be the fuzzy weighting exponent, the index is defined as follow [23, 30]:

$$V_{XB} = \frac{\sum_{i=1}^{c} \sum_{j=1}^{n} \mu_{ij}^{m} \| x_j - v_i \|^2}{n \, \min_{i,j} \| v_i - v_j \|^2} \qquad (4)$$

As stated earlier, XB validity index considers both of the validation criteria in its computation, which are compactness and separation. Eq. (4) is formed by the clustering's degree of compactness as the numerator, and separation degree as the denominator [23]. The equation produces the optimal number of clusters by solving $min_2 \leq c \leq n-1$. Hence, the optimal clustering scheme for the set of data points is represented by the minimum value of the index. In this work, the clustering validity computations were executed using Fuzzy Clustering and Data Analysis Toolbox on Matlab.

3   Results

3.1   Clustering Validation

Prior to the running of clustering validity process based on Eq. (4), two values had to be identified, namely the maximum number of clusters, c and fuzzy weighting exponent, m. The number of clusters, c used in the validity process were in the range of $c=2$ to c_{max}, hence $c_i \in \{2,3,...,c_{max}\}$. In general, the value of c_{max} can be determined either based on pre-knowledge [30] or using certain formula such as proposed by Wu and Yang [31], $c_{max} \approx \sqrt{n}$, where n is the number of data points. Thus, for 1500 data points, 38 clusters have to be considered for each QoS parameter. This number is considered as too huge. Rationally, the number of clusters should not be too huge as it will be inconvenienced for requestors to utilize the quality models with too many categories. Thus, this work applied the pre-knowledge method, and set $c_{max} = 5$.

As for m value, it is evident that the mean value of the range of 1.5-2.5, i.e. $m=2$ is normally chosen in the objective function-based method such as FCM [28]. Therefore, in this work, the authors also used $m=2$ for fuzzy weighting component value.

The results of the experiment using XB validity index upon 1500 data points, where $c = 2, 3, 4$ and 5 and $m=2$ are shown in Table 1. The data sets contained values measured in milliseconds (ms) for response time and latency, and percentage (%) for availability. Based on the results shown in Table 1, it can be concluded that the optimal number of clusters for response time, availability and latency are 4, 3 and 3 respectively.

Table 1. XB validity index scores

QoS Parameter	Number of Clusters			
	2	3	4	5
Response time	2476.1000	263.2772	148.4363	2362.9000
Availability	18.8142	15.8146	59.2782	16.0355
Latency	6535.2000	537.7606	958.0795	26506.0000

3.2   Clustering with FCM

The results presented in Table 1 were used to produce the fuzzy-represented clusters. For response time, the four clusters were known as Good, Moderate High, Moderate Low and Poor. On the other hand, availability and latency comprised 3 clusters; Good, Moderate and Poor. The clustering processes were conducted using FCM that successfully generated the cluster centers of each QoS parameter as shown in Table 2.

Then, the data set of each QoS parameter was used in the process of producing the clustering models. The models, which were named as fuzzy quality models, are shown in Fig. 1, Fig. 2 and Fig. 3. In the models, each curve represents a cluster. The peak of each curve represents the cluster center, which has the degree of membership of 1. Moreover, it can be seen that the QoS data point at each peak is correspondingly similar to the cluster centers shown in Table 2.

Table 2. Cluster Centers

Cluster	Cluster center	Cluster	Cluster center	
	Response Time (ms)		Availability (%)	Latency (ms)
Good	174.456	Good	90.695	27.373
Moderate High	491.460	Moderate	65.435	490.459
Moderate Low	1438.556	Poor	28.126	2156.339
Poor	3516.607			

Based on the fuzzy quality model shown in Fig. 1, it can be inferred that the good web services take below than 350 ms of response time. In contrast, response time's values of more than 2100 ms are categorized as poor. Besides, the fuzzy quality model presented in Fig. 2 shows that the good web services have more than 78% availability, while the poor ones have less than 48% availability. Finally, the fuzzy quality model shown in Fig. 3 reveals that the good web services take less than 350 ms of latency. It also shows that the poor web services are those with the latency of more than 960 ms. However, another important point that has to be taken into consideration is that the generated quality models are based on the fuzzy concept, which means each data point was assigned to more than one cluster with different degrees of membership. Referring to the model in Fig. 3, a web service with 300 ms latency can also be considered as moderate to a certain degree of membership. As a result, the particular web service may be viewed differently in different fuzzy-based applications based on the imposed rules upon its clusters' membership. For this reason, data clustering as proposed by this paper is important as it becomes the fundamental element of fuzzy-based applications.

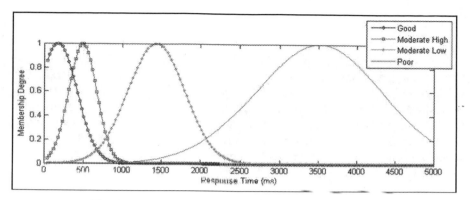

Fig. 1. Fuzzy quality model of web services' response time

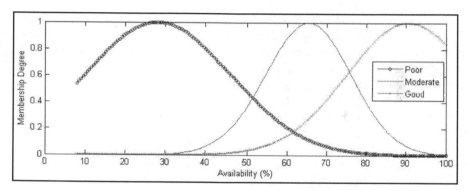

Fig. 2. Fuzzy quality model of web services' availability

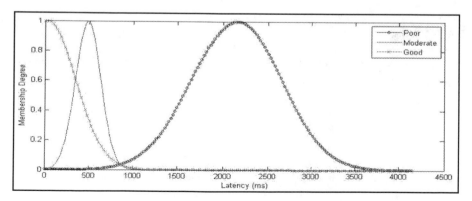

Fig. 3. Fuzzy quality model of web services' latency

Overall, the significance of the generated fuzzy quality models can be viewed in two perspectives. Firstly, the earlier part of the previous paragraph describes how they can be used as the reference for the requestors to choose suitable web services. Secondly, the fuzzy quality models can be utilized in the development of fuzzy-based web services' applications, as illustrated in the later part of the previous paragraph.

4   Conclusion

In addition to its ability to produce efficient web services' delivery, the proposed fuzzy clustering of web services' QoS could overcome two main issues. Firstly, it helps the requestors who have limited technical knowledge about the web services to understand what the realistic QoS is so that they can subscribe services that give value for their money. The outcomes of the clustering, which are the fuzzy quality models can also be used as the references for the requestors in the process of requirements negotiation and specification. Secondly, the produced fuzzy quality models also contribute towards the development of fuzzy-based web services. As fuzzy inference using knowledge from the experts requires more time, and may suffer from less accuracy and unavailability, the automatic inference like the proposed model can be considered as the better option.

In this work, the clustering of web services' QoS was performed using FCM algorithm. It was carried out using the data gathered from the real web services' implementations which comprised three QoS parameters namely response time, availability and latency. Furthermore, this work used the XB validity index to measure the clustering's optimization in terms of compactness and separation.

This paper outlines two directions for future works. Firstly, the implementation of the proposed fuzzy clustering of QoS in web services' QoS monitoring application. Secondly, the enhancement to the proposed fuzzy clustering of web services' QoS by employing an adaptive mechanism. The adaptive behavior can give significant advantages to web services due to its dynamic nature, particularly for collecting and using the updated QoS data from the Internet, and dealing with the change of requestors' fuzzy requirements during runtime.

References

1. Kumar, S., Mishra, R.: Semantic Web Service Composition. IETE Technical Review 25, 105-121 (2008)
2. Saha, S., Murthy, C.A., Pal, S.K.: Classification of Web Services Using Tensor Space Model and Rough Ensemble Classifier. Foundations of Intelligent Systems Lecture Notes in Computer Science 4994, 508-513 (2008)
3. Venketesh, P., Venkatesan, R.: A Survey on Applications of Neural Networks and Evolutionary Techniques in Web Caching. IETE Technical Review 26, 171-180 (2009)
4. Liu, H., Yu, L.: Toward Integrating Feature Selection Algorithms for Classification and Clustering. IEEE Transactions on Knowledge and Data Engineering 17, 491-502 (2005)
5. Rose, K.: Deterministic Annealing for Clustering, Compression, Classification, Regression, and Related Optimization Problems. In: Proceedings of the IEEE, pp. 2210-2239. (Year)
6. Chifu, V.R., Pop, C.B., Salomie, I., Dinsoreanu, M., Acretoaie, V., David, T.: An Ant-inspired Approach for Semantic Web Service Clustering. In: 9th RoEduNet IEEE International Conference, pp. 145-150. (Year)

7. Liang, Q., Li, P., Hung, P.C.K., Wu, X.: Clustering Web Services for Automatic Categorization. In: IEEE International Conference on Services Computing, pp. 380-387. (Year)

8. Liu, J.-x., He, K.-q., Wang, J., Ning, D.: A Clustering Method for Web Service Discovery. In: IEEE International Conference on Services Computing, pp. 729-730. (Year)

9. Crasso, M., Zunino, A., Campo, M.: AWSC: An approach to Web service classification based on machine learning techniques. Inteligencia Artificial, Revista Iberoamericana de Inteligencia Artificial 37, (2008)

10. Ladner, R., Warner, E., Petry, F., Gupta, K.M., Moore, P., Aha, D.W., Shaw, K.: Case-Based Classification Alternatives to Ontologies for Automated Web Service Discovery and Integration. In: Proceedings of SPIE Defense & Security Symposium, pp. 620117-620111 – 620117-620118. (Year)

11. Wang, H., Shi, Y., Zhouy, X., Zhou, Q.: Web Service Classification using Support Vector Machine. In: 22nd International Conference on Tools with Artificial Intelligence, pp. 3-6. (Year)

12. Mobedpour, D., Chen, D.: User-centered design of a QoS-based web service selection system. Serv Oriented Comput Appl 1–11 (2011)

13. Bacciu, D., Buscemi, M.G., Mkrtchyan, L.: Adaptive fuzzy-valued service selection. In: 2010 ACM symposium on applied computing, pp. 2467–2471. (Year)

14. Sherchan, W., Loke, S.W., Krishnaswamy, S.: A fuzzy model for reasoning about reputation in web services. Proceedings of the 2006 ACM symposium on Applied computing, pp. 1886-1892. ACM, Dijon, France (2006)

15. Zadeh, M.H., Seyyedi, M.A.: QoS Monitoring for Web Services by Time Series Forecasting. In: 3rd IEEE International Conference on Computer Science and Information Technology (ICCSIT), pp. 659-663. (Year)

16. Guillaume, S.: Designing Fuzzy Inference Systems from Data: An Interpretability-Oriented Review. IEEE Transactions on Fuzzy Systems 9, 426-443 (2001)

17. Jang, J.-S.R.: Self-Learning Fuzzy Controllers Based on Temporal Back Propagation. IEEE Transactions on Neural Networks 3, 714-723 (1992)

18. Vega-Pons, S., Ruiz-Shulcloper, J.: A survey of clustering ensemble algorithms. International Journal of Pattern Recognition and Artificial Intelligence 25, 337-372 (2011)

19. Wang, L., Wang, J.: Feature Weighting fuzzy clustering integrating rough sets and shadowed sets. International Journal of Pattern Recognition and Artificial Intelligence 26, (2012)

20. Guldemır, H., Sengur, A.: Comparison of clustering algorithms for analog modulation classification. Expert Systems with Applications 30 30, 642-649 (2006)

21. Al-Masri, E., Mahmoud, Q.H.: Discovering the best web service. In: 16th International Conference on World Wide Web (WWW), pp. 1257-1258. (Year)

22. Al-Masri, E., Mahmoud, Q.H.: QoS-based Discovery and Ranking of Web Services. In: IEEE 16th International Conference on Computer Communications and Networks (ICCCN), pp. 529-534. (Year)

23. Wang, W., Zhang, Y.: On fuzzy cluster validity indices. Fuzzy Sets and Systems 158 158, 2095 – 2117 (2007)

24. Berry, M.J.A., Linoff, G.: Data Mining Techniques For Marketing, Sales and Customer Support. John Wiley & Sons, Inc., USA (1996)

25. Halkidi, M., Batistakis, Y., Vazirgiannis, M.: On Clustering Validation Techniques. Journal of Intelligent Information Systems 17, 107–145 (2001)

26. Pal, N.R., Bezdek, J.C.: Correction to "On Cluster Validity for the Fuzzy c-Means Model". IEEE Transactions on Fuzzy Systems 5, 152-153 (1997)

27. Rezaee, M.R., Lelieveldt, B.P.F., Reiber, J.H.C.: A new cluster validity index for the fuzzy c-mean. Pattern Recognition Letters 19, 237–246 (1998)

28. Pal, N.R., Bezdek, J.C.: On cluster validity for the fuzzy c-means model. IEEE Transactions on Fuzzy Systems 3, 370-379 (1995)

29. Tang, Y., Sun, F., Sun, Z.: Improved Validation Index for Fuzzy Clustering. In: American Control Conference, pp. 1120-1125. (Year)

30. Xie, X., Beni, G.: Validity measure for fuzzy clustering. IEEE Transactions on Pattern Analysis and Machine Intelligence 3, 841–846 (1991)

31. Wu, K.-L., Yang, M.-S.: A cluster validity index for fuzzy clustering. Pattern Recognition Letters 26, 1275–1291 (2005)

Thai Related Foreign Language Specific Web Crawling Approach

Tanaphol Suebchua, Bundit Manaskasemsak, and Arnon Rungsawang

Massive Information & Knowledge Engineering
Department of Computer Engineering, Faculty of Engineering
Kasetsart University, Bangkok 10900, Thailand.
`job,un,arnon@mikelab.net`

Abstract. National web archives have been successfully made available through domain and language-specific web crawlers for years. We here propose another focused web crawler for collecting foreign language web pages that are also related to a nation. Rather finding the most relevant web pages, an ensemble machine learning has been trained with selective features to find relevant clusters of unvisited web pages, called website segments. During consecutive crawling cycles, the machine will be retrained with features extracted from new found website segments. Preliminary experiments in the real web space on Thai-tourism related topics show that this approach can take advantage of recent crawling experiences to produce more promising harvest rates than traditional breadth- and best-first baselines.

Keywords: web archive, topical crawler, focused crawler, website segment, ensemble machine learning

1 Introduction

National web archives preserve national knowledge and cultural information for generations to come. To build one, we have to gather web pages which are related to a specific nation as many as possible [1–5]. Those target web pages can roughly be categorized into 3 groups, i.e., (1) web pages which belong to a national domain name, (2) web pages which are written in a national language, and (3) web pages whose contents are written in other foreign languages, but related to the nation. For the first two groups, researchers successfully utilized a domain-specific web crawler [3, 5–7], and a specific type of focused crawlers, called the language-specific web crawler [8–11], to gather the web pages. However, the third group of foreign language web pages are uncovered by those two formers. The missing web pages may contain informative data, such as thought, aspect to a country, or useful information for foreigners. For example, English web pages which contain information about Thai tourism attractions would be beneficial for all foreign travelers who interest to visit Thailand.

In this paper, we rather consider to localize a website segment, i.e., the subset of web pages, than an individual one. We hypothesize that each already downloaded (or source) segment can give helpful clues to predict the relevancy of

T. Herawan et al. (eds.), *Proceedings of the First International Conference on Advanced Data and Information Engineering (DaEng-2013)*, Lecture Notes in Electrical Engineering 285, DOI: 10.1007/978-981-4585-18-7_72,

unvisited (or destination) ones. The set of selective features are extracted from
the source segments to train an ensemble classifier to prioritize the destination
segments in the current crawling frontier. During consecutive crawling cycles,
the machine updates its knowledge with the new found segments. Preliminary
results on the set of Thai-tourism related topic web pages written in English
show that this approach provides better harvest rate than the traditional ones.

2   Related Work

At early day, web engineers simply used domain specific web crawlers to download web pages within the corresponding country code top-level domain to build
their national web archives [1–5]. However, those simple crawlers miss many targets ending with .com, .net, .org, etc. Researchers then used another type of
web crawler, called the focused crawler, to collect national related web pages.
For example, Somboonviwat et al. [9] proposed a focused crawler, called the
language-specific crawler. They first detected the language of web pages, and
then developed a set of heuristic rules concluded from the observation of link
characteristics in a small sample set of Thai web graph to direct their crawler
to the unvisited targets. Srisukha et al. [10] proposed to use the machine learning to build the language-specific crawler, while Tadapak et al. [12] proposed a
framework to predict the relevant website rather than an individual web page.

3   Relevant web page identification

In this paper, we define a target web page as a web page whose textual content
is related to a Thai-tourism topic, and is written in English. To build a training
dataset, we manually select some seeds from the ODP [13] of following categories:
(1) Thai tourism, (2) foreign country tourism, and (3) non-tourism. We then
launch a breadth-first crawler to collect at most 300 pages per website, and
use the LangDetect library [14] to select only English web pages. We extract
word-based feature vectors; ones from the first category are marked as positive
examples, the remainders are marked as the negative ones. To train a Naïve Bayes
classifier for relevant web page identification, we run 10-fold cross validation 10
times, and finally choose the best classifier with 95.78% geometric mean value.

4   Thai-related foreign language specific web crawler

Our crawling architecture, depicted in figure 1, composes of three main components: a Segment Identifier, a Segment Predictor, and a Segment Crawler.

4.1   Segment Identifier

The observation concluded from the training data set reveals that many target
websites host only a small cluster of relevant web pages, and their URLs mostly

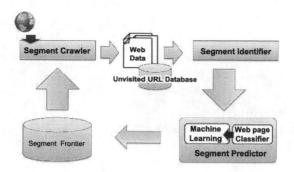

Fig. 1. The architecture of the Thai-related foreign language web crawler.

share the common prefixes. We then design our Segment Identifier to group web pages which share the same longest logical directory path, i.e., the website segments. To give an example of website segment notion, figure 2 illustrates a sample set of URLs from http://www.kpnews.org in which we can later group them into four segments.

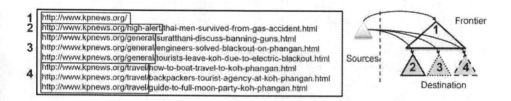

Fig. 2. Sample website segments in http://www.kpnews.org

4.2 Segment Predictor

Segment Predictor is a machine learning based classifier which will be trained to predict the relevancy of the destination website segments. At present, we observe the website segment's characteristics, and extract six features from the source segments in which there is at least one link pointing to the destination segment under consideration as follows.

Geolocation feature: We hypothesize that a destination segment should be relevant if it has been referenced by the source segments whose geolocations are often found in a certain set of countries. We then construct a geolocation feature vector by (1) counting the number of relevant and irrelevant source segments whose geolocations locate in the top-5 and the other countries of which the

source segments are often found to link to the relevant destination segment,
(2) the geolocation of the destination segment.

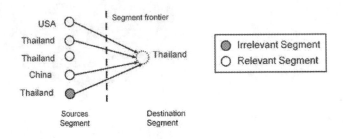

Fig. 3. Example of a geolocation feature extraction.

Suppose that, the top-5 geolocations are USA, Thailand, Japan, France, Russia, respectively, a geolocation vector feature extracted from a sample graph in figure 3 could be written as 1|2|0|0|0|1|0|1|0|0|0|0|TH. This vector means that the destination segment is cited by 1 relevant segment from USA, 2 relevant and 1 irrelevant segments from Thailand, 1 relevant segment from the other country, and the destination segment is located in Thailand, respectively.

Domain name feature: Following the same idea of the geolocation feature, we construct the domain name feature by (1) counting the number of relevant and irrelevant source segments whose the country code top-level domain names locate in the top-5 and the other domains of which the source segments are often found to link to the relevant destination segment, (2) the domain name of the destination segment.

Relevance degree feature: Using the classifier explained in section 3, we can average the percentage of relevant degree of web pages residing within a segment. We here hypothesize that a target segment should be recommended by many high relevant degree source segments. We then construct this feature vector by using the top-5 relevant degree values of the source segments. For example, if the relevant degree of the source website segments in figure 3 are 86%, 78%, 85%, 73.2% and 67% respectively, therefore, the extracted feature vector can be written as 86|85|78|73.2|67.

In-degree feature: We also hypothesize that a target segment would be linked by many relevant source segments. We then construct the in-degree feature by counting the number of relevant and irrelevant source segments.

Anchor text feature: Anchors and their surrounding text in a web page can provide contextual clue about the destination segment. In this work, we propose two methods to construct this feature. The first one is to use word occurrence extracted from the anchor texts and their 100-characters surroundings. The second method rather extracts N-gram feature from the anchor texts and their surroundings.

6) URL feature: Following the same idea of the anchor text feature, we also construct the URL feature vector by counting either word or N-gram occurrences extracted from the destination segment URL, excluding special characters, "http://" and "www".

4.3 Segment Crawler

Segment Crawler is responsible for downloading web pages within an assigned segment. Website segment with higher relevant score will be first dequeued from the segment frontier database. Then, web pages within that segment will be downloaded. From a downloaded web page, new URLs are extracted and stored in the Unvisited URL Database (cf. figure 1). To avoid downloading too many irrelevant pages from a low relevant segment, we define a discard segment threshold, S. When the Segment Crawler consecutively downloads S irrelevant web pages, it will stop downloading further web pages from that segment.

In order to help the crawler adapt to the new found web segments, we also select all false positive website segment samples and sampling some true positive samples equally at the end of each crawling iteration. Those selected samples will be used to retrain the classifier model of the Segment Predictor later.

5 Experiments

We choose the Thai-tourism related topics as the relevant target web pages. We will first explain how we prepare the training dataset, and then show the crawling result concluded from the Internet setting.

5.1 Training dataset

From around seventy unseen Thai-tourism seed URLs manually selected from the ODP [13], we first use the Google to find their backlinks. We then group those backlinks into website segments, and use the Segment Crawler to download them. From the new list of downloaded URLs, we regroup them into segments and use the Segment Crawler to download web pages in those segments within 2-hops range. Finally, all download web pages are regrouped into segments, and the feature vectors are extracted.

Setting the relevant degree threshold to 50%, i.e., a relevant website segment must compose of at least 50% relevant web pages, we obtain 1,264 relevant and 2,950 irrelevant website segments for the training dataset. We then use the under-sampling method [15] to build the 10-fold cross validation training sets, and explore all simple combiner functions [16] with the ensemble machines. We finally obtain the best predictor model (i.e., 94.8% geometric mean value) using

the Average combiner with the following setting; the link-based features which
have been trained with the Naïve Bayes classifier, the word-based anchor text
and URL features which have been trained with the Naïve Bayes Multinomial
classifiers. Figure 4 depicts the internal architecture of our Segment Predictor.

Fig. 4. The internal architecture of a Segment Predictor.

5.2   Internet Evaluation

To evaluate our crawling approach, we compare the following crawlers in the real
web space.

- Breadth-first crawler.
- Best-first crawler [17] which follows the destination URL whose parent web
  page is the most related to a Thai-tourism topic first.
- Our Thai-related Foreign Language specific web Crawler (TFLC-L) which
  employs only link-based classifier and retrains with the new found segments
  during each crawling cycle.
- Thai-related Foreign Language specific web Crawler with classifier ensemble-
  based Segment Predictor (TFLC-ENS) which retrains with the new found
  segments during each crawling cycle.
- Simple Thai-related Foreign Language specific web Crawler (TFLC-S) which
  also employs an ensemble-based Segment Predictor but it will not retrain
  with the new found segments.

We first manually select eight relevant Thai-tourism URLs from the Google,
and launch the crawlers from those seed set. We set the discard segment threshold, mentioned in section 4.3, $S = 2$. All crawlers have been restricted to download at most 300 web pages per website.

Figure 5 shows the harvest rate graph concluded from each crawler within
one hundred thousand downloaded pages. It can be seen that our proposed approaches provide much better performance than breadth- and best-first crawlers.
In other words, our approaches better focus themselves on the relevant web pages
region than the baselines.

It can also be seen that there are many ripples in the harvest rate graph
produced by our crawling approach. This is because our crawlers always visit
the website segments which has the high probability to host relevant web pages

first. Thus, at the beginning of the crawling cycle, many relevant web pages found from those website segments cause the harvest rate to increase. After crawling for a while, crawler will find more irrelevant web pages from the lower probability website segments. This will cause the harvest rate to decline and cause a ripple down in the graph.

When comparing our proposed methods, the crawling performance of TFLC-ENS is slightly better than TFLC-L and TFLC-S. This is because our classifier ensemble-based Segment Predictor can predict the relevant website segments more accurately than using only a link-based classifier. Furthermore, it can also be seen that the harvest rate of TFLC-ENS improves gradually after several crawling iterations too. This shows that the TFLC-ENS can better learn from the crawling experiences than the others. Therefore, the TFLC-ENS would be preferable to use for the large-scale crawling. For the small-scale crawling, the TFLC-S may be much preferable since it consumes less resource than TFLC-L and TFLC-ENS during their classifier update in each crawling cycle.

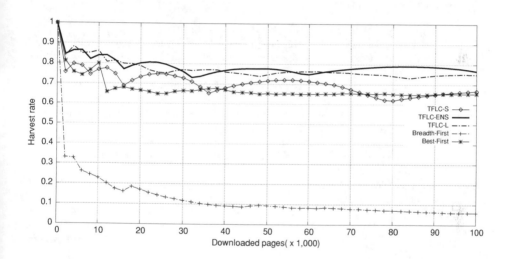

Fig. 5. Harvest rate result.

6 Conclusions

In this paper, we propose another crawling approach for collecting Thai-related web pages which are written in English. We extract several features from downloaded sources website segment to train an classifier ensemble-based Segment Predictor to predict whether the destination website segment could host relevant web pages. Web pages within the website segment with the highest probability values will be later downloaded by the Segment Crawler. Furthermore, in order to help the crawler adapt to the new environment, new found website segments will be used to retrain the Segment Predictor at the end of each crawling cycle. According to the experimental result on a Thai-tourism web dataset extracted from the real web space, this proposed crawling strategy provides better harvest rate than the breadth-first and best-first baselines. For the future work, We plan to observe the performance on other Thai-related topics, e.g., Thai food, Thai education, etc. We anticipate to exploring the effect of our Segment Crawler parameters to the harvest rate of the crawler too. For the Segment Predictor, we also plan to find more pertinent features and test them with more advance combiner functions in order to archive better harvest rate.

Acknowledgment

The first author thanks the JSTP-NSTDA Thailand for the funding support.

References

1. British Library: UK web archive. http://www.webarchive.org.uk (2011)
2. National Diet Library: Web archiving project. http://warp.ndl.go.jp (2011)
3. Baeza-Yates, R., Castillo, C., López, V.: Characteristics of the web of spain. Cybermetrics **9**(1) (2005)
4. Christensen, N.H.: Preserving the bits of the danish internet. In: Proc. of the 5th IWAW. (2005)
5. Gomes, D., Nogueira, A., Miranda, J., Costa, M.: Introducing the portuguese web archive initiative. In: Proc. of the 8th IWAW. (2008)
6. Baeza-Yates, R., Castillo, C., Marin, M., Rodriguez, A.: Crawling a country: Better strategies than breadth-first for web page ordering. In: Proc. of the 14th WWW. (2005)
7. Bordino, I., Boldi, P., Donato, D., Santini, M., Vigna, S.: Temporal evolution of the uk web. In: Proc. of the 8th ICDMW. (2008)
8. Alabbad, S.H., Alanazi, S.: Language based crawling: Crawling the arabic content of the web. In: Proc. of the ICOMP'09. (2009)
9. Somboonviwat, K., Tamura, T., Kitsuregawa, M.: Finding thai web pages in foreign web spaces. In: Proc. of the 22nd ICDEW. (2006)
10. Srisukha, E., Jinarat, S., Haruechaiyasak, C., Rungsawang, A.: Naive bayes based language-specific web crawling. In: Proc. of 5th ECTI-CON. (2008)
11. Tamura, T., Somboonviwat, K., Kitsuregawa, M.: A method for language-specific web crawling and its evaluation. Systems and Computers in Japan **38** (2007)
12. Tadapak, P., Suebchua, T., Rungsawang, A.: A machine learning based language specific web site crawler. In: Proc. of the 13th NBiS. (2010)
13. DMOZ: Open directory project (ODP). http://www.dmoz.org (2011)
14. Nakatani, S.: Language detection library for java. http://code.google.com/p/language-detection/ (2010)
15. Garcia, S., Herrera, F.: Evolutionary undersampling for classification with imbalanced datasets: proposals and taxonomy. Evolutionary Computation **17-3** (2009)
16. Ranawana, R., Palade, V.: Multi-classifier systems: Review and a roadmap for developers. International Journal of Hybrid Intelligent Systems **3** (2006)
17. Menczer, F., Pant, G., Srinivasan, P.: Topical web crawlers: Evaluating adaptive algorithms. ACM Transactions on Internet Technology **4**(4) (2004)

Web Discussion Summarization: Study Review

Ibrahim Almahy[1], Naomie Salim[1]

[1] Faculty of Computing, Universiti Teknologi Malaysia
81310, Johor, Malaysia

Email: abu_alabbas@hotmail.com

Abstract. The ease of publishing content on the internet has produced new applications that did not exist before. Discussion websites are one of these and have become a common tool between internet users. Discussions on the internet have conversation structure, in which there is asynchronous communication between participants on multiple interleaved topics, thus resulting in difficulty for readers to come up with the big picture regarding content. Producing a summary of these discussions is a real challenge. This paper presents the motivations for discussion Summarizations and why it is important. It also discusses the different methods used on thread Summarization. Some researchers consider the thread as single documents while others implement the multi-documents technique, by taking advantage of available thread features that can be exploited in producing the summary. We also point out some of the peculiarities of the task of thread Summarization which have brought about many challenges, and suggest a number of solutions.

Keywords: Thread Summarization, Web discussions, Forums, social media

1 Introduction

The ease of publishing content on the internet has produced new applications that did not previously exist. Social media applications and websites are one example of such. Nowadays, Social media has become an important and common tool used by internet users. According to Forrester Research, 75% of internet surfers used "Social Media" in the second quarter of 2008 by joining social networks, reading blogs, or contributing reviews to shopping sites; this represents a significant increase from 56% in 2007 [1]. Social media interaction methods vary in terms of the form of data used. Some websites contain only videos, some have images and text, and others contain all of these forms.

Discussion sites such as forums and blogs are an important part of social sites. It has become popular for collaboration among people as they discuss various topics online. These discussions either exist in specific sites that develop for the purpose of

T. Herawan et al. (eds.), *Proceedings of the First International Conference on Advanced Data and Information Engineering (DaEng-2013)*, Lecture Notes in Electrical Engineering 285, DOI: 10.1007/978-981-4585-18-7_73,
© Springer Science+Business Media Singapore 2014

discussion, such as forums and blogs, or can take part in many fields; overall, it is very helpful. For instance, in tourism, users can present their opinions and experiences about certain locations; hence, others can get the benefits [2]. Discussions website exist in education [3] and clearly exist in customers reviews.

Discussions in the web have conversation structure, in which there is asynchronous communication between participants on multiple interleaved topics [4], thus resulting in the difficulty for readers coming up with the big picture of the content. Consequently, there is a great demand for discussion Summarization in order to provide users with the important parts of the discussion.

Text Summarization has been widely studied for many years in many domains such as news articles, scientific papers and journals, medical files, emails and web discussion. The latter does not receive the same amount of attention, but a few people have looked at Summarization of online discussions [5].

This paper presents the motivations for discussion summarizations and why it is important, and presents the methods that people use to produce a summary from a discussion; this also addresses challenges that are faced by this domain.

1.1   Importance of discussion summarization

As presented earlier, discussions are now involved in many applications related to social media. For example, customers discuss products, but it is not always easy for a customer to read all of the reviews in order to decide on whether to buy a product or not. If the customer reads only a few reviews, then their opinion might be biased. Manufacturers also find it difficult to analyse all of the reviews and interpret the customer's opinions [6]. More importantly, discussion sites are important sources for search engines to retrieve information that is relevant to a user's queries; users find it difficult to catch up with all of the available information. This is even true for those individuals who participate in threaded discussions to deliver knowledge and ask questions. Unfortunately, they have to face an increasing amount of redundant threads when browsing web forums. Generally, people often read only a few posts ranked ahead in a thread with large post quantities, whereas this scanning style only reflects incomplete views of the conversation. Hence with an increase of the thread volume, the requirement for summarising each thread to assist comprehension is becoming more and more urgent [7].

1.2   Structure of Forums

Firstly, there are three main definitions in this area[8]:
Forum: Discussion board with a specific theme for discussion.
Thread: Series of posts which are mutually connected by reply-to relations.
Post: Message written by a participant.

A forum has conversation structure, in which there is asynchronous communication between participants on multiple interleaved topics. This structure differs from one forum to another. In its simplest form, a forum consists of an initial post written by the user who opened the topic, and comments provided by other users. The initial post with comments is called a thread. This simple form existed in the oldest discussion site, newsNet. Later on, a significant development has taken place on the form of the thread. Instead of only the initial post and replies, some have the feature of "reply of reply" which enables user to reply to other replies (rather than to the initial post). "Quotation" is another feature; it enables users to quote text from others. A few forums also have a new feature which is ranking or scoring the comments. With this, the user can give their opinion regarding the comment by choosing either a positive or negative rank.

Despite the changes that have occurred in the thread form, numerous studies have dealt with the simple form, initial post and replies, and have ignored other modern features (put some references). The idea is to make the suggested approach for solution suitable to work with other conversation forms such as blogs and emails.

From the prospective of the forums view, posts are typically organised into either sequential or tree-structured threads. Although the former has a simpler structure, the latter allows the division of many topics into smaller branches. For this reason, the tree-structured thread has been adopted in many large discussion forums (e.g. Slashdot) as well as in internal discussions in enterprises [8].

2 Thread Summarization Methods

The internet in general has two types of forums: technical and non-technical. The nature of a discussion differs in these forms. In technical forums such as the APPLE forum, the conversation often inquires about a solution to a specific problem facing the user and others answer by adding their experience in that field. Posts here, usually, are short like the initial post. Less work in previous studies has been focused on this type; most works have been performed on the second type. Non-technical forums have long discussions in various fields like politics, markets, news …etc. Users express their ideas and their opinions through these posts. Most research has addressed the second type because of the shortness of the post lengths.

From the literature, all works that have been done in this area are extraction-based summary. Generally, In order to generate a summary, different natural language processing techniques must be used. Most Summarization systems produce a summary based on the selection of key sentences. There are three different approaches for scoring and selecting sentences: Statistical approaches, Linguistic approaches and Rhetorical approaches. Recent studies implement the mentioned Summarization approaches in addition to their attempt to exploit the characteristics of conversation structure to obtain a summary.

From the view of number of documents, Summarization systems can work on single documents or multi-documents, each of which has a particular method. One of the important differences between the works in forum Summarization is in answering the question: how to deal with the thread, deal with it as a single document or a multi-document? A considerable amount of research [8],[9],[10] have taken the thread as a single document by combining the initial post with all of the comments and then implementing one of the single document approaches. On the other hand, some works

[11],[12] have tried to implement multi-document approaches by taking the initial post and replies separately from the other. The authors of [7] claim that, intuitively, a smooth transition from multi-document Summarization to web forums seems to be reasonable. However, as complicated and dynamic topics exist in each on-going thread, traditional multi-document Summarization methods fail to capture the following three characteristics that are crucial to summarising the content of one thread, especially in those with large quantities of posts:

Topic Dependencies: Thread is a type of asynchronous conversation based on temporal topic dependencies among posts. When one participant A replies to another post's author B, we consider a topic dependency has been built from user B to user A. Yet there are other factors in reality, it is believed that the reply relations among posts dominant the topic dependencies [13].

Topic Drifting: A conversation always results in various sub-topics. As the post conversation progresses, the semantic divergence among these subtopics will be widened.

Text Sparseness: Most posts are composed of short and elliptical messages. As short texts do not provide sufficient term co-occurrence information, traditional text representation methods, such as "TF-IDF", have several limitations when directly applied to the mining tasks[14] .

For the best of our knowledge, [4] is the first work tried to have produced a summary from a discussion. They investigated how to adapt extraction techniques to solve the problem of producing summaries from discussions. The challenges that they addressed include the shortness of the posts, meaning that there is no way to generate a summary for each post. Second is the reduced coherence among sentences due to the diversity of the users who write the posts. Their work focuses on how to exploit the explicit discourse structure provided by the threads (name of author and linking between posts). They focused on how to use the structural discourse relationships between postings in producing summary, and developed an algorithm called hierarchical discussion Summarization. After they had combined all posts with the initial one, sentence extraction and Summarization was recursively performed at multiple levels of a discussion hierarchy.

Their algorithm relies on determining the salient sentence in each post. Selecting the important sentence is done by the simple method that computes the frequency of terms in the sentence and the sentence position. Creating a new level above the posts involves combining the most salient sentences from all posts. Recursively, another level is generated from the previous level in the same way. This is repeated until the objective number of summary sentences is reached. Their main approach is to summarise each message individually and include every message in the summary. However, this approach ignores many important features of the thread.

[15] discusses the issue of the subtopics that are generated from the main topic. Their implementation on a technical forum shows that it is necessary to recognise subtopics in order to produce summaries. [10] extends the work which is concerned with the ordering of the comments. The comments are related to each other in some way; they found that interactions among participants can be modelled as a pair of actions: 1) Seek help or advice and 2) Give advice or an answer. Instead of making a

summary for each comment, summaries should be made and each reply should be connected to the comment.

[16] was the first study to suggest producing an indicative summary of the discussion. In fact, summaries are distinguished by their content to indicative and informative summaries. The indicative summary aims to suggest the content of the original document, and to describe the text in other words. The main approach is to select important messages and ignore the rest. This technique is called catalyst, and proposes choosing which message should be in the summary instead of creating a summary for all posts in the thread, as in [4].

The main idea is to give a score to each message in the thread and then choose the most important ones. To this end, the author proposed many factors that could be used, such as length factor, which computes the length of the message and compares it to the average length. Another factor is uniqueness, which checks the most quoted messages among all posts. Moreover, the term frequency is added to other factors to give a score to each message. Their system seems to work well, but it ignores an essential issue which is called topic drifting [7].

[17] proposed a method to extract the representative sentences from all comments. Choosing representative sentences is done by three factors, which include the username that appears in the comment, quotations made by users and the topic similarity among comments. Based on these three factors, they believe that a word is representative if it is written by authoritative readers, appears in widely quoted comments, and represents hotly discussed topics.

[8] discusses the problem of topic drifting [7], which means that the conversation thread always drifts from the main topic to subtopics. They stated that "in larger threads, it is even common that multiple topics are discussed alternately in the same sequential branch. This multi-topicality of texts is a challenge for both participants and systems to comprehend the whole content of the discussion." Their approach has two factions, the first is recognising the topics being discussed, which can be called topic extraction, and the second is sentence extraction. They used the lexical chain approach in order to perform topic extraction.

Lexical chain was first introduced by [18]. Basically, it exploits the cohesion among an arbitrary number of related words. Specifically, it tries to pass through the document and put the words that are semantically related into groups. [19] proposed lexical chains as a way to summarise texts. Topics are usually expressed by using different words rather than just one word; therefore, the lexical chains try to find the relationship between nouns in the text. For example, the occurrence of the words "car", "wheel", "seat", and "passenger" indicates a clear topic, even if each of the words is not very frequent in itself [20]. The study by [8] takes the idea of lexical chains to detecting the topic in the discussion.

2.1 Graph-Based in discussion Summarization

A graph model is a one of the approaches that is commonly used in Summarization by representing the text in the form of a graph. Vertices represent sentences, while edges between these vertices carry a weight that is equal to the similarity between the two sentences [20]. The approach reported by [8] has three parts. A graph-based model is used to represent the text and determine the important sentences by implementing PostRank [21] for the vertices in the graph. It is worth mentioning that they used an enhanced graph-based model reported by [22] which proposed working with multi-documents. The enhancement was related to the assumption which claims that given a number of documents, different documents are not equally important. [22] developed various methods to evaluate the "document-level information" and the "sentence to document relationship". Their assumption is that the sentences which belong to an important document and are highly correlated with the document should be more likely to be chosen in the summary. [8] implemented this work by considering the posts as documents in order to gauge the importance of each post.

They used cue words as an indicator of an important or unimportant sentence. Words like 'important,' 'should,' and 'I propose' are examples of bonus words (phrases), whereas those such as 'for instance' and 'example' are considered to be stigma words. Their approach seems to gives promising results in the topic detection of a thread.

While a thread is full with scattered topics, this is one of the challenges in producing the summary; therefore, [7] designed a model to solve the conflict of topic drifting. The Post Propagation Model follows the topic in the replies. They address three characteristics of the thread that traditional Summarization methods fail to capture: Topic Dependences, Topic Drifting and Text Sparseness. They propose a model named the Post Propagation Model (PPM) to handle these three challenges.

Characteristics	News articles	Social media text
Text formality	Text written in a formal way	Replies and comments are not subjected to any writing rules
Text Features (such as position of sentence, or similarity with the first sentence)	Applicable	Non-applicable
Writers	Educated and specialized persons	Users are different in education and backgrounds
Sentence boundaries	Clear	Difficult to identify

Table 1: Differences between formal text and social sites text

3   Challenges

Discussion Summarization has been attractive area for researchers, but there is still a lot to do. There is an urgent need to establish a standard dataset to enable the evaluation and comparisons between different approaches. Existing works use different forums in experiments, this diversity makes it difficult to gauge. Table 1 shows some datasets and their sources. What can be seen is that themes are various, and that writers who participate in politics forums, for instance, are very different from those who write in games forums, and so on. The average length of sentences, diversity of topics in the thread etc. In constructing the dataset, it is more useful to create it for a specific task; most tasks for recent works aim to produce a summary from the thread. However, this mission for some is not efficient because the threads often have many topics and issues discussed, and there might be no clear relationship between them. Even for humans, how can we create a summary from a number of topics? Consequently, task-oriented is better; for example, an indicative summary or a query-based summary.

Due to the nature of users and the way they write in forums, a large amount of noise exists, and this noise differs from that which exists in formal documents such as news articles (table 1). A coherent summary requires removing any unnecessary words.

Researchers are constantly evaluating their methods by using the same evaluation tools used in news article Summarization, such as the ROUGE tool. However, it may not be accurate to use the same tool with different things [8], special tools for discussion evaluation are needed.

4 Acknowledgements

This work is supported by Ministry of Higher Education (MOHE) and Research Management Centre (RMC) at the Universiti Teknologi Malaysia (UTM) under Research University Grant Category (VOT Q.J130000.2528.02H99).

5 References

1. [1] A. M. Kaplan and M. Haenlein, "Users of the world, unite! The challenges and opportunities of Social Media," *Business horizons,* vol. 53, pp. 59-68, 2010.
2. [2] Z. Xiang and U. Gretzel, "Role of social media in online travel information search," *Tourism management,* vol. 31, pp. 179-188, 2010.
3. [3] A. Carbonaro, "Towards an Automatic Forum Summarization to Support Tutoring," in *Technology Enhanced Learning. Quality of Teaching and Educational Reform*, ed: Springer, 2010, pp. 141-147.
4. [4] R. Farrell, *et al.,* "Summarization of discussion groups," in *Proceedings of the tenth international conference on Information and knowledge management*, 2001, pp. 532-534.
5. [5] A. Nenkova and K. McKeown, *Automatic summarization*: Now Publishers Inc, 2011.
6. [6] S. Hariharan, *et al.,* "Opinion mining and summarization of reviews in web forums," in *Proceedings of the Third Annual ACM Bangalore Conference*, 2010, p. 24.
7. [7] Z. Ren, *et al.,* "Summarizing web forum threads based on a latent topic propagation process," in *Proceedings of the 20th ACM international conference on Information and knowledge management*, 2011, pp. 879-884.

8. [8] J. Hatori, *et al.*, "Multi-topical discussion summarization using structured lexical chains and cue words," in *Computational Linguistics and Intelligent Text Processing*, ed: Springer, 2011, pp. 313-327.

9. [9] M. Hu, *et al.*, "Comments-oriented blog summarization by sentence extraction," in *Proceedings of the sixteenth ACM conference on Conference on information and knowledge management*, 2007, pp. 901-904.

10. [10] L. Zhou and E. Hovy, "On the summarization of dynamically introduced information: Online discussions and blogs," in *Proceedings of AAAI-2006 Spring Symposium on Computational Approaches to Analyzing Weblogs, Stanford, CA*, 2006.

11. [11] M. Sayyadiharikandeh, *et al.*, "PostRank: a new algorithm for incremental finding of persian blog representative words," in *Proceedings of the 2nd International Conference on Web Intelligence, Mining and Semantics*, 2012, p 17.

12. [12] Z. Yang, *et al.*, "Social context summarization," in *Proceedings of the 34th international ACM SIGIR conference on Research and development in Information Retrieval*, 2011, pp. 255-264.

13. [13] C. Lin, *et al.*, "Simultaneously modeling semantics and structure of threaded discussions: a sparse coding approach and its applications," in *Proceedings of the 32nd international ACM SIGIR conference on Research and development in information retrieval*, 2009, pp. 131-138.

14. [14] X. Hu, *et al.*, "Exploiting internal and external semantics for the clustering of short texts using world knowledge," in *Proceedings of the 18th ACM conference on Information and knowledge management*, 2009, pp. 919-928.

15. [15] L. Zhou and E. Hovy, "On the summarization of dynamically introduced information: Online discussions and blogs," in *AAAI Symposium on Computational Approaches to Analysing Weblogs (AAAI-CAAW)*, 2006, pp. 237-242.

16. [16] M. Klaas, "Toward indicative discussion fora summarization," *UBC CS TR-2005-04*, 2005.

17. [17] M. Hu, *et al.*, "Comments-oriented blog summarization by sentence extraction," in *Proceedings of the sixteenth ACM conference on Conference on information and knowledge management*, 2007, pp. 901-904.

18. [18] J. Morris and G. Hirst, "Lexical cohesion computed by thesaural relations as an indicator of the structure of text," *Computational linguistics,* vol. 17, pp. 21-48, 1991.

19. [19] R. Barzilay and M. Elhadad, "Using lexical chains for text summarization," in *Proceedings of the ACL workshop on intelligent scalable text summarization*, 1997, pp. 10-17.

20. [20] A. Nenkova and K. McKeown, "A survey of text summarization techniques," in *Mining Text Data*, ed: Springer, 2012, pp. 43-76.

21. [21] L. Page, *et al.*, "The PageRank citation ranking: bringing order to the web," 1999.

22. [22] X. Wan, "An exploration of document impact on graph-based multi-document summarization," in *Proceedings of the Conference on Empirical Methods in Natural Language Processing*, 2008, pp. 755-762.

Part IX
Recent Advances in Information Systems

A Quantitative Study for Developing A Computerized System for Bone Age Assessment in University of Malaya Medical Center

Marjan Mansourvar[1]*, Maizatul Akmar Ismail[1], Sameem Abdul Kareem[2], Fariza Hanum Nasaruddin[1], Ram Gopal Raj[2]

[1]Department of Information System
[2]Department of Artificial Intelligence
[1,2]Faculty of Computer Science and Information Technology
University of Malaya
50603 Pantai Valley, Kuala Lumpur, Malaysia

marjan2012@siswa.um.edu.my, maizatul@um.edu.my, sameem@um.edu.my, fariza@um.edu.my, ramdr@um.edu.my

Abstract. A quantitative study was conducted to direct the design and development of a computerized system for bone age assessment (BAA) in University of Malaya Medical Center (UMMC). Bone age assessment is a clinical procedure performed in pediatric radiology for evaluation the stage of skeletal maturation. It is usually performed by comparing an x-ray of a child's left hand with a standard of known samples. The current methods utilized in clinical environment to estimate bone age are time consuming and prone to observer variability. This is motivation for developing a computerized method for BAA. A primary analysis shows the current method used by UMMC radiologists for bone age assessment, their feedbacks, problems encountered and their opinions about new approach for BAA. Our study also extracts user requirements for designing and developing a computerized method for BAA.

Keywords: Computerized Bone Age Assessment, Quantitative Study, Bone Age, Radiography, Bone Age Assessment.

1 Introduction

Bone age assessment (BAA) is a routine radiological process utilized in pediatrics to evaluate the difference between the chronological age and skeletal age in children [1]. This discrepancy indicates the abnormality in skeletal growth or in hormonal growth. Assessment of bone age plays an important role in monitoring and controlling endocrine disorders and hormone therapy that is one of the most expensive treatments in the world [2]. Bone age assessment is commonly carried out by comparing a radiograph of left hand-wrist with a reference of known bones called atlas, or by examining specific regions of bones like carpal bone and analyzing the overlap degree with other bones. This is a manual process and is prone to observation variability [3]. Therefore, this forms the motivating factor for this study to develop a computerized method for bone age assessment in the clinical environment. Our long term research goal is to address the question of how new techniques can better benefit radiologists and experts in their daily tasks in hospitals. We have conducted a quantitative study to identify the

T. Herawan et al. (eds.), *Proceedings of the First International Conference on Advanced Data and Information Engineering (DaEng-2013)*, Lecture Notes in Electrical Engineering 285, DOI: 10.1007/978-981-4585-18-7_74,
© Springer Science+Business Media Singapore 2014

main requirements and the motivation factors for developing an automated system for BAA in UMMC. We selected postgraduate radiologists from the biomedical imaging department as respondents as they are deemed to be experienced and knowledgeable people in BAA field.

Two main methods for bone age assessment are the Greulich and Pyle (GP) method and the Tanner and Whitehouse (TW2) method in clinical environments [4]. The GP approach is easier and faster than the TW2 method and is the most preferred method used in the Netherlands. However, the TW2 method is more reliable and accurate [5]. Both approaches are based on left hand X-ray images. The GP method is an atlas based method where radiologists compare whole bones in the hand with the main reference. TW2 method is a score based method that utilizes a series of bones in the left hand. TW2 method uses twenty regions of interests (ROIs) for evaluating bone age. ROI is separated into different stages and each stage is labeled with the letter A, B, C, D, E, F, G, H and I. A numerical score is assigned to each stage. By adding these scores in the ROIs, bone age is predicted.

2   Quantitative Study

A questionnaire survey is designed to investigate the current methods used in UMMC for bone age assessment. With regard to the difficulty of contacting medical personnel which is time consuming, questionnaire survey is considred as the easiest method to collect data. Generally, directing questions to the expert respondents is useful to overcome the limitations of single source and data collection through the survey saves time in achieving the main objective. The questionnaires were distributed to postgraduate students in the biomedical imaging department who are experts in radiology. The survey aims are:

i.   To identify the method used for BAA in University of Malaya Medical Center (UMMC).
ii.   To identify the factors that affect the determination of bone age.
iii. To investigate the motivating factors in using an automated system for bone age assessment by radiologists.

2.1   Respondents

A total of 55 questionnaires were distributed to selected respondents in the UMMC biomedical imaging department. Out of these 55 questionnaires, 36 questionnaires were returned. This showed a 72% rate of response that can be considered as an adequate rate as it was very difficult to contact potential respondents and ask them for their time. Kearns and Ledere [6] cited that surveys among senior respondents generally achieve a lower rate of reply.

2.2   Data Analysis

Expreinece Level of Respondents
The majority of respondents (80.6%) have between two to three years experience in BAA. This was followed by 13.9% with below one year experience and just two respondents (5.6%) with 4 years experience. When the respondents were asked how

long it takes to be an expert in assessing bone age from X-ray images, 33.3% confirmed they need two to four months and an equal percentage indicated four to six months. Meanwhile, 16.7% of the respondents stated they only need one to two months and another 16.7% needed between six to twelve months. The answers to this question indicate that length of time to gain enough experience and skill in BAA varies from a minimum of one month to one year among the respondents. The last question in this section was about the atlas which the radiologists used for BAA. All the respondents selected the Greulich and Pyle (GP) method.

Evaluation of Current Method

Regarding the number of cases of bone age assessment the respondents had to deal with in a week, 78.9% of respondents stated between five to ten cases and 21.1% had 10 to 20cases. This result shows assessment of bone age is part of the daily task for the respondents at UMMC. Bogdanowicz M et al., [7] classified the reasons for BAA as shown in Table 1. Diagnosis of growth disorder is the main reason for BAA in UMMC, followed by the estimation of height and control on treatment for growth hormone (with 11.1% and 8.3%, respectively).

Table 1. The Main Reasons for BAA

Main Reasons for BAA	Ferequency	Percentage (%)
Diagnosis of growth disorders	29	80.6
Estimation of height	4	11.1
Control on treatment for growth hormone	3	8.3
Total	36	100

The significance of BAA is identification of growth disorders and can be categorized into two global groups. Firstly, growth deficiency is due to an inherent defect in the skeletal process. Secondly, growth deficiency is about criteria outside the skeletal process, which damage epiphyseal or even osseous maturation. These criteria can be nourishing, metabolic or unknown such as the syndrome of idiopathic growth postponement. The height of children who grow up under normal life style situations is to a large extent influenced by heredity. Hence, the final height of the children may be postulated from parental heights. Actually, different techniques in final height estimations, which consider parental height, have been identified [8]. Another issue regarding the status of bone age assessment in UMMC is time taken for evaluation of each X-ray image. The data analysis shows that 55.5% of respondents selected the timeframe of between five to 10 minutes, 41.7% claimed that they spend 10 to 15 minutes and just 2.8% needed between 15 to 20 minutes to assess the bone age from images of normal quality. Based on this result, it can be concluded that BAA is a time consuming daily task in UMMC especially when the respondents have to deal with more than one case at any one time.

Using GP atlas for BAA has a number of advantages as well as problems. The respondents were asked about the advantages and problems of GP atlas. The respondents expressed their feedback in using GP method by: easy to use, saving time and accuracy of results as advantages and time consuming, low precision and subjective decision as the main problems. 69.4% of the respondents selected easy to use as the main advantage of GP atlas followed by 22.2% for accuracy of results and just a few respondents (less than 10%) chose time or other options. Subjective decision (52.8%) and time consuming (36.1%) were chosen by respondents as the main problems in using the GP method for BAA. Furthermore, 8.3% agreed that GP method has low precision. In terms of normal error rate in reading BAA using the current method, 36.1% claimed one to two years, 30.6% stated six to 12 months, 19.4% selected three to six months, 8.3% chose two to three years and 5.6% agreed with two to three years. The difference in the response to error rate could confirm the problem of subjective decision selected as the main problem in the use of GP method by the respondents [9].

Effective Factors

Although the assessment of age is dependent on the knowledge and experience of the people assessing the image, there are some factors that can affect the estimation such as gender, race, socioeconomic situation, systemic illness, nutritional status, constitutional retardation, hypothyroidism, adrenal and hypoplasia. In this survey, 52.8% of the respondents confirmed that race or ethnicity of patients could affect the diagnosis of bone age. 88.9% of respondents confirmed the effect of gender in their assessment of bone age and only around 11% were uncertain that gender difference could be an influence on the estimation of skeletal age. The respondents were asked to give their opinions on another criteria factor for the evaluation of X-ray radiographs. This factor is called noise on the images. Abbey and Barrett [10] explained noise as variation in the brightness of images even though no image detail is present. 80.6% of respondents confirmed that noise of images could be considered as a factor that could affect the assessment of bone age from radiographs.

Motivation factors for developing a computerized BAA system

Second section in the survey questionnaire included statements on the significance and motivation for designing and developing a computerized system for bone age assessment. In order to overcome the difficulties in manual approaches in BAA, computerized systems are being developed that are based on measures of some ossification centers like carpal or epiphyses bones [11]. The questions in this section intend to find out the expectations of the radiologists of an automated BAA system in order to replace of current manual method. The respondents expressed their opinions by using the scale range of 1 for Strongly Disagree to 5 for Strongly Agree. The responses based on the highest mean are shown in Table 2.

The acceptance of a computerized system for BAA to save manpower is absolutely welcomed by the respondents. Hence the first statement had a high mean of 4.31 out of 5.00 which reflects that an automated system for BAA is considered a preferred method among the radiologists in UMMC. One of the main significant factors in bone age assessment is time spent which is considered a crucial factor in a clinical environment. The respondents ranked this factor as the most important factor with a score of 4.42 and a standard deviation of 0.554. Accuracy is measured in terms of the difference between the estimated age by the system with the chronological age of patient. Accuracy is called validity or trueness in BAA system and is expressed with standard deviations. The principal factor in the development of different methods for BAA over these years is to improve accuracy and preciseness [12]. The respondents have ranked this factor as one of the significant factors in a BAA system with a mean score of 4.17 and standard deviation of 0.775.

Table 2. Motivation factors for developing a computerized BAA system

Motivation factors	Mean	Std Deviation
Do you agree to have a computerized system for BAA in UMMC?	4.31	0.710
An automated system would help the radiologist to speed up the process of BAA.	4.42	0.554
An automated system should increase the accuracy of the assessment of bone age.	4.17	0.775
An automated system in BAA should help to eliminate the observation variability.	4.36	0.593
An automated system should be able to solve the problem for noisy images.	3.86	0.683
An automated system to record the patients' age is able to better manage toward efficiency and effectiveness in other next reference.	4.33	0.632

Determination of bone age by a computerized system could remove the manual involvement of people and therefore eliminate the subjectivity of the estimation [13]. The variability of observations is considered as the second most important factor for the development of a computerized system for BAA by respondents with a mean score of 4.36. A successful information system needs to study thoroughly end users' requirements to identify where their problems are, so that planning can be formulated. If problems are determined then there is a need to find a permanent solution. The implementation of a BAA system that could address noisy image problems could solve the limitations in manual methods. However this factor only scored a mean value of 3.86 out of 5 by respondents, but could still be considered as an important factor. Among the reasons for an automated BAA system to replace the current manual method was to solve the problems encountered with duplication of work. The system should contain a large database of patients' information for future references. The system will be wasteful of time and power if it is unable to keep the records for future reports. The computerized BAA system should not only provide accurate results for assessment of bone age, but also should keep the information in a more efficient and effective manner in order to the increase end users' satisfaction. A mean score of 4.33 out of 5 was given by the respondents to this factor.

3   Results and Discussion

The sample involves radiologists in University of Malaya Medical Center. A systematic sampling was utilized. Our research disclosed that BAA is a routine task in UMMC and about 79.8% of respondents assess at least 5 to 10 cases of bone age in a week. As far as assessment of normal images is concerned, 55.5% of respondents spent an average five to 10 minutes, where else, 47% of respondents needed to spend 10 to 20 minutes for noisy images. However, 33% claimed that it takes them 2 to 6 months to become an expert in BAA. 52.8% of respondents agreed that subjective decision is the biggest problem arising from using manual method in BAA. By considering the huge volume of cases that need to utilize BAA and with the present limitations in the manual method, a big shift towards a computerized system is necessary. Therefore it is not surprising that the answer to the question "Do you agree to have a computerized system for BAA in UMMC?" obtained a mean score of 4.31 out of 5 as shown in Table 2. The research was applied to identify the critical variables and factors that affect bone age assessment from the perspective of medical personnel in UMMC. Analysis of the survey indicated that 88.9% of respondents confirmed that gender is one the variable that affects BAA, 80.6% claimed noise and 52.8% chose race of patients. Hence, these factors should be considered in the implementation a computerized system for BAA to achieve more accurate results. The main features are congruent with the hypotheses of the study which include:

- A computerzied BAA system should be able to speed up the process of bone age assessment.
- A computerzied BAA system should be able to increase the accuracy of bone age assessment.
- A computerzied BAA system should be able eliminate the observations variability in bone age assessment.
- A computerzied BAA system should be able solve the problem of noisy images.
- A computerzied BAA system should be able save patients' information towards more efficiency and effectiveness.

Based on the results of the survey questionnaire we proposed a frame work for an automated BAA system that will discussed further in the next section.

4   Design and Implication

This research develops a computerized system to assess the bone age using hand-wrist radiographs to eliminate the variability of observations and increase the accuracy of estimation. Figure 2 shows the general workflow for the proposed system of bone age assessment.

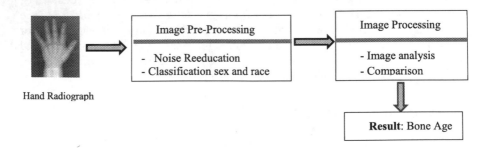

Fig. 2. Computerized process for bone age assessment

The proposed system is based on three main procedures as following:

Graphical User Interface (GUI)
This study applied a web-based interface for GUI of the proposed system that is publicly available [14]. PHP is used to interface a MYSQL database with an Apache web server on a windows-based server PC running a 64 bit Operating System. A user can

upload a radiograph involving a hand-wrist image to the BAA system. After the pre-processing and processing are completed and bone age estimated, the result is returned to the user interface.

Image Pre-Processing

Image Pre-processing, is the important task of reducing the noise of images and classifying the images based on gender and race. The pre-processing level contains the various levels, starting after uploading the image. Pre-processing stage is preparing the images for subsequent process in bone age assessment. The requirements of the pre-processing stage are based on the subsequent techniques utilized in the bone age assessment method [15]. The pre-processing provides the primary screening of the hand radiograph. This level plays a very significant role to standardize radiographs, as the images may include some irregularity features. The outputs of the pre-processing task are used for the processing stage for estimation of the bone age. The proposed pre-processing stage for this system involves two main steps: noise reduction and classification of images based on gender and race.

Image Processing

Image processing task is the crucial in bone age assessment in order to achieve the most accurate result in predicting bone age. This process needs the images in standardized content and correct orientation. The bone age is calculated using comparison techniques and similarity of images in the database. The method used for analysis of images is content based image retrieval (CBIR). CBIR method is defined as the well-known method to retrieve medical images in large databases. The system fetches the highest matched score and predicts the bone age based on following definition [16]:

$$\text{Predicted Bone Age} = \frac{\sum_{i=1}^{n} x}{n}$$

Where $x = Age\ of\ highest\ ranked\ retrieved\ images$,
$n = Total\ number\ of\ highest\ ranked\ retrieved\ images$

The tag value of similar images includes bone profiles of gender and ethnicity. The profile can be extracted after the most similar image is determined. The result of the image processing task is shown to the user as the predicted bone age in the GUI.

5 Conclusion

This study was carried out to verify the critical factors that contribute to the successful implementation of a computerized system for bone age assessment in a hospital environment. As it is an exploratory study, a quantitative approach consisting of a questionnaire survey was used for data collection. The targeted respondents were radiologists in UMMC. Based on the results of the survey the main features in BAA were identified which were also successfully verified from the research hypotheses. We proposed a frame work for an automated BAA system according to the users' requirements.

Acknowledgments. This paper is a part of PhD project in the Faculty of Computer Science and Information Technology (FCSIT), University of Malaya (UM), Kuala Lumpur, Malaysia. This project is under the Grant Number. FL012/2011.

References

[1] Thangam, P., T. Mahendiran, et al. (2012). "Skeletal Bone Age Assessment–Research Directions." Journal of Engineering Science and Technology Review 5(1): 90-96.

[2] Gertych, A., A. Zhang, et al. (2007). "Bone age assessment of children using a digital hand atlas." Computerized medical imaging and graphics: the official journal of the Computerized Medical Imaging Society 31(4-5): 322.

[3] O'Keeffe, D. (2010). "Denoising of Carpal Bones for Computerised Assessment of Bone Age."

[4] Tanner, J. M., Whitehouse, R. H.: Assessment of Skeletal Maturity and Prediction of Adult Height (TW2 Method).London, U.K.: Academic, 1975.

[5] Greulich, W. W. , Pyle, S. I.: Radiographic Atlas of Skeletal Development of Hand Wrist, 2nd ed. Stanford, CA:Standford Univ. Press, 1971

[6] Kearns, G. S., and Lederer, A.L. (2004). The impact of industry contextual factors on IT focus and the use of IT for competitive advantage. *Information & Management, 41*(7), 899-919.

[7] Berst MJ, Dolan L, Bogdanowicz M et al (2001) Effect of knowledge of chronologic age on the variability of pediatric bone age determined using the Greulich and Pyle standards. AJR 176:507–510.

[8] Rosenfeld RG, Cohen P (2002) Disorders of growth hormone / insulin-like growth factor secretion and action. In: Sperling MA, ed. Pediatric Endocrinology. Philadelphia, PA: Saunders;pgs 211–288.

[9] Unrath, M., H. H. Thodberg, et al. (2013). "Automation of Bone Age Reading and a New Prediction Model Improve Adult Height Prediction in Children with Short Stature." Hormone research in paediatrics 78(5-6): 312-319.

[10] Abbey, C. K. and H. H. Barrett (2001). "Human-and model-observer performance in ramp-spectrum noise: effects of regularization and object variability." JOSA A 18(3): 473-488.

[11] Zhang, A., A. Gertych, et al. (2007). "Automatic bone age assessment for young children from newborn to 7-year-old using carpal bones." Computerized medical imaging and graphics: the official journal of the Computerized Medical Imaging Society 31(4-5): 299.

[12] Patil, S. T., M. Parchand, et al. (2011). "Applicability of Greulich and Pyle skeletal age standards to Indian children." Forensic Science International.

[13] Mansourvar, Marjan, et al. "A Computer-Based System to Support Intelligent Forensic Study." Computational Intelligence, Modelling and Simulation (CIMSiM), 2012 Fourth International Conference on. IEEE, 2012.

[14] Muller H, Despont-Gros C, Hersh W et al. (2006) Health care professionals' image use and search behaviour. Proc MIE;24-32.

[15] O'Keeffe, D. (2010). "Denoising of Carpal Bones for Computerised Assessment of Bone Age."

[16] Mansourvar, M., et al., Automated Web based System for Bone Age Assessment using Histogram technique. Malaysian Journal of Computer Science 25(3), 107, 2012.

Customer E-loyalty: From an Estimate in Electronic Commerce with an Artificial Neural Fuzzy Interface System (ANFIS)

Nader Sohrabi Safa, Maizatul Akmar Ismail

Department of Information Science
Faculty of Computer Science & Information Technology
University of Malaya, 50603 Kuala Lumpur, MALAYSIA

sohrabisafa@yahoo.com,maizatul@um.edu.my

Abstract. Companies lose their online customers due to the competitive business environment. Customer loyalty is one of the important topics in the Electronic Commerce (E-commerce) domain. Gaining new loyal customers requires extensive expenditure of time and money. In addition, loyal customers are an important asset for a company, which brings long-term benefits. In this research, a comprehensive conceptual framework is presented that shows E-loyalty based on E-trust and E-satisfaction. The critical factors which influence E-trust and E-satisfaction are classified in organizational, customer and technological groups. Statistical analysis is applied for validity and reliability of the model. Another important method for estimation of uncertain measures is Artificial Neural Fuzzy Network System (ANFNS). E-trust and E-satisfaction data were used as inputs of the ANFIS and the output utilized E-loyalty. The result demonstrated, there is no difference between the aforementioned and the ANFIS model can be used for estimation of E-loyalty in E-commerce.
Keyword: E-commerce, E-loyalty, E-trust, E-satisfaction, Artificial Neural Network.

1. Introduction

Constant business growth is one of the important issues that are guaranteed with loyal customers. Some experts believe that loyal customers are an important asset for every company[1-3].In this research, E-loyalty which is comprised of two important components, E-trust and E-satisfaction, has been presented. Literature review and interviews with experts revealed that customer, organizational and technological factors influence E-trust and E-satisfaction. Data collection tools such as user and expert interviews and questionnaires by means of the *Likert scale* were used in this regard. This conceptual framework is accepted based on the results of data analysis. In the next part of this research, the application of ANFIS predicted E-loyalty based on two inputs of E-trust and E-satisfaction. Moreover, the usage of the Artificial Neural Network (ANN) predicted E-loyalty based on membership functions [4]. The results show that there is no difference between the results of the ANFIS model and E-loyalty calculated data.

T. Herawan et al. (eds.), *Proceedings of the First International Conference on Advanced Data and Information Engineering (DaEng-2013)*, Lecture Notes in Electrical Engineering 285, DOI: 10.1007/978-981-4585-18-7_75,
© Springer Science+Business Media Singapore 2014

2. Research framework

Different experts have paid attention to loyalty in varying aspects. According to Fang, Chiu [5], they discussed E-satisfaction through information, service and system quality. Though, as considered by Chang and Chen [6]. they put forth customization, convenience, interactivity and characters in customer interface quality. These subjects have been considered as factors within the technological group. A model was suggested for E-satisfaction by Palvia , which discussed the ease of use, perceived usefulness, competence, integrity, and belief in benevolence as important factors [7]. These subjects have been placed in the customer group factors. Assurances, empathy of organization and responsiveness have been discussed in the framework set by Lai, which were placed in the organizational category [8]. In this research framework, ten technological factors influence E-satisfaction and are listed as such; buying at 24 hours and 7 days a week, expedited paying with ease, analysing customer information, using complementary systems, providing service and product information, comparing and searching facilities, providing language options, personalization web features, information, and system quality. Other technological factors such as security of information and privacy, customer bulletin boards, complaint and following facilities, and customer feedback facilities are believed to influence E-trust. Table 1shows the classification of factors in a concise form based on literature review.

Table 1: Factors that influence E-satisfaction and E-trust

	Technology Factors	Organization Factors	Customer Factors
E-satisfaction	System Quality(9 Items) Information Quality(5 Items) Personalized Web Feature Language Options Search and Comparing Facilities Product and Service Information Using other Systems(5 Items) Collecting and Analyzing Customer Information Fast and Easy Payment Buying and Selling 24 hours and 7 days	Customer Segmentation Customize Products Fast Response to Customer Inquiries Variety of Goods and Services Rewards and Discounts(2 Items)	Perceived Site Quality Customer Experience in E-commerce Less Time Transaction Perceived Usefulness Perceived Ease of Use
E-trust	Customer Bulletin Board Security of Information and Privacy Customer Feedback Facility Complaint and Follow up Facility	Clear Shopping Process Money Back Warranty Contact Interactivity Organizational Reputation Guaranty Policy Selling High Regarded Brands Contribution with Well Known Company Tailored Advertisement and Promotion Fast and Safe Delivery	Perception of Hardware and software Reliability Perception of Risk(10 items) Perceived Market Orientation Positive Referrals from Friends Belief in Integrity Belief in Competence

The research questioner created based on the above classification and the research conceptual framework has been presented in Figure 1.

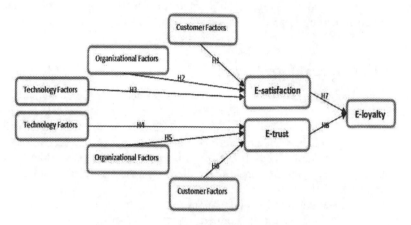

Figure 1: E-loyalty Conceptual Framework

Table 2: Correlation Results of Hypotheses

Hypothesis	Independent Variable (IV)	Dependent Variable (DV)	Pearson Correlation
H1	Technology Factors	E-Satisfaction	0.868**
H2	Technology Factors	E-Trust	0.610**
H3	Organization Factors	E-Satisfaction	0.498**
H4	Organization Factors	E-Trust	0.822**
H5	Customer Factors	E-Satisfaction	0.791**
H6	Customer Factors	E-Trust	0.771**
H7	E-Satisfaction	E-Loyalty	0.955**
H8	E-Trust	E-Loyalty	0.746**

3. Research Methodology

The Kaiser-Meyer-Olkin (KMO) measure of sampling adequacy test was used for data suitability. The outcomes from KMO test in all constructs were more than 0.5. KMO and Bartllet's test confirms our structure and data in terms of adequacy and sphericity [9]. Reliability shows the stability of measure over variety of conditions[10]. Cronbach's Alpha determines the amount of error made by any measure. Cronbach's Alpha for customer, organization and technology factors are 0.835, 0.724 and 0.845 respectively. There is no particular standard for interpreting Cronbach's Alpha. Brown suggests 0.8 as a minimum value for test measure[11]. More generally, Nually proposed that the minimum measure is 0.7 or above [10]. In this research the measure of Cronbach's Alpha for all constructs are more than Nunally's standard and close to Brown's recommendation, thus reliability of the measure is satisfactory. Correlation coefficients between customer, organization and technology factors with E-trust and E-satisfaction and between E-trust and E-satisfaction with E-loyalty have been presented in Table 2. These data show that the correlation coefficients between E-satisfaction with technology factors and E-trust with organizational factors and E-satisfaction with E-loyalty are strong (more than 0.8). These data also show the relationship between technology and customer factors with E-trust, customer and organizational factors with E-satisfaction, and E-trust with E-loyalty.

4. Artificial Neural Fuzzy System

The Artificial Neural Network (ANN), which is defined as a connected group of artificial neurons that use a mathematical or computational model for data processing according to the connection approach in calculation. In most cases ANN is an adaptive system that changes its structure based on internal or external information that flows through the network. Neural fuzzy discusses combinations of fuzzy logic and artificial neural networks. Connections called Synapses are usually formed with Axons to Dendrites [12].

Fuzzy regressions, statistical regressions, quantification analysis, neural networks and fuzzy rule base modelling have been applied in previous studies to develop customer satisfaction and loyalty models. Several experts used neural networks to consolidate the relationship between customer satisfaction and design attributes[13]. Some scholars applied neural network based on fuzzy reasoning algorithms to establish relations between a series of adjectival image words and the input from parameters in order to produce image evaluation [14]. Variety of usability dimensions, including both objective and subjective aspects are used to evaluate product usability based on statistically regressed models as discussed by [15].

5. Modelling E-loyalty based ANFIS

ANFIS can be a neural network with several layers of feed-forward which is regarded as the learning mechanism and fuzzy reasoning is used for mapping of inputs into outputs [16]. The aim of ANFIS is to create a model, which will simulate properly the inputs to the outputs [17].

 In this research, ANFIS was applied for modelling customer E-loyalty based on E-trust and E-satisfaction [18]. The steps of modelling are as follows:

The first step of customer E-loyalty modelling is, creating a conceptual framework which explains customer E-loyalty based on E-trust and E-satisfaction. This part of research had been carried out and explained in the previous parts. The validity of the model has been confirmed by experts in this area and the results of data analysis have confirmed   the validity of the model.

The second step is data collection based on the conceptual framework which was carried out by questioners from the E-customers of the largest retail chain stores in Iran.

In the third step, E-trust and E-satisfaction data can be put into the ANFIS model in order to generate fuzzy rules [19]. An ANFIS with five layers, two inputs and one output have been presented in figure 2.

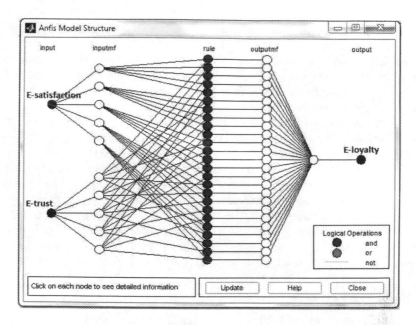

Figure 2: An ANFIS with four layers, two inputs and one output

The fuzzy modelling procedure uses neuron-adaptive learning techniques to gain information about a data set, in order to compute the membership function parameters that allow the fuzzy interface system to track the given input/output data [20].The learning process and the construction of the rules were provided by a hybrid algorithm. The finest fit model structure was determined according to criteria of performance evaluation.

The common practice is to commence learning with a high training rate such as 0.7 and decrease the rate as learning proceeds for setting the training rate and momentum[21]. For terminating network learning, illustrate there are three ways:

1) Learning can be stopped when the root mean square error (RMSE) between the expected value and network output value has reduced to a pre-set value.

2) When the pre-set number of training repetitions has been reached.

3) When the RMSE of a validation sample has started to rise. The first two conditions are based on the pre-set values.

Fuzzification, inference and defuzzification are the main parts of every fuzzy system [22].

1) Input membership functions are created based on input characteristics.

2) Rules created based on input membership functions.

3) The set of characteristics are created based on rules.

4) Output membership functions are created based on output characteristics.

5) An output single-value based on the output membership function.

Feed forward back propagation neural network consists of several layers of computational units which are interconnected in a feed-forward way. Neurons in one layer have a direct connection to the neurons of the subsequent layer. In the input layer data are introduced to the network. The data is processed in the hidden layer (that can be one or more). The results for given inputs are produced in the output layer.

Five rules have been considered for every input in the ANFIS system. Thus the inputs of the system have ten rules and twenty five rules for processing of the data to calculate the outputs. Membership functions for input variables act in layer two. Every fuzzy rule relates to one neuron which has a connection with each other. Each neuron in the third layer presents one fuzzy rule based on the precondition of the previous layer and presents output conditions. This system has one output.

MTLAB software is used for modeling of E-loyalty based on E-trust and E-satisfaction. E-trust and E-satisfaction data are the first and in the second column and E-loyalty data is in the third column of the matrix which is applied as operational data for ANFIS. Eighty percent of the data is used for training, thirty percent applied as testing and checking data. Finally, the predicted data for E-loyalty is extracted from the ANFIS model. C programming was used for this process and the data in any step is saved. Comparing the ANFIS output and observed data showed that there is a difference which is very close to zero between results.

6.   Validity of results

Root Mean Square Error (RMSE) is the difference between values actually observed and values estimated or predicted by the model. RMSE is a suitable measure of accuracy [23].

θ_1: Estimated data by ANFIS

θ_2: E-loyalty calculated data

$$\theta_1 = \begin{bmatrix} x_{1,1} \\ x_{1,2} \\ . \\ . \\ . \\ x_{1,n} \end{bmatrix}, \theta_2 = \begin{bmatrix} x_{2,1} \\ x_{2,2} \\ . \\ . \\ . \\ x_{2,n} \end{bmatrix}$$

$$\text{RMSE}(\theta_1, \theta_2) = (\text{MSE}(\theta_1, \theta_2))^{1/2} = \sqrt{\sum_{i=0}^{n} (x_{1,i} - x_{2,i})^2 / n}$$

RMSE = 0.000000001653

MSE= 0.0

One of the methods to quantify the difference between values implied by true value of the quantity and an estimator is Mean Squared Error (MSE). This error is the amount by which the value implied by the quantity to be estimated from the estimator differs [24]. The results show that there is a difference close to zero between calculated data by ANFIS and data which came out from the data analysis.

7.   Results and Discussion

The difference between the ANFIS output model and observed E-loyalty data is very close to zero. To confirm this issue the Root Mean Square Error (RMSE) and Mean Square Error (MSE) calculated, which showed there is no difference between output data from ANFIS and the observed data[25]. Figure 3 shows the same trend in the variation data. Both data change between zero and one, but in this figure to clear the trend view, E-loyalty has drawn between zero and one and ANFIS output has been drawn between one and two.

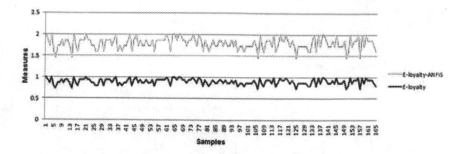

Figure 3: Comparing the prediction trend of ANFIS with real data

The measure of RMSE and MSE are 0.000000001635 and 0.0 respectively that show that there is no difference between the two outputs and ANFIS, which can be used for predicting E-loyalty.

8. Future Research

Critical factors of E-trust and E-satisfaction collected and classified for the three groups of customer, organization and technology. A conceptual framework showed the relationship between E-trust and E-satisfaction with E-loyalty. These reviews can be a valuable resource for future research of customer E-loyalty. The present study screened design variables by evaluating the model for their statistical significance. Other researchers can study these subjects from other perspectives such as international, legal or even government policy issues and so on. Another aspect for new research relates to comparing the results of different prediction tools such as ANN, ANFIS and Regressions. In this way the best method for forecasting which has the least error can be selected for predicting E-loyalty in E-commerce.

References

1.   Jacoby, J. and R.W. Chestnut, *Brand Loyalty.* 1978.
2.   Oliver, *Whence consumer loyalty?* Journal of Marketing, 1999. **63(4):** p. 12.
3.   Reichheld, F.F., R.G. Markey, and C. Hopton, *The loyalty effect - the relationship between loyalty and profits.* European Business Journal, 2000a. **12(3):** p. 7.
4.   Deng, W.J. and W. Pei, *Fuzzy neural based importance-performance analysis for determining critical service attributes.* Expert Systems with Applications, 2006. **36**: p. 3774-3784.
5.   Fang, Y.-H., C.-M. Chiu, and E.T.G. Wang, *Understanding Customers' Satisfaction and Repurchase Intentions: An Integration of IS Success Model, Trust, and Justice.* Emerald Group Publishing Limited, 2011. **21** (4): p. 479-503.
6.   Chang, H.H. and S.W. Chen, *The impact of customer interface quality, satisfaction and switching costs on e-loyalty: Internet experience as a moderator.* Computers in Human Behavior, 2008. **24**(6): p. 2927-2944.

7.   Palvia, P., *The role of trust in e-commerce relational exchange: A unified model.* Information & Management, 2009. **46**(4): p. 213-220.
8.   Lai, J.Y., *Assessment of employees' perceptions of service quality and satisfaction with e-business.* International Journal of Human-Computer Studies, 2006. **64**(9): p. 926-938.
9.   Habibpor, K. and R. Safari, *SPSS Comprehensive Guide for Research.* 2008.
10.   Nunally, J.C., *Psychometric Theory.* McGraw-Hill, 1978.
11.   Brown, F.G., *Principles of Educational and Psychological Testing.* 1983.
12.   McCulloch, W.S., *A Heterarchy of Values Determined by the Topology of Nervous Nets.* Bulletin of Mathematical Biophysics, 1945. **89**(93).
13.   Chen, C.-C., C.-S. Wu, and R.C.-F. Wu, *e-Service enhancement priority matrix: The case of an IC foundry company.* Information & Management, 2006. **43**(5): p. 572-586.
14.   Hsiao, S.-W. and H.-C. Tsai, *Applying a hybrid approach based on fuzzy neural network and genetic algorithm to product form design.* International Journal of Industrial Ergonomics, 2005. **35**(5): p. 411-428.
15.   Althuwaynee, O.F., B. Pradhan, and S. Lee, *Application of an evidential belief function model in landslide susceptibility mapping.* Computers & Geosciences, 2012. **44**(0): p. 120-135.
16.   Chang, H.H. and I.C. Wang, *An investigation of user communication behavior in computer mediated environments.* Computers in Human Behavior, 2008. **24**(5): p. 2336-2356.
17.   Keskin, M.E., D. Taylan, and O. Terzi, *Adaptive neural-based fuzzy inference system approach for modelling hydrological time series.* . Hydrological Sciences Journal, 2006. **51**(4): p. 588-598.
18.   You, H., et al., *Development of customer satisfaction models for automotive interior materials.* International Journal of Industrial Ergonomics, 2006. **36**(4): p. 323-330.
19.   Kwong, C.K., T.C. Wong, and K.Y. Chan, *A methodology of generating customer satisfaction models for new product development using a neuro-fuzzy approach.* Expert Systems with Applications, 2009. **36**(8): p. 11262-11270.
20.   Jang, S., *ANFIS: Adaptive-network-based fuzzy inference system.* IEEE Trans. Syst. Man Cybern, 1993. **23**(3): p. 665-685.
21.   Hush, D.R. and B.G. Horne, *Progress in supervised neural networks.* IEEE Signal Processing Magazine Vol. 10. 1993. 32.
22.   A. El-Shafie, O.J.a.S.A.A., *Adaptive neuro-fuzzy inference system based model for rainfall forecasting in Klang River, Malaysia.* International Journal of the Physical Sciences, 2011. **6**(12): p. 2875-2888.
23.   Armstrong, J.S. and F. Collopy, *Error measures for generalizing about forecasting methods: Empirical comparisons.* International Journal of Forecasting, 1992. **8**(1): p. 69-80.
24.   Lehmann, E.L. and G. Casella, *Theory of Point Estimation (2nd ed.).* New York: Springer, 1998.
25.   Armstrong, S. and F. Collopy, *Error Measures For Generalizing About Forecasting Methods: Empirical Comparisons.* International Journal of Forecasting, 1992. **8**: p. 69-81.

DMM-Stream: A Density Mini-Micro Clustering Algorithm for Evolving Data Streams

Amineh Amini, Hadi Saboohi, Teh Ying Wah, and Tutut Herawan

Faculty of Computer Science and Information Technology
University of Malaya (UM)
50603 Kuala Lumpur, Malaysia

Abstract. Clustering real-time stream data is an important and challenging problem. The existing algorithms have not considered the distribution of data inside micro cluster, specifically when data points are non uniformly distributed inside micro cluster. In this situation, a large radius of micro cluster has to be considered which leads to lower quality. In this paper, we present a density-based clustering algorithm, DMM-Stream, for evolving data streams. It is an online-offline algorithm which considers the distribution of data inside micro cluster. In DMM-Stream, we introduce mini-micro cluster for keeping summary information of data points inside micro cluster. In our method, based on the distribution of the dense areas inside the micro cluster at least one representative point, either micro cluster itself or its mini-micro clusters' centers, are sent to the offline phase. By choosing a proper mini-micro and micro center, we increase cluster quality while maintaining the time complexity. A pruning strategy is also used to filter out the real data from noise by introducing dense and sparse mini-micro and micro cluster. Our performance study over real and synthetic data sets demonstrates effectiveness of our method.

Keywords: Density-based Clustering, Micro Cluster, Mini-Micro Cluster

1 Introduction

Recently, a huge amount of data have been generated from various real-time applications such as monitoring environmental sensors, social networks, sensor networks, and cyber-physical systems [14]. In these applications, data streams arrive continuously and evolve significantly over time.

Mining data streams is related to extracting knowledge structure represented in streams information. Clustering is a significant data streams mining task [17, 12, 6, 2, 1]. However, clustering in data stream environment needs some special requirements due to data stream characteristics such as clustering in limited memory and time with single pass over the evolving data streams [13, 15, 19].

Traditional clustering algorithms can not deal with evolving data streams. In last few years, many data stream clustering algorithms have been proposed [20, 2, 5, 3, 4, 7, 8, 10, 19, 13, 12].

T. Herawan et al. (eds.), *Proceedings of the First International Conference on Advanced Data and Information Engineering (DaEng-2013)*, Lecture Notes in Electrical Engineering 285, DOI: 10.1007/978-981-4585-18-7_76,
© Springer Science+Business Media Singapore 2014

In this paper, we propose a density-based clustering algorithm over evolving data streams which considers distribution of data inside micro clusters. The algorithm, named as DMM-Stream, uses fading window model to deal with cluster evolution. DMM-Stream introduces the mini-micro cluster concept which is similar to micro cluster with a smaller radius (which was introduced in DenStream [10]). Our algorithm's features are described as follows. DMM-Stream:

— Considers the distribution of data points inside micro cluster.
— Increases quality by sending multiple representative points to the offline phase.
— Introduces dense and sparse mini-micro clusters to recognize real data from noise.
— Uses mahalanobis distance instead of Euclidean distance for identifying correct cluster center. This increases the quality of clustering as well.

The remainder of this paper is organized as follows: Section 2 surveys related work. Section 3 introduces basic definitions. In Section 4, we explain in details the DMM-Stream algorithm. Section 5 shows our experimental results. We conclude the paper in Section 6.

2   Related Work

Most of the clustering algorithms over evolving data streams have two phases that was firstly introduced by CluStream [2]. CluStream has online and offline phases. The online phase keeps summary information, and the offline phase generates clusters based on synopsis information. However, CluStream, which is based on the k-means approach, finds only spherical clusters. Density-based clustering can overcome this limitation. Therefore, density-based clustering is extended in two-phase clustering [11, 10, 19, 18].

DenStream [10] is a clustering algorithm for evolving data stream. The algorithm extends the micro cluster [2] concept, and introduces the outlier and potential micro clusters to distinguish between real data and the outliers. DenStream is based on fading window model in which the importance of micro-clusters is reduced over time if there are no incoming data points.

MR-Stream [19] is an algorithm which has the ability to cluster data streams at multiple resolutions. The algorithm partitions the data space in cells and a tree like data structure which keeps the space partitioning. The tree data structure keeps the data clustering in different resolutions. Each node has the summary information about its parent and children. The algorithm improves the performance of clustering by determining the right time to generate the clusters.

D-Stream [11] is a density grid-based algorithm in which the data points are mapped to their corresponding grids and the grids are clustered based on the density. It uses a multi-resolution approach to cluster analysis.

We compare time complexity and clustering quality of D-Stream, DenStream and MR-Stream algorithms. In terms of time complexity, D-Stream has the lowest time complexity; however, it has low quality since the clustering quality

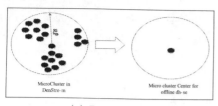 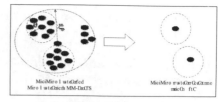

(a) DenStream   (b) DMM-Stream

Fig. 1: Distribution of data inside microcluster

depends on the granularity of the lowest level of the grid structure. DenStream has a higher time complexity in comparison with D-Stream; however, it has a better memory usage and quality. MR-Stream has the highest time complexity and memory usage while it has good quality.

We present a new method in which the granularity of micro cluster based on the distribution of data inside it by introducing new concepts called mini-micro cluster and micro cluster. This has not been considered in any of aforementioned algorithms. For example, in DenStream only the center of potential micro clusters are sent to offline phase. However, if the data points are not distributed uniformly inside micro cluster, sending only one representative point for each micro cluster leads to a low accuracy. Therefore, using mini-micro cluster increases the quality and using micro cluster prevents high time complexity (Figure 1, in which ϵ and θ are the radius of mini-micro and micro clusters respectively). We also use mahalanobis distance instead of Euclidean distance for identifying correct cluster center which increases the quality of clustering as well.

3   Basic Definitions

In this section, we introduce the basic definitions which form DMM-Stream algorithm.

Definition 1. *The Decaying Function*: *The fading function [16] used in DMM-Stream is defined as* $f(t) = 2^{-\lambda t}$, *where* $0 < \lambda < 1$. *The weight of the data stream points decreases exponentially over time, i.e the older a point gets, the less important it gets. The parameter* λ *is used to control the importance of the historical data of the stream.*

Definition 2. *MiniMicroCluster* (mmc): *A mmc for a group of data points with time stamp at time t is defined as* $\{CF^1, CF^2, W_{mmc}, C_{mmc}, r\}$. $CF^1 = \sum_{j=1}^{n} f(t - T_j)p_j$: *weighted linear sum of the points,* $CF^1 = \sum_{j=1}^{n} f(t - T_j)p_j$: *weighted linear sum of the points,* $W_{mmc} = \sum_{j=1}^{n} f(t - T_j)$: *mmc weight,* $C_{mmc} = \frac{CF^1}{W_{mmc}}$: *mmc center,* $r_{mmc} = \sqrt{\frac{CF^2}{W_{mmc}} - (\frac{CF^1}{W_{mmc}})^2}$: *mmc radius,* $r_{mmc} < \theta$.

Definition 3. *MicroCluster* (mc): *A mc for a number of its mini-micro is defined as* $mc = \{\{mmc\}, W_{mc}, C_{mc}, r_{mc}\}$. $\{\{mmc_0, mmc_1, \ldots, mmc_n\}\}$ *mini-micro Clusters list,* $W_{mc} = \sum_{i=1}^{|\{mmc\}|} W_{mmc}$ *Microcluster weight,* $C_{mc} = \frac{\sum_{i=1}^{|\{W_{mc}\}|} C_{mmc}}{|\{mmc\}|}$ *MicroCluster radius,* $r_{mmc} < \epsilon$ *MicroCluster radius.*

Definition 4. _DenseMiniMicroCluster(DMMC)_: _a mmc with a weight more than maximum threshold._ $W_{mm} > \frac{h_{mm}}{1-2^{-\lambda}} = D_{mm}$

Definition 5. _SparseMiniMicroCluster(SMMC)_: _a mmc with a weight less than maximum density threshold._ $W_{mm} \leq \frac{k_{mm}}{1-2^{-\lambda}} = S_{mm}$

h_{mm} and k_{mm} are controlling the threshold since the density cannot exceed $\frac{1}{1-2^{-\lambda}}$ (according to Lemma 1).

Definition 6. _DenseMicroCluster(DMC)_: _a mc which any of its {mmc} is dense._ $DMC = \{\{DMMC\}\}$

Definition 7. _SparseMicroCluster(SMLC)_: _a mc which all its mini-micro clusters are sparse._ $SMC = \{\{SMMC\}\}$

Definition 8. _CenterList_: _set of centers of DMMC and DMC which are sent to the offline phase:_ $CenterList = \{DMMC\} \cup \{DMC\}$

Definition 9. _MiniMicroCluster(MMC) Maintenance_: _if we have a MMC at a time t and a point p arrives in_ $t+1$ _then the statistics become_ $MMC_{t+1} = \left\{2^{-\lambda}.CF^1 + p, 2^{-\lambda}.W_{mm} + 1\right\}$

Lemma 1. _The maximum weight of the mini-microCluster(mmc) is_ $\frac{1}{1-2^{-\lambda}}$

Proof. If we assume that the data point in the data stream is added to the same mini-microCluster(mmc), the weight is equal to $W_{mmc} = \sum_{t'}^{t} 2^{-\lambda(t-t')}$ which can be converted to the following equation: $W_{mmc} = \sum_{t'}^{t} 2^{-\lambda(t-t')} = \frac{1-2^{-\lambda(t+1)}}{1-2^{-\lambda}}$ the maximum weight is defined when $t \to \infty$ therefore the maximum is defined as follows: $W_{mmcmaximum} = \frac{1}{1-2^{-\lambda}}$

Lemma 2. _The minimum time for converting the DMMC to SMMC and vise versa is:_ $t_{min} = \log_{\lambda}^{(\frac{S_{mmc}}{D_{mmc}})}$

Proof. proof is shown in [10] and [11].

4   DMM-Stream Clustering Algorithm

We now describe the key components of DMM-Stream outlined in Algorithm 1. When a new data record x arrives, we add it to the mini-micro or micro cluster depending on the distribution of data in merging algorithm. Then, we periodically prune in every gap time (which is the minimum time for converting a dense mini-micro cluster to sparse mini-micro cluster and vise versa). We remove the sparse mini micro and micro clusters in pruning algorithm.

Our clustering algorithm is divided into two phases: a) Online phase: keeping mini-micro and Micro clusters, b) Offline phase: generating final clusters.

Algorithm 1 DMM-Stream(DS, ϵ, θ)

1: Input: a data stream
2: Output: arbitrary shape clusters
3: t=0;
4: **while** not end of stream **do**
5:    Read data point x from Data Stream
6:    Merge(x,ϵ,θ);
7:    **if** t mod t_{min} == 0 **then**
8:       Pruning(MMC,MC);
9:    **end if**
10:   t=t+1;
11: **end while**
12: **if** the clustering request is arrived **then**
13:    Generate clusters
14: **end if**

4.1 Keeping mini-micro and Micro clusters

When a data point is arrived from data streams. The procedure is described as follows (Algorithm 2: Merge):

1. we try to find the nearest micro cluster to the data point
2. if we find such a micro cluster we try to find nearest mini-micro cluster to the data point. if there is such a mini-micro cluster then merge the data point to the nearest mini-micro cluster, otherwise form a new mini-micro cluster with x as a center of new mini-micro cluster.
3. otherwise, if there is not such micro cluster, form a new micro cluster with x as a center of new micro cluster.

Furthermore, we prune the mini-micro and micro cluster in the gap time in Algorithm 3: Pruning. In the pruning time, all the micro clusters are checked. We keep the list of micro and mini-micro cluster in the tree structure to make it more easier for searching and updating. For each micro cluster, its mini-micro cluster lists are checked.

We have three different situation for the mini-micro cluster list:

- if all of the mini-micro clusters are dense: micro cluster's center is kept for offline phase
- if all of the mini-micro clusters are sparse: mini micro clusters are removed as well as the micro cluster.
- if some of the mini-micro clusters are dense and some of them are sparse:
  - remove the sparse mini-micro clusters
  - keep center of the dense mini-micro clusters for offline phase

Initialization: we apply DBSCAN algorithm to the first initial points to initialize the online phase. we initialize the group of micro and mini-micro clusters by scanning data points. For each data point, if the total weight in its θ neighborhood is above its threshold , then we create a mini-micro clusters and

---

Algorithm 2 Merge(x, ϵ, θ)

1: Input: a data point from data stream
2: Output: list of MicroClusters MC with their MiniMicros
3: $mc = \{\{mmc_0^1, mmc_1^1, \ldots, mmc_n^1\}, \ldots, \{mmc_0^n, mmc_1^n, \ldots, mmc_n^n\}\}$
4: find the nearest microcluster center C_{mc} to x
5: **if** $Distance(x, C_{mc}) < \epsilon$ **then**
6:     find the nearest mmc center C_{mmc} to x
7:     **if** $distance(x, C_{mmc}) < \theta$ **then**
8:         Merge x to the mmc;
9:     **else**
10:         create a new mmc with x;
11:     **end if**
12: **else**
13:     create a new mc by x
14: **end if**

---

remove data point from data point list. furthermore, we check the aggregation of mini-micro cluster weights in the ϵ neighborhood of a microcluster. If it is above its threshold then a micro cluster is formed for these minis.

4.2 Generating final clusters

The online phase maintained micro and mini-micro clusters. However, we need to use a clustering algorithm to get the final clusters. When a clustering request arrives, DBSCAN algorithm is used on the micro and mini-micro cluster centers to get the final results. Each mini-micro and micro cluster center is used as a virtual point to perform clustering.

5   Experimental Evaluation

We implemented DMM-Stream in MOA [9]. We use the KDD CUP'99 Network Intrusion Detection data set and compare the performance and quality of the DMM-Stream with DenStream. The efficiency is measured by the execution time. The clustering quality is evaluated by the average purity of clusters. We make different sizes of data set from KDD CUP'99, and evaluate the clustering quality. In most of the cases the quality is better than DenStream. However the best answer is the KDD Cup 99 sub dataset with 17843 numerical records with 6 different classes. Our result only 1% improved the cluster quality with same time complexity in DenStream. We are trying to improve purity more than this value with bigger data set. We also evaluate our algorithm on the simple synthetic data set with arbitrary shapes. In this situation we get better purity but with higher time complexity.

6   Conclusion

In this paper, we have proposed DMM-Stream, an algorithm for density-based clustering of evolving data stream. The algorithm has two phases. The method

---

Algorithm 3 Pruning($\{mmc\},\{mc\}$)

1: Input: list of MiniMicroClusters and MicroClusters$\{MMC\},\{MC\}$
2: $mc = \{\{mmc_0^1, mmc_1^1, \ldots, mmc_n^1\}, \ldots, \{mmc_0^n, mmc_1^n, \ldots, mmc_n^n\}\}$
3: Output: Center List$\{CL_{centers}\}$
4: **for all** microclusters $\{mc\}$ **do**
5:    check all its minimicros $\{mmc\}$;
6:    $\{mmc_{initial}\} = \{mmc\}$
7:    **for each** mmc **do**
8:       **if** mmc is sparse **then**
9:          remove mmc from its mc list
10:       **end if**
11:    **end for**
12:    **if** $\{mmc\} = \{\}$ **then**
13:       remove its related microcluster mc
14:    **end if**
15:    **if** $\{mmc_{initial}\} = \{mmc\}$ **then**
16:       add mc center C_{mc} to center list
17:       $CenterList = CenterList \cup \{C_{mc}\}$
18:    **else**
19:       add all the mmc center C_{mmc} to the CenterList
20:       $CenterList = CenterList \cup \{C_{mmc}\}$
21:    **end if**
22: **end for**

---

determines the centers for offline clustering based on the distribution of the data inside the micro clusters. If the data is uniformly distributed, it only sends the micro cluster centers. However, if the data is non uniformly distributed instead of micro cluster center, its dense mini-micro cluster centers are kept for the offline phase. The pruning strategy is designed to delete the sparse mini-micro and micro clusters and to keep the dense one for the offline phase. Mini-micro and micro clusters are used in terms of increasing cluster quality and decreasing the time complexity. As a future work we want to automate the parameters of DMM-Stream and examine our algorithm in a sliding window model.

Acknowledgement

This paper is supported by High Impact Research (HIR) Grant, University of Malaya No UM.C/625/1/HIR/196.

References

1. Aggarwal, C.C. (ed.): Data Streams – Models and Algorithms. Springer (2007)
2. Aggarwal, C.C., Han, J., Wang, J., Yu, P.S.: A framework for clustering evolving data streams. In: Proceedings of the 29th international conference on Very large data bases. pp. 81–92. VLDB Endowment (2003)
3. Amini, A., Teh Ying, W.: Density micro-clustering algorithms on data streams: A review. In: International Conference on Data Mining and Applications (ICDMA). pp. 410–414. Hong Kong (2011)

4. Amini, A., Teh Ying, W.: A comparative study of density-based clustering algorithms on data streams: Micro-clustering approaches. In: Ao, S.I., Castillo, O., Huang, X. (eds.) Intelligent Control and Innovative Computing, Lecture Notes in Electrical Engineering, vol. 110, pp. 275–287. Springer US (2012)

5. Amini, A., Teh Ying, W.: DENGRIS-Stream: A density-grid based clustering algorithm for evolving data streams over sliding window. In: International Conference on Data Mining and Computer Engineering (ICDMCE). pp. 206–210. Bangkok, Thailand (2012)

6. Amini, A., Teh Ying, W.: Requirements for clustering evolving data stream. In: 2nd International Conference on Power Electronics, Computer and Mechanical Engineering (ICPECME). Cambodia (2013)

7. Amini, A., Teh Ying, W., Saybani, M.R., Aghabozorgi, S.R.: A study of density-grid based clustering algorithms on data streams. In: 8th International Conference on Fuzzy Systems and Knowledge Discovery (FSKD11). pp. 1652–1656. IEEE, Shanghai (2011)

8. Amini, A., Wah, T.Y.: Adaptive density-based clustering algorithms for data stream mining. In: Third International Conference on Theoretical and Mathematical Foundations of Computer Science. pp. 620–624. IERI (2012)

9. Bifet, A., Holmes, G., Pfahringer, B., Kranen, P., Kremer, H., Jansen, T., Seidl, T.: Moa: Massive online analysis, a framework for stream classification and clustering. In: Journal of Machine Learning Research (JMLR). vol. 11, pp. 44–50 (2010)

10. Cao, F., Ester, M., Qian, W., Zhou, A.: Density-based clustering over an evolving data stream with noise. In: SIAM Conference on Data Mining. pp. 328–339 (2006)

11. Chen, Y., Tu, L.: Density-based clustering for real-time stream data. In: Proceedings of the 13th ACM SIGKDD international conference on Knowledge discovery and data mining. pp. 133–142. KDD '07, ACM, New York, NY, USA (2007)

12. Guha, S., Meyerson, A., Mishra, N., Motwani, R., O'Callaghan, L.: Clustering data streams: Theory and practice. IEEE Transactions on Knowledge and Data Engineering 15(3), 515–528 (June 2003)

13. Guha, S., Mishra, N., Motwani, R., O'Callaghan, L.: Clustering data streams. In: Proceedings of the 41st Annual Symposium on Foundations of Computer Science. p. 359. IEEE Computer Society, Washington, DC, USA (2000)

14. Han, J., Kamber, M., Pei, J.: Data Mining: Concepts and Techniques Third edition. Morgan Kaufmann Publishers Inc., San Francisco, CA, USA (2011)

15. Kranen, P., Assent, I., Baldauf, C., Seidl, T.: The clustree: indexing micro-clusters for anytime stream mining. Knowl. Inf. Syst. 29(2), 249–272 (2011)

16. Ng, W., Dash, M.: Discovery of frequent patterns in transactional data streams. In: Transactions on Large-Scale Data- and Knowledge-Centered Systems II, Lecture Notes in Computer Science, vol. 6380, pp. 1–30. Springer Berlin / Heidelberg (2010)

17. OĆallaghan, L., Meyerson, A., Motwani, R., Mishra, N., Guha, S.: Streaming-data algorithms for high-quality clustering. In: International Conference on Data Engineering. pp. 685–694. IEEE Computer Society, Los Alamitos, CA, USA (2002)

18. Tu, L., Chen, Y.: Stream data clustering based on grid density and attraction. ACM Transactions on Knowledge Discovery Data 3(3), 1–27 (2009)

19. Wan, L., Ng, W.K., Dang, X.H., Yu, P.S., Zhang, K.: Density-based clustering of data streams at multiple resolutions. ACM Transactions Knowledge Discovery Data 3(3), 1–28 (2009)

20. Zhou, A., Cao, F., Qian, W., Jin, C.: Tracking clusters in evolving data streams over sliding windows. Knowledge and Information Systems 15, 181–214 (May 2008)

Failure Recovery of Composite Semantic Services using Expiration Times

Hadi Saboohi, Amineh Amini, Tutut Herawan, and Sameem Abdul Kareem

Faculty of Computer Science and Information Technology
University of Malaya (UM)
50603 Kuala Lumpur, Malaysia

Abstract Composite services are examples of volatile processes, which are prone to failures due to several problems that may occur during the executions. The recovery of their failure at execution time must be done efficiently to survive the system from deviation of its quality of service. Replacing a failed service with another similar service is not always reliable. Substituting a subgraph of the directed graph which represents the composite service with another sequence of services shows a major step in increasing the likelihood of the system's failure recovery. In this paper, we propose the use of expiration times of provided services to lower the time complexity of subdigraph identifications as well as other steps of a recovery approach. We evaluated our work, which shows a significant improvement to the similar approaches.

Keywords: Semantic Web Service, Composite Service, Failure Recovery, Subdigraph Replacement, Expiration Time

1 Introduction

Web services prominently substantiated the sources for Web processes. During the last decade, semantics lifted up the progress of Web services. Semantic Web services are the current phenomenon for automating the interaction between agents, and machines in general. There are two main categories of Web services, namely *information-providing*, and *world-altering* services [10]. World-altering services have one or more effects in addition to usual input, output, and precondition properties of information-providing services [14,18].

Web services are likely to fail due to their dynamic features. A mediator which acts between service providers and service consumers inspects well-execution of the services. In case of a failure during an execution of services, the *adapter* component of the mediator is triggered. The adapter repairs the process to ensure the delivery of the requested service to the consumer.

There are two major solutions to survive a service based system. First, a re-composition of the whole process. However, the re-composition is a very time-consuming task, and it delays the system's final response. Second, the mediator adapts the structure of the failed composite Web service so that the new composite service fulfills the user's goal without using the failed Web service.

T. Herawan et al. (eds.), *Proceedings of the First International Conference on Advanced Data and Information Engineering (DaEng-2013)*, Lecture Notes in Electrical Engineering 285, DOI: 10.1007/978-981-4585-18-7_77,
© Springer Science+Business Media Singapore 2014

Adaptation of Web processes is categorized into five groups: Perfective, Corrective, Adaptive, Preventive, and Extending Adaptations [6]. Our scope of study is the Corrective approach, in which a faulty behavior is removed.

An ideal adaptation is to replace a failed Web service with an exactly similar Web service (1:1 replacement) which its functional and non-functional properties are matched to the failed service. However, the opportunity of finding an equivalent service, which is called atomic-to-atomic replacement or briefly atomic-replacement, is not reliable [16].

An alternative approach is to notify the composer to compose a new composite Web service to act as a replacement for the failed Web service (1:n replacement). The approach improves the probability of recovery. Nonetheless, the time complexity is high. Recently, it has been proposed to use the composite-to-composite (n:m) replacement [19,12]. In these methods multiple services are selected, and they will be replaced by another matched set of services at the failure time. However, the problem is the high time complexity of calculations. In this paper, an approach based on expiration time is presented.

The remaining part of the paper is structured as follows. Section 2 overviews some of the similar works in the field of failure recovery of composite semantic Web services. We elaborate our model in Section 3. In Section 4 we evaluate our work. The last section concludes the paper.

2   Related Work

The problem of recovery of composite Web services has been investigated in the literature. Researchers mostly try to adapt the composite Web service to hinder from failure.

Two algorithms for dynamic adaptation of composite Web services presented in [20]. The first solution is to reroute a business process using a backup path, and the second is to reconfigure new processes in case of a failure. The replacement path is for abstract services.

The works of Canfora et al. [2,3] focused on a QoS-aware runtime binding of the abstract services to the concrete services. They re-bind the services during execution. The approach re-binds a slice of composite service from the point of the failure, i.e. non-executed services.

In [19] composite services are represented as directed graphs (hereafter digraph [1]). Potential subdigraphs of services are calculated offline and at the failure time the best subdigraph is replaced. A major bottleneck in this method is the time-consuming sub-task of subdigraph pre-calculations. The work was extended in [15,12].

Liu et al. [9] used *Alternate* exception handling strategy in their FACTS framework for composition of Web services (CWS) which is similar to our first method, i.e. Atomic Replacement.

Lin et al. [8] investigated the dynamic reconfiguration of service processes. They identify faulty regions of services, calculate constraints for each region, and recompose regions. This work can be extended using expiration times to lower the complexity of region identifications. Li et al. [7] uses the region reconfiguration and enhances it by ensuring correctness using the services behavioral type (SBT).

Researchers in [11] dynamically modified a composite service's structure at runtime.

In conclusion, approaches such as [11], and [15] show the best needed features for a failure recovery approach. However, adding the extra expiration time aspect [5], we step forward to a better model in terms of time complexity.

3 Failure Recovery Approach using Expiration Times

We have extended the work in [12] in terms of the time complexity of some of the calculations. The mentioned approach proposes to automate the recovery in two distinctive phases, i.e. offline and online.

The semantic specifications of the atomic Web services are stored in a registry called Atomic Service Registry. The structures of all available composite services (using mentioned atomic services) are represented as digraphs and stored in another registry called Composite Service Registry.

In the offline phase, possible subdigraphs of every composite semantic Web service is calculated. The approach calculates the best subdigraph to be replaced for every "Assumed Failing Web Service" (AFWS). We called the best ranked subdigraph as "Original Subdigraph". Then, searching in both atomic and composite service registries, we discover the replacements for every "Original Subdigraph". Ranking multiple choices, we find the topmost replacement which we called it "Replacement Subdigraph".

In the online phase, having the best "Original Subdigraph" for every AFWS and its related finest "Replacement Subdigraph" the replacement is done in no time. The approach shows a significant improvement in the probability of recovering the composite services from a failure [15,16,12].

The problem which we encountered is the high time complexity of finding all possible subdigraphs of all composite services. In this paper, we propose the use of Expiration Time ($\mathcal{ET}$) of provided Web services, which was initially proposed in [5] in order to accelerate the process.

Service providers may guarantee that their services will function for a specified duration, hence current reliability rate of the service. In other words, the parameters of the services will remain unchanged. The service parameters are usually provided as the services' specifications, and include functional and non-functional properties of Web services. The claim of "remaining unchanged" may not always be reliable; however, it is a criterion to lower the complexity of the calculations.

3.1 Expiration Time ($\mathcal{ET}$)

For the purpose of storing the expiration time we add an extra feature to every atomic and composite service in the Atomic and Composite Service Registries respectively. This feature shows the expiration time of the Web service as follows:

Atomic Web services: There are two possibilities.

– First, the provider of the service may specify an exact time (t) until when
the functionality of the service is guaranteed not to be changed. In this case,
the Expiration Time $(\mathcal{ET})$ is set to that specific time, i.e. $\mathcal{ET} = t$

– Second, the expiration time is unknown. In this case, we set the expiration
time to a pre-defined point of time, i.e. a bit after a *threshold* time $(\mathcal{ET} = Threshold + \epsilon)$. In other words, we assume that the expiration time of the
Web service is not close to the execution time. The *threshold* is governed by
an administrator of the mediator of the framework.

A naive approach might be to update the expiration time of the services
after (or during) every execution of the services. However, the performance is
increased if it is done based on an interval.

Composite Web services: The expiration time for every composite service is calculated based on the expiration times of its constituent atomic Web
services. The minimum expiration time of all atomic Web services, i.e. the earliest expiration time, is used for the expiration time of the composite service as
follows: $\mathcal{ET}_{Composite} = min\{\mathcal{ET}_{WS_1}, \mathcal{ET}_{WS_2}, \ldots, \mathcal{ET}_{WS_n}\}$. n is the number of
services in the structure of the composite service:

3.2 Subdigraph Renovation Plan for Failure Recovery using Expiration Time (SRPFR-$\mathcal{ET}$)

We make use of the stored expiration times in every step of our approach to
accelerate the calculations and speed-up the failure recovery process. In the
following sections, we elaborate the use of expiration time.

Subdigraph Calculations In order to lower the time complexity of the calculation of subdigraph we modify the previous subdigraph calculation algorithm
[12]. We stop calculations by reaching a vertex (Web service) which its expiration
time is later than the *threshold*, which decreases the number of subdigraphs, so
the time complexity is improved.

Ranking the Calculated Subdigraphs There might be several subdigraphs
having AFWS in their structure. We rank them for a AFWS aiming for finding
the best one to be eliminated, which we call it "Original Subdigraph".

We alter the ranking formula to get a more precise ranking. In the ranking, we
use the Expiration Time with a coefficient to increase/decrease the rank based on
the closeness to the expiry time of the guaranteed service functionality (without
any change in functional, and non-functional properties). This new statement for
the formula of ranking ensures that ideally an "Original Subdigraph" is chosen
that there is at least a Web service in its structure which its expiration time is
before the specific *threshold*.

A major advantage of this selection is that other Web service(s) with upcoming expiry time is selected to be replaced before it stops the execution of the
composite service. Hence, we may predict a probable future failure of a second

service. In other words, we replace more than one Web service which are all prone to failures.

Eqn. 1 shows a new ranking measure (OSRM) based on several criteria.

"Original Subdigraph" Ranking Measure =

$$\alpha_1(1 - |\mathcal{W}|) + \alpha_2 \mathcal{E}xec\mathcal{T} + \alpha_3 \mathcal{E}\mathcal{C} +$$
$$\alpha_4(1 - Undo_{Count}) + \alpha_5(1 - \mathcal{U}\mathcal{C}) + \alpha_6(1 - \mathcal{R}) + \alpha_7(1 - \mathcal{E}\mathcal{T}) \quad (1)$$

$$\sum_{i=1}^{7} \alpha_i = 1, \ \ 0 \leq \alpha_i \leq 1 \quad (2)$$

In every "Original Subdigraph", $|\mathcal{W}|$ is its number of Web services, $\mathcal{E}xec\mathcal{T}$ and $\mathcal{E}\mathcal{C}$ are its execution time and cost, $Undo_{Count}$ is the number of needed undo operations to compensate the effects of well-executed Web services preceding AFWS in the subdigraph, $\mathcal{U}\mathcal{C}$ is the cost of all required undo actions, $\mathcal{R}$ is the reliability of the subdigraph, and $\mathcal{E}\mathcal{T}$ is the expiration time of the subdigraph. The coefficients in Eqn. 1 are constrained as in Eqn. 2. They are either specified by the administrator of the mediator or calculated by a neural network algorithm.

Replacement Discovery In this step, we are aware of the set of services to be eliminated from the structure of the composite service. In order to select a replacement which can similarly act as the "Original Subdigraph" we discover an atomic or a composite Web service ensuring the correctness of the matchmaking.

We narrow down the number of discovered replacements using the extra feature of expiration time. The replacements are filtered with the condition of their expiration time being after a specified *threshold*: $ET_{Composite} > threshold$

Ranking the Replacements The final step of the offline phase is to rank the discovered replacement subdigraphs if there are multiple replacements.

We use a ranking formula to rank replacements based on a combination of functional and non-functional properties of the services. We consider the number of constituent services ($|\mathcal{W}|$), their total execution time and costs ($\mathcal{E}xec\mathcal{T}$ and $\mathcal{E}\mathcal{C}$), and reliability of the services ($\mathcal{R}$) as measures to rank the replacements.

Moreover, expiration time of the services ($\mathcal{E}\mathcal{T}$) is considered in the altered formula (Eqn. 3) of *"Replacement Subdigraph" Ranking Measure* and ranks the replacements higher if they have later expiration times. This will ensure that the adapted composite service may not fail because of reaching its expiration time.

"Replacement Subdigraph" Ranking Measure =

$$\beta_1(1 - |\mathcal{W}|) + \beta_2(1 - \mathcal{E}xec\mathcal{T}) + \beta_3(1 - \mathcal{E}\mathcal{C}) + \beta_4 \mathcal{R} + \beta_5 \mathcal{E}\mathcal{T} \quad (3)$$

$$\sum_{i=1}^{5} \beta_i = 1, \ \ 0 \leq \beta_i \leq 1 \quad (4)$$

Similar to Eqn. 1, weights of Eqn. 3 are constrained as in Eqn. 4.

Finally, we rank all the "Replacement Subdigraphs" for a AFWS by combining the two rank measures, i.e. OSRM, and RSRM using Eqn. 5.

$$Final\ Ranking\ Measure = \delta OSRM + \zeta RSRM \quad (\delta + \zeta = 1) \tag{5}$$

The topmost finest "Replacement Subdigraph" is indexed for every AFWS.

Online Phase (Actual Replacement) At the failure time of a Web service, the first choice is to re-execute the failed service (*forward approach*). In case of an abortive attempt of the forward approach, the alternative approach which is a *backward approach* is chosen. For the backward approach, not only the optimum "Original Subdigraph" to be eliminated, but also the finest "Replacement Subdigraph" to be implanted is already indexed, and close at hand.

To consider the world-altering services, if the order of the "Original Subdigraph" is greater than 1, and among its well-executed services, there exists world-altering services, the adapter first undoes their effects [12].

Finally, the adapter implants the "Replacement Subdigraph" and triggers the executor component of the framework to enqueue its executions. Therefore, the execution of the composite service continues.

4 Evaluation

We synthetically generated a test collection of composite semantic Web services. The atomic Web services of the test collection is created mostly like SWS-TC [4] in terms of the ontology features such as a unified single ontology, and one to two inputs and outputs for every Web service [13]. The composite services are generated using digraphs of the atomic Web services (as their vertices). The test collection included thousands of composite Web services with different structures.

We simulated the execution of a mediator for numerous times. Every composite service was executed several times, and each time a constituent Web service was marked as "failed". Then, the applicability of various studied replacement methods were evaluated. The average values of iterations of the whole method were measured and illustrated in Figure 1.

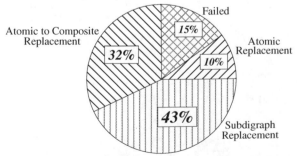

Figure 1. A Comparison of the Replacement Approaches

In 85% of the executions a replacement is possible to recover the failure of the composite Web service. In other 15% situations, there is no replacement available and the failure is declared. Using multiple strategies for recovery, i.e. the atomic-to-composite and composite-to-composite (Subdigraph Replacement) increases the probability of recovering the system from 10% (using 1:1 replacement only) to 85%. Moreover, a comparison of our previous method, and the new method in this paper shows improvements and fulfills the requirements [17].

5 Conclusion

In this paper, we have proposed an automatic failure recovery approach for composite semantic Web services using expiration time. Expiration time is an extra feature which providers can specify in the semantic service descriptions to guarantee the unchanging provision of the service until when.

The main contributions of our work are as follows. First, the subdigraph pre-calculations are significantly enhanced using expiration time. Second, using expiration time in all steps of offline calculations, we improve the probability of finding a better replacement. Hence, the deviation of quality of service is minimized and online replacement of subdigraphs is yielded in no time with a higher accuracy.

Acknowledgement

This paper is supported by High Impact Research (HIR) Grant, University of Malaya No UM.C/625/1/HIR/196.

References

1. Bang-Jensen, J., Gutin, G.Z.: Digraphs: Theory, Algorithms and Applications. Springer Monographs in Mathematics, Springer London, 2nd edn. (2009)
2. Canfora, G., Di Penta, M., Esposito, R., Villani, M.L.: QoS-aware replanning of composite web services. In: IEEE International Conference on Web Services (ICWS). pp. 121–129 (July 2005)
3. Canfora, G., Di Penta, M., Esposito, R., Villani, M.L.: A framework for QoS-aware binding and re-binding of composite web services. Journal of Systems and Software 81, 1754–1769 (October 2008)
4. Ganjisaffar, Y., Saboohi, H.: Semantic web services' test collection SWS-TC. Available online at: http://www.semwebcentral.org/projects/sws-tc/ (2006)
5. Harney, J., Doshi, P.: Speeding up adaptation of web service compositions using expiration times. In: 16th International Conference on World Wide Web (WWW). pp. 1023–1032. ACM, New York, NY, USA (2007)
6. Kazhamiakin, R., Benbernou, S., Baresi, L., Plebani, P., Uhlig, M., Barais, O.: Adaptation of service-based systems. In: Service Research Challenges and Solutions for the Future Internet, Lecture Notes in Computer Science, vol. 6500, pp. 117–156. Springer (2010)

7. Li, Y., Zhang, X., Yin, Y., Lu, Y.: Towards functional dynamic reconfiguration for service-based applications. In: IEEE World Congress on Services (SERVICES). pp. 467–473 (July 2011)

8. Lin, K.J., Zhang, J., Zhai, Y., Xu, B.: The design and implementation of service process reconfiguration with end-to-end QoS constraints in SOA. Service Oriented Computing and Applications (SOCA) 4(3), 157–168 (June 2010)

9. Liu, A., Li, Q., Huang, L., Xiao, M.: FACTS: A framework for fault-tolerant composition of transactional web services. IEEE Transactions on Services Computing 3(1), 46–59 (January-March 2010)

10. McIlraith, S.A., Son, T.C., Zeng, H.: Semantic web services. IEEE Intelligent Systems 16(2), 46–53 (2001)

11. Möller, T., Schuldt, H.: OSIRIS Next: Flexible semantic failure handling for composite web service execution. In: Fourth International Conference on Semantic Computing (ICSC). pp. 212–217. Los Alamitos, CA, USA (September 2010)

12. Saboohi, H.: An Automatic Failure Recovery Method for World-altering Composite Semantic Web Services. Ph.D. thesis, University of Malaya (2013)

13. Saboohi, H., Abdul Kareem, S.: A resemblance study of test collections for world-altering semantic web services. In: International Conference on Internet Computing and Web Services (ICICWS) in The International MultiConference of Engineers and Computer Scientists (IMECS). vol. I, pp. 716–720. International Association of Engineers, Newswood Limited, Hong Kong (March 2011)

14. Saboohi, H., Abdul Kareem, S.: World-altering semantic web services discovery and composition techniques - a survey. In: The 7th International Conference on Semantic Web and Web Services (SWWS). pp. 91–95. Las Vegas, Nevada, USA (July 2011)

15. Saboohi, H., Abdul Kareem, S.: Failure recovery of world-altering composite semantic services - a two phase approach. In: 14th International Conference on Information Integration and Web-based Applications & Services (iiWAS). pp. 299–302. ACM, Bali, Indonesia (December 2012)

16. Saboohi, H., Abdul Kareem, S.: Increasing the failure recovery probability of atomic replacement approaches. In: Asia-Oceania Top University League on Engineering Student Conference (AOTULE). p. 123. Kuala Lumpur, Malaysia (November 2012)

17. Saboohi, H., Abdul Kareem, S.: Requirements of a recovery solution for failure of composite web services. International Journal of Web & Semantic Technology (IJWeST) 3(4), 15–21 (October 2012)

18. Saboohi, H., Abdul Kareem, S., Ahakian, G.: World-altering features of semantic web service description languages. In: The Second International Conference on e-Technologies and Networks for Development (ICeND). pp. 132–136. Kuala Lumpur, Malaysia (March 2013)

19. Saboohi, H., Amini, A., Abolhassani, H.: Failure recovery of composite semantic web services using subgraph replacement. In: International Conference on Computer and Communication Engineering (ICCCE). pp. 489–493. Kuala Lumpur, Malaysia (May 2008)

20. Yu, T., Lin, K.J.: Adaptive algorithms for finding replacement services in autonomic distributed business processes. In: The 7th International Symposium on Autonomous Decentralized Systems (ISADS). pp. 427–434 (April 2005)

Improving Dynamically Personalized E-Learning by Applying a Help-seeking Model

Yousef Radi Fares[1], Maizatul Akmar Ismail[1]

Department of Information Systems, Faculty of Computer Science & Information Technology,
University of Malaya, 50603 Kuala Lumpur,
Malaysia
Yousef Radi Fares, Maizatul Akmar Ismail: youseff_89@hotmail.com, maizatul@um.edu.my

Abstract. This paper describes Yourbook, a personalized blended web-based e-learning application being developed in the University of Malaya to assist students completing introductory programming course for undergraduate studies in computer science faculty. The name is inspired by the famous social website Facebook since students are very familiar with the user experience (UX) of this famous website. In our web-based solution we are using Representational State Transfer (REST) software architecture. We are building our services on the back-end using Hypertext Preprocessor (PHP), and front-end using open-source JavaScript library Yahoo! User Interface Library (YUI). We are introducing a new model to enhance the current adaptive e-learning systems by taking in consideration some learning theories which have been introduced many times in educational psychology. Yourbook is mainly considering help-seeking strategy which is identified as a very important strategy in self-regulated learning (SRL). Through the paper we are arguing why we have chosen this specific educational theorem to serve in adaptive systems, and how the system will be designed to achieve our goals in enhancing the educational process.

Keywords: Adaptive learning, Overlay student model, Help-seeking, Concept network, Self-regulated.

1 Introduction

The importance of e-learning is increasing with time, and the realization of this importance is being noticed by users. For a long time there has been a need to create more dependable systems that can help in running the educational process, and move it forward. Many systems were created to support learning; one that has evolved and contributed to e-learning systems is adaptive e-learning system, which has played an important role among e-learning systems. The source of power in adaptive learning that it exploits the findings from different aspects related to computer science, psychology, and education to create systems that take in consideration student's needs and preferences [1]. The differences between students can be viewed from different

T. Herawan et al. (eds.), *Proceedings of the First International Conference on Advanced Data and Information Engineering (DaEng-2013)*, Lecture Notes in Electrical Engineering 285, DOI: 10.1007/978-981-4585-18-7_78,
© Springer Science+Business Media Singapore 2014

point of views, that is, knowledge differences, different characteristics; different learning styles, and different cognitive styles [2]. Many adaptive systems created to accommodate different adaptation perspectives [27][28][29][30] .This can influence the effectiveness and efficiency of learning in a big way. Programming concepts are required to be adaptive. Concepts can vary in perspective, and each concept can have different levels. For example, the concept of defining variables is usually easy to be grasped by students, but a bit more advanced concept is harder to be perceived. A closer look will reveal that each concept can have different levels. For example the concept of "if" statement can be extended to include more complicated structures as "if else". These difficulty levels can be viewed as levels of knowledge that goes vertically (.i.e. from defining variables to if statement and so on), and horizontally (.i.e. from simple if statement to nested statements). Although adaptive systems have proved to be a very effective way for learning, we will pay more attention in this paper on psychological aspects in help-seeking to improve the current systems. Considering student knowledge level adaptive solutions, how can we help the student to move from one level to another? We don't want the targeted student to be stuck in one level. It is more important to make him move faster from one level to another. Especially since courses being delivered has limited time frame.

It has been shown that moving from one level of task to another requires more information to be understood . Apparently adaptive systems deals with this aspect partially by providing information to student according to his/her level of knowledge, characteristics, and preferences. Unfortunately many systems pay more attention to technical aspects. In this paper psychological part will be mainly taken into consideration. More specifically we will consider help-seeking [8] as an effective way to support and enhance the learning process. There have been very rare systems that take this psychological aspect in consideration. Bringing this aspect into e-learning systems is not the major issue, but we need to identify where this aspect can be beneficial, and when. In our research we are investigating help-seeking from different perspectives, and our research questions are:1- Can we employ the strategy of help-seeking to increase the interaction between students, and improve the learning experience once applied to adaptive e-learning systems?, 2- Can we classify students to avoidant help-seeking students, and adaptive help-seeking students by mining their behaviors on the system and employ educational psychology of help-seeking to develop tools that can reveal their help-seeking patterns?. Section 2 reviews some work related to adaptive systems, and help- seeking. Section 3 describes our system. Finally section 4 concludes the paper.

2   Research Background

There have been many proposals that consider adaptability to build more effective systems. Some of them paid attention to adaptive navigation [5], which helps the user to locate relevant information in the context of hypermedia [6][7]. Another system on the other hand discusses the student's characteristics aspect such as cognitive style [4][26], where user interface change dynamically according to the cognitive style of the learner. And lately some of systems created to deliver materials according to student's knowledge level [3], where the contents of the system changes according to

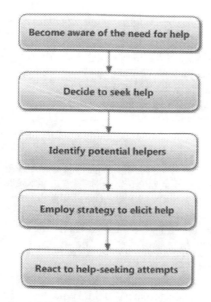

Fig. 1. Representation of Help-Seeking Process Model (Nelson-Le Gall 1981).

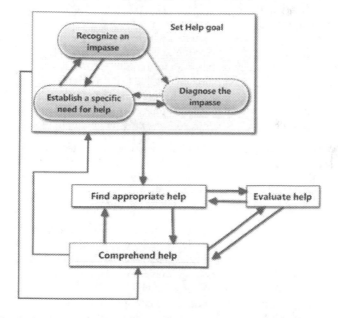

Fig. 2. Cognitive Model of help-seeking (Mercier and Frederiksen 2007).

the learner knowledge level. These systems were developed to support personalizing the system according to user preferences and needs. In the following paragraphs we review some important studies about help-seeking from the educational psychological perspective, mostly. Since these models are rarely utilized in e-learning systems, this review will help us to clarify the system design and will show the importance of this aspect.

Help-seeking in early researches is considered an activity that causes a loss in self-respect and should be avoided. It can be used as an index of dependency [9]. A conceptual model was suggested which changed the view of help-seeking from self-threatening behavior to effective alternative for dealing with current obstacles [10]. Illustration in Fig. 1 shows the conceptual model suggested by (Nelson-Le Gall 1981). The first step obviously shows when the help-seeker decides to ask for help, which is after being aware of the need for help and follow that with the corrective action.

Learning patterns are discussed in the help-seeking context and are classified mainly into avoidant help-seeking and strategic help-seeking. Avoidant help-seeking learners do not approach others to ask for help even though they realize that they need help. On the other hand strategic help-seeking learners actively use opportunities around them to reach their goals, advices and hints are adapted easily by this latter type. It is essential to mention that adaptive help-seeking can be in two forms "instrumental" and "executive" [9].

Patterns of help-seeking are associated with the characteristics of the learner. For better understanding of these characteristics, achievement goal orientation should be clarified. Achievement goal orientation is classified into mastery goal orientation and performance goal orientation [12][13][14][15]. Connecting the characteristics of learners to the patterns of help-seeking, mastery goal orientation is correlated to adaptive help-seeking [16]. Usually these learners are not feeling shy or embarrassed asking for help. They are confident enough to concentrate on mastering the skills rather than thinking of what the society think about them, in contrast to performance goal oriented people are tend to avoid help-seeking [17]. Performance goal orientation can be either performance-approach goal or performance-avoid goal [18]. Performance oriented learners in general tend to hear complements about their work, and they avoid hearing criticism. As a result they avoid help-seeking.

The spot light over the cognitive skill of help-seeking originated from the fact that it is an essential self-regulated strategy, which is important in maintaining and activating thoughts, behaviors, and affects in order to achieve goals [11]. The most important characteristic of self-regulated learners is that they can employ available resources around them to eliminate obstacles slowing down their improvement.

Fig. 2 shows a suggested and used help-seeking model designed to measure the individual differences in a graduate students' help-seeking process in using a computer coach in problem-based learning [19]. The process explained in the model is axiomatic for self-regulated learners which takes place in 5 steps [20]: (1) Analyzing the current task and interpreting the requirements according to their knowledge and beliefs, (2) Setting task-specific goals, selecting, adapting, and even inventing strategies to reach their goal, (3) Monitoring their achievement toward goals, and generating feedback about effectiveness of their strategies, (4) According to their assessment of their progress, they can adjust their strategies and efforts, (5) Motivational strategies being adopted to keep themselves on track. It has been revealed that self-regulated learners are instrumental adaptive-help seekers, with the ability to use others as resources to overcome troubles in their learning [21][22]. Hence, we want this cognitive model to also support non-regulated learners by adding new sub-processes for supporting help-seeking; identifying help-seeking patterns, provide

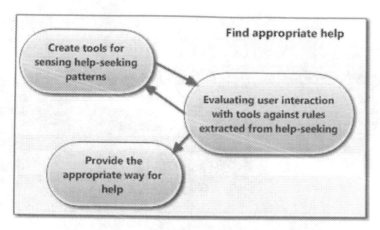

Fig. 3. Sub-processes added to find appropriate help.

appropriate help, and by running the whole of this model automatically by the system. Different characteristics for students should be considered when help is provided. Providing help at the correct time is not the only concern.

3 Suggested System

Adaptive systems that concentrate on student knowledge always try to deal with the current difficulties faced by the student. There is a high possibility that this is the time for student to seek help.

Help at the correct time is not the only concern. Learner characteristics play a main role. Advices for self-regulated learners about the best resources for supporting their learning process can be adopted easily by them. Regarding avoidant help-seeking we can run the model in Fig. 2 automatically so that we can stimulate this cognitive process for them, by taking in consideration the possible characteristics that may hold them back from asking for help. But we cannot make sure that the help-seeking process will be stimulated. Therefore there is a need to advise the teachers to pay attention to these students. In this research we are not arguing about the best way the teacher can deal with avoidant help-seeking type, but it is worth mentioning that there have been many researches to deal with this, for example help-seeking among peers taking in consideration goal structure and peer climate [23].

To plug in the functionalities we are expecting from the system, we have added extra sub-processes in the cognitive model in Fig. 3 as the second step "Find appropriate help". In this paper we are introducing a few tools to serve our second research statements, but research for this model will be open to choosing the best tools that can reveal help-seeking patterns by exploiting help-seeking psychology findings. Also research will be open to the best ways to evaluate user interactions to decide what the learner's help-seeking patterns are.

Regarding the first sub-process we have defined, we need to create tools that can help us to determine the help-seeking attitudes, and these tools will be inspired from the

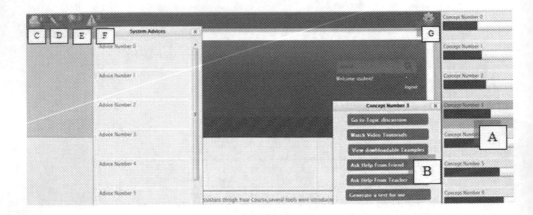

Fig. 4. Yourbook student home page.

psychology point of view. Let's start by defining instrumental help-seeking type as (HS), and avoidant help-seeking type (HA), performance goal orientation as (PGO), Mastery goal orientation as (MGO). Beside that we will divide (PGO) to (PGOpap) which belongs to performance-approach goal and (PGOpav) which belongs to performance-avoid goal. We will consider these two types mainly (.i.e. HS and HA). In our context we do not apply the model for example to problem solving, where executive help seeking is a very valid case, help will be on concepts understanding. First rule we can extract from the definition of avoidant help-seeking that if the learner has a low score in specific concept, and he/she is not turning into help (.i.e. by using the asynchronous tools provided with friends and teacher, or by posting questions to concept forum, etc..), then there is a high possibility that this is type HA. On the other hand more active learners on the system have a high possibility to be HS type. PGO as stated before is an indication for HA type, and MGO is an indication for HS. We will create a tool that can be integrated with any action in the system to give the choice for the learner to conceal his/her name for that action. For example, if a student wants to make a comment or post a question he/she will be given the choice to use a nickname, use his real name or skip this step, given that skipping will be the default choice. What is good about this tool is that student can use as many nicknames as he/she wishes. We are expecting that MGO will skip this step, while PGOpap will chose to show their real name, and PGOpav to hide their names. We expect also the latter type to be encouraged to seek for help, since hiding the name will lessen the worry about self-esteem which is a significant reason for help avoiding.

A tool for creating contests by students also implemented and we are expecting PGOpav not to participate since they avoid challenges and obstacles to maintain their self-perception of ability relative to others [18]. By contrast we expect PGOpap and MGO to participate, because the former care a lot about looking smart, showing good ability, and outperforming others, while the latter tend to be deeply engaged in the learning activities and development of competence.

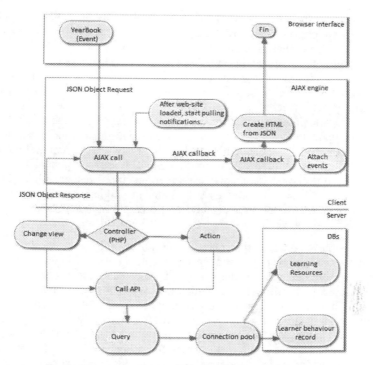

Fig. 5. Communication model between client and server.

All what have been mentioned so far in this section are hypothesis extracted from the help-seeking context, and help-seeking patterns for students will be hard coded in the system to serve our first goal in stimulating the instrumental help-seeking process. But later to evaluate and regulate our hypothesis students behavior mining will be performed to reveal the correctness of the tools we have presented and discover new behaviors that can determine online student help-seeking pattern.

Considering the student home page shown in Fig. 4, where main components are shown from A to F, component A shows the concepts being offered in the class. These concepts are offered according to the schedule being determined by the teacher. So concepts are locked according to the schedule plan, but still active student can move from one concept to another by taking automatically generated exams. He/she can notice the concepts scores shown by the graphical progress bars. We believe that this will help the student to assess his/her ability in some certain concept and set his/her help goal according to his score. Also this progress bars can help the system to assess student progress and his need for help at specific level. It is essential to mention that scores in this progress bars have specific meanings. For example it can tell if the student has problems in basic structure or more complicated structures. This serves as a hint for the user so that he/she can create sub-goals for learning the concept, and it directs the system to build the correct advice, consequently student then can establish a specific need for help, and at the same time the system is realizing this and moving to another step for defining what is the best way to deliver the help by going through the sub-processes we have defined before. The system keeps circulating between giving advices and checking if assessments on tests are satisfying. The items shown on component A are clickable, and by clicking one of these items you can popup a menu (component B), which has helping tools for that specific concept. We have separated the forum and all other facilities to measure the interactions for each concept separately. As illustrated, student can navigate to the forum, and post or view questions and answers being posted there. Also there are tools for using asynchronous communication with teacher or specific student. Videos and downloadable examples can be viewed, which can show different contents according to the student level. The last tool in component B is for generating automatic questions. Each time it is used will generate a test according to student's level, which helps the student and the system to determine his/her level of progress. A Student can use component C to generate competitions with other students, and see the competition' tests results. Components from C to F subsequently show asynchronous messaging, tests generated by the lecturer, invitations to contests, and advices generated automatically using the sub-processes discussed previously.

Fig. 5 shows the communication between the client and server to always view the corresponding learning resources that accommodate the learner knowledge, keep track of user interactive behavior on the system, and provide advice to students. Concept network is used to present the domain knowledge [24], and concepts are locked according to the course sequence. Each main concept contains sub-concepts which are dependent on each other. The Leaner can progress vertically in main concepts, and horizontally in sub concepts according to learner knowledge. For student modeling, we are using overlay student model [25], where the presentation utilized for the domain knowledge is used for student modeling. All the behavior of the student is stored in the student model.

4 Conclusion and Future Work

The system suggested previously has two main goals. First is to stimulate the cognitive skill for instrumental help-seeking, and help the students to find the appropriate help accommodates their help-seeking pattern. Part of the first goal is to investigate the significance in concepts learning progress before and after applying help-seeking system features. The Second goal is to create tools to help in identifying student characteristics automatically and utilize it to serve the first goal. In addition, our second goal includes mining the tools implemented to check if it correctly predicts student help-seeking patterns, and mine other student behaviors on the system that may give an indication about student help-seeking pattern. In the future we are planning to develop more tools that may give us a better feedback about the student help-seeking pattern, and will implement help-seeking in other e-learning contexts.

References

1. Jeremić, Z., J. Jovanović, and D. Gašević. "Student modeling and assessment in intelligent tutoring of software patterns." Expert Systems with Applications39.1 (2012): 210-222.
2. Graf, S., and T- C. Liu. "Analysis of learners' navigational behaviour and their learning styles in an online course." Journal of Computer Assisted Learning26.2 (2010): 116-131.
3. Chrysafiadi, Konstantina, and Maria Virvou. "Dynamically Personalized E-Training in Computer Programming and the Language C." 1-1.
4. Lo, Jia-Jiunn, Ya-Chen Chan, and Shiou-Wen Yeh. "Designing an adaptive web-based learning system based on students' cognitive styles identified online."Computers & Education 58.1 (2012): 209-222.
5. Hsiao, I- H., Sergey Sosnovsky, and Peter Brusilovsky. "Guiding students to the right questions: adaptive navigation support in an E- Learning system for Java programming." Journal of Computer Assisted Learning 26.4 (2010): 270-283
6. Brusilovsky, Peter. "Adaptive hypermedia." User modeling and user-adapted interaction 11.1-2 (2001): 87-110..
7. De Bra, Paul, Jill Freyne, and Shlomo Berkovsky. "Introduction to the Special Issue on Adaptive Hypermedia." New Review of Hypermedia and Multimedia19.2 (2013): 81-83.
8. Gourash, Nancy. "Help-seeking: A review of the literature." American journal of community psychology 6.5 (1978): 413-423.
9. Gall, Sharon Nelson-Le. "Help-seeking behavior in learning." Review of research in education 12 (1985): 55-90.
10. Gall, Sharon Nelson-Le. "Help-seeking: An understudied problem-solving skill in children." Developmental Review 1.3 (1981): 224-246.
11. Schunk, Dale H., and Barry J. Zimmerman. "Social origins of self-regulatory competence." Educational psychologist 32.4 (1997): 195-208.
12. Ames, Carole. "Classrooms: Goals, structures, and student motivation."Journal of educational psychology 84.3 (1992): 261-271.
13. Covington, Martin V. Making the grade: A self-worth perspective on motivation and school reform. Cambridge University Press, 1992.
14. Dweck, Carol S. "Motivational processes affecting learning." American psychologist 41.10 (1986): 1040-1048.
15. Wolters, Christopher A., Shirley L. Yu, and Paul R. Pintrich. "The relation between goal orientation and students' motivational beliefs and self-regulated learning." Learning and individual differences 8.3 (1996): 211-238.
16. Butler, Ruth, and Orna Neuman. "Effects of task and ego achievement goals on help-seeking behaviors and attitudes." Journal of Educational Psychology 87.2 (1995): 261-71.
17. Ryan, Allison M., and Paul R. Pintrich. "" Should I ask for help?" The role of motivation and attitudes in adolescents' help seeking in math class." Journal of educational psychology 89.2 (1997): 329.
18. Middleton, Michael J., and Carol Midgley. "Avoiding the demonstration of lack of ability: An underexplored aspect of goal theory." Journal of educational psychology 89.4 (1997): 710.
19. Mercier, Julien, and Carl H. Frederiksen. "Individual differences in graduate students' help-seeking process in using a computer coach in problem-based learning." Learning and Instruction 17.2 (2007): 184-203.
20. Zimmerman, Barry J. "A social cognitive view of self-regulated academic learning." Journal of educational psychology 81.3 (1989): 329-339.
21. Newman, Richard S. "Adaptive help seeking: A strategy of self-regulated learning." (1994).
22. Zimmerman, Barry J., and Manuel Martinez-Pons. "Construct validation of a strategy model of student self-regulated learning." Journal of educational psychology 80.3 (1988): 284-290.

23. Shim, Sungok Serena, Sarah M. Kiefer, and Cen Wang. "Help Seeking Among Peers: The Role of Goal Structure and Peer Climate." The Journal of Educational Research ahead-of-print (2013).

24. Tsiriga, Victoria, and Maria Virvou. "Evaluation of an Intelligent Web-Based Language Tutor." Knowledge-Based Intelligent Information and Engineering Systems. Springer Berlin Heidelberg, 2003.

25. Sison, Raymund, and Masamichi Shimura. "Student modeling and machine learning." International Journal of Artificial Intelligence in Education (IJAIED) 9 (1998): 128-158.

26. Lo, Jia-Jiunn, and Ya-Chen Chan. "Design of adaptive web interfaces with respect to student cognitive styles." Education and Educational Technology. Springer Berlin Heidelberg, 2012. 331-338.

27. Lazarinis, Fotis, Steve Green, and Elaine Pearson. "Creating personalized assessments based on learner knowledge and objectives in a hypermedia Web testing application." Computers & Education 55.4 (2010): 1732-1743.

28. Lee, Young-Jin. "Developing an efficient computational method that estimates the ability of students in a Web-based learning environment." Computers & Education 58.1 (2012): 579-589.

29. Muñoz-Merino, Pedro J., et al. "An adaptive and innovative question-driven competition-based intelligent tutoring system for learning." Expert Systems with Applications 39.8 (2012): 6932-6948.

30. Pornsawan, Insorn, and Sanrach Charan. "Designing of Adaptive Coaching System to Enhance the Logical Thinking Model in Problem-based Learning."Procedia-Social and Behavioral Sciences 46 (2012): 5265-5269.

Part X
Mining Educational Data for Effective Educational Resources

A Comparison of Three Documentation Styles for Educational Data Analysis

Sin-Ban Ho[1,1], Ian Chai[2], and Chuie-Hong Tan[3]

[1] Faculty of Computing and Informatics, Multimedia University, 63100 Cyberjaya, Malaysia
[2] Faculty of Engineering, Multimedia University, 63100 Cyberjaya, Malaysia
[3] Faculty of Management, Multimedia University, 63100 Cyberjaya, Malaysia
{sbho, ianchai, chtan}@mmu.edu.my

Abstract. Frameworks are increasingly employed as a useful way to enable object-oriented reuse. However, understanding frameworks is not easy due to their size and complexity. Previous work concentrated on different ways to document frameworks, but it was unclear which ones actually were better. This paper presents results in investigating the different philosophies for framework documentation. The philosophies include minimalist, patterns-style and extended javadoc (Jdoc) documentation. Using a survey of 90 intermediate undergraduates engaged in Command and Adaptor design patterns coding work, this exploratory study discovered that minimalist documentation has positive impact in encouraging knowledge acquisition, significantly in terms of the framework functional workings. This concludes that documentation solutions with the minimalist principle can lead intermediate undergraduates to faster growth in learning two of the design patterns.

Keywords: Educational Data Analysis, Learning Analytics, Knowledge Management, Empirical Research Result, Knowledge Surveying Work.

1 Introduction

One of the key challenges to object-oriented frameworks is introducing design patterns to intermediate undergraduates. Intermediate undergraduates are those who have already had some experience with the framework in question but are not yet experts, i.e. they are between the novice and advanced levels. This research work on online documentation adapts the philosophy of pair programming in agile development [1]. The subjects would perform the coding details of a particular portion of the code while the instructor ensures that the coding exercise is being followed with the help of time check-point in the documentation. The scope of this research project is to tackle intermediate undergraduate documentation or tutorials. This is a very important part because once past the beginner stage, one often has the familiarity to figure out the details.

---

[1] International Conference on Data Engineering (DaEng), p. 1-8, 2013.
© Springer-Verlag Berlin Heidelberg 2013

2 Motivation of the Study

Studies in pedagogical documentation show that the behavior in organizing a programming guide is a domain that has been used to describe the way beginners learn how to use a framework. For some time, studies have reported behavior differences in pedagogical framework documentation. The three philosophies being evaluated in this study include minimalist [2], patterns-style [3, 4] and extended javadoc documentation [5, 6]. Each is compatible with the idea of mixing texts, examples and diagrams.

Our main research question is to empirically test whether minimalist, patterns documentation or Jdoc presentation would give better performance in teaching intermediate undergraduates how to use design patterns. This question is indeed the main concern that is being challenged. In this paper, we use the Command and Adaptor design patterns [7] as the basis of study on the impact of the documentation philosophies. This study is a follow-up of the earlier study of beginners [8, 9]. In this paper, we intend to find the impact of intermediate undergraduates after they have undergone the beginning stage of programming within the Swing framework context.

3 Experiment Description

This research work used an exercise-based research typically used in empirical software engineering. One of the main components of the research methodology is exercise-based investigation, which was preceded with the presentation of a certain documentation set. The formulated hypotheses were used to design the documentation sets and the respective exercise, which were pre-tested for usability, soundness, and readability before it was rolled out for collecting data from the field. The data collected were then statistically analyzed using suitable data analysis techniques.

3.1 Documentation Procedure

The participants would follow the documentation and create java source code that imports the main Swing package i.e. javax.swing.* and two AWT (Abstract Windowing Toolkit) packages i.e. java.awt.event.* and java.awt.*. The expected result from these tasks is to have an outcome of running Command and Adaptor (CmdAdp) programs. An example of the CmdAdp documentation is shown in [10], which is organized into pieces to formulate the minimalist documentation.

The background information section is added to the top of each piece in order to form the patterns style [11]. For Jdoc, the background information is replaced by the output of the javadoc tool, which comprises of the extracted information from the source code about interfaces, methods and data-fields [12].

To provide a picture of the relative total length of the documentation, the documentation size is measured in kilobytes, as proposed by Beizer [13]. Through this approach, we can quantitatively characterize the documents. Table 1 gives quantitative information about the character of the documents used in this experiment.

Table 1. Characterize the relative documentation quantitatively.

Quantitative characterization	Minimalist	Patterns	Jdoc
1. Relative total length (in kilobytes, KB)	244 KB	293 KB	340 KB
2. Information that is relatively available	Short overview list of work tasks	Background information	Classes, method and interface information
3. Number of document files	10 files	13 files	22 files
4. Total sections	9 sections	14 sections	11 sections
5. Total paragraphs	17 paragraphs	27 paragraphs	24 paragraphs

3.2 Hypotheses

Standard significance testing is used to clearly specify the effects of the three documentation philosophies. The null hypotheses are stated as follows.

$E1H_0$ - There will be no difference between patterns and minimalist documentation for the intermediate undergraduates in doing the same exercise.

$E2H_0$ – There will be no difference between patterns and Jdoc documentation for the intermediate undergraduates in doing the same exercise.

$E3H_0$ – There will be no difference between minimalist and Jdoc documentation for the intermediate undergraduates in doing the same exercise.

The interpretations are derived from the rejection or non-rejection of these hypotheses for each expectation.

3.3 Experimental Design

Our experimental design uses one independent variable (factor) and five dependent variables. The independent variable consists of the documentation group. The dependent variables are the completion time, number of difficulties faced, semi completion time, workings and comprehension (understanding of the exercise).

Independent variable:

Documentation style: We use three documentation philosophies, as described in section 2, each with a similar purpose: to complete the given work task.

Dependent variables:

Semi Completion time: Time taken for the subjects to do their first compilation.

Completion time: The time taken to finish the entire exercise.

Comprehension: The subjects have to identify the method, procedure, line of the code, and constants that perform the given task. There are a number of questions to test their understanding of the code.

Workings: This is to test how well the subjects are able to follow the instructions for assigning default settings to the CmdAdp components.

Number of difficulties faced: Instead of giving all the detailed steps, some parts of the documentation allow the learners interact with the system. The subjects are to record and accumulate the number of problems they encounter.

3.4 Participants

There are 90 participants in this study. 33 (36.7%) are female and 57 (63.3%) are male, with the mean years in the university of 2.97, and SD of 0.436, a minimum 2 years and maximum 4 years in the university. Participants are all information technology undergraduates who undergo the object-oriented programming course at the university. The normal age of the students at this level is 22 years old.

To be able to test the hypotheses of our experiment, three different groups of the CmdAdp documentation are required. We arrange the participants into three different groups, according to their tutorial sections. Table 2 shows more detailed information about the groups.

Table 2. The detailed information and ANOVA tests results of years in the university (year) and previous achievement of C Language course (CLang), C++ (CPP), Data Structures and Algorithms (DataStruct) grades, and Cumulative Grade Point Average (CGPA).

Documentation style:	Minimalist	Patterns	Jdoc		F	p-value
N (participants)	26	26	38			
Mean (year)	3.08	2.92	2.92		1.175	0.314
Std. deviation (year)	0.077	0.110	0.058			
Mean (CLang)	3.16	2.95	3.16		1.015	0.367
Mean (CPP)	3.08	3.08	3.17		1.101	0.337
Mean (DataStruct)	3.04	1.58	1.74		7.843	0.001*
Mean (CGPA)	3.12	3.08	3.20		0.499	0.609

Note: * Statistically significant at 0.05 level

During the lectures, the students are taught basic object-oriented programming (OOP) principles. The lectures are supplemented by practical tutorial sessions where the students have the opportunity to make use of what they have learned through the completion of various java coding exercises using the assigned on-line documentation. Prior to this experiment, the preliminary stage of the on-line documentation presents the Swing framework [8]. The second stage discusses five of the design patterns [9]. This experiment focuses on the third stage of the intermediate undergraduates learning, which is on the CmdAdp design patterns. The participants in this experiment are regarded as intermediate undergraduates since they have attempted the prior two stages. They are not advanced users since they have not completed the OOP course yet.

3.5 Validity

To see whether the groups differ significantly, we perform ANOVA tests on the three groups of participants. In Table 2, with all the p-values > 0.05, except for the Data Structures and Algorithms course that they took in the prior semester, there is no major significant difference detected. The random assignments of the three tutorial groups are balanced in terms of their years in the university, the courses like C and C++ language, and Cumulative Grade Point Average (CGPA).

Furthermore, the total completion time of the participants shows an almost perfectly symmetric distribution. Thus, there is no evidence that slower participants hurried because of others having finished before them, in spite of the particular participant group working in the same laboratory at the same time. A final consideration is the precision and accuracy of time stamps recorded by the participants. Although the participants are informed that they have at most two hours to complete the work task, by cross checking, we discover that their responses in the time stamp to be highly reliable.

4 Data Analysis, Results and Discussion

Statistical analyses are conducted using Statistical Package for Social Science (SPSS). The results are based on the sample of 90 responses. The data is analyzed to see if one of the documentation sets let the participants compile (**Semi-Completion**) and finish the fastest (**Completion**) with the number of difficulties recorded by the subject at these intervals (**Number of difficulties**), as well as understand the most (**Comprehension**). We also check for test scores on how well their knowledge in the inner workings of the framework (**Workings**). Since we do not want to rely on the assumption of normal distribution, we test for the normality of the dependent variables. From the normality test, we discover that all dependent variables except **Number of difficulties** are normally distributed for each participant group. Thus, for this dependent variable, medians will be used as the expected values, rather than the means.

In order to determine whether any of the categories differed on any of the scales for the dependent variables, mean scores (and standard deviations) are computed for each category on each scale. Using the documentation type as the independent variable and the four dependent measures, the data are subjected to an analysis of variance. In Table 3, the minimalist column is bold-faced to indicate this documentation style has the best performance. Table 4 presents the results of the separate multivariate tests. Multivariate F-tests are conducted to determine which of the dependent variables differ across the various categories. These values are obtained via tests of between-subjects effects using Multivariate Analysis of Variance (MANOVA) with a Scheffe test adjustment [14]. We choose this test to examine the sample sizes, since the three documentation groups in this experiment are unequal. From these results, we observe that one out of four independent variables is significant.

Table 3. The means of all categories

Category (Dependent variable)	Mean		
	Minimalist	*Patterns*	*Jdoc*
1. Semi-Completion (hh:mm:ss)	**0:31:06**	0:33:59	0:36:56
2. Completion Time (hh:mm:ss)	**0:58:11**	1:04:29	1:07:03
3. Comprehension (Scale:0-18)	**14.69**	13.31	14.08

Category	Mean		
(Dependent variable)	*Minimalist*	*Patterns*	*Jdoc*
4. Workings (Scale: 0-4)	**3.42**	2.81	2.87

Table 4. Multivariate effects of the documentation type on dependent variables.

Category (Dependent variable)	F	Significance
1. Semi-Completion time	1.657	0.197
2. Completion time	2.305	0.106
3. Comprehension	1.077	0.345
4. Workings	4.639	0.012*

Note: * Statistically significant at 0.05 level

In terms of **Semi-Completion** and **Completion** in Table 3, the subjects who use minimalist documentation complete their first compilation and complete the experiment faster than the ones using the other two documentation styles. When looking for the standard significance level of 0.05 (i.e. 95% probability) in Table 4, there is evidence that the patterns group are not significantly slower. Therefore, we conclude that there is no significant difference between patterns and the other two documentation styles as to how long it takes the subjects to complete the experiment. Subjects using minimalist are faster than both of the others perhaps because there is less text to read, while subjects using patterns style are faster than subjects using Jdoc perhaps because it is not cluttered with too much class information such as inheritance and subclasses.

As for **Comprehension**, there is no significant difference between how well the subjects understand the materials. This might be because the students are still able to understand the CmdAdp code in the end, irrespective of the document styles. Their learning may reach a maturation effect after going through the four work tasks of documentation. Furthermore, this can be due to the experiment being conducted at the end of the semester. The participants learn enough from the prior eleven weeks of tutorials and lectures on object-oriented programming to bias their performance in the final stage of the experimental run.

Regarding **Workings**, the subjects in the minimalist documentation group exhibit significantly better workings scores than the other documentation styles at the 5 per cent level. Interestingly, this indicates that the $E1H_0$, $E2H_0$ and $E3H_0$ hypotheses in section 3 are rejected. These rejections show that the patterns documentation and the other two styles are not the same in teaching the subjects about completing the work tasks with the designated settings. Spending more time in directly instructing the coding of the CmdAdp can be more beneficial in having the default result rather than flooding the intermediate undergraduates with too much background information. Too much background information may motivate intermediate undergraduates to try something different. They are more confident to differ since they are equipped with the additional background.

Table 5. Kruskal-Wallis test on the number of difficulties.

Chi-square	Degree of freedom (DF)	Asymptotic significance
2.502	2	0.286

Since the **Number of difficulties** is not normally distributed over the comparison of the three groups, we use the Kruskal Wallis test [15]. With the two-sided asymptotic significant value in Table 5 more than 0.05, the number of difficulties faced by the subjects has no significant difference among the three groups. The participants might not record fully the number of difficulties they have solved the task. In summary, among the strong proxies that confirm minimalist advantages include the fastest semi completion time, the fastest completion time, the highest comprehension and workings scores. Hence, we conclude that minimalist documentation is relatively superior to others in encouraging the positive knowledge transfer strategies of intermediate undergraduates.

5 Conclusion

In this work, a set of philosophies for organizing pedagogical textual and graphical information on the CmdAdp documentation has been proposed. From the results, we realize that the effects of the patterns style documentation are not supreme all the time. Perhaps, for intermediate users, patterns are not always the best. Furthermore, Pressman [16] suggested that patterns are not suitable for every situation. Interestingly, minimalist documentation shows an overwhelming advantage in terms of the intermediate users' completion speed and comprehension in fulfilling requirements.

In order to remove any variation between groups, each group is exposed to the three documentation styles in three different stages. For instance, if a group is given minimalist documentation in the first stage, the group uses patterns-style in the second stage before proceeding with Jdoc in the final third stage of CmdAdp exercise. This provides the opportunity for each group to attempt the three techniques.

The quantitative results show that minimalist documentation did not have a significant impact on the time and comprehension that it took to perform the programming tasks. Nevertheless, in terms of the functional workings of the framework, minimalist documentation had a practically and significantly positive impact, in spite of the fact that the participants were not experts in applying design patterns into programming tasks. The aim of using the most effective documentation is to provide intermediate users with a good process that will lead to faster growth in learning the CmdAdp design patterns. All these results demonstrate the behaviors of intermediate users in using the documentation solutions.

References

1. Martin, R.C., Newkirk, J.W., Koss, R.S.: Agile Software Development: Principles, Patterns, and Practices. Pearson Education Int'l, Upper Saddle River, NJ, pp. 13, pp. 43-84 (2012)
2. Carroll, J.M.: Minimalism beyond the Nurnberg Funnel. MIT Press, Cambridge, MA (1998)
3. Chai, I.: Pedagogical framework documentation: how to document object-oriented frameworks: an empirical study. PhD dissertation, University of Illinois at Urbana-Champaign, IL, http://www.cs.uiuc.edu/research/techreports.php?report=UIUCDCS-R-99-2077 (2000)
4. Johnson, R.: Documenting frameworks using patterns. In: Proc. ACM Object-Oriented Programming, Systems, Languages and Applications (OOPSLA'92), pp. 63-76. ACM Press, Vancouver, British Columbia, Canada (October 1992)
5. Berglund, E.: Designing electronic reference documentation for software component libraries. J. Systems and Software, vol. 68, no. 1, 65-75 (2003)
6. Cockburn, A.: Supporting tailorable program visualisation through literate programming and fisheye views. J. Information and Software Technology, vol. 43, no. 13, 745-758 (2001)
7. Gamma, E., Helm, R., Johnson, R., Vlissides, J.: Design Patterns: Elements of Reusable Object-Oriented Software. Pearson Addison-Wesley, Reading, MA (1994) [Commonly called the "Gang of Four" or "GoF" book]
8. Ho, S.B., Chai, I., Tan, C.H.: Comparison of different documentation styles for frameworks of object-oriented code. Behaviour and Information Technology, vol. 28, no. 3, 201-210 (2009)
9. Ho, S.B., Chai, I., Tan, C.H.: An empirical investigation of methods for teaching design patterns within object-oriented frameworks. International Journal of Information Technology and Decision Making, vol. 6, no. 4, 701-722 (2007)
10. Example of the documentation fragment which was presented in all the three documentation groups, http://pesona.mmu.edu.my/~sbho/Swing/Jdoc/t10jDoc/ 302greetAct.html
11. Example of the documentation fragment that is available in the patterns style documentation, but not available in the minimalist and Jdoc documentation, http://pesona.mmu.edu.my/~sbho/Swing/Composite/t12Pat/302greetAct.html
12. Example of the documentation fragment that is available in the Jdoc documentation, but not available in the minimalist and patterns style documentation, http://pesona.mmu.edu.my/~sbho/Swing/Pattern/t10Pat/GreetingAction.html
13. Beizer, B.: Software is different. In: Patel, D., Wang, Y. (eds.) Comparative Studies of Engineering Approaches for Software Engineering, vol. 10, pp. 293-310. Baltzer Science Publishers, Norwell, MA (2000)
14. Neter, J., Kutner, M.H., Nachtsheim, C.J., Wasserman, W.: Applied Linear Statistical Models. McGraw Hill, Boston, MA (1996)
15. Field, A.: Discovering Statistics Using SPSS, 3rd ed. SAGE Publications Ltd, London, pp. 560-567 (2011)
16. Pressman, R.S.: Software Engineering: A Practitioner's Approach, 7th ed. McGraw Hill, New York, pp. 347-354, pp. 835-837 (2010)

Educational Data Mining: A Systematic Review of the Published Literature 2006-2013

Muna Al-Razgan[1], Atheer S. Al-Khalifa[2], Hend S. Al-Khalifa[3]

[1,3] Information Technology Department, College of Computer and Information Sciences,
King Saud University
[2] Computer Research Institute, King Abdulaziz City for Science and Technology
Riyadh, Saudi Arabia
[1,3]{malrazgan, hendk}@ksu.edu.sa, [2] aalkhalifa@kacst.edu.sa

Abstract.
Educational Data Mining (EDM) is a multidisciplinary field that covers the area of analyzing educational data using data mining techniques. Since 2008 the first annual educational data mining conference has been established. Many articles have been published in the field of EDM due to the eager interest in improving teaching practices for both the learning process and the learners. This paper presents a systematic review of the published EDM literature during 2006-2013 based on the highly cited paper in this domain. More than three hundred papers were collected through Google scholar index, then they were classified according to the application domains, while also providing quantitative analysis of publications according to publication type, year, venue, category and tasks and contributors.

Keywords: educational data mining, data mining, systematic review

1   Introduction

Due to the growing availability of educational data and the need to analyze huge amounts of data generated from the educational ecosystem, Educational Data Mining (EDM) field has emerged. EDM is a multidisciplinary field that covers the area of analyzing educational data using data mining techniques (Crist´ obal Romero & Sebasti´ an Ventura, 2010). EDM is the field that "concerned with developing methods for exploring the unique types of data that come from educational settings, and using those methods to better understand students, and the settings which they learn in" as defined in the known community website [Educationaldatamining.org].

The use of e-learning and educational software in educational systems generate huge amount of data found in web servers and access logs and collected automatically as data repositories (C. Romero & Ventura, 2007). To a large degree, EDM is the analysis of student-computer interaction using tuned data mining techniques. EDM is the

T. Herawan et al. (eds.), *Proceedings of the First International Conference on Advanced Data and Information Engineering (DaEng-2013),* Lecture Notes in Electrical Engineering 285, DOI: 10.1007/978-981-4585-18-7_80,
© Springer Science+Business Media Singapore 2014

modification of data mining techniques to fit the need of educational setting to help explore and discover student learning and the setting in which they learn. (Crist´ obal Romero & Sebasti´ an Ventura, 2010)

The data repository is collected from various resources such as log files, quizzes, interactive exercise, discussion forum, demographic data (gender, age, and student's grades), student's behaviors, administrative data (school, teacher, region), and many other. In addition, these data have hierarchy level such as course, subject, topics, time of access, time of observation (semester, year), level (school, college), etc. This raw educational data needs to be converted into useful information; If handled very well, it will help the educational institute improve the teaching for both teachers and students and it will have great impact on the educational research (Cristobal Romero & Ventura, 2013).

EDM seeks to use educational data repositories in understanding the learning techniques, thus building computational models that combines both the data and the theory to improve the teaching practice (Crist´ obal Romero & Sebasti´ an Ventura, 2010). In addition, EDM can find interesting information from educational data that can guide educators to established pedagogical basis when designing or modifying the course material or the learning environment to better suit the learners (C. Romero & Ventura, 2007). EDM extends the ability of traditional teacher observation to enable academics to build models based on students' attributes in real-time (R. S. J. D. Baker & Yacef, 2009). Moreover, EDM should capitalize on the known fact of Data Mining (DM) that turns data into golden knowledge, which guides academic to reach the correct decision for the learning environment (C. Romero & Ventura, 2007).

Research and contributions in the area of EDM have grown from workshops in various conferences to the establishment of EDM communities. Moreover, in 2008, the first annual international conference on EDM was held and extended to a Journal of EDM (R. S. J. D. Baker & Yacef, 2009). In the following years, EDM articles have increased rapidly, therefore, several surveys were conducted to monitor the progress of this domain, among them is a recent survey by (Cristobal Romero & Ventura, 2013) to review the state of the art in this field.

Our contribution in this paper is in twofold: (1) we propose an empirical method for surveying the literature based on the highly cited paper in Google scholar, and (2) we classify academic literature relating to EDM using a systematic review. Our paper will address the following research questions:

1. What types of applications were conducted in the field of EDM from 2007 until now?
2. What is the distribution of publications following the first survey paper by (C. Romero & Ventura, 2007)?
3. What are the classifications of published papers following (Crist´ obal Romero & Sebasti´ an Ventura, 2010) educational tasks?
4. What are the characteristics of the current research on EDM?

To answer these questions our paper is organized into two parts: the first part presents our method of data collection and selection criteria as well as data classification and analysis. The second part concludes the paper by answering our research questions and discussing the study limitations.

2 Method

2.1 Data Collection and Selection Criteria

The area of EDM has been studied extensively in many surveys published in journals and book chapters. The first published survey was published by (C. Romero & Ventura, 2007), then another theoretical paper was presented by (R. S. J. D. Baker & Yacef, 2009), afterwards an extended review paper of Romero's work was published recently (Cristobal Romero & Ventura, 2013).

In order to contribute to the area of EDM surveys, we applied a different approach for analyzing the literature. Basically, we analyzed different emerging trends in EDM using the chaining technique of (Gao, et al, 2012). Previous survey papers such as (R. BAKER & K. YACEF, 2009), followed another approach for distilling trends by referring to the most influential papers mentioned in (C. Romero & Ventura, 2007). Their selection was based on the citations number of each paper, which was categorized to different types of EDM methods. Then they compared the classification with new types' distribution showed solely in the publications of the following two years (2008 and 2009) of the EDM conference proceedings.

However in this work we pursue a different approach, we selected the most cited paper in EDM literature based on Google Scholar's search engine using the term "educational data mining", which revealed Romero and Ventura's (2007) survey paper. This survey gave us a comprehensive list of papers, published between 1995 and 2005, which are considered as educational data mining publications by a prominent pair of authors in EDM (Baker 2009). Moreover, it is the most influential paper as being cited by 410 other papers at the time of conducting this study.
A bibliometric and qualitative analysis have been done on Romero and Ventura's paper based on the following criteria:

1- Use forward chaining of publication referencing (C. Romero & Ventura, 2007) to identify all related papers.
2- Select English academic papers published in journals, conferences, workshops, book chapters, etc.
3- Select papers with research focus on data mining in educational settings, excluding any paper on education or data mining alone, and classifying them as relevant if their primary focus was using data mining techniques in the context of educational setting.

For each paper in the citation, the authors had to read the Title and Abstract and sometimes skim the paper in order to decide whether to include it or not. Papers that were written in another language rather than English or did not have EDM as their main focus were excluded. The data collection process resulted in 281 unique papers out of 410 and was used in this systematic review.

In the following sections, we consider how EDM has recently evolved, and investigate some of the major trends in EDM research. In order to investigate what the trends are, we analyze what researchers were studying previously, and what they are studying now, towards understanding what is new and what emerged attributes in current EDM research.

2.2   Data Classification and Analysis

After we had collected the papers, we had distributed the work among the research team as follows: The first author looked for the citation of the papers and prepared the list of statistics (number of papers published yearly, active authors, and active publications venues); this will be presented in the following sections. The second author identified the common application domains across the 281 published papers.
Then the second author carried the task of classifying the papers according to identified application domains. Application domains and their definitions will be illustrated next. Due to the large number of papers, the authors split the process of classification among them based on (Crist´ obal Romero & Sebasti´ an Ventura, 2010). The eleven categories mentioned in (Crist´ obal Romero & Sebasti´ an Ventura, 2010) were followed to classify the repository of papers collected. All of the above analysis will be presented in the following sections.

Quantitative Analysis

Articles were classified quantitatively according to the publication year, venue and type of publication. The first survey paper on EDM was published in 2006, and became since then a foundation stone to other works in the field. The number of papers about EDM has grown greatly, which can be seen in (figure 1) exceeding 70 published papers in the year 2012. Most of the publications were in the form of journal articles (121 articles), followed by 118 conference proceedings (including two posters), 14 dissertations and 28 other types of publications.

Table 1 presents the five most published authors in the repository; Sebastián Ventura has the highest number of publications with 19 published works, followed by Cristóbal Romero (16 publications). Ventura and Romero have shared many publications as co-authors along with different researchers, and it can be seen that the following three authors: Valsamidis, Kontogiannis and Kazanidis had also been co-authors in many of the listed publications.

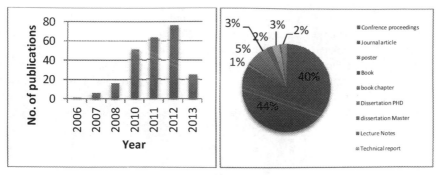

Fig. 1. Articles by publication year (left), Types of publications (right)

Table 1. Most Contributing authors

Author	No. of Articles	Details (Co-authors, year)
Sebastián Ventura	19	Porras, Romero Hervás & Zafra 2006; Romero, Antonio & De Bra 2007; Romero, Espejo & Hervás 2008; Romero & García 2008; Santos, Freire & Romero 2008; García, Romero & De Castro 2009; Romero, Zafra & De Bra 2009; García, Romero & De Castro 2009; González, Romero, del Jesús & Herrera 2009; Romero 2010; Carmona, González, Romero & del Jesús 2010; Zafra 2010; Romero & Zafra 2011; García, Romero De Castro 2011;Carmona, González,Del Jesus 2011; Zafra 2012; Espejo, Zafra, Romero & Romero 2013; Romero 2013; Márquez & Romero 2013
Cristóbal Romero	16	Porras, Ventura, Hervás & Zafra 2006; Ventura, Antonio & De Bra 2007; Ventura, Espejo & Hervás 2008; Ventura & García 2008; Santos, Freire & Ventura 2008; García, Ventura & De Castro 2009; Ventura, Zafra & De Bra 2009; García, Ventura & De Castro 2009; González, Ventura, del Jesús & Herrera 2009; Ventura 2010; Carmona, González, Ventura & del Jesús 2010; Ventura & Zafra 2011; García, Ventura, De Castro 2011; Espejo, Zafra, Romero & Ventura 2013; Ventura 2013; Márquez & Ventura 2013
Stavros Valsamidis	8	Kazanidis, Theodosiou & Kontogiannis 2009; Kazanidis, Kontogiannis & Karakos 2010; Kazanidis, Kontogiannis & Karakos 2010; Kontogiannis & Karakos 2011; Kazanidis, Kontogiannis & Karakos 2011; Kazanidis, Kontogiannis & Karakos 2011; Kazanidis, Kontogiannis & Karakos 2012; Kazanidis, Kontogiannis, Theodosiou & Karakos 2012
Sotirios Kontogiann	8	Valsamidis, Theodosiou & Kazanidis 2009; Valsamidis, Kazanidis & Karakos 2010; Valsamidis, Kazanidis & Karakos 2010; Valsamidis,

is		Valsamidis & Karakos 2011; Valsamidis, Kazanidis & Karakos 2011; Valsamidis, Kazanidis & Karakos 2011; Valsamidis, Kazanidis & Karakos 2012; Valsamidis, Kazanidis, Theodosiou & Karakos 2012
Ioannis Kazanidis	7	Valsamidis, Theodosiou & Kontogiannis 2009; Valsamidis, Kontogiannis & Karakos 2010; Valsamidis, Kontogiannis & Karakos 2010; Valsamidis, Kontogiannis & Karakos 2011; Valsamidis, Kontogiannis & Karakos 2011; Valsamidis, Kontogiannis & Karakos 2012; Valsamidis, Kontogiannis, Theodosiou & Karakos 2012

Publications' Application Domains

The articles embraced a wide-range of applications relating to EDM, the authors had identified the most used applications among the 281 articles and came up with the following list: Educational games, learning object, Mobile learning, personalized learning, scientific research into learning and learners, and intelligent tutoring system.

Classifying the articles in the previous list was done following this approach: reading the title of the paper and the listed keywords for each article then assigning the article to the appropriate application domain. Given the fact that some articles will fall in multiple application domains, we tried to match it and assign each article to one application domain.

Table 2. Types of application domains

Application Domain	Definition	# of assigned papers	% of application domains
Educational games	"Are games explicitly designed with educational purposes, or which have incidental or secondary educational value. All types of games may be used in an educational environment. "[2]	7	2%
Learning object	"a collection of content items, practice items, and assessment items that are combined based on a single learning objective"[3]	84	30%
Mobile Learning	"Learning that happens when the learner takes advantage of the learning opportunities offered by mobile technologies."[4]	5	2%

Personalized Learning	"is the tailoring of pedagogy, curriculum and learning environments to meet the needs and aspirations of individual learners"[5]	71	25%
scientific research into learning and learners	Applying educational data mining to answer questions in any of the three areas of: student models, domain models, and pedagogical support that have broader scientific benefits.(R. S. J. Baker, 2010)	95	34%
Intelligent tutoring system	"is a computer system that aims to provide immediate and customized instruction or feedback to learners"[6]	19	7%

From table 2, we notice that the majority of classified articles are found under the "scientific research into learning and learners" application domain with a total of 95 articles. Following is the "learning object" application domain with 84 articles, and "personalized learning" with 71 articles. This could be due to the fact that the main concern of educational data mining as an emergent discipline for exploring data is to explore methods to support learning and teaching processes. Nevertheless, other application domains were less common such as educational games, intelligent tutoring, and mobile learning, which suggest that more research is needed in these areas.

Publications' Tasks
In table 3, we followed the articles' categorization based on (Crist´ obal Romero & Sebasti´ an Ventura, 2010). The categories represent examples of applications or tasks in education that can be solved using data mining. Among these tasks: "Predicting student performance" is the main focus of 41 articles, followed by "Analysis and visualization of data" with 35 articles; whereas, the following categories "Detecting undesirable student behaviors" and "Developing concept maps" appeared to be the least desired tasks in the repository. However, it is important to note that many of the selected papers are included into one or more different categories. Therefore, to better highlight different emerging categories, authors were compelled to choose the most noticeable choice of category in the paper. Table 3 below lists the categories, along with their description and number of articles assigned to each category.

Table 3. Publications categories was borrowed from (Cristobal Romero & Ventura, 2013)

Tasks	Description	# of Articles
Predicting student Performance	To estimate the unknown value of a student's performance, knowledge, score or mark	41
Providing feedback for supporting in-structors	To provide feedback to support educators in deci-sion-making about how to improve students' learning and enable them to take appropriate proactive and/or remedial action	30
Recommending to students	To make recommendations to students with re-spect to their activities or tasks links to visits, problems or courses to be done, and so forth.	28
Student modeling	To develop and tune cognitive models of human students that represent their skills and declarative knowledge	33
Grouping students	To create groups of students according to their customized features, personal characteristics, students personal learning data, and so forth	14
Constructing courseware	To help instructors and developers to carry out the construction/development process of courseware and learning content automatically	32
Planning and schedul-ing	To plan future courses, student course scheduling, planning resource allocation, admission and counseling processes, developing curriculum, and so forth	18
Detecting undesirable student behaviors	To monitor students' learning progress for detect-ing in real time undesirable student behaviors such as low motivation, playing games, misuse, cheating, dropping out, and so forth	11
Social network analy-sis	Aims at studying relationships between individu-als, instead of individual attributes or properties	19
Developing concept maps	A concept map is a conceptual graph that shows relationships between concepts and ex-presses the hierarchal structure of knowledge	13
Analysis and visuali-zation of data	Provide basic information directly from data (re-ports, statistics, etc.)	35

3   Discussion and Limitations

The aim of this study is to answer the research questions related to the state of EDM research based on a systematic review of the highly cited paper in the academic literature. Using forward chaining of publication referencing of the most influential paper in the field i.e. (C. Romero & Ventura, 2007) to answer our research questions, we can report the following findings.

The types of applications conducted in the field of EDM from 2007 until now, cir-culate around: Educational games, learning object, Mobile learning, personalized learning, scientific research into learning and learners, and intelligent tutoring system. Our findings suggested the EDM research community focus more on educational games, intelligent tutoring, and mobile learning.

In terms of publications distributions, the published articles have increased since 2008 and the establishment of the EDM conference; as identified previously the number of published papers has exceed 70 papers in 2012.

Also, we had classified the articles using the same categorization used by (Crist' obal Romero & Sebasti' an Ventura, 2010). The categories represent examples of applications or tasks in education that can be solved using data mining. The most area that has many articles was scientific research into learning and learners, learning object, and personalized learning.

As for the characteristics of current research on EDM, we can say it is similar to the finding suggested by (Crist' obal Romero & Sebasti' an Ventura, 2010), it would be recommended if the tools developed for EDM can be embedded automatically in any learning management system or learning process. Moreover, the tools should be easy and attractive to educators to use, and institutes have to encourage their faculty to adapt it.

Our approach of using forward chaining of publication referencing (C. Romero & Ventura, 2007) might limit the study to a selected portion of papers. Thus, it would be interesting to investigate further on using other early important works that influenced EDM field. Another limitation in our study is not considering other non-English publications; we believe that they would have been an informative addition to the study.

Finally, EDM research became an area of concrete and solid science that will evolve and will contribute to the adaption of new teaching practices.

4   References:

Baker, R. S. J. (2010). Data Mining for Education. In *International Encyclopedia of Education , 7, 112-118.*.

Baker, R. S. J. D., & Yacef, K. (2009). The State of Educational Data Mining in 2009   : A Review and Future Visions. *Journal of Educational Data Mining, 1(1), 3-17.*

Crist' obal Romero, & Sebasti' an Ventura. (2010). Educational Data Mining   : A Review of the State of the Art. *IEEE TRANSACTIONS ON SYSTEMS, MAN, AND CYBERNETICS, 40*(6), 601–618.

Romero, C., & Ventura, S. (2007). Educational data mining: A survey from 1995 to 2005. *Expert Systems with Applications, 33*(1), 135–146. doi:10.1016/j.eswa.2006.04.005

Romero, Cristobal, & Ventura, S. (2013). Data mining in education. *Wiley Interdisciplinary Reviews: Data Mining and Knowledge Discovery, 3*(1), 12–27. doi:10.1002/widm.1075

Learning Analytics : definitions, applications and related fields

A study for future challenges

Entesar A. Almosallam, Henda Chorfi Ouertani

King Saud University, Information Technology Department
{ealmosallam, houertani}@ksu.edu.sa

Abstract. In the last few decades, the number of people connected online for educational purpose is increasing dramatically and consequently a huge quantity of data is being generated. This data is mainly "traces" or "digital breadcrumbs" that students leave as they interact with online learning environments. Confident that this data can teach us about learners' behaviors and help us enhancing learning experience, there has been a growing interest in the automatic analysis of such data. A research area referred to as Learning Analytics (LA) is identified. It is considered by many researchers as a strategic trend in education. Nevertheless, LA cannot be considered as a new field, it actually derives from different related fields such as Educational Data Mining, Academic Analytics, Action research, Personalized Adaptive Learning.

In this paper, we begin with an examination of the educational factors that have driven the need and the development of analytics in education. We study connections between LA and its most related fields (Educational Data Mining and Academic Analytics). We summarize this interconnection in a table showing for each field the objectives, the stakeholders, the methods and the initial trigger behind the analysis actions. After that we study and run through LA applications presented in the International Learning Analytics & Knowledge Conferences during the three last years. Finally, we conclude by identifying some challenges in the area of LA in relation to the driven factors related to Educational Data.

Keywords: Learning Analytics, Educational Data, Analytics Applications.

T. Herawan et al. (eds.), *Proceedings of the First International Conference on Advanced Data and Information Engineering (DaEng-2013)*, Lecture Notes in Electrical Engineering 285, DOI: 10.1007/978-981-4585-18-7_81,
© Springer Science+Business Media Singapore 2014

1   Introduction

In the last few decades, the number of people connected online for educational purpose is increasing dramatically and consequently a huge quantity of data surrounding these interactions is being generated. In higher education, most of the learning systems used by colleges and universities usually collect comprehensive log data associated with learners' behaviors and actions. However, the reports generated based on this data provide very general and limited information [1]. Recently, there has been a growing interest in the automatic analysis of educational data and how this data can be used to improve teaching and learning. This interest has seen increasingly rapid advances in the emerging field of Learning Analytics (LA).

LA is emerging as a way for educational institutions to use the data they usually collect and save for more than just reporting – to gain a better understanding of what they are doing, and so gain a strategic improvement. Generally, LA deals with the development of methods that harness educational data to support the learning process [2]. It is part of a growing academic trend towards big and open data: LA is identified in the 2013 Horizon Report [7] as a key future trend for education, and as having real potential to improve learner experience.

Although LA is still an emerging discipline; it is not new. It has roots in various fields, involving machine learning, artificial intelligence, information retrieval, statistics, visualization, and research models in general [2]. However, what is new is the rise of quantity and quality of captured data about the learning processes [4]. As a consequence, analytics have gained attention in education.

LA is a multi-disciplinary field in which several related areas of research converge. These include academic analytics, action research, educational data mining, recommender systems, and personalized adaptive learning [5]. LA borrows from these different related fields and applies several existing techniques.

The remainder of this paper is structured as follows: the next section provides some definitions of LA. How LA can change education is presented in Section 3. Section 4 identifies the related areas overlapped with LA. A through up LA applications during the last 3 years   is discussed in section 5. Conclusions and challenges are presented in Section 6.

2   What is Learning Analytics?

Different definitions have been provided for the term 'Learning Analytics'. Learning Analytics is defined on LAK11 website [1] - the first conference on LA - as "the measurement, collection, analysis and reporting of data about learners and their contexts, for purposes of understanding and optimizing learning and the environments in which it occurs".

According to [6], LA "refers to the use of intelligent data, learner-produced data, and analysis models to discover information and social connections, and to predict and advise on learning". EDUCAUSE's Next Generation learning initiative [2] describes LA as: "the use of data and models to predict student progress and performance, and the ability to act on that information" [6].

---

[1] https://tekri.athabascau.ca/analytics/
[2] http://nextgenlearning.org/

Another definition that experts in this field provide is: "Learning Analytics refers to the interpretation of a wide range of data produced by and gathered on behalf of students in order to assess academic progress, predict future performance, and spot potential issues" [7]. Even these definitions are different; it is noticeable that they focus on changing educational data to insights and actions aiming to improve learning [6].

3 How can LA changes education?

The 2013 Horizon Report [7] includes LA in the 2-3 years of adoption. It is predicted that LA will be considered as a foundational tool for informed change in education within that time [8]. Evidence inferred from LA tools can support us to make better decisions. We can better understand learning behaviors and how the learning process outputs are produced. LA helps in identifying learners who are at risk and provides the guide to the decision makers to choose and apply the best actions that will help learners to success. Techniques used in LA tools are very compelling for adoption in education [10]. Classical methods of student assessment and evaluation can report and show students results collected generally. However, L.A. permits to track student performance by integrating their data that is collected through a variety of sources. It gives a clear idea of what students are doing and their gained learning in the whole learning process. Moreover, these evaluations are usually done after the completion of the course letting impossible the interventions while the course is in progress. Also, it is difficult to find out some important information about how students use the educational content, the time consumed, and interaction with other learners. However, all these benefits and more could be accomplished using LA.

LA could offer a holistic view of the learning experience for both learners and teachers to discover the data, depending on the needs of their educational context [9]. LA can provide dramatic change in education, by altering how educational institutions develop curriculum, present it, evaluate student learning, provide learning feedback, and even allocate resources [8].

However, a combination of technological and educational factors drives the need and the development of analytics in education. Some of these factors are related to the *educational data*, the *educational task*, or the *stakeholders* [11].

3.1 Educational data

LA is data-driven approach [14], as it is based on the collection, process, and analysis of collected educational data.

In educational environments, there are many different types of *educational data* available for analyzing. These data are specific to the educational area, and therefore have essential meaning, and relationships with other data. For example, it is more difficult when dealing with *distributed* data from multiple sources with different formats rather than dealing with data that is *central* in one source [12]. Also, the *scope* and the *size* of the data to analyze is important factor. As most of the educational institutions move to online learning, this means that they deal with increasingly large sets of data [11]. Furthermore, it is also necessary to take privacy aspects of the learners and the educational institutions into consideration. The capabilities of LA tools will be limited when the data is *private* in contrast with *public* and *open* data [13].

3.2   Educational Tasks

In terms of educational tasks, there are many factors to the development of LA depending on the objective of applying LA in each application area. Although the objectives and stakeholders of all application areas overlap, they require analytics using different metrics and work on different scales. On one hand, there are many applications or tasks in educational environments that have been resolved through LA.

Most of the possible objectives of LA were stated in table 2 such as: monitoring, analysis, prediction, intervention, tutoring/mentoring, assessment, feedback, adaptation, personalization, recommendation, and reflection. On the other hand, the values that LA applications extract can be very helpful for different stakeholders mainly: learners, teachers and administrators [9, 10]. In the table 1, we present the most common educational tasks that have employed LA techniques grouped by stakeholders.

Table 1. educational tasks that employed LA techniques grouped by stakeholders

Learners	Teachers	Administrators
• Self-reflection • Assessment and feedback • Adaptation and Personalization	• Curriculum development • Interventions and Recommendations	• Monitoring • Prediction • Making decisions

3.3   Stakeholders

Learners. LA can be a valuable tool to promote self-reflection. Reflective learning offers the chance of learning by evaluating past work in order to improve future experiences and promote continuous learning. Students can benefit from data compared within the same course, across classes, to draw conclusions and self-reflect on the progress of their learning practice. [11]

In addition, by using LA, the process of evaluation of learners can be enhanced to produce a *real-time assessment* rather than the assessment at the end of a course [8]. LA supports the assessment to improve the efficiency of the learning process by providing *feedback* to different stakeholders. Feedback provides interesting information generated based on data about the user's interests and the learning context [5].

In terms of *Adaptation and personalization*, LA would be able to adapt and provide personalized improvements that adjust to the needs of each learner based on the previous academic performance [8]. The objective of LA here is to help learners decide what to do next by organizing learning resources and activities according to the preferences of each learner in adaptive way. This adaptation and personalization of the content of learning process, ensuring that each learner receives resources and teaching that reflect their current knowledge state [9].

Teachers. Teachers can benefit from LA to determine the appropriate curricula for students that adapt with their interests and preferences. When using LA, teaching components could be sharable and reusable between teachers easily. In addition, some teachers employ LA to identify specific parts in a course that cause students failures. Then, teachers can adjust curriculum or modify learning activities to enhance learning process. It helps teachers to know which academic practices need to be curbed and which need to be encouraged [9].

Indeed, the most common use of LA is to identify students who may need additional support based on their past and current activities and accomplishments and to provide *early interventions* to help them improve their performance. The effective analysis and *prediction* of the learner performance can help the teacher in intervention by suggesting appropriate actions that should be taken to assist learners achieve better outcomes [9].

In terms of *recommendations*, LA can provide valuable insight into the factors that influence learners' success. It can provide early alerts or indications of which students are at risk in their learning behavior. By recognizing these students and offering early interventions, educational institutions can translate that data into actionable insights to reduce related problems dramatically [8].

Administrators. Today, LA is most often used in higher education for *administrative decisions* [9]. There is an increasing emphasis on the use of metrics and analytics for higher *education* in order to improve learning outcomes [10]. As a result, the majority of colleges and universities are now placing a heavy emphasis on the use of LA. This will encourage a more rapid adoption of LA as a way of meeting those needs.

LA has the ability to track student activities and generate reports in order to support decision-making by the educational institution. LA monitor and evaluate the learning process with the purpose of continuously improving the learning environment. Examining how students use a learning system and analyzing student accomplishments can help administrators find patterns and make decision on the future design of the learning process.

4 Classification of Learning Analytics and their related areas

In this section, we studied connections between LA and its different related fields (Educational Data Mining, Academic Analytics). To do that, we provide a brief definition for each field then we summarize them in table 2 showing for each field the objectives, the stakeholders, the methods and the initial trigger behind the analysis actions. While LA focused on the educational issues and how can we optimize opportunities for improve learning [11], Academic Analytics (AA) focused on the administrative concerns and how to help administrative users in higher education to make better learning opportunities and decisions [5]. It combines data with statistical techniques and predictive modeling to help determine which students may face academic difficulty, allowing interventions to help them succeed. The examples in the AA literature refer mostly to the problem of detecting "at risk students", that is, those students that might drop out of a course or abandon their studies [5]. AA usually focused on enrollment management and the prediction of student academic success. AA were restricted to statistical software. On the other hand, Educational Data Mining (EDM) focused on the technical challenges and how can we extract value from the large volumes of learning-related data [13]. From a technical perspective, EDM is the application of data mining techniques to educational data, and so, its objective is to support teachers and students in analyzing the learning process [27].

The objectives, analysis domain, data, and process in LA, AA and EDM are quite similar. All of them focus on the educational domain, work with data originating from educational environments, and convert this data into relevant information with the aim of improving the learning process. However, the techniques used for LA can be different from those used in other fields. In addition to data mining techniques, LA further includes other methods, such as statistical and visualization tools or social network analysis (SNA) techniques, and puts them into practice for studying their actual effectiveness on the improvement of teaching and learning [11].

Table 2. Classification of LAs and other related research areas

Field	Stakehold-ers	Objectives	Methods	Data
AA	Educational institutions	Enrollment management, prediction Support decision making	Statistical methods	Educational environments data
EDM	Teachers-Students	Convert data into relevant information to improve learning process	Data mining techniques (clustering, classification, association rules)	Educational environments data
LA	Learners - Teachers - Educational institutions	Enrollment, prediction, Reflection, Adaptation, Personalization, recommendation	Quantitative methods Data mining techniques (clustering, classification, association rules)	Educational environments data

5   Learning Analytics Applications

The number of publications around LA and its related areas has grown rapidly in the last few years. In this section, we review the most relevant studies in this field from 2011 to 2013. We mainly reviewed studies from the LAK11, LAK12, LAK13 conferences, which discussed concrete LA applications. To note that LA publications from 2011 to 2013 are not limited solely to the aforementioned ones. Several other conferences have included special tracks addressing LA and its related topics. However, the publications at LAK conferences presented a compelling picture of the potential of LA. Therefore, we restricted our literature review to these three conferences as they represent the rich diversity of research and project directions currently underway in the field of LA. We summarize just a few of the papers to try to capture a sense of this. An overview of these papers and related learning application is presented in table 3. We summarize characteristics of 10 LA applications: their educational environments, tasks and features, for whom they are intended, what data they track and how they have been facilitated with LA in practice.

Table 3. An overview of LA applications

Tool Name	Edu. Environment	Edu. Tasks	Edu. Data	Stakeholder	LA Metric	Visualization
Video-games [15] – 2012	Educational video games	Task: Assessment	In-game situations and interaction traces	Learners	Tracking techniques, rule-based system	Human-readable reports
AAT [16] – 2011	Within the Moodle Analytics project	Task: access and analyze student behavior data in learning systems	Data from one or a set of courses hosted in one or several databases	Learning designers teachers.	Statistics, customized queries and activity reports	Graphical user interface (GUI)
CAFe [17] – 2011	eTwinning-an online community for European teachers (digital mediated learning network)	Task: Self-monitoring tool for teachers • monitor their positions in the eTwinning network, • knowing their achievements • Recognizing their weaknesses and shortages.	Data dumps gathered from relational database in the eTwinning project - anonymous data sets	Teachers coordinators researchers	Implicit assessment methods of some indicators stored in XML files	Graphs, time series charts, and bar charts
SNAPP [18] – 2011	learning management systems	Task: to visualize the network of interactions resulting from discussion forum posts and replies	Student behavioral patterns and learning activities	Teachers	Thematic Analysis	Social network visualization
Course Signals [19] - 2012	Learning Management Systems	Task: Prediction - real-time feedback to students	Edu. data collected by instructional tools	Learners	Statistical & mining techniques - predictive student success algorithms	Traffic signal indicators
GLASS [20] – 2012	Web-based visualization platform	Visualization	Datasets stored using the schema that allows to capture events occurring during the use of various computer applications.	Learners and teachers		Time lines and charts of activity events
Learn-B [21] - 2012	Self-regulated workplace learning in organizational context	Tasks: Reflective learning & collaboration Features: Makes use of ontologies for data linking and annotation	Learners progress data	Learners	Statistics & Social Waves	A set of various visualization charts
E²Coach [22] - 2012	Adaptive- recommender system for physics education	Task: Adaptive recommendations and interventions Features: Provides each student with an individualized coach.	Real time, content specific student performance data	Learners	Actionable intelligence	Dashboards

HOU2LEARN [23] - 2013	Personal Learning Environment	**Tasks:** To encourage social networking and collaboration **Features:** Open platform	E-portfolios of learners	University professors and high school teachers	Assessment method based on rubrics	Dashboards
StepUp! [24] - 2013	Visualization tool	Awareness – reflection - track student activity	Traces of learning activities collected automatically by software trackers	Learners	Statistics - tracking techniques	Dashboards

6 Conclusion and challenges

Learning Analytics promises to harness the power of advances in data analysis techniques to improve understandings of teaching and learning, and to tailor education to individual learners more effectively. However, LA requires key researchers to address a number of challenges, including questions about privacy, heterogeneity and integration, appropriate visualization, data structure, meaningful indicators and Costs issues. **Table 4** recapitulates some of the challenges and educational issues that need to be addressed during the LA development.

Table 4. Educational challenges and issues surrounding LA development

Challenges	Description
Privacy	Challenges to the ownership and use of the data - Who gets access to the data?
Heterogeneity	Different sources and formats
Intended stakeholders	The kinds of data and analysis employed depend on the intended audience and stakeholder.
Visualization	Appropriate and understandable information visualization for the different stackeholders
Data structure	The data can be structured (logged data) or unstructured (interaction data)
Lack of unique identifiers and meaningful indicators	Different stakeholders use different technologies in different ways. Stakeholders have different point of views.
Costs issues	Costs to store big data and producing LA tools

References

1. Elias, T. 2011. "Learning Analytics: Definitions, Processes and Potential".
2. George Siemens, "Structure and Logic of the Learning Analytics Field", 2013, Retrieved June 3, 2013, from http://www.learninganalytics.net/?p=185
3. Maurizio De Rose, " Future trends on data analytics to support your learners" , 2013, Retrieved May 15, 2013, from http://www.jisc.ac.uk/inform/inform35/FutureTrendsDataAnalytics.html
4. George Siemens , "Sensemaking. Beyond Analytics as a Technical Activity", April 11, 2012
5. Chatti, Mohamed Amine, et al. "A reference model for learning analytics."International Journal of Technology Enhanced Learning 4.5 (2012): 318-331.
6. Siemens, G. (2010). What are Learning Analytics? Retrieved May 14, 2013, from http://www.elearnspace.org/blog/2010/08/25/what-are-learning-analytics/
7. Johnson, L.,AdamsBecker, S.,Cummins,M., Estrada,V., Freeman,A., and Ludgate, H. (2013). NMC Horizon Report: 2013 Higher Education Edition. Austin, Texas: The New Media Consortium
8. How data and analytics can improve education, Retrieved May 14, 2013, from http://strata.oreilly.com/2011/07/education-data-analytics-learning.html
9. Jie Wu, "Revolutionize Education Using Learning Analytics", Retrieved May 11, 2013, from http://www.aboutjiewu.com/Resources/projects/INFO203/LearningAnalytics.pdf
10. Malcolm Brown, "Learning Analytics: The Coming Third Wave", April 2011, Retrieved May 12, 2013, from http://www.educause.edu/library/resources/learning-analytics-coming-third-wave
11. Ferguson, Rebecca. "Learning analytics: drivers, developments and challenges." International Journal of Technology Enhanced Learning 4.5 (2012): 304-317.
12. Suthers, Daniel, and Devan Rosen. "A unified framework for multi-level analysis of distributed learning." Proceedings of the 1st International Conference on Learning Analytics and Knowledge. ACM, 2011.
13. Romero, Cristóbal, and Sebastián Ventura. "Educational data mining: a review of the state of the art." Systems, Man, and Cybernetics, Part C: Applications and Reviews, IEEE Transactions on 40.6 (2010): 601-618.
14. Bader-Natal, Ari, and Thomas Lotze. "Evolving a learning analytics platform."Proceedings of the 1st International Conference on Learning Analytics and Knowledge. ACM, 2011.
15. Serrano-Laguna, Ángel, et al."Tracing a Little for Big Improvements: Application of Learning Analytics and Videogames for Student Assessment."Procedia Computer Science 15 (2012).
16. Graf, Sabine, et al. "AAT: a tool for accessing and analysing students' behaviour data in learning systems." Proceedings of the 1st International Conference on Learning Analytics and Knowledge. ACM, 2011.
17. Song, Ergang, et al. "Learning analytics at large: the lifelong learning network of 160,000 European teachers." Towards Ubiquitous Learning. Springer Berlin Heidelberg, 2011. 398-411.
18. Bakharia, Aneesha, and Shane Dawson. "SNAPP: a bird's-eye view of temporal participant interaction." Proceedings of the 1st International Conference on Learning Analytics and Knowledge. ACM, 2011.
19. Arnold, Kimberly E., and Matthew D. Pistilli. "Course Signals at Purdue: Using learning analytics to increase student success." Proceedings of the 2nd International Conference on Learning Analytics and Knowledge. ACM, 2012.
20. Leony, Derick, et al. "GLASS: a learning analytics visualization tool."Proceedings of the 2nd International Conference on Learning Analytics and Knowledge. ACM, 2012.
21. Siadaty, Melody, et al. "Learn-B: a social analytics-enabled tool for self-regulated workplace learning." Proceedings of the 2nd International Conference on Learning Analytics and Knowledge. ACM, 2012.

22. McKay, Tim, Kate Miller, and Jared Tritz. "What to do with actionable intelligence: E 2 Coach as an intervention engine." Proceedings of the 2nd International Conference on Learning Analytics and Knowledge. ACM, 2012.
23. Koulocheri, Eleni, and Michalis Xenos. "Considering formal assessment in learning analytics within a PLE: the HOU2LEARN case." Proceedings of the Third International Conference on Learning Analytics and Knowledge. ACM, 2013.
24. Santos, Jose Luis, et al. "Addressing learner issues with StepUp!: an Evaluation." Proceedings of the Third International Conference on Learning Analytics and Knowledge. ACM, 2013.
25. Bauer, Florian, and Martin Kaltenböck. "Linked Open Data: The Essentials."Edition mono/monochrom, Vienna (2011).
26. The Open Data Handbook, Retrieved June 19, 2013, from http://opendatahandbook.org/
27. Siemens, George, and Ryan SJ D. Baker. "Learning analytics and educational data mining: towards communication and collaboration." Proceedings of the 2nd International Conference on Learning Analytics and Knowledge. ACM, 2012.

Printed in the United States
By Bookmasters